ECONOMIC MINERAL DEPOSITS

The Lornex copper-molybdenum mine and plant, Highland Valley, British Columbia. (By permission of Rio Algon Ltd.)

ECONOMIC MINERAL DEPOSITS

THIRD EDITION

REVISED PRINTING

Mead L. Jensen
Professor of Geology and
Geophysics
University of Utah

Alan M. Bateman
Late Silliman
Professor of Geology
Yale University

JOHN WILEY & SONS

New York · Chichester · Brisbane · Toronto · Singapore

74329

Library of Congress Cataloging in Publication Data:

Jensen, Mead LeRoy, 1925–
  Economic mineral deposits.

  Second ed. by A. M. Bateman.
  Bibliography: p.
  Includes index.
  1. Geology, Economic. 2. Mines and mineral
resources. I. Bateman, Alan Mara, joint author.
II. Bateman, Alan Mara. Economic mineral deposits.
III. Title.
TN260.B3   1979   553   78-9852
ISBN   0-471-09043-3

Printed in the United States of America
10   9   8   7   6   5   4   3

PREFACE

The first edition of this book appeared at the end of 1942 and the second edition was published in April 1950. Shortly before Alan Bateman died, on May 11, 1971, he asked if I would prepare a third edition.

More than a generation has passed since the last edition, a period of time during which the peoples of the world have come to realize finally, especially because of the energy crisis, that nonrenewable energy and mineral resources are limited in amounts and localized in geographical areas or countries that are not as exploitable by industrial nations as they once were. Never before has there been a greater need for effective exploration geologists than there is at the present time. Even so, the number of universities in the United States offering courses in economic geology is decreasing, partly as the result of decisions made by geoscientists not experienced in field or exploration studies. In the USSR, in contrast, there are more than ten times as many exploration geologists as there are in all of North America.

The organization of the first two editions has been retained in this edition. Instead of following a classification of mineral deposits, processes of mineral formation and concentration are emphasized.

Part I of this text pertains to *Principles*; Part II delves into *Processes of Formation of Mineral Deposits*; Part III considers *Metallic Mineral Deposits*; and Part IV pertains to *Nonmetallic Mineral Deposits*.

Those specific chapters in the second edition pertaining to structural geology, geophysical techniques, extractive metallurgical processes, ground water supplies, and others have been omitted from this edition because these topics are extensively covered in more detail in other specific texts.

Several chapters are entirely new, including Chapter 1 on Mineral Economics and Exploration and Chapters 10 and 11, which pertain to newly recognized processes, namely, Bacteriogenic and Submarine Exhalative Volcanogenic processes. Many other chapters have been rewritten extensively while only a few chapters have received modest changes. No chapter, however, remains the same as in the second edition. Attempts have been made to update reserve, resource, and production figures, a noteworthy attribute of this text that few other modern economic geology texts include. More than half of the tables and figures are new. Graphs and tables pertaining to changing prices, reserves, and resources of all of the major metals are included in the text. This book is not only recommended as a graduate and undergraduate textbook, but it is also a reference and source text that many nongeologists may find informative.

The metric system is used predominately and metric units follow all measurements where used. English measurements have been retained in specific cases such as mine levels when measured in feet, values of gold and silver in troy ounces, and some tonnage figures.

I would be grievously remiss if I did not express my gratitude to those students, colleagues, and especially Alan Bateman with whom I spent 14 memorable years, for the wisdom, thoughts, and experiences they offered me. The institutions of MIT, Yale University, and the University of Utah, where I have spent most of my educational and teaching career, have provided leaves, funds, and time to visit mineral deposits on every inhabited continent. Also, to those exploration and mining geologists, and the personnel of foreign geological surveys and universities in different parts of the world, I extend my gratitude, for without exception, they have always made me feel a welcome guest, rather than a nuisance, and they have spent more time with me than they could afford.

Finally, but most important of all, errors and omissions in the manuscript have been spotted by former and present students, critics, colleagues, friends and enemies. I gratefully acknowledge their time, corrections, and contributions.

I cannot express my gratitude sufficiently to Don Garlick, who performed the task of critically reading parts of the manuscript. Not only did he suggest many important major changes, but he also provided many and better figures. His father, W. G. Garlick, graciously corrected and updated the material on the Zambian Copper Belt.

In addition, detailed critical reading of the galley proofs was done by Hayat Qidwai. Hasan Mohammed, and N. Carouso, all of whom made significant changes and spotted numerous errors. Robert Wiese compiled a list of corrections in the first printing that have been corrected in this revised edition.

I also express my gratitude to the following critical readers who read portions of the manuscript: B. Sharp, E. Callaghan, E. Cheney, F. W. Christiansen, W. J. Christiansen, G. de V. Klein, E. Erickson, M. P. Erickson, R. R. Shrock, and many graduate students who have used the text in manuscript form. Robina Bullock and Sylvia Howells completed the onerous task of typing by deciphering my cryptic handwriting.

Finally, I am grateful to those members of the John Wiley staff, especially Don Deneck who provided keen editorial advice and encouragment when it was greatly needed. Others are too numerous to list but Deborah Herbert, Linda Indig, Maddy Lesure, Malcolm Easterlin, Vivian Kahane, and others must be thanked.

Mead L. Jensen
Salt Lake City, Utah

CONTENTS

"The mining industry is infinitely worse off today than it ever was. We don't have access to land—due to landwithdrawals. We have inadequate financial resources—because of inflation and changes in taxation and price controls. We are cursed by the politicians' failure to recognize that the mining industry is a highly risky business."

James Boyd, 1976

ECONOMIC
MINERAL
DEPOSITS

PRINCIPLES

1

MINERAL ECONOMICS AND EXPLORATION

It is to steel and oil and uranium, not to martial ardor, that modern nations must look for victory in war.

Bertrand Russell
The Impact of Science on Society

MINERAL RESOURCES AND THEIR INTERNATIONAL IMPACT

Geologists distinguish between *mineral reserves* and *mineral resources*. The term *resource* refers to hypothetical and speculative, undiscovered, subeconomic mineral deposits or an undiscovered deposit of unknown economics. Reserves are concentrations of a usable mineral or energy commodity, which can be economically and legally extracted at the time of evaluation. (Fig.1-1.)

Today, mineral resources have become almost synonymous with industrial power, and industrial power is in turn dependent upon ownership of, or access to, large quantities of mineral resources that have become the backbone of the industrial way of life. Mineral resources have much to do with man, and man with mineral resources. It is startling to realize that the rise of the industrial age has so accelerated the demand for minerals that the world has dug up and consumed more of its mineral resources in the period shortly before and after World War II than in all preceding ages. This insatiable demand for minerals to feed industry has made some sources of supply that we used to think were adequate now look rather small, and sources capable of meeting large demands are being depleted rapidly.

Prior to and less than a century and a half ago, the rate of increase in the consumption of copper and iron was directly proportional to the increase in population. If the population doubled, the consumption of these metals doubled. In contrast, during the first hundred years of the Industrial Revolution, roughly from 1812 to 1912, the population increased threefold whereas the consumption of copper increased 80 times and iron 100 times!

The quest for the wealth of minerals wrested from the earth for man's vanity, necessities, or comforts has ever been a powerful incentive to discovery,

exploration, and trade. Their search has given rise to voyages of discovery and settlement of new lands. Their ownership has resulted in industrial development and in commercial or political supremacy, and has also caused strife and war. In the ancient country of Saomes, the winter torrents brought down gravels containing gold, which the barbarians passed through inclined troughs lined with sheeps' fleeces to catch the gold. The fleeces that were hung on trees to dry, so that the fine gold could be beaten out of them, spurred Jason and the Argonauts in the ship *Argo* to seek the Golden Fleece near the shore of the Euxine. This is the earliest record of a placer gold rush and a poetic expression of an early mining venture. It was tin that apparently drew the Phoenicians and Romans to Britain; it was gold and silver that lured the Spanish Conquistadores to the settlement of the New World. The gold rush of 1849 led to the settlement of California and then the acquisition of the western part of the United States from Mexico and Spain.

A glance at the history of development of leading industrial nations reveals that their rise has coincided with the utilization of their mineral resources, initially coal and iron, for coal is one energy that makes the wheels go around and the wheels are made of iron. Those countries supplied with diversified mineral resources are the ones that became the great industrial nations and the politically and militarily strong ones. Additional nations are joining this group based on their petroleum reserves such as the Middle East countries; or by purchasing new resources, such as Japan; or by the development of vast mineral resources, such as Russia, Australia, and Brazil.

It is no accident that great industrial centers sprang up in central England, in France, and in the Great Lakes and Appalachian regions of the United States. For there iron and coal were brought to-

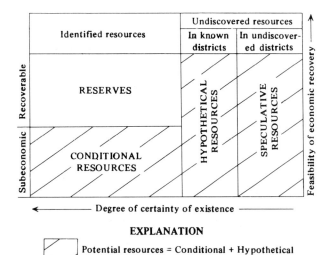

Figure 1-1 Terms used in Mineral reserves and resources.

gether; they spawned the first iron and steel industries that spelled industrial growth. Nations lacking an adequate supply of minerals either became agricultural or aggresive, such as Argentina, Germany, and Japan, respectively. Many less developed countries (LDC's) blessed with mineral resources have had difficulties in the Industrial Revolution. Brazil has much iron ore but little coal. India has much manganese, coal and iron. It not only has five major producing steel plants but is also exporting iron ore to Japan and Iran. China and Russia have immense resources of coal and iron ore and are or will become major industrial nations.

With the waning of iron ore in Great Britain came a change in Britain's economy and a plummeting of the pound's value. Of course, the loss of the British colonies and the necessity of heavy mortgaging during World War II added to Britain's problems.

Probably, few people in this industrial age realize the extent to which we have become dependent upon minerals. In our homes, our light, power, heat, stoves, TV sets, washers, windows, and all appliances are made of mineral products. Our various means of transportation (automobiles, ships, and air craft) and of communication (telephones, radio, and television) all require the use of metals.

UNIQUENESS OF MINERAL ECONOMICS

The values, costs, long-term capital investment, reserves, distribution, ownership, and international flow of minerals are some of the factors involved in the subject of mineral economics. Factors that are almost economically unique in reference to minerals are:

Occurrence

Limited and Localized. Economic minerals are limited in occurrence or distribution. Their occurrence shows no relationship to international borders; many minerals of economic value are highly localized in only a few deposits, and these may be quite limited in area.

Uncertainty of Reserves and Discoveries

Reserves of minerals are difficult to determine as the value and costs of extraction and metallurgical treatment and transportation costs determine whether the resources are potentially economic. Nevertheless, and fortunately, even though estimates of future consumption are generally underestimated, potential resources are almost invariably greatly underestimated. Because of these uncertainties, mineral exploration is a world-wide program that raises even more uncertainties with the validity of title or ownership of properties in other countries, especially in LDC's.

Another problem contributing to the reluctance to invest more in politically uncertain countries is the huge increase in the capital cost of new mineral development. In the United States, the cost of developing an annual ton of new copper capacity from mine through the refinery without intrastructure costs (town sites, transportation, water supply, etc.) has risen from $3500 in 1965 to $6500 in 1976. The value of copper has risen less, from about $0.45 per pound to less than $0.60 per pound, during the same period. The price of copper would have to exceed $0.80 per pound to offset the inflationary rate of production costs.

Depletion, Exhaustion, and Costs of Mining

The exhaustion of an ore deposit may occur suddenly or slowly as the result of price changes, costs, market, exhaustion of ore, and so forth. At the present time, inflation has raised many costs to new heights and even with the increased value of most minerals the profit margin is still narrow. Even ignoring inflation, copper is a specific example of a case in which the price was historically high but is now dropping because of increased world production, stock piles, and new deposits; even per capita demand has not increased as rapidly, which is resulting in curtailed production.

The depletion allowance in the mining industry is badly misunderstood by the general population, which is quite evident because the depletion allowance may be facing extinction. As an ore deposit is being depleted, a portion of the proceeds should be allotted for the cost of exploration in locating other deposits or the mining company will go out of business when the deposit is exhausted. In the United States the depletion allowance before taxes of 15 percent of the proceeds is allowed for many metal deposits. Some nonmetallic deposits, such as sulfur, receive more; petroleum concerns were allowed 27.5 percent, but this has been reduced to zero percent for the major companies. Depletion differs little from depreciation, which is allowed in all other industries. Depreciation is readily determined, as it is based on costs while depletion is difficult to determine, as it is based on the amortization of reserves and the expected life of an ore deposit.

Scrap Return

The recycling of minerals is becoming much more effective as an efficient process of using metals. Of course, it is not possible to recover scrap from some of the uses of minerals, such as lead in paint and tetraethyl gasoline, tin plating of cans, and especially fossil fuels. Even so, there are many uses that are not yet being recognized for their scrap return potential. In contrast, however, almost 5 billion (Fig. 1-2) aluminum cans were recycled during 1976. Not only did this save aluminum, it also reduced the amount of bauxite needed, the cost of shipping that amount of bauxite to hydroelectric sites, and the conservation of 4 kilowatts per kilogram of aluminum that was recycled.

Environmental Requirements

During the past decade, the mineral industry has faced more severe restrictions than it ever had before on the effect of ground prospecting, development, mining, refining, and smelting on the environment. Begrudgingly initially, the industry is now spending thousands of millions of dollars in complying with these requirements. Nevertheless, the subject is volatile and emotional but understandable and desirable when rational.

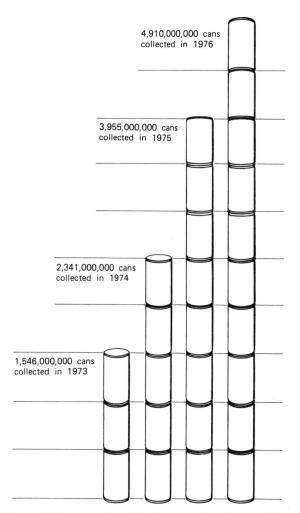

4,910,000,000 cans collected in 1976

3,955,000,000 cans collected in 1975

2,341,000,000 cans collected in 1974

1,546,000,000 cans collected in 1973

Figure 1-2 The recycling of aluminum cans has increased three fold in three years.

As an example, the "Surface Mine Control and Reclamation Act," which was initially vetoed by President Gerald Ford was thought to be not so much regulatory as it was prohibitory in both spirit and purpose. The vetoed bill did not meet the essential requirement of providing adequate coal to satisfy our immediate fuel needs, or permit substitution of coal to replace oil and gas inputs. Conflicting views have been expressed on the subject. Representative Hechler, for example, stated: "We are going to get my bill before too many years have passed. I intend to keep fighting for the phasing out of strip mining so we can turn to where the vast preponderence of our energy supplies are available." His example of "vast preponderence of energy supplies" needs further data.

In contrast, Representative A. G. Gilbert, presents the following

"When all of the facts and costs of providing a

pristine environment for modern man are clearly in focus, one wonders if there is not too much "overkill" in the Environmental Protection Agency environmental regulations. The writer, as a committee of one, prefers: (1) greater use of U.S. coal reserves to provide an adequate supply of electricity and thereby reduce our oil and gas imports to a manageable level; (2) as much improvement to the environment as we can afford in this decade; (3) an all-out research program now for better utilization of all of our natural energy resources in the future."

In 1977, a surface coal mine regulation that greatly restricts the areas of strip mining for coal became law. The Act requires 13 cents/ton of coal mined for the cost of reclamation. With coal selling between $20 to $40/ton, a cost of less than one percent of the mine production of coal may seem inadequate or reclamation is not costly.

ENERGY AND MINERAL RESERVE "CRISES"

Petroleum Reserves

The energy "crisis" has brought home not only to Americans, but also to the people of all industrial countries, and some not so industrialized, that oil and gas reserves are finite. Such a situation was given little recognition by laymen even a few years ago, and the mineral economist feared overstating the case because of geological success in locating new petroleum deposits. They had cried "Wolf" too many times in the past. Finally, however, the wolves are with us.

The first successful oil well in the United States was drilled near Titusville, Pennsylvania, in 1859. Although it was a producer, and still is from a stratigraphic trap, it was an economic failure to the New Haven, Connecticut, initial investors. They would have given up months before the well reached 69 feet, at which depth oil was found, if it had not been for the persistence of their employee at the well, "Col." Edwin L. Drake, an unemployed conductor for the New York–New Haven–Hartford Railroad. Since the nineteeth century, proven reserves of petroleum never have been sufficient to last more than 15 to 20 years at the given rate of consumption. Now, even including the reserves of Alaska and the U.S. continental shelves, proven reserves are about 50 billion barrels and consumption is 7 billion barrels per year. The United States must import more than 40 percent of its petroleum demand, and at a price now established by OPEC, at about $32 a barrel. Domestic oil is about $12 per barrel but is being allowed to increase about 10 percent per year.

Actually, a geologist, M. K. Hubbert, had predicted scarcities years before the crisis. He realized that the actual amount of the original resource is of surprisingly little importance with regard to its lifetime. The exponential increase in production and consumption speedily disposes of whatever may exist. Hubbert has taken two estimates of the original world oil resource and used discovery data plus the logistic-curve analysis to calculate lifetimes (Fig. 1-3). If the resource was originally 1350 billion barrels, 90 percent of it will be gone by about A.D. 2020. If it was about 50 percent greater, it will be gone by about A.D. 2030.

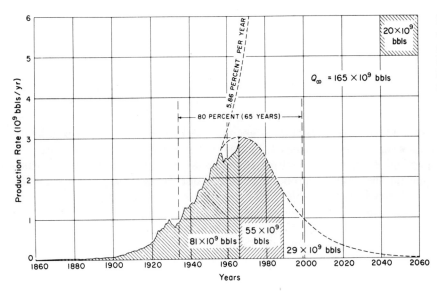

Figure 1-3 Complete cycles of world crude-oil production for two values of ultimate quantity of oil produced (from Hubbard. "Energy Resources" in Cloud, *Resources and Man*, W. H. Freeman and Co. Copyright, 1969)

The petroleum resources of the United States are, of course, greater than the reserves. Offshore petroleum resources of the eastern seaboard may be large, but environmental impact studies delayed the drilling of this immense offshore area until 1978.

President Carter has announced a program to decrease the present imported 9 million barrels of daily petroleum consumption in the United States to about 6 million barrels per day by 1985. His desire will require an increase in the cost of petroleum in the United States to provide the exploration and drilling funds for deeper drilling and extensive offshore drilling of the eastern U.S. continental shelf.

In contrast, the petroleum reserves of the Middle East and northern Africa are as follows in billions of barrels:

Saudi Arabia	163
Kuwait	65
Iran	58
Iraq	31
Neutral Zone	65
Libya	27
Algeria	8
Oman, Egypt, and other Middle East countries	18
Total =	435

If there is another Middle East anywhere in the world, the chances are that it is in Western Siberia. Russia and China are estimated to have 2500 billion barrels, with a large range of uncertainty. Only 150 billion barrels of the Russian and Chinese petroleum potential have been discovered so far. Less than 50 billion barrels has been produced.

The remainder of the world, excluding the USSR and China, has proven reserves of about 600 billion barrels. As extensive as the reserves of the Middle East are, they may be largely depleted by the turn of the century. Reliance on liquid petroleum as an energy source is a relatively short-term solution. The development of other energy sources is a necessity.

Mineral Resources

Now that an energy "crisis" is realized, the next crisis is the apparent dwindling of the economic mineral reserves of many industrial nations. Prior to World War II, for example, the United States was rather self-reliant for its needs of mineral resources. Only about a dozen materials had to be imported, such as nickel, manganese, chromium, other ferroalloys, asbestos, and industrial diamonds.

World War II, however, brought an awful shock. The United States did not have the mineral reserves to supply the military requirements of ourselves and our allies. The United States had to turn to 53 countries for imports of 65 critical minerals, of which 27 came exclusively from foreign sources. In other words, the United States is not self-sufficient to maintain its vast industry during wartime; this also has a bearing on future international relations. Peacetime requirements are also awesome.

Of the 32 minerals most important for industry, the United States is self-sufficient in 9, deficient in 18, and lacking in 5. Furthermore, it is expected that U.S. consumption of these 32 minerals will double before the twenty-first century. A Nonfuel Minerals Policy Coordinating Committee was established in 1978 by President Carter with instructions to submit its policy recommendations and options to the White House within 15 months. The major focus of the study will be on those minerals regarded as most critical to the U.S. economy, such as copper, aluminum, iron, zinc, manganese, chromium, lead, nickel, tungsten, cobalt, titanium and platinum.

Figure 1-4 exhibits the tremendous growth of the steel industry in Europe and especially Russia and Japan with growth rates of 7.5 and 16 percent, respectively, for the latter two. The U.S. growth rate in steel production is about 1.5 percent, and the United States imports 40 percent of its iron ore. Of course, Japan imports all of her iron ore but has a major available source in western Australia.

The Mining Congress Journal has indicated (Fig. 1-5) the demand, consumption, and deficit of energy, metals, and nonmetallics, for the United States from 1950 to predictions for the year 2000. In 1950, the deficits were zero to $1 billion, respectively. The predicted deficiencies in the year 2000, are $22, $36, and $6 billion, respectively. This is a total of $64 billion in the year 2000. These deficits were estimated in 1972, and the situation has severely worsened since then.

Only a few years ago, the value of metals in the United States at the adit or shaft, petroleum and gas at the well, coal at the mine, and nonmetallics at the crusher or bin had a total primary value of about $20 billion but provided the raw materials for manufacturing products that gave rise to a Gross National Product of more than $1000 billion. Now those minerals excluding fuel resources cost $40 billion and may exceed $100 billion in less than 10 years with much less change in the GNP.

Finally, environmental requirements and the creation of Wilderness and Primitive areas are limiting

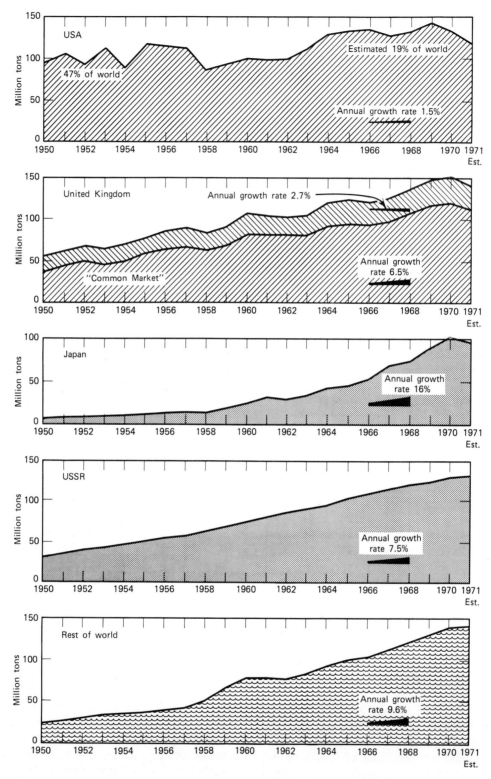

Figure 1-4 World Steel Production (From: Mining Congress Jour. July, 1972)

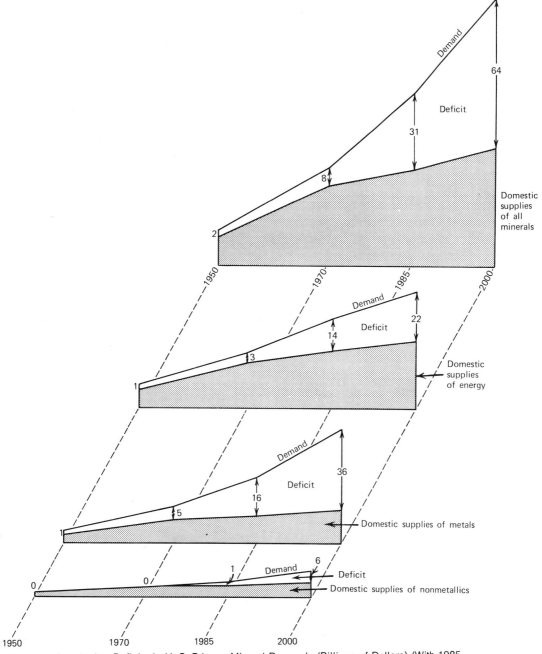

Figure 1-5 Developing Deficits in U. S. Primary Mineral Demands (Billions of Dollars) (With 1985 and 2000 at 1970 prices) (From Mining Congress Journal, July 1972).

the domestic discovery and development of potential mineral resources. When the Wilderness Act was passed in 1964, there were 12.6 million acres including an additional 3.8 million acres classed as "primitive." On top of this, 274 more areas were "being studied." Now two-thirds of the public domain is limited to mineral entry. Ohle states it succinctly, "Meanwhile we are forced to spend billions of U.S. dollars annually in importing what we could

produce domestically. Wilderness areas are fine and we all applaud and enjoy them; but when they become so numerous that they begin to unduly sap our national supply of natural resources, they are an extravagance we can ill afford."

On the other side of the coin, when 5 percent of the world's population consumes 30 percent of the world's mineral resources, wilderness areas may be more precious than more Cadillacs.

SCIENCE AND TECHNOLOGY—POSSIBLE ANSWERS

Dire predictions of doom are less exasperating than the lack of suggested actions that may resolve problems. What follows are some suggestions and hopeful thoughts pertaining to solutions expressed with the knowledge that simple solutions do not solve complex problems.

Energy

There will be no immediate relief from the energy gap. Figure 1-6 illustrates the situation that with conventional sources of energy such as oil and gas, the "crunch" will gradually increase. Coal may be the nearest immediate relief for some forms of energy. Transition sources such as coal and possibly ^{235}U nuclear fission reactors may peak before the twenty-first century. The increasing price of uranium, now approaching $60 per pound of U_3O_8, the apparent lack of extensive domestic reserves, and environmental protests of the use of nuclear energy may prevent nuclear energy from reaching the potential expected.

Only the "ultimate sources" of energy, such as geothermal and solar energy, will provide for the future beyond the twenty-first century. Oil shale resources in Colorado and Utah are admittedly massive (Table 1-1) even though the reserves approach zero. The most optimistic view of synthetic oil produced from kerogen in oil shale is about 1 million barrels per day. This was not expected until about 1980 by which time this would have provided less than 5 percent of the U.S. demand. Now the development of these resources has been curtailed or stopped, with little hope that development will be resumed during the present U.S. economic conditions of inflation and high interest rates. The Alaskan pipeline will provide about 1.2 million barrels per day or about 6 percent of the U.S. demand. With additional 48 inch-diameter pipe lines, production could obviously be more, but the Alaskan reserves are estimated at about 20 billion barrels, less than three times our present total annual consumption. There are hopes, of course, that larger tankers (Fig. 1-7) will haul more oil at less cost from foreign ports but such ships lack deep port channels, and there is presently a surplus of smaller oil freighters. Because of these and other factors, tankers of the *Nisseki Maru* and *Globtik Tokyo* type are not being built. Only three ports in the United States, all on the West coast, can handle up to

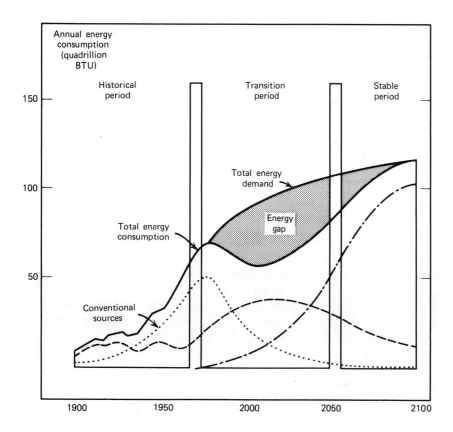

Figure 1-6 A possible scenario for U.S. energy: poorly managed transition to ultimate energy sources. The evolution of the U.S. energy system will likely consist of three phases: a growth period with major dependence on conventional sources consisting of coal, domestic oil, gas, and hydropower; a transition period, with major shifts to dependence on oil and gas imports, nuclear power, and coal; and a period of stable demand, with primary dependence on ultimate sources such as solar power, fusion, bioconversion, and geothermal energy. Without careful planning of transition sources, the U.S. could suffer a serious energy gap during the transition period. (From: Technology Review, Ed. by MIT Press, p. 20, Oct-Nov, 1975).

Table 1-1 Shale-Oil Resources of the United States, in Billions of Barrels, By Grade (Oil Yield) of Oil Shale

[All resource figures except those for the Green River Formation are adapted from Duncan and Swanson, 1965, table 2; Ne, not estimated]

Deposit	Identified[a] 25-100 gal/ton	Identified[a] 10-25 gal/ton	Hypothetical[b] 25-100 gal/ton	Hypothetical[b] 10-25 gal/ton	Speculative[c] 25-100 gal/ton	Speculative[c] 10-25 gal/ton
Green River Formation, Colorado, Utah, and Wyoming	418[d]	1,400	50	600	—	—
Chattanooga Shale and equivalent formations, Central and Eastern United States	—	200	—	800	—	—
Marine shale, Alaska	Small	Small	250	200	—	—
Other shale deposits	—	Small	Ne	Ne	600	23,000
Total	418[d]	1,600	300	1,600	600	23,000

Source: U. S. Geol. Survey Prof. Paper 820, p. 500, 1973.

[a] Identified resources: Specific, identified mineral deposits that may or may not be evaluated as to extent and grade, and whose contained minerals may or may not be profitably recoverable with existing technology and economic conditions.

[b] Hypothetical resources: Undiscovered mineral deposits, whether of recoverable or subeconomic grade, that are geologically predictable as existing in known districts.

[c] Speculative resources: Undiscovered mineral deposits, whether of recoverable or subeconomic grade, that may exist in unknown districts or in unrecognized or unconventional form.

[d] The 25-100 gal/ton category is considered virtually equivalent to the category "average of 30 or more gallons per ton."

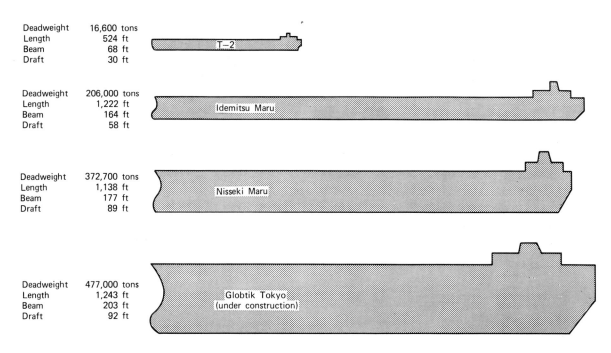

Deadweight	16,600 tons
Length	524 ft
Beam	68 ft
Draft	30 ft

T-2

Deadweight	206,000 tons
Length	1,222 ft
Beam	164 ft
Draft	58 ft

Idemitsu Maru

Deadweight	372,700 tons
Length	1,138 ft
Beam	177 ft
Draft	89 ft

Nisseki Maru

Deadweight	477,000 tons
Length	1,243 ft
Beam	203 ft
Draft	92 ft

Globtik Tokyo (under construction)

Figure 1-7 Comparative sizes of yesterday's, today's, and possibly tomorrow's tankers. Fewer larger ships means fewer collisions—but a supership collision could result in a catastrophic oil spill. (From: Marcus, H. S., Tech. Rev., March/April, 1973).

100,000 dead weight ton (dwt) tankers, namely, (1) Puget Sound, (2) San Francisco, and (3) Long Beach.

Certainly, *nuclear reactors* are increasing in numbers and megawatt output despite strong opposition. At the end of June, 1977 (Fig. 1-8), 68 *fission reactors* were operating within the United States, 94 are under construction, and plans exist for 60 more; that is a total of 222 reactors with a capability of 219 million kilowatts. There remains, however, an increasing concern that there is insufficient fissionable material to provide the fuel for any more reactors. An average nuclear fissionable power plant, producing a thousand megawatts of electricity will consume 3 kilograms (6.6 pounds) of ^{235}U each day. Figure 1-9 illustrates the large drop in uranium production from 1958 until 1966 with only a gradual increase since then, even though uranium has increased almost three-to fourfold in price.

The large nuclear reactors presently being built for electric power generators are known as *"thermal"* reactors because the fission of ^{235}U nuclei within them is caused by neutrons that are moderated in energy so as to be nearly in thermal equilibrium with their environment. These conventional fission reactors, which are increasing in number, use light water (LWR) for thermalizing neutrons and removing the heat energy released by fission. A similar reactor that is now in operation under the CANDU project in Canada uses ^{232}Th that is converted to ^{233}U, which is fissionable. Heavy water (D_2O) is used for thermalizing neutrons because it is a more efficient moderator that allows the use of less enriched fuel. The advantage of this HWR is that little ^{239}Pu is produced.

An attractive alternative to the conventional (LWR) fission or thermal reactor is the *breeder reactor,* which produces more fuel than it consumes. Of course, the LWR does breed some fissionable ^{239}Pu because even enriched ^{235}U fuel for the LWR contains about 97 percent of nonfissionable ^{238}U in the initial fuel rods, and in some reactors, a surrounding "blanket" of ^{238}U or ^{232}Th. These isotopes are transformed partially into fissile ^{239}Pu or ^{233}U respectively; ^{239}Pu is readily removed chemically from spent or residual ^{238}U. Most of the ^{239}Pu produced so far has been used in the manufacture of A-bombs. As little as 12 pounds of ^{239}Pu plus some engineering ability has been cited to be sufficient to produce a "small" A-bomb.

Because of the possibility that ^{239}Pu will be used in the construction of A-bombs, either by responsible governments, or irresponsible governments, or terrorists, the United States is attempting to curtail the domestic production and recovery of ^{239}Pu with the altruistic hope that other governments will do likewise. There seems little hope that they will do so, and it is fully expected that President Carter may have to rescind his order, especially as the limited reserves of uranium are increasingly being depleted by the continuing construction of many more (LW) reactors.

Much more ^{239}Pu is produced by the liquid-metal-cooled fast breeder reactor (LMFBR) that produces a shower of neutrons sufficient to convert a nonfissionable isotope of uranium (^{238}U) into a fissionable isotope of ^{239}Pu, as well as causing the fission that ultimately generates electricity. The LMFBR has been developed to the point of operation in the United States England, and the USSR, and efforts around the world have been increasingly focusing on this important concept. Its development in the United States, however, may almost cease because of President Carter's desire to avoid the production of ^{239}Pu.

More advanced breeder concepts include the gas-cooled fast breeder (GCFBR), which was under development (now ceased) by the Gulf Corporation, the light-water breeder (LWBR), and the molten salt breeder (MSBR) only one of which was developed at Oak Ridge. These systems are all in an early research phase, and prior to President Carter's desire to discontinue production of ^{239}Pu, they offered little likelihood of having any impact on the LMFBR program.

Uranium reserves in the United States are about 920,000 tons of U_3O_8, or about 13 million pounds of fissionable $^{235}U_3O_8$ at a price of $50 per ton. Identified uranium resources of the world were estimated in 1971 by the U.S. Geological Survey to be about 1.6 million tons of U_3O_8 (Table 1-2). In 1973, according to the Organization for Economic Cooperation and Development (OECD), estimated free-world reserves of 2.3 million tons of U_3O_8 were available at a price of $8 to $18 per pound. Actually, even more ore was needed because reactors were only about 50 percent as efficient as expected and required, therefore, more fuel per reactor. At prices of U_3O_8 given below marketable reserves calculated by DOE in 1980 are as follows:

$/lb U_3O_8	Tons U_3O_8
$ 30	645,000
$ 50	936,000
$100	1,120,000

It is estimated that at least 2.2 million tons of U_3O_8

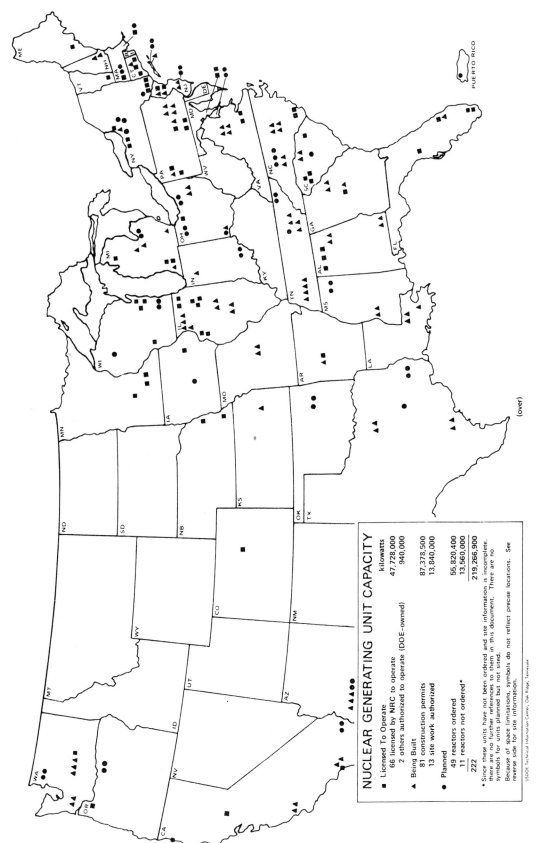

NUCLEAR GENERATING UNIT CAPACITY

		kilowatts
■	Licensed To Operate	
	66 licensed by NRC to operate	47,728,000
	2 others authorized to operate (DOE-owned)	940,000
▲	Being Built	
	81 construction permits	87,378,500
	13 site work authorized	13,840,000
●	Planned	
	49 reactors ordered	55,820,400
	11 reactors not ordered*	13,560,000
	222	219,266,900

* Since these units have not been ordered and site information is incomplete, there are no further references to them in this document. There are no symbols for units planned but not sited.

Because of space limitations, symbols do not reflect precise locations. See reverse side for site information.

USDOE Technical Information Center, Oak Ridge, Tennessee

(over)

Figure 1-8 Nuclear Power Reactors in the United States. (From: Energy Research and Development Administration, June 30, 1976).

13

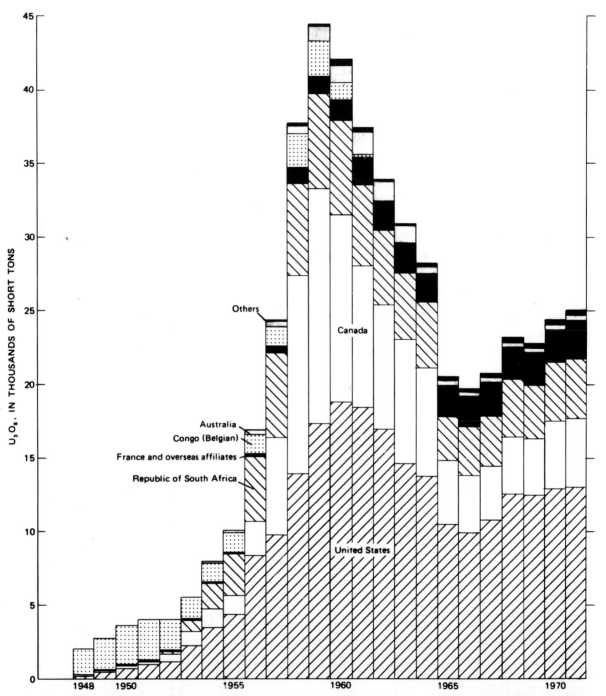

Figure 1-9 Uranium production, 1948–71, from the United States, Canada, Republic of South Africa, France and African affiliates, Belgian Congo (now Zaire), Australia, and other countries including Argentina and Sweden in most years, and Finland, West Germany, Portugal, and Spain in some years after 1956. Data from U.S. Atomic Energy Commission (1959), European Nuclear Energy Agency and International Atomic Energy Agency (1969, table 2), U.S. Bureau of Mines (1949–72), and Williams (1971). (From: U.S. Geol Survey Prof. Paper 820, p. 463, 1973).

14

Table 1-2 Identified Uranium Resources in the United States and Other Countries That Have Reported Resources

[Identified resources are specific, identified mineral deposits that may or may not be evaluated as to extent and grade, and whose contained minerals may or may not be profitably recoverable with existing technology and economic conditions]

Locality	Age of Deposit	Grade (percent U_3O_8)	Short Tons U_3O_8	Source of Data
United States: western States	Peneconcordant in sandstone, vein.	0.22	273,000	U.S. Atomic Energy Comm. (1972b).
Canada:				
Blind River-Elliot Lake, Wollaston Lake, Bancroft, Beaverlodge.	Precambrian conglomerate, veins, and other types.	>.10	230,000	Sherman (1972).
Elliot Lake	Precambrian conglomerate.	<.10	130,000	Sherman (1972).
Mexico: Chihuahua (Aldama) and other States	Peneconcordant in sandstone, vein.	>.10	2,380	Sherman (1970).
Argentina: Salta, Mendoza	Sandstone.	>.10	9,000	European Nuclear Energy Agency and International Atomic Energy Agency (1969).
France: Vendée, Limousin, Forez, Morvan	do	>.10	66,400	Sherman (1972).
Portugal: Viseu (Urgeirica), Nisa, Guarda	do	>.10	9,500	European Nuclear Energy Agency and International Atomic Energy Agency (1969).
Spain: Ciudad Rodrigo (Salamanca), Andujar	do	>.10	11,000	European Nuclear Energy Agency and International Atomic Energy Agency (1969).
Sweden: Västergötland (Billingen), Närke	Marine black shale.	.03	350,000	Sherman (1972).
Other Europe: West Germany, Italy, Turkey, Yugoslavia	Vein, peneconcordant in sandstone.	>.10	6,500	European Nuclear Energy Agency and International Atomic Energy Agency (1969).
Central Africa Republic: Bakouma	Phosphatic-filling karst in Precambrian dolomite.	~.10	19,500	Sherman (1972).
Gabon: Mounana (Franceville), Mikouloungou	Structural controlled dissemination in Precambrian sandstone.	>.10	19,500	Sherman (1972).
Niger: Arlit	Peneconcordant in sandstone.	.29	26,000	Sherman (1972).
Azelick, Madaouela	do	>.10	13,100	Sherman (1969).
South Africa, South West Africa: Witwatersrand, Rössing	Precambrian conglomerate, igneous rock.	<.10	300,000	Sherman (1972).
Australia: Mary Kathleen, Alligator Rivers area (Nabarlek, Jim Jim, Ranger), Frome Lake.	Vein and other types.	>.10	300,000	Skillings Mq, Rev.
India: Jaduguda, Bihar	Vein.	.06-.07	68,000	IAEA Vienna (1977).
Japan: Ningyo-Toge	Peneconcordant(?) in Miocene and Pliocene conconglomerate.	.06	4,000	European Nuclear Energy Agency (1965).
Total			1,637,880	

Source: U. S. Geol. Survey Prof. Paper 820, 463, 1973.

will be needed to last until the year 2000. At the present time, nuclear reactors are providing about 12 percent of the energy requirement of the United States. Some European countries have sufficient reactors to provide more than 25 percent of their energy requirements.

If, however, the costs of production and the price of U_3O_8 allows lower-grade ore to be mined, the reserves will increase further. Of course, estimates of lower-grade material are less precise than the higher-grade reserves. It is reasonable to expect that all of these reserve estimates are conservative. The OECD study reports that known worldwide reserves of high-grade ore increased by one-third between 1970 and 1973. And lower-grade reserves could increase by much higher factors.

With the higher price of U_3O_8, exploration is increasing again after the 1969 peak. Drill hole exploration in the search for deeper deposits is increasing rapidly (Fig. 1-10) with some deposits being found at depths exceeding 2000 feet, especially in the Ambrosia Lake area of New Mexico.

Geothermal power is limited but its potential is so good that it might be considered to be an ultimate source. There are few hazards or environmental problems with geothermal power. The noise "pollution" is controllable with mufflers, but the reinjection of condensed steam back into the ground or forcing even cold water into a subterraineous heat source may trigger earthquakes in seismic areas. It is preferred if not necessary that the steam be dry; when wet, it is capable of carrying dissolved corrosive salts. Natural steam reservoirs are not abundant. As an ultimate source, hot dry rock resources will have to be developed.

The rate of growth of geothermal sources is developing rapidly. A list of those in operation or at an advanced stage of development is shown in Table 1-3. The number in the United States is not impressive, especially with only one operating, namely, the Geysers, in California. This facility, however, is the largest in the world with plans to increase its capacity by 50 percent from 600 megawatts to 900 megawatts.

Solar Energy

This, of course, is the ultimate source of energy, which incidentally was also required for the development of all hydrocarbon resources. Without light, no animal or plant life would have existed. Photosynthesis uses less than 1 percent of solar energy reaching the earth with about 30 percent used to heat the land and seas. The remainder is reflected back into the atmosphere. There is an abundance, therefore, yet available to man.

Solar energy is being used today in experimental buildings and especially in spacecraft where solar panels are used to provide electricity. The present cost, however, of solar panels exceeds other sources of energy. Still, research is being done to develop cheaper sensors or even construct home residents with roofs containing heat absorbing panels or circulating tubes filled with an antifreeze solution and oriented to absorb more solar radiation. This source of energy creates few environmental problems and will be available for the life of the sun, roughly another 10 billion years or so. Lower-cost solar panels may become increasingly attractive in the future when conventional sources of energy become increasingly expensive.

Not all environments are suitable for solar energy. Desert regions of comparatively little cloud cover are certainly suitable. Temperate latitudes are well situated; equatorial zones may have too much cloud cover; and near polar regions are excluded because of the low sun angle. It may be that solar energy developed in space or remote areas could be converted to microwave energy and beamed, or reflected, from directional antennas located in space to receiving antennas located in areas lacking solar energy. This possibility is highly speculative and the energy loss at the antennas would be considerable.

Energy Summary

Conventional energy sources including oil shale are obviously becoming inadequate for the demands now existing. Coal and nuclear reactors can provide increased energy needs but still not enough. Ultimate sources such as geothermal and solar energy probably will not be sufficient until well into the twenty-first century (Fig. 1-6).

MINERAL EXPLORATION

The development of mineral deposits is dependent upon sound geological, geochemical, geophysical, and remote sensing exploration techniques. Political and economic factors and labor conditions must also be considered. Innovations in extractive processes, both at the mine and the refining sites, to

DRILLING, RESERVES, AND PRODUCTION IN ORE COMPARED TO VARIOUS HISTORICAL FACTORS

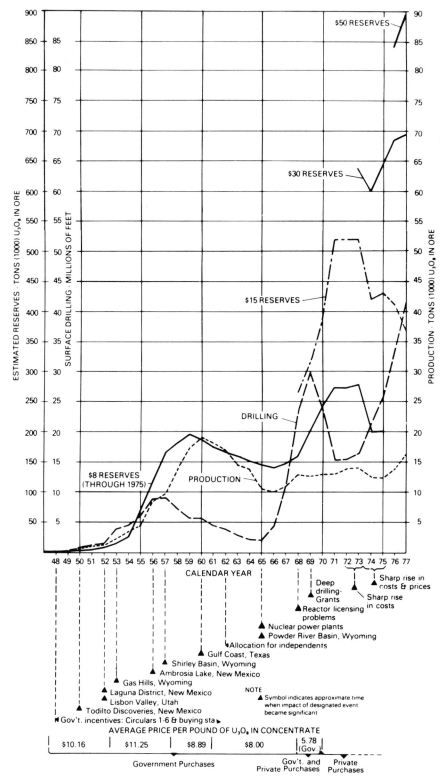

Figure 1-10 History of Uranium Exploration, 1948–1978.

Table 1-3 Geothermal Power Installations in Operation or at an Advanced Stage of Development

	Geological Situation	Average (max) Drillhole Depth (m)	Average (max) Temperature (°C)	Discharge Type (S = steam; W = water)	Total Dissolved Solids in Water (g/kg)	Total Generating Capacity (MW) Installed	Total Generating Capacity (MW) Planned Addition
Chile							
El Tatio	Quaternary and Tertiary rhyolite, andesite; Mesozoic sediments	650 (900)	230 (260)	S + W	15	—	15
El Salvador							
Ahuachapan	Quaternary andesite	1000 (1400)	230 (250)	S + W	20	30	50
Iceland							
Namafjall	Quaternary basalt	1000 (1400)	250 (280)	S + W	1.0	2.5	—
Italy							
Larderello region	Triassic-Jurassic sediments	600 (1600)	200 (260)	S	—	406	—
Mount Amiata	Triassic-Jurassic sediments; Quaternary volcanics	750 (1500)	170 (190)	S(+W)	—	25	—
Japan							
Matsukawa, N. Honshu	Quaternary andesites; Miocene sandstones	1000 (1500)	220 (270)	S	—	20	—
Otake, Kyushu	Quaternary andesites	500 (1500)	230 (250)	S + W	2.5	11	—
N. Hachimantai	Quaternary andesites, dacites	800 (1700)	— (>200)	S + W	—	10	—
Hatchobaru, Kyushu	Quaternary andesites	1000 —	250 (300)	S + W	5.5	—	50
Onikobe, Honshu	Quaternary andesites, dacites, granite	300 (1350)	— (288)	S(shallow) + W (deep)	1.5	—	20
Mexico							
Cerro Prieto	Sandstone, shales, granite	800 (2600)	300 (370)	S + W	17	75	75
New Zealand							
Wairakei	Quaternary rhyolite, andesite	800 (2300)	230 (260)	S + W	4.5	192	—
Kawerau	Quaternary rhyolite, andesite	800 (1100)	250 (285)	S + W	3.5	10	—
Broadlands	Quaternary rhyolite, andesite	1100 (2420)	255 (300)	S + W	4	—	100
Philippines							
Tiwi, S. Luzon	Quaternary andesites	920 (2300)	—	S + W	—	—	10.5
Turkey							
Kizildere	Pliocene-Miocene sandstones limestones; Palaeozoic schists	700 (1000)	190 (220)	S + W	5	—	10
USA							
The Geysers, Cal.	Jurassic-Cretaceous graywackes and shales, basalt	1500 (2900)	250 (285)	S	—	600	300
Roosevelt, Utah	Tertiary breccia	250 (400)	260	S + W	6	—	240
USSR							
Pauzhetsk	Quaternary andesite, dacite, rhyolite	— (800)	185 (200)	S + W	3	5	7

remove metals or materials from rock, such as leaching and bacterial methods, will allow the development of lower-grade deeper deposits.

Hidden Mineral Deposits

Those areas with post-ore covers have been prime areas for exploration for some time. The discoveries of the Pima, Mission, and Twin Butte deposits in the pediment of the Sierrita Mountains, Arizona, are prime examples. Glacial moraine covers have hidden many deposits and post-ore volcanic covers are common. Perspicacious geoscientists are needed in the search for such "blind" deposits.

Geologic studies indicate hopefully the best-mineralized areas using advances in hydrothermal alteration studies, mineral zonation understanding, mapping, and other geotechniques. Exploration geochemistry may aid in detecting mineral anomalies at the surface, such as mercury halos, for example, and geophysical "tools" are increasingly effective in testing potentially promising anomolous areas, especially using the induced polarization method for the location of disseminated sulfide deposits. Improved drilling techniques, which even now penetrate many times the depth of most geophysical determinations, and innovative mining methods will be needed. The primary responsibility, however, is that of the exploration geologist in this age of the search for hidden ore bodies.

Dramatic evidence of the significance of strictly geological prospecting in mineral discovery is shown by Canadian exploration results. In the decade 1946 to 1956,

77 mines were brought into production:

53 (69 percent) were found by conventional prospecting

17 (22 percent) were found by geological methods

7 (9 percent) were found by geophysical prospecting

In the decade 1955 to 1965,

175 mines were brought into production:

87 (50 percent) were found by conventional prospecting

49 (28 percent) were found by geological methods

28 (16 percent) were found by geophysical methods

11 (6 percent) were found by other methods

Geochemical prospecting was a supporting method during this decade.

Figure 1-11 indicates the areas of potential exploration in the northern portion of the Basin and Range province. All known ore deposits in this area are located in bedrock, consisting predominantely of sediments and intrusive bodies. Few exist in preore volcanics; exceptions are Goldfield, Tonopah, and the East Tintic district. All of the alluvial covered areas and most of the volcanics, more than 60 percent of the area of Fig. 1-11, cover surface evidence of mineralization. Of course, the deeper valleys presently prohibit exploration to their depths, but pediments and shallow volcanic covered areas are more numerous than thought. An example of the development of pediments in the San Francisco mining district, Millard County, Utah, is shown in Fig. 1-12. The areas of shallow alluvial cover have not yet been adequately explored.

Eugeosynclinal rocks of western Nevada have been thrust over the more favorable carbonate host rocks of central Nevada by the Roberts Moutain overthrust (Fig. 1-13). Where intrusive rocks or other sources of mineral concentration appear in windows in this thrust, mineralization is not uncommon. The Carlin gold mine is such an example. Determining the depth to the thrust is a major problem, but small mineralized veinlets that have penetrated the unfavorable silicic and volcanic allochthonous rocks are suggestive of more concentrated mineralization in the autochthonous carbonate rocks below the thrust fault.

Remote Sensing

This subject is generally limited to airborne and space craft sensors and images acquired from low level to space craft altitudes. Surface geochemical and geophysical techniques attempting to determine geologic features hundreds and sometimes thousands of meters below the earth's surface should also be included in the subject of remote sensing, but such is not generally accepted.

Comparatively low level air photos are commonly used for geological mapping and are used for field and base maps. These photos or images, however, do not show major geostructural features at their scale in contrast to space craft images that exhibit major structural features, and because of their altitude, have very little distortion. One land-

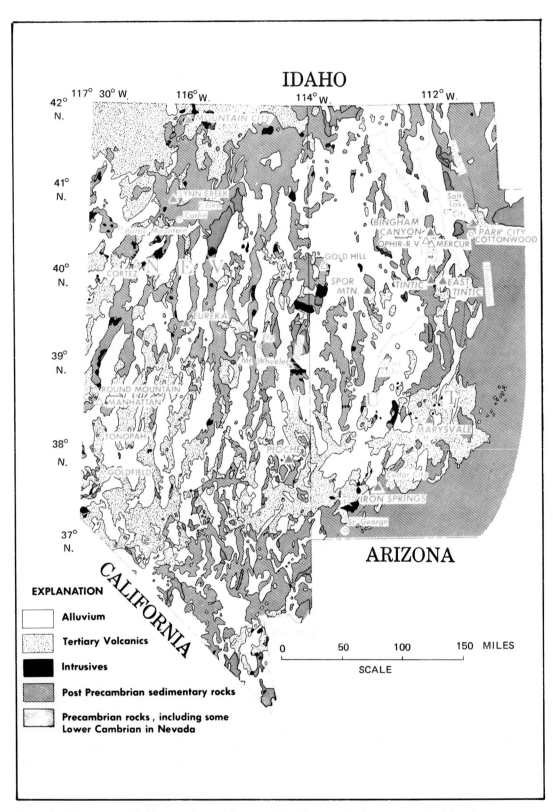

Figure 1-11 Generalized Geology: western Utah and eastern Nevada showing post-ore alluvial and volcanic cover. (From: Hewitt, Ore Deposits of the U.S., Ed. Ridge., 1967).

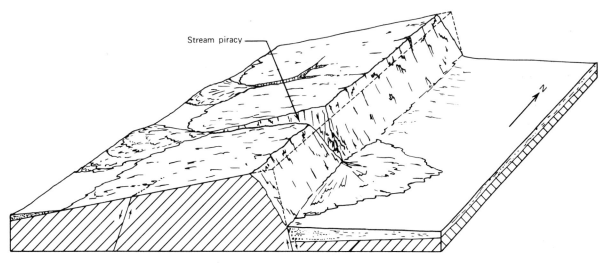

Stream piracy

N

Past (mid—tertiary) physiography.

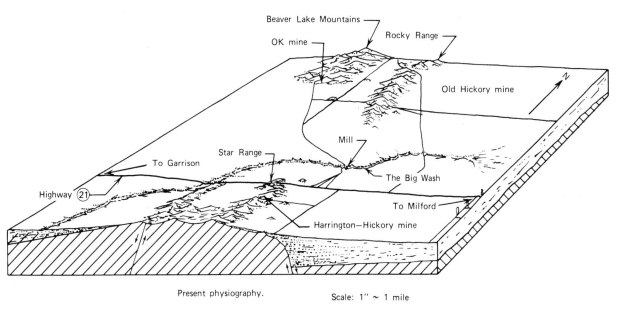

Beaver Lake Mountains

OK mine

Rocky Range

Old Hickory mine

N

To Garrison

Star Range

Mill

The Big Wash

Highway ㉑

To Milford

Harrington—Hickory mine

Present physiography.   Scale: 1″ ~ 1 mile

Figure 1-12 Pediment formation in the San Francisco mining district, Utah.

sat or skylab image, for example, covers an area of more than 180 × 180 kilometers.

Figure 1-14 is an oblique image, obtained by astronauts in the skylab satellite, of the Bonneville flats and the Great Salt Lake. The San Rafael swell in the upper portion of the image is obvious as is Utah Lake and snow on the Wasatch Range and the western portion of the Uintah Mountains. The contrasting reflection of the Great Salt Lake is the result of different bacteria on each side of the Southern Pacific railroad earthen-fill causeway across the lake. Eighty percent of the freshwater streams that enter the lake flow into the southern portion of the lake raising its elevation about one foot higher than the northern portion of the lake. Two small culverts prevent the more rapid flow of water into the northern lake, which is near a supersaturation content of salt. Obviously, brine evaporation ponds are located better around the banks of the northern lake.

Other space imagery that is of aid in determining water or brine content is shown in Fig. 1-15. This

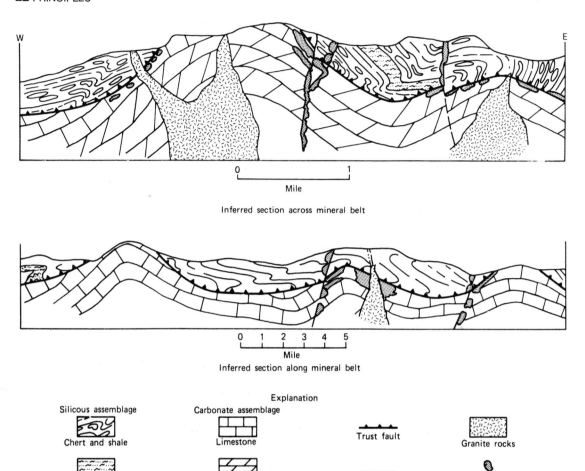

W   E

0   1
Mile
Inferred section across mineral belt

0 1 2 3 4 5
Mile
Inferred section along mineral belt

Explanation

Silicous assemblage   Carbonate assemblage

Chert and shale   Limestone   Trust fault   Granite rocks

Silt   Dolomite   Steep fault   Ore deposit

Figure 1-13 Inferred sections across and along a mineral belt in north-central Nevada showing possible sites for ore deposition. (After R. Roberts, 1968).

is an infrared, RB-57 (reconnaissance, fixed-wing, high-altitude aircraft) image of the Kaiser Industries plant near Wendover, Utah. The airfield, Interstate 80 Highway, the railroad line, and Lake Bonneville shore lines are quite obvious. Only on the near infrared image (Fig. 1-15), is the brine solution evident.

Landsat (formerly ERTS, Earth Resources Technical Satellite) images have the advantage of repetitive coverage of the same area of the earth every 18 days. Figure 1-16 is a landsat 1 image of north-central Nevada with the town of Battle Mountain located in the upper central portion of the image. Southwest of Battle Mountain is the Antler Peak Range in which two of Duval's porphyry copper deposits are located. This image was acquired on August 9, 1973, when the sun angle declination was 56°. Compare this figure with Fig. 1-17 taken on December 14, 1972, when the sun angle declination

was 22°. Structural trends, fractures, or linaments are much more evident in Fig. 1-17. Those fracture trends that are mineralized in bedrock could be traced into pediment and alluvial covered areas with ease by this image, especially when enlarged about 4X, with the potential of determining the sites of hidden mineralization.

Figure 1-18 is an image of the Ruth porphyry mining district and environs acquired during a U2 flight. The image can be enlarged many times with little loss of detail. Incidentally, this district is often used as an example of an elongate E-W trend, but the Permian Arcturus Formation exhibits, as is evident, an abrupt swing to a northerly trend. Not only the strike but the dip of the beds can be readily ascertained on this U2 image.

For further comparative purposes, Fig. 1-19 is a Skylab S192B image of the Bingham porphyry copper deposit. Indications of hydrothermal alteration

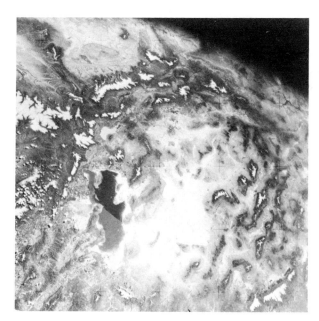

Figure 1-14 Oblique image of Bonneville Salt Flats, Great Salt Lake, and portions of Utah, Idaho, Nevada, and Arizona from Skylab Mission SL2.

and gross structural features can be discerned on this image. The digital tape, used to produce Landsat images, and computer aids can provide images of structural features, hydrothermal alteration, extraordinary detail of rock types, and other unexpected geological features. Such studies have been applied to the Goldfield district, Nevada.

Remote sensing is no panacea for mineral exploration. It is just one more "tool" of which the exploration geologist should be aware and knowledgeable.

Geophysical and Geochemical Methods

Geophysical and geochemical instrumentation and techniques are indispensable in the search for specific types of mineral deposits. Even minor summaries of both techniques are beyond the scope of this text. The geologist should, however, understand the uses and limitations of these exploration aids.

An example of a geochemical prospecting tool is the measurment of mercury escaping to the surface of the earth in soil-gas that almost invariably contains mercury in minute quantities. The mercury soil-gas method is a valuable exploration method for the following reasons:

1. Mercury is associated as a trace element with the majority of mineral deposits.
2. Mercury's high vapor pressure allows it to continually diffuse from mineralized zones, even from considerable depths.
3. Mercury is readily collected in soil-gas by a suction apparatus, where it is deposited as an amalgam on silver or gold screens.
4. The mercury present on the screens can be detected in concentrations of less than one part per billion by the precision AA (Atomic Absorption) method.

An example of this technique is shown in Fig. 1-20. The large circles of >150 ppb Hg follow the southward extension of the Comstock Lode even into the alluvial cover. The smaller circles, of <150 to >100 ppb and <100 to >50 ppb are located farther from the vein system.

One of the prime geophysical techniques used to detect massive, and especially disseminated, sulfide deposits is the IP (Induced Polarization) technique. One way to measure IP is to compare ground resistivity as determined by direct current with the apparent impedence determined by alternating current having a frequency of between 0.1 to 0.5 cycles per second. Any difference in the two measurements suggests the presence of conductors, such as sulfide minerals, some of which may be detected almost as

Figure 1-15 Infrared image of Kaiser Industries brine ponds near Wendover, Utah. RB-57 aircraft image from 58,000 feet altitude.

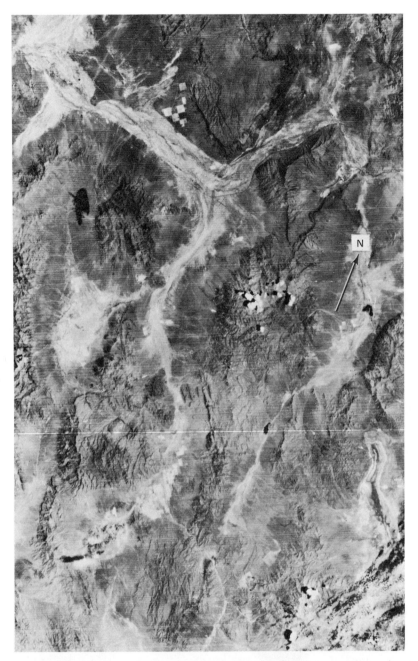

Figure 1-16 Summer scene, Aug. 5, 1973, of ERTS E-1378-17591-7 with sun declination angle of 56 degrees. Compare to Fig. 14. Area is northern Lander County, Nevada.

deep as 1000 feet if a broad electrode spread is employed. Some clay minerals, magnetite, and carbonaceous or graphitic formations yield similar effects, and the technique is not usually able to determine the sulfide species that are present, leading some to suggest that IP means "Indicates Pyrite." This, however, is one of the prime sulfide explo-

ration tools, and possibly the only geophysical technique used for detecting disseminated sulfides that occur in porphyry copper deposits.

Geological methods

Geology is the basic foundation of mineral exploration. Reconnaissance to detailed geologic mapping is indispensable. The identification of rock units, structural and stratigraphic relationships, and correct recognition of specific types of hydrothermal alteration, a primary exploration aid that is covered in a later section of this book, lead to recommendations as to which other specific geotools should be used and when to test the geologist's hypotheses by drilling.

An example of various exploration steps and techniques is provided by the Safford property in the Lone Star district of southeastern Arizona. This area had undergone extensive study over many years that finally culminated in the discovery of a deep ore body by the Kennecott Cooper Corporation. Geologic mapping on air photos at a scale of 1 inch = 0.5 miles was initiated in 1955. During the

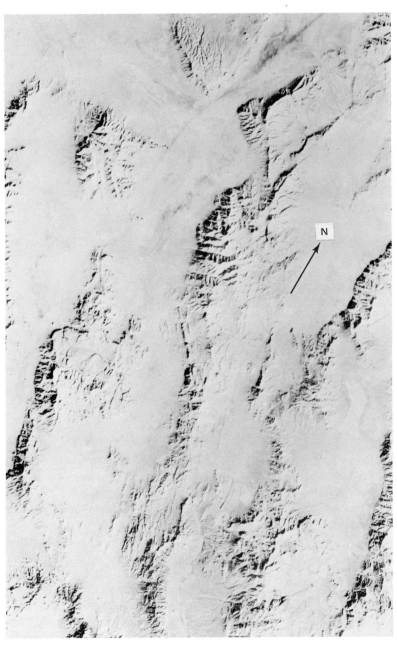

Figure 1-17 Winter snow covered scene of area shown in Fig. 1-15, Dec. 14, 1972, of ERTS E1144-18001-7 image with sun declination angle of 22 Degrees.

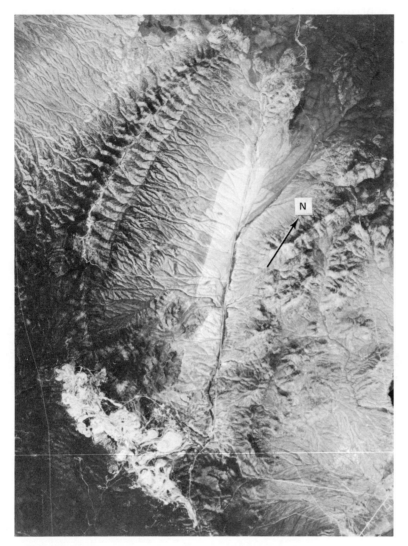

Figure 1-18 Enlarged U2 photograph of the Ruth Porphyry copper deposits and the structural trend towards the north of Paleozoic limestones to where the Tertiary dacite volcanics cover the Paleozoic sediments.

following year, a geochemical survey was made that consisted of 476 rock-chip samples collected on a 100 foot grid over an area of 3000 by 3000 feet that covered a pyritized area adjoining the buried ore body. During the following year, soil samples were collected over the entire Lone Star district at 200- or 400 feet intervals on traverse lines 1200 feet apart. Both semiquantitative X-ray fluorescent anlayses and colorometric tests of total readily soluble heavy metals were performed (Fig. 1-21).

Numerous geophysical techniques were used, but only the induced polarization and self potential geophysical results were evaluated as useful. Both techniques provided significant anomalies over the pyritized near surface area as shown by Figs. 1-22 and 1-23. The I.P. anomoly map indicates the true areal distribution of significant quantities of sulfides more conclusively than does the S.P. data and yields better indications of their minimum depths, which are about 700 feet in the area of unaltered volcanics. Diamond and rotary drilling was being done before and after the geophysical studies that led to the discovery of the deep ore body. By 1958, 76,000 feet of drilling had been completed. A development shaft was sunk, and cross-cutting and drifting were done on the 3900-foot level, 754 feet below the collar. Underground geologic mapping was completed in detail on a scale of 10 feet to the inch. Samples of 5-foot channel cuts were collected underground for assay. The deposit is presently

Figure 1-19 Bingham Porphyry copper pit by Skylab photo. S190B image.

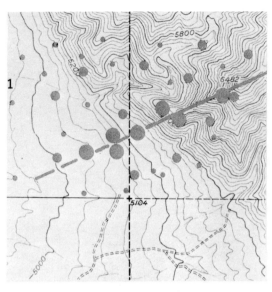

Figure 1-20 The tracing of narrow mineralized structures form exposed outcrops on to and across alluvial covered pediments based on Hg soil-gas analyses. Structure is southern end of Comstock Lode, Nevada.

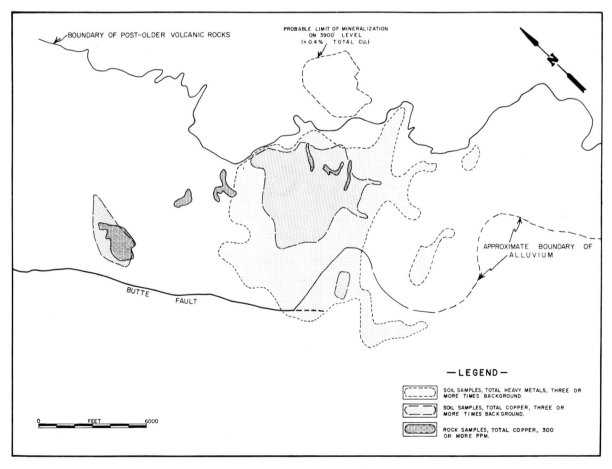

Figure 1-21 Map showing geochemical anomalies of Safford, Ariz. (From: Geology of the Porphyry Copper deposits, Ed. by Titley and Hicks, 1966).

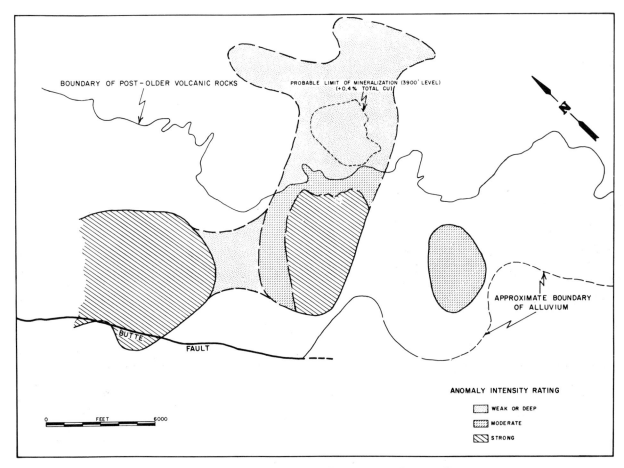

Figure 1-22 Map showing S.P. anomalies of Safford, Ariz. (From: Geology of the Porphyry Copper Deposits, Ed. by Titley and Hicks, 1966).

being considered for an in-place leaching of the composite oxide, composite oxide-sulfide, and sulfide ore.

The leaching of sulfide and oxide deposits, in situ or on surface dumps or leach pads is rapidly becoming a common metallurgical process. Various chemicals and processes, even bacteria, are used to increase the leaching rate (Fig. 1-24).

CONCLUSION

In order to improve the rate, volume, and value (Table 1-4) of exploration successes, including the prolonging of the availability of mineral resources, the following goals must be reemphasized:

1. Train and better educate perspicacious students as exploration geoscientists.

2. Foster new techniques for exploration, mining, and metallurgical processes.
3. Develop uses for abundant underutilized minerals.
4. Apply the principle of recycling more broadly.
5. Be able to develop lower-grade deposits (Fig. 1-25).
6. Develop and use the available resources more economically.
7. Reduce the demand for specific minerals through the substitution of other minerals, reduction of waste, or actual elimination of some uses.
8. If economically possible, import raw materials or refined minerals from available foreign sources.

None of these factor is especially new. They only need renewed emphasis in order to increase the successes (Table 1-4) of the past 30 years.

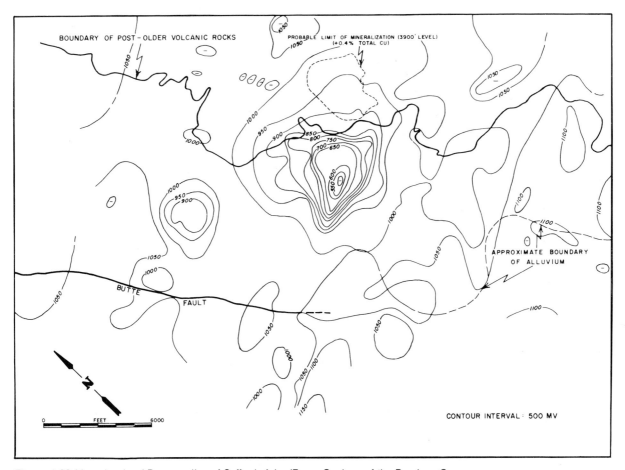

Figure 1-23 Map showing I.P. anomalies of Safford, Ariz. (From: Geology of the Porphyry Copper Deposits, Ed. by Titley and Hicks, 1966).

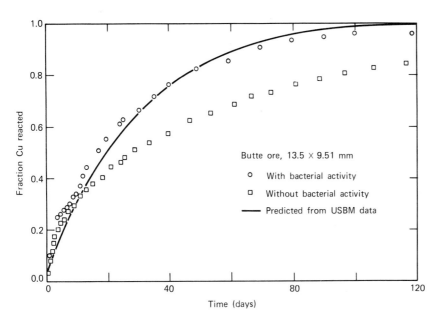

Figure 1-24 Difference in extraction of Cu from Butte ore with and without bacteria. (From: Averill and Wadsworth, Univ of Utah, 1976).

Butte ore, 13.5 × 9.51 mm

○   With bacterial activity

□   Without bacterial activity

—   Predicted from USBM data

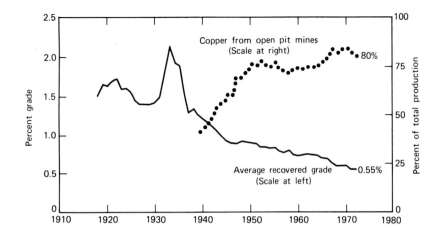

Figure 1-25 Recovered grade of U.S. copper ore and relative production from open pit mines. (From: Skillings Mg. Rev., Oct. 11, 1975).

Table 1-4 Major Mine Discoveries of the Last 20 Years with Gross Mine Production Value for 1973 (L. Miller)

1948	Knob Lake	Iron	$312,000,000
1949	Mexican Sulphur	Sulfur	140,000,000
1951	Pima	Porphyry copper	120,000,000
1952	Granduc	Volcanogenic	29,000,000
1953	Gas Hills-Shirley Basin	Uranium	193,000,000
1953	Mission	Porphyry copper	63,000,000
1953	Brunswick	Volcanogenic	53,000,000
1954	Heath Steele	Volcanogenic	12,500,000
1954	Cerro Bolivar	Iron	227,000,000
1954	Pine Point	Carbonate lead-zinc	143,000,000
1954	Gaspé	Porphyry copper	40,000,000
1955	Viburnum Trend	Carbonate lead-zinc	143,000,000
1955	Thompson	Nickel	150,000,000
1955	Ambrosia Lake	Uranium	105,000,000
1955	Bethlehem	Porphyry copper	45,620,000
1956	Palabora	Proterozoic	150,000,000
1956	Craigmont	Porphyry copper	30,104,000
1956	Mattagami Lake	Volcanogenic	64,000,000
1957	Mt. Whaleback	Iron	316,000,000
1959	Kidd Creek	Volcanogenic	280,314,000
1960	Weipa	Bauxite	103,500,000
1961	Lee Creek	Phosphate	20,000,000
1961	Lake Dufault	Volcanogenic	14,000,000
1962	North Florida	Phosphate	28,000,000
1962	Blackwater	Coal	74,000,000
1962	Mt. Tom Price	Iron	352,800,000
1963	Carlin	Gold	20,000,000
1964	Bougainville	Porphyry copper	307,530,000
1964	Twin Buttes	Porphyry copper	91,000,000
1965	Prieska	Volcanogenic	28,600,000
1965	Kambalda	Nickel	109,500,000
1965	Anvil	Proterozoic lead-zinc	104,000,000
1966	Gilbralter	Porphyry copper	83,000,000
1967	Mattabi	Volcanogenic	37,444,000
1967	Rustler Hills	Sulfur	60,000,000
1968	Lornex	Porphyry copper	70,000,000
1968	Island Copper	Porphyry copper	63,000,000

Source: L. Miller, Econ. Geol., v. 71, 1976.

SUMMARY

There is no geotechnique that is a panacea for mineral exploration. Each specialization of methods, techniques, or equipment is indispensable to mineral exploration. The geoscientist should, therefore, be knowledgeable about the applications, limitations, and ramifications of each tool, whether it be alteration studies, remote sensing, geophysics, geochemistry, and even the application or limitations of laboratory investigations.

Successful mineral exploration within the borders and offshore areas of the United States is more difficult today, even with the myriad of aids, knowledge, and technology available to the versatile geoscientist. The search for mineral deposits now focuses more on depth, rather than the lateral or area search. As a result, the professional geologist-geophysicist must "look" to greater depth and below postore coverings or at those targets for which there is no visual surface evidence.

Miller has compiled a list of Major Mine Discoveries made between 1948 to 1968 that indicates that "elephant ore deposits" are still being found (Table 1-4).

Selected References for Mineral Economics

The Mineral Position of the U.S. 1975–2000. 1972. E. Cameron, ed. Soc. Econ. Geol. Foundation, University of Wisconsin Press, 159 p. *Series of papers on various mineral economic topics by different expert authors.*

The transition to coal. 1975. R. F. Nail, et al. *Technology Review,* v. 78. MIT, p. 18–29.

Nuclear power, 1973–2000. 1972. U.S. Atomic Energy Comm., WASH-1139.

Energy Self-Sufficiency and an Economic Evaluation. 1974. MIT Policy Study Group, *Technology Review,* v. 76, no. 6.

General References 19. *Numerous chapters on the subject of mineral economics, mineral reserves, etc.*

Remote sensing of earth resources. 1972. F. Shahrokhi, ed. Univ. of Tenn. Space Inst., 183 p. *Numerous papers presented at conference.*

Minerals in World Affairs. 1972. A Sutulov. University of Utah Press, Salt Lake City, 200 p.

Affluence in Jeopardy. 1968. C. F. Park. Freeman, Cooper and Company, San Francisco, 368 p. *Mineral economics by an authority.*

Energy Resources. 1969. K. Hubbard. In *Resources and Man.* P. Clard, ed. W. H. Freeman and Company, Publishers.

2

BRIEF HISTORY OF THE USE OF MINERALS AND THE DEVELOPMENT OF ECONOMIC GEOLOGY

"The evolution of science is, in fact, not a progression from the simple to the complex, but quite the opposite. It is a progression from the complex to the simple."

M. K. Hubbert,
G.S.A. Presidential Address, 1963

ANCIENT TIMES

Economic geology probably began with tne ancient utilization of mineral products. Long ages must have passed, however, before the early crude knowledge became a craft, later to develop into a science. The early incentive for the acquisition of such knowledge was undoubtedly utilitarian, but eventually it was raised to an intellectual plane by the Greek philosophers.

The first earth materials used by primitive man were nonmetallic substances—flint, chert, quartz, obsidian, quartzite, soapstone, and limestone— sought for their use in weapons, implements, and utensils, and for carving. Clay was widely and extensively used, first for pottery and then for bricks. Unquestionably, clay represents the first large-scale mineral industry, an industry that has persisted continuously through the ages. Burned clay figures believed to be Aurignacian (30,000-20,000 B.C.) have been discovered in Moravia, and excellent Paleolithic pottery of the Solutrean period (10,000 B.C.) has been found in Egypt. Brick, tile, and clay tablets were extensively used by the Chaldeans, Babylonians, and early Egyptians for building their cities, for irrigation, and for writing materials. Early Asiatic and African dwellings were built with bricks made of clay. But, even earlier, building stones were extensively used. During the building of the pyramids this extractive industry must have been on a grand scale, as the Pyramid of Gizeh contains 2,300,000 blocks of stone averaging $2\frac{1}{2}$ tons apiece.

Paleolithic man, beginning 400,000 years ago, used numerous varieties of minerals: chalcedony, quartz, rock crystal, serpentine, obsidian, pyrite, jasper, steatite, amber, jadeite, calcite, amethyst, and fluorspar. Later ochers of mineral paints were

utilized. At about the time Neolothic man became acquainted with gold and copper, he also used nephrite, sillimanite, and turquois. These nonmetallic materials are mostly common substances that probably were found by accident; the search for them neither greatly stimulated human curiosity nor created specialized knowledge regarding their occurrence. They were accepted as found and utilized. Economic geology had not yet arisen; it was the predawn stage.

Egyptian, Greek, and Related Cultures

As the desire for gemstones and metals became more urgent, however, economic geology probably had its inception. Facts of occurrence were noted and recorded; crude theories of origin were evolved; expeditions were organized for the discovery and exploitation of deposits; and ownership and barter of these substances became an important part of the life of the people. The use of gemstones and the mining of them reached a high art among the early Egyptians, Babylonians, Assyrians, and Indians. Gemstones were greatly prized, and, living or dead, the Egyptian was bedecked with jewels, which attained important significance among a people obsessed with mysticism. In pre-Dynastic times (+ 3400 B.C.), it was the color rather than the substance that the Egyptian prized most. The Theban craftsmen created pleasing color schemes, utilizing the azure of the lapis lazuli, the red of the carnelian, the purple of the amethyst, the green of the malachite, the yellow of the jasper, and the blue of the turquois. They also used agate, beryl, chalcedony, and garnet and shaped and polished hard stones, producing not only ovoid but also faceted beads. All these stones except lapis have come from Egypt itself. Even in those remote times there must have been international barter, since the lapis was probably obtained from Afghanistan. about 4000 kilometers away.

The oldest form of mining was for gems and decorative stones, and for over 2000 years the Pharaohs dispatched expeditions including engineers and prospectors to the Sinai Peninsula for turquois, and into the Sudan. Ball identifies as the first economic geologist the Egyptian, Captain Haroeris, who about 2000 B.C. led an expedition to Sinai and after 3 months' prospecting discovered and extracted large quantities of turquois. The ancient Egyptians (from 1925 B.C.) sank hundreds of shafts for emeralds on the Egyptian coast of the Red Sea; certain workings are said to have been 250 meters deep and sufficiently large to permit 400 men at a time to work therein.

The Stone Age changed to the Copper Age when man began hammering tools out of native copper. This age long preceded the Bronze Age when Mesopotamians learned that instead of adding arsenic to copper to form the earliest bronze, adding a little tin to a lot of copper made a much harder metal, which we call bronze.

When the Mesopotamians produced enough bronze to equip their army with bronze swords, helmets, and armor, their armies conquered central Asia and the Middle East because the older copper weapons were no match for the new slashing swords and hard armor of the invaders, nor were the bronze swords any match for later iron swords. The Egyptians certainly had the ability to smelt copper and also produce bronze at early dates. Bronze, however, did not become common until about 2500 B. c. in Egypt, possibly because of the relative scarcity of copper in that land.

The Hoover's have compiled a chronology of the uses and mineral operations developed during human civilization as shown in Table 2-1.

Gold is presumed to have been used before copper, and copper is considered by some to have been discovered 18,000 years before Christ; certainly copper was known to the Egyptians in 12,000 B.C. and was widely used in Europe about 4000 B.C.

At the ancient mines of Cassandra, Greece, which Sagui estimates to have been mined from about 2500 to 356 B.C., the skillful extraction of the gold-silver ores was based upon a knowledge of their localization at the intersections of fissures, toward which tunnels were run below the oxidized zones. Also, the complications of faulting were sufficiently understood to trace the displaced end of a lode beyond an important fault. A beginning had been made in understanding the occurrence of ores. Mine development existed before 1500 B.C. as indicated by a multilevel mine discovered in Sinai, where malachite was mined.

A knowledge of the occurrence of ores and the beginning of curiosity regarding their genesis is in the writings of the Greek and Roman philosophers. Herodotus (484–425 B.C.) told of the occurrence of gold in quartz veins in the Drissites district, Greece, later described by Diodorus. Theophrastas 372–287 B.C.), a pupil of Aristotle, in his *Book of Stones,* the first mineralogy textbook, described 16 minerals, grouped as metals, stones, and earths. Strabo, writing in A.D. 19, says in reference to alluvial mining in Spain, "Gold is not only dug from mines, but likewise collected; sand containing gold being washed down by the rivers and torrents . . . at the present day more gold is produced by washing than by digging it from the mines" (H. C. Hamilton and W. Falconer). Many descriptions of ore occurrences in Spain are given in Pliny's elaborate technical descriptions of ore occurrences in Spain. He also tells us that Hannibal had a silver mine, named the Baebulo, in southern Spain, in a mountain that had been penetrated 1500 paces. Pliny said it yielded 300 pounds of silver daily. The production of silver-lead ores was an important industry in Attica at a remote period, the famous mines of Laurium having been worked long before the day of Xenophon, who wrote a report upon them in 365 B.C. The ancients sank here more than 2000 shafts, one of which is 115 meters deep, and their locations disclose an accurate knowledge of the occurrence of the ores. Throughout the Dark Ages little appears to have been added to the knowledge of the early philosphers, except by Avicenna (A.D. 980–1037), the Arabian translator of Aristotle, who grouped minerals as stones, sulfur minerals, metals, and salts (Crook), thus definitely recognizing the sulfide group of minerals.

COMMENCEMENT OF THE SCIENTIFIC ERAS

The first reasonable theory of ore genesis was formulated by Georgius Agricola (Bauer) (1494–1555). Born in Saxony amidst the mines of the Erzgebirge, he became a keen observer of minerals and a careful recorder. Although some of his views were fantastic, he showed in his *De Re Metallica* that lodes originated by the deposition of minerals in "can-

Table 2-1 Approximate Times That Particular Mineral Operations Appear on the Historical Horizon

Gold washed from alluvial deposits	Prior to recorded civilization
Copper reduced from ores by smelting	Prior to recorded civilization
Bitumen mined and used	Prior to recorded civilization
Tin reduced from ores by smelting	Prior to 3500 B.C.
Bronze made	Prior to 3500 B.C.
Iron reduced from ores by smelting	Prior to 3500 B.C.
Soda mined and used	Prior to 3500 B.C.
Gold reduced from ores by concentration	Prior to 2500 B.C.
Silver reduced from ores by smelting	Prior to 2000 B.C.
Lead reduced from ores by smelting	Prior to 2000 B.C. (perhaps prior to 3500 B.C.)
Silver parted from lead by cupellation	Prior to 2000 B.C.
Bellows used in furnaces	Prior to 1500 B.C.
Steel produced	Prior to 1000 B.C.
Base metals separated from ores by water concentration	Prior to 500 B.C.
Gold refined by cupellation	Prior to 500 B.C.
Sulfide ores smelted for lead	Prior to 500 B.C.
Mercury reduced from ores by (?)	Prior to 400 B.C.
White-lead made with vinegar	Prior to 300 B.C.
Touchstone known for determining gold and silver fineness	Prior to 300 B.C.
Quicksilver reduced from ore by distillation	Prior to Christian Era
Silver parted from gold by cementation with salt	Prior to Christian Era
Brass made by cementation of copper and calamine	Prior to Christian Era
Zinc oxides obtained from furnace fumes by construction of dust chambers	Prior to Christian Era
Antimony reduces from ores by smelting (accidental)	Prior to Christian Era
Gold recovered by amalgamation	Prior to Christian Era
Refining of copper by repeated fusion	Prior to Christian Era
Sulfide ores smelted for copper	Prior to Christian Era
Vitriol (blue and green) made	Prior to Christian Era
Alum made	Prior to Christian Era
Copper refined by oxidation and poling	Prior to A.D. 1200
Gold parted from copper by cupelling with lead	Prior to A.D. 1200
Gold parted from silver by fusion with sulfur	Prior to A.D. 1200
Manufacture of nitric acid and *aqua regia*	Prior to A.D. 1400
Gold parted from silver by nitric acid	Prior to A.D. 1400
Gold parted from silver with antimony sulfide	Prior to A.D. 1500
Gold parted from copper with sulfur	Prior to A.D. 1500
Silver parted from iron with antimony sulfide	Prior to A.D. 1500
First textbook on assaying	Prior to A.D. 1500
Silver recovered from ores by amalgamation	Prior to A.D. 1500
Separation of silver from copper by liquation	Prior to A.D. 1540
Cobalt and manganese used for pigments	Prior to A.D. 1540
Roasting copper ores prior to smelting	Prior to A.D. 1550
Stamp-mill used	Prior to A.D. 1550
Bismuth reduced from ore	Prior to A.D. 1550
Zinc reduced from ore (accidental)	Prior to A.D. 1550

Source: Hoovers, in De Re Metallica by Agricola, p. 354.

ales'' (fissures) from circulating underground waters, largely of surface origin, that had become heated within the earth and had dissolved the minerals from the rocks. He made a clear distinction between *homogeneous minerals* (minerals) and *heterogeneous minerals* (rocks); the former he divided into *earths, salts, gemstones, metals, and other minerals.* He also classified ore deposits genetically into veins (vena profunda), beds, stocks, and stringers. Prior to Agricola, most writers thought that lodes were formed at the same time as the earth, but he recognized clearly that they were of a different age from the enclosing rocks, as he states: ''To say that lodes are of the same age as the earth itself is the opinion of the vulgar'' (*De Re Metallica,* 1556). This knowledge of earth materials led him to be the first to refute vigorously the efficacy of the forked hazel stick commonly used at that time in attempting to find metals and water. He made accurate observations on the weathering of rocks and the surface decomposition of metallic sulfide ores. Many of his careful observations are quaintly portrayed by interesting woodcuts (Fig. 2-1). His description and cuts of veins were drawn freely from von Kalbe's *Bergbüchlein* of 1518.

Agricola's writings are among the most original contributions to the study of ore genesis. They were a marked advance in scientific thought and influenced greatly the thought of later writers.

Contributions of the Seventeenth and Eighteenth Centuries

Although knowledge of minerals and rocks must have continued to accumulate in the extraordinary

Figure 2-1 Medieval Miners operating mine, sluice box, sieving, and making love. (Agricola, *De Re Metallica*)

mining atmosphere of the Erzgebirge and the Harz Moutains, little information was recorded from the time of Agricola to that of Descartes, whose *Principia Philosophae* was published in 1644. His conception of the earth as a cooled star with a hot interior led him to suggest that the ore minerals were driven upward from a deep metalliferous shell by interior heat in the form of exhalations and resurgent surface waters, to be deposited as lodes in the fissures of the outer stony crust. This conception is clearly the forerunner of some of the ideas held today.

In the eighteenth century, the accumulated factual knowledge further incited human curiosity about the genesis of ore materials. Under the stimulus of inspiring leaders, hypotheses of origin burst forth, which at the end of the century led to vigorous controversies. Most of these theories emanated from the German mining districts, but the Swedes also made early contributions. Becher (1703) and Henkel (1725) attributed the origin of ore veins to the action on stony materials of vapors arising from "fermentation" in the bowels of the earth. Henkel's idea of "transmutation" had in it the germ of modern metasomatism. In 1749, Zimmermann also anticipated the idea of metasomatic replacement when he ascribed the origin of lodes to the transformation of rocks into metallic minerals and veinstones by the action of solutions that entered through innumerable small rents and other openings in the rocks. This idea had in it also a suggestion of the subsequent lateral-secretion hypothesis. To von Oppel (1749) belongs the credit for having shown that veins were mainly the filling of fault fissures whose formation preceded the circulation of the ore-depositing solutions. His ideas, however, escaped attention for a long time. Lehman, in 1753, explained that the upward branching of veins indicated deposition from exhalations and vapors that emanated from the earth's interior and rose through the crust, like sap rising from the roots into the branches of a tree. Such were the theories of the origin of mineral deposits before 1756.

In that year an event occurred that profoundly affected the subsequent development of economic geology. The famous Mining Academy of Freiberg, Germany, was founded in the midst of the varied mineral deposits of the Erzgebirge. Its famed teachers conducted excursions to study the nearby ores and enclosing rocks; extensive mineral collections were made and studied in the Academy. Scholars flocked to study under its masters. Here the newer geology flourished, and for over a century and a quarter its teachings in ore deposits influenced world thought.

Throughout most of the eighteenth century the prevalent view was that mineral deposits were formed by exhalations from the earth's interior that brought up the metals from depth and deposited them in fissures, or by substitution of rock matter. Lassius, however, in 1789, following Delius (1770) and Gerhard (1781), explained ore solutions as diffused ascending water that dissolved scattered grains of metals from the rocks through which it passed. Thus most of the germs of theories of mineral deposits had been considered, even if unscientifically, before the controversial time of Hutton and Werner.

When Abraham Gottlob Werner (1749-1807) became the professor of mineralogy and geology at the Freiberg Mining Academy in 1775, he discarded the theories of an interior source for the metals and became an insistent advocate of the theory that mineral veins were formed by descending percolating waters derived from the primeval universal ocean, from which, according to his views, not only sediments but also all the igneous and metamorphic rocks were precipitated. These waters, he thought, descended from above into fissures and there deposited the vein materials by chemical precipitation. His stimulating personality and fiery lectures caused students from all Europe to flock to him and to return as zealous disciples to defend his Neptunist views. His thought then dominated in all matters relating to ore genesis, particularly after the publication, in 1791, of his classic treatise on the origin of veins. Werner's enthusiastic lectures perhaps carried conviction to his hearers more by his personality and oratory than by the soundness of his facts. He published little but his students spread his views. In one sense his leadership retarded the advancement of thought regarding mineral genesis, but at the same time his dogmatic statements aroused vigorous opposition and thus stimulated wider consideration of other ideas, of which the most noteworthy was the Plutonist or Vulcanist school headed by Hutton.

Hutton (born in 1726), a quiet Scotchman, also adverse to publishing, was a careful observer and investigator. In his *Theory of the Earth* (1788) he first defined the true origin of plutonic and metamorphic rocks, and his proponents waged bitter controversy with the Neptunist school. Hutton also applied his theory of the magmatic origin of igneous rocks to all mineral deposits. He claimed that ore minerals were not soluble in water but were igneous injections. In the words of his advocate, Playfair (1802), "The materials which fill the mineral veins were melted by heat and forcibly injected into the clefts and fissures of the strata." Hutton's observations were confined largely to rocks rather than veins, and with Werner it was the reverse. Hutton's correct conclusions of the origin of igneous rocks made him go too far in attributing all veins to melted injections and discarding water as a possible agent. Werner's correct conclusions regarding the role of water in the formation of veins made him go too far in ascribing granites and basalts to water deposition. The Plutonists won out with respect to the rocks; the Neptunists prevailed with respect to the dominance of water in the formation of mineral veins although, owing to the disrepute of Werner's explanation of rocks, his ideas regarding ores were overlooked for some time.

Nineteenth and Twentieth Centuries

The Plutonist and Neptunist controversy quickened observations regarding the occurrence of minerals and rocks, and many data, particularly regarding ore veins, became available. Hutton's igneous injection theory of mineral veins was quickly forgotten. The early nineteenth century writers reverted to the pre-Werner ideas of mineral formation by exhalations from the interior of the earth, recognizing more, however, the significance of water in the formation of veins. Gradually, water of igneous derivation was considered to play the important role. This view received confirmation by the work of Necker (1832), who, believing that the intrusions generated the vein materials, demonstrated the close relationship, in various regions, between igneous rocks and mineral deposits.

The connection of the mineral-forming solutions with magmas was given further emphasis by the French geologists Daubrée, Scheerer, and Elie de Beaumont. The brilliant Daubrée introduced the first experimental methods into the study of mineral deposits. In 1841 he produced artificial cassiterite (tinstone) from stannous chloride and inferred that vapors or mineralizers containing water, fluorine, and boron generated at depth had deposited the tin ores and associated minerals.

Scheerer in 1847, following Scrope, who concluded that magmatic water played a part in the formation of igneous rocks, stated clearly that water was an important constituent of granite magmas and that mineral veins were formed by exudations of aqueous solutions from granite intrusions

A few months later an important paper by Elie de Beaumont appeared, which Thomas Crook stated was perhaps the "most important and influential paper ever published on the theory of ore deposits." Elie de Beaumont might well be called the father of our modern thought on the formation of mineral deposits. He was the first to show that most mineral deposits must be regarded as just one phase of igneous activity. He recognized that water vapor was an essential feature in volcanic activity and that metalliferous veins were formed as incrustations upon the walls of fissures from hot waters of igneous origin. He distinguished these veins from dikes injected in a molten condition. He cited occurrences of segregations of magnetite and chromite in basic igneous rocks, which he considered had crystallized out during the cooling of the intrusive. Elie de Beaumont thus recognized clearly the igneous affiliation of many types of mineral deposits—that some were formed as segregations during crystallization of the magma and that others were formed from hot aqueous emanations that escaped upward from the igneous intrusion. Essentially similar views are held today, but for many years the clear statements of Elie de Beaumont were overlooked.

The conflicting opinions of these times were carefully weighed by Von Cotta, whose excellent treatise on ore deposits appeared from Freiberg in 1859 (in English in 1870) and remained a standard textbook on mineral deposits for two decades. He presented concise information regarding the content, character, and structural and textural features of mineral deposits and careful descriptions of the outstanding mineral districts. He examined judiciously the various theories of mineral genesis and correctly concluded that no one theory was applicable to all deposits. In his concluding observations he remarks (Prime's translation, 1870) "Thus the formation of lodes shows itself to be . . .very manifold; and appears to have always stood in some connection with neighboring. . . eruptions of igneous rocks. The local reaction of the igneous-fluid interior of the earth created fissures, forced igneous-fluid masses into many of the same, caused gaseous emanations and sublimations in others; and in addition, during long periods of time impelled the circulation of heated water, which acted, dissolving at one point and again depositing the dissolved substances at another, dissolving new ones in their stead. The whole process is not confined to any particular geological period, or any particular locality." Such statements might well be a part of a

modern textbook. He shows clearly that most mineral deposits were the products of deep-seated igneous action. He recognized not only a definite zonal arrangement of minerals dependent upon temperature and pressure conditions of deposition but also certain superficial changes imposed upon mineral deposits by weathering. Von Cotta's balanced treatise exerted a profound influence upon the subject of economic geology.

A trend back toward the Hutton views of a straight igneous origin for mineral deposits is to be noted in the writings of Fournet (1844, 1856) and Belt (1861), who considered that veins and many mineral deposits were the result of an igneous injection into fissures in a molten state, and were thus *ore magmas,* to use the term later proposed by Spurr (1923). Belt, however, believed that water played a part in helping to lower the temperature of the "liquefaction" (fusions) of such minerals as quartz.

In the meantime these views of hydrothermal and igneous action in the formation of mineral deposits were partly obscured during the advocacy of another startling hypothesis on the origin of ores.

The earlier assumptions of Delius, Gerhard, and Lassius, that water percolating through the rocks had dissolved out certain ingredients and afterward precipitated them in fissures, was later taken up by Bischoff (1847) and put forward as a reasoned theory of lateral secretion by waters of meteoric origin. This theory was supported by data regarding the dissemination of lode minerals in superficial rocks and their significance in ore genesis. Somewhat similar ideas were advanced in America by T. Sterry Hunt in 1861 and in England by J. A. Phillips in 1875, but they were advocated in an extreme form in Germany in 1882 by Sandberger, who sought to establish two facts: (1) that the gangue of ore veins corresponded with the wall rocks, and (2) that traces of the heavy metals occurred in the wall rocks. Meteoric waters were supposed to search out these ingredients from the surrounding wall rocks and deposit them in the fissures. Daubrée in his later comprehensive study of underground waters (1887) concluded that hot waters were the most important agent in the formation of mineral deposits and that these waters, for the most part, were not magmatic but were meteoric waters that had become heated in depth and had risen again. These were the views of Hunt and Phillips, and likewise later of S.F. Emmons, who in 1886 explained the origin of the ore deposits of Leadville, Colorado, by surface waters that had leached metallic ingredients from the neighboring rocks. Similar views were elaborated in 1901 by C. R. Van Hise, who concluded that magmatic waters played a minor role and that most mineral deposits resulted from surface waters that had descended to depths where they became heated, dissolved metals from the rocks, and rose again to deposit their metallic content in fissures or other openings—a circulation resembling that of a hot-water heating system

There then followed animated discussions on the respective merits of the *descensionist-, ascensionist-,* and *lateral-secretionist* theories.

The supporters of lateral secretion, however, lost ground rapidly under the attacks of Stelzner (1879), Patera (1888), Pošepný (1894), De Launay (1893), and others who contended that the mineral substances were not dissolved from the surrounding rocks by meteoric waters but rather were deposited there by ascending hot waters that had carried the materials up from the deep-seated sources. Both Pošepný and De Launay, following the earlier ideas of Elie de Beaumont, sought a source for the metals and waters in deep-seated eruptive rocks or still deeper sources in the barysphere. They were hot-water ascensionists, although Pošepný clearly distinguished certain types of deposits formed by the circulation of meteoric waters in the vadose zone. The vigorous discussions that followed the presentation of Pošepný's classical paper in Chicago in 1893 directed attention once more to the close association between mineral deposits and igneous rocks, and an igneous origin for mineralizing solutions was urged by J. F. Kemp (1901), Waldemar Lindgren (1901), W. H. Weed (1903), and others whose ideas were rather generally accepted but are now modifed by $\delta^{18}O$ and D analyses that indicate the importance of meteoric water drawn into igneous magmas.

In the meantime J. H. L. Vogt of Norway had been laying the foundations of our present-day conceptions of the origin of some mineral deposits. Starting from the ideas of Elie de Beaumont, he delved further, utilizing physicochemical principles, into the source of ascending hot mineralizing solutions. On the basis of careful field studies of magnetite, chromite, nickel, pyrrhotite, and pyrite, he concluded that such substances were igneous injections of material derived from their igneous source by processes of magmatic differentiation (1893). He also concluded that the hot mineralizing waters were similarly derived by magmatic differentiation. Thus he recognized magmatic differentiation as a process of ore formation that gave rise to

(1) ore segregations or injections, (2) mineralizing gases and vapors, and (3) hot mineralizing waters. In past years these processes have been elaborated, clarified, and experimentally verified. Such conceptions are generally current today to explain the origin of most primary mineral deposits. It is recognized, however, as was pointed out by Von Cotta long ago, that there are many types of mineral deposits, which have been formed by different processes.

A later contribution to the theories of the genesis of mineral deposits is the magmatic view of J. E. Spurr (1923) who, following the earlier conceptions of Thomas Belt (1861), postulated that many, or most, ore deposits have resulted from the injection and rapid freezing of highly concentrated magmatic residues, for which he proposed the terms *ore magmas* and *vein dikes.* Spurr's ideas are not generally accepted.

Waldemar Lindgren (1860-1939) is recognized as one of the world's foremost economic geologists for his perspicacious knowledge, a major textbook still valued for succinct and time-tested genetic theories, for the Lindgren classification of mineral deposits that still requires only minor changes, the establishment of the Annotated Bibliography of Economic Geology, and many other contributions. (Fig. 2-2.)

Lindgren was born in Sweden; his interest in geology began at the age of 10. He studied at the Royal Mining Academy, Freiberg, Saxony, following which he sailed for the United States. He obtained a field assistant position in Montana at $25 a month with William Morris Davis. Later he joined the U.S. Geological Survey where he spent a quarter of a century. He had the almost unparalleled experience of visiting, studying, directing, and publishing reports pertaining to hundreds of mineral deposits primarily located in the western United States.

The Massachusetts Institute of Technology (MIT) invited Lindgren to give a series of lectures during 1909-1911. Afterwards, he accepted the position of Professor of Economic Geology and Chairman of the MIT Department of Geology. Almost immediately, the first edition of his textbook (*Mineral Deposits*) appeared; it was completed during his first three years of lecturing. His influence and writings have been major milestones pertaining to mineral genesis.

The role of surface waters in the oxidation and secondary enrichment of ore deposits has long been recognized, and circulating meteoric waters have

Figure 2-2 Waldemar Lindgren, 1860–1939, Geologist, Teacher, Author, and Editor. (From: R. R. Shrock provided by Bachrach).

also been called upon to account for the origin of uranium deposits in sandstone.

There has also been a recent trend toward a variation of the old lateral secretion theory in that an increasing number of authors consider that certain mineral deposits have originated by the mobilization of metals from underlying or adjacent rocks, which were then carried upward in solution to sites of deposition. Others have suggested a regeneration of older deposits below to new sites above.

The role of meteoric waters in the formation of mineral deposits has been receiving increasing attention. Mineralizing solutions, formerly thought to be entirely of magmatic hydrothermal origin, are now believed to have been composed predominantly of meteoric water. The evidence is based upon fluid inclusions, geothermal waters, and the hot metal-bearing brines of the Salton Sea and Red Sea but mostly upon the stable isotopic compositions of oxygen and hydrogen.

Submarine exhalative processes accompanying submarine volcanic eruptions have lately been called upon to account for many massive sulfide deposits in volcanic host rocks including the Ku-

roko deposits of Japan and a growing number of other deposits, many of which were classified otherwise. This process is recognized in many massive sulfide deposits that are sandwiched between older andesite rocks and overlying rhyolitic rocks, or their metamorphic equivalents.

Biogenic processes have long been recognized as factors in the formation of some mineral deposits (e.g., sedimentary iron deposits and Gulf Coast sulfur deposits), but lately they have been more broadly applied to the formation of other specific sulfide ore deposits.

These genetic ideas are dealt with further in later chapters. Succinct abstracts of the major changes in concepts of ore genesis between 1933 and 1967 have been compiled by J. D. Ridge in general reference 11.

The advance in the study of ore deposits has in large part centered around the development of theories of origin, since each new theory quickened field observations directed to prove or disprove it. Consequently, a great mass of material regarding the character, distribution, and localization of mineral deposits has accumulated.

Selected References on Development of Economic Geology

A review of the development of ideas concerning the genesis of mineral deposits is enlightening because it presents the background of the current prevailing ideas and discloses that most of them have been advanced in previous generations. At present we are only elaborating upon, amplifying, revising, and revitalizing earlier conceptions.

History of the Theory of Ore Deposits. 1935. Thomas Crook. Thos. Murby and Co., London. *An interesting, brief, historical treatise.*

General Reference 10, Chap. 1. Historical Review of Geology as Related to Western Mining. F. L. Ransome. *Development of geological thought as related to Western United States mineral deposits.*

The Birth and Development of the Geologic Sciences. 1938. F. D. Adams. The Williams & Wilkins Company, Baltimore. *An exhaustive treatment dealing with prenineteenth century development of ideas; particularly chap. 9, "The Origin of Metals and Their Ores."*

Fiftieth Anniversary Volume, 1888–1938. Geological Society of America, 1941. *Development of progress of thought in the various branches of geology until 1938 by 21 authors.*

The Romance of Mining. 1944. T. A. Rickard. Macmillan Publishing Co., Inc., New York. *Many references to historical development. General Reference 17,*

Chap. 3. Minerals in history before the Industrial Era. *Early use, distribution, and trade in minerals.*

Some cultural contributions of the geological arts and sciences. 1966. T. S. Lovering. Bull. New Mexico Academy of Science v. 7. *History of the early use, development and significance of metals.*

Bergwerk- und Probierbuchlein. 1949. Translation of U. R. von Kalbe's German of 1518 by A. G. Sisco and C. S. Smith. AIME, New York. *Translation of first-known treatises on technology of minerals, metals, and veins.*

Theory of the Earth with Proofs and Illustrations. 1795. Hutton, James, Edinburgh.

De Re Metallica. Georgius Agricola. 1912. Transl. from the Latin ed. of 1556 by Herbert Hoover and Lovit Hoover. Mining Magazine, London. reprinted by Dover Publications, Inc., New York, 1950. *Early ideas of ore genesis.*

Fiftieth Anniversary Volume. 1955. Chap. 1, Economic geology by Alan M. Bateman. *Review of ideas appearing in Economic Geology 1905-1955.*

General Reference 11, v.2.

3

MATERIALS OF MINERAL DEPOSITS AND THEIR FORMATION

Science, ever since the time of the Arabs, has had two functions (1) to enable us to know things, and (2) to enable us to do things.

Bertrand Russell,
The Impact of Science on Society

Mineral deposits, whether metalliferous or nonmetallic, are accumulations or concentrations of one or more useful substances that are for the most part sparsely distributed in the earth's outer crust. They are relatively scarce in igneous rocks and in sedimentary rocks, although igneous and sedimentary deposits are among the largest known. A few deposits, however, consist of concentrations of common rock-forming minerals such as feldspar or mica (Table 3-1).

The elements that enter into the materials of mineral deposits have been derived either from the rocks of the earth's crust and upper mantle or from molten bodies (*magmas*) that have cooled to form igneous rocks. Originally, all the elements except those that may have persisted from the primitive atmosphere were derived from magmas or igneous rocks. Of the 106 known elements, only 8 according to S. R. Taylor are present in the earth's crust in amounts exceeding 1 percent, only 4 of the economically important elements exceed 1 percent (Table 3-2), and 99.5 percent of the earth's crust is made up of the following 13 elements: oxygen, silicon, aluminum, iron, calcium, sodium, potassium, magnesium, titanium, hydrogen, phosphorus, manganese, and fluorine. The remaining economically important elements, constituting only 0.5 percent, include all the precious and useful substances such as platinum, gold, silver, copper, lead, zinc, tin, nickel, and others (Table 3-2). It is thus clear that geologic processes of concentration are necessary to collect these diffuse elements into workable mineral deposits, and it will be our purpose to study some of these exceptional and interesting processes.

The minerals that constitute the bulk of the earth's crust are also few in number. About 2500 mineral species are known; about 50 of these are rock-forming minerals, of which only 29 are common ones. About 200 are classed as economic minerals. The remainder come mostly from mineral deposits. A compilation of some of the common minerals that constitute the bulk of the earth's crust are listed in Table 3-3.

MATERIALS OF METALLIFEROUS DEPOSITS

Metalliferous deposits represent, in general, extreme concentrations of formerly diffuse metals. The desired metals are generally chemically bound with other elements to form *ore minerals*. These in turn are commonly interspersed with nonmetallic minerals or rock matter, called *gangue*. The mixture of ore minerals and gangue constitutes *ore*, which generally is enclosed in *country rock*. Some metalliferous deposits, however, may lie upon the surface and thus are not enclosed within country rock.

Most of the ore minerals are metallic minerals, such as galena, which is mined for its lead. A few are nonmetallic minerals, such as malachite, bauxite, or cerussite, ore minerals of copper, aluminum, and lead. The ore minerals occur as native metal, of which gold and platinum are examples, or as combinations of the metals with sulfur, arsenic, oxygen, silicon, or other elements. Combinations are the most common.

A single metal may be extracted from several different ore minerals. Thus there are several ore minerals of copper, such as chalcocite, bornite, chalcopyrite, cuprite, native copper, and malachite; one or more of these may occur in an individual deposit. Also, more than one metal may be obtained from a single ore mineral; stannite, for example, yields both copper and tin. An individual ore de-

Table 3-1 Common Minerals of Earth's Crust in Percent

Mineral	Lithosphere	Igneous Rocks	Sedimentary Rocks
Feldspar	49	50	16
Quartz	21	21	35
Pyroxene, amphibole, and olivine	15	17	—
Mica	8	8	15
Magnetite	3	3	—
Titanite and ilmenite	1	1	—
Others	3	—	3
Clay	—	—	9
Dolomite	—	—	9
Chlorite	—	—	5
Calcite	—	—	4
Limonite	—	—	4
	100	100	100

Table 3-2 Crustal Abundance of Economically Important Elements

Name	Chemical Symbol	Atomic Number	Crustal Abundance (Percent by weight)
Aluminum	Al	13	8.00
Iron	Fe	26	5.8
Magnesium	Mg	12	2.77
Potassium	K	19	1.68
Titanium	Ti	22	0.86
Hydrogen	H	1	0.14
Phosphorus	P	15	0.101
Manganese	Mn	25	0.100
Fluorine	F	9	0.0460
Sulfur	S	16	0.030
Chlorine	Cl	17	0.019
Vanadium	V	23	0.017
Chromium	Cr	24	0.0096
Zinc	Zn	30	0.0082
Nickel	Ni	28	0.0072
Copper	Cu	29	0.0058
Cobalt	Co	27	0.0028
Lead	Pb	82	0.00010
Boron	B	5	0.0007
Beryllium	Be	4	0.00020
Arsenic	As	33	0.00020
Tin	Sn	50	0.00015
Molybdenum	Mb	42	0.00012
Uranium	U	92	0.00016
Tungsten	W	74	0.00010
Silver	Ag	47	0.000008
Mercury	Hg	80	0.000002
Platinum	Pt	78	0.0000005
Gold	Au	79	0.0000002

Source: Press and Siever, *Earth,* W. H. Freeman and Co., San Francisco.

posit, therefore, may yield several metals from different ore minerals.

Most of the world's gold has come from native gold; consequently, its removal from admixed minerals is a relatively simple process and offered no serious problems of extraction, even to the ancients. Silver, on the other hand, is derived not only from the native metal, but also from combinations with sulfur and other elements. This is also true of copper, lead, zinc, and most of the other metals. The vast quantity of iron used in industry is obtained almost entirely from combinations with oxygen. It is from such simple combinations that the human race has been supplied with desired metals for over 4000 years. In addition, there are more complex metallic combinations that yield considerable quantities of the common metals as well as many of the minor metals.

Ore minerals are also classed as *primary* or *hypogene,* and *secondary* or *supergene.* The former were deposited during the original period or periods of metallization; the latter are alteration products of the former as a result of weathering or other surficial processes resulting from descending surface waters. The term *primary* has also been used to designate the earliest of a sequence of ore minerals as contrasted with later minerals of the same sequence, which some writers have called secondary. This gave rise to some confusion, and to avoid this, Ransome proposed the terms *hypogene* and *supergene.* Primary and hypogene are generally considered synonymous, but hypogene, as the word implies, indicates formation by ascending solutions. All hypogene minerals are necessarily primary, but not all primary ore minerals are hypogene; sedimentary hematite, Kuroko ores, and other submarine exhalative ores, for example, are of primary deposition but have not necessarily been formed from ascending solutions. Similarly, confusion has arisen with the use of the word *secondary,* which is eliminated by the better term *supergene.*

Gangue Minerals. Gangue minerals are the associated, usually worthless, nonmetallic materials of a deposit. They may be introduced minerals, or the enclosing rock, and are usually discarded in the treatment of the ore. The *gangue,* in customary usage, includes only nonmetallic minerals, but in technical usage it also includes some metallic minerals, such as pyrite, which are usually discarded as worthless. Certain gangue materials, however, may at times be collected as by-products and utilized. For example, rock gangue may be utilized for

Table 3-3 List of the Common Ore Minerals

Metal	Ore Mineral	Composition	Percent Metal	Primary	Super-gene
Gold	Native gold	Au	100	X	X
	Calaverite	$AuTe_2$	39	X	
	Sylvanite	$(Au,Ag)Te_2$	—	X	
Silver	Native silver	Ag	100	X	X
	Argentite	Ag_2S	87	X	X
	Cerargyrite	$AgCl$	75		X
Iron	Magnetite	$FeO \cdot Fe_2O_3$	72	X	
	Hematite	Fe_2O_3	70	X	X
	"Limonite"	$Fe_2O_3 \cdot H_2O$	60		X
	Siderite	$FeCO_3$	48	X	X
Copper	Native copper	Cu	100	X	X
	Bornite	Cu_5FeS_4	63	X	X
	Brochantite	$CuSO_4 \cdot 3Cu(OH)_2$	62		X
	"Chalcocite"	Cu_2S	80	X	X
	Chalcopyrite	$CuFeS_2$	34	X	X
	Covellite	CuS	66	X	X
	Cuprite	Cu_2O	89		X
	Digenite	Cu_9S_5	78	X	X
	Enargite	$3Cu_2S \cdot As_2S_5$	48	X	
	Malachite	$CuCO_3 \cdot Cu(OH)_2$	57		X
	Azurite	$2CuCO_3 \cdot Cu(OH)_2$	55		X
	Chrysocolla	$CuSiO_3 \cdot 2H_2O$	36		X
Lead	Galena	PbS	86	X	
	Cerussite	$PbCO_3$	77		X
	Anglesite	$PbSO_4$	68		X
Zinc	Sphalerite	ZnS	67	X	
	Smithsonite	$ZnCO_3$	52		X
	Hemimorphite	H_2ZnSiO_5	54		X
	Zincite	ZnO	80	X	
Tin	Cassiterite	SnO_2	78	X	?
	Stannite	$Cu_2S \cdot FeS \cdot SnS_2$	27	X	?
Nickel	Pentlandite	$(Fe,Ni)S$	22	X	
	Garnierite	$H_2(Ni,Mg)SiO_3 \cdot H_2O$	—		X
Chromium	Chromite	$FeO \cdot Cr_2O_3$	68	X	
Manganese	Pyrolusite	MnO_2	63	X	X
	Psilomelane	$Mn_2O_3 \cdot xH_2O$	45	X	X
	Braunite	$3Mn_2O_3 \cdot MnSiO_3$	69	?	X
	Manganite	$Mn_2O_3 \cdot H_2O$	62		X
Aluminum	Bauxite	$Al_2O_3 \cdot 2H_2O$	39		X
Antimony	Stibnite	Sb_2S_3	71	X	
Bismuth	Bismuthinite	Bi_2S_3	81	X	X
Cobalt	Smaltite	$CoAs_2$	28	X	
	Cobaltite	$CoAsS$	35	X	
Mercury	Cinnabar	HgS	86	X	
Molybdenum	Molybdenite	MoS_2	60	X	
	Wulfenite	$PbMoO_4$	39		X
Tungsten	Wolframite	$(Fe,Mn)WO_4$	76	X	
	Huebnerite	$MnWO_4$	76	X	
	Scheelite	$CaWO_4$	80	X	
Uranium	Uraninite	Combined UO_2	50–85	X	
	Pitchblende	and UO_3		X	
	Coffinite	$USiO_4$	75		X
	Carnotite	$K_2O \cdot 2U_2O_3 \cdot V_2O_5 \cdot nH_2O$	60 U_3O_8		X

"road metal", fluorspar for flux, quartz for abrasive or concrete, pyrite for sulfur, and limestone for fertilizer or flux. Some gangue minerals considered worthless today may prove of value tomorrow. The gangue, even though worthless, may so influence the method or cost of treatment that it determines the value of the ore. Some of the common gangue minerals are listed in Table 3-4.

Ore. The term *ore* is often loosely used to designate anything that is mined. Technically, it is an aggregation of ore minerals and gangue from which one or more metals may be extracted at a profit. To be ore, material must, therefore, be payable, and this involves economic considerations as well as geologic. Obviously, a body of valueless pyrrhotite devoid of gold would not be ore, even though pyrrhotite is a metallic mineral, and a body of pyrite containing gold may or may not be ore, depending upon the amount of gold present and whether the value of the recoverable gold is greater than the cost of extraction.

The question of profit depends upon the amount and price of the metal and upon the cost of mining, treating, transporting, and marketing the product. This in turn depends in part upon the geographic location of the deposit. A high-grade hematite body located in the Arctic, for example, would not be iron ore because the cost of extraction and transportation would be greater than the value of the iron. However, it might become ore in the future. Likewise, increased efficiency and lower cost of metallurgical technique or mining practice may enable present worthless material to be classed as ore in the future. Also, new uses may transform worthless materials into valuable ones. The discovery that beryllium makes a fatigue-resisting alloy with copper has changed beryl from a mineralogical curiosity to a much-sought, valuable mineral.

What constitutes ore may also depend upon the gangue or upon minor constituents. Certain materials may be profitably worked for their metallic content only if some part of the gangue can be utilized. The presence of small quantities of bismuth, cadmium, or arsenic may change the value of deposits of lead, zinc, or copper.

The relative proportions of ore minerals and gangue vary enormously. In average ores, gangue greatly predominates. Because it is costly to smelt valueless gangue in order to obtain the enclosed metal, it is customary to subject the ores to ore-dressing processes (milling) whereby the ore minerals are concentrated and the waste gangue discarded. Thus from 5 to 30 tons of crude ore will yield 1 ton of concentrates containing most of the metallic content of the original lot. This is then smelted for its metallic content, thereby saving the

Table 3-4 List of Common Gangue Minerals

Class	Name	Composition	Primary	Supergene
Oxides	Quartz	SiO_2	X	X
	Other silica	SiO_2	X	X
	Bauxite, etc.	$Al_2O_3 \cdot 2H_2O$		X
	"Limonite"	$Fe_2O_3 \cdot H_2O$	X	X
Carbonates	Calcite	$CaCO_3$	X	X
	Dolomite	$(Ca,Mg)CO_3$	X	X
	Siderite	$FeCO_3$	X	X
	Rhodochrosite	$MnCO_3$	X	
Sulfates	Barite	$BaSO_4$	X	X
	Gypsum	$CaSO_4 + 2H_2O$	X	X
Silicates	Feldspar	—	X	
	Garnet	—	X	
	Rhodonite	$MnSiO_3$	X	
	Chlorite	—	X	
	Clay minerals	—	X	X
Miscellaneous	Rock matter		X	
	Fluorite	CaF_2	X	
	Apatite	$(CaF)Ca_4(PO_4)_3$	X	
	Pyrite	FeS_2	X	X
	Marcasite	FeS_2	X	X
	Pyrrhotite	$Fe_{1-x}S$	X	
	Arsenopyrite	$FeAsS$	X	

cost of treating the 4 to 29 tons of discarded gangue. For example, if a gold-copper ore contains $9 worth of metal per ton and freight and smelting charges are $9 per ton, there would be no profit in smelting the crude ore, but if it were concentrated 10:1 (the ratio of concentration), then there would be freight and smelting charges for only 1 ton of concentrates instead of for 10 tons of ore, equivalent to about 90 cents per ton of original ore. Consequently, the ratio of concentration is of vital importance in determining whether a material is ore; the gangue may thus play fully as important a part as the ore minerals.

The amount of metalliferous minerals present varies greatly in ores of different metals and also in ores of the same metal. High-grade iron ores may consist of 100 percent hematite; copper ores range from less than 0.5 percent to 60 percent of metalliferous minerals, those with low percentages being concentrated, and those with higher percentages being smelted directly. In contrast, gold ores may contain only an infinitesimal amount of gold. For example, the Alaska Juneau gold mine mined with profit ores that contained only 0.00016 percent of gold. The amount of metal that must be present to constitute ore obviously depends upon the price of the metal. During periods of low metal prices, only ore with a metallic content higher than normal can be classed as ore. Conversely, as the price of gold has increased, much material that previously was valueless has become good gold ore.

Associated Metals in Ores. Ores may yield a single metal (simple ores) or several metals (complex ores). Ores that are generally worked for only a single metal are those of iron, aluminum, chromium, tin, mercury, manganese, tungsten, and some ores of copper. Gold ores may yield only gold, but silver is a common associate. Much gold, however, is extracted as a by-product from other ores. Ores that commonly yield either two or three metals are those of gold, silver, copper, lead, zinc, nickel, cobalt, antimony, and manganese. Some complex ores may yield four or five metals, such as copper-gold-silver-lead, silver-lead-zinc-copper-gold, tin-silver-lead-zinc, or nickel-copper-gold-platinum. Many minor metals are not won directly from their ores but are obtained as by-products from ores of other metals during smelting or refining operations. This is true of arsenic, bismuth, cadmium, selenium, and others. Where precious metals accompany base metals, their presence may make good ore of otherwise uneconomic material.

Some of the common associations of metals in ores are gold and silver; silver and lead; lead and zinc; lead, zinc, and copper; copper and gold; iron and manganese; iron and titanium; nickel and copper; nickel and cobalt; chromium and platinum; tin and tungsten; molybdenum and copper; and zinc and cadmium.

Tenor of Ores. The metal content of an ore is called the *tenor* or *grade,* which is generally expressed in percentage or units, or, in the case of precious metals, in ounces per ton. The tenor varies with the price of a metal, with the cost of production, with ores of different metals, and also with ores of the same metal. The higher the price of a metal the lower the metal content necessary to make it profitable. Most iron ore to be profitable must have a tenor of 25 to 50 percent of iron from which higher grade pellets are made, whereas copper ore need contain only 0.4 percent copper, or gold ore, less than 1/1000 of 1 percent of gold. The tenor of ore need have no upper limit; the richer, the better. The lower limit, however, is fixed by economic considerations and varies according to the nature and size of a deposit; the feldspar, barite, or fluorspar deposits include considerable waste that must be removed by "processing." Gemstones, asbestos, or graphite generally constitute only a small part of the deposit.

The nonmetallic materials consist of a vast array of substances utilized in modern civilization. These include fuels, rocks, earthy materials, sand, salts, minerals of pegmatite dikes, and many other nonmetallics such as asbestos, gypsum, fluorspar, mica, barite, graphite and sulfur. The associated gangue or waste consists mostly of enclosing rock, or parts of the nonmetallic products themselves that are discarded as unfit for use because of physical or chemical defects. The separation of the two, called "processing," consists of hand-sorting, simple mechanical concentration, flotation, or washing. Since smelting, or involved metallurgical treatment, is unnecessary, the determination of what is economic material is not so dependent upon the associated gangue as in the case of ores. Rather, it is dependent upon the price and the physical and chemical properties of the products themselves. A metal is just a metal, but clay, for example, to be usable must meet definite specifications regarding plasticity, specific gravity, fusibility, shrinkage, and tensile strength. Its value depends more upon its physical than its chemical properties, and not all clays fulfill the requirements. Different requirements exist for each nonmetallic product. Consequently, the properties that determine the commercial use of nonmetallic products are multitudinous as compared with the few that determine ore.

DETERMINATION OF MATERIALS

The materials that make up mineral deposits can, for the most part, be determined visually. However, for more exact determination, precise methods are necessary, such as assaying, chemical analyses, microscopic examination, atomic absorption, X-ray fluorescence and diffraction, spectroscopic examination, differential thermal analyses, electron probe analysis, and colorometric tests.

In ores, the metallic content of the desired constituents is determined usually by atomic absorption, or by wet (chemical) assaying, and the results are expressed in terms of weight percent (wt %), $ per ton, $ per cubic yard, ounces per ton, ounces per cubic yard, units, or parts per million (ppm).

As these methods express only the metallic content, without regard to the mineralogical constituents, a supplemental microscopic and electron micro probe examination is usually desirable to reveal the identity of the minerals and their relation to each other. Important minute metalliferous constituents can thus be detected. Partial or complete chemical analyses are sometimes made in order to determine the quantities of other ingredients, particularly those that affect the metallurgical treatment. Thus, for smelting, it is necessary to know the proportions of the oxides of silicon, calcium, magnesium, and iron in order that correct proportions of fluxes may be added to the ore or concentrates. Also, the quantities of undesired ingredients, such as arsenic, are determined by chemical analyses. In some cases minute quantities of rare elements are determined spectroscopically.

In nonmetallic products, the materials are determined by chemical analyses, physical tests, visual inspection, and microscopic examination. For such fine materials as clays, X-ray examinations and thermal analyses are employed. Different methods are employed for different products. Thus physical characters, such as strength, grain size, hardness, specific gravity, plasticity, fusibility, and electrical conductivity, are determined for such materials as building stones, sands, clays, abrasives, mica, or asbestos. On the other hand, partial analyses are necessary for such materials as fuels, fertilizers, bauxite, cements, magnesite, or lime. But for such subtances as gemstones, roofing slates, quartz, or barite, generally visual inspection alone is necessary. Microscopic examinations supplement the other tests.

THE FORMATION OF MINERALS AND MINERAL PRODUCTS

An understanding of mineral deposits necessitates a knowledge of the manner in which the constituents have been formed. Much information has accumulated in recent years regarding the chemical and other conditions of formation, particularly temperature, pressure, pH, Eh (redox potential), p_{O_2} (partial pressure), and other factors. In consequence, the study of minerals has taken on a new and wider geological significance in that the presence of certain minerals may supply the information as to temperature, pressure, or chemical character of the mineralizing agencies, which thus aids in deciphering the origin of the deposits that contain them. The importance of two of these factors, namely, pH and f_{O_2} (fugacity) as determined by the environmental deposition and its bearing of such on $\delta^{34}S$ and $\delta^{13}C$ values is shown in Fig. 3-1. Obviously, the mineralogy of the deposits is also influenced by these factors; in fact, it is the differing of similar mineralogy of the different deposits that indicated the approximate environments shown in Fig. 3-1.

Because it is the intent to consider here the formation only of those minerals that are of economic importance, the following discussion deals with the formation of the *materials* of mineral deposits and not with the deposits themselves.

Temperature and Pressure

The formation of a mineral generally indicates a change from a mobile to a solid state. As most minerals have been precipitated from solutions, many physiochemical factors play important roles. The constituents of mineral deposits are formed in the different ways to be discussed here. Temperature, pressure, pH, p_{O_2}, and various solutions play an important part in the deposition of the vast majority of minerals.

Temperature is one factor affecting the solubility of materials in solution, and therefore their precipitation. Usually, a decrease of temperature and/or pressure such as that which occurs when solutions ascend in the earth promotes precipitation from

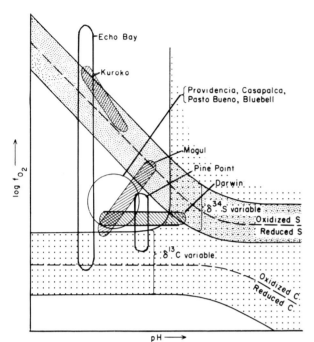

Figure 3-1 Schematic pH-fo₂ diagram showing reconstructed depositional environments of nine hydrothermal deposits. Dashed lines separate environments of predominantly reduced and oxidized carbon and reduced and oxidized sulfur. The stippled zones indicate environment where carbon and sulfur isotope values in carbonates and sulfides will be strongly influenced by change in pH and fo₂. Encircled areas indicate environments of ore deposition. (From: Econ. Geol. v. 69, 1974)

hydrogen sulfide to form sulfides; in fact, the precipitation of metals from complex ions can take place by many different processes.

Crystallization from Magmas. Since a magma is a molten or fluid silicate solution, crystallization from it follows the same laws as with an aqueous solution. When a magma cools and the saturation point of the solution is exceeded for any given mineral, that mineral will crystallize, provided that the temperature at the existing pressure is below the fusion point of the mineral. Thus, from certain magmas, economic minerals such as apatite, magnetite, or chromite have formed by crystallization. The processes involved are discussed in Chapter 5.

Reactions between solutions and solids are probably the most important natural processes in the formation of both hypogene and supergene minerals. Hypogene and supergene fluids are continuously in contact with rocks and minerals; chemical action takes place, and ore and gangue minerals are precipitated. Several processes are involved, namely, metasomatism or replacement, precipitation, reduction or oxidation, direct deposition, catalytic action, adsorption, base exchange, chemical complexes, and others.

Metasomatism or *metasomatic replacement* or simply *replacement*, as it is generally called, is a process of essentially simultaneous capillary solution and deposition by which new minerals are substituted for earlier minerals or rocks. A single mineral may replace and retain the exact form and size of a replaced mineral (**pseudomorph**); or silica may replace wood, retaining the woody structure (**petrification**); or a large body or ore minerals may take the place of an equal volume of rock. The replacing mineral (**metasome**) need not have a common ion with the replaced substance. The metasomes are carried in solution, and the replaced substances are carried away in solution: It is an open circuit, not a closed one. If, in a brick wall, each brick were removed one by one and a silver brick of similar size substituted for each, the end result would be a wall of the same size and form, even to the minutiae of brick pattern, save that it would be composed of silver instead of clay. This is how replacement proceeds, except that the parts are infinitesimally small—of molecular size. Consequently, the shape, size, structure, and texture may be faithfully preserved in the replacing substance.

Replacement is one of the important processes in the formation of epigenetic minerals and mineral deposits, or those formed later than the rocks en-

aqueous solutions (or magmas). The more soluble salts will tend to stay in solution longer and be precipitated later than the less soluble, thereby supplying one explanation of the sequences of minerals in mineral deposits and mineral zoning although the latter is the result of several other processes. Minerals once precipitated may be redissolved and reprecipitated.

Changes in pressure are probably more important than previously thought in promoting precipitation. Gases in solution are very sensitive to a change of pressure. For example, carbon dioxide dissolved in water promotes the solubility of calcium carbonate, and the release of carbon dioxide by agitation of the water and loss of carbon dioxide or lessened pressure causes precipitation of calcium carbonate from solution. Precipitation of solutes in solution can be affected isothermally by decrease in pressure alone. Likewise, the escape of fluids from magmatic solutions under decreasing pressure promotes crystallization of the magma.

Probably one of the most prevalent processes of deposition is by reduction of metals in solution. For example, complex metal chlorides may react with closing them. The ore minerals of contact-metaso-matic deposits have been emplaced by this process. Replacement is the controlling or dominating process in the formation of most supergene mineral deposits. It plays a major role in the extensive rock alteration that accompanies most epigenetic metallization. Replacement may occur through the action of hot vapors and gases, or by either hot or cold solutions. Cold copper sulfate solution may replace limestone to yield copper carbonate, or pyrite to yield chalcocite.

The *relative solubility* of solid and solute determines the precipitation of many minerals from solution. For example, if a copper sulfate solution meets sphalerite, which is more soluble, copper sulfide will be deposited at the expense of the sphalerite, which will go into solution; should the copper sulfate meet cinnabar, which is less soluble, no deposition will occur.

Oxidation and *reduction* involve the loss and gain, respectively, of electrons to specific ions that is determined by the environment, especially by the pH and Eh factors. Pyrite, for example, would be expected to form in a reducing environment, whether acid or basic, while hematite is formed in an oxidizing environment that is basic, such as fresh oxygenated seawater.

Direct deposition in open space, without involving replacement, may result by change of temperature and pressure, and by mineralizing solutions coming in contact with solids. The country rock traversed may change the solutions from acid to alkaline; certain wall rocks of fissures favor deposition of vein filling opposite them; and materials dissolved from such solids may bring about deposition.

Catalytic action, in which certain substances cause precipitation from solution without themselves entering into such solution, is another cause of mineral deposition. Many catalysts, such as platinum, are used chemically, and on a large scale; pyrolusite has been effectively employed by Zapffe as a catalytic agent to precipitate manganese and iron from the municipal water supply of Brainerd, Minnesota.

Adsorption is the taking up of one substance at the surface of another. For instance, kaolin adsorbs copper from solution to form much of the so-called chrysocolla, and silica gel adsorbs ferric oxide. In part, the change involves chemical reaction between the substances.

Base exchange occurs between solids and liquids whereby cations are exchanged, producing changed characteristics of both.

Precipitation by Bacteria. This is exemplified by the well-known precipitation of iron by bacteria. Harder recognizes three types of such bacteria; that is, those that precipitate (1) ferric hydroxide from ferrous bicarbonate solutions by absorbing carbon dioxide, (2) ferric hydroxide without requiring ferrous carbonate, and (3) ferric hydroxide or basic ferric salts from iron salts of organic acids. They are mostly thread and soil bacteria, of which *Crenothrix* is a common form. Such bacteria are considered by some geologists to have caused the precipitation of extensive deposits of iron ore (Chapter 10). Zapffe found that soil bacteria are effective precipitants of manganese from the subsurface waters of Brainerd, Minnesota. Algae precipitate travertine and in some places silica.

Sulfate-reducing bacteria are well recognized for the prominent role they play in producing native sulfur deposits, including the sulfur in the cap-rocks of salt domes by reducing $CaSO_4$ to produce bacteriogenic hydrogen sulfide, which is then partially oxidized to native sulfur.

Sulfate-reducing anaerobic bacteria liberate H_2S that may react with any metal to produce a sulfide of that metal. Syngenetic pyrite, which form at the same time as the enclosing rock, is the result of bacteriogenic H_2S reacting with indigenous iron by the following reaction:

$$9H_2S + 2Fe_2O_3 + 3O_2 \rightarrow 4FeS_2$$
$$+ 8H_2O + 2H^+ + SO_4^=$$

These bacteria, even though anaerobic, are almost ubiquitous, as they have adapted to fresh and saline waters, muds, and organisms; even a barophilic variety has been recovered from the Philippine trench at a depth of about 12 km. The most common species are *Desulfovibrio desulfuricans* and *Clostridium nigrificans*. The latter is a thermophilic variety.

In other cases, bacteriogenic hydrogen sulfide may act as the reductant that causes the deposition of specific minerals even though no sulfur is included in the reduced mineral. The reduction of U^{6+} to U^{4+} is an example.

Unmixing of Solid Solutions. Natural solutions of one solid in another are well known. Much gold contains silver in solid solution, and the ease with which gold unites with mercury to form amalgam is commercially utilized in the extraction of free gold from its ores. Solid solutions of magnetite and ilmenite, chalcocite and covellite, and other mineral pairs, are common although more recent studies suggest that some of these pairs may be the result of intergrowths. Some solid solutions form, and remain stable, at low temperatures. Others form only at high temperatures and become unstable at lower temperatures. When these cool slowly, one mineral may separate out of the other at a certain point or zone in the cooling-temperature curve. This is known as unmixing, or *exsolution.* Thus ilmenite plates separate out from solid solution in magnetite, covellite laths from chalcocite, argentite from galena, chalcopyrite from sphalerite and stannite, and cubanite from chalcopyrite. The minerals that form by exsolution remain as inclusions in the host and generally are visible only with a microscope.

Colloidal Deposition. Colloids are precipitated from natural solutions as flocculent or gelatinous masses because such waters contain dissolved salts, which are electrolytes. These may harden to rounded or colloform masses, but such forms as botryoidal, reniform, mammillary, nodular, or pisolitic are rarely thought to have resulted from colloidal deposition. Lindgren considered it proof of colloidal origin if single radial crystals cross colloform banding. Kalliokoski has recently shown that the framboidal texture that is commonly observed is the result of inorganic colloidal deposition. Others, however, suggest that this texture is the result of bacterial action.

In the past, there was a tendency among geologists to attach much importance to colloidal phenomena in mineral formation. Many hypogene and supergene minerals were considered to have had a colloidal origin, such, for example, as bauxite, chrysocolla, malachite, native arsenic, wurtzite, and others. Lindgren also believed that colloid minerals may replace older minerals. There is, however, little belief at the present time, in the action of colloids in mineral deposits.

Weathering Processes. Weathering is much more important in the formation of economic minerals than is generally realized. It is a complex operation that involves several distinct processes, such as disintegration, oxidation, hydration, reactions of solutions and gases with other solutions, gases, and solids, and evaporation—processes that may operate singly or jointly. Weathering is generally subdivided into mechanical and chemical action; usually both operate together. Mechanical action, although important in yielding valuable placer deposits, does not create the useful minerals; it merely frees and concentrates minerals already formed.

However, it facilitates chemical weathering by reducing the size of the materials, thus creating more specific surface available for attack. Chemical weathering, however, actually creates useful minerals by acting upon (1) preexisting economic mineral deposits, (2) submarginal mineral bodies, (3) gangue minerals and, (4) rocks (see also Chapter 5).

1. *Pre-existing mineral deposits* subjected to weathering yield new minerals that are stable under surface or near-surface conditions. These changes are brought about by the action of surface waters and atmospheric water, oxygen, and carbon dioxide. Some minerals are altered in situ; others are taken into solution, carried below, and there reprecipitated as new minerals. Common sulfides are particularly susceptible to attack, and in the zone of oxidation alter to "limonite," native metals, and the familiar oxides, carbonates, silicates, sulfates, and chlorides of the common metals. Below the zone of weathering, supergene sulfides, such as chalcocite, covellite, argentite, and others, are precipitated, or replace prexisting minerals, producing an entirely new group of valuable ore minerals.

2. *Submarginal bodies* of low-grade disseminated minerals, such as pyrite and chalcopyrite, are similarly converted into commercial deposits of chalcocite and covellite. Many copper deposits have been formed in this manner.

3. *Gangue minerals,* such as rhodochrosite and siderite, the carbonates of manganese and iron, and feldspar, weather to usable oxides of manganese and iron and to china clay.

4. Rocks weather into newly formed minerals, of which some form valuable mineral deposits. Feldspathic igneous rocks and shales yield clay deposits. Aluminous rocks in warm, moist climates yield deposits of bauxite, the ore of aluminum. Serpentine in Cuba yields iron laterite, which forms extensive iron and nickel deposits. Manganese laterites may have formed in India by weathering but this origin is less accepted at the present time.

Metamorphism. The agencies of metamorphism, mainly pressure, heat, and water, act upon rocks and minerals, causing recombination and recrystallization of the ingredients into new minerals that are stable under the imposed conditions. Many of them are of economic value. For example, garnet, graphite, and sillimanite minerals are created by metamorphism.

STABILITY OF MINERALS

Minerals, like life, are responsive to their physiochemical environment. They are for the most part stable under the conditions under which they were created but become unstable with different environment. Most of those formed under conditions of high temperature and pressure perish at the surface, and, vice versa, many of those formed near the surface change to more stable forms under high temperature and pressure. A few less-sensitive minerals such as diamond or cassiterite persist under changed conditions. Others called *persistent minerals*, such as pyrite or gold, form under many different conditions. Such sensitivity to conditions of formation or environment is utilized in interpreting the history of the mineral deposits that contain them. It is the earmark of progressive or retrograde metamorphism, of high or low temperature of formation, of hydrothermal alteration, of surface weathering, or other earth processes. Subjected to temperature changes or solvent action, minerals change to other substances more stable under the new conditions. A bit of iron, cradled in a fiery furnace, if thrown out, succumbs quickly to mere moisture and oxygen; it rusts, or changes to the more stable form of "limonite." So does pyrite.

The most noticeable changes of stability occur during weathering. Under the relentless attack of atmospheric water, oxygen, and carbon dioxide, few hypogene minerals survive. They are altered to native metals, oxides, carbonates, sulfates, chlorides, silicates, and other forms. Those minerals that already are oxides suffer least; sulfides suffer most. Some of our most beautiful mineral specimens are thus formed from more somber predecessors, and they are generally stable under their new environment. Their presence denotes the action of surface agencies. They presage a change in the nature of the mineral deposit below the depth of supergene alteration.

Common substances, such as clays or shale, reared on the earth's surface and later buried deeply beneath a pressing load of sediments will, under the new environment of increased pressure and temperature, change to more stable forms of muscovite or other silicates. Mica schists or garnets may thus be formed, and most of the metamorphic minerals belong to this group.

A monzonite, freshly consolidated, fissured, and coursed by mineralizing fluids arising from below, may have its feldspar and biotite changed to clay and sericite and kaolinite—minerals more stable under these conditions. This tells a story of hydrothermal alteration as convincingly as though the change had actually been observed. The feldspar is not stable in such an environment, but the alteration product minerals are.

Some minerals respond to a change in environment by undergoing a molecular change. Copper sulfide, for example, may form as many as six varieties depending upon different conditions of temperature; common ones are chalcocite, djurleite, and digenite.

Pyrite and marcasite also have the same composition but different crystallographic structure representing stabilities for different conditions. Also high- and low-quartz undergo displacive transformation at their inversion point of 573°C. Similarly, sillimanite, andalusite, and kyanite are triplets of the same composition, and with increased temperature and pressure change to mullite. Fig. 3-4 shows the pressure-temperature stability fields for carbon and the higher pressures needed to compress hexagonal graphites with layers of carbon held together by Van der Waals bonding, to isometric crystals of closely packed carbon atoms forming a diamond. The fields of stability of many minerals are now known; thus broad generalizations can be made concerning the weathering, hydrothermal alteration, metamorphism, and temperatures of formation of mineral deposits.

GEOLOGIC THERMOMETERS

Minerals that yield information as to the temperatures of their formation, and of the enclosing deposits, are termed *geologic thermometers*. They are of scientific and practical importance for a proper understanding of the origin of mineral deposits and their classification. This information has been obtained by direct observation, by laboratory experiments, and from the repeated observations of association of certain minerals with other previously determined diagnostic minerals. Some of the methods by which geologic thermometry has been determined follow.

Direct Measurements. The measurement of the temperatures of lavas, fumaroles, and hot springs yields maximum temperatures of formation for the minerals contained therein. Temperatures in slight excess of 1200°C have been recorded for basaltic lava at Kilauea, and up to 1140°C at Vesuvius. Washington estimated the more acidic lava of Santorini between 800 and 900°C. In general, the earliest minerals of the less basic rocks, form in part

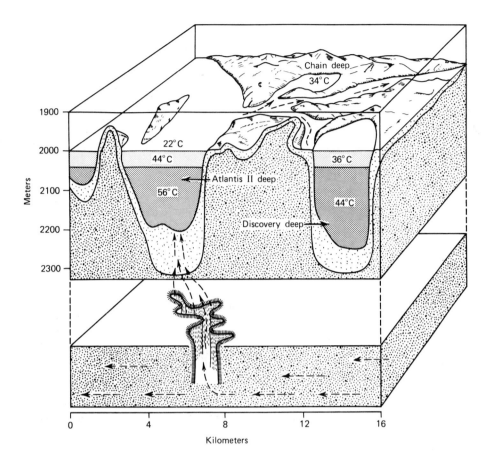

Figure 3-2 Bottom topography in the area of the three deeps in the Red Sea is seen near the top of this explored cross-section of the basement underlying the spreading center. The view is from the west; north is to the left. The long axis of the block measures 16 kilometers, but depth indications below 2300 meters are arbitrary. Top arrows show how overflow of the Atlantis 11 Deep can periodically replenish the two other pools. The winding "canyon" visible in the lower block represents zones in the basement rock where the percolating brine, cooling as it rises toward the floor of the Atlantis 11 Deep, has probably deposited rich veins of various metallic ores in the rock below. (From: Degens and Ross, "The Red Sea Hot Brines." copyright 1970 by Scientific American, Inc. [all rights reserved.].)

above 870°C but principally between 870° and 650°C, decreasing with increasing silica content. Pyrogenic ore minerals, such as chromite, for example, form within the range of magma consolidation. Also, contact-metamorphic minerals would not ordinarily form at temperatures higher than those of the magmatic emanations that produce them. The gas temperatures of fumaroles likewise indicate maximum temperatures for fumarole minerals. Temperatures in excess of 700°C have been measured at Showashinza, Japan. A temperature of 645°C has been measured in the fumaroles of Katmai, and magnetite and other minerals have been deposited about their conduits. Lava flow fumaroles reach 800°C. With waning fumarolic activity, lower temperatures occur.

The temperatures of shallow hot springs extend downward from the boiling point of water, and maximum temperatures of formation can be assigned to opal, gypsum, cinnabar, stibnite, and many others that have been observed in spring deposits.

Variations in the temperature of the Red Sea, in which mineral-bearing brine solutions have been sampled and analyzed, are shown in Fig. 3-2. Different portions of the Red Sea have different geo-

thermal gradients, and temperatures below the bottom of the sea are considerably higher.

Melting Points. The melting points of minerals indicate maximum temperatures at which they can crystallize from a melt. In a supersaturated solution, they may precipitate at considerably lower temperatures. In melts, the presence of other substances, however, greatly lower the liquid-solid temperature point. Examples of melting points are albite at 1104°C, stibnite at 546°C, bismuth at 271°C.

Dissociation. Minerals that lose volatile constituents at certain temperatures serve as poor geologic thermometers because the temperature of dissociation is increased by pressure. Most zeolites indicate low temperatures of formation because when heated they lose their water content, provided that the pressure is not too high. Calcite dissociates under atmospheric pressure at 885°C, but according to Smyth and Adams, only 40 meters of rock pressure is required to prevent dissociation at 1100°C. This dissociation is also affected by the mole fraction of CO_2 in a CO_2-H_2O environment. Furthermore, silica available for combination with the calcium oxide

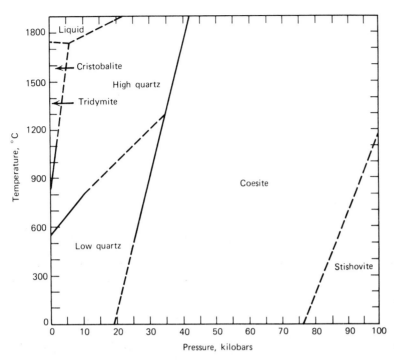

Figure 3-3 Regions of stability of the various forms of SiO_2. The $\alpha\beta$ transitions are rapid, but all other types are sluggish or have to form from a ma. (From: Dana, Hurlbut, & Klein, *System of Mineralogy*, 19th Ed.)

lowers the dissociation temperature. Thus the presence of intermixed calcite, CO_2-H_2O, and quartz indicates a temperature below their combining point, giving consideration to pressure.

Inversion Points. Some useful temperature indicators are inversion points even though they are affected by pressure. Many inversion points are known within the temperature range prevalent in the formation of most mineral deposits. Silica is erroneously readily utilized. It is of widespread occurrence and has many crystalline modifications whose ranges of stability are known. Even so, the utilization of tridymite and cristobalite is attended with complications because both occur in volcanic amygdules having formed at temperatures below their inversions at 970 and 1470°C at 1 atmosphere. The noticeable effect of pressure on the inversion temperatures of SiO_2 is shown in Fig. 3-3. At about 573°C and 1 atmosphere, high quartz inverts displacively to low quartz (and vice versa), with recognizably different symmetry. Thus low quartz may have been formed originally below 573°C or it may originally have been high quartz that has inverted to the low form. The two forms can be distinguished. Wright and Larsen found that geode quartz and much vein and pegmatite quartz were formed below 573° and that the quartz of many igneous rocks was originally high quartz. Much pegamtite quartz, formerly thought to be high quartz has been

found by Ingerson, using fluid inclusion geothermometry, in many to have been formed below 250°C.

N. W. Buerger concluded that orthorhombic chalcocite inverts at 105°C to a hexagonal form but reverts on cooling as the result of a displacive transformation. He also concluded that what was formerly called "isometric chalcocite" is digenite (Cu_9S_5), which above 78°C dissolves covellite. From this it follows that if digenite contains de-

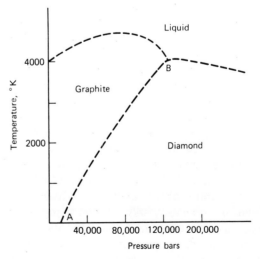

Figure 3-4 Regions of stability of the various forms of carbon. (From: Fyfe, W. S., *Geochemistry of Solids*. McGraw-Hill Co., 1964)

monstrably unmixed covellite, it was formed above 78°C and was hypogene rather than supergene. However, Roseboom has shown that there are three low-temperature phases of copper sulfide, namely chalcocite, djurleite, and digenite. Chalcocite changes to hexagonal chalcocite at 103.5°C and this to high digenite at 420°C. Djurleite cannot exist above about 93°C, but its presence in ores does not necessarily mean that it was deposited below 93°C because it can form by solid state transformation from phases stable above 93°C. Hexagonal chalcocite has not been observed in nature. Roseboom's work throws doubt upon the previous identification of copper sulfides and upon their use as geologic thermometers as previously reported.

Argentite (isometric) and acanthite (orthorhombic) represent, respectively, the high- and low-temperature forms of Ag_2S, with an inversion point of 175°C. The external form of argentite crystals is isometric. Therefore, it follows that they were formed above 175°C, and the anomalous anisotropism commonly ascribed to argentite indicates that such argentite was originally isometric and later inverted to the orthorhombic acanthite. Unquestionably, much Ag_2S that is called argentite is really acanthite. Ramdohr has shown that cubanite I inverts to cubanite II at 235°C. Many other examples are known.

Exsolution. Minerals that form natural solid solutions in each other, and at determined lower temperatures, unmix to yield distinguishable mineral intergrowths, serve as geologic thermometers, indicating a temperature of formation above that at which exsolution takes place. For example, Borchert has shown that chalcopyrite and bornite unmix at 500°C, cubanite and chalcopyrite at 450°C, cubanite and pentlandite at 400°C, and bornite and chalcocite at 175 to 225°C. Borchert has shown that chalcopyrrhotite exsolves below 255°C into chalcopyrite, cubanite, and pyrrhotite. Similarly, allemontite exsolves into arsenic and antimony at 200 to 250°C. Unmixing of magnetite and ilmenite, and hematite and ilmenite, have been demonstrated by Ramdohr, although his temperatures of exsolution, 600°C to 700 and 675°C, respectively, have been shown by Greig to be too high. The effect of rate of cooling, or time, and impurities on exsolution temperatures have usually been ignored which cast some doubt on exact temperatures of exsolution.

Recrystallization. This change is somewhat similar to inversion and exsolution but applies more spe-

cifically to native metals. Carpenter and Fisher have found that native copper undergoes a recognizable recrystallization at about 450°C. Microscopic examination reveals that most native copper has been formed well below this temperature, in fact, much native copper is supergene. Similarly, they found that native silver recrystallizes at about 200°C and that the silver ores of Cobalt, Ontario, were formed above this temperature.

Fluid Inclusions. Sorby long ago showed that fluid inclusions in cavities of crystals indicate the approximate temperature of formation of the crystals by the amount of contraction of the liquid, assuming that the liquid originally filled the cavity. W. H. Newhouse applied this method to determine the temperature of formation of various sphalerites, which were heated on a microscope stage until the liquid filled the cavities, when the temperature was read. He found that the sphalerite of the Tri-State District, for example, had been formed at temperatures of 115 to 135°C. W. S. Twenhofel similarly determined that a fluorite crystal from New Mexico started growth at 202°C and continued to 150°C. Determinations by Earl Ingerson on pegmatite quartz disclosed surprisingly low temperatures of formation, none more than 250°C, even with pressure corrections. H. S. Scott developed a procedure for opaque minerals by which mineral powders are heated until the inclusions burst or decrepitate thereby giving the upper temperature of formation.

Fluid inclusion studies have been greatly refined and enlarged upon by the careful work of E. Roedder. He has not only determined with some precision the temperature of formation of minerals, but also the density, salinity, rate of movement, and composition of the solutions that deposited the minerals and has thereby added greatly to the understanding of the origin of some ore deposits.

Even though fluid inclusions have been noted for almost two centuries, the newer methods of analysis using micro analytical techniques, mass spectrometric analysis of the δD and $\delta^{18}O$ values of the water in the fluid, and especially heating and freezing techniques have indicated the significant importance of fluid inclusions to geothermometry and their value as microsamples of the fluids from which the mineral containing the inclusion formed.

Figures 3-5 and 3-6 are photomicrographs of fluid inclusions studied by Roedder. The large primary inclusions of Fig. 3-5 are enclosed in a striated surface of a yellow sphalerite crystal from Creede, Colorado. The vapor phase bubbles form a small

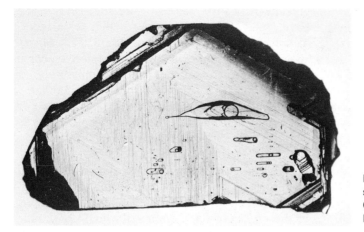

Figure 3-5 Large primary inclusions just under striated surface of clear, yellow, zoned sphalerite crystal from Creede, Colorado. (From: Prof. Paper 440 JJ, Plate 1, Fig. 6 From: E. Roedder)

portion of the inclusions. With heating of the crystal, the vapor phase will gradually decrease in size until it disappears, which is inferred to be close to the temperature at which the mineral formed. Figure 3-6 illustrates many daughter crystals in an inclusion in a *Volynian* topaz crystal from a chambered pegmatite specimen from Russia.

The validity of fluid inclusion studies is based on three major assumptions as given by Roedder,

1. If the ore or coprecipitated gangue mineral crystallizes from a fluid medium, the growing crystal is surrounded by the fluid from which it is crystallizing, that is the "ore fluid."

2. If a reentrant imperfection of any kind forms on the surface of the growing crystal, this reentrant is filled with the ore fluid, and if trapped by

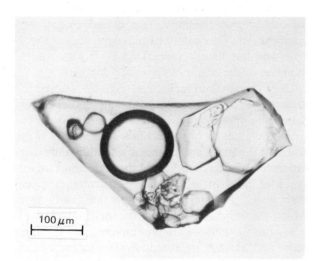

100 μm

Figure 3-6 Many daughter crystals in inclusion in Volynian topaz. (Chambered pegmatite from USSR; see plate 12 in Data of Geochemistry Prof. Paper 440 JJ. From: E. Rodder)

additional crystal growth, the resulting primary fluid inclusion is a representative sample of this ore fluid at the moment of trapping.

3. Significant quantities of material are neither lost nor gained subsequent to this trapping.

Although certainly each of these assumptions may not be strictly valid in *every* case, the available evidence indicates that if care is used in sample selection, the exceptions to these assumptions will be few and of minor significance.

A geothermometer method suggested by Fournier and Truesdell is based on the Na-K-Ca concentration of fluid inclusions. The test of this technique was done on fluid inclusions of various quartz and one fluorite sample from the Climax, Colorado, mine. The K/Na ratios of the hydrothermal fluids were controlled by exchange reactions with alkalai feldspars of the Climax stock and are a function of temperature. Calcium concentrations, however affect the temperature estimates. Comparison of the temperatures between 280 to 348°C obtained (Table 3-5) by this geothermometry technique that average about 321°C does not agree well with the temperature estimated for the late barren pyrite-quartz stage and flourite of about 230°C.

In earlier fluid inclusion studies, a heating stage attached to the microscope was used. As the thin section containing fluid inclusion was heated, the vapor phase decreased in size until the liquid phase filled the inclusion. This filling temperature was assumed to be the temperature of formation with further refinement attempted for pressure corrections.

At the present time, freezing temperatures are used as PT conditions of pure water differ from a

Table 3-5 Chemical and Isotopic Compositions of Fluid Inclusions, Climax Mine

| Host Mineral | Location and Ore Body | Chemical Composition | | | | | | | | Na-K-Ca Temp. °C[a] | Isotopic Composition δD_{SMOW} per mil |
		Na (ppm)	K (ppm)	Ca (ppm)	Mg (ppm)	Cl (ppm)	Zn (ppm)	Cu (ppm)	Total (ppm)		
Quartz	Ceresco level Ceresco ore body	64,000	13,000	26,000	2,400	70,000	200	4,000	180,100	280	−88
Quartz	600 level, roots Ceresco ore body	20,000	10,200	16,000	2,000	14,000	500	80	62,780	348	−118
Quartz	929 level Lower ore body	38,600	12,800	14,600	2,300	43,600	1,800	1,300	115,000	315	−101
Quartz	929 level Lower ore body	43,600	17,100	16,400	2,200	51,800	4,600	—	135,700	342	−102
Quartz	Storke level Lower ore body, high silica	16,000	5,000	8,000	600	16,000	300	300	46,200	290	−126
Quartz	600 level Late barren mineralization	22,000	1,400	8,000	900	14,400	1,000	600	48,300	193	−140
Quartz	600 level Late barren mineralization	14,500	1,900	5,000	500	1,400	1,200	—	24,500	230	−134
Fluorite	Phillipson level Late barren mineralization	2,100	700	1,500	200	2,300	40	10	6,850	264	−138

Source: From: Hall, et al., Econ. Geol., v. 69, 1974.
[a] Based on the Na-K-Ca geothermometer of Fournier and Truesdell, 1973.

saline solution which is prevalent in fluid inclusions. Roedder, for example, shows that at a filling temperature of 350°C, if the pressure is 2000 bars, the actual temperature is 530°C if the fluid contains 20 percent NaCl; it would have formed at 615°C if the fluid had been pure H_2O. At temperatures above the critical temperature, the pressures increase rapidly with increasing salinity.

Changes in Physical Properties. Some minerals at certain temperatures undergo recognizable changes in certain of their physical properties. The pleochroic haloes in mica are destroyed at 480°C; smoky quartz and amethyst lose color between 240 and 260°C; and the color disappears from fluorite at around 175°C. None of these changes is reliable enough to be of significant use as geothermometers, in fact, there is some evidence that radiation damage is the cause of coloration in quartz and fluorite.

Mix-Crystals. An indication of temperature is given when the composition of mix-crystals is determined by the temperature of formation. For example, Buddington has shown a temperature determination from the amount of TiO_2 in magnetite

associated with ilmenite as the TiO_2 content increases with increasing temperature of formation.

A sphalerite-pyrrhotite-geothermometer was proposed by Kullerud based on the amount of FeS in sphalerite. This was widely used as an ore geothermometer by many geologists in recent years but the variable substitution of other elements for Fe and Zn has decreased the precision of this technique.

Other pairs, such as Fe_2O_3 in Mn_2O_3, and others have been similarly used for geothermometry but the substitution of other similar ionic radius and valence change minerals limits the use of this technique.

Mg/Fe Substitution in Biotite. A possible geologic thermometer for use in the potassic zone of porphyry copper deposits is one that pertains to the differences in molecular ratios of Mg:Fe in the primary biotite of typical granitic-granodioritic host rocks in contrast to the hydrothermal biotite. The former exhibit Mg:Fe ratio < 1.0 while the latter are characterized by Mg:Fe ratios of > 1.5.

The substitution or mixing of Mg^{2+}-Fe^{2+}-Fe^{3+} provide a geothermometer for the potassic zone which is based upon the composition of the biotite coexisting with magnetite and K-feldspar.

Using this suggested technique, temperatures of potassic alteration at Santa Rita, New Mexico; and Ray and Safford, Arizona, lie in the range of 350 to 450°C. Bingham, Utah, and Galore Creek, British Columbia, which contain significant amounts of bornite, are characterized by slightly higher temperatures ranging from 450 to 550°C. These temperatures are in general agreement with temperatures determined by other techniques.

More recent studies of this geothermometry method have cast substantial doubt on the technique. Using the substitution of specific ions into given minerals at specific temperatures is fraught with the problems of similar sized and ionic charged elements that will readily substitute for the specific element ''one expects but doesn't always get.''

Conductivity. F. G. Smith devised a ''pyrite geothermometer'' for measuring the thermoelectric potential of pyrite against a metal to give the temperature of formation of the pyrite, which under certain circumstances appears to provide reasonable temperatures.

Pressure.
The effect of pressure on geothermometry and indirect determinations of temperature has often been ignored. In some cases, this may be valid as in some systems and especially in isotopic geothermometry. In other cases, however, pressure can change temperature measurements and estimates drastically. The Clausius-Claperon equation provides a means of determining the change of temperature of a phase boundary with pressure. This requires, however, knowledge of factors of the material being studied such as the change in volume accompanying the phase change with temperature as shown by the equation:

$$\frac{dt}{dp} = \frac{\Delta V}{\Delta S} = \frac{T \Delta V}{\Delta H}$$

where dt/dp = the change of temperature with pressure, ΔV = the change in volume at temperature T. ΔH is the enthalpy or heat content per mole at temperature T. Also, ΔV is almost always positive as materials expand at their phase boundaries with heat, with the important exception of ice and the lesser importance of Bi and Sb that shrink when they melt.

An example of the effect of pressure in increasing the temperatures of the phase boundaries of SiO_2 is shown by Fig. 3-3. Figure 3-4 shows the regions of stability for different forms of carbon. Diamonds are not readily formed by placing carbon in the diamond stability field. Certain catalysts are re-

quired which have allowed the formation of man-made diamonds.

Stable Isotope Geothermometry. One of the more reliable geothermometers of increasing use is based upon the measurement of stable isotopic ratios of low atomic weight elements. Oxygen-bearing minerals have been used in geothermometry with impressive results. Even though pressure increases the rates of isotopic reactions, it is of little significance in isotopic equilibrium reactions because crustal pressures have little or no effect on the nucleus of an isotope. It is possible, therefore, to determine the pressure effect on other geothermometers, such as fluid inclusions, by determining the difference in temperature of formation by both techniques on the same or similar sample.

Although there are several pairs of stable isotopes to which geothermometry studies have been applied, $\delta^{18}O$ and δD studies surpass all others. Some measurements can be done using $\delta^{34}S$ studies but not with the reliability obtained by $\delta^{18}O$ applications.

Some time ago Craig noted that fresh meteoric waters showed a correlation between their $\delta^{18}O$ and

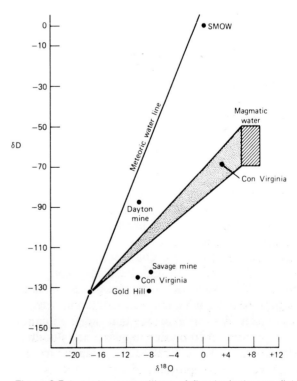

Figure 3-7 Isotopic compositions of five hydrothermal fluids from various mines in the Comstock Lode District. Stippled area represents isotopic compositions which would result from mixtures of local meteoric water and a hypothetical magmatic water. (From: O'Neill, Econ. Geol., v. 69, 1974)

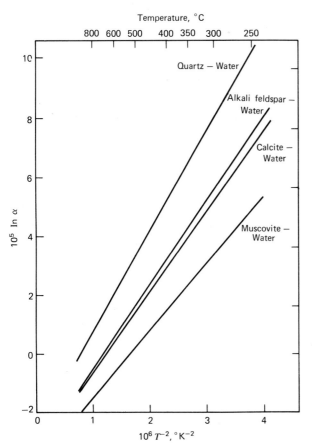

Figure 3-8 Experimentally determined isotope calibration curves. These curves are plotted as 1000 ln α versus $10^6/T$, where α is the mineral-water fractionation factor and T is the absolute temperature. (From: Taylor, Econ. Geol., v. 62, 1967)

δD values as indicated on Fig. 3-7. In contrast, magmatic water has a restricted isotopic composition as indicated in Fig. 3-7. The oxygen in unaltered igneous silicates will generally fall within this range making it possible, therefore, to determine the source of oxygen or hydrous phase sources in a variety of minerals. Hydrothermal minerals, or hydrothermally altered minerals, often have significantly lower $\delta^{18}O$ and δD values indicating the involvement of much meteoric water.

Equilibrium oxygen isotopic fractionations versus temperature have been determined for various mineral-H_2O pairs. Some of the determinations are shown in Fig. 3-8. It is evident that if the $\delta^{18}O$ value of calcite and the $\delta^{18}O$ value of quartz were determined, where both minerals formed contemporaneously and isotopically exchanged oxygen under equilibrium conditions, the absolute temperature of formation of the mineral pair could be determined, and incidentally the $\delta^{18}O$ of the water in which these minerals formed could also then be determined.

Examples of such measurements and studies have been presented by James, Clayton, O'Neill, Shepherd, Garlick, Epstein, and others on regionally metamorphosed minerals. Even though the metamorphic rank in the minerals indicated is generally well known (Fig. 3-9), the isotopic geothermometer determined the absolute temperature environments of the minerals.

In the study of mineral deposits, Taylor has shown that isotopic geothermometry has not only been an exceptional "tool" in determining temperatures and the sources of hydrothermal water, but it has also been used to study hydrothermal alteration by determining better the reactions that have occurred and also the temperature conditions under which such reactions took place. Examples are shown on Fig. 3-10 and 3-11.

There are other kinds of geologic thermometry, such as thermoluminescence, metamorphism of minerals, and phase equilibrium in silicate melts, which, however, with the exception of the first, apply to the high temperatures of formation of igneous rock but are seldom evident in mineral deposits.

General Considerations. The previous discussion shows that there are several classes of geologic thermometers, very few of which record accurately

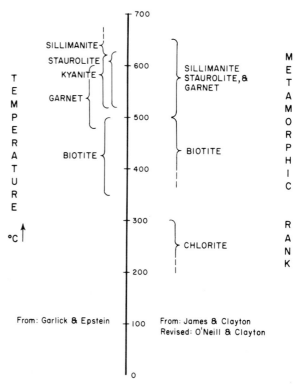

Figure 3-9 Oxygen isotopic application to determination of temperatures of regionally metamorphosed suites of minerals.

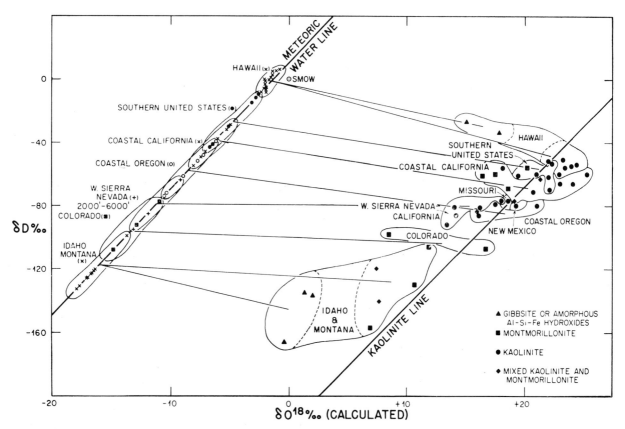

Figure 3-10 Plot of δD versus δ¹⁸O for clay minerals and hydroxides from modern soils formed on igneous parent rocks in the United States, after Lawrence and Taylor (1971). The δ¹⁸O values are only approximate because they are calculated from the bulk soil analyses by subtracting out the isotopic contribution due to parent-rock minerals. Also shown is the kaolinite line of Savin and Epstein (1970a) and some typical meteoric waters form various regions. Most of the meteoric waters shown were analyzed only for δD or δ¹⁸O, not both; they are, however, plotted on the meteoric water line. (From: Taylor, Econ Geol. v. 69, 1974)

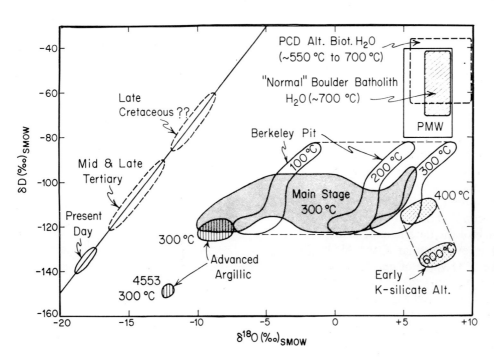

Figure 3-11 δD versus δ¹⁸O diagram showing the fields for the calculated isotopic composition of H_2O that would be in equilibrium at the stated temperatures with the alteration minerals from the pre-Main Stage, and the Central, Intermediate, and Peripheral Zones of the Main Stage at Butte. The Main Stage envelope includes calculated waters based on data for both quartz and hydrous minerals. Also shown are the fields for primary magmatic water (PMW), normal Boulder batholith H_2O, and water in equilibrium with alteration biotite from porphyry copper deposits PCD (Sheppard et al., 1971), as well as the estimated isotopic compositions of meteoric waters in the Butte area during the Late Cretaceous, Tertiary, and present. (From: Sheppard and Taylor, Econ. Geol., v. 69, 1974)

specific temperature conditions of formation. Some provide an upper or a lower temperature above or below which they do not form; others provide a range of temperature within which they may form; still others serve only as rough indicators. It should be understood that the temperature figures are subject to variations due to pressure or other factors. The presence of two or more of the less precise geologic thermometers in a mineral deposit may establish a fairly narrow range of temperature of formation for the deposit as a whole; one may fix a minimum temperature of formation, another a maximum. In summary, however, the isotopic study of the stable isotopes seems to be the most valuable. The information obtained of the temperatures that existed during the formation of the Butte mineral deposit (Fig. 3-11) is evidence enough of the value of this technique.

Selected References

General Reference 9, chaps. 1 and 2. *Relate to distribution* and formation of minerals.

General Reference 1, v. 69, no. 6. *Stable isotopes as applied to problems of ore deposits.* A volume devoted entirely to the use of stable isotopes.

General Reference 14, chap. 7. Recent summary on sulfide mineral stabilities.

Studies of Fluid Inclusions I and II. 1962. E. Roedder. Econ. Geol. v. 57, p. 1045, 1061; Econ. Geol., v. 58, 1963, p. 167–211. Classic papers on Fluid inclusion methods of analyses and results.

Methods and Problems of Geologic Thermometry. 1955. Earl Ingerson. Econ. Geol. 50th Anniversary Volume, pt. 1, p. 341–410. A thorough review of all kinds of geologic thermometry, with an extensive bibliography.

An investigation of the system Cu-S and some natural sulfides between 25 and 700°C. 1966. E. H. Roseboom, Jr. Econ. Geol., v. 61, p. 641–672. *Clarification of copper sulfides in geologic thermometry.*

Dana's System of Mineralogy. 1944, 1951. Palache, Berman, and Frendel. Vols. 1, 2. John Wiley & Sons, Inc., New York. *A general textbook for references on mineralogy.* Also *Dana's Manual of Mineralogy,* 17th ed. 1959. C. S. Hurlbut. John Wiley & Sons, Inc., New York. *A more modern treatment of selected minerals.*

Textures of the Ore Minerals, and Their Significance. 1947. A. B. Edwards, Aust. Inst. Min. and Met. Melbourne, Australia. Chaps. 1, 4, and 6. *Deals with several geologic thermometers based upon recrystallization, solid solution and exsolution, inversion points, and melting points.*

Progress in the study of exsolution in ore minerals. 1942. G. M. Schwartz, Econ. Geol. V. 37, p. 345–364. *Criteria and examples of exsolution.*

The pyrite geo-thermometer. 1947. F. G. Smith, Econ. Geol., *v. 42 p. 515–523. Determination of temperature of formation of pyrite by measuring the thermoelectric potential.*

Decrepitation method applied to minerals with fluid inclusions. 1948. H. S. Scott. Econ. Geol. v. 43, p. 637–654. *Opaque minerals with fluid inclusions are heated until they burst, thus giving an audible record of the temperature of formation.*

4

PETROLOGY OF MINERAL DEPOSITS: MAGMAS, SOLUTIONS, AND SEDIMENTS

Thoughts to unlock the fettering chains of things.

Robert Underwood Johnson,
"In Tesla's Laboratory"

It is now generally considered that many metalliferous deposits and also many nonmetallic deposits result from igneous activity. Magmas are the parents, mineral deposits the offspring. Many are immediate descendants that originated soon after the consolidation of the parent magma; others, immature at birth, have been built up during one or more periods of growth by diverse processes to yield subsequent, fully developed deposits. A placer gold deposit, for example, may have been originally a lean magmatic offspring that first underwent disintegration and then was later concentrated by moving waters into an economic workable deposit. Another intermediate stage may have intervened; the earliest surficial concentration may have collected the gold into a sedimentary formation, still too lean to be worked, and then later reworking may have further concentrated it into eagerly sought placer deposits. The progenitor of the gold, however, was some parental hydrothermal dike or intrusion. The wily prospector, aware of this, has traced the errant placer gold upstream to its source in order to search for the "mother lode," as he terms it. Many profitable gold mines attest to his success. Still other metalliferous lean magmatic offspring have not matured until they have been exposed by erosion to the beneficent activity of the atmosphere whose leaching action has removed the metals from undesired companions and concentrated them into cleaner, fatter, richer deposits of supergene origin but of magmatic ancestry nevertheless.

There have also been widely noted associations of certain metals and minerals with certain kinds of igneous rocks, for example, chromite with peridotite or tin with granite. Here the kinship is intimate. The offspring have remained within the body of the parent or close to its edge. The repeated association cannot be coincidence; it must be genetic. Slightly more wayward offspring escape from the parent body but nestle close to its fringe, and their high-temperature birth clearly indicates a source from adjacent igneous masses. The most wayward offspring wandered far, perhaps many thousands of meters, but also contain minerals that point to moderate temperatures of formation and igneous affiliations. Thus the valuable materials of many mineral deposits, whether hypogene or supergene, have directly or indirectly come from magmas. It is, therefore, of the utmost importance for a proper understanding of the origin and nature of mineral deposits to consider briefly some features of magmas and their crystallization.

MAGMAS

Magmas are masses of molten matter plus their dissolved fluids derived from the earth's crust and the upper mantle. Their composition, however, is not the same as that of the rocks from which they were derived. Magmas may be derived from the more readily mobilized constituents of the original rock. In addition, magmas contain water (Fig. 4-1) and other small but important quantities of volatile substances that escape or react with minerals already formed before complete consolidation occurs. In many cases, the water is not entirely indigenous but may be derived from groundwater or crustal formations as shown in Figs. 4-2 and 4-3. Strictly speaking, magmas are high-temperature mutual solutions of silicates, silica, metallic oxides, and always variable amounts of fluid substance. Their highest temperatures range from about 625°C for felsic magmas up to (more than 1200°C) for mafic magmas.

Their composition is as variable as the vast array of rocks that they yield; they range from felsic to mafic. The volatiles consist chiefly of predomi-

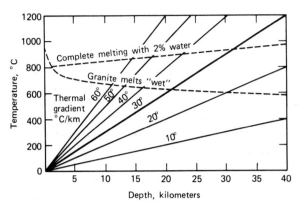

Figure 4-1 Graph showing the relations between the melting of granite and various geothermal gradients, assuming the gradients are linear. (From: Tuttle, D. F. and Bowen, N. L., Geol. Soc. Am. Memoir 74, 1958)

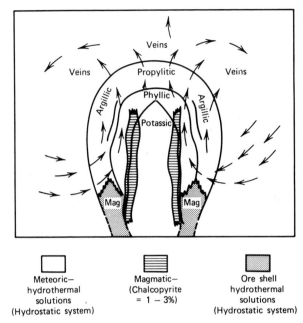

Figure 4-2 Schematic cross-section through a typical porphyry copper deposit (modified after Lowell and Guilbert, 1970), showing the two types of hydrothermal fluids generated in such environments, as determined from δD measurements (Sheppard et al., 1969, 1971). (From: Econ. Geol., v. 64, 66)

nantly water, carbon dioxide, and compounds or ions of sulfur, oxygen, hydrogen, chlorine, fluorine, and boron. Although minor in amount, the volatiles play an important role in decreasing viscosity, in lowering the melting point, in collecting and transporting metals, and in the formation of mineral deposits. These constituents are largely expelled but are also trapped in fluid inclusions, or absorbed or substituted in other minerals as the magma consolidates.

Magmas are temporary features within the earth's crust and upper mantle. Most magmas presumably originate by partial or selective melting of the upper mantle or the crust (Fig. 4-5) promoted by the presence of fluxes or mineralizing fluids.

Figure 4-5, for example, suggests that calc-alkalic magmas may form by partial melting, presumably of eclogite, in the upper mantle. Granitic magmas may form in a similar fashion or by the partial melting of sedimentary rocks. Primordial sulfur isotopic ratios and low Sr/Rb ratios of minerals in many

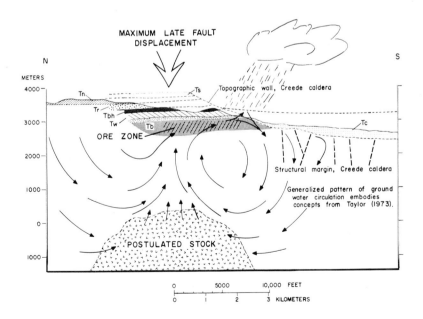

Figure 4-3 Idealized north-south section through the Creede mining district. Existing rocks are patterned; restored rocks, open. Tc, Creed Formation; Ts, Snowshoe Mountain Tuff; Tn, Nelson Mountain Tuff; Tr, Rat Creek Tuff; Tbh, andesite of Bristol Head; Tw, Wason Park Tuff; Tb, Bachelor Mountain Member, Carpenter Ridge Tuff. Coarse stipple in the Rat Creek Tuff and at the top of the Bachelor Mountain Member indicates soft, relatively impermeable tuff. (From: Steven and Eaton, Econ. Geol., v. 70, 1975)

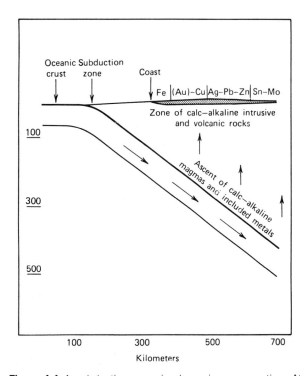

Figure 4-4 A subduction zone is shown in cross-section. At the right, magmas created by the sinking plate are rising toward the surface, and may ultimately create mineral deposits like those in the Andes. The illustration is taken from "Relation of Metal Provinces in Western America to Subduction of Oceanic Lithosphere." (Richard H. Sillitoe. Geological Society of America Bulletin, March, 1972).

granites indicate that the former is much more likely than the latter. Nevertheless, the felsic magmas presumably derive most of their water from crustal sediments as indicated by $\delta^{18}O$ and δD values. The melting is local; no continuous molten layer underlies the solid crust.

Basaltic magmas are derived from upwelling upper mantle. As illustrated in Fig. 4-5, when ultramafic periodotite reaches point A, the temperature of the partly molten mass changes within the crystal and liquid zone because of the consumption of heat consumed by the enthalpy of melting. The dotted line, therefore, shows the P-T changes of the partially melted mass as it continues to rise. When within or nearing the crust, additional heat is consumed through the pyroxenite region as the remaining peridotite minerals are incorporated in the phases in basalt. The mass may then reach the surface of the crust within a temperature at or near the liquidus of basalt. If the mass remains below

the surface with very slow cooling, differentiation takes place as shown in Fig. 4-5.

Reaction Series. Crystallization is not in all cases a simple formation of minerals that persist as such. Bowen suggested that certain minerals, once formed, may continue to react with the enclosing liquid magma, with the result that their composition is continually being modified and new minerals or solid solutions result. He described a definite sequence (Fig. 4-6) of reactions, which he called the reaction series. Later studies, such as those of the Skaergaard, Greenland, intrusion, indicate that differentiation of a mafic magma at shallow depths is not toward a residual calc-alkalic magma but actually toward a ferrogabbro (Table 4-1). Bowen's reaction series is only a generalization that is modified in detail by specific studies of magmatic differentiation.

The minerals precipitated in a magma are not simply the result of the source of the magma. Temperature, pressures, and even the amount of water in the magma will determine, for example, the abundance of biotite, hornblende, and other hydroxyl-bearing minerals. Hornblende and biotite crystalliz-

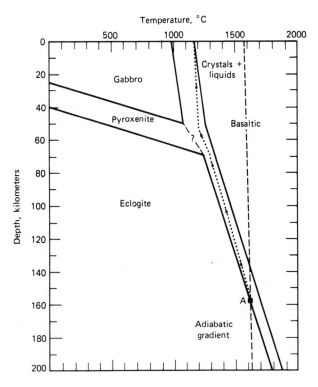

Figure 4-5 Generation of basaltic magma. (From: Tuttle, *Generation of Basaltic Magma*, Nat'l Acad. of Sci., 1976.)

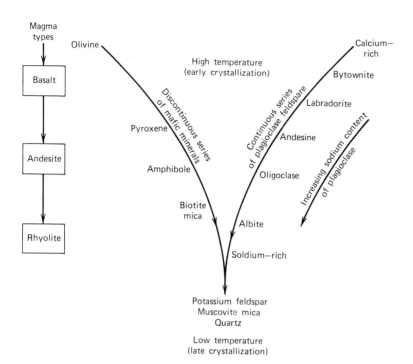

Figure 4-6 Bowen's reaction series, showing how the sequence of fractional crystallization of a melt leads to the formation of differentiated magmas. (From: Press and Siever, Earth, W. H. Freeman and Co., San Francisco, 1974.)

ing from magmas with varying temperatures and water pressures (Figs. 4-1 and 4-2) indicate the increased amount of original water present.

Upward movement may be induced by less overburden, by earth movements that tend to squeeze the magma out; or the magma may melt and assimilate some of the overlying cover rock and move upward aided by gas pressure within itself. Assimilation is unlikely because of insufficient heat needed to fuse host rocks. Nevertheless, in its upward movement, the magma may pry off blocks of

roof rock that sink into the liquid, giving rise to magmatic stoping and the formation of *xenoliths*. It may reach the surface, giving rise to volcanic extrusions; or it might solidify at depth, forming large intrusives of stocks or batholiths.

CRYSTALLIZATION

Inasmuch as magmas are solutions, crystallization or precipitation of the constituent minerals will

Table 4-1 Variation of Mineral Composition with Depth in the Skaergaard Intrusion

| Height above lowest exposed level, m. | Mean composition of minerals | | | | |
	Plagioclase	Clinopyroxene	Ortho-pyroxene	Olivine	Rock
2,500[a]	An_{30}	$Wo_{30}En_2Fs_{68}$	Absent	Fa_{96}	Quartz ferrogabbro
1,800	An_{40}	$Wo_{27}En_{23}Fs_{50}$	Absent	Fa_{60}	Ferrogabbro
1,200	An_{45}	$Wo_{32}En_{32}Fs_{36}$	Fs_{57} [b]	Absent	Gabbro
500	An_{56}	$Wo_{42}En_{40}Fs_{18}$	Fs_{45}	Fa_{37}	Olivine gabbros
0	An_{61}	Not given	Fs_{41}	Fa_{34}	

[a] Roof.
[b] A minor constituent.
Source: Turner & Voerhogan, *Igneous Metamorphic Petrology.*

commence when their concentrations exceed their solubilities. The crystallization of minerals is not determined simply by their temperatures of fusion. The solution of one constituent lowers the melting point of another constituent. Consequently, magmas remain fluid at temperatures well below the melting points of all its minerals.

Crystallization commences below this point; as the temperature lowers and the solution becomes supersaturated in specific minerals, their crystallization eventually ensues, giving rise to the paragenises of individual minerals or isomorphous mixtures.

Order of Crystallization. The lower solubility substances crystallize first, and these include some of the accessory minerals, such as apatite, zircon, ilmenite, magnetite, and chromite. Olivine and orthorhombic pyroxene are among the earliest essential minerals to crystallize, followed by clino-pyroxenes, Ca-plagioclase, hornblende, Na-Ca plagioclase, Na-plagioclase, orthoclase, mica, and quartz (Fig. 4-6). This is the normal succession, but there are exceptions. In fact, the final rock product is more often a ferrogabbro, rather than granite.

With the subtraction of the most mafic minerals from the magma, the residual magma in general becomes progressively more felsic. Granitic residual magmas are solutions rich in silica, alkalies, water, and mineralizing fluids; some of this may be squeezed out into fissures to form pegmatites. Volatile substances or mineralizers, such as fluorine, boron, chlorine, along with tin, concentrate in the mother fluid or felsic magmas.

With progressing crystallization, the residual volatile fluids gather the metals possibly as complex chlorides that originally were either sparsely contained in the magma or existed within the connate fluids that were drawn into the magma cupola, along with the rare elements, the rare earths, and chlorine, boron, fluorine, hydrogen, sulfur, arsenic, and other substances. These solutions become expelled or are deposited in the shattered cupola (Fig. 4-9B) of the intrusion upon final crystallization and constitute the hydrothermal solutions that give rise to some economic mineral deposits. Consequently, they are a part of the magma of particular interest to economic geologists.

THE IGNEOUS ROCKS

As a result of crystallization and differentiation, associations of minerals are formed that yield many kinds of igneous rocks (Table 4-2); the varieties far exceed the number of the relatively few rock-making minerals that enter into their composition. Different sources of the magma could also provide the variety of igneous rocks as an alternative to differentation. Figure 4-7 illustrates different sources of rock and different processes that could produce a variability of igneous rock compositions.

The rock texture is determined chiefly by the silica content, the rate of cooling, and also by the amount of volatiles present during consolidation. Slower cooling gives opportunity for large crystals to grow and results in a coarse or *granular texture.* If cooling is rapid, the crystals are microscopic and the texture is *aphanitic;* if very rapid, no crystallization occurs and glass (obsidian) forms, as in the case of some lavas. Interrupted crystallization may give a *porphyritic* texture consisting of large crystals (*phenocrysts*) in a finer-grained matrix. This texture was probably caused by the initial crystallization of early formed minerals, followed by movement of the magma into another place where the remaining liquid underwent more rapid complete crystallization. The viscosity of rhyolitic magma with >70 percent silica content is about 100×10^6 times greater than that of a basaltic magma with about 50 percent silica. The rate of crystal growth, therefore, is impeded in rhyolitic magmas that often results in obsidian while even rapid cooling of a basalitic magma still allows growth of small crystals.

Thus different varieties of igneous rocks result from both compositional and textural differences. A granitic source, depending upon the rate of cooling, which may be dependent on the depth of burial, may yield granite, granite porphyry, rhyolite porphyry, rhyolite, obsidian, or tuff.

Many of the igneous rocks are in themselves economic mineral deposits, as they are used for building stones, road and building construction, and other purposes. Also, certain ore deposits of magnetite, chromite, and other minerals are merely unusual varieties of igneous rocks. A grouping and classification of the more common igneous rocks is shown in Table 4-2.

Pegmatites. With progressing crystallization the late residual liquid of a granite, for example, is made up principally of low-melting silicates and considerable water, along with other low-melting compounds and fluids, and a relative concentration of many of the substances that enter into mineral deposits of igneous origin. In addition to water, the fluid substance consists of liquid CO_2 and compounds of

Table 4-2 Classification of Igneous Rocks

Color	L e u c o c r a t i c			Mesocratic	Melanocratic	Hypermelanic
Mineralogic	Felsic			Intermediate	Mafic	Ultra Mafic
Chemical	Intermediate		Acid	Intermediate	Basic	Ultra Basic

Mineral diagram labels: K – Spar, Orthoclase, Microcline, Perthite, Sanidine; Quartz; Labradorite Anorthite; Olivine; Andesine; Nepheline; Pyroxene; Oligoclase; Albite; Hornblends; Biotite

				GRANITE		Diorite	Gabbro	Peridotite
		Coarse Grained	Nepheline Syenite	Syenite	Granite — Granodiorite; Qtz. Granite — Monzonite — Granodiorite	Diorite	Gabbro	Peridotite
T	F	Few Phenocrysts	Phonolite	Trachyte	RHYOLITE — Rhyolite — Dacite; Qtz. Rhyolite — Latite — Dacite	Andesite	Basalt	N
E	I N E							
X	G R A I N E D	Prominent Phenocrysts	Phonolite Porphyry	Trachyte Porphyry	RHYOLITE PORPHYRY — Rhyolite Porphyry — Qtz. Latite Porphyry — Dacite Porphyry	Andesite Porphyry	Tachylyte and Sideromelane	O
T U R E		Fragmental	Phonolite Tuff	Trachyte Tuff	Rhyolite Tuff	Andesite Tuff	Rare or Unknown	N
	G L A S S Y	Massive	O B S I D I A N					N
		with Phenocrysts	V I T R O P H Y R E					E
		Perlitic cracks	P E R L I T E					
		Vesicular	P U M I C E			Scoria		

65

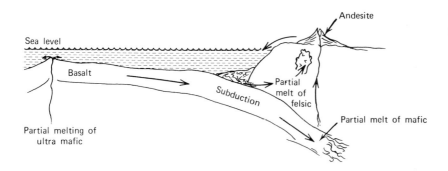

Figure 4-7 Suggested origins of igneous rocks of different compositions.

boron, fluorine, chlorine, sulfur, phosphorus, and other more rare elements. They aid crystallization by decreasing the viscosity of the magma and by lowering the freezing point of minerals. This is an aqueo-igneous stage—a transition between a strictly igneous stage and a hydrothermal stage, leaning more to the igneous—and is referred to as the *pegmatitic stage*. Withdrawals of the early liquid yield simple pegmatite dikes that are varieties of igneous rocks; later withdrawals of a more aqueous stage yield complex pegmatites, commonly characterized by druses, and may contain compounds of tungsten, tin, uranium, titanium, beryllium, phosphorus, chlorine, fluorine, and other elements, and some of the minerals of ore deposits; sulfides are uncommon. These pegmatites have often been referred to as pegmatite veins rather than dikes. Some, indeed, are valuable mineral deposits and are discussed later.

FLUIDS

A magma consists of molten material plus its dissolved volatiles, which subsequently are partially expelled giving rise to postmagmatic fluids. These fluids are predominantly water, carbon dioxide, complexes of chlorides, and possibly sulfides. Such complexes as $PbCl^+$ and $PbCl_4^{2-}$ could result in deposition of galena while $CuCl^+$ and $CuCl_4^{2-}$ would result in copper sulfides. These complexes make up dilute solutions in the aqueous phase. These hot water solutions constitute the *mineralizing solutions* or *hydrothermal solutions* to which many metallic minerals deposits owe their formation.

The fluids were originally in solution in the magma, and as long as the magma remained liquid and the pressure unchanged, they remained in solution. But the rise of the magma to levels of lesser pressures, or crystallization of anhydrous minerals, brings about an increase of fluid or hydrothermal

phases. With final consolidation, the hydrothermal fluids are expelled under great pressure (except as they may enter into the composition of the rock or become imprisoned in the rock minerals) toward and in the zones of greater porosity and permeability. A part may actually reach the surface in the form of fumaroles and hot springs, after undergoing change in the passage through the rocks. Studies of the constituents of fumaroles and hot springs have contributed toward an understanding of the genesis of mineral deposits.

Gas and Vapors. Enormous quantities of gas and vapor escape from the magma into the atmosphere during volcanism; fumaroles also represent discharges from beneath the surface. Deep-seated intrusives must similarly yield large emissions, of which only the effects become visible. Of the observed emissions, water constitutes 90 percent or more of the volume; CO_2, H_2S, and S are abundant; and CO, HCL, HF, H_2, N, Cl, F, B, and others occur. In addition, there are many other compounds present, such as the chlorides of H, Na, K, Mg, Ca, NH_4, and the metals, fluorides, tellurides, arsenides, and sulfur dioxide.

The fumaroles of the Valley of Ten Thousand Smokes gave much information not only of the quantities and composition of the gases and compounds emitted by volcanic ash flows (tuffs), but also of the vast quantities of metallic minerals dispersed in the porous tuffs. Such minerals as magnetite, specularite, molybdenite, pyrite, galena, sphalerite, covellite, and others were found around the fumarole conduits, where no liquid solution had at any time been evident. Zies found that the magma also contained lead, copper, zinc, tin, molybdenum, nickel, cobalt, and manganese. Also deposited were NaCl, KCl, S, Se, Te, and evidences of As, Bi, Th, and B_2O_3. Zies estimated that these fumaroles yielded annually 1,250,000 tons of hydrochloric acid and 200,000 tons of hydrofluoric acid.

Copper chloride was detected by Murata in flames from the Kilauea volcanic eruption of 1960.

Gilligham, Morey, Krauskopf, and especially Kennedy have shown that silica and other essentially nonvolatile substances can be dissolved and transported by water at temperatures below and above the critical temperature of water (Fig. 4-8). Pressure greatly increases the solubility of silica at temperatures above the critical temperature of water, and upon condensation, the mineral load is transferred from the gaseous to an aqueous state. Figure 4-8 illustrates the efficacy of precipitation of silica under isothermal conditions with only a decrease in pressure. At a temperature of 400°C, for example, a decrease in pressure from 1000 bars to 300 bars results in an isothermal decrease in the solubility of SiO_2 from about 0.175 percent to 0.04 percent.

Daubree, curious about the origin of tin deposits, introduced in 1841 the first experimental methods into the study of ore deposits by producing cassiterite from volatile tin halides. As a result, the terms *pneumatolytic* action and *pneumatolysis* arose.

These terms were later redefined by Fenner as "pertaining to processes or results effected by gases evolved from magmas or entrained with gases of that origin." He considered that gaseous emanations were the agents best adapted to effect the primary separation and transportation of materials from the magmas and that pneumatolytic processes were important in producing contact metasomatism. The term *pneumatolytic* became widely used and is even today incorporated as one division in some European classifications of ore deposits. The term is now used rarely because it is realized that the physical states of liquids and gases are essentially the same under specific conditions of high temperature and pressure.

Residual Fluids. It is generally considered that many mineral deposits of igneous affiliations result from hot waters (hydrothermal solutions) of predominantly connate and meteoric derivation and such hydrothermal solutions spring, directly or indirectly, from the magma in consequence of crystallization and differentation (Fig. 4-9). Burnham,

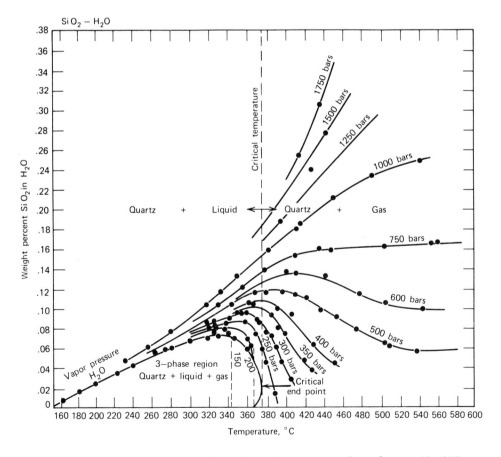

Figure 4-8 Isobaric solubility curves of SiO_2 in H_2O. (From: Kennedy, Econ. Geol., v. 45, 1950)

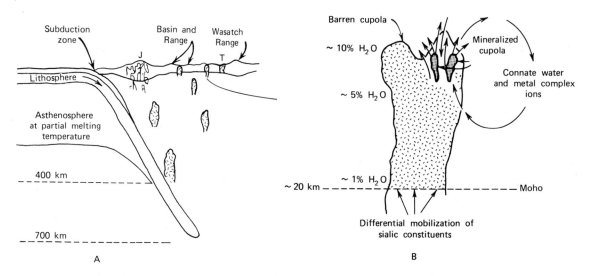

Figure 4-9 A&B Suggested derivation of some magmas from the Upper Mantle in the Basin and Range Province and the addition of connate brine waters to a magma. The cupola zone could contain two cupolas, one barren (lacking water) and the other mineralized such as the Last Chance and Bingham stocks respectively, Bingham Canyon, Utah.

Jahns, Holland, and others have pointed out that magmas, depending on pressures, may contain from 1 to 10 percent of water but that the consolidated igneous rock contains less than 1 percent. It is also evident that there has been an enrichment of soluble metal ions in the hydrothermal solutions. Thus hydrothermal fluids become containers and carriers of metals to form mineralizing solutions. Possibly, magmas are derived by differential mobilization of constituents from the upper mantle (Figs. 4-9 and 4-10) and obtain their water from mobilized connate

brine solutions in the crust with the latest magmatic crystallization stage carrying most of the hydrothermal waters (Fig. 4-9B). The metals may have been collected as complex chlorides and are included in the high-vapor-pressure stage of the hydrothermal solutions. The model shown in Fig. 4-9B may explain the mineralized Bingham stock in juxtaposition to the barren Last Chance stock, Bingham Canyon, Utah.

The nature of the residual fluids has been determined only by inference because they cannot be

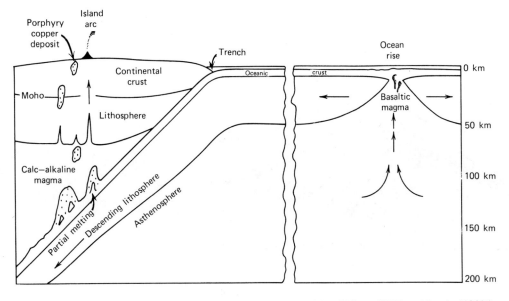

Figure 4-10 Plate Tectonic origin of porphyry copper deposits (after Sillitoe [1972] and Issaks [1968]).

observed directly, as can gaseous emanations at the earth's surface. Hot springs have little indication of the original fluid composition because their waters may have condensed partially from volatiles to start with; undoubtedly they must have undergone chemical reactions with each other and other changes in passing through the rocks. Recent analyses of fluid inclusions, brine solutions collected from mile-deep drill holes near Niland, California, and analyses of Red Sea brine solutions are great aids to our understanding of the composition of some hydrothermal solutions.

Bowen and other geologists suggested that the magmatic hydrothermal solutions are the residue of pegmatite injections left after the pegmatite constituents have crystallized. Most economic geologists, however, now believe that they represent a stage beyond the pegmatitic liquids—the last stage of differentiation in the magma chamber—and that the residual fluids have been greatly modified by

the crystallization of the last rock minerals. The withdrawal and crystallization of pegmatitic liquid would unquestionably leave a tenuous liquid residue outside of the magma chamber, from which quartz and other minerals may be precipitated. We do not accept, however, that this residual fluid is the ore-forming fluid. The pegmatitic fluid must have been expelled from the magma chamber while some residual magma was still present and when quartz and alkalic feldspar were still forming in the parent magma. After withdrawal of the pegmatitic fluid, there would still remain some residual fluid undergoing differentiation and enrichment in mineralizers and metals until the cessation of crystallization; this fluid most probably was an important ore-forming fluid. Kennedy has suggested that as water will diffuse and distribute itself in a magma according to the equal chemical potential of the water throughout the magma, the water will tend to concentrate in the magma chamber in the regions

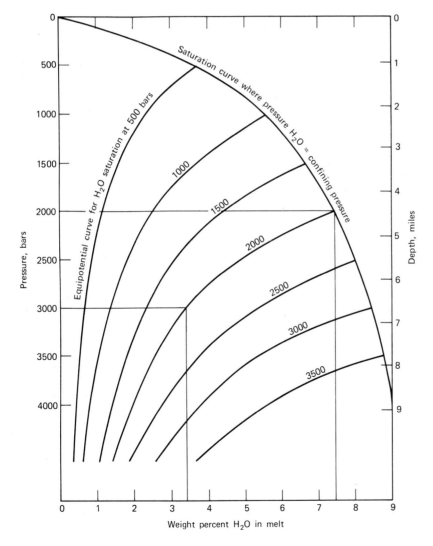

Figure 4-11 Equilibrium distribution of water in a melt of albite composition in the gravitational field of the earth. (From: Poldervaart, Ed., 1955, Crust of the Earth. GSA Special Paper 62, G. C. Kennedy)

of lowest pressures and temperatures (Fig. 4-11). These regions might not be at the bottom of the chamber or even the core of the magma where crystallization may begin and continue upward and outward, respectively, which is the opposite of general assumptions.

This suggestion substantiates the observation that hydrothermal alteration and associated mineralization generally occur in the cupola or upper portion of an intrusion. Kennedy has illustrated these conditions diagrammatically in Fig. 4-11. His data are based on some of Goranson's studies, and his values are plotted for the 1000°C isotherm.

As an example, Kennedy suggests that the top of an intrusion extends to within 4.5 miles of the earth's surface. The confining pressure caused by the overlying weight of rock will be about 2000 bars (Fig. 4-11), and if the magma is saturated with water, there will be 7.4 percent water at the top of the intrusion. At a greater depth, where the confining pressure has increased to 3000 bars, the water content of the melt will be only 3.3 percent and the melt will be far from saturated. The diagram suggests, therefore, that the deeper the burial of the magma chamber, the stronger the tendency for most of the water to be concentrated in the uppermost portion or cupola zone of the intrusion.

Ingerson and Morey consider that there is a late orthomagmatic liquid stage devoid of critical phenomena and a coexisting vapor phase of different composition that would exhibit critical phenomena. Morey believes that the magma (or orthomagmatic-liquid phase) remains liquid until it reaches the hydrothermal or hot-spring stage, but he points out that a volatile stage would probably be present if the external pressure were less than the vapor pressure and the enclosing rock were somewhat porous. Generally similar conclusions are expressed by Krauskopf.

Field evidence indicates that in many types of deposits there may be several phases of mineralization. The earlier phases are barren of metals; the last phase is the ore carrier. In some contact-metasomatic deposits, for example, the barren aureole of alteration that has universally preceded the metallization may have been caused by an escaping and penetrating fluid phase, and the introduction of late metals of lower temperature of formation may, in some cases, represent an invasion of the coexisting orthomagmatic residual liquid. At Butte, Montana, in the central zone of intense mineralization, the widespread and pervasive barren rock alteration that preceded the vein filling may possibly have been caused by a penetrating low-temperature fluid phase as an advancing surge of hydrothermal liquids from which the metals were deposited. It is believed that the similar widespread pervasive, barren, rock alteration of "Porphyry copper" deposits has universally preceded the copper introduction by a pervasive hydrothermal solution.

MAGMAS AND MINERAL DEPOSITS

So far, the relationship between magmas and mineral deposits has been assumed. The evidence consists of the occurrence of igneous rocks that are themselves ores; of the relations of specific igneous rocks and metals; of the depositions from volcanoes, fumaroles, and hot springs; of mineral zoning about igneous centers; and of the character of mineralizing solutions as evidenced by the minerals deposited.

Igneous Rocks as Ores. Some igneous rocks are themselves bodies of ore, such as some magnetite deposits and bodies of chromite, ilmenite, corundum, or diamonds. Since these are merely igneous rocks, although of unusual composition, they indicate a direct relationship between magmas and mineral deposits.

Relationship Between Certain Metals and Specific Rocks. Field observations disclose an association of certain ore minerals with specific rocks that is so general and widespread as to preclude coincidence. This association establishes a definite relationship between the ore minerals and rocks, indicating thereby a magmatic source for both. For example, primary platinum deposits occur only in ultrabasic rocks, such as dunite or peridotite; diamonds in kimberlite; chromite in peridotite or serpentine; ilmenite in gabbro or anorthosite; titaniferous magnetite in gabbro or anorthosite; corundum in quartz-free rocks, such as nepheline syenite; nickeliferous sulfides in norite or gabbro; tin in silicic granites; and beryl in granite pegmatite. These associations are so universal as to render inescapable the conclusion that the rock and associated ore issued from the same magma

Relationship Between Plate Tectonics and Meso-Cenozoic Magmas. The resurgence of the belief in sea floor spreading and continental drift and the resulting subduction zones has suggested the cause of

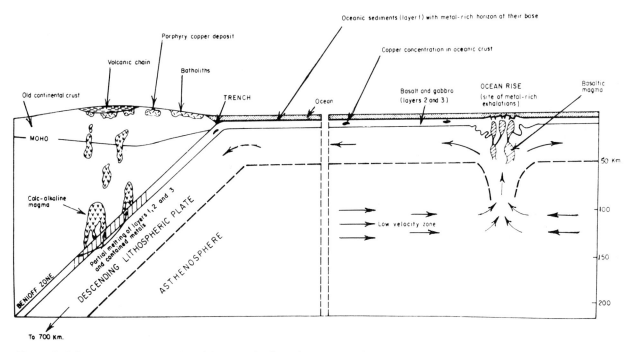

Figure 4-12 Schematic representation of the genesis of porphyry copper deposits in the context of plate tectonics. (Modified from: Sillitoe, Econ. Geol., v. 67, p. 188, 1972)

compositionally, similar intrusions derived from calc-alkalic magmas. These magmas may be mobilized from the upper mantle as suggested by Fig. 4-12, which also offers an explanation for the specific geographical occurrence of porphry copper deposits (Fig. 4-13). Incidentally, from such a plot of subduction zones, there seems little reason why the Aleutian Islands should not contain porphyry copper deposits.

Composition of Mineralizing Fluids. In 1942, a new volcano began forming in the southern portion of Hokkaido, Japan. It was named Showashinzan and offered an excellent repetitive opportunity to study the composition, both chemically and isotopically, of the materials collected from numerous fumaroles varying in temperatures between 187 to 722°C. The study was done by students at Nagoya University under the direction of Professor Shinya Oana. A summary compilation for one year of some of the constituents from some of the fumaroles is given in Table 4-3.

It is evident from these analyses that water vapor is the most abundant gas, but the significance of the increase in H_2S with decreasing temperature agrees well with the reaction $H_2S + 2H_2O \leftrightarrows SO_2 + 3H_2$. The direction of the reaction is temperature-dependent favoring SO_2 and H_2 at high temperatures and conversely H_2S and H_2O at low temperatures. Table 4-3 indicates these results with the H_2S/SO_2 ratio being lowest at the high temperature of 722°C while the H_2S/SO_2 ratio is highest at the

Table 4-3 Fumarolic Gases from Showashinza Volcano, August 1960. (Shinya Oana).

	A-1	A-6	B-lb	B-5	B-6	C-2	C-3
Temperature °C	722	695	260	446	347	605	187
H_2O 10001	1.66	1.27	0.67	0.55	0.66	3.2	1.7
HCl 10001	0.42	0.60	0.13	0.27	0.25	0.65	0.12
H_2S 10001	0.008	0.009	0.10	0.11	0.12	0.08	0.17
SO_2 10001	0.14	0.10	0.003	0.004	0.004	0.047	0.002
H_2S/SO_2	0.057	0.090	33	27.5	30	1.7	85

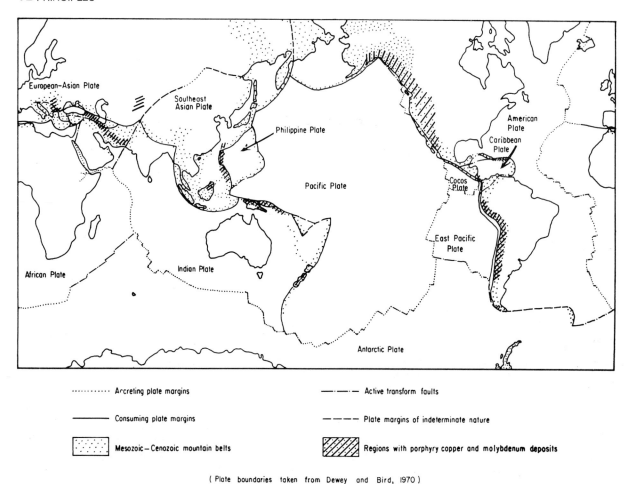

.... Accreting plate margins

——————— Consuming plate margins

▢ Mesozoic—Cenozoic mountain belts

———·——— Active transform faults

———— Plate margins of indeterminate nature

▨ Regions with porphyry copper and molybdenum deposits

(Plate boundaries taken from Dewey and Bird, 1970)

Figure 4-13 The western Americas, southwest Pacific and Alpine porphyry belts in relation to Mesozoic-Cenozoic orogenic belts and accreting and consuming plate boundaries. (From: Sillitoe, Econ. Geol., v. 67, 1972)

lowest temperature of 187°C. The significance of this observation is that H_2S becomes more and more abundant as temperatures drop allowing it to be available to react with complex metal chlorides. Thus metal sulfides form and release the chloride ion to react with sodium to form NaCl, some of which is captured in fluid inclusions, and release [H+] ions to form HCl. Some of the HCl escapes from the hydrothermal fluids while some of the [H+] ions react with the intrusive minerals resulting in hydrothermal alteration, much of which, is the result of hydrogen ion metasomatism. At higher temperatures, HCl is relatively undissociated and may allow a mineralizing fluid to transit carbonate host rocks with little chemical reaction, an observation often noted whereby ore-bearing fluids have passed through some carbonate formations before forming a metal deposit in a later-reached, quite similar carbonate zone.

In regard to H_2S and SO_2, they readily exchange isotopes and, under isotopic equilibrium conditions, H_2S is enriched in ^{32}S relative to SO_2. Figure 4-14 illustrates the contrasting extent of ^{32}S enrichment in H_2S in comparison to SO_2. The contrast between $\delta^{34}S$ values of H_2S and SO_2 is the greatest at low temperatures and the least at higher temperatures. Figure 4-15 schematically suggests some of the chemical reactions that may occur in a fumarole or hot spring systems with some of the reasons for $\delta^{34}S$ variations.

Volcanism offers an opportunity for direct observation of relationships between mineral deposits and magmas, although the deposits formed are relatively unimportant economically. Scientifically, however, they suggest a magmatic derivation for many minerals and metals. Deposits of native sulfur found in and about volcanic craters are well known. Among the sublimates from Vulcano have been

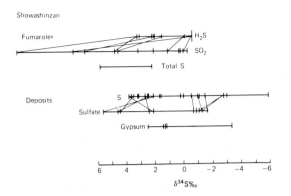

Figure 4-14 $\delta^{34}S$ values of H_2S-SO_2 pairs collected at varying temperatures.

noted sulfur, realgar, glauber salt, tellurium, cobalt, tin, zinc, lead, copper, bismuth, and phosphorus. Table 4-3 indicates the composition of Showashinzan fumaroles. From Vesuvius specular hematite, tenorite, sodium, iron, copper, boracic acid, and others have been recorded; CuCl was detected at Kilauea and Nyirogongo. Skinner and Peck found an immiscible sulfide melt in a Hawaiian lava.

Samples of gases from Kilauea consisted mostly of water, CO_2 and SO_2 with lesser CO, H_2, H_2S, S, SO_3, Cl_2, and $CuCl_2$. Samples from Kamchatka show chiefly water, carbon gases, sulfur gases, and halogen compounds. Sulfur again at high tempera-

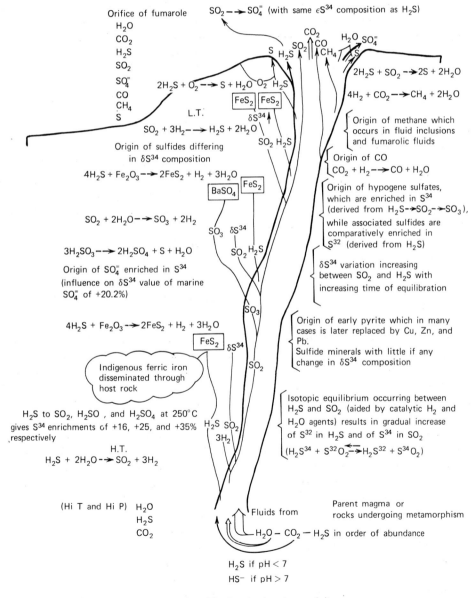

Figure 4-15 Chemical and isotopic characteristics of an idealized volcanic conduit.

tures is SO_2 and at low temperatures is H_2S as noted previously.

Hot Springs. Hot springs associated with volcanism have long enticed the geologist and chemist because they contain many dissolved substances and demonstrate that such hot waters dissolve, transport, and deposit minerals.

Hot spring waters may be meteoric, magmatic, or metamorphic in origin. The distinction is difficult without δD and $\delta^{18}O$ isotopic analyses (Fig. 3-7). Temperature alone is no criterion because hot springs may originate without volcanic heat, even those in areas of volcanism may be meteoric waters heated by hot volcanic rocks, geothermal gradient, or by small additions of volcanic gases. Probably few, if any, surface hot springs consist predominantly of magmatic waters because of inescapable dilution by connate or near-surface meteoric waters.

Lindgren divides "juvenile" springs into sodium chloride–silicate waters and sodium carbonate waters. Sodium carbonate springs are widely distributed, with many representatives in western North America, France, and Germany. The Carlsbad Springs of Germany are examples of sodium carbonate springs containing arsenic, antimony, boron, cobalt, copper, chromium, fluorine, gold, iodine, lithium, nickel, phosphorus, selenium, strontium, tin, and zinc.

The chloride-silicate waters are characteristic of the three great geyser regions of the world. They all deposit abundant silica or geyserite.

The New Zealand geyser waters at Wairaki also carry gold, silver, and mercury. The Steamboat Springs of Nevada have deposited chalcedony; quartz; calcium carbonate; precious metals and sulfides of the base metals; antimony; mercury; arsenic; and traces of cobalt, nickel, and manganese.

In both types of springs, the rare metals are prominent, and borates, arsenates, and fluorides are relatively abundant. Both types are clustered about recent volcanic centers, and the contents suggest expiring magmatic activity. These waters may well be the type that produce shallowed-seated veins. Indeed, the issuance of such thermal waters from mineral veins at Plombieres, France; and the Comstock Lode, Nevada, suggest a direct connection between the present thermal waters and the underlying ore deposits.

Submarine Volcanism. The actual physiochemical conditions of volcanic or submarine exhalative solutions before they reach the sea floor is a subject of increasing study. Haas has provided information on densities and vapor pressure of heated brine solutions under differing hydrostatic pressures; Ridge and Phillips have amplified upon these basic data.

In the study of fluid inclusions, Roedder has noted that massive stratiform deposits contain more NaCl (~20 percent) in their minerals than do "normal" hydrothermal solutions that contain about 5 percent NaCl. The maximum boiling temperature of an aqueous solution containing 20 percent NaCl is about 600°C at the critical pressure of about 922 atmospheres. The submarine mineral-bearing solutions would be cooled rapidly and undergo pressure decrease as they rise and approach the sea floor, resulting possible in the rapid deposition of massive sulfide deposits.

Sources of such solutions have been summarized by Ridge as having been derived below the sea floor by (1) volcanic emanations that carried their mineral load in the vapor state, having obtained their mineral content either from magmatic sources or from the rocks through which they passed; (2) heated solutions that may have been hydrothermal in the classic sense, having brought their mineral content with them from their magmatic source; and (3) solutions of near-surface origin that picked up heat and their mineral material from the rocks through which they passed on their subsurface path to the site where they deposited ores.

SUMMARY

The magma is the direct source of much of the materials of hypogene mineral deposits. Through crystallization and differentiation, some constituents collect as crystal aggregates or molten liquids, before final consolidation, to form magmatic oxide and sulfide deposits. Progressive differentiation of magmas results in a residual liquid that becomes more and more enriched in the volatiles and other constituents. They include possibly metal complexes that formerly were dispersed in country rock and/or the magma. Before final consolidation, some or all of this liquid may be tapped off to form pegmatites, whose crystallization leaves an aqueous residue that may form one class of mineralizing solutions. When consolidation is complete, or nearly complete, this mobile, nonviscous, aqueous residue may be expelled as hydrothermal mineralizing fluids that later deposit their load to form various types of mineral deposits.

If the consolidating magma is rather close to the surface, under slight pressure, the volatiles will escape from the residual aqueous liquid and may reach the surface directly as acid gases, forming fumaroles, such as those of Katmai. They transport and dissipate some of the metals that they formerly collected; no appreciable metallization results.

At less-shallow depths, boiling may occur, giving rise to an acid distillate that rises through the fractured rocks. Reactions with the wall rocks will result in hydrothermal alteration and probably cause deposition of some ore minerals. The continued reactions will render the solutions more neutral, in which condition, after mingling with more meteoric waters, they may emerge as juvenile hot springs still carrying part of their mineral load.

If the external pressure is greater than the vapor pressure of the residual fluids, no vapor phase will result, and the fluids may be expelled as such to form rising hydrothermal metallizing solutions.

Under deep-seated conditions and high pressure, a vapor phase may be absent, and the last residual liquids, which may or may not contain sufficient concentrations of metals to form economic mineralization may be tapped off as hydrothermal solutions, or they may be excluded by final crystallization of the residual magma, or they may be absorbed in the formation of the last minerals to crystallize.

It seems evident, in summary, based on fluid inclusion studies, stable isotopic studies, Niland geothermal well fluid analyses, the composition of Red Sea hydrothermal solutions, and a more complete understanding of complex metal chloride ions and the physical and thermal chemistry of hydrothermal solutions at several hundred degrees centigrade, that mineralizing solutions may be slightly acid to near neutral at high temperatures. As the solution cools, dissociation of HCl begins, which leads to more acid conditions as the dissociated hydrogen ions react with host rocks, resulting in the metasomatic process of hydrothermal alteration. As this process continues with decreasing temperatures, it may be followed at a later time, by the metal chlorides reacting with H_2S to form the last-stage metallization deposition.

Selected References

General Reference 10, chap. 3, pt. 1: Pneumatolytic Processes in the Formation of Minerals and Ores, by C. N. Fenner. Magma and phases and role of gaseous emanations. Pt. 2: *Magmatic Differentiation Briefly Told*, by N. L. Bowen. *A concise summary of differentiation.* Pt. 3: Differentiation as a Source of Vein and Ore-Forming Materials, by C. S. Ross. *Hydrothermal mineralizers leave the magma as liquids.*

General Reference 9, chaps. 7,8, and 10. *Mineral springs, underground water, and magmas.*

Nature of the Ore-Forming Fluid. 1940. L. C. Graton. Econ. Geol., 35, Supp. to no. 2. *Classical but early views on the origin, state, composition, and migration of fluids; ore fluids leave magma as alkaline liquids.*

Environments of generation of some base metal deposits. 1968. D. E. White. Econ. Geol., v. 63, p. 301–335. *Nature of solutions that formed some ore deposits.*

Mercury and base-metal deposits with associated thermal and mineral waters. 1969. D. E. White. *In* Geochemistry of Hydrothermal Ore Deposits, chap. 13.

Hydrothermal Fluids at the Magmatic Stage. 1958. C. Wayne Burnham. *In* Geochemistry of Hydrothermal Ore Deposits, chap. 2.

Solubilities and transport of ore minerals. 1968. H. L. Barnes and G. K. Czamanske, *In* Geochemistry of Hydrothermal Ore Deposits, chap. 8.

Fluid inclusions as samples of ore fluids. 1968. Edwin Roedder. *In* Geochemistry of Hydrothermal Ore Deposits, chap. 12.

Introduction to Geochemistry. 1957. K. B. Krauskopf. Chaps. 14 and 15, Crystallization of magmas; chap. 17, Ore-forming solutions; chap. 16, Volcanic Gases.

Theoretical Petrology. 1952. Tom F. W. Barth. John Wiley & Sons, Inc., New York. *Fundamental petrology; magmas, their differentiation and products.*

Occurrence of CuCl emission in volcanic flows. K. J. Murata. Am. Jour. Sci., v. 258, p. 769–772.

Magmatic, connate, and metamorphic waters. 1957. Donald E. White, Bull. Geol. Soc. Am., v. 68, p. 1659–1682. *Character of some hydrothermal waters.*

A portion of the system silica-water. G. C. Kennedy. Econ. Geol., v. 55, p. 629–653.

Complexing and Hydrothermal Ore Deposition. H. C. Helgeson. Pergamon Press, Inc., Elmsford, N.Y. *Treatise on theory, methods, and resulting efficacy of complex ions as mineral transporters in hydrothermal fluids.*

Ore Petrology. 1972. R. C. Stanton. McGraw-Hill International Series. McGraw-Hill Book Company, New York. *A treatise concerned with the study of ores as rocks.*

The tops and bottoms of porphyry copper deposits. 1973. Richard H. Sillitoe, Econ. Geol., v. 68, p. 799–815.

PROCESSESS OF FORMATION OF MINERAL DEPOSITS

A theory is a tool—not a creed.
J. L. Thompson

GENERAL REMARKS

The formation of mineral deposits is complex. There are many types of deposits, generally containing several ore and gangue minerals. No two are alike; they differ in mineralogy, texture, content, shape, size, and other features. They are formed by diverse processes, and more than one process may enter into the formation of an individual deposit. The modes of formation of minerals discussed in Chapter 3, although many, are but part of the larger processes that build up of mineral deposits of igneous derivation; there are also many others. Diagrammatic views that summarize these processes and the more common metals concentrated by these processes are shown in Figs. A and B.

Among the agencies that enter into the formation of mineral deposits, water plays a dominant role. It may be water vapor; meteoric connate water; magmatic water; metamorphic water; or ocean, lake, or river water. Temperature also plays an important part, but many processes operate at surface temperature and pressure. Other agencies are pressure magmas, gases, vapors, solids in solution, the atmosphere, organisms, and country rock.

VARIOUS PROCESSES

The various processes that have given rise to mineral deposits are

1. Magmatic concentration.
2. Sublimation.
3. Contact metasomatism.
4. Hydrothermal processes.
   (a) Cavity filling
   (b) Replacement
5. Sedimentation
6. Bacterial processes
7. Submarine exhalative and volcanic processes
8. Evaporation
9. Residual and mechanical concentration
10. Oxidation and supergene enrichment
11. Metamorphism

The various processes of formation of mineral deposits are considered in detail, starting with original magmatic materials, through those of lower-temperature conditions of formation, to secondary processes.

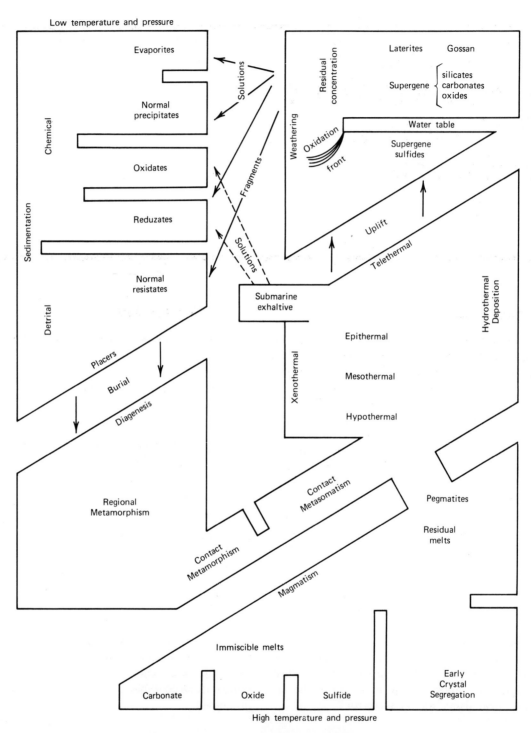

Figure A Simple graphical classification of mineral deposits (D. Garlick).

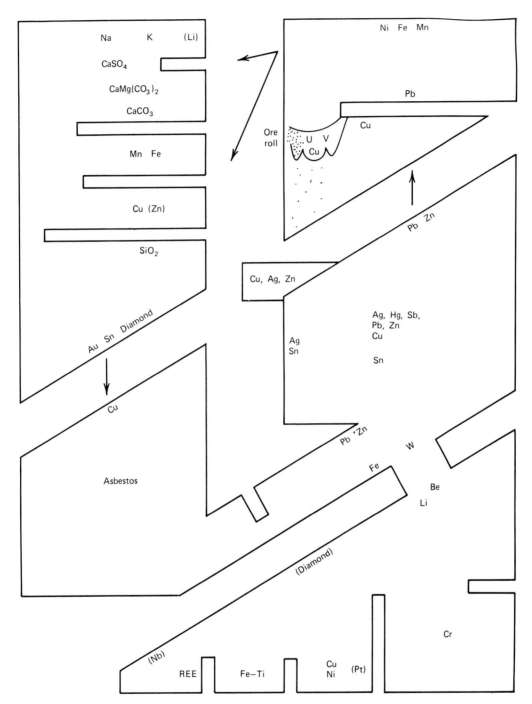

Figure B Simple graphical classification of elements correlated with Fig A (D. Garlick).

5

MAGMATIC CONCENTRATION

I hope you are not intending to publish your conclusions; indeed I cannot suppose this, as you know how dangerous it is to work out geological problems with three thousand miles between you and the region to be investigated.

J. D. Dana letter to
Sir Archibald Geikie, 1886

Certain accessory or uncommon constituents of magmas may become concentrated into bodies of sufficient size and richness to constitute valuable mineral deposits. Some of these are large and rich; some, such as chromite and platinum, are the sole source of these minerals; many, however, are of greater scientific than commercial interest. Representatives of magmatic concentration are many and widespread, but there are relatively few types; their mineralogy is simple, and the products yielded are not numerous. Although individual deposits are of great value, as a whole they are overshadowed in importance by those resulting from other processes.

Magmatic ore deposits are characterized by their close relationship with intermediate or deep-seated, intrusive igneous rocks. Actually, they themselves are igneous rocks whose composition happens to be of particular value to man. They constitute either the whole igneous mass or a part of it, or form offset bodies. They are magmatic products that crystallize from magmas. They are also termed *magmatic segregations, magmatic injections,* or *igneous syngenetic deposits.*

MODE OF FORMATION

Magmatic deposits result from simple crystallization, or from concentration by differentiation, of intrusive igneous masses. They were firmly established by Vogt as one of the major types of mineral deposits, at which time they were considered to be simply an ore facies of igneous rock, formed by earliest crystallization of the ore minerals. Subse-

quently, it was learned that in many of the deposits the ore minerals actually crystallized later than the rock minerals, and for such deposits the former simple concept had to be modified. It is now realized that there are several modes of formation of magmatic deposits and that they originate during different periods of magma crystallization. In some, the ore minerals crystallized early, in others late; in still others they remained as immiscible liquids until after crystallization of the host rock. Graton and McLaughlin proposed the term *pneumotectic,* now abandoned, for deposits in which mineralizers played a part and *orthotectic* for the early straight magmatic ones, which is the same as Niggli's term *orthomagmatic.*

In the formation of magmatic concentrations the processes of differentiation discussed in Chapter 4 apply.

Much confusion exists in the literature on magmatic deposits, in part because of fuzzy terminology. "Magmatic segregation" has often been assumed to be the only kind of magmatic deposit; early and late magmatic too often have not been distinguished. As a result, some authors have considered that certain deposits could not be magmatic simply because they were not segregations or because ore minerals later than rock silicates precluded separation by crystallization. Unnecessary conflict of interpretation arose and still exists.

Some of this confusion need not occur if it is realized that there can be more than one period of magmatic ore formation within the magmatic period and that there may be more than one process of differentiation by which magmatic deposits may result.

Table 5-1 Classification of Magmatic Deposits

Type		Examples
I. Early Magmatic:		
A. Dissemination	Disseminated crystallization without concentration	Diamond pipes; some corundum deposits, Muskox chromite
B. Segregation	Crystallization differentiation and accumulation	Bushveld chromite; Stillwater, Mont.; Selukwe chromite; and Muskox chromite
II. Late Magmatic:		
A. Gravitative liquid accumulation		
(1) Residual liquid segregation	Crystallization differentiation and residual-magma accumulation	Allard Lake, Que.; Bushveld titanomagnetite; Tabeg (?); Duluth Complex; Bushveld platinum; and Allard Lake, Que.
(2) Residual liquid injection	Same, with filter pressing and/or injection	Adirondack magnetite; Kiruna, Sweden (?); Pea Ridge, Mo.; pegmatites; and Allard Lake, Que.
B. Immiscible liquid		
(1) Immiscible liquid segregation	Immiscible liquid separation and accumulation	Insizwa, S. Africa
(2) Immiscible liquid injection	Same, with injection	Klackfontein, Bushveld, S. Africa

Table 5-2 Classification of Ultramafic and Related Mafic Bodies

Class	Examples	Remarks
A. Bodies emplaced in active orogenic areas		
1. Bodies contemporaneous with eugeosynclinical volcanism		
(a) Tholeiitic suite		
(i) Picritic subtype		
	Munro-Dundonald area (Naldrett and Mason, 1968; McRae, 1969; Arndt, 1975, 1976)	Gravity-differentiated flows, capped by hyaloclastite with 13–15 wt% MgO. Gravity differentiated sills.
	Kakagi Lake (Ridler, 1966)	
	Pechenga (Gorbunov, 1968)	
(ii) Anorthositic subtype		
	Doré Lake Complex (Allard, 1970)	Some examples of this class are conformable, others appear to be discordant. They are possibly differentiation chambers for overlying volcanism.
	Bell River Complex (Sharpe, 1965)	
	Kamiskotia Complex, Timmins, Ont.	
(b) Komatiitic suite		
	Munro-Dundonald area, Ont. (Naldrett and Mason, 1968; Pyke et al., 1973; Arndt et al., in prep.)	Simple flows, spinifex-capped flows, differentiated flows, and differentiated sills. Composition of flows ranges from peridotite to basalt and of sills from dunite to anorthositic gabbro.
	Eastern Goldfields, Australia (McCall and Leishman, 1971; Nesbitt, 1971; Hallberg and Williams, 1972; Williams, 1972; Naldrett and Turner, 1977)	
2. Alpine-type bodies		
(a) Large obducted sheets		
	New Caledonia (Guillon, 1975)	
	Papua-New Guinea (Davies, 1968, 1971)	

Table 5-2 (Continued)

Class	Examples	Remarks
	(b) Ophiolite complexes	
	Vourinos (Moores, 1969)	
	Troodos (Gass, 1967, 1968; Moores and Vine, 1971)	
	Bay of Islands (Dewey and Bird, 1971; Irvine and Findlay, 1972)	
	Canyon Mountain (Thayer and Himmelberg, 1968)	
	Eastern Townships, Quebec (Lamarche, 1972; Laurent, 1975)	
	(c) Deformed ophiolite complexes and clastic blocks in mélange terranes	
	TwinSisters,Washington(Ragan,1967)	
	Vermontserpentinites(Jahns,1967)	
	Iran(Gansser,1959)	
	FranciscanSeries(Hamilton,1969;Hsu,1968)	
	(d) Possible diapirs	
	Mt. Albert, Quebec (MacGregor, 1962)	
3.	Alaskan-type complexes	
	Intrusions of Alaska and British Columbia, including Duke Island (Irvine, 1974), Union Bay (Ruckmick and Noble, 1959), Tulameen (Findlay, 1969)	
	Intrusions of Urals (Vorobeyeva et al., 1962)	
B. Bodies emplaced in nonorogenic areas		
4.	Large stratiform layered complexes	
	Bushveld (Hall, 1932; Wager and Brown, 1968; Visser and von Gruenewaldt, 1970)	
	Stillwater (Hess, 1960; Jackson, 1961; Bowes et al., 1973)	
	Muskox (Irvine and Smith, 1967)	
	Duluth (Taylor, 1964; Phinney, 1970)	
	Kiglapait (Morse, 1969)	
	Sudbury (Naldrett et al., 1970, 1972)	
5.	Sills and sheets equivalent to flood basalts	
	Palisades sill (F. Walker, 1940; K. Walker, 1969)	Generally occur in areas in which extrusion of flood basalts has occurred. The sills are chemically similar to the extruded basalts.
	Insizwa-Ingeli intrusion, South Africa (Maske, 1966)	
	Dufek Intrusion, Antarctica (Ford and Boyd, 1968; Ford, 1970; Himmelberg and Ford, 1976)	
6.	Medium- and small-sized intrusions	
	Skaergaard (Wager and Brown, 1968)	
	Rhum (Wager and Brown, 1968)	
	Noril'sk-Talnakh area (Godlevskii, 1959; Zolutuchin and Vasil'ev, 1967)	
7.	Alkalic ultramafic rocks in ring complexes and kimberlite pipes	

Source: By Naldrott and Cabri.

In an attempt to clarify the matter, we propose the classification shown in Table 5-1 for magmatic deposits and for the processes that give rise to them, retaining familiar terminology.

A tectonic classification of ultramafic and related mafic bodies has been used recently to indicate the environment in which the ultramafic and related mafic bodies have been emplaced. This classifica-tion proposed by Naldrett and Cabri including the deposit characteristic of each group is shown as Table 5-2.

Early Magmatic Deposits

The early magmatic deposits resulted from straight magmatic processes—those that have been called orthotectic and orthomagmatic. These deposits have been formed by (1) simple crystallization without concentration, (2) segregation of early formed crystals, and (3) injections of materials concentrated elsewhere by differentiation. The ore minerals have crystallized earlier than the rock silicates and in part presumably have separated by crystallization differentiation.

Dissemination. Simple crystallization of a deep-seated magma in situ will yield a granular igneous rock in which early formed crystals may be disseminated throughout it. If such crystals are valuable and abundant, the result is a magmatic mineral deposit. The whole rock mass, or a part of it, may constitute the deposit, and the individual crystals may or may not be phenocrysts.

The diamond pipes of South Africa are examples. The diamonds are sparsely disseminated throughout kimberlite rock and the whole kimberlite pipe is the mineral deposit. The diamonds are phenocrysts, and in this example, as far as known, they crystallized in a former deep magma chamber and were transported with the enclosing magma and perhaps even continued to grow before final consolidation occurred in the present pipes. The disseminated corundum in nepheline syenite in Ontario is another example.

The resulting deposits of this class have the shape of the intrusive, which may be a dike, pipe, or small stocklike mass. Their size is large in comparison with most mineral deposits.

The same process may also yield noncommercial bodies of valuable minerals which, however, may undergo later concentration by mechanical or residual processes to yield economic placers or residual concentrations. Placer deposits of ilmenite, monazite, and some gemstones are examples.

Segregation. The term *segregation* has often been loosely used to designate magmatic deposits as contrasted with those formed by solutions or other means. In the sense here used, however, following the original meaning, it is restricted to concentrations of early crystallizing minerals in place and is to be distinguished from an *injection,* where the differentiate has undergone a change in position before consolidation. Early magmatic segregations are early concentrations of valuable constituents of the magma that have taken place as a result of gravitative crystallization differentiation. Consequently, they have often been referred to as "liquid

magmatic." Such constituents as chromite may crystallize early and become segregated in bodies of sufficient size and richness to constitute economic deposits. The segregation may take place by the sinking of heavy early formed crystals to the lower part of the magma chamber, by marginal accumulation, or by constrictional flowage, as explained by Balk.

The early demonstration of magmatic ore deposits at Taberg by Sjögren, Törnebohm, and Igelström, and their later establishment by J. H. L. Vogt in 1891 to 1893 led to the conclusion that all magmatic ore deposits were segregations formed as described above. Then it was found that in many deposits, particularly those of iron, the ore minerals were later than the silicates and, therefore, could not have originated by the separation of early formed crystals. This led to a restudy of certain Adirondack titaniferous magnetite deposits which, since the work of Kemp in 1897, had been considered type examples of segregations. The work was undertaken by F. F. Osborne, who found that those deposits are injections and not segregations and are either concordant or discordant with the primary structure of the host rock. He found that the ore minerals are definitely later than the silicates and, therefore, could not be segregations of early crystals.

The restudy of other magnetite deposits has led to similar conclusions, and today few representatives of early magmatic segregations, as defined above, remain. Perhaps some of these will prove to be injections instead of segregations; others may prove to be late magmatic deposits.

Chromite deposits have long been considered unimpeachable illustrations of early magmatic segregation. Field and microscopic evidence testify that in many places this is the case. However, Diller; J. T. Singewald, Jr.; Sampson; C. S. Ross; E. Cameron; and others cite evidence of some late magmatic or even hydrothermal chromite. Similarly, several magmatic ilmenite deposits, long considered typical early segregations, have been shown to be late magmatic. Some nickeliferous sulfide deposits were also thought to be early magmatic segregations, but the sulfides are now known to be later than the silicates and, therefore, cannot be considered as early magmatic. They are either late magmatic or hydrothermal.

Few new deposits have been added to the class of early magmatic segregations. Many have been removed and few remain. It is probable that with further detailed work even more examples will be moved into other classes and that this group will hold only a small number of chromite deposits.

The Stillwater Complex of Montana, although less than half is exposed, is considered to be an early magmatic segregation-type complex. It has been studied extensively but has had little economic success although it has had sporadic attempts at production. One of the more promising recent economic developments is on Johns-Manville Corporation's platinum-palladium mining claims on the north wall of the West Fork of the Stillwater River, where geologists have found, in 23 sample locations, a new zone of 0.5 ounces per ton combined platinum-palladium. The samples also contain Cu-Ni values of 0.15 percent over an averaged width of two meters. The samples were taken at irregular intervals in the zone over a length of 1200 meters and over 300 meters in a vertical distance. Minor amounts of gold and silver were also indicated.

The mineral deposits formed by early magmatic segregation are generally lenticular and of relatively small size. Commonly, they are disconnnected pod-shaped lenses, stringers, and bunches. Less commonly, they form layers in the host rock. This is notably the case in the Bushveld Igneous Complex (BIC) of South Africa, where stratiform bands of chromite (Fig. 5-1) of remarkably uniform thickness lie parallel to the pseudo-stratification of the enclosing mafic igneous rocks and can be traced for several kilometers (Fig. 5-2). Even more remarkable are other thin bands of chromite in the "Merensky Reef" horizon that contain economic quantities of platinum. Some of the South-African

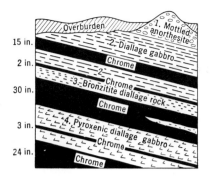

Figure 5-2 Upper chromite horizon in Bushveld Complex, Pritchard's workings, Kroonendal, Transvaal. (Kupferburger et al., South Africa Geol. Survey)

geologists consider the Bushveld chromite and chromite-platinum reefs to be early magmatic segregations formed by the sinking of early formed crystals. Also, Cameron and Emerson came to a similar conclusion for the Eastern Bushveld chromite.

With the increase in suggestions that hypervelocity impacts have formed large geologic features, suggestions arise that the BIC may have formed as the result of four simultaneous hypervelocity impacts modified by large-scale endogenic processes triggered by the impact. Rhodes, for example, has described the results of such an event to explain the origin of the magma as the result of sudden pressure release, the origin of the Rooiberg felsite, the chaotic attitude of the congolomerates and sandstones lying in moats of the impact craters, and even the origin of the overlying red granites.

Figure 5-1 Chromitite bands (black) in stratified anorthosite. Dwars River, Transvaal. Note convergence of some chromite bands and inclusion of anorthosite surrounded by chromite.

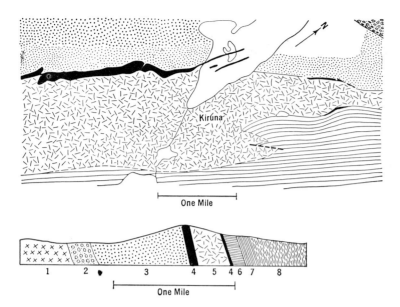

Figure 5-3 Plan and cross-section of magnetite deposit, Kiruna, Sweden. 1,Greenstone; 2, conglomerate; 3, syenite porphyry; 4, magnetite; 5, quartz porphyry; 6, 7, 8, sediments. (Geijer, Econ. Geol.)

One Mile

One Mile

Injections. Many magmatic ore deposits were formerly thought to belong to this group. The ore minerals presumably have been concentrated by crystallization differentiation (see Chapter 4) and are earlier than or contemporaneous with the associated rock silicate minerals. They have not remained at the place of accumulation, however, but have been injected into the host rock or surrounding rocks, perhaps as a mush of crystals of oxides fluidzied by residual magma. The structural relations of the deposits to the enclosing rocks suggest that they have been injected; they transect enclosing rock structures, include rock fragments, or occur as dikes or other intrusive bodies; they may even metamorphose the wall rocks. They resemble an intrusion of igneous rock, which they are. Tentative examples of such injections are the titaniferous magnetite dike of Cumberland, Rhode Island; perhaps the largest body of magnetite in the world, at Kiruna, Sweden, that has recently been suggested to have an exhalative-sedimentary origin (Fig. 5-3); the platinum pipes of South Africa and possibly some Bushveld chromite; and especially the ilmenite of Allard Lake, Quebec. Figures 5-4*A* and *B* illustrate a few aligned plagioclase subhedral to euhedral crystals enclosed in massive, late-forming, residual liquid titanomagnetite accumulation of the Bushveld Igneous Complex.

If such large concentrations of massive magnetite at Kiruna, or ilmenite at Allard Lake, are injected bodies, the differentiation and concentration must have occurred elsewhere before injection. If the concentration occurred by early crystal settling in a magma chamber, one is faced with the problem of injection, since solid crystals can hardly be injected. Vogt and Lindgren have maintained that the crystals were remelted and then injected. But Bowen has pointed out that there is insufficient heat to remelt magnetite, and further, to obtain remelting, the high temperatures required would also remelt the surrounding silicates, and miscibility might be evident. Others have suggested that the iron oxides and silicates form immisible liquids, but experimental investigations do not support this. The Kiruna and Luossavaara bodies, if magmatic concentrations, are extreme differentiations, as the ore is almost pure iron oxide. The difficulty of separating iron oxides from silicates should show some evidence of remaining silicates with the several occurrences of Kiruna-type deposits existing in Sweden; it would seem more likely, therefore, that they have not formed by magmatic segregation but by a sedimentary exhalative process that could explain the monomineralic origin of the deposits.

Some geologists have suggested that ultramafic intrusives have been emplaced as a mush of early formed crystals lubricated by minor residual liquid or as solid tectonic slabs (*obduction*). Possibly some of the magmatic ore deposits may have been similarly emplaced. We suggest elsewhere that in differentiation of mafic igneous magmas rich in iron, titanium, volatiles, and fluxes, these substances form a residual magma, which would be available for injection. If so, there may be no rep-

Figure 5-4a Aligned plagioclase crystals enclosed in a late liquid accumulation of titano magnetite, Bushveld Igneous Complex.

Figure 5-4b Aligned plagioclase crystals enclosed in a late liquid accumulation of titaniferous ore, Bushveld Igneous Complex.

resentatives of this class of early magmatic injections.

Late Magmatic Deposits. Late magmatic deposits consist of igneous ore minerals that have crystallized from a residual magma toward the close of the magmatic period. They are the consolidated fractions of the magma left after crystallization of the early formed rock silicates and in this respect differ from concentrations of early formed ore minerals mentioned previously. Consequently, the ore minerals of late magmatic deposits are later than the rock silicates and cut across them, embay them, and yield *reaction rims* of alteration products around the margins of enclosed silicate minerals. These changes, termed *deuteric alteration* occurred before the final consolidation of the igneous body and are not to be confused with somewhat similar effects produced in later rock alteration by hydrothermal solutions. Some of these effects, however, may have been caused by solutions excluded from the consolidating end products.

The late magmatic deposits are always associated with mafic igneous rocks. They have resulted from variations of crystallization differentiation, gravitative accumulation of heavy residual liquid, and liquid separation of sulfide droplets (called *liquid immiscibility*), like drops of oil out of water, or other modes of differentiation.

This late magmatic group now includes most magmatic deposits of iron and titanium ores that formerly were considered to have been formed in an early magmatic stage. It also includes some chromite deposits showing crosscutting relationships and probably the platinum pipes of the Bushvelt Complex of South Africa

A residual magma liquid enriched in iron, titanium, and volatiles could collect in the interstices of previously formed rock grains. If freezing occurred at this state, the result would be late disseminated grains of magnetite such as have repeatedly been reported in mafic igneous rocks. Or the residual ore liquid may be forcibly squirted out, or it may quietly drain out to settle on an underlying solid layer. There it may solidify, or while still fluid it may be subjected to pressure that may cause it to be injected as an intrusion into the fractures, the outer solidified part of the parent igneous body, or even into the surrounding rocks. Its mobility permits concentration and also injection. Thus there may be residual liquid segregation of injection. Figure 5-5 illustrates the steps in this process.

The essential difference between early magmatic and late magmatic deposits is that the early deposits must lie within the igneous body at the place of settling, and since the ore minerals accumulate as solids, there is no mobility after accumulation. In the late deposits, however, accumulation is through mobility, and the deposit may lie snugly and conformably enclosed within its host rocks or cut across its internal structures or those of other rocks, in disrupting fashion.

Residual Liquid Segregation. In Chapter 4 differentiation by crystallization differentiation was traced from the initial stage to final consolidation. It was pointed out that in most magmas the more mafic minerals crystallize first and the residual magma becomes progressively enriched in silica, alkalies, and water, with the silicic minerals the last to crystallize. However, in certain mafic magmas, the residual magma becomes enriched in iron, titanium, and volatiles. As mentioned earlier in discussing types of mafic rocks, magnetite was the last, or almost the last, mineral to crystallize; it lies moulded around, or fills the interstices between, early formed grains of calcic plagioclase or dark rock minerals. A restudy of many titanium-bearing magnetite deposits has also disclosed that the iron oxides crystallized last and surround, cut across, or embay the earlier formed rock minerals. The factual evidence is indisputable.

Given a somewhat mafic magma, perhaps with a higher than normal content of iron, undergoing differentiation at depth, the early formed crystals will be calcic plagioclase and mafic minerals. The first crystals in such liquid will tend to sink because their specific gravity will be greater than that of the liquid. Calcic plagioclase may be jostled upward by the heavier minerals. With advancing crystallization there will be a mush of crystals in diminishing liquid in which the crystals will be growing larger and new ones forming. Meanwhile, the residual liquid will be growing richer in excess iron and its density may come to exceed that of the crystals.

At this stage two possibilities may happen: (1) If the silicate crystals grow to touch each other and freezing occurs, the result would be a granular igneous rock with interstitial magnetite. The mafic sills of Connecticut and the Palisades of the Hudson river, for example, contain from 2 to 10 percent of such late interstitial iron ore. (2) The heavy iron-rich residual liquid may drain downward to collect below as a segregation resting on a solid floor of

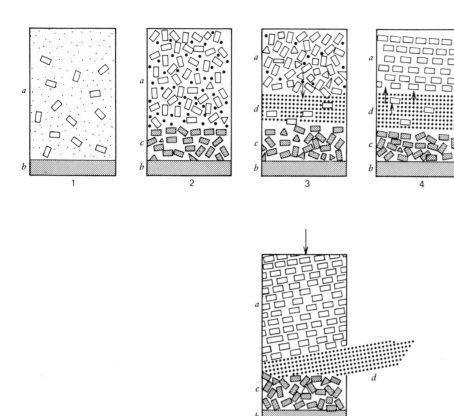

Figure 5-5 Idealized diagrammatic representation of late gravitative liquid accumulation. 1, Early stage of crystallization of basic magma a, after formation of chill zone b; 2, layer of sunken early-formed ferromagnesian crystals c, resting on chill zone b, with mesh of later silicate crystals above, whose interstices are occupied by residual magma enriched in ore oxides; 3, mobile, oxide-rich, residual liquid draining down to layer d, and floating up later silicate crystals; 4, formation of concordant oxide ore body in which a few late silicate crystals are trapped as mobile, enriched gravitative accumulation d squeezed out or decanted to form the magmatic injections.

early formed sunken crystals. To make room for itself, it would float up the enclosed lighter silicates, which would gradually be pushed upward in the magma chamber (Fig. 5-5). Under quiescent conditions this heavy liquid would form a layer concordant with the grain structure of the host rock. During its cooling it would expel upward the silicate crystals still contained in it. If the remaining silicate were abundant, the resulting ore would be low grade; if sparse, the ore would be high grade.

On the bottom would be a considerable thickness of early formed sunken crystals. On this would settle the oxide liquid, kept fluid by mineralizers and volatiles; above it would be upfloated, calcic plagioclase crystals in layered pattern. Thus the concordant layer of oxides would lie not at the bottom of the intrusive but in a midsection.

Concordant bodies of iron oxides entirely enclosed in mafic igneous rocks have previously been an enigma. The explanation just given could adequately account for them. The process may be termed *late gravitative liquid accumulation*. Representative examples of this process are the exten-

sive titaniferous magnetite layers of the Bushveld Igneous complex in South Africa, concordant magnetite deposits of the Adirondack mountains, and some platinum deposits.

Residual Liquid Injection. The iron-rich residual liquid accumulated in the manner described above may, before consolidation, be subjected to movement. A gentle tilting of the host rock may cause it to be decanted and move laterally, perhaps intersecting slightly the grain structure of the host rock. Or it may be subjected to pressure by disturbances such as often accompany magma formation and igneous intrusion and be squirted out to places of less pressure into the overlying consolidated portions of the host rock, or into adjacent rocks to form *residual liquid injections*.

The place of repose of the iron-rich residual liquid may be (1) in the interstices between the previously consolidated silicate grains, or (2) as an unconsolidated liquid segregation in the lower part of the igneous body.

The squeezing out of a residual liquid from the

silicate interstices like water out of a sponge has aptly been termed *filter pressing*. It could take place only by mashing together the silicate crystals, but even then some of the residual liquid must remain. The silicate crystals would show the effect of such mashing in the form of broken corners and edges, and strained, bent, cracked crystals, and undulatory extinction under the polarizing microscope. This is just what is found in the crystals of plagioclase in many anorthosites where filter pressing is thought to have taken place. Osborne has described such effects of mashing in anorthosite rocks associated with titaniferous magnetites in the Adirondack region, which he considered had resulted from filter pressing. In a large igneous body, however, filter pressing would not be likely to yield large concentrated bodies of oxides; rather there would be numerous avenues of escape of the fluid oxides, which would give rise to many smaller bodies of oxides.

However, if a late gravitative liquid accumulation were subject to pressure, the mobile liquid already collected as a concentrated body could be forced out en masse to a place of less pressure. It might be injected along the primary structure of the already consolidated parent intrusive to give rise to concordant bodies of oxides. More likely, it would be injected across the primary structure giving discordant relations along the contact, or out into the invaded rock along shear zones or fissures to form a sill or a dike, such as the dike of Cumberlandite (magnetite, silicates, apatite) in Rhode Island. If a high degree of purification had taken place before injection, mineral bodies of high purity might result. The injected nature of such deposits distinguishes them from segregations. They exhibit the intrusive relations of normal igneous intrusives. The ore minerals surround, cut across, corrode, and react with earlier-formed rock minerals. If the injected iron-rich fluids are rich in volatiles, hydrothermal reactions might be evident.

Some deposits of magnetite and ilmenite, which were formerly called early segregations, are now suggested to be of the *late, gravitative, liquid injection* type. This has been suggested for many of the titanomagnetite deposits of the Adirondack region of New York and the Allard Lake deposit. The huge magnetite deposits of Kiruna, Sweden, with its hundreds of million of tons of high-grade ore, is probably an injection, even though it encloses fragments of the country rock. Several recent studies, however, have recognized that the banded nature of the Kiruna iron is not the result of magmatic

segregation but actually sedimentary layering resulting from a submarine exhalative origin. In these deposits, the ores and the composition of the rocks above and below the ore indicates that the ores were deposited in a volcanic-marine environment suggesting a proposed classification of "Kiruna-type exhalative-sedimentary apatite iron ores." The Kiruna-type magmatic injection genetic type may be better illustrated by southeast Missouri; Cerro Mercado, Durango, Mexico; and Algarrobo, Chile, deposits. Some geologists also consider that the titanomagnetite deposits of Iron Mountain, Wyoming, are late magmatic injections. The huge ilmenite deposits of Allard Lake, Quebec, exceeding 300 million tons of ore, intrude the enclosing anorthosite, cross the grain structure, and include blocks and fragments of coarse anorthosite. This also appears to be a late magmatic injection into the parent host rock. The chromite "dike" at La Coulee, New Caledonia, looks like an injection, as do also some of the Turkish chromite deposits. The chromite at Dwars River Bridge in South Africa (Fig. 5.1) angles across the grain of the rock and includes fragments of parent anorthosite. It might be a decanted late magmatic accumulation. The platinum pipes of South Africa, actually olivine rock containing minute platinum, cut vertically across the pseudo-stratification of the BIC for hundreds of meters. It is difficult to escape the conclusion that these pipes and accompanying dikes are late magmatic injections.

Irvine has suggested that fractional crystallization by itself cannot yield a concentrated deposit of chromite once olivine or pyroxene or other silicates have begun to form. If, however, siliceous material is added to the magma on the olivine-chromite side of the phase diagram, the composition would shift to the chromite primary-phase field. It is possible that if added silica content were provided from granite melt derived, for example, from the roof of the intrusion, chromite separation from the magma might occur. This suggested process is substantiated by an experimental study conducted by Irvine at the Carnegie Geophysical Laboratory.

Residual Liquid Pegmatitic Injection. The formation of pegmatites, mentioned in Chapter 4, results from the injection of late magmatic fluids containing the ingredients of the late rock-forming minerals, along with much water, carbon dioxide, concentrations of rare elements, mineralizers, and metals. Many such pegmatites are valuable mineral deposits and are mined for their industrial minerals and metals.

These form dikes and irregular bodies of intrusives. Rarely, noninjected pegmatitic segregations occur within an igneous body.

Pegmatites of economic importance are mostly associated with felsic igneous bodies such as granitic or quartz diorite rocks. The simple types consist dominantly of quartz, feldspar, and mica; these minerals are apt to be arranged zonally from the walls toward the cores, generally with feldspar and mica dominant in the wall or intermediate zones and quartz in the cores. The less common pegmatitic minerals may be concentrated in one or other of these zones. The larger crystals in the core of zoned pegmatites are the result of the higher fluidity of the magma (lower viscosity) allowing more rapid movement of ions to growing crystal faces. Interestingly enough, the predominant mineralizer in the core of many pegmatites is liquid carbon dioxide, with water being more abundant at the outer fringe zone. Such is indicated by fluid inclusion studies.

The chief industrial minerals in complex pegmatites are feldspar, mica, quartz, corundum, cryolite, gemstones, rare earths, and minerals containing beryllium, lithium, cesium, and rubidium. The chief metals of economic interest are tantalum, niobium (columbium), tin, tungsten, molybdenum, and uranium.

The chief economic minerals of pegmatites are

feldspar	spodumene	tantalite
mica	lepidolite	columbite
quartz	amblygonite	wolframite
beryl	petalite	ferberite
pollucite		scheelite
bertrandite		molybdenite
corundum		uraninite
gemstones		

Immiscible Liquid Segregation. Skinner and Peck convincingly proved that sulfide immiscibility actually occurs. They found an immiscible sulfide melt in a late stage of a cooling Hawaiian lava, saturated in sulfide sulfur at a temperature of 1065°C. There was separation of two sulfide-rich phases—an immiscible sulfide-rich liquid followed by a copper-rich pyrrhotite solid solution. The sulfide-rich liquid quenched to a mixture of pyrrhotite, chalcopyrite, and magnetite. The liquid contained a significant oxygen content, which depressed the sulfide liquidus surface. They concluded that the first phase to crystallize is a copper-nickel-rich pyrrhotite solid solution. Thus the early phase of crystallization of the immiscible sulfide melt concentrates the copper and nickel. This may occur in sufficient quantity to form commercial ore bodies.

Although metallic oxides apparently do not or rarely form immiscible phases in silicate magmas, Vogt has shown that iron-nickel-copper sulfides are soluble up to 6 or 7 percent in mafic magmas and that upon cooling they may in part separate out as immiscible sulfide drops, which may accumulate at the bottom of the magma chamber to form liquid sulfide segregations. In this respect it would resemble the accumulation of molten copper (*matte*) that collects at the bottom of a smelting furnace by draining down through the overlying molten slag. The sulfides remain liquid until after the silicates crystallize; hence they penetrate and corrode the silicates and crystallize around them. Such sulfides are the latest pyrogenic minerals to crystallize, and because they penetrate and corrode the earlier silicates they give rise to relationships that have often been interpreted as hydrothermal.

The accumulated sulfide need not necessarily be a pure sulfide melt. Rather it must be thought of as an enrichment of sulfide in the lowest part of the magma, which upon consolidation gives rise to a mafic igneous rock with sulfide segregations. Rarely, an almost pure sulfide melt might occur, giving rise to bodies of massive sulfide.

The deposits formed in this manner consist of pyrrhotite-chalcopyrite-pentlandite nickel-copper ores, with accompanying platinum, gold, silver, and other elements. They are confined to mafic igneous rocks of the gabbro family that generally display pronounced differentiation. The deposits occur commonly as disconnected bodies along the lower margins of the differentiated intrusives, notably where there are depressions in the floor. Their size is proportionate to that of the mother intrusive. The mineralogy is simple and monotonously uniform. Pyrrhotite, chalcopyrite, and pentlandite are the chief metallic minerals; sperrylite and lead and zinc sulfides are rare; the platinum metals and gold, tellurium, and selenium are generally present. Vogt noted a relatively uniform ratio of nickel to copper. Examples of this class of segregation are the nickel-copper sulfide deposits of the *Insizwa* type, South Africa (Fig. 5-6), the nickeliferous sulfide deposits of the Bushveld, South Africa, and of Norway; and possibly some of the marginal deposits of Sudbury, Ontario. The breccia ore and the massive sulfide ore of Sudbury are considered by Hawley and by Naldrett and Kullerud to be definitely magmatic in origin and to have been injected late in the sequence

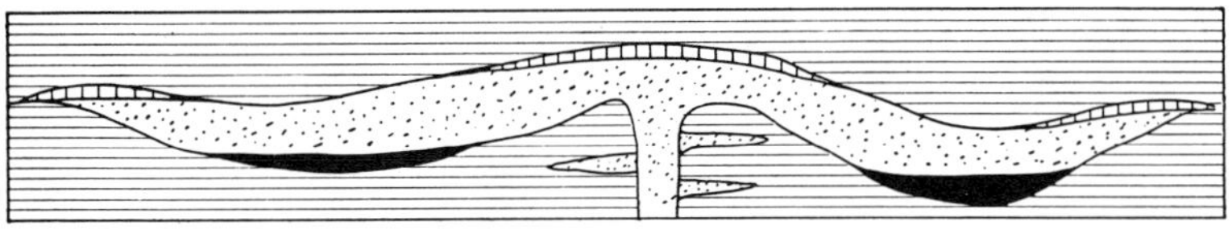

Figure 5-6 Generalized diagram of Insizwa type intrusion. Black, differentiated basal zone carrying nickel-copper sulphides. (After D. L. Scholtz, Geol. Soc. South Africa, 1936).

of intrusives. Other origins for the Sudbury ores have been suggested.

Recent geological studies of the layered *Duluth Complex* have resulted in a resurgence of exploration activity and success in discovering mineralization and ore deposits. The *Minnamax* copper-nickel deposit is a prime example that occurs in a relatively uncomplicated southeasterly dipping series of formations that make-up a portion of the Complex. The deposit is located 8 Km south of Babbitt, Minnesota. It is localized along the base of the Gabbro Complex and extends along a 5 Km length of the gabbro contact and is about 2 km miles in width.

Mineralization consists of disseminated copper and nickel extending from the surface downdip to a depth of 750 meters. Higher grade portions of the deposit of >0.60 percent copper through a minimum thickness of 15 meters are the Bathtub deposit on the west side that extends from 365 to 490 meters in depth and the Tiger Boy deposit that extends from 425 to 680 meters in depth. The sulfides are principally cubanite, nonmagnetic pyrrhotite, chalcopyrite, and pentlandite. Some massive sulfides are localized at the base of the mineralized intrusive and also in the underlying hornfels.

Immiscible Liquid Injection. If the sulfide-rich fraction accumulated in the manner described above, and should be subjected to disturbance before consolidation, it might be squirted out toward places of less pressure, such as sheared or brecciated areas along the margins of the consolidated mother rock or in the enclosing rocks. There it will consolidate to form immiscible liquid injections. Such deposits give unmistakable evidence of late magmatic age. They intrude older rocks and enclose brecciated fragments of foreign rock. They exhibit the intrusive relations of dikes. The ore minerals penetrate, corrode, alter, and even replace the silicates. The remnants of consolidation produce some hydrothermal alteration of the surrounding silicates. The deposits are irregular or dikelike in form. If the residual fraction is rich in volatiles, the resulting deposits might display transitions into hydrothermal types.

Examples of this class of deposit are the Vlackfontein mine of South Africa and probably some of the Norway nickel deposits.

ASSOCIATION OF ROCKS AND MINERAL PRODUCTS

Definte associations exist between specific magmatic ores and certain kinds of rocks. Platinum occurs only with mafic to ultramafic rock, such as the varieties of norite, peridotite, or their alteration products. Chromite, with rare exceptions is found only in peridotite, anorthosite, and similar mafic rocks. Titaniferous magnetite and ilmenite are mothered by gabbro and anorthosite, and magmatic magnetite deposits occur with syenite. Nickel-copper deposits are associated universally with norite, and magmatic corundum with nepheline seyenite. Diamonds occur in commercial quantities only in kimberlite, a variety of peridotite. Pegmatite minerals such as beryl, cassiterite, lepidolite, scheelite, and niobium-bearing minerals occur chiefly with granitic rocks. It is thus seen that the deep-seated mafic rocks are the associates of most of the important magmatic mineral deposits, which indicates a genetic relationship, probably during the early magmatic history of mafic rocks.

Selected Readings on Magmatic Concentration
Bushveld igneous complex—Magmatic ore deposits. 1969. J. Willemse. Econ. Geol. Mono., no. 4, p. 1–22. *Chromite, magnetite, and platinum deposits. Also vanadiferous magnetites*, p. 187–208.
Occurrence and characteristics of chromite deposits—Eastern Bushveld complex. 1969. Eugene N. Cameron and A. Desborough. Econ. Geol. Mono., no. 4, p. 23–40; also by E. N. Cameron, Econ. Geol., v. 54 p.1151–1213, 1959. Chromite deposits of this classical area.
Stillwater, Montana, chromite deposits. 1969. Edward Sampson. Econ. Geol. Mono., no. 4 p. 72–75.
Chromite seams—Great Dyke, Rhodesia. 1969. R. Bicham. Econ. Geol. Mono., no. 4, p. 95–113. *Magmatic concentration in separate intrusions.*
Gravity Differentiation and Magmatic Podiform Chromite Deposits. 1969. T. P. Thayer. Econ. Geol.

Mono., no. 4, p. 132–146. Gravitational segregation and reemplacement.

Symposium on chrome ores. 1962. C.T.O., Ankara, Turkey. *Review and occurrence of Asian deposits.*

Origin of chromite deposits, Bay of Islands, Newfoundland. 1953. Charles H. Smith. Econ. Geol., v. 48, p. 408–415. A different idea of formation of chromite deposits.

General Reference. 1. 1977. 72, p. 1279–1284, Manwaring and Naldrett. *Paper on an intrusion into the Duluth Complex.*

Duluth complex, history and nomenclature. 1972. In Geology of Minn: A centennial vol., W. C. Phinney, ed. Minn. Geol. Survey, p. 333–334.

Chromite deposits of selukwe, Rhodesia. Peter Cotterill. 1969. Econ. Geol. Mono., no. 4, 154–186. *Magmatic segregations on definite horizons in ultrabasic intrusions.*

Merensky Reef of the Bushveld complex. 1969. C. A. Cousins. Econ. Geol. Mono., no. 4, p. 239–251. *Concentration of sulfides of nickel, copper, and iron, along with platinum at the base of a mafic rock layer.*

The sulfide ores of Sudbury. E. Souch, et al. 1969. Econ. Geol. Mono., no. 4, p. 252–261. *Nickel-copper ores crystallized from an inclusion and sulfur-rich magma.*

The Sudbury ores: Their mineralogy and origin. 1962. J. E. Hawley. Minerals Assoc. of Canada, *Review of earlier works; new mineralogical data; magmatic origin.*

An immiscible sulfide melt from Hawaii. B. J. Skinner and D. L. Peck. 1969. Econ. Geol. Mono., no. 4, p. 310–322. *An immiscible sulfide melt in the late stage of a cooling lava.*

Notes on the Muskox intrusion, Coppermine River area district of Mackenzie. 1962. C. H. Smith. Geol. Survey Canada, Paper 61–25. *Unique complex of almost complete exposure.*

Phase relation in the Cu-Fe-Ni-S system and applications to Magmatic ore deposits. 1969. J. R. Craig and G. Kullerud. Econ. Geol. Mono., no. 4, p. 344–358. *Application of phase relations to Ni-Cu magmatic ores.*

Magmas and igneous ore deposits. 1926. J. H. L. Vogt. Econ. Geol., v. 21, p. 207–233, 309–332 and 469–467. *Early discussions of magmatic processes.*

Certain magmatic titaniferous ores and their origin. 1925. F. F. Osborn. Econ. Geol., v. 23, p. 724–761, 895–922. *Filter pressing applied to magmatic ores.*

Formation of late magmatic oxide ores. Alan M. Bateman. 1951. Econ. Geol., v. 46, p. 404–426. *Origin and classification of different kinds of magmatic oxide ores.*

Sulfides in the Skaergaard intrusion, Greenland. 1957 L. A. Wager, et al., Econ. Geol., v. 52, p. 855–903. *Immiscible droplets of sulfides form in magma and collect below.*

Geology of titanium and titaniferous deposits of Canada. E. R. Rose. Geol. Survey of Canada. Econ. Geol. Rept. 25, 1969. *A broad survey of titanium deposits associated with anorthesite; late magmatic injections.*

Therometric and petrogenetic significance of titaniferous Magnetite. A. F. Buddington, 1963. Am. Jour. Sci., v. 253, No. 19, p. 497–532.

Geology and genesis of ultrabasic nickel-copper-pyrrhotite deposits, Pacific nickel Property, 1956. British Columbia, by A. E. Aho. Econ. Geol., 51, p. 444–481. *Late magmatic origin.*

6

SUBLIMATION

Where there is smoke, or rather steam, there is fire, or rather magma.

Sublimation, a very minor process in the formation of mineral deposits, is included here for the sake of completeness of processes. Sublimation applies only to compounds that are volatilized and subsequently redeposited from vapor at lower temperature or pressure. It involves direct transition from the solid to the gaseous state, or vice versa, without passing through the liquid state, which usually intervenes between the two. It does not include minerals formed by reactions of gases and vapors. Many compounds cannot be sublimed in the presence of oxygen. The process is associated with volcanism and fumaroles.

There are many sublimates deposited around volcanoes and fumaroles, but they are seldom in sufficient abundance to make workable mineral deposits. Noticable concentration have formed in some geothermal cells. Sulfur, supposedly of this origin, has been mined in Italy, Japan, and elsewhere, and sodium chloride or common salt has been extracted locally around volcanoes. Common, but noncommercial sublimates are the chlorides of iron, copper, zinc, oxides of iron and copper, boracic acid, and various salts of the alkali metals and ammonium. Most of these, however, are quickly blown away or washed away.

7

CONTACT METASOMATISM

"If a man does not keep pace with his companions, perhaps it is because he hears a different drummer. Let him step to the music which he hears, however measured or far away."

Henry David Thoreau

The great explosive volcanic outbursts that occur from time to time are most impressive and terrifying. They testify to the vast volume of fluids that are released. These constituents originally were in solution in the magma, and as long as the magma remained liquid and the pressure unchanged, they remained in solution. With the release of pressure they escape. Most magmas, however, do not reach the surface but are enclosed in the earth's crust and their gaseous constituents do not directly reach the surface. However, during or shortly after the last stages of magmatic consolidation, the high-temperature fluid emanations may escape into the surrounding rocks and produce pronounced effects upon them. They also collect and carry with them other substances from the magma. With progressing crystallization of the magma there would tend to be an increasing concentration of the volatiles into the diminishing residual liquid. Thus, during the later stages of crystallization, a greater volume of volatiles would be available, and they would have greater opportunity to react with metals, host rock, and parent rock.

The effects of igneous emanations upon the surrounding rocks have been divided by Barrell into two types: (1) the effects of heat, alone, without appreciable accessions, giving rise to *contact metamorphism;* and (2) the effects of heat combined with accessions from the magma chamber, giving rise to *contact metasomatism.* The two are to be sharply distinguished, since contact metamorphism does not give rise to mineral deposits except in a few rare cases of nonmetallic deposits such as sillimanite, and contact metasomatism may give rise to valuable and distinctive mineral deposits.

Contact metamorphism manifests itself by (1) *endogene* or internal effects upon the margins of the intrusive body itself, and by (2) *exogene* or external effects upon the rocks invaded by the igneous mass. The endogene effects consist chiefly of textural and mineral changes in the border zone; pegmatite minerals such as tourmaline, beryl, or garnets may be present.

The exogene effects of large intrusive bodies are generally pronounced. They consist of a baking or hardening of the surrounding rocks and generally their thorough transformation. Old minerals are broken up, and their ions recombine to form new minerals that are stable under the changed conditions. For example, original minerals AB and CD may recombine to AC and BD. In an impure limestone consisting of carbonates of calcium, magnesium, and iron and quartz and clay, the calcium oxide and quartz may combine to form wollastonite; dolomite, quartz, and water form tremolite, or, by addition of iron, actinolite; or calcite, clay, and quartz form grossularite garnet. These, and other changes, yield simple recrystallized rocks such as quartzite from sandstone or marbles from limestone or dolomite, and more highly metamorphosed rocks, such as hornfels, from shale or slate and complex silicate rocks from impure limestones. In such alteration the chemical composition of the rock undergoes little change, except for the loss of some volatiles and the addition of magmatic volatiles such as boron or fluorine. The alteration is most intense nearest the intrusive and less intense farther away. The result of these changes is to form around the intrusive a contact-metamorphic *aureole* that varies in shape and size according to the size and shape on the intrusive and the character and structure of the invaded rocks.

Contact metasomatism differs from contact metamorphism in that important accessions from the magma are involved, which by metasomatic reaction with the contact rocks form new minerals under conditions of high termperature and pressure. To the effects of the heat changes of contact metamorphism are added those of metasomatism, by which the replacing minerals may be composed in part or in whole of constituents added from the magma. The resulting mineralogy is thus more varied and complex than that by heat metamorphism alone. If the magmatic emanations are highly charged with the constituents of mineral deposits, *contact-metasomatic deposits* result, particularly in the favorable environment of calcareous rocks. Such deposits have been frequently designated as "contact-metamorphic deposits"; but, as pointed out by Lindgren, they are not metamorphic—they are metasomatic, and their substance has largely been derived from the magma and not the invaded rock. Consequently, the use of the term *metasomatic* instead of *metamorphic* will eliminate some of the confusion that has existed. Lindgren proposed the term *pyrometasomatic deposits* and defined that as "formed by metasomatic changes in rocks, principally in limestone, at or near intrusive contacts under influence of magmatic emanations." Therefore, contact-metasomatic deposits and pyrometasomatic deposits are essentially the same, except that Lindgren includes under pyrometasomatic numerous deposits remote from intrusive contacts. The term *contact-metasomatic deposits* is favored, because it connotes the much-described relationship with intrusive contacts and a class of highly characteristic deposits recognized by their distinctive assemblage of high-temperature minerals so widely referred to in the older literature as contact-metamorphic deposits.

The process was first realized by Von Cotta in 1865, from a study of the deposits at Banat, Hungary, and was definitely recognized as a distinct ore-forming process by Von Groddeck in 1879. By 1894, Vogt established it firmly as the important ore-forming process in the formation of the Kristiania ores. The main concept was familiar to Lacroix by 1900. Lindgren, the first to recognize the type in the United States, in 1899, at Seven Devils, Idaho, redefined it. Since the nineties the work of Emmons, Kemp, Blake, Weed, Barrell, Lindgren, Goldschmidt, and others led to the realization of the importance of this type of deposit and helped unravel the intricacies of the mode of formation.

PROCESS AND EFFECTS

General Comments. The high-temperature effects of deep-seated magma intrusions upon the invaded rocks result from the heat transferred by the magmatic emanations and to a very minor extent by conduction. Entire beds of carbonate rocks may be changed to complex silicate rocks called *tactite,* or to *skarn* with the addition of iron oxides, or to garnet rock. Huge additions and subtractions of materials are involved. The emanations may carry the constituents of mineral deposits that replace the invaded rock to form metallic and nonmetallic mineral deposits distributed spasmodically within the contact aureole. Not all magma intrusions yield mineral deposits. For their formation, certain types or environments of magma are essential: The magma must contain the ingredients of mineral deposits; it may be intruded at deep to shallow depths and even in submarine depths, and it must contact reactive rocks and presumably invade zones of connate or meteoric water.

Temperature. The temperature at the immediate contact must have been that of the intruding magma, which in felsic to mafic magmas ranges from 600 to 1100°C, respectively. Outward from the magma contact the temperature would decline. Also, there would be a gradual lowering of the temperature during the slow cooling of the intrusive. Consequently, there may have been a decreasing temperature of the magma of several hundred degrees and more in the more distant zones of contact minerals. Mineral zones are observed in many metasomatic contacts. The Cornwall, England, district is a prime example of mineral zoning as indicated in Fig. 7-1.

On the other hand, the Donut scheelite ore body of the *Pine Creek mine* located near Bishop, California, has contact from top to bottom and had the most complete contact metasomatic zoning observed in the mine. Gray, Hoffman, Bagan, and McKinley state that there was no significant temperature gradient produced during the formation of the ore zone; the zoning was not, therefore, the result of temperature but must be the result of a chemical gradient. They conclude that the tactite

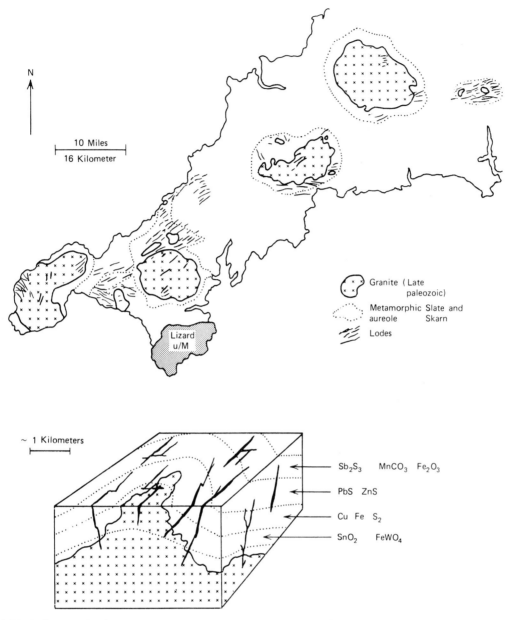

Figure 7-1 Block diagram showing ore zones relation to granite-stock contacts, Cornwall, England. (From K. Hosking)

formed under pneumatolytic conditions at a temperature just under 700°C. Hence isothermal conditions can result in contact metamorphic zones of different mineral or rock facies by reactions in different rocks, differing diffusion rates, or a varying chemical gradient.

One of the newer and better geothermometers that indicates the gradual decrease of temperature in contact-metasomatic deposits with increasing distance from the intrusive source involves the sta-

ble oxygen isotopic study. If isotopic exchange approached equilibrium between the invading hydrothermal water of uniform $\delta^{18}O$ content and the carbonate host rock, contours of equal $\delta^{18}O$ value could be plotted as isopleths as shown in Fig. 7-2. An actual plot of such a study of the Bohemia mining district, Oregon, is shown in Fig. 7-3.

When a coexisting pair of oxygen-bearing minerals has undergone isotopic equilibrium exchange of oxygen, the absolute temperature values can be

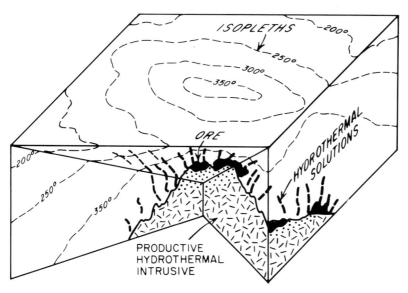

Figure 7-2 Idealized potential application of $\delta^{18}O$ studies to hydrothermally altered contact zones near a mineralizing fluid source.

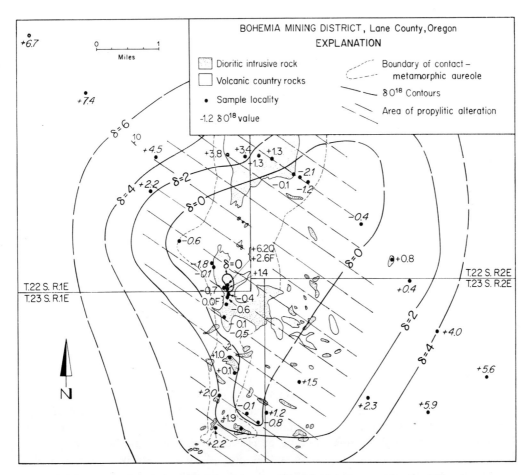

Figure 7-3 Generalized geologic map of the Bohemia mining district, Oregon, showing $\delta^{18}O$ values of volcanic country rocks (italic numbers) and intrusive bodies (modified after fig. 4, Taylor, 1971). The Brice Creek stock is at the top and the Champion Creek stock is in the center of the map. Q = quartz, F = plagioclase. (From: Taylor, Econ. Geol., v. 69, 1974)

obtained. With only one mineral the isopleths represent relative temperature changes. The isotopic composition of the water can also be calculated from the $\delta^{18}O$ compositions of two coexisting minerals.

The presence of wollastonite in carbonate rocks with indigenous or introduced silica indicates that the temperature did not exceed about 900°C at extremely high pressures. In the same system at higher temperatures, spurrite and larnite form. Wollastonite begins to form between 380 and about 900°C, depending on the pressure (Fig. 7-4). A large number of additional silicate minerals form in this temperature range but at more narrow temperature ranges as mentioned below. Similarly, the presence of andalusite indicates a temperature of less than 525°C, since above this temperature it changes to mullite or sillimanite and kyanite with increased pressure (Fig. 15-2). Merwin showed that contact-metamorphic andradite commonly exhibits anomalous birefringence and heating to 800°C makes it isotropic. A temperature below 800°C, therefore, is indicated. Evidence of high quartz identified in a contact-metamorphic zone indicates a temperature of formation above 573°C. Lindgren and Whitehead

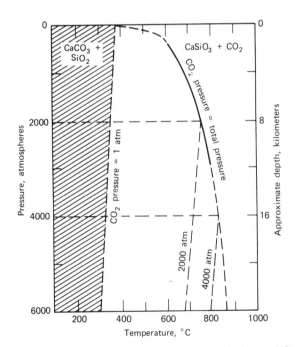

Figure 7-4 Equilibrium curves for the reaction $CaCO_3 + SiO_2 \leftrightarrows CaSiO_3 + CO_2$ at $Pco_2 = P$ total (heavy line) and $Pco_2 = 1$ atm (light line). Heavy solid line experimentally determined (Harker and Tuttle, 1955). Dashed lines extrapolated or theoretical. At P-T values between the two lines, calcite and quartz are stable if $Pco_2 = P$ total, and wollastonite is stable if $Pco_2 = 1$ atm. (After Barth 1962)

concluded that the contact-metasomatic deposits at Zimapan, Mexico, had been formed between 400 and 500°C. Thus contact metasomatism takes place between temperatures of about 300 to 800°C, perhaps, but rarely, even higher.

Recrystallization, Recombination and Accessions. The Recrystallization and recombination of rock minerals take place in the alteration halo. Limestone and dolomite are recrystallized to marble; carbonaceous impurities may form graphite; sandstone may be converted to quartzite, and shale to hornfels. Recrystallization alone generally indicates mild contact action in the outer zone of alteration. Usually, recombination of rock ions occurs also. With additions of materials from the magma, the original minerals AB and CD that recombined to AC and BD may become ACX and BDY, where X and Y represent accessions. Heating dolomite and quartz gives rise first to tremolite and, in the order of rising temperature, forsterite, diopside, periclase, wollastonite, monticellite, okemanite, spurrite, merwinite, and larnite. Thus many complex calcio-silicates are formed such as amphiboles, pyroxenes, vesuvianite, tourmaline, or garnets. The ore minerals, of course, result almost entirely from accessions from the magmas.

The ore minerals and those with constituents foreign to the invaded rocks give abundant testimony of noteworthy additions to the contact zone from the magma.

The magmatic accessions consist chiefly of the metals, silica, sulfur, boron, chlorine, fluorine, potassium, magnesium, and some sodium.

Volume Changes. Lindgren, in his classical work on the contact-metasomatic deposits at Morenici, Arizona, has shown that vast quantities of material have been added to, and subtracted from, the invaded rocks. He shows that if all the CaO in 1 cc of $CaCO_3$ were converted to andradite garnet, the volume would become 1.40 cc, or a volume increase of nearly one-half, but the field relations show no volume change. He concludes that for every cubic meter of the altered limestones 460 kilograms of CaO and 1190 of CO_2 have been carried away, and 1330 kilograms of SiO_2 and 1180 of Fe_2O_3 have been added. These astonishing figures of 1.8 tons of material removed and 2.8 tons added per cubic meter attest to the vigorous wholesale transfer of material and the quantity of accessions during contact metasomatism. Additional quantities of metallic minerals were also introduced. Some studies have indicated that volume expansion may have actually occurred.

Stages of Fromation. Contact metasomatism apparently commences shortly after intrusion and continues until well after consolidation of the outer part of the intrusive. In general, the first stage, which is one of heat, consists of recrystallization and recombination with or without accessions from the magma. This gives rise to many of the silicates. Magnetite and hematite form with the silicates and later than them, but they generally precede the formation of the sulfides, although Butler found magnetite later than chalcopyrite in Utah. The sulfides mostly form later than the silicates and oxides. A common order is pyrite and arsenopyrite followed by pyrrhotite, molybdenite, sphalerite, chalcopyrite, galena, and sulfo-salts last. In some places, however, the sulfides form contemporaneously with the silicates, but the older idea that the silicates, oxides, and sulfides were formed simulanteously no longer seems to hold, and successive stages of development have generally been noted. The stages of development, however, seem to vary considerably with different magmas.

Mode and Timing of Transfer. The recrystallization and part of the recombination may have been achieved by heat alone immediately after the intrusion. The main transfer of materials by magmatic fluids, however, must have occurred at a later period, after the freezing of the border or chill zone of the intrusive and during the accumulation of the restmagma in which the mineralizers were being concentrated.

Opinion is not yet unanimous in regard to the mode of transfer of the magmatic materials. It is shown in Chapter 4 that emanations can and do transfer materials outward from magmas. Ingerson and Morey, and Kennedy have demonstrated the effectiveness of water vapor in transporting silica. If the pressure is such that a vapor phase is formed, high-temperature gaseous emanations are to be expected. If no vapor phase is formed, magmatic fluids may emanate as high-temperature liquids and the contact metasomatism and accompanying mineral deposition may be produced by them.

RELATION TO INTRUSIONS

Contact metasomatism that gives rise to mineral deposits does not occur indiscriminately with all magmas. It is restricted exclusively to some intrusive magmas; it appears to depend upon the composition of the magma, the host rock, and it is related to the size and depth of formation of the intrusive body. Extrusive bodies may produce a little baking, hardening, or other effects at the contact but rarely mineral deposits. Some volcanic rocks, however, are excellent host rocks for the formation of mineral deposits.

Composition of Intrusions. Although widespread contact-metamorphic effects may be produced by magmas that produce rocks of wide variation in composition, those that yield mineral deposits are mostly felsic ones of intermediate composition, such as quartz monzonite, monzonite, granodiorite, or quartz diorite. Some deposits are associated with syenite as the Marmoraton, Ontario, iron mine. Highly silicic rocks, such as normal granite, rarely yield contact deposits. Neither are contact-metasomatic deposits found with ultrabasic rocks, and only rarely with mafic rocks. Two noted examples of such deposits in mafic rocks are deposits with quartz diabase at Cornwall, Pennsylvania, and with a white gabbro at Hedley, British Columbia. The lack of contact-metasomatic deposits around the great Bushveld Igneous Complex of South Africa is noteworthy, especially since so many ore deposits are included within the intrusive.

The Brownstone tungsten mine near Bishop, California, has a tactite zone in contact with a quartz monzonite intrusive. Successive zones of marble (that contain the scheelite), and hornfels are located farther from the contact (Fig. 20-39). Alaskite, however, cuts all of the rock units, but no contact metasomatic features are associated with this intrusive rock that presumably lacked water.

The reason contact-metasomatic deposits are produced from some felsic intrusives and seldom from mafic intrusives is probably because the felsic material has a higher content of fluids than the mafic. Schwartz has shown that greenstone in Minnesota develops hydrous minerals where intruded by granite and nonhydrous minerals where intruded by gabbro, indicating that the gabbro was deficient in water.

Size and Form of Intrusive. Most contact-metasomatic deposits are associated with stocks, batholiths, and intrusive bodies of similar size; they are rarely associated with laccoliths and large sills and are absent from small sills and dikes. Intrusive bodies whose flanks dip gently produce wider zones of contact metasomatism than those with steeply dipping flanks. Also, those whose roofs are irregular in form and are marked by cupolas and roof pendants give rise to extensive zones of pronounced contact metasomatism. The Pine Creek tungsten deposit near Bishop, California, is a prime example of a roof pendant extending from an elevation of 4150 meters, to an unknown depth below 2,130 meters.

Depth of Intrusion. The depth of intrusion appears to be an important factor in the formation of contact-metasomatic deposits, because deposits are found only with rocks of granular groundmass, which generally indicates relatively slow cooling at considerable depth. The absence of deposits with rocks of glassy and aphanitic texture, indicative of rapid cooling at shallow depths, shows that near-surface conditions are not favorable for the formation of contact-metasomatic deposits. This is probably because of rapid loss of the magmatic emanations at shallow depth.

The actual depth of consolidation of granular intrusives is known only imperfectly. At Philipsburg, Montana, the depth of crystallization was ~2100 meters. Brögger showed that near Oslo, Norway, some intrusives had crystallized at depths of less than ~1000 meters. Barrell concluded that at Marysville, Montana, the batholith had reached within 1200 meters of the surface, and similar depths were ascertained for the intrusives associated with contact-metasomatic deposits in Utah and New Mexico.

Alteration of Intrusive. In general, the intrusive itself has been little affected during contact metamorphism. Rarely, its margins may be so altered as to obscure the exact boundary between intrusive and altered intruded rock. Epidote, the chief mineral formed in the intrusive, presumably results from the absorption of CaO and CO_2 released from the invaded rock. Less commonly, garnet, vesuvianite, chlorite, diopside, and other minerals occur. Sericitization of the intrusive is common.

RELATION TO INVADED ROCKS

The character and extent of the alteration of the invaded rocks depend upon their composition and in part upon their structure. Certain rocks are more susceptible than others and exhibit extreme effects; some are only slightly changed.

Relation to Composition. Carbonate rocks are the ones most affected by intrusion of magma. Pure limestone and dolomite readily recrystallize and recombine with introduced elements. Impure carbonate rocks are affected even more, since such impurities as silica, alumina, and iron are ingredients available to enter into new combinations with calcium oxide. The entire rock adjacent to the intrusive may be converted to a mass of garnet rock, silicates, and ore.

Sandstones are but little affected. They recrystallize to quartzite and may contain sparse contact-metasomatic minerals.

Shales and slates are baked and hardened, or altered to hornfels, generally with andalusite, sillimanite, staurolite, and garnet. The degree of alteration varies with their purity, being greatest with carbonate varieties. The argillaceous rocks seldom contain important mineral deposits.

In some districts, volcanic rocks form better host rocks for mineralization than do carbonate formations. Invaded igneous rocks do not contain contact-metasomatic deposits. Their slight alteration is most pronounced if their composition is markedly unlike of the intrusive. For example, granodiorite may intrude granite, with little effect, but a mafic rock such as gabbro might undergo considerable change. Likewise, the metamorphic rocks are uncongenial for ore and undergo little change. This is generally true of the crystalline schists, whose constituents have already been metamorphosed to minerals that are stable under conditions of fairly high pressure and temperature. A noteworthy exception is the Pinal schist of the Miami, Arizona, district in which several major ore bodies have formed.

In general, then, the rocks most susceptible to contact metasomatism and most favorable for the formation of ore deposits are the sediments, particularly the impure calcareous ones, and some volcanic rocks.

Relation to Structure. The structure of the invaded rocks, and faults, affect the extent and position of

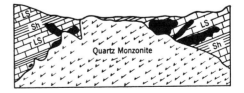

Figure 7-5 Relation of contact-metasomatic deposits (black) to monzonite intrusive and structure of limestone.

the contact-metasomatic zone. Where sedimentary beds dip into the intrusive, as in Fig. 7-5, good up-dip channelways are provided for escaping emanations in contrast to the slower migration across the bedding; ore deposits are likely to be larger and more widely distributed up the dip than across it.

Faults that extend outward and upward from the intrusive serve as through channelways that concentrate and conduct the emanations far from the intrusive, and localize the metamorphism and mineralization in tabular form.

RESULTING MINERAL DEPOSITS

The mineral deposits that result from contact metasomatism constitute a distinctive class characterized by an unusual assemblage of ore and gangue minerals. They contribute to the world's mineral production and supply many of the uncommon mineral products. The deposits generally consist of several disconnected bodies. They are mostly small as compared with "porphyry coppers," or sedimentary deposits. They are vexatious deposits to exploit because of their relatively small size, their capricious distribuiton within the contact aureole, and their abrupt terminations. Like the scattered plums in a plum pudding, they are difficult to find; they exhibit few "signboards" pointing to their presence, and costly exploration and development are necessary to discover and outline them. Their development must be undertaken with caution, and the optimism attendant upon mining such concentrated and often rich bodies frequently gives way quickly to disappointment upon the sudden termination of the ore body.

Position. The ore bodies occur within the contact aureole and generally relatively close to the intrusive contact. J. B. Umpleby, however, has pointed out that in several sulfide deposits the ore bodies lie on the outer or limestone side of the contact aureole, perhaps because of the neutralizing effect of the limestone upon acid generated as HC1 dissociates with decreasing temperature resulting in the lower pH.

The deposits are generally scattered irregularly around the contact but tend to be concentrated upon the side of the intrusive that has the gentler dip. If the dip of the intrusive is low, the deposits may lie a considerable distance horizontally from the contact and still be relatively close to it. Roof pendants (Fig. 7-6) and large inclusions are especially favorable loci, as at Mackay, Idaho. Where prominent faults extend outward from the contact, deposits may be aligned along them and extend hundreds of meters or more from the contact. This is the case at Bisbee, Arizona, and Knopf describes contact-metasomatic deposits at Rochester, Nevada, along a fault zone at a distance of 600 m from the intrusive.

Form and Size. Contact-metasomatic deposits are notably irregular in outline. They may have almost any shape. Generally, they are more or less equidimensional, but ramifying tongues may project outward along bedding planes, joints, or fissures enhancing the irregularity of outline. Those more irregular in shape occur chiefly in thick beds of limestone. Roughly tabular bodies are formed where deposits are aligned along the contact for a distance, or along fault zones, or where they have been formed by selective replacement of individual susceptible limestone beds. Just why certain beds prove more congenial to ore formation than others is uncertain.

In general, contact-metasomatic deposits are of comparatively small size, with dimensions of 30 to 120 meters and contain from a few tens of thousands to a few hundreds of thousands of tons of ore; a few deposits contain several million tons.

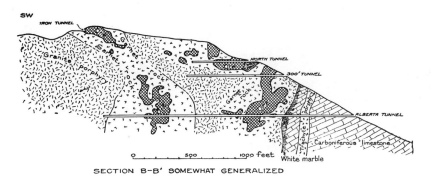

SECTION B-B' SOMEWHAT GENERALIZED

Figure 7-6 Cross-section of contact-metasomatic deposits at Mackay, Idaho, showing ore (cross-hatched in garnet rock developed from huge limestone inclusions. (After J. B. Umpleby, U.S. Geol. Survey)

Table 7-1 Types of Mineral Deposits Formed by Contact Metasomatism With the Chief Constituent Minerals and Examples

Deposit	Chief Minerals	Examples of Deposit
Iron	Magnetite and hematite	Cornwall, Pa.; Fierro, N. Mex.; Banat, Hungary; Gora, Urals; Elba, Italy; Iron Springs, Utah.
Copper	Chalcopyrite and bornite with pyrite, pyrrhotite, sphalerite, molybdenite, and iron oxides	Some deposits of Morenci and Bisbee, Ariz.; Bingham, Utah; and Ruth, Nev.; Cananea, Matehuala, Mexico; Susan, Korea
Zinc	Sphalerite with magnetite, sulfides of iron and lead	Hanover, N. Mex.; Long Lake, Ontario; Kamioka, Japan
Lead	Galena, magnetite, and sulphides of iron, copper, and zinc	Magdalena, N. Mex.; Inyo County, Calif.
Tin	Cassiterite, wolframite, magnetite, scheelite, pyrrhotite	Pitkäranta, Finland; Saxony; Dartmoor, England; Yak, Alaska
Tungsten	Scheelite and minor sulphides, or wolframite with molybdenite and minor sulfides	Mill City, Nev.; Inyo County, Bishop, Calif.; King Id; Australia; Tempiute, Nevada.
Molybdenum	Molybdenite, pyrite, garnet	Yetholm, Australia; Azegour, Morocco; Buckingham, Quebec
Graphite	Graphite and contact silicates	Adirondacks, N. Y.; Buckingham, Quebec; Sri Lanka; South Australia
Gold	Gold with arsenopyrite, magnetite, and sulfides of iron and copper	Cable, Mont.; Hedley, B. C.; Suan, Korea
Silver	Argentite, native, argentiferous galena	Bingham district-Lark and U.S. Mines
Manganese	Manganese and iron oxides and silicates	Längban, Sweden
Emery	Magnetite and corundum, with ilmenite and spinel	Virginia; Peekskill, N. Y.; Turkey; Greece
Garnet	Garnet and silicates	
Corundum	Corundum with magnetite, garnet, and other silicates	Peekskill, N.Y.; Chester, Mass.

Texture. Commonly, the ores are coarse in texture, containing large crystals or clusters of crystals. Few of the minerals, except garnet, show cyrstal outlines, and even garent appears often in sugarlike masses of irregular outline. Columnar and radiating minerals may show crystal faces several centimeters in length; calcite is generally in coarse grains. Magnetite may occur in bunches composed of large grains of shapeless outline. The individual minerals appear to be closely interlocked; open space is rare. Crustification and banding are absent, but orbicular structures have been noted. The metallic minerals, except pyrite and arsenopyrite, generally lack crystal outlines.

In some cases, however, as in parts of the deposits of Hanover, New Mexico, and magnetite at Iron Springs, Utah, the minerals are extremely fine grained and the ore has a flinty appearance, resembling dense hornfels; the individual grains of the intimately admixed gangue and metallic minerals cannot be distinguished by the naked eye.

Mineralogy. The outstanding feature of the mineralogy is the distinctive assemblage of gangue minerals characteristic of high-temperature formation. These include grossularite and andradite garnet, hedenbergite, hastingsite, tremolite, actinolite, wollastonite, epidote, zoisite, vesuvianite, ilvaite, diopside, forsterite, anorthite, albite, fluorite, chlorite, and micas. Quartz and carbonates are generally present. In addition, silicates containing the mineralizers, such as tourmaline, axinite, scapolite, ludwigite, chondrodite, and topaz, may be present.

The ore minerals consist of oxides, native metals, and sulfides, arsenides, and sulfo-salts. The oxides are represented by magnetite, ilmenite, hematite (specularite), corundum, and spinels. Magnetite is particularly abundant. Graphite, gold, and platinum represent the native minerals, but the last two are rare. The sulfides consist chiefly of base-metal sulfides. Sulfo-arsenides and antimonides are rare, as are the tellurides. In addition, scheelite and wolframite occur.

Mineral deposit formed by contact metasomatism include the minerals listed in Table 7-1.

Selected Reference on Contact Metasomatism

Clifton-Morenci district, Arizona. 1905. Waldemar Lindgren. U.S. Geol Survey Prof. Paper 43. *Outstanding work on contact metasomatism, particularly the transfer of materials from magma to invaded rocks.*

Occurrence of ore on limestone side of garnet zones. 1916. J. B. Umpleby. University of Calif. (Berkeley) Pub., Vol. 10, p. 25-37. *Review of contact-metasomatic deposits showing that ore bodies lie on the outside of the garnet zone.*

Contact metamorphic reactions and processes in the Mt. Talläc roof remnant, Sierra Nevada, Calif. 1966. A. A. Loomis. Jour. Geol., v. 7, p. 221-245.

Contact metasomatic tungsten deposits in tungsten mineralization in the United States. 1946. Paul F. Kerr. Geol. Soc. Am. Mem. 15, p. 41-46. Some excellent examples.

Structural control of contact metasomatic deposits in the Peruvian Cordellera. 1958. A. J. Terroves. Trans. AIME, v. 211, p. 365—372. *Geology of several contact metasomatic deposits. Contact metasomatism and ore deposition, Conception del Oro, Mexico. 1966. P. R. Buseck. Econ. Geol. v. 61, p. 97—136.*

Equilibrium temperature during formation of Marmoraton pyrometasomatic iron deposit. 1965. F. B. Park. Econ. Geol. v. 60, p. 1366-1379.

Triasssic magnetite and diabase, Cornwall, Pa. 1968. D. M. Lapham, Graton-Sales AIME, p. 1532—1554.

Bishop Tungsten district, California. 1968. R. F. Gray, et al. Graton-Sales AIME, p. 1532—1554.

The contact metasomatic magnetite deposits of S.W. British Columbia. 1970. D. F. Sangster. Geol. Survey, Canada Bull. 172, Ottawa.

General Reference 10, chap. 11, pt. 3, pp. 537—556. *Brief review of western occurrences illustrating Lindgren's classification.*

General Reference 9, chap. 28.

8

HYDROTHERMAL PROCESSES

"I propose to show how this seeming extravagance of poetry is actually sober scientific fact."

J. D. Dana, 1855

Thus far, ore-forming processes within the magmatic chamber have been considered first, then those operating just outside of the magma chamber during magma consolidation. This chapter pertains to a further step in magma consolidation that gives one variety of *hydrothermal solutions,* namely, hydrothermal mineralizing solutions, some of which are associated with magmas and others are not. These may be the direct source of epigenetic mineral deposits or they may be derived from meteoric or connate waters or with water of metamorphic origin and still be hydrothermal solutions. (Fig. 8-1).

The term *hydrothermal* means hot water, and hot waters may originate by other than magmatic processes. They may be a mingling of waters as already mentioned or they may be meteoric, or connate, or the water content of minerals released during metamorphism that have become heated within the earth and thus become hydrothermal solutions. Recent isotopic studies of hydrothermal solutions indicate that connate and meteoric waters play a more prominent role than hitherto recognized (Fig. 8-1). Also, Roedder, who has made extensive studies of the character and temperature of formation of fluid inclusions in ore deposit minerals, suggests that the Mississippi Valley type of deposits have been formed from modified, deep circulating, heated connate brine solutions carrying trace amounts of lead and zinc.

As a definition, Helgeson has suggested that "*hydrothermal solutions* are concentrated, weakly dissociated, alkali chloride-rich electrolyte solutions" (which is somewhat more definitive than that of mere "hot water"). With the high chloride content of hydrothermal solutions and the presence of H^+ ions, it has been expected that such solutions would be highly acidic. This depends, however, upon the degree of dissociation of HCl into H^+ and Cl^- ions. At temperatures of 100°C and less, HCl is almost completely dissociated and the pH is consequently low.

Hydrothermal solutions give rise to high-temperature hydrothermal deposits nearest the intrusive, intermediate-temperature deposits at some distance outward, and low-temperature deposits farther outward. Lindgren has designated these three groups as *hypothermal, mesothermal,* and *epithermal* deposits, according to the temperatures, pressures, and geologic relations under which they were formed as indicated by the contained minerals. Lindgren's temperature ranges, however, are now believed to be too low, especially for the hypothermal deposits that evidently have reached 600°C, and possibly even higher. Butte, Montana, and Bingham, Utah, minerals exhibit evidence of such temperatures. Figure 8-2 illustrates the isotherms at the Wairakei geothermal area with a narrow conduit, wider spread temperatures near the surface and flow lines of the water.

Deposits formed by solutions that migrate far from the intrusive, or possibly are not derived from the intrusion at all, may approach the temperature of the host rock. They generally produce only weak reactions and are referred to as *telethermal* deposits. Some would include the Mississippi-type deposits in this category. On the other hand, near surface solutions that are under high initial pressure and high initial temperature would result in rapid reactions and rapid deposition of an unusual variety of minerals. Such deposits are named *xenothermal* as suggested by Buddington. Xenothermal deposits may exhibit hypothermal to epithermal mineralogy but in overlapping or telescopic configuration.

Recent significant studies by Helgeson and Korzinskii indicate that HCl is a stable complex[1] at high temperatures and high chloride content. In a

---

[1] A complex is a chemical species formed by the association of two or more simpler species, each capable of independent existence, e.g., H^+ and CO_3^{2-} to form HCO_3^-.

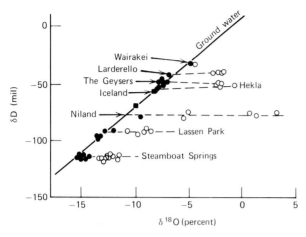

Figure 8-1 Variations in δD and $\delta^{18}O$ values of water from different areas. Variations in essentially $\delta^{18}O$ from the meteoric water line are presumably the result of hydrothermal reactions with rocks of lower $\delta^{18}O$ values.

chloride-rich solution, above the critical temperature of water, HCl forms to a high degree at the expense of H_2S and HS^- because it is a far more stable species than the two latter complexes at high temperatures. The pH of such a solution is near neutral or slightly acidic because of the comparative lack of free H^+ ions, which aids in explaining the not uncommon occurrence of how such a solution might transit carbonate beds without chemical reaction. Furthermore, such a solution would lack the ability to enter specific types of hydrothermal alteration reactions because of the lack of free hydrogen ions that allow hydrogen ion metasomatism.

Both the Red Sea and the Salton Sea brines have salinities that are almost 10 times that of seawater. Hydrogen and oxygen isotopic studies indicate that the leaching by surface waters that moved downward to geothermal sites dissolved chlorides from sediments and provided a heated brine solution that is possibly the precurser of hydrothermal solutions. At the same time, the chloride-rich hydrothermal solutions would form metal chloride complexes if metals were present. Helgeson has calculated the increased solubility of metal chloride complexes as being sufficient to explain how metals are transported in solution by near-neutral or slightly acid hydrothermal solutions. Seward, as an example, has plotted (Fig. 8-3) the increased solubility of silver chloride at various temperatures and Cl^- concentration.

As the temperature of the solution decreases, HCl begins to dissociate and the increasing acidity

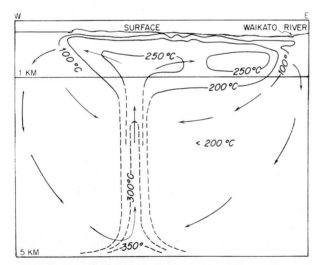

Figure 8-2 Vertical section through the modern geothermal system at Wairakei, New Zealand, showing actual measure (solid lines) and estimated (dashed lines) isotherms down to a depth of 5 kilometers beneath the surface (modified after Banwell, 1961, and Elder, 1965). The approximate flow lines of the meteoric water are shown with arrows. (From Econ. Geol., v. 69. 1974.)

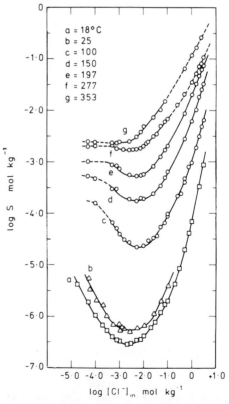

Figure 8-3 The solubility, S, of silver chloride (solid) as a function of sodium chloride concentration, $[Cl^-]_{in}$, from 18 to 353°C; the data at 18 and 25°C are from Mironov (1962) and Jonte and Martin (1952); the solid lines are the solubility curves calculated using the values of Ks_0 and β_n obtained in this study. (From Seward, 1976, Geochim. et Cosmochim. Acta, v. 40.)

initiates specific types of hydrothermal alteration. As a result, the pH of the cooling alkali, chloride-rich hydrothermal solution may remain near neutral. The increased degree of formation of such complexes as H_2S, HS^-, and $H_2CO_3^=$ (expected at low temperatures) opposes the trend toward acidity resulting from dissociation of HCl with decreasing temperatures.

With the increase in the formation of H_2S as temperatures drop, it is expected that the reaction of H_2S with indigenous iron will result in early pyritization of the deposit. Finally, when H_2S encounters metal complexes, precipitation of metal sulfides results.

It has been shown by Kennedy that water is quite soluble in silicate melts (Fig. 4-11) and such a magma entering a zone highly saturated with meteoric water under high pressure could pick up the meteoric water and expel it later as a hydrothermal solution. Burnham has indicated that felsic melts may have some difficulty in absorbing water. Shallow-seated intrusives, however, do seem to discharge mineralized gases and vapors into groundwater to create hydrothermal solutions.

In their journey through the rocks, the hydrothermal solutions may lose their mineral content by deposition in the various kinds of openings in the rocks, to form *cavity-filling deposits,* or by metasomatic replacement of the rocks to form *replacement deposits.* The filling of openings by precipitation may at the same time be accompanied by some replacement of the walls of the openings. Thus there may be a gradation between these two types of mineral deposits. In general, in favorable host rocks, replacement dominates under the conditions of high temperatures and pressures nearer the intrusive where *hypothermal* deposits are formed and cavity filling dominates under the conditions of low temperatures and pressures where the *epithermal* deposits are formed; both are characteristic of the *mesothermal* zone.

PRINCIPLES OF HYDROTHERMAL PROCESSES

Geologists attribute to hydrothermal processes that vast array of metallic mineral deposits that supply part of our useful metals and minerals. From such deposits are won most of the gold and silver, copper, lead and zinc, mercury, antimony and molybdenum, most of the minor metals, and many nonmetallic minerals. Consequently, it is these deposits

that have been mined, investigated, and written about far more than any other group. They have given rise to many of the great mining districts of the world; part of the lore of mining sprang from them.

Essentials for the formation of hydrothermal deposits are: (1) available mineralizing solutions capable of dissolving and transporting mineral matter, (2) available openings in rocks through which the solutions may be channeled, (3) available sites for the deposition of the mineral content, (4) chemical reactions that result in deposition, and (5) sufficient concentrations of deposited mineral matter to constitute workable deposits.

Openings in Rocks

The movement of hydrothermal solutions from source to site of deposition is dependent largely upon available openings in the rocks. The deposition of large bodies of extraneous minerals involves the necessity of continuing supplies of new material, and this means that through channelways must be available. The openings must be interconnected. Furthermore, cavity-filling deposits obviously cannot form unless there are available cavities to be filled.

A characteristic of many porphyry types of deposits is a shattered cupola zone where even small hand specimens exhibit planer surfaces or breaks, commonly referred to as the "plumbing system," that have allowed the passage of mineralizing fluids. A typical rock specimen showing these characteristics is shown by Figs. 8-4 and 8-5.

The various types of openings in rocks that may serve as receptacles for ore or permit the movement of solutions or their constituents through the rocks may be classified as shown in Table 8-1.

Most of the kinds of cavities listed in Table 8-1 may in rare cases under special conditions become filled to form various types of mineral deposits. Some serve only as conduits for mineralizing solutions; some permit the ingress of solution for replacement; others serve as receptacles or conduits for water, oil, and gas; still others such as pore spaces, shear zones, and rock alteration openings play a more important role in replacement than in cavity filling. The formation of epigenetic mineral deposits is dependent upon them.

Pore Spaces. Rock pores are interstitial openings between grains, capable of absorbing fluids. They make rocks permeable and serve as containers for

cμ

Figure 8-4 Typical shattered hand specimen from a porphyry copper deposit (Bingham) illustrating the planar fractures that allow the movement of hydrothermal solutions through the rock.

Figure 8-5 Different view of Fig. 8-4 showing large veinlet that has allowed the passage of siliceous and metallic solutions.

ores, petroleum, gas, and water. The oil and gas pools of the world are contained for the most part in rock pores, as are also subsurface water supplies.

Porosity. The porosity of a rock is the volume or pore space measured in percentage of the rock volume. It ranges from practically nothing in hard indurated rocks to a maximum of more than 50 per-

Table 8-1 Types of Openings in Rocks

Original Cavities	
Pore spaces	Cooling cracks
Crystal lattices	Igneous breccia
Vesicles or "blow	cavities
holes"	Bedding planes
Lava drain channels	

Induced Cavities	
Fissures, with or	Volcanic pipes
without faulting	Tectonic breccias
Shear-zone cavities	Collapse breccias
Cavities due to folding	Solution caves
and warping	Rock alteration
Saddle reefs	openings
Pitches and flats	
Anticlinal and	
synclinal	
cracking and	
slumping	

cent in loose-textured soils. With spheres of uniform size, the minimum porosity is 25.95 percent and the maximum is 47.64 percent, but rocks are never perfect spheres. The maximum occurs when the arrangement of the spheres is in the unstable form, with the center of one directly above the center of another. The minimum occurs with the stable form of packing, similar to the arrangement obtained when cannon balls are stacked in a pyramid. The same maximum and minimum porosities are obtained regardless of the diameters of the spheres. The spheres in a container arranged in unstable packing approach a stable arrangement when jarred. Angular materials have greater porosity than spherical ones, and finer angular materials have considerably greater porosity than coarser angular materials. In other words, an ore bin filled with fine ore will contain less ore than one filled with coarse ore.

Permeability. Permeability is a measurement of the ease by which fluids transit material. The permeability of a rock exists by virtue of its porosity, but a rock may be porous and not permeable. Permeability does not increase in direct proportion to the porosity but is dependent upon the size of the pore, the total amount of pore space, and particularly the interconnection of pore spaces. Some vesicular lava has high porosity but if the pores do not connect, the permeability is nil. The smaller the pores, the greater the surface exposed, the greater the friction, and the less the flow. If the same sand is packed to give 26 percent and 47 percent porosity, the flow through the latter will be 7 times greater than through the former. Thus a coarse sandstone with only 15 percent porosity may be much more permeable than a shale with 30 percent porosity. Wet clay, because of its very fine pores, exerts a very tight hold upon the contained water. Consequently, water-wet clays and shales are essentially impermeable. Coarse-pored rocks, on the other hand, even if of low porosity, are quite permeable if the pores are connected. The percentage of porosity of some common rock samples are as follows:

	Average	Maximum	Minimum
Granites (25)	0.369	0.62	0.19
Building limestones (25)	4.88	13.36	0.53
Building sandstones (35)	15.9	2.8.28	4.81
Oil sands (29)	19.4		
Clays (14)	28.43		

Bedding Planes. These are well-known features of all sedimentary formations. They permit the ingress of hydrothermal solutions and replacement by ore of adjacent walls.

Vesicles or "Blow Holes." Vesicles or "blow holes" are openings produced by expanding vapors typical of the upper part of many basaltic lava flows. They are tubular in shape, roughly circular in cross section, and may be spaced from $\frac{1}{2}$ to 2 inches apart. If the vesicles are filled, the rock is called an *amygdaloid*. If they are closely crowded together, they form a cellular rock like a sponge, called *scoria*.

Volcanic Flow Drains. Volcanic flow drains form in lava flows when the outside of the lava has solidified and liquid lava in the center drains out leaving a pipe, or tunnel.

Cooling Cracks. Formed as a result of contraction in cooling igneous rocks, cooling cracks may be regularly spaced joints that divide the rock into blocks, or parallel platy partings, or irregular cracklings.

Igneous Breccias Cavities. Igneous breccias are of two types: volcanic breccias forming agglomerates, and intrusive breccias. Both consist of coarsely and angularly fragmented igneous rocks with finer interstitial material. They may be quite permeable.

Fissures. Fissures are continuous tabular openings in rocks, generally of considerable length and depth. They are formed by compressive, tensile, or torsional forces operating on rocks and may or may not be accompanied by faulting. Thus faults are fissures, but all fissures need not be faults. They may constitute long and continuous channelways for solutions. When occupied by metals or minerals, they form *fissure veins*.

Shear-Zone Cavities. Shear zones result where fractures, instead of being concentrated in one or two single breaks, are expressed in innumerable closely spaced and more or less parallel, discontinuous surfaces of deep-seated rupture and crushing. Faulting is present generally. The thin, sheetlike openings, mostly of infinitesimal size, make excellent channelways for solutions, as is evidenced by the copious water flows where they are cut in tunnels and mines. Because of the minute openings, only minor open-space deposition can occur, but the large specific surface available makes shear zones favorable localizers of replacement deposits.

Folding and Warping. Flexing and folding of sedimentary strata gives rise to: (1) *saddle reef* openings at the crests of closely folded, narrow anticlines; (2) *pitches*, which are highly inclined, and *flats*, which are openings formed by the parting of beds under gentle slumping; and (3) longitudinal cracks along the crests of anticlines and synclines.

Volcanic Pipes. When explosive volcanic activity bores pipelike openings, the material blown out may fall back or be washed back into the opening, forming an angular breccia with spaces between fragments. These form excellent confined conduits for mineralizing solutions from which cavity filling or replacement deposits may form.

Breccias may be formed by the crushing of any brittle rock caused by folding, faulting, intrusion, or other tectonic forces, forming tectonic breccias; or by the collapse of rock overlying an opening, giving rise to collapse breccias. As with other breccias, the openings between the angular fragments provide space for the circulation of solutions, cavity filling, or replacement.

Solution openings, such as caves and enlarged joints or fissures in soluble rocks, supply channelways and open spaces for cavity filling.

Rock Alteration Openings. Wall rocks that have been altered by solutions are found by tests to be generally more porous than unaltered rocks and permit the ingress of mineralizing solutions.

Dolomitization of limestone, for example, results in greater porosity and permeability of the hydrothermal dolomite. Pinckney and Rye noted an increase in fracturing and jointing in Mississippian limestone that had been dolomitized. It has been suggested that the substitution of magnesium for calcite during dolomitization results in a reduction in volume. Hewett has noted that dolomitized limestone near ore deposits is commonly slightly more porous but states that the porosity is rarely more than five percent.

Silicification also results in an increase in porosity and permeability. The processes of silicification are the result of introduced silica and its production by specific processes of hydrothermal alteration. Being a more brittle rock, and producing more brittle rocks by silicification, it results in more fractures and openings.

Figures 8-4 and 8-5 are different views of the same shattered hand specimen of a typical porphyry copper (Bingham) hydrothermally altered rock that exhibits the numerous planes and openings that allowed for the passage of mineralizing fluids.

Movement of Solutions Through Rocks

The movement of hydrothermal solutions through rocks appears to take place most readily where long continuous openings, such as fissures, are available, or where smaller openings are interconnected as in shear zones, vesicular lava beds, or permeable porous sediments. Since many deposits contain millions or even hundreds of millions of tons of ore, vast quantities of solutions must have been necessary to transport this substance. Hence fairly large confined channelways must have been available. Diffusion through crystal lattices is known to be too slow and too incapable of acting in large volume over long distances. However, diffusion is an important factor in permitting the ingress of replacing ions into crystals at the actual foci of replacement where voluminous moving solutions could not gain entry. Moreover, the flow of solutions, particularly prior to reaching the site of deposition, must have been relatively confined or their mineral matter would have become too dispersed. A widespread volcanic breccia, for example, might be quite permeable throughout, but, just because of this widespread permeability, mineralizing solutions

could be so spread out that the minerals deposited from them would be too diffuse to constitute ore.

To supply a large amount of new mineral matter at the point of deposition, fissures, channelways, and permeable beds serve as the main freight lines, and small fractures (Figs. 8-4 and 8-5), cleavage planes, and pore spaces further distribute it to the front lines, where diffusion acting over short distances rarely more than a few meters, may deliver it to the point of deposition. As is shown under "Replacement," diffusion is essential for the entry of ions through an already formed wall of replacement ore, but it is not essential for cavity filling. Thus rock openings are essential to transmit large volumes of mineral matter to the site of deposition.

Rock-Particle Size and Specific Surface

The size of rock particles is important not only in the flow of solutions through rocks but particularly in chemical reactions between rock matter and solutions. The surface area of a unit mass of a finely divided substance is termed the *specific surface*. With large surface area and small particle size, pores are small and permeability is low. But these features are opportune for mineral deposition, since large specific surface permits greater contact between rock and solution and, therefore, greater opportunity for reaction between them. An angular open breccia, for example, offers to mineralizing solutions simultaneously many times greater rock surface than confining walls of a fissure vein and, hence, greater opportunity for deposition or replacement. A permeable bed, such as an arkose with connected but small pores, presents large specific surface for reaction between rock grains and solutions, and in addition the small pores retard the rate of flow and afford more time for reactions.

Effect of Host Rocks

Field evidence indicates that in many cases the world over, reactive wall rocks not in equilibrium with solutions exert a profound effect upon hydrothermal mineral deposition, more particularly in replacement than in cavity filling. Certain carbonate beds, for reasons not previously known, are congenial to ore deposition whereas others are not. It is evident now, however, that those solutions in which HCl has not dissociated could transit carbonate beds with little reaction of the weakly acid solution with the carbonate rock.

Metallizing solutions seem almost to have been empowered with peculiar discrimination in passing by certain carbonate beds and replacing others at Leadville, Colorado; Bisbee, Arizona; Bingham, Park City, or Tintic in Utah; or Kennecott, Alaska. The preference of ore for dolomitic limestone over pure limestone in lead-zinc-silver ores of Mexico has been pointed out by Hayward and Triplett. However, at Santa Eulalia, Mexico, the reverse is true, and massive sulfide ore in limestone terminates abruptly at a dolomite contact. Likewise, the selective mineralization of "greenstone" in contrast to adjacent granite and "porphyry" has been demonstrated repeatedly in the gold camps of northern Canada. Wandke and Hoffman point out that at Guanajuato, Mexico, bonanza silver ores occur in fissures opposite andesite or porphyry but not opposite schist. Weston-Dunn shows that the Giblin tin lode at Mt. Bischoff, Australia, widens in slate but pinches in quartzite. Similarly, other rocks are known definitely to favor ore deposition, and innumerable examples can be drawn from all over the world indicating that ore deposition, particularly replacement, appears to be influenced or localized by the character of the host rock. With the present knowledge of undissociated acids in hydrothermal solutions, it should be more readily understood why certain beds have been more selectively mineralized. The selective replacement of specific formations is a subject of vast study, observation, and speculation.

Mineral Deposition

Hydrothermal solutions gradually lose their heat to the rocks through which they pass. The temperature drop depends upon the rate of loss of heat to the wall rocks, which in turn depends upon the amount of solution moving past, and particularly the capacity of the wall rock to conduct heat away. In the initial stages of circulation with cool wall rocks, the temperature drop will be relatively rapid, but the continued flow of solutions will heat the wall rocks to the temperature of the solutions.

The nature of the rock openings also affects the heat loss. Rapid flow through a straight-walled open fissure (Fig. 8-2) would entail less loss of heat than flow through the intricate openings of a breccia with large specific surface, where the initial drop in temperature would be rapid. Once heated up, however, the breccia would not remove much heat from the solution. The greater the volume of new solution passing a given point, the greater would be the supply of heat and the slower the drop in the temperature of the solutions. Thus, in a fissure with characteristic constrictions and open spaces, the temperature of the solutions would drop less in the constricted than in the wider areas. Such features are also important in causing and localizing the deposition of mineral content.

The solutions are initiated under the high pressures that prevail at the depths where they originate. Their upward journey, through zones of lower pressure, normally is accompanied by a drop in pressure, which promotes precipitation. But other factors may also result in changes in pressure. Constrictions in channelways, or partial filling by mineral deposition, or barriers may build up excess pressure below them. The escape of the solutions to more open areas above the constrictions would lower the pressure and promote deposition. Thus many changes in the physical character of the openings through which solutions pass play an important role in causing and localizing deposition of minerals from hydrothermal solutions.

Kennedy has proved, for nonvolatile substances in supercritical steam, that at constant temperature there is an increase in solubility with an increase in pressure; at a constant pressure below 700 bars there is a decrease in solubility with an increase of temperature and an increase of solubility at any given pressure above 700 bars; and the solubility of some nonvolatile substance in a gas is more dependent on pressure than on temperature (Fig. 4-8). These startling conclusions lead to the interesting corollary that in contrast to the precipitation of minerals from a liquid by decrease in temperature, precipitation *can be effected by decrease in pressure alone.* This means that at the depths of vein formation where temperature is relatively constant, if a mineral-laden fluid enters a porous rock where pressure release takes place, a sudden dumping of mineral content can occur. It also follows that a fluid at constant pressure could dissolve substances from the rock walls while a drop in temperature occurs through the loss of heat to the wall rocks.

HYDROTHERMAL ALTERATION

The recognition and application of the various types of *hydrothermal alteration or wall rock alteration* has provided a valuable prime exploration tool that must be used by the exploration geologist because the alteration halo is much more widespread than the smaller target of a hidden ore deposit (Fig. 7-

2). The subject has mushroomed during the last score of years not only with the study of the recognition of alteration types and processes in the field but also with a profuse amount of laboratory studies.

Field studies and laboratory experiments have dovetailed remarkably well in understanding the stability conditions of minerals under dehydration, temperature, and pressure variables, and determining reaction curves and divariant equilibrium areas. A detailed summary of such applicable studies chiefly by Hemley, Burnham, Roy, Yoder, Meyer, and others no less important but too numerous to list has been summarized by Creasey as it applies to Arizona porphyry copper deposits.

The factors that determine the types and intensities of hydrothermal alteration are

1. Characteristics and composition of the host rock.
2. Composition of the hydrothermal fluid.
3. Temperature and pressure conditions and changes in phase of the hydrothermal fluid.
4. Changes in certain constituents such as early released H_2S that may become a strong acid.

Variations in these factors give rise to a large variety of hydrothermal alteration types. Comments on five of these types follow.

Dolomitization. Variable amounts of magnesium substituting for calcium in carbonate rocks commonly occurs in sedimentary processes. These primary dolomites are sometimes difficult to distinguish from dolomitization resulting from hydrothermal solutions transporting Mg^{2+} ions, which replace Ca^{2+} in $CaCO_3$ beds. Although primary dolomite formations contain a comparably more uniform amount of Mg^{2+} throughout the areal extent of a formation, hydrothermal dolomitization is more erratic, leaving portions of the preexisting carbonate rock unaltered. The extent of magnesium replacement also diminishes in distance from the mineralizing source; however, certain formational members may be selectively replaced much more extensively with little or no dolomitization in juxtaposed members. This makes it difficult to always distinguish between primary and hydrothermal dolomite.

One means of distinguishing between primary dolomite and the replacement of calcium by magnesium during hydrothermal alteration is to determine if the aqueous solution exchanged oxygen atoms from its H_2O with the original carbonate, especially when done under isotopic equilibrium conditions with an abundance of H_2O such that there is no significant change in the $\delta^{18}O$ composition of the H_2O. As the oxygen of the hydrothermal H_2O replaces oxygen of a different $\delta^{18}O$ composition in the CO_3^{2-} ion, the $\delta^{18}O$ composition of the dolomite may increase with distance from the source because of decreasing temperature (Fig. 7-2), according to the isotopic exchange equilibrium reaction; changes in $\delta^{13}C$ values have also been noted. Isotope analyses of the carbonate sample can, therefore, allow the construction of an isopleth map of the surface of a hydrothermally dolomitized area similar to that drawn on the surface of the block diagram shown in Fig. 7-2. This procedure could be a major aid in exploration for deposits enclosed or adjacent to the much larger target of widespread hydrothermal dolomitization. In addition, however, dolomitization releases the preexisting Ca^{2+} ion and may redeposit it in veins, by replacement of other minerals, or reacting with constituents occurring in some of the following types of alteration.

Silicification. This variety of hydrothermal alteration is not only quite common but an obvious ingredient in many of the other types of alteration discussed here. The addition of quartz or opaline silica to the other rocks in deposits is well displayed. Cinnabar deposits are commonly formed in fractures in opaline deposits. Of course, there are many siliceous sinter or opalized surface deposits devoid of economic minerals.

The Mother Lode, California; Yellow Knife, N.W.T.; and other quartz-gold vein deposits have formed by the leaching of silica from adjacent host rocks with redeposition in dilatant fractures that have resulted in gold-quartz veins or fissures at these locals. On a much smaller scale, a similar process is shown in Fig. 8-5.

In many deposits and in the adjacent country rock, hypogene solutions carrying silica have dispersed the silica throughout the area and have sometimes "overwhelmed" or "flooded" the deposit. Such siliceous bodies of this type are referred to as *jasperoid*. It is often colored or darker as the result of "limonite," or ferric oxide, hydrated or not, impregnated in the jasperoid.

Silicification of carbonate rocks is not uncommon, a process that occurs so carefully that the sedimentary features are remarkably preserved. As a cold water process, silicification is well known in the formation of petrified wood, but silicification is also well known under hydrothermal processes.

The solubility of SiO_2 has been thoroughly studied in a quantitative manner by Kennedy. Figure 4-8 illustrates the increasing solubility of SiO_2 with increasing temperatures and especially pressures above the critical temperatures of water. A decrease in pressure from 1250 bars to 400 bars, but isoclinally at 440°C, for example, results in the deposition of more than two-thirds of the SiO_2 dissolved at that temperature. Such an isothermal process may better explain how some quartz veins form by filling dilatant zones rather than by the process of just decreasing temperature. In additional support of the loss of silica (Figure 4-8) from host rocks during hydrothermal alteration, note the decrease by leaching of the SiO_2 content of different host rocks Fig. 8-6.

Figure 8-6 is a summary of the chemical changes from fresh wall rock (No. 1) to increasing alteration of the wall rock. The results were determined by Tooker in a study of hydrothermal alteration of different rocks in the Colorado Mineral Belt and illustrate the metasomatic variable changes in different host rocks.

Argillic Alteration. This term was suggested by Lovering in 1940 to emphasize the prevalence of clay minerals formed by hypogene solutions reacting with wall rocks concordant with the leaching of lime. Weak to intermediate argillic alteration is made manifest with the dispersed formation of kaolin, montmorillonite, dickite, halloysite and illite.

Advanced argillic alteration is commonly found adjacent to veins where it grades outward to intermediate and weak argillic alteration. Argillic alteration halos of one vein may merge with that of another, encompassing a large area and leaving no unaltered rock between the closely spaced veins. This occurs in the central zone at Butte, where the veins are so closely spaced that they are referred to as "horsetail." Many porphyry copper deposits, for example, San Manuel, Globe-Miami, Ray, Bisbee, Silver Bell, and Morenci, Arizona; Bingham, Utah; and others, show these mergings of alteration.

The absence of minerals in the epidote and carbonate groups (note granodiorite in Fig. 8-6) is indicative of the loss of lime. Amphiboles are also unstable under argillic alteration conditions.

Propylitic Alteration. This type of hydrothermal alteration has been less commonly recognized in the past, especially in porphyry copper deposits, but the large assemblage of minerals included in this type of alteration makes its recognition of significant importance. It is generally a fringing type of alteration common in epithermal or hot spring deposits.

Basically, propylitic alteration differs from the argillic type by the addition of lime-bearing minerals such as calcite, chlorite, and epidote. Albite, some sericite, zoisite, leucocene, and pyrite are common associates (Fig. 8-6).

A typical reaction of propylitic alteration of biotite to chlorite is as follows:

$$2K(Mg, Fe)_3AlSi_3O_{10}(OH)_2 + 4H^+ \rightarrow$$
$$Al(Mg, Fe)_5AlSi_3O_{10}(OH)_8 + (Mg, Fe)^{2+} + 3 SiO_2 + 2K^+$$

The only constituent needed to bring about this reaction is the H^+ ion. Note the free SiO_2 that is available for silicification. This is such a typical type of hydrothermal reaction that it is common to define hydrothermal alteration as "hydrogen ion metasomatism" whereby H^+ ions replace other cations. Of course, this is an accurate definition for some types of alteration but it would exclude certain types such as dolomitization, calcic alteration, pyritization, and others.

Sericitization. Sericitic or potassic alteration is probably the most abundant, more readily recognized, and widespread variety of alteration noted. Sericite is a fine-grained, white mica, commonly associated with ore deposits. It is readily observed in thin sections under crossed nichols of the microscope because of its birefringent appearance. Once again, it is the result of hydrogen ion metasomation as, for example, the alteration of orthoclase to sericite, K ions, and silica, respectively, as follows:

$$3KAlSi_3O_8 + 2H^+ \leftrightarrows KAl_2AlSi_3O_{10}(OH)_2 + 2K^+ + 6 SiO_2$$

In this type of alteration, assemblages of sericite, quartz, and pyrite are common. The reaction is reversible and does result in the change of muscovite to K feldspar over a large temperature range as shown by Hemley and Jones in Fig. 8-7. Creasey refers to this as *potassic* alteration and notes that it is distinguished by the assemblage of muscovite-biotite-K feldspar or any two of the three phases.

Diagrams referred to as AKF and ACF are commonly used to illustrate the composition of various types of hydrothermal alteration. A set of such diagrams for different assemblages of the San Manual-Kalamozoo (the recently discovered, deeper faulted extension of the San Manual ore deposit) is shown in Fig. 8-8. These diagrams illustrate the chemical

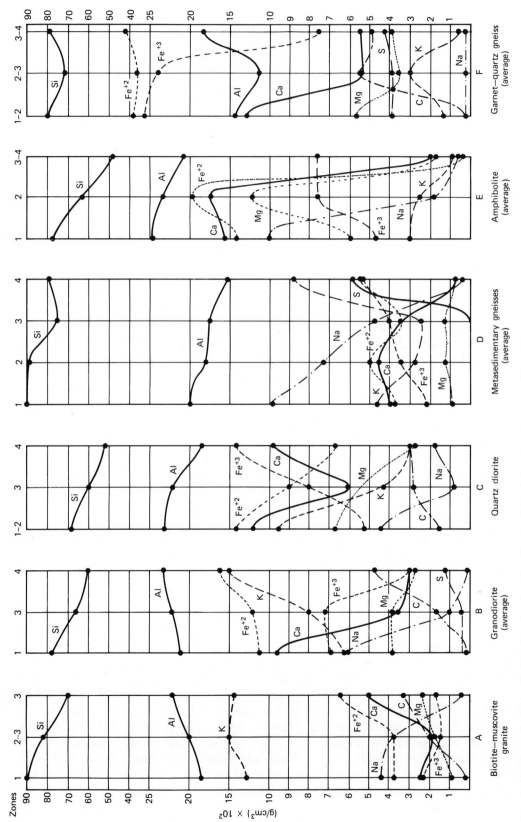

Figure 8-6 Summary of ion distributions in altered wallrocks of the Colorado mineral belt. (From E. Tooker, 1963, U.S. Geol. Survey Prof. Paper 439.)

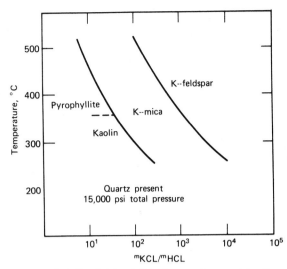

Figure 8-7 Stability fields of K-feldspar, mica, and kaolinpyrophyllite. A similar diagram relates sodium feldspar (albite) and sodium mica (paragonite) stabilities to temperature and alkali-to-hydrogen ion ratios. Clearly, both feldspars and micas can be stable over a wide temperature range, even down to surface temperatures in alkaline environments. Also, one can cross stability fields without necessarily changing temperature. (From Hemley and Jones, 1964, Econ. Geol., v. 59.)

and mineral changes that occur during hydrothermal alteration.

In order to detect alteration in the field, the exploration geologist often resorts to simple tests and the hand lens. The degree of hydrothermal alteration can be estimated in the field by recovering material from, for example, an altered feldspar phenocryst with a pocket knife and then *carefully* grinding a small portion of it between one's teeth to determine the extent of plasticity or unaltered gritty feldspar remaining.

This discussion of the various types of hydrothermal alteration is by no means complete. Other minerals commonly formed during alteration stages lead to other categories such as alunitization, pyritization, and jasperoidization. Some geologists would even include ore mineralization as the last stage of hydrothermal alteration although the economic stage is not always present. The fact that ore deposits are commonly associated with hydrothermal alteration, which is generally much more widespread than the ore deposits, make the study and utilization of hydrothermal alteration a prime exploration tool.

With the use of instrumental techniques, such as petrographic microscopes, differential thermal analysis (DTA), x-ray diffraction, and electron probe, the rapid recognition and better identification of alteration products and their chemical composition is more readily determined.

CAVITY FILLING

The filling of open spaces or cavities in the rocks was early thought to be the only mode of formation of mineral deposits and was long considered the most important. Particularly, a filled fissure, called fissure vein was almost synonymous with ore deposit and is still considered the representative type of mineral deposit. More mineral deposits are formed by the filling of cavities than by any other process, and of these the fissure vein is by far the most common and important. However, as stated previously, fissure filling is generally accompanied by replacement.

The Process and Characteristic Features

Cavity filling consists in deposition from solutions of minerals in rock openings. The solutions may be dilute or concentrated, hot or cold, and of magmatic or meteoric derivation. Mostly they are hot and dilute. Precipitation of the minerals is brought about by the processes considered in Chapter 3, of which changes in chemical character and temperature and pressure of the mineralizing solutions are the chief ones.

The first mineral to be deposited lines the walls of the cavity and grows inward generally with the development of crystal faces pointed toward the supplying solution. In some cases the same mineral or minerals may be deposited continuously on both walls until the cavity becomes filled or nearly so. Such a type of filling gives rise to homogeneous or *massive* ore. Generally, however, successive crusts of different minerals are deposited upon the first one, perhaps with repetition of earlier minerals, until the filling is complete, and this gives rise to *crustification;* if the cavity is a fissure, a *crustified vein* results (Fig. 8-9). If the crusts surround breccia fragments, *cockade ore* may result. If prominent crystals project from the walls, as in Fig. 8-10, it forms *comb structure.* Commonly, the filling may not be complete and open *vugs* (Fig. 8-10) remain in the center; some are large enough to admit a man. The vugs may contain one or more sequences of crystals perched on the walls and are eagerly sought by mineral collectors because they often prove to be treasure houses of the beautiful and

SHALLOW–MODERATE DEPTH ASSEMBLAGES					
FRESH QM, PORPHYRIES	**PROPYLITIC ZONE**	**ARGILLIC ZONE**	**PHYLLIC ZONE**	**POTASSIC ZONE**	
Quartz	No Change	Augmented	Augmented	Augmented	
Orthoclase–Microcline	No Change	Flecked with Sericite	Sericitized	Recrystallized, in part replaced by alteration K-feldspar–quartz	
Plagioclase (An35-45)	Tr. Mont, flecks & granules ep, zois, car, chlorite, kaol.	Montmorillonite → Kaolin	Sericitized	Fresh to completely replaced by brn-grn alt'n biotite, K-spar, ser.	
Biotite	Chlor, zois, car, leucoxene	Chloritized, + leucoxene, qtz	Sericite, pyrite, rutile	Fresh or recrystallized to sucrose brn-grn granules, ± chlorite	
Hornblende	Ep, car, mont, chlor (2 types)	Chloritized	Sericite, pyrite, rutile(?)	Biotite, ± chlorite, rutile	
Magnetite	trace pyrite	Pyritized	Pyritized	Pyritized	
A–K–C–F A = Al K = K, Na C = Ca salts F = Fe, Mg					
Veinlet Fillings	Q-cal- K-spar-chlor-rare ab–rt	Q–ser–py–chlor	Q–ser–py	Q–K-spar–bi–ser–anhy–cal–ap	

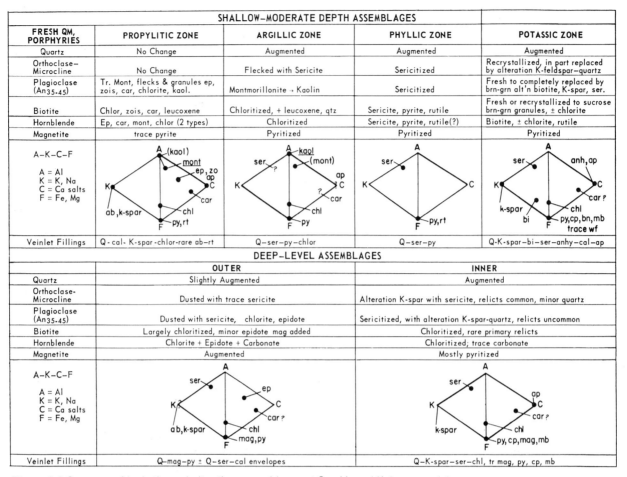

DEEP–LEVEL ASSEMBLAGES		
	OUTER	**INNER**
Quartz	Slightly Augmented	Augmented
Orthoclase-Microcline	Dusted with trace sericite	Alteration K-spar with sericite, relicts common, minor quartz
Plagioclase (An35-45)	Dusted with sericite, chlorite, epidote	Sericitized, with alteration K-spar-quartz, relicts uncommon
Biotite	Largely chloritized, minor epidote mag added	Chloritized, rare primary relicts
Hornblende	Chlorite + Epidote + Carbonate	Chloritized; trace carbonate
Magnetite	Augmented	Mostly pyritized
A–K–C–F A = Al K = K, Na C = Ca salts F = Fe, Mg		
Veinlet Fillings	Q–mag–py ± Q–ser–cal envelopes	Q–K-spar–ser–chl, tr mag, py, cp, mb

Figure 8-8 Summary of hydrothermal alteration assemblages at San Manuel-Kalamazoo, Arizona. (From Lowell and Guilbert, Econ. Geol., v. 65, p. 378.)

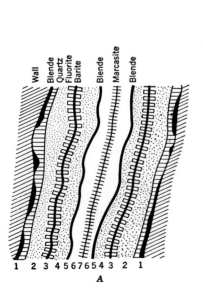

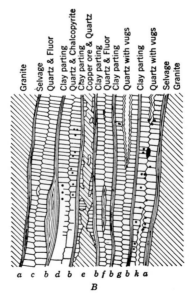

Figure 8-9 Crustified veins. A, Symmetrical crustification, Freiberg, Germany (after Von Wissenbach); B, asymmetrical crustification in tin vein, Redruth, Cornwall (after Phillips).

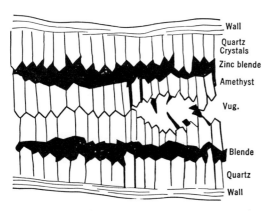

Figure 8-10 Growth of quartz crystals from fissure walls toward the center, forming a comb structure, and a vug in the center containing latest crystals.

rare crystals that adorn mineral museums. Such crustified deposits and vugs permit the paragenesis of the minerals to be worked out.

Vein crustification may be *symmetrical* with similar crusts on both sides (Fig. 8-9A and 8-10) or *asymmetrical* (Figs. 8-9B and 8-11) with unlike layers on each side. The latter is generally caused by the reopening of the fissure, permitting further deposition, as between layers *a* and *b* of Fig. 8-9. Several reopenings may occur. Many quartz veins are characterized by *ribbon structure,* consisting of narrow layers of quartz separated by thin dark seams of smeared altered wall rock, which presumably represent successive reopenings and movement. Angular fragments of wall rock partly enclosed by the filling may occur.

Cavity filling involves two separate processes: (1) the formation of the opening to start with, and (2) the deposition of the minerals. The two may operate almost simultaneously, but generally they are independent processes separated by an interval of time.

RESULTING MINERAL DEPOSITS

The process of cavity filling has given rise to a vast number of mineral deposits of diverse form and size, and such deposits have yielded a great assemblage of metals and mineral products. Much of the literature on economic geology relates to them.

For convenience, the deposits resulting from cavity filling may be grouped as follows; discussions of each follow in sequence.

1. Fissure veins.
2. Shear-zone deposits.
3. Stockworks.
4. Saddle reefs.
5. Ladder veins.
6. Pitches and flats; fold cracks.
7. Breccia-filling deposits: volcanic, collapse, and tectonic.
8. Solution cavity fillings: cave, channel, and gash veins.
9. Pore-space fillings.
10. Vesicular fillings.

1. Fissure Veins

A fissure vein is a tabular ore body that occupies one or more fissures; two of its dimensions are much greater than the third. Fissure veins are the most widespread and most important of the cavity fillings and yield a great variety of minerals and metals. There are several varieties that differ from each other chiefly in form. They are the earliest described type of bedrock deposit, around which the lore of mining has grown.

Formation. The formation of a fissure vein involves (1) the formation of the fissure itself, and (2) the

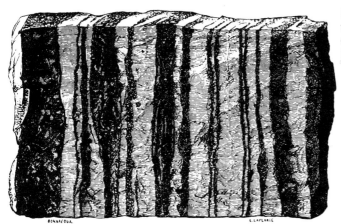

Figure 8-11 Drawing of Zellerfeld vein, Harz, Germany, with crusts of galena, blende, and calcite. (Burat, 1870.)

ore-forming processes—a distinction often over-looked. The two may have been separated by a long interval of time. Neither can result in the formation of a fissure vein by itself; a coincidence of the two is necessary.

Fissures may be formed by stresses operating within the earth's crust and may or may not be accompanied by faulting. Also, they may be formed or enlarged at the time of mineralization by the intrusive force of the mineralizing solutions, which acts as a wedge from below and spreads the rocks apart along some crack or line of weakness. It has been contended that the force of growing crystals may wedge apart the walls of a crack to make a wider fissure. Growing crystals do exert some pressure, within the limit of their strength, upon confining substances, but it seems extremely doubtful that this force is sufficient to form fissures.

Varieties. The varieties of fissure veins are simple, composite, linked, sheeted, dilated, and chambered; each of these may be massive or crustified.

The simple fissure vein occupies a single fissure whose walls are relatively straight and parallel. Where the walls are irregular and brecciated, particularly the hanging wall, owing to formation under light load near the surface, it is often called a *chambered* vein (Fig. 8-12).

Dilation or *lenticular veins* (Fig. 8-12) are fat lenses in schists. Generally, several occur together, like a string of sausages, or they may be disconnected *en echelon* lenses (Fig. 8-12). They are thought to be caused by the bulging or dilation of schistose rocks due to pressure transmitted by the mineralizing solutions. Some are due to the pulling apart of a preexisting vein during later metamorphism of the enclosing rock. In width they range from a few centimeters to several tens of meters.

A group of closely spaced, distinct, parallel fractures is a *sheeted vein* (Fig. 8-12C). Each fracture is filled with mineral matter and is separated by layers of barren rock, and the whole is mined as a single deposit lode whose width may be several tens of meters. If individual fractures are linked by diagonal veinlets, a *linked vein* is formed (Figs. 8-12E).

A *composite vein* or *lode* is a large fracture zone, up to many tens of meters in width, consisting of several approximately parallel, ore-filled fissures and connecting diagonals, whose walls and intervening country rock have undergone some replacement. The Comstock Lode of Nevada is an example.

Physical Features. Most fissure veins are narrow and range in length from a few hundreds of meters to a few kilometers. Few are vertical; most are highly inclined and apex at the surface. Consequently, their outcrop is the trace of an approximate plane upon the surface. If the surface is flat, the outcrop will be relatively straight; if irregular, the outcrop will be irregularly curved, depending upon the relief and the relation of the dip to the surface slope, as in Fig. 8-13. Two parallel veins may thus form seemingly divergent outcrops. Greater disparity results if two veins have parallel strikes but different dips. To determine the true relation of the veins to each other it is necessary to plot their strike and dip, or project them to a horizontal plane. Misconceptions of the true strike and dip in regions of pronounced topographic relief have given rise to many erroneous deductions regarding the extensions of veins in strike and dip. Many a tunnel has been projected to intersect a vein at depth but has missed it because the outcrop has been confused with the strike.

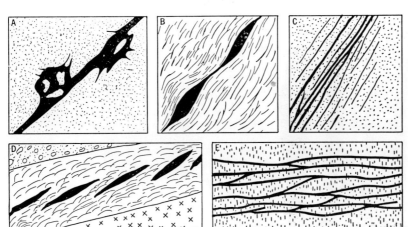

Figure 8-12 Varieties of fissure veins. A, Chambered vein (after Becker); B, dilation veins in schist; C, sheeted vein, Cripple Creek, Colorado; D, en echelon veins in schist; and E, linked vein.

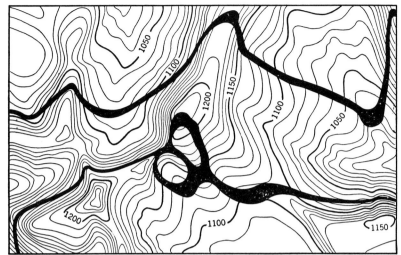

Figure 8-13 Relation of outcrop of 10-foot southerly dipping inclined veins to topography, accurately plotted. The two veins are parallel in strike but of different dip. One would not suspect that the peculiar outcrop of the lower vein could result from the intersection of an inclined plane with an irregular surface.

Veins are seldom planes, notwithstanding the customary textbook illustration. Most of them curve, along both strike and dip, but the dip is apt to be straighter than the strike.

Most veins exhibit irregularity in width, or *pinches* and *swells,* owing to movement of one wall past the other (Fig. 8-14*A*). A swell followed by a pinch is apt to be followed by another swell. Ordinarily, it is difficult to conceive of wide, straight-walled, inclined veins having existed as open yawning cavities before being filled, but the support of juxtaposed protuberances, as in Fig. 8-14*C* may allow wide premineral spaces to persist.

Displacement along fissure veins is mostly small, although generally present. It is noteworthy that large faults are seldom mineralized although there are exceptions, such as the Comstock Lode. Ransome has pointed out that in the Coeur d'Alene district the veins exhibit small displacement and the large faults of the region are barren. Most fault fissure veins have displacements of only a few centimeters to a few tens of meters, rarely a few hundred meters, and only exceptionally over a kilometer.

Fissure veins branch, divide, and join again, enclosing *horses* of country rock, split into stringers, and form *brecciated* zones.

The *walls* of fissure veins are commonly marked by a band of *selvage* or *gouge,* which is a claylike or gummy substance formed by the movement of one wall upon the other and subsequently altered. The vein filling often scales off cleanly from a gouge wall constituting what is called a *free wall.* In contrast, the walls may be *frozen;* that is, the vein matter, particularly quartz, adheres to the wall so tenaciously that in mining it cannot readily be separated, and the ore becomes diluted with country rock. The vein matter may be sharply delimited against the country rock or it may merge into it, forming commercial walls whose limits are determined by the economic workable ore.

The *vein matter* may consist of several minerals. Generally, both gangue and ore minerals are present. Cavity fillings, unlike most other classes of deposits, generally contain more than one gangue mineral, such as quartz, calcite, and rhodochrosite. Several metallic minerals are also commonly present.

Relation of Fissure Veins to Each Other. Fissures seldom occur alone but tend to occur in groups (Fig. 8-15), and if the fissures of a group are of the same age and have approximately parallel strike and dip they constitute a *fissure system* (Fig. 8-

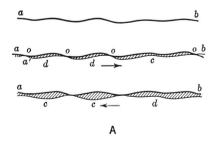

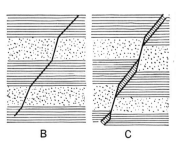

Figure 8-14 Pinches and swells in fissures, produced by movement along irregular fissures, and with walls supported at juxtaposed protuberances. (A, from De la Beche.)

A   B   C

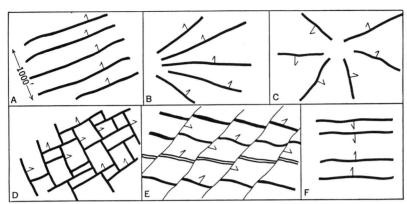

Figure 8-15 Classes of fissure systems. A, Parallel system; B, fan-shaped; C, radial; D, intersecting cognate; E, intersecting systems; and F, conjugated.

15A). Fissures formed at the same time are *cognate fissures* as in Fig. 8-15D, those with parallel strike but intersecting dip are said to be *conjugated (Fig. 8-15F)*.

There are seven systems at Butte, Montana, each of which displaces the earlier ones. Other notable examples of intersecting systems are those of Freiberg, Germany; Pachuca-Real del Monte district, and the Silverton-Telluride district, Colorado (Figs. 8-16, 8-17, and 8-18). Figure 8-18 exhibits the typical radial and intersecting veins formed as the result of a collapsed caldera.

Two intersecting systems may differ in age (Fig. 8-15E), or they may be cognate and of the same age (Fig. 8-15D). Those differing in age generally displace each other, carry different ores, and have independent legal rights; those of the same age carry similar ores, and do not fault each other. To distinguish between them is important, both geologically and legally. Either they are (1) cognate; (2) different systems that displace each other but both are premineral; or (3) different systems, and older vein fillings faulted by younger veins.

In cases (1) and (2) the intersecting systems would be filled simultaneously by similar ores, the vein matter would pass without break from one system to the other, as in Fig. 8-19, and would not be faulted; and the junctions would legally be branches and not faulted ends. In (*a*) no displaced part would be found beyond the junction, but in (*b*) the premineral displaced end is present.

In case (3) (above) the older veins (Fig. 8-20) may carry different ores than the younger veins, as in the classical example of Freiberg, Germany; since the veins themselves are faulted, they carry different legal ownership rights. The filling of the different systems may have been widely separated or the mineralization may have been a continuous event, within which different periods of fissuring and faulting occurred, as at Butte, Montana, where similar filling occurs in three fissure systems that displace each other. The relative ages of the systems may be determined from the intersections, the fillings, and their relation to rocks of known age. The older veins may be slickensided, bent, and crushed; and the younger veins may cut sharply across the older veins and may contain *drag ore* from the older veins (Fig. 8-20). The older veins are commonly displaced by the younger veins in the same direction and for similar distances, as, for example, at Butte, where

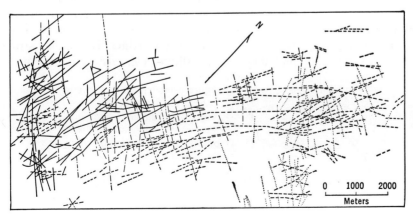

Figure 8-16 Part of mineral district of Freiberg, Germany, showing five systems of veins. 1 and 2 are rich silver-lead veins trending N-S and N-W (solid black); 3 (dash lines), quartz-lead veins; 4 (dotted), barite-lead veins; and 5 (dots and dashes), barren veins.

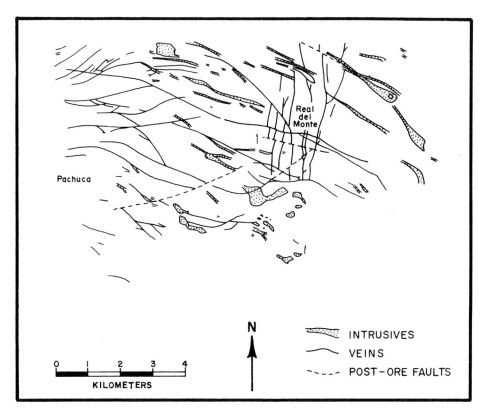

Figure 8-17 Veins, faults, and intrusives in the Pachuca-Real del Monte district, Mexico. The paired lines and spotted areas are dikes and other intrusives, mostly rhyolite but including some dacite; the solid lines are veins along faults; and the broken lines are postore faults. Intrusives are as mapped on the surface; veins and faults are projected to the common level. (After Geyne, 1956, figures 3 and 4.)

all the older veins of the east-west system are thrown to the left by members of the younger northwest system.

The legal features just mentioned relate to the apex rule, in American mining law, by which the prior locator of the top or apex of a vein is entitled to follow it down dip, within the end lines of the mining claim, beneath the land of another. If, down the dip, the vein branches, the holder of the apex is entitled to the ore beneath the junction; if, however, the junction is a faulted end, the holder of the apex may not be entitled to the ore beneath the junction. Much mining litigation has resulted in the interpretation of junctions.

Effects of Change of Formation on Fissure Veins. Fissure veins tend to change in both fissuring and filling when they pass from one rock formation into another. This is partially due to the different physical behavior of different rocks toward stresses, but the chemical composition of the rocks also influences mineral deposition.

If alternate layers of rubber and glass, as in Fig. 8-21, are subject to stress, each layer of brittle glass will fracture, but the intervening rubber will bend. Similarly, a strong, regular fissure that passes from a tough rock, such as greenstone, into a brittle rock like sandstone is likely to break up into a series of irregular, interlacing cracks as in Fig. 8-23A or in Fig. 8-23F. This feature is remarkably well shown (Fig. 8-22) by Burbank at Arrastre Basin, Colorado. Conversely, a strong fissure passing from a brittle rock to a yielding rock, such as shale, is likely to be distorted or disappear (Fig. 8-23B). Or, such a fissure may undergo constriction where it passes through a narrow band of different rock, such as a dike (Fig. 8-23C). Again, it may divide into a group of stringers, as in Fig. 8-23D. If a fissure enters or passes through a strongly schistose or sheared rock, it may break up into an en echelon arrangement of disconnected lenses.

Where a fissure enters another formation at a low angle of incidence, it may be reflected or refracted as in Fig. 8-23E, but if the angle of incidence is

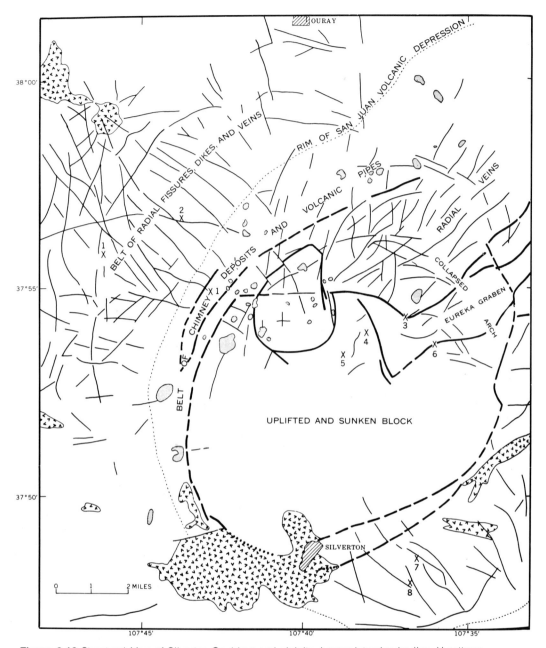

Figure 8-18 Structural Map of Silverton Cauldron and vicinity. Larger intrusive bodies, V-pattern; volcanic pipes, stippled pattern; faults, heavy lines; dikes and veins, light lines. Locations of mineral mines: 1, Idarado; 2, Camp Bird; 3, Sunnyside; 4, Gold King; 5, American Tunnel; 6, Terry Tunnel; 7, Shenandoah; and 8, Silver Lake. (From Ridge, ed., Ore deposits of the U.S., AIME.)

high, little or no change may be expected. Rarely, a fissure may undergo no change in passing from one formation to another.

Terminations of Fissure Veins. As fissures cannot enter rock formations later in age than themselves, unless there is later movement along them, but may continue into older rock formations, obviously it is essential to know their relative age in the geologic history of the region.

In general it is difficult to determine which part of a vein outcrops, but certain features may suggest it. (1) The top may be indicated if it is known that the vein is confined to a formation that has not undergone much erosion. For example, some Tertiary veins near Chinapas, Mexico, occur in an an-

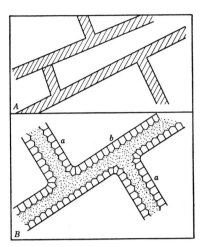

Figure 8-19 Simultaneous filling of premineral fissures. A, Member of intersecting cognate system the parts of which do not displace each other; B, fissure of system *a* displaced by *b*, both of which are premineral. Members of both A and B form junctions.

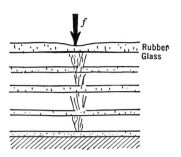

Figure 8-21 Alternate layers of glass and rubber. Under an applied force (*f*) the layers of incompetent rubber bend but the layers of brittle glass become fissured.

desite flow and die out in overlying tuff. The tuff has been eroded only slightly below the andesite floor. Therefore, the present vein outcrop is near its top. Blind veins, of course, give a definite top. (2) In rare cases of young fault veins, if fault scarps along the veins persist, they indicate scant erosion. (3) Associated placer deposits may give a clue as to the amount of erosion if it can be shown that the placers were derived from the vein or veins under consideration. Such a check in Alaska confirmed a conclusion that the roots of certain veins were exposed and only shallow depth might be expected. The placer gold ceased upstream above the vein outcrops. The total placer gold, extracted and in the ground, was estimated. Many samples gave an average value of the veins; their area was computed. To yield the amount of placer gold, thousands of meters of the veins must have been eroded even under the assumption that, higher up, their value and their area were three times as great as at the present surface. (4) The degree of supergene enrichment may give a clue. Where rich supergene sulfide ores overlie lean protore, it indicates much erosion (Chapter 14), since a high degree of enrichment is possible only by the oxidation of much of the overlying vein. For example, a copper vein with hypogene ore averaging 1 percent, overlain by 300 meters of 5 percent supergene ore, must have undergone a minimum of 1,200 meters of oxidation, provided that the tenor and width persisted unchanged above.

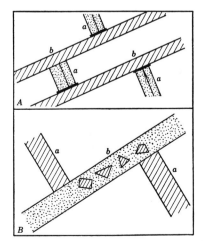

Figure 8-20 Veins of earlier system *a* displaced by those of later system *b* and containing different ores (also drag ore from *a*.) These do not constitute junctions.

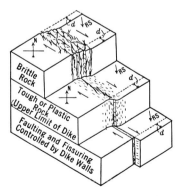

Figure 8-22 Arrastre Basin, Colorado. (From W. S. Burbank, Colo, Sci. Soc. Proc.)

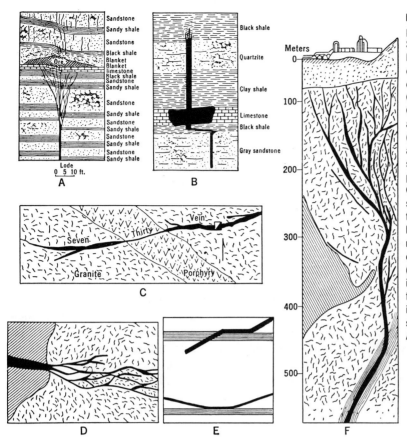

Figure 8-23 Relation of fissures to physical features of rocks. A, Strong vein in competent rocks breaking up into stringers and disappearing upon entering incompetent rocks. (Ransome, U.S. Geol. Survey.) B, Strong fissure at Ouray deflected in shale and disappearing in black shale above. (Irving, U.S. Geol. Survey.) C, Vein at Georgetown, Colorado, constricted in passage through prophyry. (Spurr and Garrey, U.S. Geol. Survey.) D, Division of strong vein into stringer upon entering schist. Gottlob Morgengang vein, Freiberg, Germany. (Beck.) E, Refraction and diversion (upper) and deflection (lower) of a vein upon encountering an incompetent rock. F, Strong fissure in schist diverging upward into minor fissures in brittle andesite, Mazarron, Spain. (After Pilz, A. prakt. Geol.)

From such data, approximations as to the depth of erosion and rough predictions as to the expected depth may be made. It should be remembered that fissure veins are essentially shallow features and that all of them will terminate at depth.

Distribution of Values. The valuable hypogene minerals are never equally distributed throughout a fissure vein. They tend to be concentrated toward the hanging or foot wall, or in the center. Horizontally, some parts may be rich or lean or barren, and, vertically, they decrease or increase with depth, or some intermediate levels may be rich or lean or barren. Where large-scale concentrations occur, they are termed *ore shoots*. These vagaries of distribution of values constitute one of the hazards of mining fissure veins.

Examples of Important Fissure Vein Deposits. Some of the world's most famed deposits, both ancient and recent, are fissure veins. Included among them are some of the deepest and richest mines of the world as well as some of the oldest worked deposits. Although they do not contain the enormous tonnages of some magmatic, sedimentary, or re-

placement deposits, vast treasures of gold and silver have been won from them. They have also contributed largely to the world's production of copper, lead, zinc, tin, tungsten, mercury, and fluorspar. They have been the chief contributors of antimony, cobalt, germanium, and much uranium. Two areas that are claimed to be the "richest hills on earth," namely Butte, Montana, and Potosi, Bolivia, are mostly fissure vein deposits.

A brief list of important fissure vein deposits is shown in Table 8-2.

2. Shear-Zone Deposits

The thin, sheetlike, connected openings of a shear zone serve as excellent channelways for mineralizing solutions, and some deposition takes place within the seams and crevices in the form of fine grains or thin plates of minerals. The open space is insufficient to accommodate enough of the nonferrous metals to constitute ore, but gold with pyrite forms workable deposits, as at Otago, New Zealand. Many lodes and "fahlbands" formerly thought to have been formed by impregnation of the shear openings are now known to have origi-

Table 8-2 Some Important Fissure Vein Deposits

Ores	Examples
Gold	Cripple Creek, Camp Bird, San Juan, Colo.; Mother Lode, Grass Valley, Calif.; El Oro, Veta Madre, Mexico; Porcupine, Kirkland Lake, Ontario; Kalgoorlie, Australia
Silver	Sunshine, Idaho; Tonopah, Nev.; Tintic, Utah; Fresnillo, Pachuca, Guanajuato, Mexico; Potosi, Huanchaca, Bolivia; Cobalt, Ontario;
Silver-lead	San Juan, Colo.; Park City, Utah; Coeur d'Alene, Idaho; US and Lark mines, Bingham district Utah; Clausthal, Frieberg, Germany; Przibram, Austria
Copper	Parts of Butte, Mont., and Cerro de Pasco, Peru; Walker Mine, Calif.
Lead	Calusthal, Freiberg, Germany; Przibram, Austria; Linares, Spain
Zinc	Butte, Mont.; Sardinia
Tin	Llallagua, Huanani, Oruro, Bolivia; Cornwall, England; Erzgebirge, Germany
Antimony	Hunan, China; Mayenne, France
Cobalt	Cobalt, Ontario; Annaberg, Germany
Mercury	New Idria, Calif.
Molybdenum	Temiskaming County, Quebec
Uranium	Great Bear Lake, Canada; Joachimsthal, Czechoslovakia; Shinkalobwe, Zaire
Tungsten	Boulder County, Colo.; Bishop, Calif.; Kiangsi, China; Chicote, Bolivia
Fluorspar	Illinois-Kentucky region; Thomas Range, Utah
Barite	Missouri; Harz, Germany; Buchans, NH
Gems	Colombia

nated by replacement of the sheared rock. Shear zones, because of their large specific surface, are particularly favorable for replacement, and this process has given rise to many large and valuable ore deposits.

3. Stockworks

A stockwork is an interlacing network of small ore-bearing veinlets traversing a mass of rock. The individual veinlets rarely exceed a centimeter or so in width or a few meters in length, and they are spaced a few centimeters to a few meters apart. The interveinlet areas may in part be impregnated by ore minerals. The entire rock mass is mined. In general, the veinlets consist of open-space fillings that exhibit comb structure, crustification, and druses.

Stockworks may occur as separate bodies, or in association with other types of deposits. As separate bodies they may attain a large size, as at Altenberg, Germany (Fig. 8-24), where the tin stockworks have a diameter of over 1000 meters. Such deposits, in general, yield low-grade ore because the entire rock mass is mined, but the large tonnages compensate for this. Outlines are irregular, and the walls merge into unprofitable country rock. Stockworks occur as large bulges on the hanging wall of the great Veta Madre gold vein in Mexico; in association with replacement deposits at Leadville, Colorado; and associated with veins at Victoria, Australia; Slocan, British Columbia; and tin-tungsten veins in Bolivia.

Stockworks yield ores of tin, gold, silver, copper, molybdenum, cobalt, lead, zinc, mercury, and asbestos. Much of the world's supply of lode tinstone in the past has come from stockworks at Altenberg and Zinwald, Germany; Mount Bischoff, Tasmania; Cornwall, England; New South Wales; and South Africa; disintegrated stockworks have yielded much placer tin in Indonesia, Thailand, and Malaysia. Gold occurs in stockworks in Victoria, Mexico, and Juneau, Alaska. Bastin describes a gold-silver-bearing stockwork at Quartz Hill, Gilpin County, Colorado, that is 150 to 250 meters across and has been traced downward for 500 meters. Large, low-grade, gold stockworks in granite porphyry have been explored at Lake Athabaska, Canada. These consist of quartz veinlets carrying pyrite, a little chalcopyrite, and free gold. Most of the gold of the old, large, low-grade and now defunct Alaska Juneau mine was contained in irregular quartz stringers. Silver-cobalt ores are obtained from a large stockwork in Schneeberg, Germany, according to Müller; and a silver stockwork is mined at Fresnillo, Mexico.

Stockwork veinlets are formed by: (1) crackling upon cooling of the upper and marginal parts of intrusive igneous rocks, and (2) irregular fissures produced by tensional or torsional forces. For example, fault movement downward along a curved fissure produces a crackling where the hanging wall

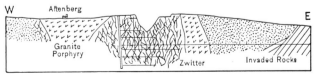

Figure 8-24 The tin stockworks of Altenberg. (After Dalmer, Z. prakt. Geol.)

moves over humps on the footwall, as shown by Wandke and Martinez for the great Veta Madre vein in Guanajuato, Mexico.

4. Saddle Reefs

If a thick stack of writing paper is sharply arched, openings form between the sheets at the crest of the arch. Similar ore receptacles are formed when alternating beds of competent and incompetent rocks, such as quartzite and slate, are closely folded. When filled by ore, they resemble the cross section of a saddle, hence their name (Fig. 8-25). The saddle reefs of Bendigo, Australia, are an example; they have yielded over $500 million in gold.

The saddles are repeated in ore bed after ore bed down the axial plane or "center-country," which may be vertical or inclined (Figs. 8-25 and 8-26). Fifteen "lines of reef" are known in Bendigo, of which five lines of superimposed saddles have been highly productive. The saddles are 6 to 15 meters across and have irregular crests, and the leg depth is mostly less than 30 meters. The cross sections are small, but one saddle has been followed horizontally for 2750 meters in Bendigo and mining has reached 1480 meters vertically. Superimposed saddles are generally connected by faults (Fig. 8-25B), which according to Stone conducted the ore solutions to the saddles; ore seldom is found where faults are lacking.

The ore emplacement has been chiefly by cavity filling, but Stillwell and Stone believe that considerable replacement has also taken place.

Similar reefs occur in other parts of Victoria and in New Zealand, and smaller pitching saddle reefs occur in Nova Scotia (Fig. 8-26).

5. Ladder Veins

"Ladder vein" is the name applied to more or less regularly spaced, short, transverse fractures in dikes. These generally extend roughly parallel to each other, from wall to wall of the dike. Their width is thus restricted, but they may extend for considerable distance along the dikes. Such openings may become filled with mineral matter to form commercial deposits, as the Morning Star gold-bearing dike in Victoria, Australia. The individual fissures may form separate veins, or if they are closely spaced the dike as a whole may be mined. Such transverse veins in a vertical dike resemble the rungs of a ladder, hence their name.

The fractures that constitute ladder veins have generally been considered to be contraction joints. However, some fissures extend beyond the dikes into the walls, which indicates that cooling contraction may not be their cause. Also, Grout found that the total width of quartz filling in ladder veins in Minnesota greatly exceeded the possible contraction through cooling. He suggested that tangential movement of weak wall rocks opened up transverse tension cracks in an enclosed or brittle dike. Tangential movement is probably more important than contraction.

Ladder veins are not numerous or important, but examples are the gold quartz ladder veins of the Morning Star, Waverly, and All Nations mines in Victoria, Australia; molybdenite veins in New

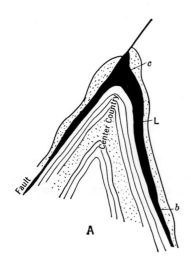

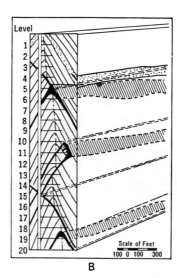

Figure 8-25 A, Typical saddle reefs of Bendigo, Australia, c, cap; L, leg. b, back; black is gold quartz ore. B, Three types of bendigo reefs and methods of mining them. (After Report by Bendigo Mines, Ltd.)

Quartz veins interbedded
Thickness of vein in inches

Quartz veins not interbedded
Called angulars or stringers

Quartz veins interbedded and corrugated

Bed of slate

Drift along veins

Dip of strata

10 0 10 20 30 40 50
Feet

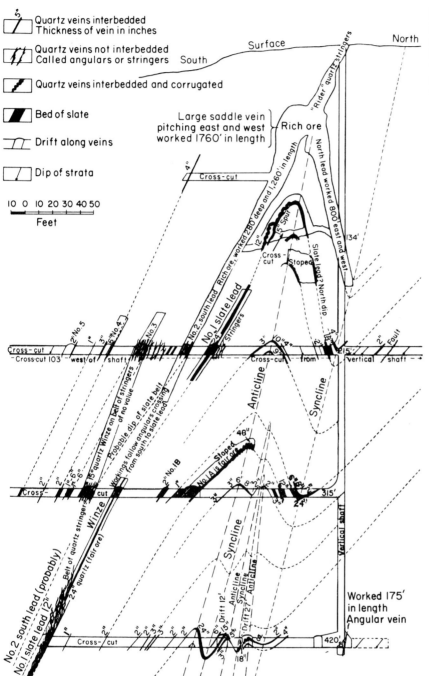

Figure 8-26 Saddle reefs in the Dufferin mine, Salmon River gold district, Nova Scotia. (After Faribault as quoted by Malcolm 1912, from Park and MacDiarmid, *Ore Deposits*, W. H. Freeman, 1964.)

South Wales; and the copper ladder veins of Telemarken, Norway.

6. Pitches and Flats—Fold Cracks

Under light load, slumping or gentle synclinal folding of brittle sedimentary beds gives rise to a series of connected tension cracks or openings collec-

tively known as *pitches and flats*. Gentle open folding also forms *anticlinal tension cracks* at the crests of anticlines (or troughs of synclines).

Fold Cracks. Anticlinal tension cracks are produced by folding under light load. The anticlinal cracks generally have small vertical extent but may extend along the axis of the fold, as a series of disconnected fractures, for considerable distances. They are generally wedge-shaped and commonly terminate abruptly, either up or down, against another bed. Such anticlinal tension cracks contain gold ores in Victoria, Australia; zinc ores in Mexico; and gilsonite in Utah. Some of the copper bodies at Kennecott, Alaska, are in synclinal tension cracks.

7. Breccia-Filling Deposits

The haphazard arrangement of the angular rock fragments in breccias gives rise to numerous openings that permit the entry of solutions and mineral deposition, forming breccia-filling deposits (Fig. 8-27 and 8-28). The breccias may result from volcanism, collapse, or shattering.

Volcanic Breccia Deposits. Explosive volcanic activity gives rise to bedded breccia deposits and breccia pipes or craters. The unconfined bedded deposits are unimportant ore loci, but the confined pipes offer an opportunity for concentrated mineralization.

The pipes are vertical or highly inclined, roughly oval-shaped, and are filled by breccias. They are ideal channelways for mineralizing solutions and for ore deposition. Beautiful crustification of gangue and ore minerals is often seen around the rock fragments; and vugs partly filled by dainty crystals delight the mineralogist and indicate the sequence of mineral deposition.

The great Braden copper deposit of Chile is considered by Lindgren and Bastin to be an old volcanic crater in part filled by breccia, but Brüggen holds that it is not a crater. The copper minerals in part fill cavities in the breccia.

Collapse Breccia Deposits. In the caving methods of mining, an excavation is started at the bottom of a block of ore; the roof rock then caves, and this caving extends itself upward until the block is a mass of jumbled, angular fragments of ore, with considerable open space. It is an artificial breccia. Similarly, in nature, the roof of a solution opening may collapse and the collapse may extend upward to yield a mass of broken rock. The Woodrow pipe

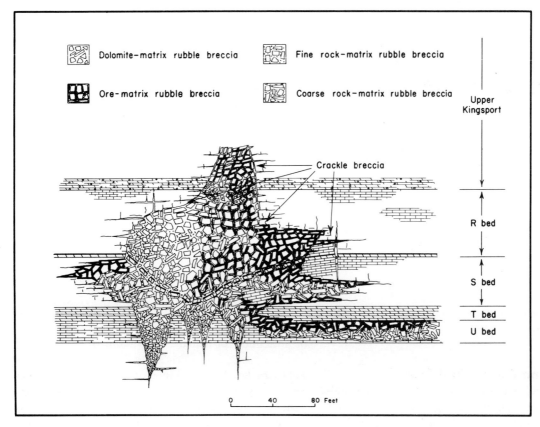

Figure 8-27 Generalized section through a portion of the Jefferson City Mine showing breccia. (From Crawford and Hoagland, J. Ridge, ed., 1968, Ore deposits of the U.S., p. 253.)

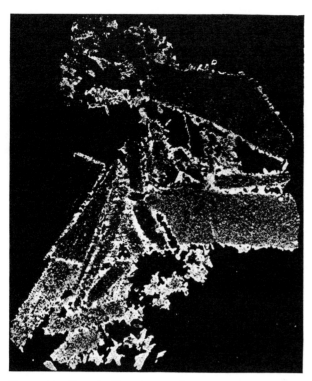

Figure 8-28 Breccia-filling deposit. Breccia cavities are lined with quartz crystals and filled with chalcopyrite. Latite ore, Pilares mine, Sonora, Mexico. (Locke, Econ. Geol.)

is a collapse breccia as the result of solution leaching of the Todilto carbonate formation allowing collapse into the void space (Fig. 8-29). Such subsidence may extend upward for hundreds of meters and may be bordered by crackled and fissured rock.

Breccias of this type (Fig. 8-30) have been described as *cave-collapse* breccias by Watson and Ulrich, as *crackle* (Fig. 8-27) and *founder* breccias by Norton in the Appalachian zinc areas, and as mineralization-stoping breccias by Locke. The *Cactus mine,* which is now a defunct open pit mine, in Utah (Fig. 8-31), is described by Butler as a "pipe" because it extends nearly 275 meters vertically. Locke believes that subsidence was initiated by solution caused by the mineralizing solutions themselves, during the earliest stages of mineralization. He cites as other examples the Southwest pipe at Bisbee, Arizona; the South Ibex and Cresson in Colorado (Fig. 8-31B); and the Pilares, Catalina, and Duluth pipes in Sonora, Mexico. Walker cites the Emma chimney in the Little Cottonwood Canyon, Utah, as an example of cave-collapse breccia; it is mineralized by silver-lead-zinc ore to a depth of over 300 meters. The ore has been emplaced chiefly by open space filling in the breccia cavities.

Wisser describes similar collpase breccias in Bisbee, Arizona, that were initiated by shrinkage attendant upon sulfide oxidation, the effects of which extend upward for over 300 meters. Some reach the surface, and such breccias and accompanying crackled areas are considered to indicate the presence of postbreccia ore below them. The breccias themselves are not filled by ore, since they are postore features.

Whether the subsidence in all cases cited by Locke was initiated by hypogene solutions is open to question; it is not established for the Pilares mine in Mexico.

Tectonic Breccia Deposits. Breccias produced by folding, faulting, intrusion, or other tectonic forces have been referred to variously as crush, rubble, crackle, and shatter breccias (Fig. 8-32). In the Appalachian zinc deposit region, Currier distinguishes rubble from shatter or crackle breccias. The rubble breccias are characterized by prominent relative displacement of fragments and by some rounding, whereas the shatter or crackle breccias are composed of angular fragments that show little rotation. The breccias localize the zinc ores, and in Currier's opinion, they were formed by warping or folding and by thrust faulting. Similar shatter breccias, according to Fowler, localize ore in the Tri-State zinc district. The zinc bodies at Mascot, Tennessee, were formerly thought to be tectonic breccias, but Crawford and Hoagland consider them to be collapse breccias due to underlying karst solution areas, and somewhat modified by later Appalachian orogeny. The deposits are limited to a stratigraphic range of about 60 meters, average 5 to 6 meters in thickness, are up to 45 meters wide, and are exposed for 100 meters in depth.

Crackle Breccia. Recent studies of the Mississippi Valley type of deposits show structures that differ from the pitches and flats that were formerly described. Irregular masses that sometimes have the shape of bioherms but consist of *crackle breccia,* as shown in Fig. 8-27, now are more commonly recognized.

8. Solution Cavity Fillings

Various types of solution openings in soluble rocks have afforded receptacles for primary and secondary mineral deposits. They occur most commonly in limestone, at shallow depth, and are generally

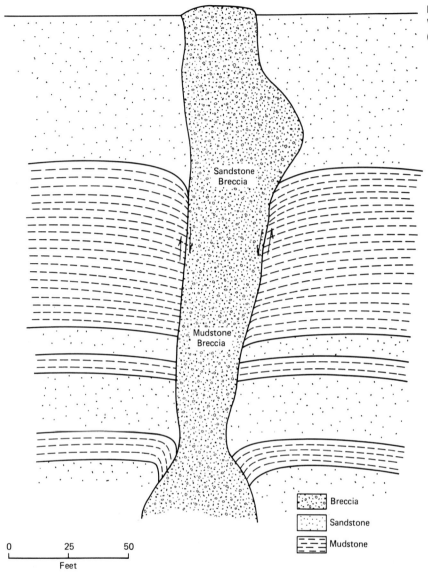

Figure 8-29 Cross section of Woodrow mine breccia pipe. (Geology after Moench, 1959.)

Sandstone Breccia

Mudstone Breccia

Breccia

Sandstone

Mudstone

0   25   50
Feet

believed to have been dissolved above the water table by surface waters charged with carbon dioxide although Davis has postulated their formation below the water table. Most solution cavities are modifications of preexisting openings, such as joints, fissures, or bedding planes, resulting in caves, galleries, and gash veins.

Cave Deposits. Caves of various shapes and sizes (Fig. 8-33) are characteristic of limestone plateau areas that have undergone prolonged erosion. They are commonly accompanied by "sinkholes" that result from roof collapse or from surface solution of joint or fissure intersections. Joint solution gives

rise to funnel-shaped sinks; collapse gives large irregular openings, and the floors of the underlying caves are strewn with broken pieces of rock and mud. Caves unaccompanied by sinkholes are generally floored with "cave earth," a residual accumulation of the insoluble materials of the limestone.

Small caves may be almost completely filled by ore minerals, but large caves generally contain only peripheral crusts of ore minerals, among which may be large and beautiful crystals. It was formerly thought that all large, massive, irregular ore bodies in limestone were cave fillings until replacement was realized. Caves are common containers of zinc and lead ores, such as those of Wisconsin and Illi-

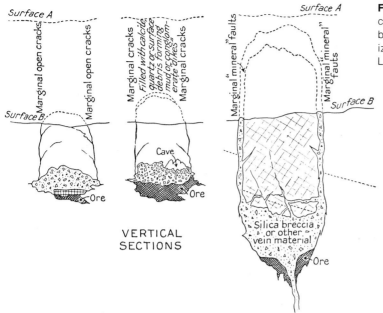

VERTICAL
SECTIONS

Figure 8-30 Vertical sections of collapse breccia pipes produced by mine oxidation and "mineralization" collapse. (Kingsbury-Locke, Econ. Geol.)

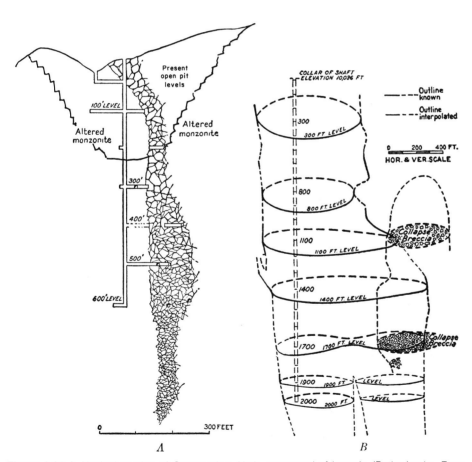

A   *B*

Figure 8-31 A, Vertical section of Cactus pipe, Utah, composed of breccia (Butler-Locke, Econ. Geol.). B, Outline of Cresson pipe of basaltic breccia, Cripple Creek, Colorado. (Loughlin, AIME.)

131

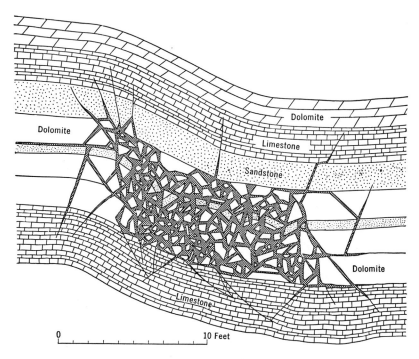

Figure 8-32 Tectonic breccia formed by the shattering of dolomite along the axis of a bend, forming an "ore run" in northern Arkansas lead and zinc district. (After McKnight, U.S. Geol. Survey.)

nois, and of oxidized ores of copper, lead, silver-lead, zinc, vanadium, and other metals. The caves of Bisbee, Arizona, have been particularly productive of oxidized copper ores in the past. Caves also yield nitrates, clays, fluorspar, barite, strontianite, guano, and other materials. In general, the individual deposits are relatively small and disconnected.

Galleries. Galleries are horizontal or gently inclined attenuated caves, which result from solution along fissures. They probably represent former underground streams. They yield ores similar to those found in caves.

Gash Veins. Gash vein is frequently erroneously employed to designate small wedge-shaped fissures in stressed, brittle rocks. The term as originally applied by Whitney, however, designates vertical solution joints in limestone (Fig. 8-33C). Solution is essential or they are not gash veins. They are confined to single formations, seldom reach 60 meters in depth, and widen and narrow conspicuously.

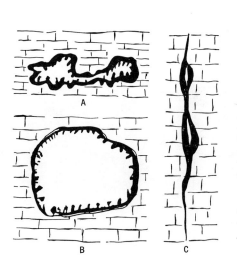

Figure 8-33 Solution caves and cavities in limestone. A, B, Open solution cavities lined with crusts of crystals, Wisconsin lead-zinc region (after Chamberlin); C, gash vein or solution enlargement along joint (after Whitney); D, solution cave occupied by ore (black) and cave breccia on bottom, overlain by later breccia and ore, and by breccia fragments (after Walker).

In some respects they resemble fissure veins. They are common in limestone regions; those in the upper Mississippi Valley were the first described.

The fillings were characterized by crustification, large vugs, and beautiful crystals and consist of lead, zinc, silver-lead, copper, fluorspar, and barite. The deposits are mostly small, although they may be rather rich.

9. Pore-Space Fillings

Pore spaces in rocks may contain ores, in addition to oil, gas, and water. Copper ores occupying sandstone pores, known as the red beds ores, occur in Texas, New Mexico, Arizona, Colorado, Pennsylvania, and Utah. The best ones are in New Mexico, but none of them is of present economic importance since the Yuba mine is now closed. Similar sandstone copper ores occur in the Permian red beds of the Pre-Ural region, where large reserves average 2 to 2.5 percent copper; in the Don Basin and Djeskasjan, Russia; in Turkestan; in Nova Scotia; Germany, and the Midlands, England. Copper in pore spaces in Triassic arkose was formerly worked at Bristol, Connecticut. Similarly, lead ores have been mined in Rhenish, Prussia; and silver ores at Silver Reef, Utah. Vanadium and uranium ores occur as impregnations in sandstone in Wyoming, New Mexico, Colorado and eastern Utah; quicksilver ore impregnates sandstone in California and Arkansas.

10. Vesicular Fillings

Permeable vesicular lava tops may serve as channelways for mineralizing solutions. In the Lake Superior region, vesicles in basalts are filled with native copper and have given rise to some of the great copper deposits in the world, which have been followed downdip for nearly 2700 meters.

Somewhat similar, but at present, noncommercial deposits occur in the White River region of Alaska and the Yukon and around the Coppermine River in Arctic Canada. They also occur in Nova Scotia, New Jersey, Brazil, Faroe Islands, Norway, Germany, Siberia, and the Russian Arctic.

Selected References on Hydrothermal Processes and Cavity Filling

General Reference 9, chaps. 12, 13, 15, and 23. *Descriptions of various types of openings.*

Structural relations of ore deposits. S. F. Emmons. In *General Reference 10. A general discussion of rock openings filled by ore.*

Structural control of ore depositsion. 1929. C. D. Hulin. Econ. Geol., v. 24, p. 15–49. *Kinds of openings that control ore deposition.*

Vein systems of Anastre Basin, Colorado. 1933. W. S. Burbank. Colo. Sci. Soc. Proc. 13, no. 5. *An interesting group of vein systems.*

General Reference 15. Brief resumes of many hydrothermal deposits that exhibit localization of ore by structural features.

General Reference 16. This volume contains many descriptions of Canadian deposits that exhibit structural controls.

Solubility of metal sulphides in dilute vein-forming solutions. 1944. R. M. Garrels. Econ. Geol., v. 39, p. 472–483, *Dilute solutions cannot carry metal sulfides; colloids play an important role.*

Some considerations in determing the origin of the ore deposits of the Mississippi-Valley type. 1959. *Ernest L. Ohle.* Econ. Geol., v. 54, p. 769–789.

Geology of the Upper Mississippi Valley zinc-lead deposits. 1960. Heyl, Agnew, Lyons, and Behre. U. S. Geol. Survey Prof. Paper 309.

Facies and types of hydrothermal alteration. 1962. C. *Wayne Burnham.* Econ. Geol., v. 57, p. 768–784.

Hydrothermal alteration at the Climax molybdenite deposit. J. W. Vandewilt and R. V. King. 1955. Trans. AIME, v. 202, p. 41–53.

Some effects of environment on ore deposition. 1962. *Pierce D. Parker.* Econ. Geol., v. 57, p. 293–324.

Breccia pipes in Shoshone range, Nevada. 1959. *Olcott Gates.* Econ. Geol., v. 54, p. 790–815.

Breccia and pebble columns associated with epigenetic ore deposits. 1961. *L. Bryner.* Econ. Geol., v. 56, p. 448–518.

Geology and origin of mineralized breccia pipes in copper basin, Arizona. 1961. W. P. Johnstone and J. D. Lowell. Econ. Geol., v. 56, p. 916–940.

Geology of the Porphyry Copper Deposits, South Western North America. 1966. Titley and Weeks, eds. University of Arizona Press, *Twenty-three chapters by different authors on Southwestern U.S. porphyry copper deposits.*

Complexing and Hydrothermal Ore Deposition. 1964. H. C. Helgeson. Macmillan Publishing Co., Inc., New York. *The role of complex ions in transporting metals in hydrothermal solutions. Application of thermal chemistry.*

The Marcot-Jefferson City zinc district, Tennessee. J. Crawford and A. D. Hoagland. Graton-Sales AIME, v. , p. 243–256.

Ore deposits in the Central San Juan Mountains, Colorado. Thomas A. Steven. T. D. Ridge. vol. ed. Graton-Sales AIME v. , p. 715–727. *Detailed description of calderas and cauldrons.*

Hydrothermal alterations. 1966. S. C. Creasey. *In Geology of the Porphyry Copper Deposits*, S. R. Titley and C. L. Weeks, eds. University of Arizona Press, Tucson, General Reference 14, chap. 6.

METASOMATIC REPLACEMENT AND FISSURE VEINS

Metasomatic replacement, or simply replacement, as it is generally called, is the most important process in the emplacement of epigenetic mineral deposits. It is the dominating process of mineral deposition in the hypothermal and mesothermal deposits, and important in the epithermal group, the ore minerals of contact-metasomatic deposits have been formed almost entirely by this process. Also, it is the controlling process of deposition in supergene sulfide enrichment and dominates in the formation of most other supergene mineral deposits. In addition, it plays the major role in the extensive rock alteration that accompanies most epigenetic metallization.

Replacement may be defined as a process of essentially simultaneous capillary solution and deposition by which a new mineral is substituted for one or more earlier formed minerals. The process, however, is not so simple as the definition implies, as will be shown later. By means of replacement, wood may be transformed to silica (petrification), a single mineral may take the place of another, retaining its form and size (pseudomorphs), or a large body of solid ore may take the place of an equal volume of rock. Thus many mineral deposits originate. The replacing mineral (metasome) need not have a common ion with the replaced substance. The replacing minerals are carried in solution and the replaced substances are carried away in solution; it is an open circuit, not a closed one.

Replacement was early recognized in the explanation of mineral pseudomorphs, particularly where the pseudomorph exhibited a diagnostic crystal form of an earlier unlike mineral. Naumann introduced the word *metasomatism,* meaning a change of body, to designate the process. It was vaguely realized as a mode of emplacement of mineral bodies by Zimmermann, who in 1749 ascribed the origin of lodes to the transformation of rocks into metallic minerals and veinstones by the action of solutions that entered through small rents and other openings in the rocks.

Hints concerning its application to ore formation were subsequently made from time to time, but only in the beginning of the present century was replacement realized to be a large-scale process of ore formation. Huge bodies of massive sulfide ore completely enclosed within limestone were early thought to be the filling of old solution caves in limestone. But, with the growing knowledge of the effectiveness and extent of replacement, and, particularly, of criteria for its recognition, it was realized that such bodies had been formed by replacement. As knowledge of this type of deposit has increased, especially in America, many deposits formerly thought to have originated by different processes are now ascribed to replacement. Even many formerly thought to be igneous injections are now conclusively recognized as replacement deposits. Also to this group now belong many deposits that formerly were thought to have been fissure fillings and impregnations of rock pores. In the case of supergene sulfide enrichment the supergene sulfides were formerly thought to have been merely deposited as coatings on earlier sulfides, but now it is generally known that they are deposited only by replacement of earlier sulfides.

The Process of Replacement

If mineralizing solutions encounter minerals that are unstable in their presence, substitution may take place and replacement ensues. The exchange is practically simultaneous, and the resulting body may occupy the same volume and may retain the identical structure of the original body.

Mode of Interchange. If, in a brick wall, each brick were removed one by one, and a silver brick of similar size substituted as each brick in the wall were removed, the end result would be a wall of the same size and form, even to the minutiae of brick pattern, except that it would be composed of silver instead of clay. This is how replacement proceeds, except that the parts interchanged are infinitesimally small—of molecular or atomic size. Consequently, the shape, size, structure, and texture may be faithfully preserved even below the visible magnifications of the microscope.

If the replacement were molecule by molecule, a simple chemical equation would express the interchange, such as

$$ZnS + CuSO_4 \rightleftarrows CuS + ZnSO_4$$

Here, dense covellite replaces less-dense sphalerite, and a unit volume of sphalerite yields a smaller volume of covellite, with consequent shrinkage. Such interchange, however, takes place only with free-growing crystals or in incoherent materials where pressure is negligible.

Extended observations show that replacement in rigid rocks is not attended by a change in volume. A cubic centimeter of sphalerite or calcite is replaced by a cubic centimeter of covellite or galena. This is the *theory of equal volumes*. The calculable shrinkage, according to the chemical reactions, does not occur. It follows from the law of equal volumes that excess covellite or galena is deposited to make up for the shrinkage that would ensue if denser covellite or galena replaced lighter sphalerite or calcite according to the customary chemical reactions. Therefore, in volume-for-volume replacement, the interchange is not molecule for molecule. A single crystal of pyrite, for example, may cut across and replace some half-dozen different rock minerals, which proves further that volume-for-volume replacement cannot be expressed by any single chemical equation. It has been repeatedly demonstrated by large-scale field relations, by crystal pseudomorphs, and by microscopic observations that replacement is definitely a volume-for-volume interchange. Consequently, ordinary balanced chemical equations do not express what actually happens in volume-for-volume replacement, and the oft-quoted equations must be considered only as indicating the trend and end products of the exchange. The process is not as yet fully understood.

Procedure of Substitution. Although the procedure of substitution is imperfectly known, certain features are recognized. The simultaneous interchange must be by infinitesimal particles of molecular or atomic size. The growing mineral is in sharp contact with the vanishing substance; between them there must be a thin film of solution that supplies by diffusion the replacing materials and removes the replaced substances. In the case of liquids such a film will be supersaturated, facilitating reaction. The rate of reaction will depend upon the rate of supply of new material and the readiness of removal of the dissolved material. The instant that space is made available by solution, some of the replacing mineral will separate out from the supersaturated film. Thus the metasome will continuously advance against the host and grow at its expense. In this,

the growth pressure of the impinging crystals will also facilitate the solution of the host. The replacing minerals thus present a constantly advancing front against the host as long as the supply of new material and disposal of dissolved material keeps up.

Where solution is supplied to a center, such as a pore space, growth may proceed outward in all directions from the center, giving rise either to discrete, shapeless grains, or to crystals with well-developed faces in sharp contact with the enclosing host. In this manner isolated, doubly terminated crystals may grow at the expense of limestone or other rock. It is obvious that such crystals could not have been formed by the filling of preexisting cavities with shapes coinciding exactly with the crystals. Therefore, it is concluded that they must have been formed by replacement, and such isolated, doubly terminated crystals are rightly considered diagnostic of replacement. If the supply of material to feed a given crystal ceases for any reason, other minerals may continue to deposit at its margins, and eventually earlier-formed euhedral minerals will be enclosed within later replacement minerals. Thus pyrite crystals are common within copper, lead-zinc, and other ores. This explanation may also account for the interesting "polar" pyrite cubes of Ducktown, Tennessee; fringes of chalcopyrite at opposite pyrite faces probably are due to the greater ease of replacement at these ends.

Entry and Exit of Solution. Replacement involves the necessity of continuing supplies of new material and removal of the dissolved material. How does the new material arrive at the point of deposition? This question becomes more pointed in the case of a visably unfractured pyrite cube that is undergoing supergene replacement by compact chalcocite from the outside toward the center, by means of a liquid solution at atmospheric temperature and pressure. First, the faces of the cube are replaced, say to a depth of one-quarter of its diameter. Then, the replacing copper must penetrate a dense layer of chalcocite in order to arrive at the interior front of replacement, and the dissolved iron from the pyrite must escape through the same layer of compact chalcocite. The chalcocite may have no determinable pore space. Obviously, the necessary quantity of solution cannot flow bodily through the dense chalcocite layer.

Diffusion is probably the answer. This is the movement of molecules or ions in a solution from a point of supply to a point of deposition or from

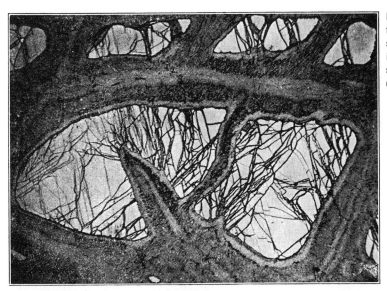

Figure 8-34 Replacement extending outward from cracks. Residual grains of pyrite isolated by invading covellite and intersected by covellite veinlets. (Photo by L. C. Graton.)

a place of higher to one of lower concentration or chemical potential. Particles of ionic or molecular size can move through a layer where a bodily flow of solution could not. But diffusion is known to be exceedingly slow and to act only over short distances. Therefore, it cannot be a means of transporting large quantities of replacing substances over long distances; it is incompetent by itself to build up large mineral deposits. However, it is an effective process for supplying and removing the products and by-products of replacement over the short distances at the actual front of replacement, whereas voluminous moving solutions could not gain entry.

Replacement will occur first along the major channelways. The walls succumb, and further advances are fed through minor openings, down to capillary size, whose walls in turn succumb. If the wall rocks or early formed replacement minerals have been minutely fractured, this provides a greater opportunity for replacement, and such openings may serve for the final stages of delivery of the replacing solutions without the necessity of having to call to any great extent upon the slow process of diffusion. However, the microscopic examination of polished ore surfaces discloses (Fig. 8-34) waves of replacement, whose growing fronts are nurtured by diffusion, extending outward from the smallest microscopic cracks.

The refuse of replacement is probably removed through conduits similar to those that permitted ingress to the new materials and is swept away to be dispersed within the mass of the groundwater.

Stages of Replacement. Those who view polished ore surfaces through a microscope become aware that replacement ore deposits are commonly built up in stages and earlier replacement minerals are themselves replaced by later minerals. Definite successions are thus recognized. In the initial stages of the replacement of rocks, ferromagnesian silicates are attacked first, followed by the feldspars and quartz. Introduced gangue minerals replace all silicates. Likewise, all sulfides replace all rock minerals and also introduce gangue minerals; gangue minerals rarely replace sulfides.

The first-formed metallic minerals, of which pyrite and arsenopyrite are common ones, may also be replaced by later sulfides. Pyrite, for example,

Figure 8-35 Camera lucida drawing of supergene chalcocite (white) replacing sphalerite (black), leaving unreplaced residuals; stippled, incompletely replaced sphalerite. Morenci, Arizona. ×40.

is often seen to be veined by minute fractures along which the later sulfides have penetrated and replaced the walls. Little islands of unreplaced sulfides left between intersecting replacement veinlets are characteristic features (Fig. 8-35). Again, the margins may be attacked, and unreplaced residuals may remain enclosed within the later sulfides. Similarly, still later metallic minerals may, in turn, replace those of the second generation, and the process may be repeated until 8 or 10 successive stages have resulted. The sequence among some common, hypogene metallic minerals generally is pyrite, enargite, tetrahedrite, sphalerite, chalcopyrite, bornite, galena, and ruby silver.

Growth of Replacement. In nonhomogeneous rocks the growth of replacement may be controlled by favorable beds, structural features, or chemical or physical properties of the host rock as already indicated.

In homogeneous rocks, replacement may advance in one or more of three ways: (1) Starting from a fissure, the walls are first replaced and the replacement then advances outward with a bold face of massive ore against unreplaced country rock to the extreme limit of mineralization (Fig. 8-36A). It is a wavelike advance, and the end product is a massive ore body consisting mostly of introduced materials, such as the massive sulfide bodies of Kennecott, Alaska, or of Bisbee, Arizona. (2) The growth may take place with a bold front, but preceding it, like skirmishers flung out in front of an advancing army, is a fringe of disseminated replacement where partial replacement is going on at many small centers (Fig. 8-36B). The latter may gradually coalesce to form a farther front of massive ore, beyond which still continues, however, the outward-flung fringe of disseminated ore. The resulting deposit is a massive or high-grade ore body flanked by disseminated or lower-grade ore. The massive part of the ore body has resulted from the continued growth and coalescence of innumerable centers of replacement. (3) The third method may be termed *multiple-center* growth (Fig. 8-36C). If the country rock becomes permeated by mineralizing solutions, replacement may start simultaneously at innumerable closely spaced centers, such as, for example, at the loci of ferromagnesian minerals in a monzonite porphyry. With arrested growth, disseminated replacement deposits result. The individual metallic mineral grains are peppered throughout the rock and range in size from specks invisible to the naked eye to a half-centimeter in diameter; they commonly constitute only a few percent of the rock mass. The porphyry coppers are examples of this mode of replacement, some of which contain several hundred million tons of copper ore.

Agencies of Replacement

Replacement deposits are produced by both liquid and gaseous solutions, and in both of them water predominates. The liquid solutions occupy the major role. Most hypogene replacement deposits are considered to have been deposited mainly from later hydrothermal solutions. Such hot waters may have been diluted by intermingling with near-surface meteoric waters. The materials carried in solution largely came from the magma, but some were dissolved from the country rock.

Cold surface or artesian waters also produce both primary and supergene replacement deposits, as, for example, some manganese deposits and many oxidized and supergene sulfide deposits. Gaseous emanations have also been demonstrated to be effective agencies of replacement in diverse types of high-temperature deposits.

Temperature and Pressure of Formation

Replacement may take place under almost any condition of temperature and pressure. It is an effective process in supergene enrichment at surface temperature and pressure; it is dominant at the high temperatures and pressures that prevail during contact metasomatism. It is, of course, most effective

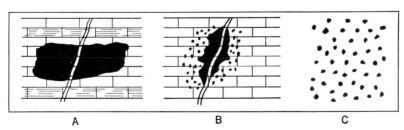

Figure 8-36 Replacement advance. A, Bold face advance; B, outer fringe of disseminated replacement (black is ore); C, multiple center.

at elevated temperatures, since heat tends to accelerate chemical activity.

The nature of replacement varies somewhat according to the conditions of temperature and pressure, pH, and partial pressures of specific constituents.

Atmospheric Temperatures. The formation of replacement deposits by cold meteoric waters is confined, mostly, to soluble rocks, such as limestones. These may be replaced by oxides of iron or manganese to form iron and manganese deposits, or by calcium phosphate to form phosphate deposits.

The most noteworthy replacement under atmospheric temperature and pressure is in the zone of oxidation and supergene sulfide enrichment. As shown in Chapter 14 near-surface limestone beds are extensively replaced by smithsonite and malachite-azurite, forming large and valuable zinc and copper carbonate deposits. Similarly, copper and zinc silicates form valuable surficial replacement deposits by means of cold solutions. In addition, vast tonnages of valuable metallic ores have resulted from the replacement of lean hypogene sulfides by supergene sulfides, such as has occurred in many of the great porphyry copper deposits. In general, the metallic minerals formed by replacement from cold solutions are of simple composition.

Some of the common ones are chalcocite, covellite, marcasite, wurtzite, native copper, native silver, and cerargerite.

Elevated Temperatures. With warmer solutions the ranges and intensity of replacement increase. More rocks are affected, and more extensive replacement occurs. The metallic minerals formed are mostly simple sulfides and sulfo-salts, and the gangue minerals are chiefly carbonates, quartz and simple silicates; many are hydrous.

With solutions at intermediate temperatures, the extensive wholesale replacement of rocks may take place and large replacement ore deposits may be formed. Many ore minerals replace carbonates, quartz, silicates, and metallic minerals.

At high temperatures hardly any rock may escape replacement. Even such relatively refractory rocks as granite may be almost completely altered to greisen. Silicates containing boron, chlorine, and fluorine are common. Sulfides and oxides (particularly magnetite) develop in coarse texture. Many tin, tungsten, and magnetite deposits have been formed by replacement at high temperatures.

Host Rocks

Every rock is susceptible to replacement, but, naturally, the readily soluble carbonate rocks are the most widespread hosts of replacement mineral deposits. Calcareous shales and sandstones are also congenial host rocks. Igneous rocks are also readily susceptible to replacement and include some of the largest known ore deposits. Among the metamorphic rocks, field studies show that crystalline schists and marble yield most readily to replacement, and gneisses, phyllite, and quartzite the least. Even these, however, are known to succumb. Pure argillaceous rocks and quartzites are the least susceptible to replacement by ore.

In the granular igneous rocks, replacement is commonly selective and differential in that certain minerals, such as the dark minerals or feldspar, may be selectively replaced by ore minerals. This gives rise to one type of disseminated replacement deposit of which the porphyry coppers are outstanding examples.

Localization of Replacement

Various physical, chemical, and structural features serve to localize hydrothermal deposits. Most of these features operate in connection with replacement deposits. Also the effects of the host rock are particularly pertinent to replacement deposits. Their chemical character alone may be the controlling factor in localizing ore, but generally structural features operate in conjunction. In addition, certain minerals of an igneous rock may localize replacement, as, for example, where ferromagnesian minerals are selectively replaced by sulfides. The chemical control of replacement is perhaps most clearly demonstrated in the selective replacement of disseminated hypogene sulfides by supergene sulfides, in which chalcopyrite is replaced in preference to pyrite, and adjacent rock silicates are untouched.

Structural Features. Of the various structural features mentioned previously, fissures are the most important ore localizers. Single fissures may have their walls replaced along favorable beds to form tabular deposits known as *replacement veins* or *replacement lodes* (Fig. 8-37A). The fissure may intersect several favorable sedimentary beds, each of which may be replaced to form a succession of replacement bodies, of which the fissure is the major locus, and the beds the minor locus. Examples of such controls are numerous and form a working basis for the mining geologist in ore hunting.

Closely spaced *sheeted fissures* and *shear zones,* because of their large specific surface, give rise to still larger replacement lodes (Fig. 8-37*B*). Deposits along sheeted zones are likely to be more regular in width than those in shear zones, and replacement of the plates of rock may give rise to a banded structure that simulates crustification. Commonly, in shear zones, the replacement is more fitfully distributed within the zone, and an irregularly shaped deposit, although roughly tabular in form, results.

Intersections of fissures may yield a type of deposit shown in Fig. 8-38. These give rise to ore bodies of limited horizontal extent, but their vertical extent may be great. Intersections of two groups of closely spaced multiple fissures may divide the rock mass into many-surfaced polygonal blocks whose large specific surface offers an opportunity for the development of large, irregularly shaped replacement deposits confined by the outlines of the intersection.

Pitching folds and drag folds that result in dilatancy have localized the great Homestake Mine, the largest gold producer of the western hemisphere.

The tin-bearing pipes of eastern Australia and South Africa are another example of structural control.

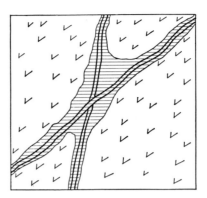

Figure 8-38 Replacement ore localized by fissure intersections.

Sedimentary Features. The features previously mentioned due to sedimentation are especially effective in localizing replacement bodies, as is shown in Fig. 8-39. In contrast, some sulfide features in sediments are not formed by replacement but are formed contemporaneously with the sediments (Fig. 10-3).

Criteria of Replacement

Replacement deposits generally exhibit one or more diagnostic features characteristic of replacement. Some can be seen only in field relationships, some in hand specimens, and others only under the microscope. Some of the criteria afford positive identification, others negative. An extraneous mineral can gain entry into a rock only by filling a preexisting cavity or by making way for itself by replacement. Cavity-filling deposition is distinguished by certain characteristic features discussed under "Cavity Filling," which are lacking in replacement deposition. Therefore, their absence constitutes dependable negative criteria for replacement.

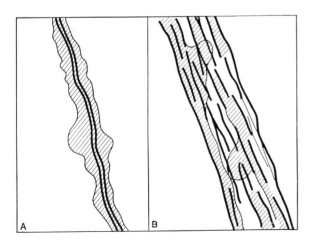

Figure 8-37 Replacement lodes. A, Developed along single fissure; B, along shear zone.

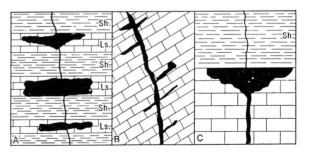

Figure 8-39 Relation of replacement to sedimentary features. Black, ore. A, relation to intercalated limestone beds; B, to bedding planes; C, to overlying impervious bed.

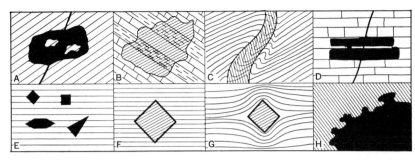

Figure 8-40 Features illustrating criteria of replacement. A, Unsupported residual nuclei; B, preserved rock strata; C, preserved folded structure; D, ore abutting bedding; E, doubly terminated crystals; F, pyrite cube truncating bedding, in contrast to G, which has grown in yield shale; H, irregular outlines of replacement contacts.

Unsupported Residual Nuclei. The frequent observations of small islands of country rock entirely enclosed within mineral bodies (Fig. 8–40) puzzled geologists for a long time, and it was not until replacement became understood that their presence could be explained. In fact, their elucidation was largely instrumental in leading to the concept of replacement deposits. They are unsupported residuals of country rock that escaped replacement while the surrounding rock was converted to ore. Obviously, if the ore body had not been formed by replacement, the space now occupied by ore must have been a preexisting opening, in which case such inclusions would have rested on the floor of the opening. Where the residuals contain bedding or other structural features that are in alignment with similar structures of the wall rock (Fig. 8-40A) they constitute the strongest evidence of replacement.

Preservation of Rock Structure. When rock is replaced by ore, particle by particle, the structure of the replaced body is commonly faithfully preserved in the ore. Where such inherited structure is diagnostic of the preexisting rock, it constitutes definite and usable criteria of replacement. Such features as stratification, cross-bedding, fossils, and dolomitization rhombs, if preserved in ore, indicate clearly that the ore has replaced the rock that formerly showed those features (Fig. 8-40B). Similarly, phenocrysts of igneous rocks and the schistosity of metamorphic rocks may be delicately preserved. In addition, larger structural features, such as folds (Fig. 8-40C), breccias, faults, and joints, are inherited. These various features are analogous to the preserved wood cells in petrified wood.

Intersection of Structural Features. Since many types of mineral deposits intersect rock structure, this criterion by itself is inconclusive. A massive sulfide body for example, may abut the ends of thin limestone beds (Fig. 8-40A,B,D). This shows that the ore is later in age than the bedding of the rock, but the ore may have filled an open cavity that abutted the beds or it may have grown by replacement into an abuting position. If, however, the bedding extends uninterrupted from rock into ore, it indicates replacement.

Complete Crystals. Crystals alien to the original rock, that grow by replacement in homogeneous rocks, commonly display freely developed faces in contrast to the imcompletely developed crystals that are attached to, and grow outward from, the walls of open spaces. Consequently, complete, doubly terminated crystals tightly enclosed in rock (Fig. 8-40E) are diagnostic of replacement, particularly when they transect several individual rock grains. Delicate crystals of barite in a fine-grained limestone are an example. Such crystals may be of megascopic or microscopic size.

Care must be used in differentiating between these crystals and crystals that may have grown by pushing aside the adjacent rock. For example, cubes of pyrite in shale may have grown by pushing aside the shale laminae (Fig. 8-40G), which continue unbroken around the cube. However, if the cube face squarely intersects the shale laminae it is a sure indication of replacement (Fig. 8-40F).

Mineral Pseudomorphs. The presence of pseudomorphs of a mineral of one composition after another of quite different composition is evidence of replacement. A cube having the form and characteristic striations of pyrite may consist of chalcocite. The chalcocite must have replaced the pyrite crystal.

Outlines. The sharp outline of a compact mineral or body of ore against the host rock may indicate replacement. In Fig. 8-40H the wavy knobby outline of the ore, particularly the unusual-shaped protuberances and embayments into the rock, is not characteristic of solution cavities filled by ore but is typical of replacement. Even more characteristic are irregular outlines formed as a result of differential replacement whereby grains or bands of one mineral are more completely replaced than adjacent minerals or bands.

Idealized drawings of some of the openings and textures used to study replacement are shown on Fig. 8-41.

Form. Extreme irregularity of form is characteristic of replacement deposits. Such shapes might be confused with those of solution cavities in limestone but not with cavities produced by fissuring or rupture. However, solution caves have characteristic walls marked by intersecting concavities and commonly have floor debris. Also, some deposits were formed at a geologic time when several thousand feet of uneroded cover overlay them—at a depth at which solution caves do not form.

Shrinkage Cavities. Irving noted irregular cavities or "volume vugs" at Leadville, Colorado, and in the siliceous gold ores of the Black Hills, which he considered characteristic of replacement. They are, however, not diagnostic, since replacement proceeds without a change of volume.

Transection of Different Crystals. Small wavy veins or veinlets of irregular width that transect several diversely oriented host crystals indicate replace-

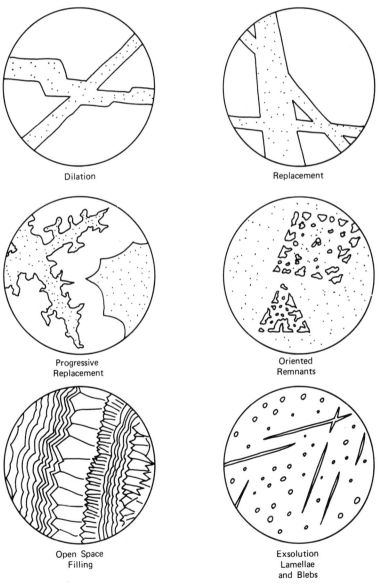

Dilation

Replacement

Progressive Replacement

Oriented Remnants

Open Space Filling

Exsolution Lamellae and Blebs

Figure 8-41 Different types of openings, replacement, and exsolution features generally observed in thin sections. (D. Garlick.)

ment. Their irregularity of width and outline distinguishes them from somewhat similar veinlets formed by the filling of small fissures.

Absence of Crustification. Crustification is absent from replacement deposits so that its presence distinguishes cavity-filled deposits, but its absence does not prove replacement.

Resulting Mineral Deposits

Mineral deposits formed by replacement may be divided into massive, replacement lode, and disseminated deposits. Except for deposits of iron ore, they are the largest and most valuable of all metallic mineral deposits.

Massive Deposits. Massive deposits are characterized by great variations in size and extremely irregular form. Bodies in limestone generally thicken and thin, display wavy outlines, and ramify irregularly in all directions. Others are great irregular massive bodies whose larger dimensions may be measured in thousands of meters. Generally, the deposit consists mostly or entirely of introduced ore and gangue minerals, and included rock matter constitutes only a small or negligible part. Some deposits consist of stupendous masses of pure yellow sulfides, such as Rio Tinto, Spain. More than 600 million tons of pyrite were originally introduced into the rocks at Rio Tinto, probably by submarine exhalative processes. Other massive sulfide deposits, such as Noranda, Quebec (Fig. 8-42), and Flin Flon, Manitoba, may be submarine volcanic exhalative deposits rather than large replacement deposits.

Massive replacement ore terminates abruptly against the country rock. Although most common in limestone, they occur widely in various kinds of rocks. There is extensive replacement over almost two kilometers in length of selective replacement of only one formation, the Bluebell dolomite, in the Chief Mine, Tintic district, Juab County, Utah (Fig. 8-43).

Replacement Lode Deposits. Replacement lode deposits are localized along thin beds or fissures whose walls have been replaced. Consequently, they resemble fissure veins in form. Many so-called fissure veins are actually replacement lodes. In general, they are wider than fissure veins, and the width varies greatly along a single lode; it may range from a few centimeters to several tens of meters. The walls are commonly wavy, irregular,

and gradational into the country rock. The ore may be massive or irregularly scattered in the rock; commonly it is flanked by a fringe of disseminated ore minerals so that the ore boundary is a commercial one. The gold veins of Kirkland Lake, Ontario; the copper veins of Kennecott, Alaska; and the lead veins of the Coeur d'Alene, Idaho, are examples.

Disseminated Replacement Deposits. In disseminated replacement deposits, the introduced material constitutes only a small proportion of the ore. The ore minerals are peppered through the host rock in the form of specks, grains, or blebs, generally accompanied by small veinlets, and represent the multiple-center type of replacement. The amount of introduced gangue is small, and the ore consists of altered host rock and the disseminated ore grains. The total content of metallic minerals may be less than one percent but commonly ranges from 5 to 10 percent of the mass. The ores are mostly low grade. The boundaries are vague; the metallized

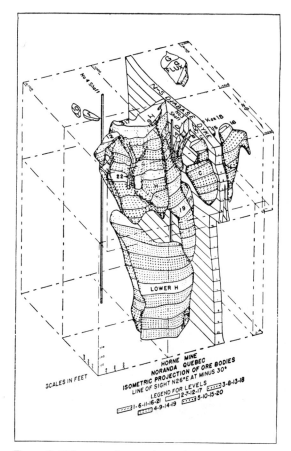

Figure 8-42 Supposedly massive replacement bodies of sulfide ore, but possibly submarine volcanic pile, Noranda mine, Quebec, in isometric projection. Line of sight N26°E at minus 30°. (Peter Price, Bull. Can. Inst. Min. and Met.)

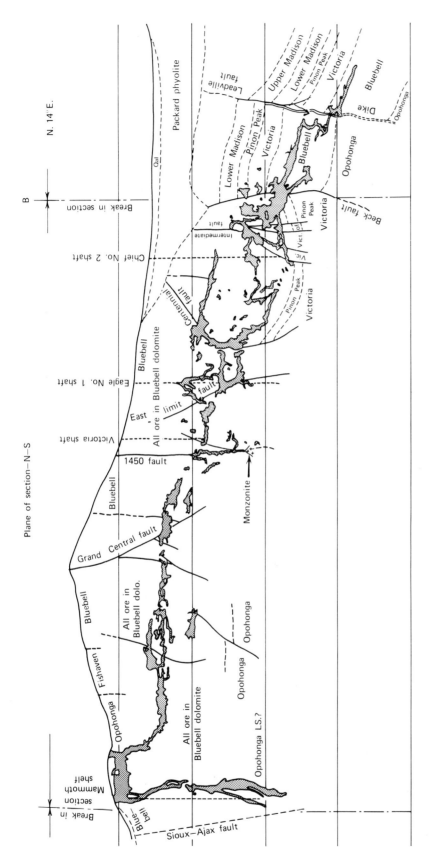
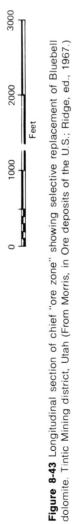

Figure 8-43 Longitudinal section of chief "ore zone" showing selective replacement of Bluebell dolomite. Tintic Mining district, Utah (From Morris, in Ore deposits of the U.S.; Ridge, ed., 1967.)

143

part fades gradually into waste rock, and the ore limits are determined by the workable grade of the ore.

Disseminated replacement deposits are generally huge, which permits large-scale mining operations and the utilization of low-grade ores. Since the workable parts of the deposits are fringed by zones of somewhat lower-grade material, a slight reduction in the grade of the rock that can be profitably treated permits the inclusion of the lower-grade marginal ore within the ore boundaries and thereby brings about a large increase in the size and tonnage of the deposits. A lowering of a workable-grade copper ore, for example by 0.25 percent, may increase the ore reserves by tens of millions of tons. The great "porphyry copper" deposits, many of which are mined from huge open pits, fall in this group of deposits. Some idea of the enormous size of disseminated deposits and the immensity of the operations may be realized from the following examples: The Chile Copper Co. mine at Chuquicamata, Chile, is reported to have reserves of more than 1 billion tons of copper ore, which to date has averaged 1.10 percent copper. The great Utah Copper mine at Bingham, Utah, has officially reported reserves of 600 million tons of ore containing about 0.65 percent copper, although known reserves and

grade are now higher and lower respectively. It has treated over 125,000 tons of ore per day. More than 1 billion tons of ore and 1 billion tons of waste have been mined from the Bingham Canyon Open pit. The Zambian copper belt has estimated reserves of 900 million tons, averaging 3.5 percent copper. The Climax Molybdenum mine at Climax, Colorado, has ore reserves of over 500 million tons averaging about 0.24 percent molybdenum and has produced more than 10^9 pounds of contained molybdenum.

The Alaska Juneau gold mine handled 12,000 tons of ore per day, which averaged about 0.035 ounce ($1.23 at $35 per ounce) of gold per ton and has mined around 88 million tons yielding about $81 million from ore averaging 0.043 ounces of gold per ton. Another example of ores and deposits that belong to this group is the disseminated lead deposits of southeastern Missouri with the greatly increased reserves of the Viburn belt.

Form and Size. The form of replacement deposits is determined largely by the structural and sedimentary features that localize them. Accordingly, they are irregular, blanket-shaped, tabular, pipe-shaped, synclinal, or anticlinal, or they may be large irregular disseminated deposits. Figure 8-44 shows representative forms.

The size varies greatly. The deposits may be

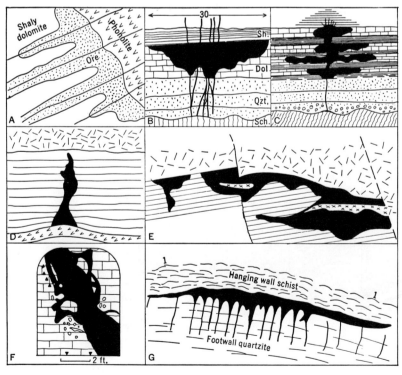

Figure 8-44 Forms of replacement deposits. A, Along fissures and where phonolite dammed solutions, Mineral Farm, Black Hills, South Dakota (after Irving); B, ore restricted below shale and abutting dolomite bedding, Union mine, South Dakota (after Irving); C, Portland mine, South Dakota (after Irving); D and E, cross and longitudinal sections of Iron Hill, Leadville, Colorado (after Irving); F, sketch of replacement vein, Jumbo mine, Kennecott, Alaska; G, relation of ore to fissures in quartzite and to overlying schist, Ferris Haggarty mine, Encampment, Wyoming. (After Spencer, U.S. Geol. Survey.)

mere cracks containing high-grade ore or they may have the dimensions of the massive Henrietta-Wolftene ore body at Leadville, Colorado, which was 1000 meters long, 500 meters wide, and 60 meters thick. Lode replacements may reach a few thousand meters in length and depth and as much as 60 meters in width. Disseminated replacement deposits range in size from small to the dimensions of the Chuquicamata deposit in Chile, whose maximum length is 3,200 meters and maximum width is 1,100 meters.

Texture of Replacement Ores. The texture of replacement ores varies considerably according to the conditions of temperature and pressure of formation and the degree of replacement. All replacement ores lack crustification, and drusy cavities are generally absent.

Dissemination ores are characterized by pepper-and-salt texture. The valuable mineral grains may be shapeless, as shown by copper ores, or crystals, as shown by disseminated lead deposits.

Massive ores may retain the texture and structure of rocks they replace, such as the texture of oölitic limestones, the rhombs of dolomite, or the phenocrysts of porphyry. Commonly, however, such original texture is wholly destroyed. With incomplete replacement, residual rock particles may constitute a small or major part of the ore. With more or less complete replacement, the texture is generally holocrystalline. At high temperatures of formation, a coarse texture is characteristic. Ores formed at intermediate temperatures are generally characterized by finer-grained texture, and those at low temperatures are mostly fine grained. Agate-like and colloform textures are not uncommon.

Under the microscope, crisscrossing and replacement of earlier minerals by later minerals are often observed (Fig. 8-45), or the replacing mineral may form rims at the expense of earlier minerals. Exsolution textures, such as dots, lenses, or plates, oriented along cleavage planes of the host mineral, may be present but are not diagnostic of replacement deposits.

Ores Formed, and Examples. Outside of some iron and some nonmetallic deposits, replacement processes have given rise to the world's largest and most important mineral deposits. Of first rank are the base metal and precious metal deposits, but the rarer metals and many nonmetals are well represented. The chief metals, the chief types of deposits, and some examples of important deposits are listed in Table 8-3.

Figure 8-45 Sphalerite (dark) cutting across and replacing pyrite (light) along fractures in pyrite. ×90 (L. C. Graton, Econ. Geol.)

Table 8-3 Examples of Important Deposits of Ores

Ores	Type	Examples of Important Deposits
Iron	Magnetite	Dover, N.J.; Lyon Mountain, N.Y.
	Hematite	Iron Mountain, Mo.
	Limonite	Oriskany ores, Va.
Copper	Disseminated	Utah Copper; Nevada Consolidated; Chino, N. Mex.; Ray, Ajo, Miami-Inspiration, and Morenci, Ariz.; Zambian copper belt; Chuquicamata, Braden, and Potrerillos, Chile; Phillipines, British Columbia; and Papau
	Lode	Kennecott, Alaska; Bingham, Utah; Magma, Morenci, and Bisbee, Ariz.; Britannia, British Colombia; Cerro de Pasco, Peru
	Massive	Bingham, Utah; Bisbee and United Verde, Ariz.; Granby, British Columbia; Boliden, Sweden
Lead	Massive	Leadville, Colo.; Bingham, Utah; Sullivan, British Columbia; Santa Eulalia, Mexico; Broken Hill, Australia
	Lode	Coeur d'Alene, Idaho; Park City and Tintic, Utah
	Disseminated	Southeastern Missouri, Pine Point, NWT
Zinc	Massive	Leadville, Colo.; Bingham, Utah; Sullivan, British Columbia; Flin Flon, Manitoba; Silesia
	Lode	Broken Hill, Australia; Part City, Utah; Trepca, Serbia
Gold	Massive	Noranda, Quebec
	Lode	Homestake, S. Dak.; Kirkland Lake and Porcupine, Ontario
	Disseminated	Juneau, Alaska; Witwatersrand, South Africa
Silver	Massive	Leadville, Colo.; Bingham, Utah
	Lode	Park City and Tintic, Utah; Coeur d'Alene, Idaho; Cerro de Pasco, Peru; Santa Eulalia, Mexico
Tin		Transvaal; Australia
Mercury	Lode	Almaden, Spain
	Disseminated	New Almaden, Calif.
Molybdenum	Disseminated	Climax, Henderson, Colo.; Utah Copper, Utah
Manganese	Massive	Leadville, Colo.; Potgietersrust, South Africa; Turkey; India; USSR.
Barite	Lode	Missouri; Nevada
Fluorite	Lode	Illinois-Kentucky field; Chihuahua, Mexico
Magnesite	Massive	Manchuria; Washington; California
Kyanite	Disseminated	North Carolina; Virginia

Selected References on Replacement

The nature of replacement. 1912. Waldemar Lindgren. Econ. Geol., v. 7, p. 521–535. *A fundamental discussion of the process of replacement.*

General Reference 9., chaps. 9, 14, 26, and 27. *Descriptions of some replacement deposits and processes operating toward their formation.*

Metasomatism. 1925. Waldemar Lindgren. Geol. Soc. Am. Bull., v. 36, p. 247–261. *Development of ideas; processes involved and problems raised.*

Metasomatic processes. 1947. W. T. Holser. Econ. Geol. v. 42, p. 384–395. *Physico-chemical processes of transport and reaction in fine-grained rocks.*

Sulfide ore deposits at Yauricocha, Peru. 1959. H. J. Ward. Econ. Geol., v. 54, p. 1365–1379.

Role of replacement in some cockade textures. 1966. J. Kutina and J. Sededikoŕa. Econ. Geol., v. 56, p. 149–176.

ORE SHOOTS

Ore deposits are rarely equally rich throughout. Generally, the valuable primary minerals tend to be concentrated in certain sections called *ore shoots*, which contrast with lean or barren portions of the deposits. Ore shoots may be present in most hydrothermal deposits, but they are most characteristic of fissure veins and replacement lodes.

An ore shoot may differ from the lean portions of a deposit by the presence of only as little as 0.0004 percent gold or 1 percent copper, or there may be a readily detectable mineralogical difference, such as the presence of galena in the ore shoot and its absence from the barren portion.

Terminology

The term *ore shoot* should apply only to concentrations of hypogene ore, and it is advisable to restrict the term to this usage in order to distinguish hypogene concentrations from supergene concentrations because the latter are produced by quite different processes and may have been formed from either the rich or lean hypogene parts of veins. The terms *pockets, nests, bunches,* or *kidneys* are variously employed in different places to designate small, irregular concentrations of ore. They refer either to hypogene or supergene concentrations, and although they fall within the definition of ore shoots, the latter term is usually applied to larger bodies. *Bonanza* is commonly used to designate an exceptionally rich shoot or bunch of ore, particularly with reference to gold and silver. Generally, it refers to rich secondary masses, and Irving has suggested its restriction to that field. *Chimneys* or *pipes* are terms used to designate vertical or highly inclined elongated ore shoots that resemble huge smokestacks. These may occur within fissure veins, but the terms are more commonly applied to any body of pipelike shape.

Shape and Size. The outline of ore shoots may be irregular, but generally they tend to be elongated bodies that extend in a vertical or highly inclined position up and down a fissure vein. This inclination within a vein is called the *pitch* or *rake* (Fig. 8-46). Commonly, several shoots of more or less similar shape, size, and pitch occur in a vein and tend to be spaced at approximately equal intervals; those in nearby parallel veins tend to pitch in the same direction. The shape of some typical ore shoots may be seen in Fig. 8-46.

In size, most ore shoots range in level-length from a few tens to a few hundreds of meters and in pitch length from a hundred or so to several hundred meters; rarely they attain lengths of 500 or 1000 meters. Notable exceptions occur, however, as in the Mother Lode of California, where shoots

attain pitch lengths of 1,500 meters or more, and at Grass Valley, where the North Star shoot is 2,700 meters in pitch length.

Grouping. Ore shoots may be grouped as follows:

1. Open-space shoots, due to available open space.
2. Intersection shoots, due to vein intersections.
3. Impounded shoots, due to damming of mineralizing solutions.
4. Wall-controlled shoots, due to the effect of wall rock upon precipitation.
5. Structure-controlled shoots, due to various structural controls.
6. Depth-controlled shoots, due to a decrease of temperature and pressure.
7. Recurrent mineralization shoots, due to successive periods of mineralization.
8. Unsolved shoots, due to unknown factors.

The last group includes the greatest number of shoots.

Open-space shoots are localized by available open spaces in fissures, such as are caused by the relative movement of opposite walls of a curved fissure (Fig. 8-14). The ore shoots of the Mother Lode gold veins of California are considered by Knopf to have been so controlled. Similarly, Wandke ascribes many of the ore shoots of Guanajuato, Mexico, to open-space control.

Intersection shoots are localized at vein intersections or cross fissures and are among the oldest known and the commonest types (Fig. 8-38). Intersections are particularly favorable because at such places different solutions meet, also the walls are more shattered and afford greater specific surface. Howe states that most of the ore shoots of the Grass Valley gold veins are at vein intersections.

Impounded shoots result from the impounding of mineralizing solutions against impervious barriers, such as shales or fault gouge (Figs. 8-39 and 8-44).

Wall-controlled shoots are those occuring adjacent to certain favorable wall rocks that presumably influence deposition from the mineralizing fluids. The well-known precipitating effect of carbonaceous rocks upon gold is an example. A fissure may intersect several favorable beds and contain ore shoots adjacent to each, alternating with barren stretches of vein opposite less favorable rocks. Wandke and Martinez point out that at Guanajuato, Mexico, bonanza silver ores occur in fissures opposite andesite or porphyry but not opposite schist.

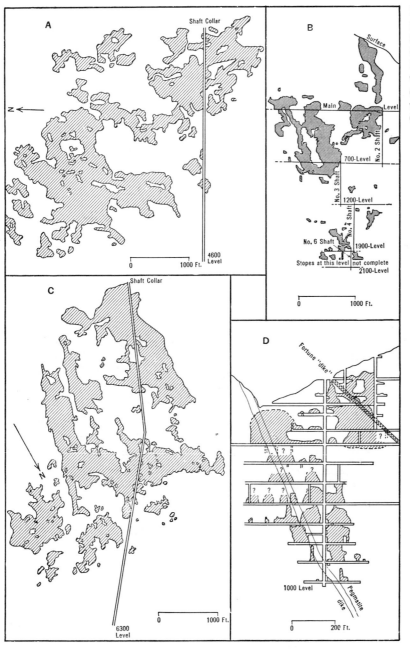

Figure 8-46 Forms of ore shoots in longitudinal section. A, North Star vein, and C, Empire vein, Grass Valley, California (after W. D. Johnston, Jr., U.S. Geol. Survey); B, Chichagoff mine, Alaska (after Reed, AIME); D, Ingram vein, Gold Hill, Colorado. (After Goddard, Colo. Sci. Soc. Proc.)

Similarly, at Porcupine, Ontario, good gold ore in greenstone gives way to barren gangue where the fissures pass into porphyry.

Structure-controlled shoots are localized by various structures. Places of change in strike and dip of fissures are favorable sites for ore shoots. Closely folded, competent beds alternating with incompetent beds form potential openings between the layers at the crests of anticlines and the troughs of synclines and localize ore deposition. Intermineral movement along veins may produce breccia-

tion of one wall or a part of the vein, forming a locus for further mineral deposition, and thus give rise to ore shoots.

Depth-controlled shoots result from the control exerted by decreasing temperature and pressure upon deposition from solutions (Chapter 3). The rapid release of pressure by near-surface, shattered rocks may cause a sudden dumping of the minerals in solution, such as described by Turneaure for the rich tin shoots of Llallagua, Bolivia. This type of shoot occurs particularly with "typomorphic" min-

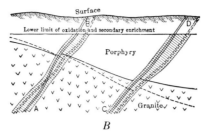

Figure 8-47 Sections along veins showing geology plotted on both walls (solid lines, far wall; dotted lines, near wall), and ore shoots coincide with the diabase dikes, indicating their control in localizing the ore and affording basis for search for other shoots. In B, intersecting fissures have localized the shoots.

erals (cinnabar or silver minerals), or those deposited within a narrow range of temperature and pressure. Similarly, near-surface chemical changes in the metallizing solutions, such as described by Ransome for Goldfield, Nevada, and by Graton and Bowditch for Cerro de Pasco, Peru, give rise to rich, near-surface ore shoots.

Recurrent mineralization shoots are due to reopenings by intermineralization movements, accompanied by recurrent mineralization during which certain parts of a vein, commonly along either wall, are enriched. Several stages of movement and recurrent mineralization may occur.

Unsolved shoots are numerous. Although many ore shoots may readily be placed in one of the above-described groups, a large number as yet defy interpretation. They do not appear to be localized by any of the conditions considered here. Probably, many are localized by chemical controls, such as the complexes described by Garrels and Helgesen.

Causes of Ore Shoots. Although separate causes of ore shoots have been indicated previously, a single shoot may be due to more than one cause. Its formation may involve the coincidence of two or more separate factors; suitable conditions of temperature, pressure, and chemical character of the solutions must prevail. Once the cause is determined,

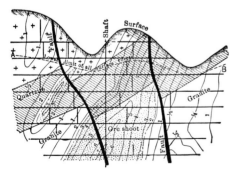

Figure 8-48 Section along vein showing geology, and vein width indicated by contours, with superimposed ore shoot (stippled). Here the ore shoot occurs in disregard of geological features but is clearly related to the wider portions of the vein.

however, a practical as well as a scientific achievement has been attained because a search for similar conditions may be rewarded by the discovery of other shoots.

Recognition and Search for Shoots. Shoots must be recognized before their cause can be determined or search directed for others. Often the two unfold themselves together. They may be recognized by visual observation of the mineralogy, by assays or analyses, and by the plotting of data on maps, particularly on longitudinal sections. The plotting of visual observations is sufficient in the case of gross ores or minerals, but assays or analyses are necessary for invisible metals. The plotting will indicate if the valuable materials are uniformly or haphazardly distributed, or if they are localized into definite shoots, whose outlines and size will also become evident.

If the geology is superimposed on a longitudinal section (Fig. 8-47), a relationship between ore shoot and kind of wall rock, or fissure intersection, or other controls, may become apparent. If the ore shoots occur in apparent disregard of such features, some other cause of ore localization must be sought. Any relationship between ore shoots and width of vein or former open space may be made apparent by plotting width contours, that is contour lines connecting parts of equal vein width, on the longitudinal assay plat (Fig. 8-48). If the vein is a fault fissure, the walls of which differ in their geology, longitudinal geological plats may be made for both walls, and the hanging wall tracing can be superimposed upon the footwall plat. An established ore-shoot control yields information for search for other shoots.

Selected References on Ore Shoots
Structural control of ore deposition. 1929. C. D. Hulin. Econ. Geol., v. 24, p. 15–49. *Excellent discussion.*
Outcrops of ore shoots. 1939. H. Schmitt. Econ. Geol., v. 34, p. 654–673. *Detection.*
Openings due to movement along a curved or irregular

fault plane. W. H. Newhouse. Econ. Geol., v. 35, p. 445–464. *One type of ore-shoot formation.*

General References 15. *Innumerable examples and discussions of ore shoots.*

Certain ore shoots on warped plane surfaces. 1943. W. H. Emmons. AIME Tech. Pub. 1545. *Comparison of ore shoots in veins; good bibliography.*

Formation of primary ore shoots (Colo.). 1946. V. C. Kelley. Colo. Sci. Proc., v. 14, no. 7, p. 318–333. *Review of ore-shoot control in general.*

9

SEDIMENTATION

The river rolls by us in silent majesty; the quiet of the camp is sweet; our joy is almost ecstasy.

John Wesley Powell during first
expedition down the Colorado River

The process of sedimentation as distinct from evaporation has resulted not only in the formation of common sedimentary rocks but also in valuable mineral deposits of iron, manganese, copper, phosphate, coal, oil shale, carbonates, cement rocks, clay, diatomaceous earth, bentonite, fuller's earth, magnesite, sulfur, and, less directly, uranium-vanadium deposits. These substances may be regarded as merely exceptional varieties of sedimentary rocks that happen to be valued because of their chemical or physical properties. Their mode of formation, therefore, is that of sedimentation with special variations to account for the special materials. They are composed of inorganic and organic materials; their source, like that of any sedimentary rock, is from other rocks that have undergone disintegration, the ultimate source, of course, being the igneous rocks. Some of the materials, such as oxygen and carbon dioxide, have been obtained from the atmosphere, and a few have been derived from former deposits.

The formation of sedimentary deposits involves, first, an adequate source of materials; second, the gathering of the materials by solution or other processes; third, the transportation of the materials to the site of accumulation if that is necessary; and fourth, the deposition of the materials in the sedimentary basin. Subsequent compaction, chemical alteration, or other changes may take place.

The source materials, solution, mode of transport, and the nature of the deposition are generally similar for each product involved. The variations of some of the sedimentary cycles will be considered separately.

SOURCE OF MATERIALS

The materials that enter into sedimentary mineral deposits have been derived chiefly from the weathering of rocks. Occasionally, materials have come from the weathering and oxidation of former mineral deposits such as iron, manganese, and copper. Others have passed through an intermediate organic stage. The rocks, however, constitute an adequate source for most of the sedimentary iron, manganese, and copper ore that is known.

Krauskopf has shown that in the earth's crust the average content of iron is 5.6 percent. Eckel has calculated that the portion of the earth's crust beneath the United States to a depth of 300 meters contains over 275,000 billion tons of iron, of which only about 0.01 percent has been concentrated into commercial deposits, in a ratio of about 80,000:1. Clearly, one need not look beyond the ordinary rocks for an adequate source of the iron in deposits. The iron comes from the weathering of iron-bearing minerals of igneous rocks such as hornblende, pyroxene, or mica, from the iron-bearing minerals of sedimentary and metamorphic rocks, and from the red coloring matter of sedimentary rocks.

Similarly, the manganese of sedimentary deposits has been derived chiefly from the weathering of manganese-bearing minerals in the rocks and, to a minor extent, from former sedimentary concentrations and epigenetic lode deposits. Hewett states that there are over 200 minerals that contain manganese as an essential element, and Krauskopf estimates that manganese makes up 0.095 percent of the earth's crust, there being about 50 times as much iron as manganese. Using this proportion, and applying it to Eckel's estimate for iron, there should be 5,400 billion tons of manganese beneath the United States to a depth of 300 meters—an adequate source of supply.

The source of sedimentary phosphate is phosphorus-bearing rock minerals, among which apatite is the most common. Some is also derived from the weathering of collophanite and dahllite in sedimentary rocks.

The constituents of sedimentary carbonate deposits such as the industrial limestones, dolomite, and magnesite are derived from the sea or saline waters to which they are largely supplied by rock weathering; also, the constituents of the numerous types of industrial clayey deposits such as clays, bentonite, and fuller's earth originate in rock weathering.

SOLUTION AND TRANSPORTATION

Solution of the constituents of economic sedimentary deposits in large part goes on during weathering. This is true of iron, manganese, phosphates, carbonates, copper, and some other metals, but, of course, does not apply to clays. The chief solvents are carbonated water, humic and other organic acids, and sulfate solutions.

Carbonated waters are very effective solvents of limestone, iron, manganese, and phosphorus. Where iron is present in the ferrous state, its solution offers no difficulty, since in that form it is unstable and soluble. But ferric iron is almost insoluble in most surface waters and to undrgo solution must first be changed to the ferrous state. Organic matter aids this. The Precambrian iron ores that were formed before organic matter or vegetation became very abundant were probably transported as ferrous bicarbonate or in the colloidal state. Vast quantites of calcium carbonate, as well as other salts, are readily removed in solution and transported to bodies of standing water of higher pH, where precipitation may occur to form sedimentary deposits.

Humic and other organic acids derived from decomposing vegetation are considered effective solvents by Harrar. The hydroxyl acids dissolve large quantities of iron, but the weak organic acids dissolve remarkable quantities and are the most effective of all natural solvents. Moore and Maynard's experiments on the solution and precipitation of iron indicate that carbonated water is the most effective solvent of iron and silica from norite and diabase and that peat solution is next. They concluded that iron is not carried as bicarbonate in surface waters high in inorganic matter, but that the main part entering sedimentary iron formations was probably transported as ferric oxide hydrosol stabilized by organic colloids.

Garrels considers that iron is soluble in river waters at pH 7 or lower. The solution of manganese has not been investigated so thoroughly as that of iron, and knowledge of its solution by organic compounds and its transportation in colloidal form is inadequate. It will probably be learned, as in the case of iron, that organic compounds are important solvents and that manganese is removed largely in the colloidal state.

Sulfate solutions are effective solvents of iron and manganese but are rarely abundant enough to effect large-scale solution and transportation. The oxidation of pyrite yields sulfuric acid and ferric sulfate. At Rio Tinto, Spain, for example, the oxidation of the huge pyritic ore bodies yielded vast quantities of iron sulfate, some of which was precipitated nearby to form thick deposits of bog iron ore, but much of it has been carried down the Rio Tinto (hence the river's name) to the sea and deposited near the shore.

Most of the substances that make up sedimentary mineral deposits (coal excepted) are transported by means of rivers and subsurface waters. For the most part the substances reach the sea, but some are arrested en route or find a resting place in inland bodies of water or interior land basins.

The sedimentary substances will remain in solution as long as the solution does not undergo any appreciable physical or chemical change. Some or all of the iron or manganese, however, may be lost during transportation if the solutions traverse limestone or are subjected to other agencies of deposition. If the iron or manganese in solution escapes these hazards, it may be transported to bogs, lakes, playas, or the sea, where quantity concentration can take place.

DEPOSITION

The materials that form economic sedimentary beds are deposited mechanically, chemically, or biochemically. The manner of deposition depends upon the nature of the solvent, the place of deposition, and the pH and redox (Eh) conditions as, for example, whether in the sea or in a swampy basin. The types of sediments and their resulting classification of rocks are shown by Table 9-1.

Garrels and others have aided the understanding of solution and deposition at low temperatures through the study of mineral equilibria under varying pH and Eh conditions. This subject is treated in more detail in the section on "Oxidation and Supergene Processes." A simple fence diagram of the equilibrium pH-Eh conditions is shown in Fig. 9-1, where the areas bounded by "fences" indicate

Table 9-1 Economic Products of Sedimentation

Substance	Composition
Coal (nonanthracite)	Carbonaceous material
Oil shale	Shale and bitumens
Diatomaceous earth	Chiefly diatoms
Fuller's earth	Clay minerals
Bentonite	Clay minerals
Tripoli	Silica
Sand and sandstone	Sand grains
Greensand	Sand, glauconite

the environmental relations of chemical sediments in normal seawater. As an example of the relationships, note that pyrite is deposited in a reducing environment under both acid and alkaline conditions while, in contrast, hematite forms in an alkaline oxidizing environment. The Clinton hematite ores, therefore, could have readily formed by the transport of iron to the seas where oxidizing shallow seawater with a pH of about 7.8 would allow precipitation of iron as hematite with some replacement of carbonates.

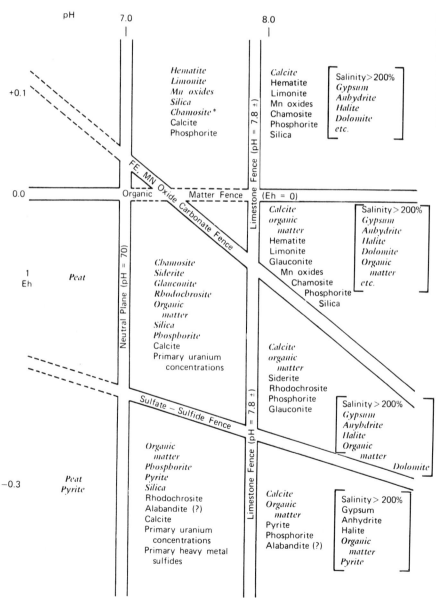

Figure 9-1 Fence diagrams showing relations of chemical sediments to Eh and pH controls for normal sea-water conditions. Associations in brackets refer to hypersaline solutions. (Adapted from Krumbein and Garrels (1952). Copyright by the University of Chicago.)

*Chamosite as used here is representative of the sedimentary iron silicates.

Products of Deposition. Iron is commonly precipitated as (1) ferrous carbonate (siderite); (2) hydrous ferric oxide, goethite (limonite); (3) ferric oxide (hematite); and (4) minor basic ferric salts. It seems probable that most of the marine hematites were deposited directly as ferric oxide. Glauconite, chamosite, and greenalite are less common forms. In the presence of air, ferric oxides form; at lower pH and Eh, siderite, the ferrous carbonate, forms. Manganese is deposited largely as oxides. The carbonate may be deposited in the absence of air and in the presence of excess carbon dioxide.

Depositional Separation of Manganese and Iron. An interesting feature in connection with the formation of manganese ore is its separation from iron during deposition. In the case of chemical precipitation from carbonate solutions, the separation occurs because manganese carbonate is more stable in solution than iron carbonate; hence it is carried farther and is thus separated from the iron. Krumbein and Garrels have shown (Fig. 9-1) that the separation of iron and manganese in an oxidizing environment takes place because the iron oxides precipitate at a lower oxidation potential than the comparable manganese compounds at any given pH. Similarly, under fixed Eh, iron starts precipitating as an oxide at a lower pH than manganese, but in a neutral environment both iron and manganese may precipitate together as carbonates.

Conditions of Deposition

The conditions under which deposition occurs determine in large part the mineralogical composition of the resulting deposits; their size; purity; and distribution, both areal and stratigraphic. Sedimentary iron and manganese ores are deposited in both fresh-and marine water, in bogs, swamps, marshes, lakes, lagoons, and in the ocean. Phosphates and sulfur form mostly under marine conditions.

Bogs and Lakes. In restricted bogs and boggy lakes the resulting deposits are small and local. Iron is deposited as hydroxide or as ferrous carbonate, which in the absence of organic matter readily oxidize to ferric oxides. Impurities are generally present. In Quebec and Sweden iron ore is dredged from existing lake bottoms. The conditions for deposition of bog ores are widely present in glaciated regions.

Manganese may accompany bog iron or be deposited by itself. The resulting product consists of impure manganese wad. Hewett mentions that manganese oxide nodules have been found in lakes, such as those of Lake Tyne, Scotland; and Seller Sea, Austria. A bog manganese deposit 2 meters thick with an area of 17 acres occurs at Hillsborough, New Brunswick.

Swampy Basins. In areas of brackish water and in marine swamps and marshes iron is deposited in the presence of plants. Precipitation takes place from ferrous bicarbonate or organic solutions chiefly through the depletion of carbon dioxide. Decaying vegetation inhibits oxidation, and the iron is deposited as ferrous carbonate (siderite); if any were thrown down as hydroxide, reduction and carbonation to ferrous carbonate would take place.

The conditions of deposition are also those of coal accumulation; therefore, sedimentary siderite occurs in coal measures. If the siderite is deposited along with accumulating vegetation, *black band iron ore* is formed. This occurs in coallike beds, is associated with coal, looks like coal, and might even be mistaken for dull coal except for its weight. If the deposition takes place in coal measures along with clay but not with coal, *clay ironstone* results— an impure carbonate. Sedimentary carbonate ores generally contain impurities of organic matter, clay, sand, and other carbonates.

Manganese is similarly deposited, and Hewett states that carbonate deposition in sediments is more widespread than is generally realized. Residual enrichment (Chapter 13) of these carbonate beds has given rise to many economic oxide deposits.

Peneplain Depressions. Deposition might also take place in isolated basins on peneplains during eras of low seasonal rainfall, as suggested by Woolnough.

Marine Conditions. The majority of economic sedimentary deposits are formed in marine areas, mostly under shallow-water conditions but also as open-sea deposits. Manganese nodules of economic importance containing significant amounts of base and precious metals occur in deep-sea sediments, and the iron silicates have been deposited abundantly in open-sea sediments. These only rarely constitute economic deposits, but some investigators hold that some iron oxide deposits have been derived from them.

Shallow-water areas such as marine lagoons or long, narrow, epeiric seas are the sites of the greatest deposition of sedimentary deposits of iron ores, manganese, phosphate rock, sulfur, commercial carbonate rocks, and clays.

THE CYCLE OF IRON

Because of the low solubility of ferric iron, it is probably ferrous iron that is dissolved during rock weathering and moves largely in streams to favorable sites of deposition. The iron may be lost during transportation (1) if the solutions traverse limestone, where reactions cause deposition of ferrous carbonate or ferric oxides; (2) if the solutions come to rest in an enclosed basin undergoing evaporation; (3) by contact with organic matter; or (4) by decrease in carbon dioxide content of the solutions. That which is precipitated in bogs gives rise only to small, impure, low-grade deposits rarely exceeding 45 percent iron. The ore consists chiefly of limonite with some ferrous carbonate and phosphate of iron mixed with sand and clay, as at Three Rivers, Quebec; and in Maine, Sweden, Russia, and England. The iron that reaches coal basins is thrown down as impure and low-grade siderite deposits. For extensive oxide precipitation, however, the iron must reach the sea.

Open-Sea Deposition. The iron that reaches the open seas becomes deposited in vast quantities as the hydrous iron silicates, glauconite, greenalite, chamosite, or thuringite. "Greensands" are marine deposits of sand, glauconite granules, and some clay and shell matter. Eckel estimates that three Cretaceous beds in New Jersey contain about 250 thousand million tons of ferric oxide. They are not iron ore, but their content of potassium and phosphorus make them useful for fertilizer and water purification.

Glauconite is deposited in deep-sea muds, and recent glauconite favors warm, slightly shallow mud bottoms, where some organic matter is present, an environment that is neither strongly oxidizing nor reducing. It is thought that glauconite is formed from pellets of colloidal silica and clay in which colloidal iron replaces the alumina and absorbs potash from the seawater.

Greenalite is thought by Van Hise and Leith to be the iron silicate from which the Lake Superior iron ores were derived by the action of alkaline silicates on ferrous salts.

Chamosite, a hydrous ferrous silicate, occurs

mostly as oölites in the sedimentary iron ores of Newfoundland, Alabama, England, and central Europe. With thuringite, it has been worked in Thuringia.

Marine Shallow-Water Deposition. The iron solutions that reach the shallow seas give rise to the largest iron-ore deposits of the world. Apparently, the optimum conditions are where sluggish streams enter from deeply eroded, low-lying, coastal areas, with gradients too low to permit abundant suspended matter to be transported. Consequently, little sediment accumulated with the iron ore. Shallow waters are indicated, where waves, alternating with periods of quiet water, gently churned the bottom, and macerated the fossils present. Ripple and current markings also show that the depth of water was not great, and mud cracks indicate occasional elevation and exposure to the sun. The marine life was not dwarfed, which gives evidence of no unusual conditions of environment. Limestone deposition occurred, alternating with the deposition of thin beds of shale and sandstone.

Garrels and co-workers have shown that the deposition of iron in seawater in equilibrium with calcium carbonate is largely controlled by the pH and the Eh conditions. They showed that when slightly acid river water containing iron in solution entered such seawater, the iron would be precipitated as pyrite at low Eh and either acid or alkaline pH. However, with increasing Eh and pH the iron would be deposited as siderite and then as hematite, as shown in Fig. 9-2, over a broad range of Eh. Aerated seawater has a pH of about 7.8. Thus iron in solution would be precipitated as ferric oxide in aerated seawater. Colloidal iron would be deposited almost instantaneously upon contact with oppositely charged electrolytes of seawater. Extensive beds of marine iron ores, such as the Clinton iron beds, may thus be formed. Some iron was deposited as oölites; some coated or replaced shell fragments on the sea bottom; and some was deposited as an iron mud. It has also been suggested that low Eh groundwater disgorging Fe^{2+} into shallow marine seas could form the Clinton-type deposits.

A little calcium carbonate was precipitated at the same time, and variable amounts of clayey matter became admixed, relatively small amounts in the rich ores and relatively large amounts in the lean ores. Eckel points out that in the Clinton iron beds the associated clayey and shaly matter is higher in alumina and iron than normal shales, suggesting

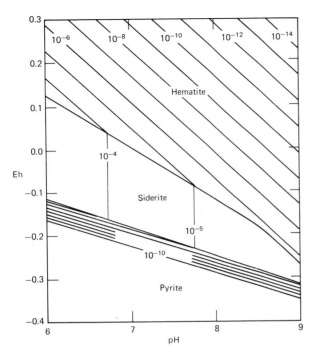

Figure 9-2 Stability fields and "solubility" (Fe^{++} + Fe^{+++}) contours for some iron bearing minerals. (From: Shepherd and Garrels, Jour. Geol., v. 60, 1956.)

tioned. Some geologists have contended that they have been formed by the replacement of limestone, and others that they are residual accumulations or derived form glauconite beds. Such conclusions, however, are not substantiated by field evidence. Some of the iron is considered to have been dissolved and transported as bicarbonate or as a colloid in organic solutions and to have been deposited directly as hematite or goethite. In some deposits, diagenetic replacement of fossil fragments is evident, and in others, such as those of Newfoundland, chamosite has been deposited in concentric shells with hematite to form the oölites; siderite is also relatively abundant there.

At Clinton, several iron-ore beds have been deposited of which from 3 to 10 have proved commercial. A long period of continuous iron deposition over a wide area is indicated by the Big Seam of Clinton iron ore at Birmingham, Alabama, which, according to Eckel, attains a thickness of 10 meters, averages over 3 meters, and has been tested over a width of 15 kilometers and a length of 75 kilometers. Outcrops of Clinton ore are almost continuous for a distance of over 1100 kilometers, so the main basin of deposition must have been at least this long and 80 kilometers wide. Eckel states that in one part of it the enormous quantity of 5000 million tons of iron oxide was laid down in continuous deposition and that the Wabana basin of Newfoundland probably had a continuous deposition of 7000 million tons. Iron accumulation was on such a vast scale that purely local conditions cannot be invoked to explain it. Examples of deposits formed follow.

that the iron was derived from deeply weathered basic igneous rocks or limestones.

The marine character of these oxides is indicated by the contained fossils, the oölites, the nature of the sedimentation, and the size of the basins of deposition. The oölites are considered to be diagnostic of marine conditions. The origin of these marine oxides, however, has not been unques-

Deposits Formed and Examples

Type	Character	Chief Localities
Bog ores	Limonite	Quebec, Maine, Sweden, Russia
Carbonate	Black band ore, clay ironstone	Pennsylvania; Ohio; Middleboro, North Staffordshire, Lowmoor, Dowlais, Clyde Basin, Great Britain; Westphalia, Saarbrücken, Germany; Russia
Iron silicates	Thuringite	Thuringia, Switzerland, South Africa, Russia
	Chamosite	Cleveland Hills, England; Newfoundland
	Greenalite-glauconite	Widely distributed
Marine oxides	Hematite or limonite	Clinton oölitic hematites of United States; "Minette" oölitic limonites of Lorraine and Luxemburg; Lake Superior area; Newfoundland; Krivoi Rog, Krusch, Russia; Brazil; Australia.

Note: These various deposits are described in Chapter 20.

THE CYCLE OF MANGANESE

Manganese may be deposited as a minor constituent of iron ores or it may be deposited separately as sedimentary manganese deposits relatively free from iron. Iron oxides precipitate at a lower pH than manganese oxides; thus iron and manganese may be separated but often are not. Neutral precipitation in a slightly reducing environment gives simultaneous precipitation of both carbonates. Also, separation of iron and manganese occurs under oxidizing conditions because iron oxide or hydroxide precipitates at a lower oxidation potential (Eh) than the comparable manganese compounds at a given pH.

Large deposits of sedimentary manganese, however, are relatively few but very important as compared with manganese deposits produced by other processes, and they are fewer and much smaller than sedimentary iron deposits.

Deposition of sedimentary manganese ore in the form of carbonate or oxides may occur in lakes or bogs, or in the sea. It parallels that of iron and so needs only brief mention. The two differ, however, in the degree of oxidation and in the oxides formed. Whereas ferrous oxide and ferrous hydroxide are unknown in nature, the equivalent manganese compounds are well-known minerals. Manganite ($Mn_2O_3H_2O$) and hausmannite (Mn_3O_4), the respective counterparts of goethite and magnetite, are common, but the dioxide (MnO_2), which is the chief ore mineral of manganese, has no iron counterpart.

Marine depositions, chiefly in the form of the dioxide, have been formed under shallow-water conditions and in deep-sea sediments, where it is widely distributed as nodules, as coloring matter, and as coatings on fossils; according to Hewett, it is greatest in amount of oceanic areas where low sedimentary rates prevail. Deep-sea manganese nodules are of scientific and increasing commercial interest. Hewett concludes that most of the stratiform manganese deposits of the world are attributable to low-temperature hydrothermal waters from a depth related to centers of volcanism. They lack the large amount of iron that should be present had the metals been derived from the weathering of continental rocks.

Under near-shore conditions, as in the case of iron, manganese oxide hydrosol, or bicarbonate solutions, precipitate oxides or carbonates or both. The oxides commonly form oölites, which are made up chiefly of psilomelane and pyrolusite. These, along with included marine fossils, indicate a marine origin of the manganese. The associated rocks are shales, limestones, and, less commonly, sandstones.

Deposits Formed and Examples

Carbonates. Hewett states that impure sedimentary manganese carbonate is widely distributed but is seldom commercial. Beds of relatively pure sedimentary manganese carbonate are reported from Newfoundland, and others occur in Arkansas, Minnesota, South Dakota, California, Nevada, the Appalachian states, Wales, Belgium, and Russia. The beds seldom exceed a meter or more in thickness or contain more than 20 percent manganese. Their chief importance is in the light they throw on the origin of manganese and in the fact that they supply preliminary concentrations of manganese which, upon weathering, may yield marketable deposits of secondary oxides. At Usinsk, Siberia, very extensive carbonate deposits contain up to 48 percent manganese oxide.

Oxide Deposits. The great manganese deposits of Russia at Chiaturi, Georgia, and at Nikopol, Ukraine, are not only the outstanding sedimentary oxide deposits but the greatest manganese deposits of the world. The ore occurs in oölites and nodules in Tertiary beds of sand and clay from 1 to 4 meters thick. Somewhat similar deposits occur in the Urals, Siberia, and Milos, Greece. Vast quanities of manganese nodules cover large areas of the ocean floor.

THE CYCLE OF PHOSPHORUS

The sedimentary cycle of phosphorus is fascinating and puzzling. Dissolved from the rocks, some of it enters the soil, from which it is abstracted by plants, from them passes into the bodies of animals, and is returned via their excreta and bones to accumulate into deposits. These in turn may undergo re-solution, reach the sea, and there the phosphorus is deposited or accumulated by sea life, embodied in sediments, and returned to the land upon uplift, when a new cycle may start.

Phosphates are soluble in carbonated water and, in the absence of calcium carbonate, will stay in solution. The phosphate in limestones resists solution. Some phosphoric acid in solution reaches the sea, where it is extracted by organisms; some is redeposited as secondary phosphates, which may be redissolved; and some is retained in the soil. Swamp waters rich in organic matter also dissolve phosphates, and some phosphorus compounds are thought to enter solution as colloids. Phosphorus is probably transported by streams as phosphoric acid and as calcium phosphate. Some is transported by birds and animals.

Special Conditions of Deposition.

Economic beds of phosphate are formed under marine conditions in the form of phosphorite. The beds range in age from Cambrian to Pleistocene and extend with remarkable uniformity over thousands of square miles. They are interstratified with other sediments and grade laterally into them. The beds

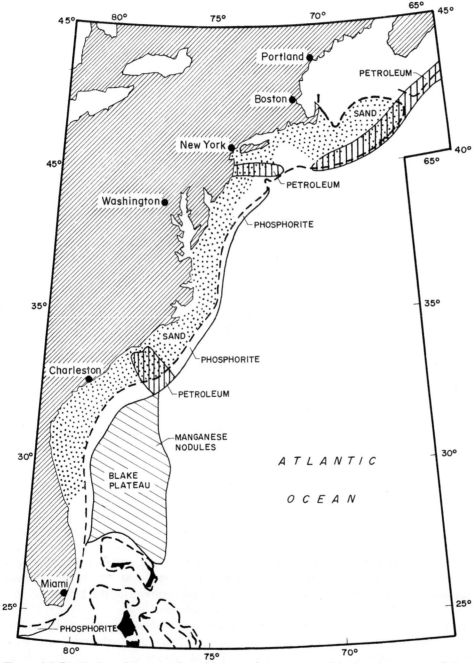

Figure 9-3 Distribution of the most favorable areas for some potential mineral resources off the Atlantic coast of the United States. The dashed line denotes the position of the edge of the continental shelf; abundant manganese nodules corresponds to the surface of the Blake Plateau. (After Emery, 1965, Figure 1.)

are sparingly fossiliferous and are interlaminated with marine fossiliferous beds. These features together with their own oölitic character prove a marine origin. Such deposits are found in the southeastern United States, the northern Rocky Mountain states, North Africa, and Russia west of the Urals.

According to Mansfield, the accumulation of sufficient phosphates to produce sedimentary beds must have required unusual marine conditions. Other sedimentation must have ceased to permit sufficient accumulation of pure phosphatic materials to form the oölites that aggregated into phosphate beds. The presence of hydrocarbons and marcasite in these deposits indicate deposition under reducing conditions, and replacement of shells by phosphate proves some diagenesis. Mansfield thinks that the material represents a slow accumulation, shut off from the sea, of phosphatic debris under anaerobic conditions and that for a long time cool temperatures were frequent during climatic oscillations. Such conditions favored the growth of life in the shallow waters and reduced the activities of denitrifying bacteria, which curtailed the deposition of calcium carbonate and favored the concentration of phosphatic solutions from which the oölites were formed.

Later work by Krauskopf on the Phosphoria formation in Utah, Idaho, Wyoming, and Colorado recognizes that the well-preserved sedimentary features suggest direct precipitation in the Permian Sea. But the high degree of insolubility of phosphate and its minor content in seawater makes this suggestion most difficult. Calcium phosphate does precipitate under the same conditions as calcium carbonate whence there may have been a conversion of the calcium carbonate to calcium phosphate. The only source of the vast quantity of phosphorus needed must be from organic activity for it is only in the proteins and hard parts of organisms that phosphorus accumulates to any natural extent.

Krauskopf describes the sinking of organic material from shallow marine seas to the deeper areas where sunlight does not penetrate and where the phosphorus from the organic material changes to dissolved phosphate. When upwelling occurs, especially where the continental shelf meets the continental slope (Fig. 9-3), the phosphate-rich water would result in increased organic activity, resulting in a concentration of phosphorite in the accumulating sediments reaching concentrations of more than 50 percent in locations that are now mined.

Direct inorganic precipitation of apatite may occur, according to Krauskopf, by the upwelling of cold water to shallow areas where greater organic activity is present. Of course, calcium carbonate would be expected to precipitate in much greater abundance except that phosphorite alone will precipitate in a very narrow pH zone lower than that for calcium carbonate. These needed critical conditions are unusual, which explains why such deposits are uncommon. The results of this upwelling and restricted environments are shown in Fig. 9-4. Some investigators (Youssef) believe that the North African phosphate beds have been formed on sea bottom depressions where decaying organisms produce ammonium phosphate and ammonia; the former reacts with calcium ions to produce calcium phosphate. It is probable that Krauskopf's explanations are more valid, as Youssef's more common conditions would have produced more phosphorite deposits of which we are not aware.

Sedimentary deposition has given rise to the great phosphate deposits of the world. Algerian, Tunisian, Moroccan, and now Egyptian deposits together yield the largest production in the world, and the marine phosphate beds of Idaho, Montana, Wyoming, and Utah are the most widespread. Sedi-

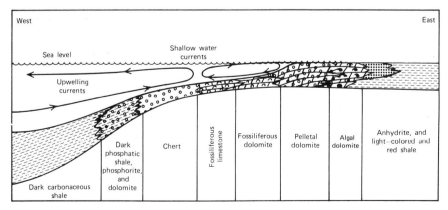

Figure 9-4 Lateral sequence of environments and sediments in the Phosphoria sea. (after Cheney and Sheldon, 1959.)

mentary phosphate beds also yield secondary ''pebble'' deposits. These and other deposits are described in Chapter 28.

THE CYCLE OF SULFUR

Sulfur is distributed in the earth's crust in the form of sulfates, sulfides, and native sulfur. It is a copious and important constituent of volcanic gases and magmatic emanations and is common in hot springs.

The sulfur of sedimentary deposits is derived from sulfates in rocks, from partial oxidation of hydrogen sulfide obtained from volcanic emanations, and anaerobic bacterial reduction of sulfates in solution.

Sulfur is deposited from sulfates and hydrogen sulfide in bodies of water where reducing conditions and anaerobic bacteria of the *Desulfovibrio desulfuricans* prevail. Sulfates are also reduced by *Clostridium nigrificans* to hydrogen sulfide, which, in turn oxidizes to sulfur and water. Hydrogen sulfide becomes so highly concentrated in some waters lacking in oxygen that it inhibits marine life.

The reduction of sulfates by sterile inorganic compounds could not be accomplished in experiments by Bastin and Allen-Crenshaw-Merwin. Experiments, however, by Thiel, Beyerinck, Van-Delden, Thode, Feeley, Kaplan, Nakai, Jensen, and others using microorganisms yielded hydrogen sulfide. Beyerinck and Van Delden's studies of Netherland problems with bacterial-produced hydrogen sulfide led to their estimate that 45 kilograms of sulfur per square meter would accumulate in 100 days from a zone of hydrogen sulfide water 10 meters deep in a lake. They were the first to recognize that the natural reduction of sulfates at low temperatures necessitates sulfate-reducing anaerobes.

Example of Deposits. Important sedimentary sulfur deposits occur near Knibyshev, Sukeievo, and Chekur, in Russia. Those deposits at Knibyshev illustrate the sedimentary process. The occurrences consist of thin gypsum beds with layers of pure sulfur, laminations of sulfur and calcite, or sulfur nodules in bituminous limestone. Celestite is an unusual associate. The sulfur is pure or bituminous; some is recrystallized and some is öolitic. A lagoonal sedimentary origin is indicated. These deposits are of Permian age, and those of Chekur occur in upper Tertiary lagoonal clay beds. Hydro-

gen sulfide springs are numerous near the deposits, probably as the result of reduction, at the present time, of sulfate by sulfate-reducing anaerobes.

The Sicilian deposits are considered by Dessau, Jensen, and Nakai, on the basis of sulfur isotope determinations and geology, to have been formed by anaerobic bacterial reduction of sulfate to hydrogen sulfide followed by exothermic oxidation to native sulfur.

An example of the $\delta^{34}S$ and $\delta^{13}C$ stable isotopic study of one of these deposits, Trabia, is shown diagramatically in Fig. 9-5. In the sulfur limestone portion of the stratigraphic section, the native sulfur is depleted in ^{34}S with $\delta^{34}S$ values of zero to 10 per mil. In contrast, the isotopic values of the original SO_4^{2-} exhibit $\delta^{34}S$ values of about 22 to 26 per mil. In addition, the $\delta^{16}C$ values of carbonates associated with native sulfur, vary from -18 to -30 per mil. These $\delta^{13}C$ values are unlike those of inorganic carbonates having $\delta^{13}C$ values from -3 to -12 per mil. These isotopic characteristics are strongly suggestive of bacteriogenic processes in the formation of these deposits. Similar isotopic results have been obtained on sulfur-bearing Gulf Coast salt domes, and sulfur deposits of Poland, namely Lubaczow and Tarnobrzeg-Stoszow, with similar bacteriogenic suggestions for the origin of sulfur.

The sulfur deposits of Sicily are an example of sedimentary sulfur. Sulfur-bearing formations lie in isolated basins up to 7.5 kilometers long and almost a kilometer across. The deposits consist of cellular limestone, interstratified with bituminous shale and gypsum, overlain by beds of marl, clay, limestone, and sandstone that are now folded and faulted. The sulfur is disseminated through the cellular limestone and occurs also as pure layers of sulfur up to a few centimeters in thickness. The sulfur content ranges from 12 to 50 percent and averages 26 percent. During the nineteenth century, Sicily provided all of Europe's sulfur. Since it held a monopoly, Sicily raised the price of sulfur, which led to the search for sulfur elsewhere. Eventually, the process for the recovery of sulfur from pyrite was developed in England; as a result, Sicily's sulfur industry never economically recovered.

Salt Domes. There is little doubt that certain mineral deposits have formed by bacteriogenic action. More than half a century ago, it was suggested that the millions of tons of sulfur in salt-dome cap rocks had been reduced to H_2S by sulfate-reducing bacteria and limited oxidation of H_2S to native sulfur had resulted in the formation of these deposits. No sul-

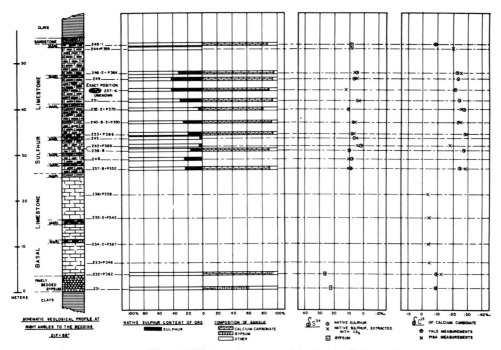

Figure 9-5 Stratigraphy and $\delta^{34}S$ and $\delta^{13}C$ isotopic results of the Trabia, Sicily, sulfur deposit.

fur exists in salt domes that contain no organic matter, specifically petroleum, associated with the cap rock. Furthermore, the cap rock environment had to be above depths of higher temperatures that would affect the viability of the anaerobes.

Although sulfur is not a metal, the association of native sulfur in the cap rock of those salt domes associated with petroleum provides an example of bacteriogenic sulfur and the use of stable isotopes in determining the genesis of the native sulfur. As the domes formed, soluble salt was leached from the cap rocks leaving behind the less soluble sulfates, predominately anhydrite. If the dome either cut oil-bearing strata or had oil rise along the up-tilted formations adjacent to the dome, an energy source was provided for sulfate-reducing bacteria. Soluble sulfate provided the oxygen for these anaerobes that reduced sulfate to H_2S. In the limited oxidation state of the environment, the H_2S was oxidized to $S°$. The hydrocarbon energy source allowed the bacteria to strip oxygen from the SO_4^{2-} and combine it with carbon from the hydrocarbons. The bacteria also yielded bacteriogenic CO_2, which reacted with the Ca of the anhydrite to produce "white calcite." The $\delta^{34}S$ and $\delta^{13}C$ isotope values confirm this genetic model including the fact that almost all sulfur-producing domes are within one or two kilometers of the surface where temperatures would permit the viability of the anaerobes. Sulfur,

therefore, in salt domes is formed by the bacteriogenic process.

Volcanic hydrogen sulfide that oxidized to sulfur is considered by Kato to be the source of sulfur of sulfur layers in a lake at Kozukie, Japan. Murzaiev, also attributed the sulfur of Kazbed, Gamur, and Kamchatka, in Russia, to volcanic sources, but they may prove to be of bacteriogenic origin.

Other examples of sedimentary sulfur deposits occur in Persia, Roumania, Croatia, Calicia, and Upper Silesia.

THE CYCLE OF COPPER

A syngenetic sedimentary origin is held for a growing number of copper deposits, and with decreasing dispute. The stability fields of copper minerals are shown on Eh-pH diagrams at 25°C in Figs. 9-6 and 9-7. It is, of course, known that copper also has a sedimentary cycle. Dissolving during oxidation, it moves to basins of fresh or salt water. It has been precipitated in sea muds as sulfides and native copper and has been thrown down by microorganisms. Oysters can absorb more than 5 mg of copper per oyster. It is a constituent of the manganese nodules that occur in vast portions of the ocean floors. Red-beds copper deposits are associated with the sandstone-type of uranium deposits of the Colorado Pla-

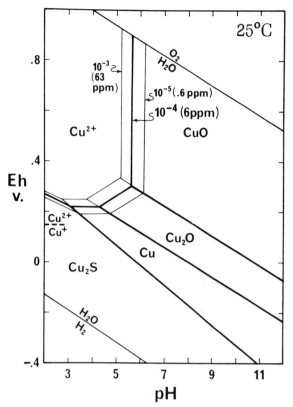

Figure 9-6 Ep-pH diagram for system Cu-O-H-S, 25°C, $\Sigma S = 10^{-4}$ m (From: Rose, Econ. Geol., v. 71, 1976.)

teau, both of which are nonmagmatic even though they form epigenetically.

A sedimentary origin has been maintained for the famed *Kupferschiefer* of Mansfeld, Germany, but an opposing school vigorously advocated a hydrothermal origin for it. These remarkable deposits, mined since A.D. 1150, have been the chief copper reserves of Germany. The Permian Kupferschiefer basin has an area of 58,500 square kilometers. Above a basal conglomerate is thin, black cupriferous shale, 1 meter thick, which Beyschlag describes as "one of the most remarkable products of the geologic ages." It is a shallow, marine, organic mud, full of land plants washed in from adjacent coasts. The syngenetic advocates believe that into this reducing bottom there were swept cupriferous solutions, probably sulfate, derived from the oxidation of distant lodes, perhaps even in the Harz Mountains. Alternatively, it is possible that the metals were derived form submarine hot springs. The metals were probably precipitated as iron-copper sulfide gels by bacteriogenic H_2S, as Schneiderhöhn believed. The chief minerals are copper sulfides, but there are also those of iron, lead, and zinc. Silver, nickel, cobalt, vanadium, molybde-

num, and other metals are also present. The arguments for a syngenetic origin are strong.

There is no adequate genetic explanation of the Kuperschiefer or any of the stratabound copper deposits that are free from controversy. Yet, stratiform metalliferous deposits underlain by red beds or other oxidized strata and overlain by evaporites account for about 30 percent of the world's copper production. Deposits included within this type are the Kuperschiefer, the Dzhezkazgan of Russia, possibly the Nonesuch, certainly the Zambian copper belt, and others.

Several investigators have suggested a so-called "Sabkha Process" during past recent years. Renfro has summarized these views in which he defines *sabkha* as the Arabic word for barren, uninhabitable, evapororite flats bordering partially landlocked seas. A coastal sabkha forms at the margin of a very large body of water where the groundwater table is very near the surface and the land has a flat surface but dips very slightly seaward. Groundwater so close to the surface evaporates leaving its dissolved salts at the surface, similar to a playa. In contrast, however, because of evaporative dis-

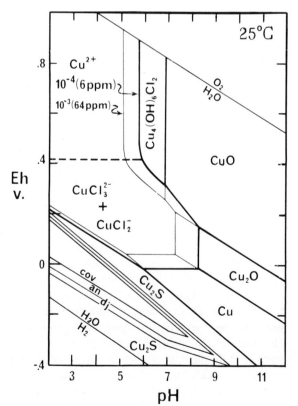

Figure 9-7 E-pH diagram for system Cu-O-H-S-Cl, 25°C, S = 10^{-4} m, $Cl^- = 0.5$ m as NaCl. Cov = covellite, an = anilite, dj = djurleite. (From: Rose, Econ. Geol., v. 71, 1976.)

charge of water through the sabkha, a subsurface hydraulic gradient toward the sabkha is created. Seawater, therefore, flows toward the sabkha. It may be augmented by storm action or even high tides. Renfro describes the resulting intertidal mud-flat that is covered with a leatherlike mat of sediment-entrapping, blue-green algae, which becomes a fetid, organic ooze containing sulfate-reducing anaerobes, such as *Desulfovibrio desulfuricans*.

If the oxygenated terrestrial water carries trace amounts of metals, they will be concentrated in the bacterial H_2S zone as their transporting fluids continue upward and are evaporated. Transgression and regression add refinements to the process allowing lateral buildup of the deposits and mineral zoning. Several of the stratabound syngenetic sulfide deposits contain evidence suggesting their formation in an ancient sabkha.

The red-beds copper deposits of the southwestern United States have often been spoken of as sedimentary in origin but not formed syngenetically. The evidence seems enticing that these deposits formed epigenetically in porous horizons where sulfate-reducing bacteria played a role in producing the H_2S reductant that precipitated the copper as sulfides from the transiting cupiferous solutions.

Red-beds copper deposits in the United States exist in Utah, Colorado, Oklahoma, Texas, New Mexico, and Pennsylvania. Nova Scotia and New Brunswick contain numerous deposits as does also Russia southwest of the Urals and Siberia at Ukokan. Deposits also exist in Bolivia at Corocoro. Few of these deposits, however, are actively being mined. Of the hundreds of deposits in the western United States, only the Nacimiento, New Mexico, deposit near Cuba was being mined, and this was being done by leaching. The Mongum, Oklahoma; and Corocoro, Bolivia, deposits are also in production.

A recent analysis by Rose indicates that red-beds copper deposits are introduced into their host rocks epigenetically by flowing subsurface waters approximately in equilibrium with hematite, quartz, feldspar, and mica at temperatures less than about 75°C. A reductant such as bacteriogenic H_2S, or a reducing environment brings about precipitation of "chalcocite" (digneite, djurleite, anilite, etc.), pyrite, chalcopyrite, native copper, covellite, and some bornite. In the western United States, sandstone-type uranium deposits are often intermingled with red-beds copper deposits such as at the Happy Jack Cu-U deposit in Utah. The reducing environmental

conditions for all of these minerals and deposits is located in the lower left-hand corner of Fig. 9-7.

Although the transport of hexavalent uranium occurs readily, Rose indicates that the solubility of copper in the system Cu-O-H-S, as illustrated in Fig. 9-6, is 6.3 ppm at pH = 5.67, 0.6 ppm at pH = 6.17, and decreases rapidly at higher pH. Because of the above low solubilities, the well-known association of evaporate beds with red-beds copper deposits, and the significance of complexing of chloride ions with metals, the possibility of increased solubility of low-temperature cuprous chloride complexes has been suggested by Rose. In contrast, Fig. 9-7 is the Eh-pH diagram for the system Cu-O-H-S-Cl at 25° C that shows a tenfold increase in the solubility of copper, at the expense of the Cu^{2+} and Cu fields, in comparision to the Cu-O-H-S system solubilities. Furthermore, the solubility is extended to pH values of 8.5 in contrast to the decreasing solubility of copper at pH values of 6 and higher as shown by Fig. 9-6. Rose has also determined the importance of increased chloride content on increased solubility of copper. In addition, his Eh-pH studies indicate the stability field of Cu_2S, which is most prevalent in red-beds deposits, to be in the reducing portion of the diagram.

Silver solubilities are not unlike copper solubilities and some native silver is associated with red-beds copper deposits. The Silver Reef mine in southern Utah, north of St. George, is most likely a red-beds silver deposit. Incidentally, this deposit also shows the characteristic wide spread in $\delta^{34}S$ values with depletion of ^{34}S, which is characteristic of sandstone-type uranium and red-beds copper deposits.

Although Kuperschiefer and the Oklahoma copper deposits are stratabound and presumably low-temperature deposits, they might not be considered to be red-beds copper deposits because of their syngenetic origin and much larger lateral extent. The former is suggested as a volcanic exhalative deposit while the latter is suggested to be bacteriogenic.

Widely mineralized copper that is stratigraphically controlled exists in the Late Proterozoic *Seal Lake* basin of Central Labrador. Studies by Brummer, Gandhi, Brown, and others of more than 250 native copper and copper sulfide showings have been done within tholeiitic basalts and associated gray shales (now metamorphosed to slates) of this area.

The occurrences resemble those of the Keweenawan copper province in host rock lithology, stratigraphic configuration, constituent metals, and mineralogic zoning. The tenors, however, rarely reach 1 percent copper; even with some silver grades of one ounce/ton, there is insufficient localized volume to form economic deposits.

The correlation of distinct litholigic units is recognized, and apparently all copper sulfide showings lie within a narrow stratigraphic zone within the Adeline Formation (Fig. 9-8). The occurrence shows similarities to the White Pine mineralization of the Nonesuch Shale with the exceptions that the grade is much lower and iron-rich sulfides are replaced by copper-rich sulfides at Seal Lake. On the other hand, mineralized showings exist over a length of more than 115 kilometers and about 42 kilometers in width. A sabkha process may be appropriate for this copper occurrence.

SEDIMENTARY URANIUM DEPOSITS

Sedimentary deposits of uranium are located worldwide even though they differ in age, mineralogy,

and occurrences. Presumably, the Blind River, Ontario; Witwatersrand, South Africa; Bahia, Brazil; and Alligator Rivers, Australia, deposits are all sedimentary deposits and presumably the uranium in these deposis is syngenetic in contrast to the epigenetic uranium in the sandstone-type of uranium deposit (Fig. 9-9). The reserves of all of these areas are large even though the grades are generally less than 0.2 percent. The exception is the Australian deposits that exhibit tenors as high as 2.3 percent U_3O_8.

Figure 9-10 shows the location of the Australian deposits, especially those in the Alligator River area of the Northern Territory. Even though Australia has some of the world's richest uranium deposits, only one small uranium deposit is producing yellowcake at the present time, namely, the Mary Kathleen mine, which has the lowest reserves and produced only 332 tons during three-quarters of 1975, and at a loss.

The known resources of U_3O_8 in Australia are slightly greater than those of the United States. But while it took the United States more than 36 years to discover 420,000 tons of U_3O_8, it took Australia

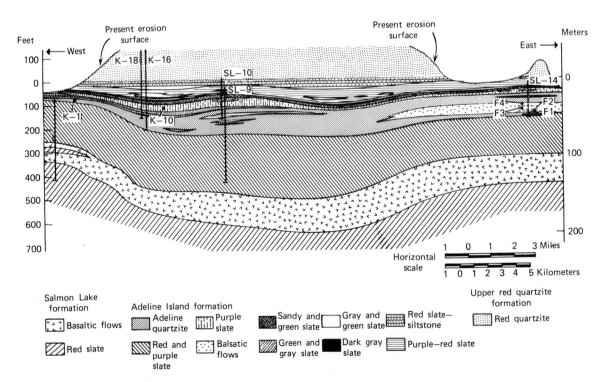

Figure 9-8 Illustration of the stratigraphy of the Adeline Island Formation along the northern limb of the Seal Lake syncline with respect to the base of the Upper Red Quartzite Formation restored to the horizontal position. Intrusive diabase omitted. Vertical bars with numbers represent selected diamond drill holes: K-1, Adeline Island Prospect (Brummer and Mann, 1961, plate 7); K-10, K-16, and K-18, Ellis prospect (ibid., p. 1375-1376); SL-9 and SL-10, Brian prospect; and SL-14, Brandy Lake Prospect. (From: Econ. Geol., v. 70, 1975.)

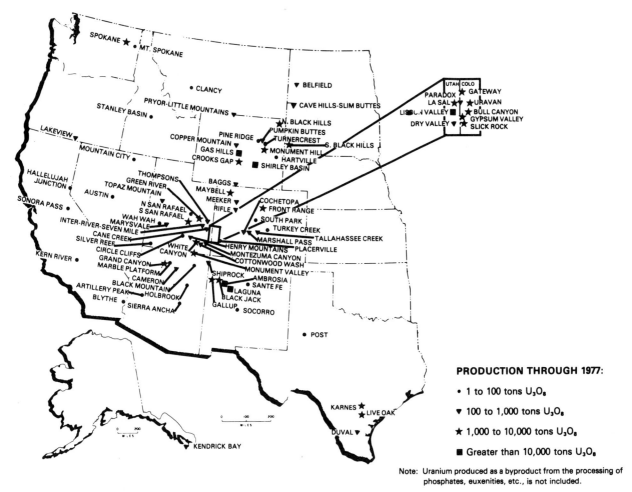

PRODUCTION THROUGH 1977:

- • 1 to 100 tons U_3O_8

- ▼ 100 to 1,000 tons U_3O_8

- ★ 1,000 to 10,000 tons U_3O_8

- ■ Greater than 10,000 tons U_3O_8

Note: Uranium produced as a byproduct from the processing of phosphates, euxenities, etc., is not included.

Figure 9-9 Index map showing the regions that contain the principal mining districts and mineral belts of uranium deposits in sandstone. Wyoming Teritary basins: 1. Powder River Basin; 2. Shirley Basin; 3. Gas Hills; 4. Crooks Gap. Colorado Plateau region: 5. Grants: 6. Monument Valley, White Canyon; 7. Big Indian: 8. Uravan; 9. Texas coastal region. (After Fischer, Econ. Geol. v. 69, p. 363.)

less than 7 years to discover a comparable amount. Australia had only 7 percent of the world's reserves in 1970; it had 20 percent of the world's reserves in 1977. Three-fourths of these reserves are in the Northern Territory where geologists have barely begun to scratch the surface in this highly forested, deep-weathered mantle, near-equitorial area.

The largest of these deposits is in Jabiluka, Northern Territory (Fig. 9-10), with reserves of 207,000 metric tons of U_3O_8, and incidentally it also contains a substantial amount of recoverable gold (about 15.3 grams per ton). The ore is contained in the Cahill Formation of Lower Proterozoic age. The formation comprises a carbonate-carbonaceous, pelitic lower unit according to Needham, and a more psammitic upper unit both containing amphibolite-grade schist as the major rocktype. The uranium is probably syngenetic having been precipi-

tated under reducing conditions in a lagoonal environment adjacent to an ancient granitic land mass. Concentration took place during regional metamorphism and deformation 1800 million years ago.

THE CARBONATE CYCLE

The solution, transportation, and deposition of calcium and magnesium carbonate give rise to deposits of commercial limestones, dolomite, and magnesite.

Limestones are of marine or freshwater origin, and magnesium may in part replace the calcium, giving dolomitic limestones even though dolomite is also of primary origin. Impurities of silica, clay, or sand are commonly present, as well as minor

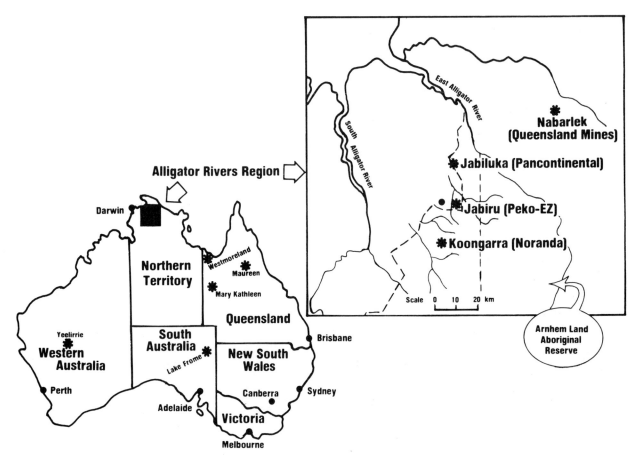

Figure 9-10 Location of Australian uranium deposits. (From: Mining Eng., v. 29, 1977.)

amounts of phosphate, iron, manganese, and carbonaceous material. The calcium is derived from the weathering of rocks and is transported to the sedimentary basins chiefly as the bicarbonate, in part as carbonate, and abundantly as sulfate.

Calcium carbonate is deposited at all Eh conditions but mostly at higher pH values (Fig 9-1). It is also deposited by organic and mechanical means. Carbon dioxide plays a dominant role in inorganic processes because the solution of the calcium carbonate in the sea is dependent upon it. If it escapes, calcium carbonate is precipitated. The amount of carbon dioxide in the sea depends upon the water temperature and the amount in the air, which is in equilibrium with that in the water. More carbon dioxide is held in cold water than in warm water. Warmed seawater loses carbon dioxide and, since it is practically saturated with calcium carbonate, precipitation ensues.

Organic deposition is brought about by algae, bacteria, corals, and Foraminifera. Calcium carbonate is also deposited by the photosynthesis of plants. Entire limestone beds may consist of Foraminifera, or nummulite shells, or of coral, or of larger shell forms (*coquina*).

Limestone may be formed mechanically through the deposition of comminuted shell matter and coral sand, which became cemented into compact limestone. Most limestones are deposited in shallow to moderately deep seawater, free from terrigenous sediments.

Marl, a friable, incoherent, pure limestone, is deposited in lakes from calcium carbonate supplied by streams or springs. It is common in glacial lakes because the glaciers that made the lakes supplied ground limestone and yielded cold water that held much carbon dioxide and, therefore, calcium carbonate in solution. The cold melt waters lost their carbon dioxide content in the warmer lake waters and calcium carbonate was precipitated. Low aquatic plants, such as *Chara,* however, probably deposit most of the marl.

Chalk, white earthy limestone, is deposited mainly in shallow waters and consists of a chemical precipitate of calcium carbonate and the minute shells of Foraminifera and other organisms.

Dolomite consists of the double carbonate of calcium and magnesium (54.35 percent $CaCO_3$ and 45.65 percent $MgCO_3$), but in dolomitic limestones the proportion of $MgCO_3$ is less than in dolomite. Most so-called dolomites are really dolomitic limestones; some of the magnesium may be replaced by iron or manganese. The three carbonates, with calcium, form isomorphous mixtures within certain limits. Many dolomites are not sedimentary but are epigenetic replacements of limestone. Under certain marine conditions the magnesium of seawater will react with $CaCO_3$ to form dolomite. It is known, however, to be abstracted from seawater into the shells of organisms, and some coral reefs consist in part of dolomite. The staining of polished surfaces shows that some dolomite is intimately intercalated and contemporaneous with limestone.

Magnesite, the carbonate of magnesium, is an important industrial mineral. The sedimentary variety occurs in association with salt and gypsum, or shales and limestones, and is formed by deposition of magnesium carbonate, along with some calcium carbonate from concentrated waters of saline lakes. Apparently, the deposition has been brought about by chemical precipitation with subsequent dehydration. Presumably the magnesium was transported as magnesium sulphate by surface or underground waters and reacted with sodium carbonate to yield insoluble hydromagnesite, which accumulated as a relatively pure precipitate, and sodium sulphate, which with other soluble salts remained in solution. Examples of sedimentary deposits occur in Kern County, California, Nevada, Idaho, British Columbia, and Germany.

Varieties of sedimentary carbonate rocks of economic interest are listed as follows.

Kind	Use	Reference Chapter
Building limestones	Building and structural	**25**[a]
Cement limestones	Hydraulic cements	**25**
Siliceous limestones	Hydraulic limes	**25**
Silico-aluminous limestones	Natural cements	**25**
Limestone	Flux, fertilizer, chemicals	**26**, 28, 29
Lime rock	Quicklime	**25**
Chalk	Cements, powders, crayons, fertilizer	25
Marl	Cement, fertilizer	**25**, 29
Lithographic limestone	Fine engraving	—
Dolomite	Cement, refractory	25, 29
Magnesite	Cement, refractory, chemical	29, **12**, 28

[a] Bold-face indicates the chapter of chief description.

THE CLAY CYCLE

The clay cycle differs from the preceding sedimentary cycles in that the constituents of clay are transported not in solution but in suspension, and their deposition is mechanical rather than by chemical or organic processes.

As shown in Chapter 13 the chemical decomposition of aluminous rocks yields clay minerals. These may become deposited in situ to form residual clay deposits or be transported and deposited as sediments. The sedimentary clays may be divided into marine, estuarine, lake, swamp, and stream clays.

Marine clays settle from mechanically transported suspensions in quiet water some distance offshore. The beds may be of great extent, of considerable thickness, and of fairly uniform composition although lateral variation is to be expected because different streams supply the materials; vertical variations are similarly introduced. The beds are finely laminated. Marine clays are widely distributed in Paleozoic and Mesozoic formations.

Estuarine clays, since they are laid down in shallow ocean arms, are restricted in extent and commonly contain many sandy laminations that increase toward the source of supply. Marsh products are also interlaminated. Examples occur along the New Jersey side of the lower Hudson River and in the Chesapeake Bay region.

Lake clays occur in restricted basins and typically alternate with beds of sand. They are common in glacial lake beds. Consequently, many of them are varved, that is, made up of thin bands of alternate coarse and fine material, each pair of bands representing seasonal accumulations of a year's growth.

Swamp clays or *fire clays* may underlie coal seams, and upright stumps of ancient trees are found in them. The beds are small, lens-shaped, and exhibit little lamination. The clays are highly plastic, generally refractory, and relatively pure. Consequently, they are eagerly sought for the host of uses to which they are put. They are thought to have originated from suspended matter carried by low-gradient streams into coal swamps whose early marginal fringe of vegetation filtered out the coarser sediment, allowing only the finest to reach and settle in the interior of the basin. The organic acids present have been held by some investigators to purify the clay sediment.

Stream clays are deposited in protected places on flood plains during periods of overflow. Consequently, the deposits are pockety and grade laterally into sandy material. The pockets, however, yield a fine plastic clay, but different pockets vary greatly in composition. Streams may also deposit delta clays in isolated basins on deltas. Clays of stream origin are widely distributed.

Selected References on Sedimentation

Solution, Transportation, and Deposition of Iron and Silica. E. S. Moore and J. E. Maynard. 1929. Econ. Geol., v. 24, p. 272-303, 365-402, *Theoretical treatment of natural solution and deposition of iron that forms ore beds.*

Wabana Iron Ore Deposits of Newfoundland. 1931. A. O. Hayes. Econ. Geol., v. 26, p. 44-64. Occurrence, mineralogy, and origin.

Potash Reserves of the United States. 1946. S. H. Dolbear. American Potash Institute, Washington, D.C.

The effect of cuprous chloride complexes in the origin of Red bed copper and related deposits., 1976, Rose, Econ. Geol., v. 71, p. 1036-1048.

Sedimentary Manganese Ores. 1964. I. M. Varentsov. American Elsevier Publishing Co., Inc., New York. *Principal types of ores and their geochemistry.*

Genesis of iron ores of the Pretoria series, South Africa. 1965. H. Schweigart. Econ. Geol., v. 60, p. 269-298.

A Sedimentary Origin by Submarine Exhalations. 1965.

M. I. Youssef. Econ. Geol., v. 60, p. 590-600. *Genesis of bedded phosphates.*

Transportation and precipitation of uranium and vanadium at low temperatures. 1962. P. B. Hostetler and R. M. Garrels. Econ. Geol. v. 57, p. 137-167.

The Chemistry and Physics of Clays, 3rd ed. 1959. A. B. Searle and Rex W. Grimshaw. Interstate Publishers, New York, Chap. 1. *Formation of clays.*

Geology and isotope studies of Sicilian sulfur deposits. 1962. G. Dessau, M. L. Jensen, and N. Nakai. Econ. Geol., v. 57, p. 410-438. *A biogenic origin of the sulfur in these deposits.*

Correlation between the uranium content of marine phosphate and other rock constituents. 1962. F. Habashi. Econ. Geol. v. 57, p. 1081-1084. *Formation of marine phosphates.*

Principles of Geochemistry. 1958. Brian Mason. John Wiley & Sons, Inc., New York, chap. 6, Sedimentation and sedimentary rocks. *Covers iron, uranium, phosphate, clays, and carbonates. Relation of pH and oxidation potential to sedimentary iron mineral formation.*

Stratigraphy and Sedimentation, 2nd ed. 1963. W. C. Krumbein and L. L. Sloss. W. H. Freeman and Company, San Francisco, chaps. 5, 6, and 13.

Submarine Geology, 2nd ed. 1963. F. P. Shepard. Harper & Row, Publishers, New York, chap. 16. *The Sea,* Vol. 2. 1963. M. N. Hill, ed. Interscience Publishers, John Wiley & Sons, Inc., New York, chap. 2.

The Geology of Uranium. 1958. V. G. Urana. (Translated from Russian.) Consultant's Bureau, Inc., New York, chaps. 3 and 4.

Origin of Uranium Deposits. 1955. V. E. McKelvey, et al. Econ. Geol. 50th Anniv. Vol.

Weathering. 1969. C. Dollier. American Elsevier Publishing Co. New York. *General text on processes of weathering.*

Uranium-vanadiaum deposits of the Thompson area, Utah. 1952. W. Lee Stokes. Utah Geol. Survey Bull. 46.

Introduction to Geochemistry. 1967. K. B. Krauskopf. McGraw-Hill Book Company, p. 721. *A comprehensive text on geochemistry.*

Sedimentary Rocks, 3rd ed. 1975. F. J. Pettijohn. Harper & Row, Publishers, New York, chaps. 9, 11, and 13. *Limestones, dolomites, and phosphate.*

Sedimentary facies of iron-formation. 1954. H. L. James. Econ. Geol., v. 49, p. 235-293. *Excellent treatise of sedimentary oxide, carbonates, and sulfides.*

Manganese ore. 1962. J. A. Dunn; rev. by S. Narayanaswamy. Geol. Survey India Bull. 20. *Comprehensive survey of manganese of India and the world.*

General Reference 1. 1974. A. R. Renfro, *Genesis of evaporite-associated stratiform metalliferous deposits—A sabkha process,* v. 69, p. 33-45.

10

BACTERIOGENIC DEPOSITS

"What an unfortunately short time you were permitted to stay in many places yet how much you managed to see."

J. D. Dana letter to Charles Darwin, 1849

The recent renovation of the two genetic hypotheses of *Bacteriogenic and Submarine Exhalative Volcanic or Endogenic processes* may, in many specific deposits, involve both processes. The former enigmatic source of the metals in these deposits seems to be resolved by the transport and escape of submarine volcanic or submarine hydrothermal fluids to the surface or near surface of the sea bottom where they may spread laterally for some distance. Their precipitation may occur as concretionary growth as with manganese modules, or by bacteriogenic production of H_2S that reacts as a most effective reductant of the metal solutions and also provides the source of the sulfur to form the insoluble sulfides.

It is difficult, therefore, to separate some of these deposits from the two processes and, in fact, to even exclude them from a hydrothermal process. Nevertheless, a few clearcut genetic classifications have been made, but the large number of remaining deposits could be considered in either Chapter 10 or 11.

Sulfur in *salt domes* is a major source of native sulfur. Production figures, reserves, and extractive processes are discussed in Part IV. The origin of the sulfur in the domes is undoubtedly from the sulfur in the gypsum and anhydrite of the cap rocks. The reduction of these sulfates to native sulfur was suggested as a bacteriogenic process as long ago as 1915. Stable carbon and sulfur isotopic studies by Thode and Feely, however, have provided convincing evidence of the sulfate-reducing bacterial role. Additional isotopic studies of the salt domes of Poland and the Sicilian sulfur deposits also indicate that their sulfur is of a bacteriogenic source.

Pyrite crystals in marine shales are so widespread that only a syngenetic or diagenetic origin is plausible, and it is reasonable to suggest that bacteriogenic sulfate reducers provided the H_2S reductant that reacted with indigenous iron to form the hydrotroilite that ultimately formed pyrite.

Copper, vanadium, uranium, and even silver (Silver Reef mine, Utah) in sandstone-type uranium deposits have formed by being concentrated in reducing environments which may have been produced inorganically or by sulfate-reducing anaerobes liberating the H_2S reductant. These deposits are not syngenetic but epigenetic. Although many bacteriogenic processes occur in sediments during diagenesis, it is important to realize that some bacteriogenic processes are epigenetic.

Sulfate-reducing anaerobic bacteria of the *Desulfovibrio desulfuricans* and *Clostridium nigrificans* genera and species reduce SO_4^{2-} to H_2S using organic material or a hydrogen acceptor as an energy source. In so doing, isotopic fractionation occurs with the $\delta^{34}S$ value of the produced H_2S being enriched in ^{32}S. The isotopic fractionation is also affected by pH and the partial pressure of oxygen. Under partial anaerobic conditions, the H_2S oxidizes to native sulfur and the CO_2 liberated by the bacteria reacts with the Ca^{2+}, released during reduction of SO_4^{2-} to form rather pure white $CaCO_3$. The $\delta^{13}C$ values are also indicative of a bacteriogenic origin of the $CaCO_3$. Only those salt domes that are near enough to the surface and at lower temperatures that allow the viability of the bacteria, and only those domes that have organic material, generally petroleum, associated with the cap rock have become sulfur-rich sources.

With increasing frequency, a bacteriogenic genesis is being suggested for many deposits.

Nairne Pyritic Deposit. This deposit is located at the base of the Cambrian succession in the Eastern portion of the Mount Lofty Ranges of South Australia (Fig. 10-1).

The pyritic horizons, of which five have been recognized, occur within the graywackes and siltstones several thousand feet above the base. The principal sulfides are pyrite and pyrrhotite, which are distributed along bedding planes and persist for such distances as to suggest strongly that they are of syngenetic origin. Within the formation as a whole, sulfides have been traced along the strike for about 100 kilometers but it is only the two lower horizons that have almost complete continuity.

The two major sulfide-bearing horizons are separated by about 450 meters of graywacke and quartzite; it is in the lower of these two horizons that the Nairne pyrite open-cut mining project has been developed.

The lower horizon consists of three sulfide-rich ore beds that are separated by two sulfide-poor waste beds. This assemblage of beds is confined within wall rocks that are virtually free from sulfides.

The composition and metamorphic modification of the pyritic sediments has been described by Skinner, who suggests that the sulfides were originally precipitated in a hydrate form and not as detrital grains derived from some adjacent source. He indicated that sulfides entering the sedimentary basin in company with detrital quartz, feldspars, and micas would have built-up rythmic, heavy mineral accumulations within the graywackes. It may be that bacteriogenic H_2S reacted with syngenetic iron oxide to produce the pyrite. Some question organic matter as an energy source because amorphous carbon is not found in the deposit. Nevertheless, appreciable amounts of very fine graphite have been observed by Whittle in polished sections of the sulfide-bearing graywackes. Further evidence of carbonaceous material is manifest in experimental ore dressing, where graphite is concentrated in flotation cells. It is probable that the small amount of carbonaceous matter in the original sediments has been transformed to graphite under the influence of metamorphism. It is, therefore, inferred that the Nairne pyrite deposit suggests a bacteriogenic origin for the sulfur of this deposit. Such a conclusion is not all inconsistent with the geological and mineralogical factors.

Mount Isa, Australia. This deposit, discovered in 1923, has been one of the world's largest lead-zinc producers and has been recently assigned a syngenetic, possibly bacteriogenic origin. The deposit is also a producer of copper for which some geologists

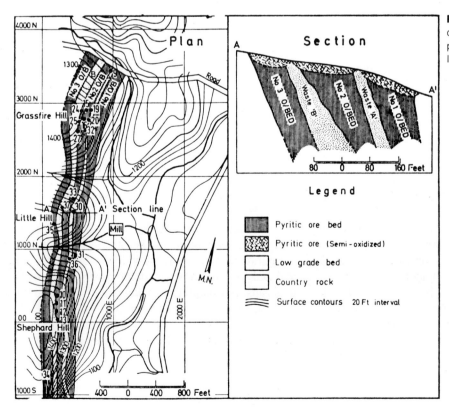

Figure 10-1 Geologic map and cross section of Nairne, Australia, pyrite deposit showing sample location sites. (From Econ. Geol.)

still clutch to a magmatic hydrothermal origin even if the lead-zinc mineralization may be syngenetic.

The deposit is named from a regional term applied to a sequence of carbonate volcanic shales, bedded carbonates, and siltstones that strike north-south along an outcrop that extends for 40 km across a width of 4 km.

A diagrammatic, composite cross section of the deposits is shown in Fig. 10-2 which exhibits the narrow width of sequential beds that are favorably mineralized. Regional metamorphism has modified the original deposits that are obviously premetamorphic. A not unusual, polished hand specimen from the deposit is shown in Fig. 10-3, where the galena is more readily deformed than the thin zones of pyrite. Although the genesis of a mineral deposit is not determined by one hand specimen, the texture of this polished rock is certainly indicative of a syngenetic origin of the sulfides.

Bacteriogenic deposits obviously dovetail with the sedimentary and submarine exhalative deposits. With the large amount of volcanic rock associated with the marine shale of the Mount Isa group, it is suggestive that the metal composed in the Mount Isa deposit may have been derived from a submarine exhaltive-volcanic source. Such deposits, especially syngenetic deposits, have large strike or lateral extent. It is not surprising, therefore, that exploration for similar deposits should have continued in the area approximately along the strike resulting in the discovery of the MacArthur River deposits and even more recent discoveries.

Mount Isa, Australia, and the associated MacArthur River deposits were thought to be magmatic hydrothermal deposits for some time after Mount Isa's discovery in 1923, with the MacArthur River deposit being developed much later. Increased studies, however, favor a syngenetic origin and the associated volcanic ignimbrites in the formations lend evidence to a submarine exhalative or volcanic

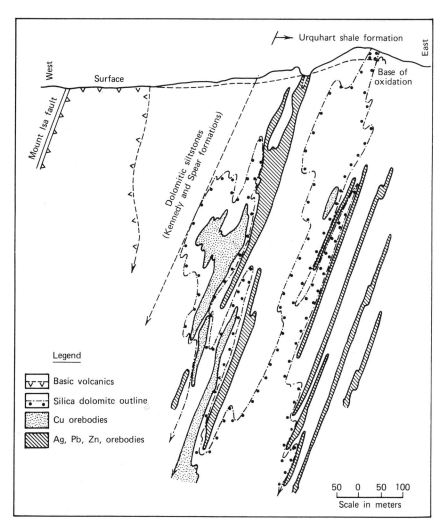

Figure 10-2 Composite cross section at Mount Isa northern mine area, Australia. (Compiled by Mount Isa Mines Limited Geology Dept.)

Figure 10-3 Polished slab of Mount Isa rock specimen showing plastic deformation of galena bands and undulations of sedimentary pyrite layers

process in the highly conformable but metamorphosed formations (Fig. 10-2).

Other deposits that might be included in this process are discussed in Part II. They include such deposits as White Pine, Michigan, and the sandstone-type uranium deposits. These latter deposits are certainly stratabound and occur as stratiform deposits but instead of being of syngenetic origin, they are decidedly epigenetic, formed by the reduction of transiting solutions carrying U^{+6} meteoric or connate water.

In the submarine exhalative deposit case, heated solvent solutions could leach metals from the volcanic rocks that form in the ocean bottom with precipitation occurring as the temperature of the solution decreases or as the chemical environment changes, even in some cases by bacteriogenic reduction. Of course, the metals might be derived from the deeper magma from which the volcanics were derived. Some such formed deposits are Kidd Creek, Ontario; stratabound sulfide deposits in the Skellefte district of northern Sweden; Buchans,

Newfoundland; Kuroko deposits of Japan; the Flin Flon massive copper-zinc-rich deposits, Saskatchewan and Manitoba; and a growing number of other deposits as geologists reconsider genetic models in contrast to volcanic, exhalative submarine-derived deposits.

Oklahoma Red-Beds Copper Deposits. The genesis of the rather unique, comparatively thin red-beds copper occurrences in Oklahoma has been a source of controversy for some, but of interest to those concerned with the exhalative and/or bacteriogenic origin of heavy metal deposits. Of particular interest are the syngenetic features of the metallic minerals "such as cross-bedding and microfaults in minute detail" as described by White and Wright. Furthermore, a few $\delta^{34}S$ values of chalcocite samples and two gypsum samples strongly corroborate the bacteriogenic origin of those sulfides that have not been oxidized.

The deposits in the Pauli district, Oklahoma, are

formed as copper-silver solution fronts in sandstone paleochannels in the Permian Wellington Formation. These ore fronts are similar to sandstone-type uranium and red-beds roll-front deposits. Ore-grade copper-silver is not continuous in the Wellington Formation solution front even though individual fronts have been traced as far as 3 kilometers. The average tenor is approximately 0.75 percent Cu and 6 grams per ton Ag. Selected individual samples assay, of course, much higher.

The genesis of these deposits is thought to be almost identical to sandstone-type uranium deposits, where oxygenated water flushing through permeable sediments deposits ore minerals at an oxidation-reduction interface in the Wellington formation. An external source of the ore metals is required because of the barren nature of the Wellington Formation. The source may be shaley red-beds surrounding the sandstone paleochannels, especially as there is evidence of volcanic constituents in these red-beds.

Lockwood has completed a geochemical and petrological study of the deposits of *Creta*, Jackson County, and Mangum, Greer County, Oklahoma, and has included a section of $\delta^{34}S$ analyses that exhibit the variable enrichment in ^{32}S.

The deposits occur in marine shales in the upper part of the Permian Flowerpot formation in the El Reno Group. The interbedded gypsum-and halite-bearing horizons indicate a brackish water lagoonal or shallow marine environment.

The $\delta^{34}S$ composition of marine water during Permian time was about +11 permil, which is not abnormally different from the two gypsum samples analyzed that gave $\delta^{34}S$ values of +4.7 and +6.3 permil. The characteristic enrichment in ^{32}S of chalcocite samples is typical of other red-beds copper deposits and sandstone-type uranium deposits:

$(\delta^{34}S‰)$ of samples

-34.6
-19.2
$+0.9$
-26.8
-19.1
-29.7
-30.3

It is reasonable to assume that the deposits are bacteriogenic-diagenetic in origin. The source of the copper is conjectural but may have been derived from an anomalous copper-rich upland, carried to the depositional site while absorbed on clay minerals, or derived from hydrothermal fluids provided by submarine volcanic exhalations, or from abnormal concentrations in seawater. It is significant, however, that the sulfides rarely exhibit pseudomorphs after pyrite but reveal their characteristic primary chalcocite crystal forms.

THE CYCLE OF URANIUM AND VANADIUM

Uranium occurs in many types of deposits many of which definitely belong to a sedimentary cycle. Possibly all sandstone-type uranium deposits (Fig. 9-4) are epigenetic and deposited from nonmagmatic solutions.

Uranium is widely distributed in the earth's crust amounting to an average of 3 to 4 ppm, and ranging from 5 to 40 ppm in some felsic igneous rocks and volcanic tuffs. During weathering it is released from rocks or former deposits, and being readily soluble, is transported in solution, as the uranyl ion$(UO_2)^{2+}$. When soluble uranium-bearing solutions enter reducing environments, precipitation occurs by reduction of UO_2^{2+} to UO_2, a valence change of U^{6+} to U^{4+}.

Some coals, formed in reducing environments, contain up to 6.4 percent uranium. In lakes, uranium is deposited in muds that form shales; in broad marine embayments where black carbonaceous shales accumulate, the uranium content is unusually high. Practically all such shales contain uranium; the Chattanooga shale contains up to 0.035 percent uranium; the black alum shale (Kolm) of Sweden contains up to 0.5 percent uranium. Marine phosphate beds are particularly rich in uranium and range up to 0.03 percent. The marine Phosphoria formation of the Western United States generally contains 0.01 and 0.02 percent uranium but reaches 0.065 percent. Where shales and phosphate rock occur together, the uranium appears to show a preference for the phosphate rock because of the ease with which phosphorus replaces calcium.

The uranium in the black bituminous shales appears to be definitely syngenetically sedimentary, as is also that in phosphate beds. The marine deposits are quite low grade. The extensive Precambrian sedimentary uranium deposits of the Alligator Rivers of Queensland Australia are discussed in Chapter 9.

SANDSTONE-TYPE URANIUM DEPOSITS AND RED-BEDS COPPER DEPOSITS

The genesis of the sandstone-type uranium deposits of the Colorado Plateau, Wyoming basins, New Mexico, and Texas (Fig. 9-5) has been an enigma to economic geologists for some time and may still be. Possibly, the most favored theory of today was suggested as early as 1920 and summarized by Warren and Granger: The deposits were formed by circulating waters that collected the metallic constituents from dispersed sources within the sediments and precipitated them in favorable environments (Figs. 10-4, 10-5, and 10-6). Prior to, and after World War II, the magmatic hydrothermal view gained recognition but weakened as geological investigators noted the close relationship of these deposits to sedimentary features that suggested erroneously a syngenetic or penecontemporaneous origin with the enclosing sediments although the primary features did have some control on the deposits, almost all of which are epigenetic.

Uranium was mined from the Colorado Plateau as early as 1896 when uranium oxide was used in the ceramic industry to produce an intense red color on ceramic ware when baked in kilns. The supply for this purpose dwindled from 1942 to 1945, during the years of the Manhattan Project that culminated with the advent of atomic bombs and a new form of energy. Interestingly enough, the source of uranium for the Los Alamos and Hiroshima atomic devices came form Shinkolobwe, Zaïre, ore that had been stockpiled in the United States prior to World War II.

With the formation of the Atomic Energy Commission and the increased price incentive for U_3O_8 in 1948, the exploration for uranium "ballooned" and thousands of penny share companies formed. During some days, more shares (penny shares of course) passed through the Salt Lake City Stock Exchange than were traded on the New York ex-

change! A large number of government geologists, many from the U.S. Geological Survey, combined with the AEC geologists to produce quadrangle maps of "uranium country."

New discoveries were made in New Mexico, Wyoming, Utah, Colorado, and Texas. Major discoveries were also made in Blind River, Ontario; and in the Witwatersrand banket ore, South Africa.

Following World War II, numerous uranium deposits were discovered as the result of an accelerated exploration program to establish adequate reserves of the fissionable material in the United States. Because of their large size, partial discordance, and common occurrence in groups, apparently related to structural features and igneous intrusives, as in the Big Indian Wash district of Utah, the syngenetic hypothesis was shunned again in favor of the magmatic hydrothermal origin. Nevertheless, the continued search for uranium in the western United States resulted in the discovery of many sandstone-type deposits that were unrelated to igneous activity. As the increased geologic evidence now seems to preclude a magmatic hydrothermal origin for these deposits, mineralization is attributed to uranium-bearing ground waters moving along permeable aquifers.

These deposits are localized in ferric-oxide-cemented Mesozoic and Cenozoic sediments. The cement has been converted to ferrous sulfide, where the uranium or copper mineralization occurs. Low-temperature hydrothermal solutions are credited with forming these deposits before, and during, early Tertiary time when more than 3 kilometers of Cretaceous sediments covered the deposits. Even with an average geothermal gradient of 30°C per kilometer, the temperature of the solutions could have exceeded 100°C.

Beginning in 1957 and since, thousands of $\delta^{34}S$ analyses have been made on sulfur-bearing constituents in these deposits. Austin and King have catalogued many of these analyses for the U.S. Atomic

Figure 10-4 Uranium country. The Morrison formation near Thompson, Grand County, Utah. In the foreground, are sandstone cliffs representing fillings of river channels. The upper part of the formation is in the smooth slopes along the upper left skyline. This area is famous for uranium deposits, petrified dinosaur bones, and other fossils. (Photograph by William Lee Stokes.)

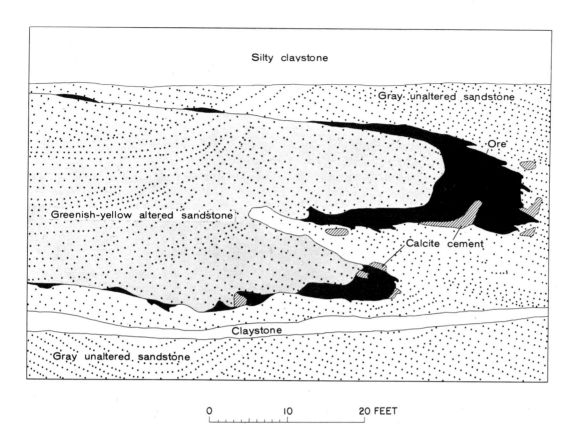

Silty claystone

Gray unaltered sandstone

Ore

Greenish-yellow altered sandstone

Calcite cement

Claystone

Gray unaltered sandstone

0   10   20 FEET

Figure 10-5 Section normal to west edge of altered sandstone tongue, showing the relation of ore to calcite cement to altered sand. (From J. Ridge, ed., *Ore Deposit of the United States*, 1968.)

Figure 10-6 Uranium ore roll and elongate roll, ore mined from the lower portion of the roll and erosion stripped ore from the upper portion. Sand lens channel in Yellowcat district, Grand County, Utah. (Photograph by William Lee Stokes.)

175

Energy Commission (now DOE Department of Energy). These analyses were done by Field, Cheney, Nakai, Jensen, and others.

The highly variable enrichment of ^{32}S in the sulfides of these deposits is strongly suggestive of sulfur derived from sulfate-reducing bacteria. It was originally suggested, therefore, that H_2S, derived from sulfate-reducing bacteria, provided the reducing environment for the reduction and concentration of soluble, hexavalent uranium to the relatively insoluble tetravalent form. The sulfate source was most likely connate water; the energy source for the bacteria was provided by the pockets of former and some still existing cellulose-bearing organic material in these nonmarine formations. The bleached effect of the original red sediments where uranium and red-beds copper mineralization has occurred is the result of the reduction of indigenous ferric oxide to ferrous sulfide, presumably by the H_2S according to the following reaction:

$$9H_2S + 2Fe_2O_3 + 3O_2 \rightarrow 4FeS_2 + SO_4^{2-} + 2H^+ + 8H_2O$$

Red-beds copper deposits have a similar origin with the exception that cupriferous solutions are as prevalent or more prevalent than the uraniferous solutions. Bacteriogenic H_2S is also efficaceous in

the reduction of copper to a sulfide. In some deposits, for example, Happy Jack, Utah, the deposit is both a uranium and a copper deposit, as are similar red-beds copper deposits in Pennsylvania, Nova Scotia, and New Brunswick. Possibly, these deposits should be considered as bacteriogenic and epigenetic but still derived from low-temperature hydrothermal solutions.

The extreme enrichment of ^{32}S in these deposits is quite comparable to the $\delta^{34}S$ results obtained by Kaplan, Emery, and Rittenberg from their study of biogenic sulfides in recent marine sediments in which ^{32}S was highly enriched.

Based upon their study of uranium ore-rolls (Figs. 10-5 and 10-6) and their spatial correlation with $\delta^{34}S$ results (Fig. 10-7) King and Austin noted that, in some ore-rolls, there are appreciable changes in the $\delta^{34}S$ values at the roll interface and at the zone of high uranium content. The idealized correlation is of highly enriched ^{32}S in the oxidized portion of the roll with an abrupt increase in ^{34}S at the interface that gradually approaches zero permil in correlation with the decreasing U_3O_8 content. Not all rolls exhibit the example of Fig. 10-7, but the generalization is apparent in most rolls.

Granger and Warren have suggested that the isotopic variations of sulfur might better be explained

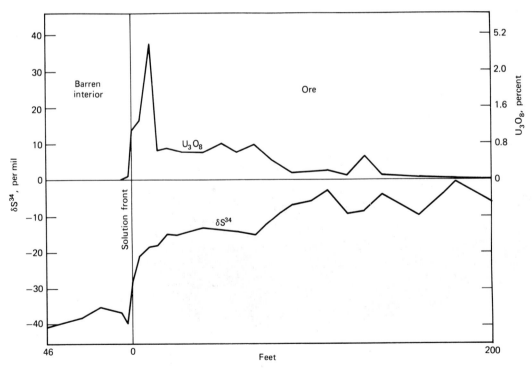

Figure 10-7 Sulfur isotope fractionation and uranium distribution across a roll front, Gas Hills, Wyoming. (From King and Austin.)

by a nonbacteriological process, by postulating a chemical origin with the formation of thiosulfates in which sulfur undergoes isotopic fractionation during the leaching in the oxidation zone and reduction at the interface of the roll. This is a totally different genetic process from the bacteriogenic process, although the initial sulfides may still be of bacteriogenic origin. The geologic evidence is inadequate to determine which of the two hypotheses is acceptable, possibly both, although the latter non-bacteriogenic process is gaining proponents. The complexities and different processes and sources of *sulfur* are shown diagramatically in Fig. 10-8. Variations in H_2S, SO_4^{2-}, and O_2 are also shown at various depths in the idealized body of water.

Other Bacteriogenic Processes. Harrar has long advocated a prominent role for bacteria in *iron* deposition. He divides the iron bacteria into three main groups: (1) those that precipitate ferric hydroxide from ferrous bicarbonate solutions, such as *Spirophyllum ferrugineum* and *Gallionella ferrugineum*, which thrive best in the absence of organic matter and the presence of carbon dioxide; (2) those that deposit ferric hydroxide from organic or inorganic iron solutions, such as *Leptothrix ochracea* and *Cladothrix dichotoma;* these need organic matter; (3) those that attack organic iron salts, using the organic radical for food, producing ferric hydroxide. Moore and Maynard also proved that ferric hydroxide is precipitated by soild bacteria of natural waters.

The chief *manganese*-precipitating bacteria are *Crenothrix polyspora, Leptothrix ochracea, Cladothrix,* and *Clanothrix.* The first two forms were found in abundance by C. Zapffe in manganese precipitates at Brainerd, Minnesota, where the water mains, supplied by well water, were being rapidly choked by a precipitate of oxides of manganese and iron, the manganese predominating. In the well water, iron predominated and had been largely removed by aeration, which, however, did not remove the manganese. After many experiments, Zapffe built a simple, effective, manganese-removal plant, in which the manganese was precipitated from the aerated water by using pyrolusite (MnO_2) as a catalyst. Deposition of manganese dioxide in the water mains was started by bacterial action, and once started it was accelerated by the earlier deposition. By catalysis, the manganese bicarbonate in solution is broken up into carbon dioxide and manganous hydroxide, and the latter is converted into manganic hydroxide and manganese dioxide.

Phosphorus is also precipitated by bacteria. Most of it, however, is probably removed from solution by vertebrates of shellfish, and deposition occurs best under reducing conditions in shallow water. Most investigators agree that the concentration of the phosphate of sedimentary beds has taken place through organic agenices. Although this is a minor process in contrast to the inorganic process of the formation of large phosphate deposits.

Bacteria are also effective in the deposition of sulfur by sulfate production and by generation of H_2S that may result in the deposition of sulfides of bacteriogenic origin.

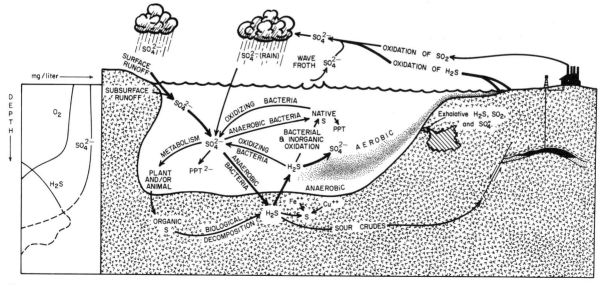

Figure 10-8 Sulfur cycle in an idealized body of water.

Oxidizing bacteria are utilized extensively in the large-scale oxidation of waste dumps, which generally contain about 0.4 percent copper. At Kennecott's Bingham and Chino mines, bacteria produce sulfuric acid, and thus supply more solvent for copper and other metals. They are utilized extensively by the Kennecott Copper Corp. in their large-scale mine dump leaching at their Utah and Chino mines, and by Anaconda on their large waste Butte, Montana, dumps (Fig. 1-24).

Many other sulfide deposits have been suggested as having a syngenetic-bacteriogenic origin with the source of the metals being controversial, or by submarine hot springs, or by volcanism. Whether bacteriogenic processes were needed for the reduction and supplying the source of sulfur by reduction of marine SO_4^{2-} is also difficult to ascertain. Some such deposits are the Zambian copper belt; Broken Hill, Australia; Sullivan mine, British Columbia; Kidd Creek, Ontario; massive sulfide deposits; and most certainly the Kuperschiefer and Mansfield deposits of Europe.

Biological processes are becoming more evident as a means by which certain mineral deposits have formed. Stratabound deposits that are thin but extend for miles have always presented problems of explaining such as an epigenetic processes. Such deposits at the *Nairne pyrite* deposits, *Mount Isa,* Australia; *Zambian copper belt,* Zambia and Zaire; *White Pine,* Michigan; *Oklahoma copper beds; Red-beds* copper and *sandstone-type* uranium deposits of the United State and Russia; the *Kuperschiefer,* Germany; and some Canadian deposits that include *Kidd Creek,* the *Sullivan* deposits and *massive sulfide deposits* have all been considered at one time or another as either bacteriogenic, or exhalative or volcanic deposits some forming syn-

genetic and others epigentic. For some deposits, postore metamorphic processes have complicated the primary evidence but generally some syngenetic evidence remains.

Selected References on Bacteriogenic Deposits

Microbiological processes in the formation of sulfur deposits. 1968. G. D. Sokolva, and Karavaiko. U.S. Dept. of Commerce. M.V. Ivanov. *Physiology and geochemical activity of Thiobacilli.* 1968. U.S. Dept. of Commerce.

Introduction to Geological Microbiology. 1963. S.I. Kuznetsov, M.V. Iranov, N.N. Lyalikora, C. H. Oppenheimer, and P. T. Broneer, McGraw-Hill Book Company, New York.

The role of microorganisms in chemical mining. 1971. E. E. Malouf. v. 23, p. 43–46.

Geochemistry and Petrology of Some Oklahoma Redbed Copper Copper Occurrences. 1972. R. P. Lockwood, Ph.D. dissertation, University of Oklahoma, Norma, Okla.

Limits of the natural environment in Terms of pH and oxidation in reduction potentials. 1960. G. M. Bass Becking, I. Kaplan, and D. Moore. Jour. Geol., v. 68, p. 243–284.

Sedimentary deposits of copper, vanadium-uranium, and silver in southwestern U.S. 1937. R. P. Fischer. Econ. Geol., v. 58, pg 447–456.

Evidence for organic complexed copper in sea water. 1967. J. F. Slowey and L. M. Jeffrey. 1967. Nature, v. 214, p. 377–378.

Syngenesis of sulfide ores: Description of adsorbed metal ions and their precipitation as sulfides. 1964. K. L. Temple and N. W. LeRoux. Econ. Geol., v. 59, p. 647–655.

General Reference 1. *Contains numerous papers pertaining to uranium deposits.* General Reference 19. *Chapters on uranium and coal.*

11

SUBMARINE EXHALATIVE AND VOLCANOGENIC PROCESSES

Black's researches gave the world a gas that differed from common air in being heavier, very poisonous, and in having the properties of an acid, capable of neutralizing the strongest alkalies; and it took the world some time to become accustomed to the notion.

T. H. Huxley

During the past score of years, following early suggestions by Schneiderhöhn, there has been a growing concern and understanding of the importance of metallic mineral deposits having formed by an exhalative or volcanic source from which the mineralizing fluids were derived. Many such sulfide deposits have formed in a submarine environment and commonly occur between generally fractured volcanic rocks below and unaltered volcanics above (Figs. 19-14, and 11-1). The deposits may be described as stratabound, lenticular bodies of massive pyritic mineralization. They contain variable amounts of chalcopyrite, sphalerite, and galena in layered volcanic rocks, overlain by thin-bedded siliceous and iron-rich sedimentary or volcanic rocks. There are some differences among the compositions; rock associations; and ages of volcanogenic, massive, base metal sulfides. The deposits may be divided into three distinct types. Generally, pyrite-sphalerite-chalcopyrite bodies are in mafic to felsic volcanic rocks; pyrite-galena-sphalerite-chalcopyrite bodies occur in more felsic, calc-alkalic volcanic rocks; and pyrite-chalcopyrite bodies are in mafic-ophiolitic volcanic rocks. Some characteristics of different volcanogenic sulfide deposits are shown on Tables 11-1 and 11-2.

Many examples of such deposits have already been referred to, such as the Noranda type or massive sulfide deposits; Kidd Creek, Ontario; the Sullivan ore body, British Columbia; Kuroko deposits, Japan; the Rammelsberg mine, Germany, the Big Mike and Rio Tinto deposits, Nevada; and the stratabound deposits of Australia. The massive Sullivan

ore body is included primarily because of its "conformable nature." Some would still favor the classical hydrothermal origin for the deposit. The Mattagami Lake deposit is a classical example of this genetic type but highly modified by metamorphism. Hutchinson has summarized the geology of Canadian massive sulfide deposits in Table 11-2.

Submarine Manganese Nodules

Nodules or concretionary layers of hydrous manganese, iron oxides, and various metals such as copper (Fig. 11-2), cobalt, nickel, molybdenum, lead, and chromium are located in widespread areas of the ocean bottoms. Such deposits have been studied extensively, and specially constructed seagoing dredges will soon attempt to recover these nodules in economic quantities. Even attempts to claim exclusive mineral rights, or "ocean mining claims," in portions of the Pacific Ocean recently have been made.

The nodules vary in approximate diameter between less than 1 mm to more than several decimeters. Some nodules weigh more than several hundred kilograms. Manganese and iron make up about 20 percent and 15 percent, respectively, of the nodules. Many contain clay, calcium carbonate, and volcanic fragments. Not uncommonly, copper, nickel, and cobalt contents range between 0.1 to a few percent.

The widespread existence of these nodules on the ocean bottom suggests that some may derive their metal content as detrital material from the conti-

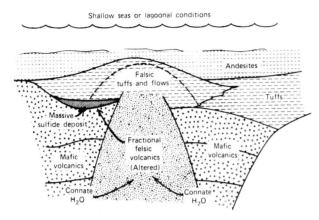

Figure 11-1 Idealized cross-section of typical Precambrian stratabound-volcanic massive sulfide deposit.

nents. Because of the widespread occurrence at the ocean bottoms at vast distances from shore (Figs. 11-3 and 11-4), a submarine volcanic source is by far the most logical source. And there is no reason why some should not be found even at some depth below the present ocean floor, even though very few are found in deep-sea cores. It is even expected

that the hot spring or volcanic sources should have localized higher concentrations of these deposits when such orifices reached the ocean floor. Apparently, however, the mineralizing fluids migrated laterally from such sources to provide widespread occurrences.

Manganese nodules are concretionary and show growth rings in cross section. Their growth rate is extremely slow averaging a few mm in 1 million years. Even small nodules, about 2 cm in diameter, have been forming for more than 10 million years. They seem to grow from a nucleus that may be a rock or volcanic fragment, an organic particle, or even a shark tooth. The nodule may remain at the surface for more than 5 million years before being trapped in the underlying sediments and buried. As W. Broecker states, "Manganese nodules are so abundant on the sediment surface compared to the underlying sediment because the product of roughly 5 million years of nodule growth is stored in the upper few centimeters of sediment. For each mature nodule that becomes trapped in the sediment, another newly arrived nucleus is beginning its long accumulation of iron and manganeze. Nodules are

Table 11-1 Some Geological Characteristics of Different Volcanogenic Sulfide Deposits

Base metal type	Precious metal association	Associated volcanic rock types	Type of volcanism	Type of sedimentation	Tectonism	Age	Examples
Zn-Cu-pyrite	Both Au (with high Cu) and Ag (with high Zn)	—Fully differentiated suites of intermediate bulk composition (?) —Tholeiitic to calc-alkaline —Basalt-andesite-dacite-rhyolite, etc.	—Initial deep, subaqueous mafic platform, with differentiation toward felsic volcanism, building domical centers	—Chemical cherts —Clastic, immature, first cycle, volcanogenic, graywackes, volcano-clastics	—Early geosynclinal orogenic stage —Major subsidence	Archean Proterozoic (?)	Timmins, Ont. Noranda, Que. United Verde, Ariz.
Zn-Cu-pyrite	ditto	ditto	ditto	ditto	ditto, early subduction	Pre-ordovician Mid-devonian	Rambler, Newfoundland W. Shasta, Calif.
Pb-Zn-Cu pyrite	Mainly Ag	—Intermediate to felsic calc-alkaline volcanic suites —Andesite-dacite-rhyolite-porphyry crystal tuff, etc.	—Felsic centers of explosive, pyroclastics and ignimbritic activity, subaqueous to subaerial	—Epiclastic predominates; immature volcanogenic graywackes, graphitic shales and argillitic, siltlines —Chemical minor, cherts, iron formations —Sulfate	—Later eugeosynclinal-orogenic stage	Proterozoic Ordovicin	Mt. Isa, Queensland Errington, Vermillion (Sudbury Basin) Bathurst, New Brunswick
Pb-Zn-Cu pyrite	ditto	ditto	ditto	ditto	ditto, later subduction	Triassic Tertiary	E. Shasta, Calif. Kuroko, Japan
Cu-pyrite	Mainly Au	—Poorly differentiated mafic suites —Tholeiitic —Basaltic pillow lavas, serpentinite, etc.	—Deep subaqueous quiescent fissure eruptions	—Chemical predominates, cherts, iron-stones —Clastic insignificant	—Early stage of continental plate, rifting, tension, separation	L-Ordovician U-cretaceous Jura-cretaceous Cret-eocene	W. Newfoundland Cyprus Islan Mountain, Calif. Phillipines

Source: Adapted from Hutchinson, 1973.

Table 11-2 The Geology of Canadian Massive Sulfide Deposits

Regional geologic setting	Keewatin eugeosynclinical belts—abundance of basic, pillowed volcanics
	Volcanic complexes and lava piles within Keewatin belts
	In extrusives, both acidic and basic, toward top of piles
Local geologic setting	At position in lava pile
	(a) a break in volcanism, transition from acidic to basic volcanics.
	(b) a thin-bedded, evenly laminated, siliceous and pyritic sediment, tuff or iron formation
Size	Widely variable: few thousand to 50 million tons (ore)
Shape	In regular, lenticular, podlike, elongated
Composition	Massive, underlying stringers sulfides
Economic metallization	0.5 to 5.0 % Cu, 0.5 to 12.0 % Zn
	0.50 to 4.0 oz/ton Ag, 0.005 to 17 oz/ton Au, minor Pb
Rock alteration	Complex, several types
Mineralogy	Pyrite, chalcopyrite, marmatitic sphalerite; pyrrhotite common
Structural geology	All rocks deformed, intruded, regionally metamorphosed. Structural control important
Attitude	Long dimension near vertical; conformable and locally cross-cutting
Wall-rock relationship	Sharp contacts, particularly above. Numerous wall-rock inclusions
Textures	Colloform structure, pumiceous, dense and compact, 8 cu ft per ton, fine to coarse, paragenetic sequence pyrr-sph-cpy, cataclastic pyrite, unmixed pyrrhotite

Source: Adapted from Hutchinson. 1965

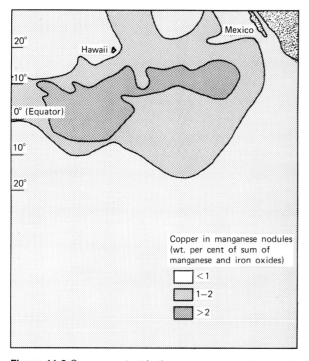

Figure 11-2 Copper content in the manganese and iron oxides of manganese nodules on the Pacific ocean floor. Optimal combinations of nodule coverage and copper-nickel content are found in the border regions of the equatorial high biological productivity zone, created by the Equatorial Undercurrent. (Data from Frazer and Arrhenius, 1972.)

born and die; at any given time, nearly the same number inhabit the sea floor.''

Imagination has led to numerous methods of recovery of these submarine mineral deposits, but a complex bottom scouring and dredging system with suction devices seems to be favored at the present time. A trailing suction hopper system is shown in Fig. 11-5, where suction draws material into the hopper and the material is then transported to the surface by a bucket-line. Such operations have been performed at depths exceeding the 12,000-foot average depth of the oceans. In 1972, Kennecott Copper Corporation conducted a research project to recover 200 tons of manganese nodules for the purpose of metallurgical testing.

"Küperschiefer," Germany and Holland. This interesting and extensively mineralized Permian formation, mined since A.D. 1150 covers more than 3550 square kilometers extending from England, through the Netherlands and northern Germany to Poland. Yet it only averages about half a meter in thickness. The sulfides consist of pyrite, chalcopyrite, "chalcocite," barite, galena, and sphalerite. The reduction of the sulfide in this Permian sea could have been brought about by bacteriogenic H_2S, but the source of the metals is the same prob-

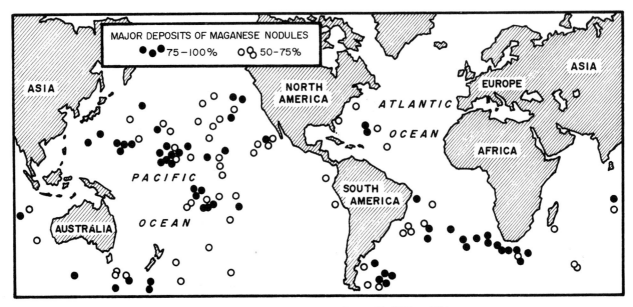

Figure 11-3 Major deposits of manganese nodules. (From Newsweek, Sept. 17, 1973, p. 85.)

lem faced by most syngenetic sulfide deposits. The prime suggestions are the sabkha process, submarine exhalative sources of mineralizing fluids, derivation from hot springs, heated mineralized ground water sources, or magmatic or volcanic fluids, all entering the sea at submarine orifices.

In order to better determine which of the above

sources might be reasonable, Marowsky analyzed isotopically 106 samples mainly from profiles of the "Kuperschiefer" in northwestern Germany and northern Holland. Samples were analyzed for $\delta^{13}C$, $\delta^{18}O$, and $\delta^{34}S$ values. The $\delta^{34}S$ values of sulfides varied between −4 to −44 permil and the "residue" sulfates varied between +11 and +36 permil. Ma-

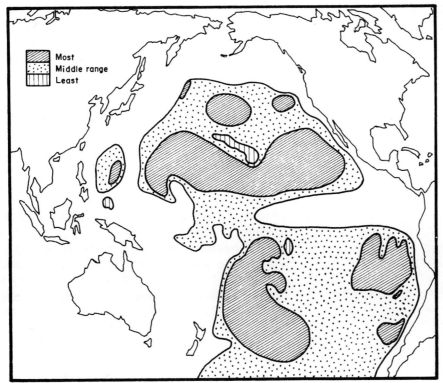

Figure 11-4 The relative distribution of manganese nodules on the Pacific Ocean floor from the estimates of Soviet scientists. A similar general pattern may exist for the Atlantic Ocean. The ridge areas, or areas under strong surface biological productivity where the sediments are rich in calcium carbonate, do not appear to have high concentrations of manganese nodules. (After Skornyakova and Andruchenko, 1964.)

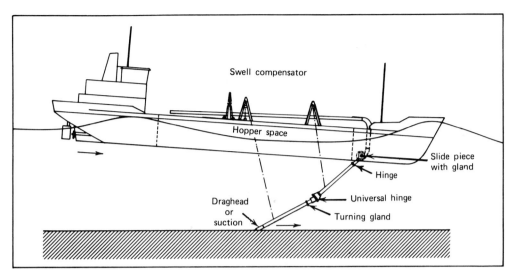

Figure 11-5 Trailing suction hopper dredge (right) permits efficient dredging of soil bottoms. (From: Donkers and Groot, 1974, Min. Eng., v. 26, p. 23.)

rowsky states that "the abundance of light carbon and sulfur as well as the correlation between the $\delta^{34}S$ and $\delta^{13}C$ values are strong indications for bacterial sulfate reduction." He further states that the "assumption of a syngenetic sulfide precipitation is confirmed by the results of the carbon isotope analyses. The trend to heavy carbonates in the sequence of the profiles can be interpreted as due to rising marine influence with time in the early Zechstein basin. A model of epigenetic metal supply forming this regular pattern of correlated sulfur and carbon distribution and composition cannot be constructed." With a syngenetic-bacterial origin for the formational deposit, the actual source of the economic metals is problematical, but it is probably a sabkha process source or a submarine hot spring or volcanic source, especially with a copper content at Mansfield varying between 2 to 3 percent.

Rammelsberg, Germany. More than 200 papers have been written on the (Altes and Neues lager) Rammelsberg county, ore deposits, not because of their economic importance, as these deposits contain only about 6×10^6 tons of zinc, lead, and copper, but because of their enigmatic but essentially stratabound character (Fig. 11-6). Obviously, it was a prime deposit for a $\delta^{34}S$ study that was done by Anger, et al., and included $\delta^{34}S$ values on 278 sulfides and 90 barite samples.

The results indicate that within a given part of the ore bodies, the $\delta^{34}S$ values of galena, sphalerite, and chalcopyrite specimens are in most cases quite similar to each other. Pyrite, some of which occurs

as concentrations, on the other hand, shows a significant spread in $\delta^{34}S$ values (between -10 and $+25$ permil but an obvious enrichment in ^{32}S in contrast to the sulfide ore minerals. Sulfur in barite, incidentally, averages about $+25$ permil, which is about the same as sulfur in Devonian seawater.

The investigators concluded that the ore sulfides contain sulfur of magmatic hydrothermal origin whereas the sulfur of pyrite is bacteriogenic in origin. "The sulfides of copper, lead, and zinc, but only about 50 percent of the pyrite must have been derived from submarine springs, while the other pyrites (especially the concretionary ones formed from bacterial reduction of marine sulfate)."

Inasmuch as the ore sulfides exhibit enrichment in ^{34}S by about 10 to 20 permil on the average, the source of this sulfur is of questionable classic magmatic hydrothermal origin but, considering the stratabound nature of the deposits, it may also be bacterial in origin with the source of the metals being by submarine exhalation or by the sabkha process source. Gehlen has considered the possibility of "very strong assimilation of unusual material or extreme isotopic fractionation in a secondary magma chamber." Neither of these two possibilities is obvious in other deposits studied.

White Pine, Michigan. More than 5 percent of domestic copper production is derived from this large stratabound deposit that measures one to 8 meters thick and several kilometers across and averages about 1.2 percent copper. It is the only known major copper shale deposit in the United States. It

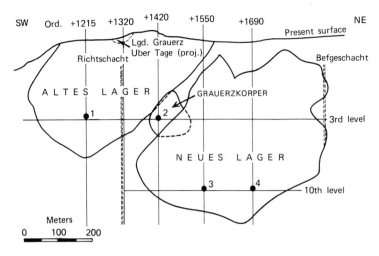

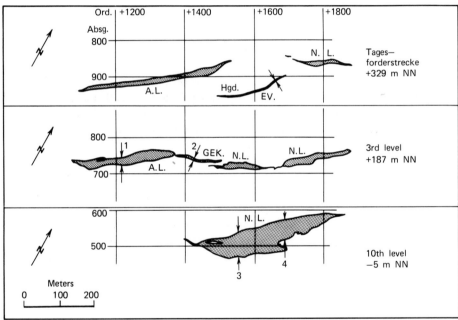

Figure 11-6 Sections through and along the Rammelsberg ore lenses in their present state (after Kraume). Dots or arrows mark the profiles that were investigated by Anger, et al. (From: G. Anger, et al., Econ. Geol.)

is, however, similar to the Küpferschiefer deposits in Germany, and especially similar, but with greater metamorphism, to the Zambian copper belt deposits.

Even though all of these deposits appear to be stratigraphically controlled, White Pine is unique in the predominance of the single sulfide, chalcocite. Although the White Pine deposit appears to be syngenetic, it has been classified at various times as epigenetic, magmatic hydrothermal, and formed either early, diagenetically, or formed later than the structures. These vastly differing genetic suggestions stem primarily from the question of the source

of the copper. The leaching of copper from the rocks of the Keweenawan Basin by saline connate or ground water has been suggested by White, Wright and Richardson. Such a source would also aid in providing the amygdaloidal copper in the Calumet-Houghton area located 70 miles northeast of White Pine.

Recent studies of the White Pine deposits by White and Wright indicate that the mineralization shows no correlation between shale ore and the abundant faults and fissures, which others have suggested as solution conduits. The "apparent pretectonic age of the mineralization and the probability that significant igneous activity in the region had ceased long before the rocks were deformed" strongly supports a syngenetic or diagenetic origin. On the other hand, the persistence of copper throughout as much as 15 meters of beds that show major ecological environmental changes, the absence of any copper in the upper parts of the Nonesuch shale, even though it exhibits favorable rocks and depositional environments, and the barrenness of specific beds in precisely the same area, but later than other beds that are well mineralized, suggest a syngenetic origin. If submarine exhalative sources for the metal occurred only during the time of mineralization, it is understandable why similar shale beds are barren.

It seems unlikely that copper was leached from the surrounding rock, transported through the sandstone that underlies the shale, and moved upward to replace pyrite, or more likely was reduced to chalcocite by H_2S, all of which occurred prior to lithification. By such a process, other shale host rocks, faults, or even diagenetic fractures would be mineralized. The possibility that the sabkha process formed the deposit is worthy of consideration.

Kidd Creek, Timmons District, Ontario. On March 3, 1959, seven years after TexasGulf, Inc., initiated a Canadian Shield exploration program, an electromagnetic anomaly was detected by Leo Miller near Kidd Creek, (Kidd Township), located about 15 miles north of the town of Timmins, Ontario. Options on the first of four half lots overlying the ore body were obtained in early 1963, followed by field geological and geophysical studies. In November 1963, the first diamond drill core "showed copper mineralization." By April 1964, eight drill holes confirmed the existence of what is now believed to be one of the most unusual and extraordinarily rich zinc, copper, silver, and tin ore bodies discovered during the century. By December 31, 1969, 5½ years after the discovery, the Kidd Creek mine had produced $460,528,993 (Canadian dollars) of ore consisting of chalcopyrite, sphalerite, pyrite, native silver, galena, and greenockite.

This discovery initiated one of the largest staking rushes in Canadian history. It created excitement among several nations and caused hysterical activity on the American and Canadian stock exchanges, and of course resulted in almost unnumbered lawsuits.

The Kidd Creek mine is located in the largest Early Precambrian greenstone belt within the Canadian Shield. This Abitibi belt has been well known for the large gold and base metal deposits associated with it. The Geology Department of Ecstall Mining, Ltd., the wholly owned subsidiary of Texas Gulf Inc., which operates the Kidd Creek mine, characterizes the ore deposits as stratiform, massive base-metal sulfides associated with felsic volcanics. The deposit is analogous to other massive, base-metal deposits of the Abitibi belt, primarily those of the Noranda camp, and is thought to be a near-surface, submarine exhalative type of deposit.

The ore deposit reaches about 170 meters wide, has a minimum length of 670 meters, and extends at least to 1220 meters depth. Ecstall reports that "since commencement of mining in 1966 until 1974, 25 million tons of ore have been milled averaging 1.52% Cu, 9.75% Zn, 0.40% Pb, and 4.30 oz/ton Ag." Significant amounts of tin and cadmium are recovered as by-products. Reserves are estimated at 95 million tons above the 2800-foot level. Drill hole exploration has extended from the 2800-foot level to more than an overall depth of 1220 meters. 230 meters of core from one hole assayed 4.16% copper and 0.63 ounces per ton Ag. This is the richest hole ever drilled at Kidd Creek and the ore structure continues beyond a depth of 1220 meters.

A cross section of the geology of the Kidd Creek ore deposit is shown in Fig. 11-7, which also shows the outline of the open pit limit and the present shaft and various Levels. Figure 11-8 is a generalized geologic map of the mines and bed of the surface below the glacial moraine veneer.

Bathurst-Newcastle Area, New Brunswick. In 1952, the Brunswick No. 6 and No. 12 sulfide deposits were discovered by diamond drilling based on electromagnetic anomalies near the Austin Brooks magnetite iron formation and an associated pyritic sulfide body. This association led to extensive exploration in New Brunswick, which resulted in the discovery of 25 or more sulfide deposits having in excess of hundreds of millions of tons of copper-lead-zinc ore (Table 11-3).

There is some structural control for some of the deposits, such as Heath Steele, that consists of thickening or dilatancy in zones around the crest of folds. Others question a structural control. The host rocks are sedimentary rocks, and iron formations and schists of Ordovician and Silurian age that are commonly interlayered with volcanic rocks (mostly green schists) and intruded by igneous rocks ranging from gabbro to granite.

Some of the deposits are fine-grained assem-

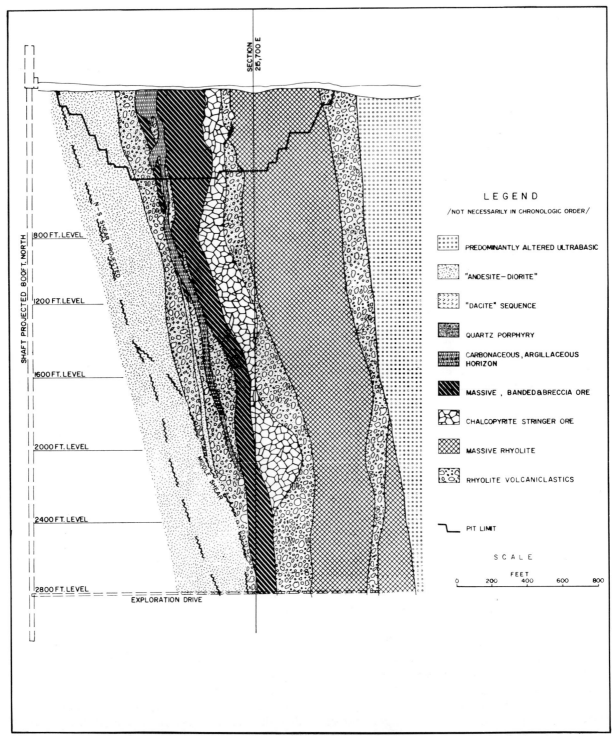

Figure 11-7 Kidd Creek Mine Section 214,600 N. (From Can. Inst. Min. and Metal.)

blages of (1) pyrite with sphalerite, galena, minor chalcopyrite, and pyrrhotite; or (2) pyrrhotite with chalcopyrite in a gangue of chlorite and quartz. Some of these deposits consist of disseminated sulfides while the majority are massive pyritic sulfide deposits that form lenticular, pipelike, or sheetlike bodies that are banded, folded, brecciated, and perhaps recrystallized.

Because of the intrusions, the deposits are thought to be epigenetic by some. Others suggest

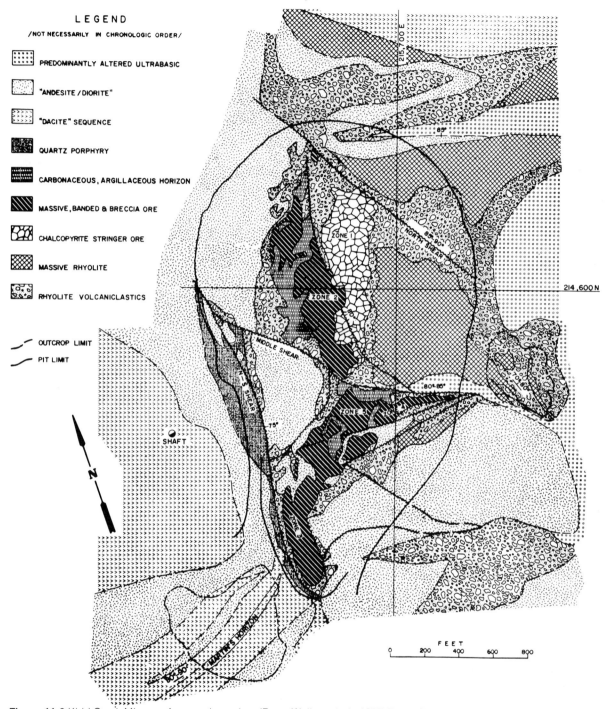

Figure 11-8 Kidd Creek Mine, surface geology plan. (From: Walker, et al., 1975, Econ. Geol., v. 70)

an original sedimentary origin except for the un-doubted fissure-vein deposits in northern New Brunswick. Lusk has suggested that these strati-form deposits have undergone low rank regional metamorphism. The structural deformations sug-gest somewhat higher rank of metamorphism but the common occurrence of quartz-sericite-schist and chlorite-sericite schist and green schists cor-roborates Lusk's suggestion. With the advent of the submarine exhalative volcanic genesis, and consid-ering the evidence of these massive sulfide depos-its, stratiformed, and associated with volcanics, this

Table 11-3 Size, Tonnage, and Grade of Some Sulfide Deposits of the Bathurst Area, New Brunswick (Davies and Smith, 1966; modified after McAllister, 1960)

Deposit	Location (approx.)	Size (Feet)			Thousand tons	Grade			
		Length	Width	Depth		% Cu	% Pb	% Zn	Ag oz./ton
Anaconda (Caribou)	47°34'N	3,800	20–130	1,200	28,000		2	5	1
Anaconda (Armstrong Brook)	47°36'N 66°17'W	A—450 B—450	50 (max) 150 (max)	800 800					
Brunswick No. 6	47°28'N 65°53'W	1,000	200	1,000	22,278 4,350 1,684	0.4 0.5 0.8	1.9 0.4 0.5	5.0 1.0 0.5	1.6 0.7 0.3
Brunswick No. 12	47°28'N 65°53'W	1,200	200	1,400	15,787 13,500	0.3 0.7	3.6	3.6 3.2	2.6 1.6
Captain mines	47°17'N 62°52'W	500	80	800	802	1.15			
Canoe Landing Lake	47°25'N 66°12'W	3,200	30	700	3,500	0.5	0.5	1.5	1.5
Chesterville mines (massive zone only)	47°07'N 66°12'W	650	650	30	650	0.92	1.62	3.4	
C.M. and S. (Nepisiquit A)	47°22'N 66°03'W	700 700	75 50	350 500					
C.M. and S. (Nepisiquit B)	47°22'N 66°03'W	900	75	300					
C.M. and S. (Wedge mine)	47°24'N 66°08'W	850	50	500					
Heath Steele mines	47°17'N 66°05'W	400–600	50	400–800	4,700	1.1 1.4	3.0 1.3	7.2 3.6	3.3 2.0
Kennco (Murray Brook)	47°32'N 66°27'W	1,000	300	800	23,000	0.44	0.86	1.95	0.91
Key Anacon mines	47°36'N 65°42'W	1,200a		1,500	3,000		2.35	6.4	2.16
Middle River Mining Co. (Texas Gulf Sulfur)	47°17'N 66°19'W	1,000	75	1,000	5,000	1.0	2	5	
New Calumet (Orvan Brook)	47°38'N 66°08'W	6,000 800c	14b 14b	500 500	200		3.25	6.3	1.0
New Jersey Zinc (Teck Corp.)	47°38'N 66°08'W	1,500	100	500	3,000		5	9	3.0
Stratmat (C.M. and S.)	47°19'N 66°06'W	600	100	500					

Source: Geology of Canada, Geol. Sur. Comm., 1968, p. 321.
a Several lenses along zone 1200 feet long.
b Maximum width.
c Ore zone.

origin has been suggested even though the epigeneticists resist this suggestion.

Mineralization has been recognized in the province since 1897, but only sporadic activity existed in the area until the early 1950s when electromagnetic surveys were initiated over the area and discovered the intimate relationship between magnetic iron formation and massive sulfides. Further surveys led to the discovery of many other disseminated massive sulfide bodies.

One of the largest massive sulfide deposits in the Bathurst-New Castle area of New Brunswick is the Brunswick No. 12 deposit. It has lead-zinc ore reserves of 98,173,000 tons grading 9.22% Zn, 3.79% Pb, 0.30% Cu, and 2.79 ounces per ton Ag. It also contains copper reserves of 14,094 tons grading 1.27% Zn, 0.45% Pb, 1.11% Cu, and 1.03 ounces per Ag.

The Brunswick No. 12 ore deposit is located about 19 kilometers southwest of Brunswick. The deposit is a stratiform, massive sulfide body occurring in the Ordovician Tetragouch group. Luff sites evidence that the massive sulfides were formed as chemically precipitated sediments. The Ordovician rocks are complex but appear to have been tuffa-ceous felsic volcanic rocks before several main pe-riods of metamorphism; the massive sulfide ore contains metamorphosed basic volcanic rocks, which are typical of massive sulfide deposits with a presumed volcanic exhalative origin.

The No. 12 ore body is being mined to a depth of more than 600 meters.

Noranda, Quebec. Prior to about 1960, the massive copper-zinc sulfide deposits in the Archean green-stone of the Canadian Shield were attributed to epigenetic emplacement of the sulfide minerals in the host rocks. The Noranda-type deposits associ-ated with felsic volcanic rocks, were cited as prime examples of the process. Now they are cited as prime examples of syngenetic volcanic exhalative deposits formed in the enclosing rocks and more particularly as end phases of mafic to felsic volcanic cycles.

Specifically, the copper-zinc massive sulfide de-posits of the Noranda, Quebec, area (Fig. 11-9) occur in a thick pile of Archean andesites and rhyol-ites. The ore deposits occur at or near the top of the rhyolitic formations. The deposits are attributed to submarine volcanogenic processes forming sul-fide sinters over hot springs according to Spena and de Rosen-Spence. Table 11-4 is a summary of pro-duction, grade, and values of ores from the major mines of the district.

Mattagami Lake Ore Body, Quebec. The Mattagami Lake ore body is a stratabound, massive Fe-Zn-Cu sulfide deposit, according to R. G. Roberts, of Ar-chean age emplaced in rhyodacitic, vitroclastic tuff overlain by effusive andesite. It is an example of the exhalative volcanic origin as shown in Fig. 11-1 but highly modified by metamorphism. It is class-ified with the important group of massive sulfide deposits of the Canadian Precambrian Shield that includes the deposits of the Noranda district (Fig. 11-9), Kidd Creek of the Timmons district (Fig. 19-16) Ontario, and the deposits of the Flin Flon dis-trict in Saskatchewan and Manitoba (Fig. 19-17).

The geology, according to Sharpe, is dominated by a westward-plunging anticline that extends across the area. The Wabasse group of volcanics (andesite, basalt, dacite, and rhyodacite) forms the outer limbs of the anticline with the older Watson Lake group, predominately rhyolite, formed in jux-taposition below the Wabasse group. Several ore bodies, such as Bell Channel, Radiore "A," Bell

Table 11-4 Production or Reserves of Base Metal Mines in Noranda Area

Mine	Years	Ore (tons)	Copper (tons)	Zinc (tons)	Gold (oz)	Silver (oz)
Copper-Zinc						
Aldermac	1931–1943	2,057,100	30,845	—	10,750	389,100
Horne[a]	1927–1970[c]	56,264,700	1,226,018	—	8,549,029	N.A.
Millenbach[a]	1971– [b]	2,415,000	3.45%	4.35%	0.018/ton	1.1/ton
Norbec[a]	1964–1970[c]	2,800,200	93,242	134,034	58,318	3,524,269
Quemont	1949–1970	15,013,000	183,801	283,991	1,895,771	8,241,356
Vauze	1961–1964	385,000	11,150	3,600	7,435	266,600
Waite Amulet	1930–1962	9,658,000	404,009	352,921	261,448	7,692,690
(A		5,872,000				
B,C,D,E, Bluff		596,000				
F		290,000				
O. Waite		1,245,000				
E. Waite)		1,655,000				
		88,593,000				
Zinc						
Delbridge	1969–1970	400,000	2,170	34,000	26,000	780,000
D'Eldona	1950–1952	86,500	14	4,360	13,282	69,024
West MacDonald		1,030,000	125	30,000	2,000	5,300
		1,516,500				
Mobrun	—[b]	3,000,000	0.69%	2.18%	0.05/ton	0.62/ton

Source: Econ. Geol., v. 70, 1975.
[a] Current producer.
[b] Reserves 1972.
[c] Last available data.

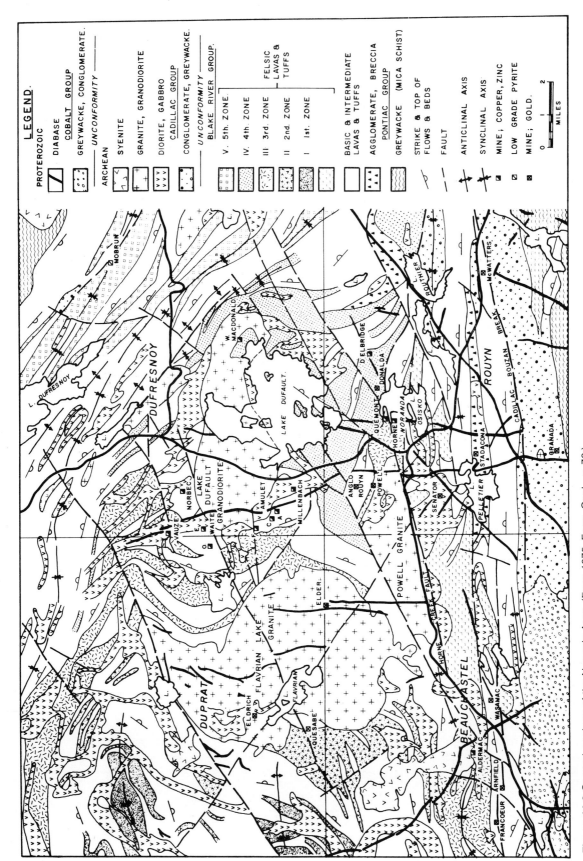

Figure 11-9 General geology of the Noranda Area. (From 1975, Econ. Geol., v. 70.)

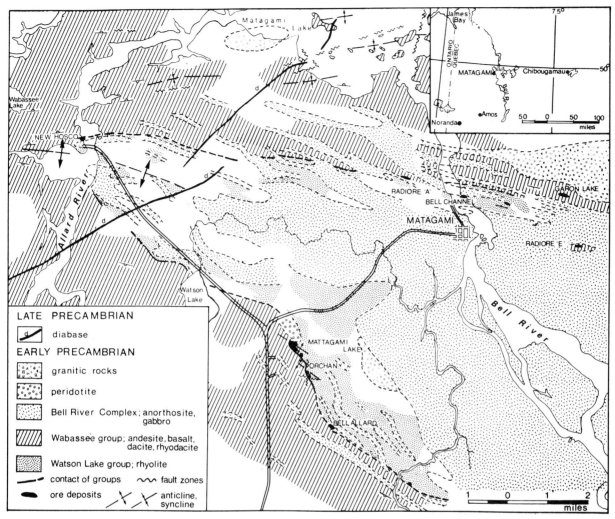

Figure 11-10 Geologic map of the Mattagami district, Quebec. (After Sharpe, 1968; Econ. Geol. v. 70, 1975.)

Allard, Orchan, New Hosco, and Mallard Lake occur between these two volcanic groups (Fig. 11-10). The ore deposits are stratabound and confined to the stratigraphic zone of felsic rocks at the top of the Watson Lake group.

Thirteen zinc-copper massive sulfide deposits have been discovered in the area since 1956, but only four have been developed into producing mines. Pyrite and pyrrhotite comprise 70 to 80 percent of the sulfides and oxides of the deposits. Field relationships, according to Sharpe, indicate that the deposits are cut by the intrusive rocks of the area and that they have undergone the regional metamorphism. The deposits are, therefore, preintrusive.

A location map of several of the Canadian mineral deposits is given in Fig. 11-11.

Central Africa Copper Deposits. The outstanding development of copper in Zambia (Table 11-5) made this country the third ranking producer of copper, and the Rhodesian or Zambian copper belt the most important single copper district in the world. This belt and the adjacent Zaire belt constitute a central African copper province. The deposits of both areas occur in parts of the same formation, but oxidized ores dominate in Zaire and sulfide ores in Zambia.

The Zambian copper belt or district has emerged, since 1927, from a forest wilderness to a center of industrial activity. Its unique disseminated copper deposits are confined to certain stratigraphic horizons of late Precambrian sedimentary rocks. The area is 225 kilometers long by 50 kilometers wide and extends from Konkola in the northwest to Roan Antelop (Luanshya Baluba) and Bwana Mkubwa in

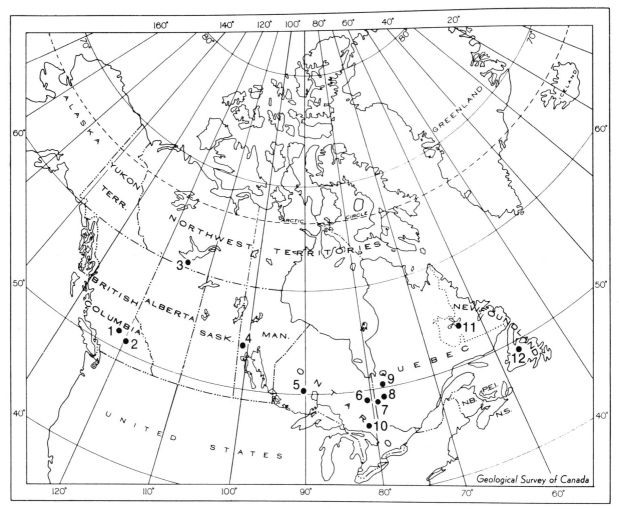

Figure 11-11 Location map of some mineral deposits or areas in Canada. 1, Highland Valley; 2, Boss Mountain; 3, Pine Point; 4, Flin Flon; 5, Mattabi; 6, Kidd Creek; 7, Vauze deposit and Noranda area; 8, Dumont; 9, Mattagami Lake; 10, Sudbury; 11, Bathurst-Newcastle; and 12, Buchans. (Adapted from Econ. Geol., v. 70, 1975.)

the southeast. It is adjacent to Zaire, the province of Shabain and includes numerous large producing mines.

The formations include about 5000 feet of continental and marine sandstones, arkoses, shales, and dolomites of the Mine series, which unconformably overlies a basement complex of ancient Precambrian schist and granite. Above them is 4,000 meters of Kundelungu sediments. The mine series has been folded into pitching anticlines and synclines, of which erosion has left only synclinal remnants below an erosional surface (Fig. 11-12). Basic sills occur in the Mine series.

Deposits. The chief deposits (Fig. 11-13) all occur in the Lower Roan group of the Mine series in a narrow belt 110 kilometers long. There are 18 to 20 proven deposits, all in operation. They are consol-

idated into six operating units: the Roan Antelope, Rokana, Mufulira (Fig. 11-14), N'Changa (five open pit mines and one underground), Chibuluma, and Chambishi. The latter is one of the latest to be exploited. The Chibuluma ore body has no outcrop or surface indication. Pits over the suboutcrop of the ore body revealed limonite-stained quartzite containing up to 0.2 percent copper. The metallized bed of the Roan Antelope deposit, discovered in 1927 and the first deposit to be developed, laps around the nose of a plunging syncline and extends along the south limb for a length of close to 15 km. The ore bed has been developed down the plunge of the syncline to 760 meters and extends into the Roan Extension deposit and Muliashi.

The Baluba copper-cobalt orebody occupies a subsidiary synclinorium north of Muliashi. In the

Table 11-5 List of Zambian Deposits

Mine	Owner	Copper (percent)	Annual Capacity (1000 tons Cu)	Ore Resources (million tons)	Cobalt (percent)
Zambia					
Bancroft	Bancroft Mines Ltd.	4.76	51	105	—
Nchanga	Nchanga Consolidated Copper Mines Ltd.	4.65	177	180	—
Lyanshya	Roan Antelope Copper Mines Ltd.	3.04	91	95	—
Chambishi	Chambishi Mines Ltd.	3.37	—	35	—
Chibuluma	Chibuluma Mines Ltd.	4.89	22	10	0.18
Mufulira	Mufulira Copper Mines Ltd.	3.35	103	178	—
Rhokana	Rhokana Copper Mines Ltd.	3.07	104	120	0.18
Baluba	Roan Antelope Copper Mines Ltd.	2.41	—	112	0.16

Roan syncline, chalcocite dominates near the nose, as that is the farthest east, near the ultimate shore line. Sulfide zoning is prominent, chalcocite, bornite, chalcopyrite, and then pyrite from east to west and down across the strata of the upper Roan.

Ores. The ore consists of minute disseminated specks, course blebs, and veinlets of sulfides in feldspathic sandstones and shales. The only gangue is the host rock. The sulfide minerals are predominately chalcocite, bornite, chalcopyrite, and pyr-

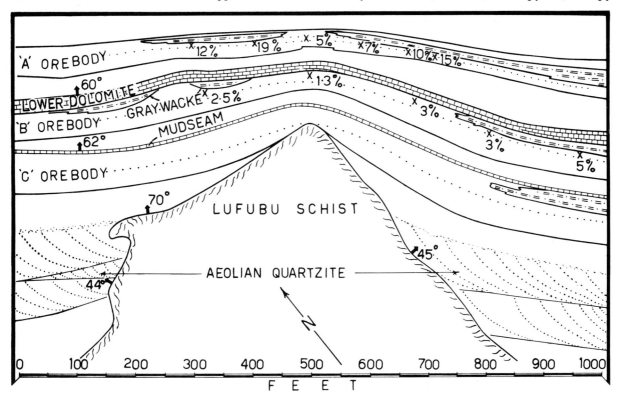

Figure 11-12 Mufulira: Plan through compaction fold showing changing ore grade. 1400 level. (From *Geology of the Northern Rhodesia Copperbelt*, Mendelsohn, ed. MacDonald and Co., London, 1961.)

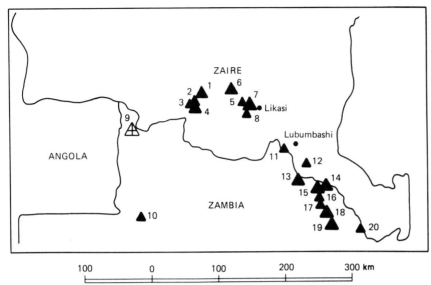

Figure 11-13 Zambian Copper Belt mines including those in Zaire. 1, Ruwe; 2, Musonoi; 3, Kamoto; 4, Mutoshi; 5, Kakanda; 6, Tenke-Fungurume; 7, Kambove-Sesa; 8, Shinkolobwe; 9, Lumwan; 10, Kalengwa; 11, Kipushi; 12, Etoile; 13, Konkola; 14, Mufilira; 15, Nohanga; 16, Chambishi; 17, Chibuluma; 18, Nkana; 19, Luanshya Baluba; and 20 Bwana Mkubwa. (From Mendelsohn.)

ite. At the Roan Antelope, chalcocite dominates nearer the nose. In general, chalcopyrite and bornite are more abundant at depth and chalcocite less so.

The deposits consist of one or more beds impregnated by specks of copper sulfides. Their outstanding features are the persistency and uniformity of the metallization over a uniform width for great horizontal distances. Being bedded deposits, their shapes are those of the parts of the synclines that are metallized.

In 1945, Garlick produced his theory of the formation of syngenetic deposition, partly based on Schneiderhöhn's ideas on the Mansfeld Kuperschiefer, that the ores were formed by the precipitation of copper and iron as sulfides by the production of H_2S produced by sulfate-reducing bacteria in stagnant water. Mendelsohn concludes that "the bulk of the evidence, therefore, supports a syngenetic origin for the valuable metals in the "Roan," and the majority of Copperbelt geologists extend the theory throughout the Copperbelt."

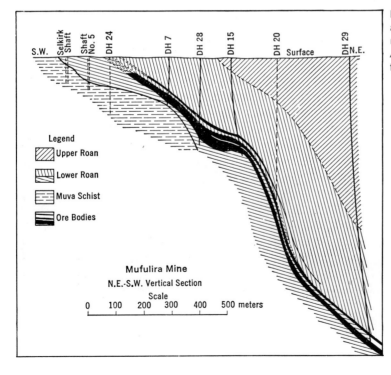

Figure 11-14 Vertical section across Mufulira ore bodies in northern Rhodesia, showing the *A*, *B*, and *C* horizons of the ore formation. (Gray, Econ. Geol.)

The syngenetic theory, as reasonable as it appears, raises questions pertaining to the source of the copper and the metals such as nickel and cobalt. It is possible that the metal source is from widely dispersed submarine exhalations. Obviously, the genesis of these deposits is still controversial. The hydrothermal origin is most unlikely, but the respective roles of bacterogenic, sedimentary, or submarine exhalative origins must still be considered especially in light of the sabkha process.

Zaire. The rich Zaire copper belt, just north of Zambia, follows the regional structure for 300 kilometers. The geology is similar to that of Zambia except that the folding has been closer, faulting more abundant, and nappe thrusting important. In addition, erosion has penetrated less deeply, and anticlines and more of the higher formations are exposed.

Deposits. There are about two-score commercial copper deposits, of which all except two are oxidized, plus a worked out uranium deposit at Shinkolobwe.

The oxidized deposits lie at or near the surface and vary greatly in size. Some of the bodies are worked by large power-shovels. Most of them are confined generally to two dolomitic shale beds in the Upper Roan dolomites of the Mine series, about 200 meters thick. The two horizons commonly mineralized throughout the Shaba belt are the siliceous shale, the "Feuilletees," and the "Schiste dolomitique," separated by the very distinctive "Roache Cellulaire," which commonly shows stromatolitic structures. The oxidized minerals are malachite and chrysocolla, with subsidiary cuprite and native copper, and these form films, veinlets, and replacement masses. Ore grades of over 4 percent copper are common.

Most of the deposits have oxidized in place. The dolomitic host rock tends to precipitate the migrating copper as carbonate and prevent supergene enrichment. However, a little supergene chalcocite and covellite occur in siliceous rocks. Where the mineralized beds persist below the oxidized zone, they contain finely disseminated chalcocite, bornite, chalcopyrite, or pyrite. Such primary mineralization has been intersected by deeper drill holes at many of the deposits and is similar to that found in Zambia. The Shinkolobwe uranium deposit, which carries copper and cobalt sulfides and also nickel, is considered to be a hydrothermal deposit. This deposit provided the first uranium used in the development of atomic bombs. It is now worked

out. The Kipushi mine is located just west of Lubumbashi and right on the Zambian border. It is an important replacement deposit in Kundelungu limestone, just above the tillite and adjacent to a cross-cutting fault. Massive bornite, chalcopyrite, passing outwords into sphalerite and galena form one of the world's richest sulfide deposits.

Japan. Copper is one of the most important mineral resources of Japan. It surpasses all other metallic minerals in the number of mines and amount of production. In the eighth century, copper was sold to China and, in the fifteenth and sixteenth centuries to the Netherlands. Prior to World War 1 Japan ranked second in the world in copper production. During World War II, 550 mines were worked that produced between 81,000 to 125,000 tons of copper per year. During the 1970s, production approached 140 million tons per year from about 30 mines.

The most important copper mines of Japan are located in Northeast Japan, Honshu and Skikoku. The principal deposits were once classified as (1) bedded cupriferous pyritic deposits, (2) fissure veins, (3) replacement deposits ("Kuroko" type), (4) stockwork deposits ("Kuroko") type, and (5) pyrometasomatic deposits. With the recent recognition of the submarine exhalative type of deposit, and the

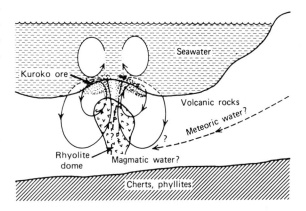

Figure 11-15 A model for the formation of the Kuroko ores. Heat associated with shallow submarine volcanic activity initiates convective circulation of seawater in the volcanic rocks. Because the solubility of gypsum decreases with increasing temperature, gypsum and anhydrite may precipitate directly from the overlying seawater (that is, formation of gypsum/anhydrite beds). Seawater that circulates through hot volcanic rocks may become an ore-forming fluid by decreasing its oxidation state and by dissolving metals from volcanic rocks. Small amounts of magmatic and/or meteoric water may mix with the circulating seawater. When a drop in temperature and change in the chemical environments occur near the sea floor, precipitation of metallic sulfides may take place. (From Ohmoto and Rye, 1974, Econ. Geol., v. 69, p. 947–953.)

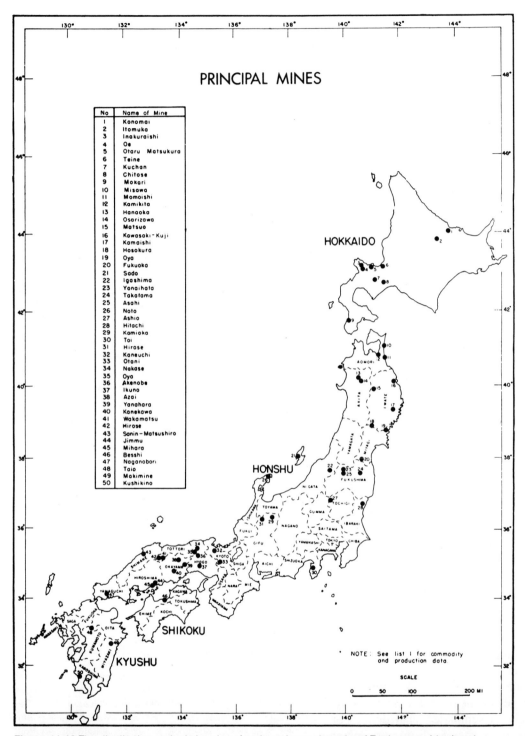

Figure 11-16 The distribution and relative size of endogenic ore deposits of Tertiary age (dots) and Quaternary age (triangles) in Honshu Island, Japan. A portion of a map compiled by Takeuchi. (From Kutina and Takeuchi, in preparation.)

geological features including the hydrothermally altered Miocene tuffs around "Kuroko"-type deposits, it is now thought that many of the Japanese base metal deposits are of this origin. This would include many of the bedded, replacement, and massive deposits.

Ohmoto and Rye have diagramatically suggested how the kuroko deposits were formed (Fig. 11-15)

Table 11-6 Major Copper Mines of Japan and Their Production in 1958

Mine	Prefecture	Type of deposit	Copper ore	
			Average grade of crude ore (%)	Production (tons) Cu content in crude ores
Besshi	Ehime	Bedded cupriferous pyritic deposits	1.3	7,396
Hitachi	Ibaraki	Bedded cupriferous pyritic deposits	1.0	5,860
Ozarizawa	Akita	Vein	1.2	5,439
Asko	Tochigi	Vein and replacement	1.1	4,158
Kamaishi	Iwate	Pyrometasomatic	0.7	3,929
Sazare	Ehime	Bedded cupriferous pyritic deposits	2.3	3,731
Hanaoka	Akita	Replacement	1.9	3,588
Kishu	Mie	Vein	1.1	2,856
Ogoya	Ishikawa	Vein	1.5	2,644
Ikuno	Hyogo	Vein	1.8	2,560

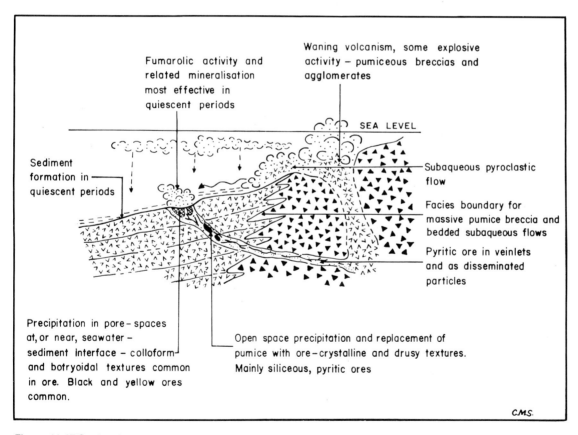

Figure 11-17 Section through the Undu mine area showing conditions at the time of formation of the Undu ore deposit, a kuroko-type deposit. Fiji (From Econ. Geol., v. 70, 1975.)

Table 11-7 Details of Kuroko Deposits in the Hokuroku District

Name of mine	Production in 1972 (per month)	Size of mine
Kosaka	Crude ore = 45,000 tons Acid leached Cu = 70 tons	25 million tons
Hanaoka Matsumine	Cu ore = 25,000 tons Pyritic ore = 10,000 tons	35 million tons
Fukazawa	Crude ore = 10,000 tons	8 million tons
Shakanai	Crude ore = 35,000 tons	10 million tons
Hanawa	Crude ore = 11,000 tons	2 million tons
Furutobe	Crude ore = 15,000 tons	7 million tons
Matsuki	Crude ore = 10,000 tons	10 million tons
Ainai	Crude ore = 9,000 tons	10 million tons

Source: Adapted from Lambert and Sato, 1974

by such a submarine volcanic process. A heat source and possibly part of the metal source is indicated as an igneous dome that acquires mineralizing fluids not only from seawater but from connate water in the sediments.

A list of the more important copper deposits of Japan in order of production in 1958 is given in Table 11-6. The locations of some of these deposits are shown in Fig. 11-16. A list of the kuroko type deposits in the Hokuroku district with summaries of production and reserves is given in Table 11-7.

Some of the terms used in referring to Japanese mineral deposits are "Kuroko" or black ore, which is an intimate mixture of sphalerite, galena, and barite with varying amounts of chalcopyrite, tetrahedrite, and pyrite. "Oko" or yellow ore is composed mainly of pyrite and chalcopyrite, with bornite, luzonite, tetrahedrite, chalcocite, and rarely digenite. "Keioko" or siliceous ore is composed of variable amounts of pyrite and chalcopyrite impregnated in light-colored silicified rocks containing some gold and silver.

Total reserves of copper ores in Japan containing more than 0.4 percent copper of 143 working mines in 1958 were estimated to be about 90 million tons with an average grade of 1.54 percent copper. Nevertheless, the number of working copper mines decreased to 69 in 1969, to 33 in 1972, and to 22 in 1975, but production exceeds 130 million tons. Yet

more than 80 percent of Japan's copper is derived now from imports.

Fiji. A diagrammatic example of a non-Japanese kuroko type submarine exhalative volcanic deposit is shown in Fig. 11-17. This deposit is the Undu mine located on the island of Fiji.

Selected References on Submarine Exhalative Volcanogenic Deposits

Massive sulfide deposits and volcanism. 1969. C. A. Anderson. Econ. Geol., v. 64, p. 129–144.

Genesis of Canadian massive sulfides reconsidered by comparison to Cyprus deposits. 1965. R. W. Hutchinson. Min. & Met. Bull., v. 56, p. 972–986.

On the Geographical Association of Strata-Bound Ore Deposits with Evaporites. 1968. Brongersma-Sanders, Mineralium Deposita, v. 3.

Association of mineralization and algal reef structures on northern Rhodesian copperbelt. 1964. W. G. Garlick, Econ. Geol., v. 59, p. 416–427.

"Küperschiefer" and the problem of syngenetic ore deposition. 1964. K. H. Wedepol. Bull. Geol. Soc. Am., v. 75, p. 187.

Cyprus sulfide deposits. 1973. B. N. Fredericton. Econ. Geol., v. 68, p. 843–858.

The Geology of the Northern Rhodesia Copper Belt. 1961. F. Mendelsohn, ed. MacDonald, London. *Comprehensive account of the Copper Belt.*

Volcanism and ore genesis. 1970. T. Watanabe. Econ. Geol., v. 65, p. 424–443.

Ore Petrology. 1972. R. L. Stanton. McGraw-Hill Book Company, New York, 713 p. *A text with many references to volcanogenic exhalative deposits of Canada.*

The environments of formation of volcanogenic massive sulfide deposits. 1973. F. H. Sillitoe. Econ. Geol., v. 68, p. 1321–1336.

The geological setting of the Mattagami Lake mine, Quebec; A volcano-genic massive sulfide deposit. 1975. R. G. Roberts. Econ. Geol., v. 70, p. 115–129.

Volcanogenic sulfide deposits and their metallogenic significance. 1973. R. W. Hutchinson. Econ. Geol., v. 68, p. 1223–1246.

Examination of volcanic exhalative and biogenic origins for sulfur in the stratiform massive sulfide deposits of New Brunswick. 1972. J. Lusk. Econ. Geol., v. 67, p.

Island-Arc Evolution and Relates Mineral Deposits. 1973. A. H. G. Mitchell and J. D. Bell. Jour. Geol., v. 81, p.

The White Pine Copper Deposit. 1960. W. S. White. Econ. Geol., v. 55, p.

The White Pine Copper Deposit, Ontonagon County, Mich. 1954. W. S. White and J. C. Wright. Econ. Geol., v. 49, p.

12

EVAPORATION

Resources are not; they become.
E. S. Zimmerman. 1951

Evaporation has been important in producing many valuable types of nonmetallic mineral deposits. Underground waters have migrated to arid surfaces, there to be evaporated, leaving behind valuable minerals once in solution. Lakes have disappeared under relentless desert suns to form playas encrusted with various usable salts, or salt layers covered later by the shifting sands of arid regions. Or, the evaporation may not have gone to dryness but has yielded concentrated liquors from which are obtained much-used household salts. The searcher for useful substances eagerly avails himself of these rich liquors and extracts them for precipitation by more rapid natural or artificial evaporation. Natural brines are pumped from great depths or surface brines (Fig. 1-15) and spread out to the sun or put in artificial evaporators so that their salts may be obtained (Fig. 12-7).

Great sections of the oceans have been cut off during slow oscillations of land or sea and gradually evaporated to the point that gypsum or common salt has been deposited in many parts of the world. Still greater concentration by evaporation has yielded rich and valuable potash deposits that have been a continuing source of wealth. A list of those evaporite deposits in the United States is given in Table 12-1.

Thus evaporation is a great mineral-forming process, simple and familiar in its operation, which supplies homely materials used by the householder, the farmer, the chemist, the builder, the engineer, the manufacturer, and even by the birds and beasts and plants.

PROCESS OF MINERAL FORMATION BY EVAPORATION

Evaporation proceeds most rapidly in warm, arid climates. In the evaporation of bodies of saline water, concentration of the soluble salts occurs, and when supersaturation of any salt is reached, that salt is precipitated.

Deposition of minerals by evaporation is dependent on factors other than solubility constants. Factors such as temperature (Fig. 12-1), pressure (Fig. 12-2), depositional environment, and seasonal and climatic changes are a few of the more important factors.

Ocean water is the prime source of minerals formed by evaporation. About 3.45 percent of seawater consists of dissolved salts of which 99.7 percent by weight is made up of only seven ions that are as listed below in weight percentages.

Na^+	30.61	Cl^-	55.04
Mg^{2+}	3.69	SO_4^{2-}	7.68
Ca^{2+}	1.16	HCO_3^-	0.41
K^+	1.10		

About 45 other elements, whose concentration in seawater is known, form definite compounds that generally occur as trace minerals in evaporites. Exceptions exist such as borate deposits formed by evaporation of playas, for example, in Death Valley, California.

The number of carbonates that exist in marine evaporites is few in comparison with those that form from terrestrial evaporites. Calcite, magnesite, and dolomite are the chief carbonates of marine evaporites.

DEPOSITION FROM OCEANIC WATERS

Materials Involved

The salts of oceanic waters are mainly the contribution of land waters; small amounts are contributed by volcanism, and some are dissolved from the ocean basins. They are the soluble products of the weathering of rocks and of solution by subsurface waters. The ocean forms a great mixing pot for the diverse contributions of rivers.

Table 12-1 Marine Evaporite Deposits of the Conterminous United States
[Thicknesses are approximate and include interbedded nonevaporitic sediments. Evaporite types are calcium sulfates, including gypsum and anhydrite, S; halite, H; and bedded bittern salts of potassium and magnesium, B. (After W. C. Krumbein, 1951, with slight modifications.)]

Age	Area	Formation or interval	Thickness of evaporite section (ft)	Evaporite type
Ordovician	Williston basin	Whitewood-Bighorn	25	S
	Illinois basin	Joachim	50	S
Silurian or Devonian	Williston basin	Niagaran	25	S
	Michigan basin	Bass Island, Salina	3000	S, H
	New York	Camillus, Syracuse	500	S, H
	West Virginia	Salina	800	S
	Iowa-Missouri	Niagaran-Cayugan?	100	S
Devonian	Williston basin	Potlatch, Jefferson	400	S
		Prairie	600	S, H, B
	Iowa	Cedar Valley—Wapsipinicon	50	S
	Michigan Basin	Detroit River	1200	H
	Southwestern Montana	Three Forks	50	S
Mississippian	Williston basin	Otter, Charles	1000	S
	Southwestern Montana	Kibbey	100	S
	Iowa	Keokuk	30	S
	Illinois basin	St. Louis	200	S
	Michigan basin	Michigan	350	H
	Western Virginia	Maccrady	1000?	S, H
Pennsylvanian or	Paradox basin, Utah	Paradox	4000	S, H, B
	Gypsum basin, Colorado	Maroon	500+	S
	Black Hills region	Minnelusa	50	S
Permian	Michigan basin	Virgilian?	50+	S
	Black Hills	Minnekahta, Opeche	100	S
	Gypsum basin, Colorado	Maroon	500	S
	Grand Canyon area	Kaibab, Toroweap	900	S, H
	Northern New Mexico	San Andres	100	S
	South central New Mexico	Yeso, Abo	2000	S, H
		Chalk Bluff, Whitehorse	1000	S, H
Permian	Southeastern New Mexico	Rustler, Salado, Castile	4500	S, H, B
	Central Texas	San Angelo, Clear Fork, Wichita	1500	S
	Texas Panhandle	Pease River, Clear Fork, Wichita	2000	S
	Oklahoma Panhandle	Dog Creek, Blaine, Cimarron	1000	S
	Western Oklahoma	Blaine, Clear Fork, Wichita	1500	S, H
	West central Kansas	Harper, Wellington, Marion	800	S, H
	Iowa	Leonardian?	50	S
or Jurassic	Gulf Coast	Louann, Werner	1500±	S, H
Triassic	Wyoming, Nebraska, North and South Dakota	Chugwater	Several hundred	S
Jurassic	Central Wyoming	Gypsum Springs	200	S
	Central Montana	Gypsum Springs	100	S
	Southeastern Idaho	Preuss	450	S, H
	Central Utah	Arapien	1000?	H
	South central Utah	Carmel	400	S
	North central New Mexico	Todilto	100	S

Table 12-1 (Continued)

Age	Area	Formation or interval	Thickness of evaporite section (ft)	Evaporite type
	East central Colorado	Summerville equivalent	200	S
	Southern Arkansas	Buckner	350	S
	Northeastern Texas	Buckner	700±	S, H
	Southwestern Alabama	Buckner	950	S, H
Cretaceous	Northwestern Louisiana	Ferry Lake	300±	S
	Southwestern Texas	Fredericksburg	150	S, H
	South central Florida	Comanchean	6000	S
	Southeastern Florida	Comanchean	4000+	S, H
Tertiary	Florida	?	Several hundred	S

Source: Fleischer, 1963, USGS Pro. Paper 440 Y.

Ocean water, dropped over the land as rain, garners a new load of soluble materials and again reaches the ocean; the cycle repeats itself. Thus salts have accumulated in the oceanic reservoir ever since streams first coursed over the earth. The total amount of salts in the ocean is calculated to be 21.8 million cubic kilometers enough to form a layer 60 meters thick over the ocean bottoms. Of this, common salt would constitute 47.5 meters; $MgCl_2$, 5.8 meters; $MgSO_4$, 3.9 meters; $CaSO_4$; 2.3 meters; and the remaining salts, 0.6 meters. The rivers of the world are estimated to contribute annually about 4 billion tons of salts to the ocean.

Ocean water also contains gold, silver, base metals, manganese, aluminum, vanadium, nickel, cobalt, as well as iodine, fluorine, phosphorus, uranium, arsenic, lithium, rubidium, cesium, barium, and strontium. Of these, iodine concentrates in sea weeds and copper in shellfish; and nodules containing manganese, copper, nickel, and other metals form on Guyots and on the ocean floor. Only the nodules have become concentrated into potentially commerical deposits.

Conditions of Deposition

The mean depth of the oceans is about 3660 meters, and the surface would have to be lowered nearly 3350 meters before salt could be deposited. Consequently, to attain the necessary concentration to induce precipitation, bodies of sea water must be-

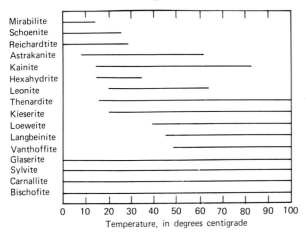

Figure 12-1 Temperature ranges of formation of oceanic salts between zero and 100°C. (From Phillips, fig. 2, 1947, reproduced by permission of the Chemical Society; M. Fleischer, U.S. Geol. Survey Prof. Paper 440-Y.)

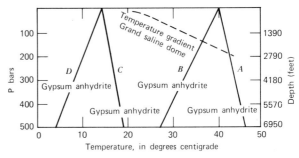

Figure 12-2 Pressure-temperature relations for the reaction gypsum = anhydrite + water. Curve *A*, Same pressure acting on all phases, in the presence of pure water. Curve *B*, Rock pressure acting on solid phases, hydrostatic pressure acting on pure water. Curve *C*, Same pressure acting on all phases in the presence of saturated NaCl solution. Curve *D*, Rock pressure acting on solid phases, hydrostatic pressure acting on saturated NaCl solution. (From MacDonald, p. 999, fig. 2, 1953.)

come isolated from the ocean in places where evaporation exceeds inflow. Such isolation may be effected by the formation of barrier reefs, which tend to restrict the continual inflow of ocean water and permit evaporation concentration. If the barriers are low, periodic, or steady, an inflow of seawater may occur. Cutoffs occur near coasts where sills or reefs isolate sinking inland basins, or result from delta building or crustal movements. Natural "salt pans" are numerous in coastal areas of arid regions. The Gulf of Karabughgaz, on the eastern side of the Caspian Sea, is a well-known example of the concentration of salt water behind a bar with a narrow, shallow inlet. Here, amid the surrounding deserts, evaporation is so rapid that a current flows continuously from the Caspian Sea into the Gulf, augmenting the Gulf waters with about 130 tons of salt daily. Gypsum and halite are deposited. If its inlet should become closed, potassium compounds and other salts would be deposited eventually.

In a cutoff body of seawater, under conditions of evaporation in excess of inflow, as the volume diminishes the original salt content becomes concentrated in the deeper portions of the basin. That which is deposited on the uncovered shores of the receding water body is mostly washed in again by infrequent rains. Consequently, the salt originally diffused throughout the entire body of water gradually becomes concentrated in a relatively small volume in a central depression. If the original cutoff body had a volume of 100 cubic kilometers of water (Karabughgaz has 183 cubic kilometers) and this became concentrated to about 50 cubic kilometers, the iron oxide and calcium carbonate present would be precipitated. If evaporated to about 20 cubic kilometers, the water would then contain about 3500 million tons of salt, of which about 2700 million tons would be common salt. At this point gypsum would be deposited. When the volume reaches about 10 cubic kilometers, common salt would be deposited. Subsequent evaporation would next bring about deposition of magnesium sulfate and chloride, followed by the bittern salts.

Calcium Sulfate Deposition

Calcium sulfate may be deposited either in the form of gypsum or anhydrite, depending upon the temperature, pressure, and salinity of the solution.

In general, the first salts to separate by the evaporation of seawater are carbonates. When the water has been evaporated to about 19 percent of its original volume, calcium sulfate starts to separate. Ac-

cording to Stewart and McDonald, experimental and thermodynamic work indicates that only anhydrite separates from seawater above about 34°C; at lower temperatures, gypsum separates first and is followed by anhydrite at a later stage of evaporation.

When the solution has evaporated to about 9.5 percent of its original volume, much of the calcium sulfate has been deposited. Halite then begins to precipitate together with gypsum at temperatures below 7°C, and with anhydrite above that temperature as determined by MacDonald.

Posnjak concluded that gypsum would be deposited from saturated calcium sulfate solutions below 42°C and anhydrite above that temperature (Fig. 12-3). Later work by MacDonald changed this figure to 40°C at 1 atmosphere and 46°C at 50 atmospheres. Posnjak and others show in Fig. 12-4 that when seawater evaporates at 30°C, calcium sulfate will always be deposited as gypsum until the salinity reaches 4.8 times normal, beyond which anhydrite is deposited. Further, about one-half of all the calcium sulfate present will be deposited as gypsum before anhydrite is deposited. This means that at the temperatures of evaporation of marine basins, much gypsum will always be deposited first if the temperature is below 34°C, and that marine beds of pure anhydrite imply either that the early deposited gypsum was converted to anhydrite or that deposition occurred above the conversion temperature of 34°C. Hardie has shown that the equilibrium temperature for the reaction $CaSO_4 \cdot 2H_2O = CaSO_4 + 2H_2O_{(liq. sol)}$ is a function of activity of H_2O of the solution. His data indicate that in sea water saturated with halite, gypsum should dehydrate above

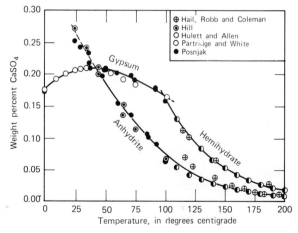

Figure 12-3 Solubility of gypsum, hemthydrate, and anhydrite in the system $CaSO_4-H_2O$ (Posnjak, 1938, p. 268, fig. 3; M. Fleischer, 1963, U.S. Geol. Survey Prof. Paper 440y.)

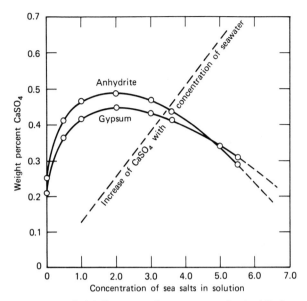

Figure 12-4 Solubility curves for gypsum and anhydrite in solutions of sea salts. Broken line shows the amount of calcium sulfate that is available when ordinary seawater is being evaporated. Concentration is expressed in units of the usual salinity of seawater. (From Posnjak, 1940, p. 565, fig. 1.)

of calcium sulfate of 75 meters are known in Nova Scotia, of 90 meters in Germany, about 600 meters in New Mexico, and more than 1,370 meters in southeastern New Mexico. Since the evaporation of 300 meters of seawater yields only 0.2 meters of gypsum, the evaporation of 130,000 meters, or a depth of 135 kilometers of seawater, would be required to yield 90 meters of anhydrite. As this is ridiculous, it follows that new supplies of seawater must have been added to the basin during evaporation, and the residual liquors became concentrated in subbasins. The depth of water in most closed basins should, upon evaporation, also yield a layer of common salt above the calcium sulfate. Therefore, in the case of gypsum beds with no overlying salt, unusual conditions must have prevailed toward the conclusion of deposition. Further complications are added by the occurrence of salt without gypsum. Any theory of origin of thick beds of gypsum or salt or both must therefore account for a method by which enormous volumes of additional seawater are added to the evaporating basins, for the cessation of deposition before the period of deposition of sodium chloride or bittern salts is reached, and must also account for the alternation of salt and gypsum.

Many theories have been proposed. Most investigators adopt a modification of the ''bar theory'' of Ochsenius, which postulates partial evaporation of an arm of the sea isolated by a barrier or sill over which new supplies of seawater are added to a subsiding interior basin. Under high evaporation, the surface layers supposedly become more dense and sink, causing bottom concentration, and with reversal of flow the residual bitterns are removed. The Gulf of Karabughgaz is used as an example. This theory does not account for the lack of sodium

18°C. This aids in understanding the scarcity of anhydrite in modern evaporite deposits. Zen and others have also determined the effect of salt solutions on the gypsum-anhydrite equilibrium at atmospheric pressures. After sodium chloride starts to precipitate, only anhydrite can be deposited if the temperature exceeds 7°C. Changes in temperature above and below 7°C could cause alterations of gypsum and anhydrite. Under hot, arid conditions anhydrite would be expected; and in temperate regions, gypsum. Either form may change into the other. The increase of NaCl and temperature determine whether gypsum or anhydrite is formed (Fig. 12-5). Gypsum subject to the pressures and temperatures accompanying deep burial is converted to anhydrite. Anhydrite is converted slowly into gypsum rinds under conditions of weathering, as is evident in old gypsum-anhydrite pits.

Most salt deposits throughout the world consist only of calcium sulphate and common salt; the general lack of bittern salts means that the sea water has not been completely evaporated. Consequently, evaporation must have been interrupted, and the bittern liquors were drained off or diluted by influx of more sea water or fresh water.

Deposition of Thick Beds. The origin and conditions of deposition of thick beds of calcium sulfate, or calcium sulfate and salt, is a problem. Thicknesses

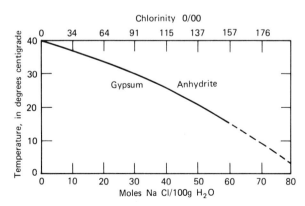

Figure 12-5 Dependence of the dehydration temperature of gypsum on the concentration of NaCl in solution at 1 bar pressure. (From MacDonald, 1953, p. 893, fig. 3.)

chloride or bittern salts in many occurrences. A modification by Branson postulates a series of semiisolated subbasins separated by low barriers so that new supplies of water first enter the outer basins and overflow into the inner basins, carrying there the salt brines of the outer basin; thus, gypsum would be deposited in the outer basins and salt in the inner basins. However, the influxes of new seawater could run over into the inner basins, and bittern salts would also be deposited in the inner basins. Other theories assume the influx of fresh supplies of salts from stream drainage.

When salt is deposited without underlying gypsum, the explanation probably is that the brines from which the gypsum has already been deposited have been decanted into another basin where later deposition of salt could occur. This could be brought about by coastal tilting.

Resulting Products. Calcium sulfate deposition occurs in (1) beds of relatively pure gypsum or anhydrite from a few meters to many hundreds of meters in thickness (gypsum beds constitute one of the most important nonmetallic resources and anhydrite finds little use); (2) gypsum beds with impurities of anhydrite; (3) alabaster, a softer and lighter variety of gypsum; and (4) gypsite, an admixture with dirt. The beds are generally interstratified with limestone or shale, and they are commonly associated with salt.

Common Salt (Halite) Deposition

The deposition of salt beds provides the source for about three-fourths of all salt used. Salt beds are generally lenticular and range in thickness from a meter to hundreds of meters. Beds 110 meters thick occur in Michigan, 210 meters thick in New Mexico, 900 meters (with K) in the Siberian platform, and the Paradox Basin in Utah has a total evaporite thickness of 1330 meters. Thicknesses of hundreds to thousands of meters have been intersected in salt-dome regions of the Gulf Coast and Germany, where salt beds have been squeezed and upthrust. The salt is commonly intricately folded in salt domes, dazzling white and pure. Salt in sedimentary formations may be underlain by gypsum, anhydrite, or shales. The beds are of many geologic ages; in North America they occur mostly in the Silurian, Permian, and Triassic formations. Salt beds are extensive in the northeastern and western states, where they extend into Canada (Fig. 28-2).

Salt is also recovered by evaporation of saline water such as from the Great Salt Lake (Fig. 12-7).

Salt Domes. Salt domes are unusual and interesting. They are known in the Gulf Coast, Mexico, Germany, Poland, Spain, Rumania, USSR, and the Middle East, but those of the United States and the Middle East are of particular interest because of the prolific oil accumulations associated with some of them. Surface expression may be absent, or they may be indicated by low mounds or depressions. Most of them have been located by geophysical methods, especially gravimetric methods, and proved by drilling. They are dome-shaped, pipe-shaped, or also mushroom-shaped (Fig. 12-6). The plug consists of salt and anhydrite, and anhydrite or gypsum generally constitutes a caprock up to 300 meters thick. The salt plug is several thousands of meters thick; the Grand Saline dome is 2700 meters in diameter and 5000 to 7000 meters long. The tops may protrude through the surface or extend to several thousands of meters beneath the surface.

No drill has yet penetrated to their bottoms. The flanks are bordered by upturned strata, which form the reservoir rocks of associated oil pools. They are gigantic plugs that have risen upward diapirically because of their low density into the overlying strata, bending the host rocks upward around the flanks and above the dome to make oil traps. The salt originally existed in sedimentary beds (usually Mesozoic?) several kilometers beneath the surface. Under pressure, the yeilding plastic salt with a density of 2.4 floated upward through sediments of greater density along some line of weakness, such

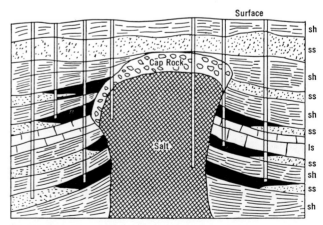

Figure 12-6 Diagram of salt dome showing salt plug, anhydrite caprock, and upturned strata containing oil and gas pools (black) tapped by wells.

as fault intersections. The great thickness of salt is thus not original but has been induced subsequent to deposition. Barton estimates that less than a dozen of the famous salt domes in the Gulf Coast region contain enough minable salt to supply the world for 4000 years and that all known salt domes would supply the world demand for 300,000 years. More than 300 salt domes are known in the Gulf Coast area.

Potash Deposition

After the deposition of common salt, chlorides and sulfates of magnesium and potassium are the other chief salts deposited. The potassium minerals result from evaporation carried almost to completion and, therefore, only rarely are they deposited. Conse-

quently they are not common, and important deposits are known only in Germany, Alsace, Russia, Spain, Poland, Great Britain, Canada, and the United States. Some of the world's supply of potash is derived from marine evaporites, but a growing quantity is won from brines of salt lakes.

The recovery of magnesium metal and chlorides from the Great Salt Lake, Tooele County, Utah, has been initiated by NL Industries. Figure 12-7 shows the solar evaporite site of the project an area formerly covered by the Great Salt Lake, but the lake brine will now be pumped into the area indicated by the numbered mile square sections. The plant is designed to reach 45,000 tons per year of magnesium metal and 80,000 tons per year of chlorine. Lithium may also be produced in the future.

Name	Composition	Percent K_2O
Anhydrite	$CaSO_4$	—
Bischofite	$MgCl_2 \cdot 6H_2O$	—
Bloedite	$Na_2Mg(SO_4)_2 \cdot 4H_2O$	—
Carnallite	$KCl \cdot MgCl_2 \cdot 6H_2O$	16.9
Douglasite	$K_2FeCl_4 \cdot 2H_2O$	—
Glaserite	$3K_2SO_4 \cdot Na_2SO_4$	37.5
Halite	$NaCl$	—
Kainite	$MgSO_4 \cdot KCl \cdot 3H_2O$	18.9
Kieserite	$MgSO_4 \cdot H_2O$	—
Langbeinite	$K_2SO_4 \cdot 2MgSO_4$	22.7
Leonite	$MgSO_4 \cdot K_2SO_4 \cdot 4H_2O$	23.3
Polyhalite	$K_2SO_4 \cdot 2CaSO_4 \cdot MgSO_4 \cdot 2H_2O$	—
Sylvite	KCl	63.2

Potassium and Associated Salts. The famous Stassfurt deposits of Germany represent the only complete sequence of deposition of oceanic salts. There, some 30 saline minerals are known, but many of them have been formed by subsequent secondary reactions. Some of the more common salts of the deposits are shown in the table above, the important potash salts are italicized.

Sylvite is the preferred salt because it has the highest K_2O content, is a simple soluble chloride, and is the most usable mineral.

Special Conditions on Depositions. The deposition of potash salts from sea water requires special conditions of formation that have not occurred with the worldwide deposits of nonpotash gypsum and halite. Their abundance means not only almost complete evaporation but also a stupendous concentration of residual bitterns. This must mean a gradual drainage of bitterns, after removal of gypsum and

much halite, into low portions of basins of evaporation. Thus a small area may come to contain the residual constituents formerly dispersed in a great volume of water spread over a large basin. In addition it seems probable that in most places there must have been some tilting of the land to drain off the bitterns into adjacent basins and aid in concentration.

Thoughout the deposition of potash salts, halite is also deposited. Consequently, beds of pure potash salts are rare; they also contain magnesium salts. Thus certain beds are richer than others. Stratigraphic repetitions are not uncommon, indicating that the cycle of evaporation and deposition was interrupted and renewed many times.

Features of the Stassfurt Deposits. The deposits of Stassfurt and the related areas, provide an illustration of potash deposition by evaporation of sea waters because they not only represent a complete

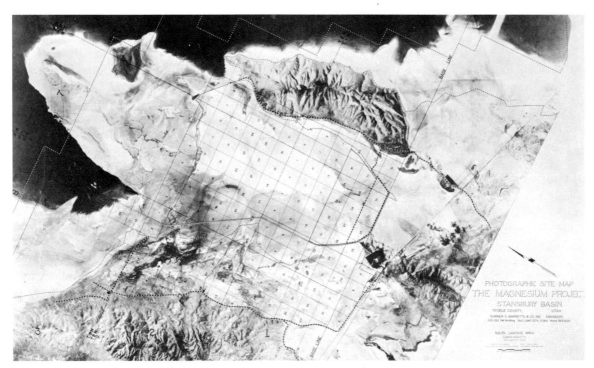

Figure 12-7 Air photograph view of National Industries, Inc. magnesium project where evaporation of Great Salt Lake brine solution will ultimately provide 45,000 tons of magnesium metal per year. (Courtesy N. L. Ind., Inc.)

sequence of oceanic salts but they have long been carefully studied (see Table 12-2).

The deposition of potassium salts closed with the older series (Zones 10 to 6), and a further influx of seawater started the deposition of the younger series (Zones 5 to 1) in which, however, no commercial potassium salts were deposited.

The potash zones also contain magnesium and calcium chlorides, and magnesium sulfate. Halite is present throughout all the zones from 10 to 2. Complex double salts also form as a result of resolution and reprecipitation due to slight changes in temperature and salinity during deposition. A yield of about 9 percent K_2O is obtained from the carnallite

Table 12-2 Vertical Section, in Meters, of Salts at Stassfurt, Germany

No.	Zones	Materials	Percent K_2O	Thickness	Depth from surface
1		Surface drift		10	10
2	Red clay	Clays, sandstone, shales, salt, and anhydrite	0	100	110
3	Younger salt	Rock salt	0	100–150	210–260
4	Main anhydrite	Massive, widespread	0	30–90	240–350
5	Saline clay	Clay, salt	0	6–10	245–360
6	Carnallite	Carnallite, halite, kieserite, sand, clay, some sylvite, kainite	9.27	15–40	300±
7	Kieserite	Kieserite + carnallite and bischofite rock salt, anhydrite	2.17	20–40	350
8	Polyhalite	Rock salt, $MgCl_2$, gypsum, etc.	1.02	40–60	410
9	Older rock salt	With anhydrite in thin annular layers	Trace	300–500	800
10	Older anhydrite	Anhydrite and gypsum	Trace	60–100	1025
11	Underlying	Zechstein marine limestone		4–10	1030

salts and about 13 percent K_2O from the sylvite-kainite salts.

Several of the Stassfurt salts have resulted from subsequent transformations due to increased temperatures attendant upon burial. Kainite and sylvite are alteration products of carnallite. Other salts represent retrograde metamorphism or reversals brought about by later rise of the beds after deep burials. Such potash minerals constitute delicate geologic thermometers. According to Van't Hoff and his associates, salts of the Stassfurt region had temperatures of formation about 10°C for loewite with glaserite and 60°C with vanthoffite; and 72°C for kieserite with sylvite. Recent synthesis by Ide suggests much higher temperatures for langbeinite, vanthoffite, and polyhalite. Investigations by Weber show that much replacement has gone on: polyhalite and kieserite replaced anhydrite; vanthoffite replaced kieserite and carnallite, and kieserite has been altered to sylvite. Some of these changes are considered to have occurred by downward percolation of potassium and magnesium chloride solutions through underlying anhydrite. Ahlborn has shown that where pressure was greatest "hartsalz" was converted to langbeinite; and where pressure was least, sylvite-rich or carnallite-rich beds were formed. It is evident that the formation of the complex salts has been a complicated geochemical process and not merely simple deposition through evaporation.

Borate and Bromine Deposition

Minor quantities of borates and bromine are obtained from marine salts. Although *borates* are mostly formed under other conditions, some are precipitated along with potassium minerals from marine residual liquors. Thus boracite and other borates occur in the potash salt of Germany in association with carnallite and the overlying potash minerals. Magnesium borates are considered to be typical of marine conditions and calcium borates of lake-bed deposits. Most borates of commerce are obtained from lakes, lake-bed deposits, or dry lakes (playas).

Bromine is also deposited from residual liquors of seawater. The carnallite of Stassfurt contains 0.2 percent bromine, which is extracted in Germany during the refining of the potash salts. Bromine is remarkably concentrated in the Dead Sea to the extent of 0.4 percent compared with 0.0064 percent in ocean water. Most bromine, however, is a by-product of salt, from salt brines and seawater.

Figure 12-8 shows a portion of the sequence of deposits idealized for a part of a closed basin where long continued evaporation has taken place. The basin may be hundreds of kilometers long but is greatly magnified vertically to illustrate the distribution of bromine and potassium. It is possible to determine the salinity of the evaporating brine at different stages by measuring the amount of bromine substituting for chlorine in the halite. As an

Figure 12-8 The distribution of bromine and potash in an idealized evaporite basin. (Based on Raup and Hite, 1965.)

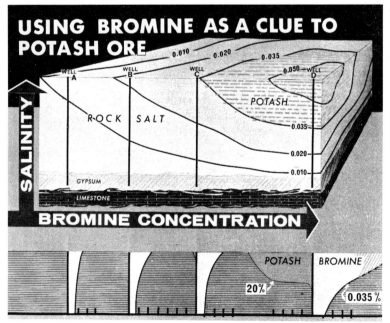

example, the first crystals of salt to precipitate from natural marine waters contain about 0.007 percent bromine. As evaporation continues, the bromine content also increases and more bromine replaces salt. The idealized increase in bromine content in the salt is shown by Fig. 12-8, and so the bromine content of salt is an index of salinity of the brine from which it formed.

Deposition from Salt Lakes

Salt lakes contain the same salts as the ocean but generally in greater proportions. The salinity may range from less than that of the ocean to that of Great Salt Lake, which has a varied composition. This depends upon snowfall in the Wasatch Range, which feeds melt water to all the streams that flow into the lake. The salinity, therefore has varied from 13.79 percent in 1877 to 27.72 percent in 1904.

A landfill railroad causeway was built across the Great Salt Lake in 1958 with only two small culverts connecting the north and south partitions of the lake. Because almost all of the fresh water rivers flow into the southern portion of the lake, the salinity of the lake north of the causeway is at the saturation level of about 27 percent; south of the causeway the salinity is considerably less. This result is of deep concern to those solar evaporation industries in the southern portion of the lake, especially at the present time of increased snowfall and greater fresh water flow into the southern part of the Great Salt Lake. This difference in salinity is shown in space photographs on images as the variety of bacteria in different salinity contents reflect light in different portions of the spectrum as shown by Fig. 1-14.

Since the Great Salt Lake is only a residual of the greater Lake Bonneville, its salts have undergone considerable concentration by evaporation, and it has lost most of its calcium carbonate. Gilbert estimated it to contain 400 million tons of common salt and 30 million tons of sulfate. The Dead Sea is an example of a similar type that has been subject to extreme evaporation and is highly enriched in magnesium chloride.

Deposits. The deposits formed from the evaporation of salt lakes are similar to those obtained from ocean water. The relatively small size of lakes, however, makes them more responsive to climatic changes, with the result that they exhibit greater fluctuations of deposition. Evaporites formed during periods of desiccation may be redissolved during subsequent periods of expansion. Moreover, lakes constantly receive new supplies of fresh water, salts, and also sediments. The resulting saline deposits, therefore, are generally thin-bedded alternations of impure salts and clays. Also, on salt playas, desert winds distribute sands and silt, upon which later salts may be deposited during subsequent lake periods. This also gives alternations of salines with sand, clay, and minor calcium carbonate.

Examples of salt-lake deposits are common throughout the Basin and Range province of the United States. Hundreds of square kilometers of level, glistening white salt lie west of Great Salt Lake and support a growing industry (Fig. 1-15). The Salton area of the Imperial Valley of California formerly yielded considerable salt from a surface crust 25 to 50 centimeters thick. Examples of salt pans are numerous throughout central Asia and northern Africa. Huntington tells of the salt plain of Lop in Asia, where choppy waves of hard salt can be seen over a length of 320 kilometers and a width of 80 kilometers. Salt lakes of the past have been disclosed by drilling in Oklahoma and elsewhere.

Deposition from Alkali and Bitter Lakes

Lakes rich in sodium carbonate and sodium sulfate are common in the arid regions of North America where igneous rocks, or sedimentary rocks rich in sulfates, are found. Many igneous rocks upon weathering yield sodium and some potassium in greater abundance than do sedimentary rocks. Thus the high content of sodium is explainable. The lakes are high in $CO_3^=$, Na^+ and $SO_4^=$, relatively high in chlorine, contain appreciable quantities of magnesium and potassium and variable amounts of SiO_2, and are low in calcium and iron. In the early stages of the formation of such lakes, calcium carbonate derived from weathering is introduced, but, with increasing concentration from evaporation, calcium is thrown down in the form of tufa, leaving sulfates and chlorides in solution.

Alkali Lakes. In alkali or soda lakes, sodium carbonate predominates, potassium carbonate may be abundant, and common salt is always present. Most of the sodium carbonate has been derived directly by decomposition of volcanic rocks, but some is also formed by slow and complex chemical reactions with other sodium and calcium salts; it may be formed also by the action of algae on sodium sulfate. The potassium carbonate is considered to be the indirect product of the work of organisms.

Examples of alkali lakes are Owens and Mono Lakes in California, the Soda Lakes of Nevada, Lake Goodenough in Saskatchewan, and the Natron Lakes of Egypt; they are numerous in the arid parts of the world. Owens Lake has high salinity that varies with seasonal and climatic cycles. The Soda Lakes of Nevada contain about one-fifth each of sodium carbonate and sodium sulfate and three-fifths of sodium chloride. The deposits of hydrated salts, however, consist of nearly one-half sodium carbonate, one-third bicarbonate, and only a little sodium sulfate and halite. The Natron Lakes of Egypt are alternately wet and dry, and evaporation leaves a layer of natron and salt, bordered by sodium carbonate.

Bitter Lakes. In bitter lakes, sodium sulfate predominates, but carbonate and chloride are present. The sulfate may be derived from the decomposition of rocks that contain sulfates, or from the leaching of buried beds of sulfates. Such lakes are common in the arid regions of America and Asia. Examples are Verde Valley Lake in Arizona; Soda and Searles Lakes in California; and numerous lakes in New Mexico, Wyoming, and Saskatchewan. Lakes Altai and Domoshakovo in Russia are Asiatic representatives. The Soda Lake playa salts contain 88 percent anhydrous sodium sulfate, and the crude salts of the Downey Lakes of Wyoming contain over 95 percent sodium sulfate. The Gulf of Karabughgaz deposits marginal sodium sulfate each winter, and this is harvested.

Deposits. The alkali and bitter lakes yield commercial sodium carbonate and sodium sulfate from brines, and from marginal and playa deposits.

Sodium carbonate is deposited around soda lakes as an efforescence, known as "black alkali" because of its effect upon vegetation. The efforescences also contain some chlorides, sulfates, or nitrate and borax. According to Chatard, fractional crystallization attendant upon evaporation yields (1) trona with some soda, (2) sodium sulfate, (3) common salt, and (4) normal sodium carbonate. Gaylussite ($CaCO_3 \cdot Na_2CO_3 \cdot 5H_2O$) is also deposited from soda lakes. The Natron Lakes of Egypt become desiccated during the dry season and leave a crust 25.4 to 68.5 centimeters thick of common salt and natron and an efforescence of natural soda.

The largest deposits of sodium carbonate and the greatest production at the present time is from the underground trona deposits in the Bridger basin of southwestern Wyoming. These evaporite deposits

provide more than 85 percent of the U.S. consumption of sodium (Refer to Chapter 28.)

Sodium sulfate is obtained from concentrated brines of bitter lakes. Harvestable crops are also deposited as marginal efforescences of "white alkali." It is deposited in the bottoms of dried-up lakes mainly as Glauber's salt or mirabilite ($Na_2SO_4 \cdot 10H_2O$) and as the anhydrous sulfate thenardite (Na_2SO_4). Minor amounts of halite, natron, and magnesium and calcium sulfates are generally present. Some of the associated sulfate salts are: glauberite, bloedite, and aphthitalite.

The deposition of sodium sulfate generally precedes the deposition of salt and may be controlled by the season of the year. Since it is much more soluble in warm than in cold water, whereas salt is not, it may be deposited as mirabilite in winter and washed up on the shores, as at Great Salt Lake and the Gulf of Karabughgaz. From warm (above 32.4°C) concentrated waters, thenardite is deposited. The mirabilite in warm dry air loses its water and changes to thenardite. The bottom deposits may be interstratified with beds of epsom salts and sodium carbonate.

The crude salts as mined, mainly Glauber's salt or thenardite, are commonly relatively pure sodium sulfate (85 to 96 percent). Glauber's salt is harvested from the Saskatchewan lakes, Great Salt Lake, Downey Lakes, and Karabughgaz; thenardite is characteristic of warm regions and is the salt obtained from Searles Lake and Verde Valley.

For uses, production, occurrence, and extraction of sodium sulfate see Chapter 28.

Deposition from Potash Lakes

Some of the alkali lakes contain potash in amounts that permit commerical extraction. The potash lakes of Nebraska, which are just hollows in sand dunes, are of interest. The potash is believed to have come from the surrounding country that formerly was burned over by the Indians, releasing plant ashes. The evaporated salts are high in potassium sulfate and carbonate and contain soda, salt, and sodium sulfates; one crust contained 21 percent K_2O.

The Great Salt Lake, Utah, is the most important lake source of potash in the United States. The Searles Lake, California, deposit is also rich in saline minerals of which some 15 are unknown in the Stassfurt deposits. The list includes six sulfates, six carbonates, four triple salts, two chlorides, and one borosilicate. The potassium is confined to the residual brine. Potash is also present in Mono Lake, California; and Columbus Marsh, Nevada.

Deposition from Borate Lakes

Borate lakes are relatively uncommon, but several are known in California, Nevada, Oregon, Tibet, Argentina, Chile, and Bolivia. Formerly, most of the borax in the United States was obtained from lake waters in California and Nevada or from playas. Subsequently, borax was made less expensively from colemanite and ulexite, and later from kernite. At present, the only lakes yielding commercial borax are Searles and Owens, in California, where it is extracted in conjunction with other salts.

Boron is a constituent of some igneous minerals, and, in the form of boric acid and calcium and magnesium borates, it is well known in the volcanic exhalations of Tuscany, where steam jets have yielded much commercial boric acid. It also occurs in the exhalations of Vulcano. Borax is a constituent of some hot springs, and it occurs in some lakes in volcanic regions, as Borax Lake, California, where it constitutes 12 percent of the evaporated salts. The borax of the lakes is considered to have been leached from surrounding igneous rocks or to have been contributed by magmatic hot springs. Clarke thinks calcium borates are indicative of lake deposits.

Deposits. Borates are obtained from brines or playa deposits. The brines of Searles and Owens Lakes yield borax, as do also those of Borax Lake, where, in addition, borax crystals are embedded in the marginal mud. Incrustations on the Searles Marsh playa have also yielded much borax. The famed dazzling white surface crusts of borax in Death Valley, however, are no longer worked. Borax crusts also occur in Chile, Bolivia, and Argentina.

The chief boron minerals of playas and brines are borax ($Na_2B_4O_7 \cdot 10H_2O$), colemanite ($Ca_2B_6O_{11} \cdot 5H_2O$), and ulexite ($Na_2 \cdot 2CaO \cdot 5B_2O_3 \cdot 16H_2O$); searlesite ($Na_2O \cdot B_2O_3 \cdot 4SiO_2 \cdot 2H_2O$) is also found at Searles Marsh. Other boron minerals and their occurrence are discussed in Chapter 28.

Deposition from Nitrate Lakes

Nitrate lakes are common in arid regions but rarely contain commercial nitrate deposits. The nitrogen of nitrates originally came from the atmosphere. Sodium and potassium nitrates are characteristic of fertile soils, but, being readily soluble, they are removed in humid regions and become concen-

trated only in dry regions. They may originate in the soil by oxidation of organic matter in contact with alkaline salts, or by the action of nitrogen fixing bacteria, which live in the roots of legumes. The existence of such deposits is rare. It is limited to extremely dry desert areas because of the high solubility of nitrogen-bearing minerals.

Soda niter, along with ammonium compounds, is known in Searles Lake and in lake deposits in Death Valley, Utah, and Nevada. Saline lakes in Argentina contain liquors from which soda niter is obtained. It is also found in playa deposits in Colombia and Bolivia, again in association with boron compounds. It is reported also from Matsap, South Africa. Deposits occur near Winkelman, Arizona; and Glenrock, Wyoming. The almost sole source of nitrogen at the present time is from the atmosphere.

DEPOSITION FROM GROUNDWATER

Evaporation of groundwater is universal, but in humid regions the evaporites are mostly redissolved and removed by rainwater. In arid regions the evaporites may accumulate as long as the climate remains arid.

The groundwater contains salts similar to those of lakes and oceans. However, the concentration is generally low, and the proportions of the individual salts vary with the character of the soil, bedrock, topography, and climate. Calcium carbonate is almost invariably present; magnesium, sodium, potassium, iron, and manganese compounds are common; silica is generally present, as is also phosphorus; and, locally, boron and iodine may be relatively abundant. The groundwater in Chile has yielded vast quantities of commercial nitrates, and some deposits of copper are due to evaporation.

Deposition ensues when evaporation occurs at or near the surface, or in caves. If the site of evaporation is fed by fresh supplies of groundwater, extensive deposits may eventually result. Evaporation will, of course, proceed most rapidly where groundwater is supplied relatively close to the surface, such as valley bottoms, slopes where hills and valleys merge, and long hill slopes interrupted by gentler or reverse grades.

Deposits. The deposits of economic importance formed by evaporation of ground water are nitrate salts with iodine, some boron, calcium carbonate, sodium sulfate, and sodium carbonate.

The processes of formation of *nitrate deposits* are illustrated by the famed *Chilean nitrate deposits,* which are the only large deposits and, until the development of nitrogen recovery from the atmosphere, formerly supplied the world with natural nitrate for fertilizer, explosives, and other chemicals. The deposits are found in the Atacama, Tarapaca, and Antofagasta deserts of northern Chile, where several discontinuous deposits occur at moderate elevations in a long narrow belt from 20 to 160 kilometers inland. The deposits, called *caliche*, lie on gentle slopes on the east side of the Coast Range. They consist of a thin bed of gravel cemented by sodium nitrate and associated salts at depths of a few centimeters to a few meters beneath the surface. About one-fourth is sodium nitrate, along with common salt, sulfates, borates, bromides, iodates, phosphates, and lithium and strontium compounds. They also yield most of the world's supply of iodine. The overburden contains a little nitrate, salt, gypsum, and sodium sulfate. (See also Chapter 28.)

The origin of these deposits is controversial. It is generally agreed that the nitrates were transported by groundwater and were deposited by its evaporation, as advocated by Miller and Singewald. The source of the nitrate, however, has been attributed to (1) bird guano, subsequently leached (Penrose); (2) nitrogen fixation by electrical discharges from thunderstorms that sweep the Andes (Pissis, Rogers, and Van Wagenen); (3) the bacterial fixation of former nitrogenous vegetable matter, and (4) a volcanic source via nearby Triassic and Cretaceous tuffs containing ammonium salts that were oxidized during weathering under unusual climatic conditions (Clark, Whitehead). The soluble nitrates were carried downslope to the site of evaporation. Whitehead estimates that the erosion of 1 meter of the volcanic rock, carrying 1 percent NH_4Cl, over the nitrate areas would supply all the nitrate of the deposits. The correspondence between volcanic rocks and nitrates can hardly be casual. Moreover, such an origin is supported by the constant association of borates, a relationship noted previously in connection with the California borates.

Graham, however, states, without elaboration, that "it seems definite that nitrate was finally collected in an inland sea, which eventually deposited part of its load along its shore lines. Finally, the sea drained into the Pacific Ocean, but the deposits laid down on the eastern shore were either dissolved or washed away by the drainage from the Andes or covered with alluvial debris, leaving the present deposits on the western shore of the inland sea or the eastern slope of the Coast Range."

Be all this as it may, there is fair agreement at the present time that the nitrogen was derived from nitrogen-fixing bacteria, a process occurring in many environments; but the nitrogen minerals are rapidly dispersed except in areas of little or no rainfall, such as the Atacoma desert. The relative lack of phosphorus in the deposits rules out guano as a source of the nitrates.

Natural brines result when sediments saturated with seawater are buried beneath the level of circulating groundwater. These are called connate waters and become sources of commercial salts. Other natural brines result from underground solution of salt beds. The salt waters of oil pools are also natural brines. The brines yield salt, bromine, lithium, iodine, and calcium and magnesium chlorides.

Artificial brines are formed by pumping water down wells drilled into rock-salt beds, and the brines so formed are the main source of common salt production in the United States. Potash is also obtained by artificial brines at great depth in Saskatchewan. The large playa, south of Silver Peak, Nevada, is the largest producer of lithium, which is pumped from brine solutions remaining in the deeper sediments below the playa.

Calcium carbonate in solution in the ground water is also deposited by evaporation at or near the surface to form "caliche" deposits in the arid regions. These deposits commonly underlie arid valleys floored by detrital matter and serve to cement the detritus, but they are most abundant on piedmont slopes, where subsurface waters from higher regions approach the place where hills and valleys merge. Rarely such deposits are locally utilized for lime, fertilizer, or cement. Cave dripstones also result from evaporation.

Other depositions consist of crusts of common salt, Glauber's salt, soda, epsom salts, and borax, deposited in arid regions by evaporation of subsurface water. Silica is also deposited by evaporation, causing surface silicification or "case hardening" of altered rocks, gossans, or fault material, which have often been mistaken for outcrops of mineral bodies.

DEPOSITION FROM HOT SPRINGS

The substances contained in solution in hot-spring waters build up deposits about their orifices, partly as a result of evaporation. A few such deposits are of commerical importance; others are scenically beautiful. Their deposition, however, is not, in all cases, the result of evaporation alone; microorganisms help deposit some substances, and the escape of carbon dioxide under the reduced surface pressure also causes deposition. The chief substances deposited in this manner are (1) calcium carbonate, in the form of tufa, travertine, or calcareous sinter; (2) silica in the form of siliceous sinter or geyserite; (3) iron oxide in the form of ocher; and (4) manganese dioxide in the form of wad. Many other nonmetallic and metallic substances are deposited from hot springs but in relatively small amounts.

The *calcium carbonate* deposited from hot carbonated waters forms deposits of tufa; the famed White Terrace of New Zealand and the Mammoth Hot Springs of Yellowstone Park, composed of almost pure calcium carbonate, are noted for their beauty and scenic interest. Other deposits of tufa are locally utilized for cement, lime, and fertilizer. One form, known as *travertine,* is much prized as an interior decorative stone because of its pleasing texture and vuggy nature; the type locality is at Travertine, Italy, where it has been quarried since the time of the Romans.

Silica in hot-spring waters is deposited by evaporation and hot-water algae as a colloidal gel, which crystallizes to chalcedony, opaline quartz, agate, microcrystalline quartz, and quartz. The famous siliceous sinters or geyserite of the Yellowstone Park geysers consist of almost pure silica. Siliceous sinters of hot springs generally contain small quantities of metals, as at Steamboat Spring, Nevada; the Yellowstone geyserite contains arsenic and pyrite. A few spring deposits have been mined for metals, for example, mercury in New Zealand.

Selected References on Evaporation

Saline deposits. R. B. Mattox, Editor, 1968. Geol. Soc. Amer. Spec. Paper 88. *A symposium on various features of world wide saline deposits and their origin.*

Salt Deposits, the Origin, Metamorphism and Deformation of Evaporites. 1964. H. Borchert and R. O. Muir. Van Nostrand Reinhold Company, New York.

Data of Geochemistry, 6th ed. Chap. Y, Marine evaporites. 1963. F. H. Stewart. U.S. Geol. Survey Paper 440Y.

Early stages of evaporite deposition. 1961. E-an Zen. U.S. Geol. Survey Prof. Paper 400B, p. 458–461.

Stratigraphy and Sedimentation, 2nd ed. 1963. W. C. Krumbein and L. L. Sloss. eds. W. H. Freeman and Company, Publishers, San Francisco, p. 567–590.

Physiochemical conditions of the formation of potassium salt deposits of the past. 1957. M. Y. Valyaskho. Geochem. Acta, v. 6, p. 553–563.

Symposium on salt. 1961. A. C. Bersticker, ed. Northern Ohio Geol. Soc., Cleveland, Section 1.

Geology and stratigraphy of Upper Silurian Cayugan evaporites.

Deposition of calcium sulphate from sea water. 1940. E. Posnjak. Am. J. Sci. v. 238, p. 559–568. *Chemical controls of deposition of gypsum and anhydrite.*

The gypsum-anhydrite equilibrium at one atmosphere pressure. 1967. L. A. Hardie. The Amer. Mineralogist, v. 52, p. 171–200.

Solubility measurements in the system $CaSO_4 \cdot NaCl\text{-}H_2O$ at 35°, 50°, and 70° and one atmosphere pressure. 1965. E-an Zen. J. Petrol., v. 6, p. 124–164.

Potash in North America. 1943. J. W. Turrentine. Am. Chem. Soc. Mon. Ser. 91. *General Survey.*

Owens Lake: Source of sodium minerals. 1947. G. D. Dub. AIME Tech Publ. 2235. *Saline minerals and occurrence.*

General Reference 9, Chap. 20, and Chapter 19 on evaporites.

13

RESIDUAL AND MECHANICAL CONCENTRATION

I seem to have been only like a boy playing on the seashore, and diverting myself in now and then finding a smoother pebble or a prettier shell than ordinary, while the great ocean of truth lay all undiscovered before me.

Sir Isaac Newton

Under the slow, unrelenting attack of weathering, rocks and enclosed mineral deposits succumb to mechanical disintegration and chemical decomposition. Weathering is a complex operation consisting of distinct processes that may operate singly or jointly. The minerals that are unstable under weathering conditions suffer chemical decay; so soluble parts may be removed and the insoluble residues may accumulate, and some of them may form residual mineral deposits. Even such a stable mineral as quartz undergoes chemical changes and may be leached from its enclosing matrix and thereby result in a residual enrichment deposit.

Weathering is generally considered to consist of mechanical and chemical action; usually both operate together. Mechanical disintegration, such as that by frost action does not create new minerals; it merely frees minerals already formed. However, it facilitates chemical decomposition by reducing the particle size and creating greater specific surface available for chemical attack. Chemical weathering, however, actually creates new minerals of which some remain stable under surface conditions.

Weathering Processes

Mechanical disintegration by itself is confined largely to arid regions, or to very cold regions where surface chemical changes proceed slowly and frost action is vigorous much of the time. Chemical weathering is most active in warm, humid regions where more rainfall containing more CO_2 allows the increased solubility of limestone. In arid regions, limestone forms rim rocks and ledges because of the lesser precipitation containing CO_2.

Rainfall supplies and nourishes plant life that yields humic and organic acids to assist chemical activity.

Deep and long-continued weathering is necessary to yield residual products. Hence such products are generally absent from glaciated regions where residual products have been removed by ice and the time since glaciation has been too brief to permit much weathering under the unfavorable cold climates that prevailed then.

The agents of decomposition that operate at the surface are water, oxygen, carbon dioxide, acids, alkalies, vegetable and animal life, and some of the soluble products of the decomposition of the rocks themselves. Without water, little decomposition occurs; hydration is general and is an important factor in both decomposition and disintegration. Eh values permit oxidation. Carbon dioxide dissolved in water is a powerful solvent, especially of limestone; acids, such as sulfuric acid, and some sulfates, generated by oxidation of sulfides, are active agents of decomposition.

Weathering commonly does not extend deeper than a meter in arid areas to a few tens of meters in areas of about 100 centimeters of precipitation, and depths of 100 to 200 meters of weathering are known in such areas as the Amazon with 750 centimeters of precipitation per year. Sulfides are known to oxidize down natural fractures to depths of more than 1000 meters.

Temperate versus Tropical Climate Weathering. There is a difference, not fully understood, in the nature of weathering in temperate climates as compared with that in tropical and subtropical climates. In temperate climates the silica of silicate rocks is not extensively removed but remains in the form of clay, along with hydrous oxides of iron and perhaps residual grains of quartz. Thus clayey soils are not common products of weathering of such rocks in all temperate climates. Tropical and subtropical climates are characterized by alternate wet and dry seasons, by hot weather and warm surface waters throughout the year, and, generally, by luxuriant vegetation with an abundant supply of bacterial life and organic compounds. Under these conditions, rock decay is carried further, leaching is more complete, the silicates are more thoroughly decomposed, but, particularly, the surface waters readily and extensively remove silica in solution. This is a very important distinction. The result is a *laterite* soil, which is a mixture of hydrous oxides of aluminum and iron with some silica and other impurities. Instead of the hydrous aluminum silicate (clay) of temperate regions, it is hydrous aluminum oxide (bauxite). Laterites may be so high in iron that they constitute a laterite iron ore, or, conversly, they may be so high in alumina and low in iron and silica that they constitute an ore of aluminum; most laterites are neither.

RESIDUAL CONCENTRATION

Residual concentration results in the accumulation of valuable minerals when undesired constituents of rocks or mineral deposits are removed during weathering. The concentration is due largely to a decrease in volume effected almost entirely by surficial chemical weathering. The residues may continue to accumulate until their purity and volume make them of commercial importance.

Process of Formation

The first requirement for residual concentration of economic mineral deposits is the presence of rocks or lodes containing valuable minerals, of which the undesired substances are soluble and the desired substances are generally insoluble under surface conditions. Second, the climatic conditions must favor chemical decay. Third, the relief must not be too great, or the valuable residue will be washed away as rapidly as formed. Fourth, long-continued crustal stability is essential in order that residues may accumulate in quantity and the deposits may not be destroyed by erosion. Given these conditions, a limestone formation free from other impurities except minor iron oxide, for example, will

slowly be dissolved, leaving the insoluble iron oxide as residue. As bed after bed of limestone disappears, the iron oxide of each bed persists and accumulates as an insoluble residue, until eventually there may be formed an overlying mantle of iron ore of sufficient thickness and grade to constitute a workable deposit.

The above example typifies one mode of formation wherein the residue is simply an accumulation of a preexisting mineral that has not changed during the process. In another mode of formation, however, the valuable mineral first comes into existence as a result of weathering processes and then persists and accumulates. For example, the feldspar of a syenite decomposes upon weathering to form bauxite, which persists at the surface while the other constituents of the syenite are removed.

Tropical climatic conditions yield residual concentrations, although tropical and subtropical conditions favor the formation of more kinds of deposits. Valuable deposits of iron ore, manganese, bauxite, clays, nickel, phosphate, kyanite, barite, ochers, tin, gold, and other substances accumulate as residual concentrations. The basic processes of weathering and concentration apply to each, but, inasmuch as the source materials, chemical changes, and other details of formation differ considerably for each substance, the cycles of the various products are considered separately below. Possibly those residual deposits now located in temperate regions, such as Arkansas bauxite, could have been leached earlier when continental drift allowed the area to be farther southeast in a tropical climate.

Residual Iron Concentration

Most rocks contain iron, and under favorable conditions a sufficient quantity may accumulate in the residue to form workable deposits of residual iron ore. This depends, in part, upon the nature of the rock that contains it, the form in which the iron occurs, and the amount of iron present. Also, the climatic and physiographic conditions must be favorable.

Source Materials. The iron or residual iron-ore deposits may come from:

1. Lode deposits of siderite or iron sulfide. The iron oxide residues of pyritic deposits, however, are only rarely used as iron ores.
2. Disseminated iron minerals contained in non-aluminous limestone.

3. Limestones that have been partly replaced by iron minerals, either before or during the period of weathering.
4. Basic igenous rocks.
5. Ferruginous siliceous sediments.

Mode of Formation. Of the source materials in the previous list epigenetic deposits of siderite, or limestone, ferruginous chert, and basic igneous rocks are the chief sources of residual iron ores. In temperate regions, limestones and siderite lode deposits are the only source rocks of residual iron ores, since all others leave a residue too high in silica or in alumina, or both, to constitute ore. The various source materials are considered in the following paragraphs.

From Iron Carbonates. Bodies of siderite or ankerite readily weather to yield oxides of iron, which may accumulate as residual concentrations. The highly ferriferous carbonates yield good clean iron ore, such as the valuable deposits of Bilbao, Spain.

From Pyrite. Massive pyritic bodies yield extensive gossans of goethite and hematite that locally are used for iron ore, but their high sulfur content makes them generally undesirable. Limonite of this type has been mined at Rio Tinto, Spain, and in Shasta County, California. Tarr described sink holes 150 meters in diameter in Missouri, occupied by marcasite that weathered to hematite nearly 30 meters thick, which was mined as iron ore.

From Limestone. The weathering of relatively pure limestone containing a little iron carbonate will result in solution of the limestone and a residuum of iron oxide. If conditions favor accumulation and if a considerable thickness of limestone is weathered, as in weathering on plateaus, then workable iron deposits may result.

E. C. Eckel points out, however, that many of the limestones underlying residual iron ores contain only a slight amount of disseminated iron and much larger amounts of silica and alumina and that their weathering would yield a clay instead of an iron oxide residue. He concludes that the decay of such limestones cannot yield deposits of residual iron ores. If a limestone contains 4 percent of insoluble material, of which one-eighth is iron, it would be incorrect to conclude that the decay of 30 meters of limestone would yield a 15 centimeter thick bed of iron ore. On the contrary, it would yield, under ordinary weathering, about a one meter thick bed of residue that would be a clay with only about 12 percent iron content. Consequently, residual iron-ore deposits resting on such a limestone must have

been derived from overlying beds in which the iron was somewhat concentrated either before or perhaps during weathering. This is the conclusion of Eckel for most residual iron deposits resulting from limestone decay in temperate regions.

During weathering and accumulation of iron residue there was unquestionably some solution and redeposition of iron. This is indicated by the common radial and colloform structure of the limonite.

Basic Igneous Rocks. Basic igneous rocks will not yield residual iron ores from normal weathering in temperate climates because clay accumulates in excess of iron oxide. Under tropical weathering, however, all the constituents except alumina and iron may be removed in solution, leaving a residue composed chiefly of either brown ore (limonite) or bauxite (aluminum hydroxide) or both. The weathering of serpentine has yielded extensive residual iron and nickel deposits in Cuba, and the formation of laterite in tropical and subtropical regions is well known. Since the weathering residuals in tropical regions are chiefly the hydrated oxides of iron and alumina, it follows that the iron ores of such regions are high in alumina and low in silica, phosphorus, and sulfur.

Summary. There are many puzzling features about the formation of residual iron deposits that are not yet understood. In the Appalachian region, carbonates were removed and silica remained; in the Lake Superior region, India, Brazil, Liberia, and many other places, silica and carbonates were removed and iron and a little clay remained; in Arkansas, silica, iron, and other substances were removed and aluminum hydroxide remained; in Cuba, silica was removed and iron and nickel remained. The large-scale removal of silica from the iron formation of the Lake Superior region has been demonstrated by Marsden and others, and it is clearly recognized that silica has been removed in the formation of bauxite deposits. The abundance of rainfall and the pH of the leaching aqueous solution determine the solubility of silica. Of course, the geological time of the leaching of these different areas must be considered as the areas could have been located in different climatic conditions of the earth because of continental drift.

Classes of Resulting Deposits

Commercial iron-ore deposits of residual origin may be divided into:

1. Those derived from preexisting masses of sid-

erite (or associatged ankerite). These are relatively few but locally important. The ores are of high grade, are valuable, and are relatively low in impurities.

2. Those derived from the oxidation of iron sulfides. They are small, impure, and rarely of even local importance.

3. Those derived from limestone that contains seams, bunches, or beds of iron carbonate. This group includes the greatest number of residual iron deposits and the most important of the limestone-derived ores. In general they contain medium-grade ores and impurities of clay and silica but they are low in deleterious sulfur.

4. Those derived from disseminated iron minerals in beds of limestone. The ores commonly have a high clay content or consist of nodular and concretionary masses of hydrous ferric oxide embedded in clay. They are apt to be of low grade and difficult to treat.

5. Those derived from basic igneous rocks. They are formed only under tropical and subtropical climates. The deposits may be rather extensive but are generally high in alumina, although low in silica, phosphorus, and sulfur.

6. Those derived from ferruginous siliceous sediments. Extensive bodies of noncommercial banded ironstone may result, or, in the case of the Lake Superior type iron ores derived from such materials; the resulting deposits may be high grade, large, and number among them some of the important commercial iron-ore deposits of the world. Those residual deposits in the United States, however, are mostly depleted.

Distribution and Examples of Deposits

Residual iron ore deposits are widely distributed except in glaciated regions; they are particularly characteristic of warm, humid regions underlain by limestone or highly ferriferous silicate rocks. Thus they are common in the southeastern United States, the West Indies, Brazil, Venezuela, southern Europe, Africa, and India. A few examples will illustrate residual concentration processes.

Valuable iron ores near Bilbao, Spain, have resulted from the weathering of large replacements of iron carbonate in limestone, yielding extensive surficial blankets of residual iron oxides of high purity that still contain some unweathered carbonate.

Southern States. The southern states include nu-

merous deposits of residual "brown ores," principally in three regions—the Appalachian, the Tennessee River, and northeastern Texas.

The best deposits occur in the Appalachian Valley, in lower Paleozoic strata near the Precambrian contact where there are several hundred million tons. The content ranges from 30 to 55 percent Fe, and the ores are high in Al_2O_3, SiO_2, P, and Mn, and low in S. Even so, they are not considered to be an important future reserve. The ores are mostly irregular pocket deposits enclosed in clay and derived from the decomposition and solution of folded Cambrian-Silurian limestones and dolomites. The "valley ores," containing from 50 to 55 percent Fe, rest on limestone and are mostly shallow; the "mountain ores," containing 35 to 50 percent Fe, rest in residual material generally above the Lower Cambrian quartzite and may extend to depths of a few hundred feet. The ores consist of lumps and nodules in clay, from which they have to be "log-washed." The low-grade "Oriskany" ores of Virginia replace fracture zones in the folded calcareous Oriskany sandstone and reach depths of 200 meters.

The Tennessee River brown ores occur as nearly horizontal mantles in lower Carboniferous limestones. They also form pockets in clay and are of somewhat higher grade than most of the Appalachian ores.

The brown ores of northeast Texas (Fig. 13-1), according to Eckel and Purcell, are in Eocene beds. The Northern Basin ore occurs in nodules, concretionary masses, or as thin lenses in a 2 to 30 meter zone of weathered greensand. The washed ore contains from 48 to 57 percent Fe, 5 to 13 percent SiO_2, 2 to 7 percent Al_2O_3, and less than 0.12 percent P. The Southern Basin ore occurs as a solid, almost continuous horizontal bed about one meter thick, resting on white clay that grades downward into weathered greensand. No washing is necessary. The ore contains from 42 to 48 percent Fe, 10 to 12

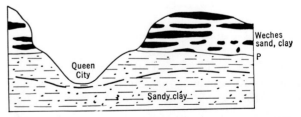

Figure 13-1 Cross section of residual brown iron ores of East Texas, showing relation of limonite (black) in Weches greensand to perched water table, *P* (dashed line), which lies above permanent water table. (After Eckel and Purcell, Tex. Bur. Econ. Geol.)

percent SiO_2, 8 to 12 percent Al_2O_3, and less than 0.24 percent P. Eckel and Purcell estimate that between 150 and 200 million tons of high-grade ore are available in this field. These ores illustrate the accumulation of alumina and silica in subtropical environments that existed during continental drift.

Cuba. Cuba contains extensive deposits of residual iron ores on the north coast, of which the chief ones are the Mayari, Moa, and San Felipe. The ores consist of goethite and hematite and occur as residual mantles overlying weathered serpentine, from which they have been derived. They rest on plateaus and are probably of Tertiary age.

The ore is a typical iron laterite from which all the magnesia and most of the silica content of the serpentine have been removed. Leith and Mead showed that 100 kilograms of the original serpentine containing 1.5 kilograms of alumina and 10 kilograms of ferrous iron shrank during weathering to a residual of 17.5 kilograms, consisting of 11.7 kilograms of hydrous ferric oxide, 3.8 kilograms of bauxite and kaolin, and perhaps 2 kilograms of other constituents. The residue carries 44 percent iron and is an iron ore. Chromium and nickel, common constituents of serpentine, are also residually concentrated in the ore. This deposit is a clear example of the removal of silica under subtropical weathering. Two major nickel deposits were developed in Cuba by Freeport Minerals and the U.S. government, but both were nationalized by the Cuban government when Fidel Castro acquired political power.

Other Localities. Little-developed residual brown ores, generally similar to those of Cuba, occur in Colombia, Venezuela, and Surinam. Extensive deposits of brown ores occur in the Kirtsch district of Russia. They are low-grade Pliocene hematites with 35 to 42 percent Fe, 1 to 8 percent Mn, and are high in phosphorus and silica. Several deposits of brown ores in Greece are similar to those of Cuba, with high chromium and some nickel content. In Bihar, Orissa, and Madhya Pradesh in India, are several deposits of high-grade residual ores, and smaller residual deposits occur in New Zealand and the East Indies.

Residual Manganese Concentration

Manganese accompanies iron and occurs under generally similar conditions but in smaller amounts. Its oxides also resist solution, and residues accumulate from the weathering of manganiferous rocks to form residual manganese deposits. The conditions of formation are similar to those of residual iron deposits, which commonly contain some manganese. Special conditions of pH and Eh, however, are necessary to give rise to residual deposits in which the manganese dominates.

Source Minerals. The sources of manganese that give rise to residual deposits are more restricted than those of iron. Manganese is always present in igneous rocks but in amounts too small to give rise to residual deposits; this is also true of many schists.

Many residual deposits of manganese result from the weathering of:

1. Limestones or dolomites low in alumina but containing disseminated, syngenetic manganese carbonates and oxides.
2. Limestones containing disseminated introduced manganese. Carbonate rocks precipitate manganese under certain conditions.
3. Manganiferous silicate rocks, such as crystalline schists or altered igneous rocks. Crystalline schists in Brazil containing rhodochrosite, spessartite, and tephroite, and in India and the Gold Coast containing rhodonite and spessartite, are sources of large residual deposits.
4. Lode deposits of manganese minerals, or ores high in manganese. Veins, replacement deposits, or contact-metasomatic deposits may contain or consist of such minerals as rhodochrosite, rhodonite, manganiferous siderite and calcite, spessartite, tephroite, alleghenite, piedmontite, hausmannite, manganosite, and others. Hewett has shown that these minerals are much more widespread than previously realized. Most residual manganese deposits, however, are derived from manganiferous rocks rather than from manganiferous lodes.

Mode of Formation. Residual manganese deposits (Fig. 13-2), like iron, are formed by the accumulation of insoluble oxides released by weathering of manganiferous rocks or lodes. They result from silicate rocks only under tropical or subtropical weathering, during which process silica is removed.

Few large deposits have been formed from the weathering of limestone. Since the amount of syngenetic manganese in limestone is, in general, much less than iron, a larger reduction in volume of limestone is necessary to yield a large deposit. Consequently, more clay and silica accumulate. Therefore, manganese deposits of limestone derivation

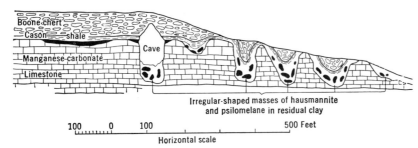

Figure 13-2 Sketch section at Clubhouse mine, Batesville, Arkansas, showing irregular bodies of residual manganese ore (black) in Fernvale limestone and masses in residual clay (white). (After Miser, U.S. Geol. Survey.)

are apt to be small, of low grade, and to have more clay per ton of ore. The few large resdiual deposits of limestone derivation have come from limestone previously enriched in manganese minerals.

From silicate rocks, the silica, magnesia, lime, and other constituents are removed under tropical weathering, and a residue of manganese oxides and bauxite results. Most of the important residual manganese deposits have been derived from crystalline schists.

Lode deposits containing manganese-bearing minerals, particularly the carbonates, yield upon weathering a residue of the oxides of manganese. Carbonated solutions or humic acids dissolve the carbonate, and, in the presence of oxygen, the higher and nearly insoluble oxides of manganese are precipitated and may accumulate as a superposed residue. If sulfides are present in the lode, the manganese may be taken into solution as manganese sulfate which, upon neutralization, breaks down and in the presence of oxygen yields the higher oxides of manganese.

In general, the formation of residual manganese deposits closely parallels that of residual iron deposits just described, but slight changes in pH and Eh are effective in separating the two elements.

Classes of Deposits and Distribution. Manganese deposits formed by residual concentration may be grouped as follows:

1. Those derived from crystalline schists in tropical regions.
2. Those derived from limestone previously enriched with manganese minerals.
3. Those derived from former manganiferous mineral deposits.

The first two groups are the most important and, together with the sedimentary deposits, yield most of the manganese of the world.

Residual deposits of manganese are found throughout the unglaciated parts of the world, but the best ones occur in tropical or subtropical regions. The large deposits of manganese in India, Ghana, Brazil, Egypt, and Morocco are of residual origin. Other important residual deposits occur in Rumania, Japan, Malaysia, and the Philippines; there are minor deposits in Hungary, Czechoslovakia, Austria, Italy, Chile, Puerto Rico, Malayasia, Zaire, and the United States.

Residual Bauxite Formation

Bauxite, the ore of aluminum, is formed in residual deposits at or near the surface under tropical or semitropical conditions of weathering. Although aluminum is the most abundant metal in the earth's crust and the third most abundant element, it occurs mainly in combinations that so far have defied commercial extraction. It is an important constituent of all clays and soil and of the silicates of common rocks, and yet until the beginning of this century it was necessary to travel to far-distant Greenland to obtain the mineral cryolite for a source of metallic aluminum, now displaced by extensive deposits of bauxite, which in the future may be replaced by aluminous igneous rocks or high aluminous clays. At the present time, the United States imports 87 percent of its bauxite from Jamaica, Surinam, and Australia. The largest producer of bauxite in the world is Australia.

Constitution of Bauxite. Bauxite was formerly the mineral name of $Al_2O_3 \cdot 2H_2O$, which is no longer recognized as a mineral species. The present usage of the term, both mineralogically and in commerce, is to designate a commonly occurring substance that is a mixture of several hydrated aluminum oxides with considerable variation in alumina content. It is a hardened and partly crystallized hydrogel that consists of variable proportions of the minerals gibbsite or hydrargillite, and boehmite [$Al(OH)_3$] and its dimorphous form, diaspore. Impurities are invariably present in the form of halloysite, kaolinite, nontronite, and iron oxides; rarely, bauxite contains octahedrite. A typical bauxite contains 35 to 65 percent Al_2O_3, 2 to 10 percent SiO_2, 2 to 20 percent Fe_2O_3, 1 to 3 percent TiO_2, and 10 to 30 percent combined water. For aluminum ore, bauxite should contain preferably at least 35 percent Al_2O_3 and less than 5 percent SiO_2, 6 percent Fe_2O_3, and 3 percent TiO_2. For the chemical industry the percentage of silica is less important, but iron and titanium oxides should not exceed 3 percent each; and for abrasive use, silica and ferric oxide should be less than 6 percent each.

Commercial bauxite occurs in three forms: (1) pisolitic or oölitic, in which the kernels are as much as a centimeter in diameter and consist principally of amorphous trihydrate; (2) sponge ore (Arkansas), which is porous, commonly retains the texture of the source rock, and is composed mainly of gibbsite; and (3) amorphous or clay ore. All three may be intermingled.

Mode of Formation. *General.* Bauxite is not a product of normal weathering in temperate regions; it is almost entirely lacking from soils formed there. It is, however, a constituent of lateritic soils formed in tropical and subtropical regions.

Bauxite is an accumulated product of peculiar weathering of aluminum silicate rocks lacking in much free quartz. The silicates are broken down; silica is removed; iron is partly removed; water is added; and alumina, along with titanium and ferric oxide (and perhaps manganese oxide), becomes concentrated in the residuum (Fig. 13-3). This is the generally accepted view, but some investigators have thought that bacteria may have played a part in bauxite formation.

Since hydrous aluminum oxides, and not the stable hydrous aluminum silicates of clays were formed, it means that conditions peculiar to tropical weathering must have prevailed during bauxite formation.

The conditions necessary for bauxite deposits are: (1) humid tropical or subtropical climate; (2) rocks high in aluminum material and susceptible of yielding bauxite under suitable weathering conditions; (3) available reagents, including abundant precipitation, to bring about breakdown of the silicates and solution of silica at specific pH and Eh conditions; (4) surfaces that permit slow downward infiltration of meteoric water; (5) subsurface conditions that allow the removal of dissolved waste products; (6) long time of tectonic stability; and (7) preservation.

Climate. Dittler has shown that a temperature above 20°C favors chemical processes by which SiO_2 goes into solution and Fe_2O_3 and Al_2O_3 remain behind. Actually, the pH of the dissolving fluid is much more important because of the insolubility of alumina between pH's of 4 to 9, but soluble at a pH of <4, with a moderate solubility of SiO_2 below a pH of 10. Actually low pH of about 3 and low Eh are necessary to achieve transport of iron with residual enrichment of alumina. Accumulation of free CO_2 on the surface is hindered during a wet season. The wet season of the tropics, during which acid solutions are diluted, is one of formation of Al_2O_3 and Fe_2O_3; the dry season, when alkaline residual solutions exist, is one of the leaching of silica from these oxides. This may supply the answer to why a tropical climate is necessary for bauxite formation. The bauxite deposits in the temperate regions of Arkansas, Georgia, and France do not contradict

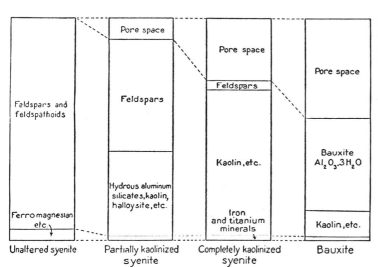

Figure 13-3 Diagram showing mineral and volume change in alteration of syenite to bauxite. (From W. J. Mead, Econ. Geol.)

this. They were formed during Early Tertiary or Late Mesozoic time when, because of continental drift, they were nearer the equator.

Source Rocks. Bauxite deposits are formed from rocks relatively high in aluminum silicates and low in iron and free quartz. Thus the Arkansas, Brazil, and Guinea deposits have been formed from nepheline syenite, the Georgia-Alabama deposits from clay, the French deposits at Baux (the source of the name *bauxite*) from limestones or clays in limestone, the Guyana deposits from residual clay derived from crystalline and metamorphosed rocks, the Jarrahdale, Coore, and Mitchell Plateau deposits of Australia are similar to the Guyana deposits although recent studies suggest that these deposits, including Weipamon are derived from a clastic source or sediment, derived by erosion of the Archean rocks, the Ghana deposits from clay shales and other aluminous rocks, the Indian deposits from basalt, and Thailand deposits from clay alluvium. Krook in 1969 concluded that the Surinam deposits were derived from feldspathic sandstones. Jamaican, Bahaman, and Cayman Islands bauxite is derived from limestone residue or the insoluble constituents of limestone.

Chemical Reagents. The chemical processes involved in bauxite formation are imperfectly understood. Because sulfuric acid or sodium carbonate decompose clays, these reagents have been mentioned as the active ones. It is likely that the ordinary reagents available in tropical weather such as plant acids and rainfall have been the effective agents of alteration.

It is probable, earlier as stated by Cooper for the Ghana deposits, that water, carbon dioxide, humic acids, and tepid rainwater are the reagents involved. Carbonic and organic acids break up silicates, and the resulting alkali carbonates are competent solvents for silica. Carbon dioxide in rainwater is an effective dissolver of limestone. Cooper mentions that in shafts sunk in bauxite in Ghana, carbon dioxide and methane are so abundant that workers have to be safeguarded. Bacteria

may also aid in bringing about solution and redeposition of alumina, as demonstrated by Thiel. He has shown that aluminum sulfate in solution is hydrolyzed to the hydrate and that natural reduction of sulfates may take place through biochemical processes. It has also been suggested that peat or lignite in overlying beds may have contributed organic compounds that effected or aided the alteration. However, not all deposits have overlying lignites, and further, the bauxite in most if not all cases was formed before the overlying lignites. Probably, the lignite resulted because of the conditions that gave rise to bauxite. The strangeness of the alteration that yields hydrous aluminum oxide instead of the usual aluminum silicate was early attributed to the action of hydrothermal solution—an origin with but few advocates at present.

Erosion Surfaces. Bauxite deposits characteristically lie upon, or are part of, gently undulating surfaces of erosion. Many are associated with old peneplains, as in the southern United States (Fig. 13-4) and Ghana; most deposits mark unconformities. Such gently sloping surfaces restrict waste removal and permit continual and slow downward seep of rainwater and a seasonal high water table. A flattish surface is also necessary to retain the accumulating residuum, which would otherwise be washed away as soon as formed.

Waste Disposal. Subsurface drainage must permit underground withdrawal, commensurate with inflow, of the rainwater charged with the silica and other substances removed during the formation of the bauxite. If the groundwater were stagnant, the waste products could not be removed, and alteration would be inhibited.

Time. The formation of bauxite apparently requires a long period of time—perhaps equivalent to the amount involved in the formation of a peneplain. Most deposits are, however, post-Cretaceous in age; many are early or middle Tertiary, probably because older deposits had a higher probability of being lost to erosion.

Preserval. Probably countless deposits of bauxite

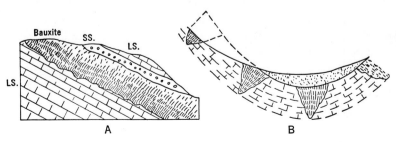

Figure 13-4 Sections of French bauxite deposits. A, bedded deposits between compact limestones at Maigret; and B, pocket deposits in Jurassic limestone at Centre de Barjol. (After de Lapparent.)

have been formed throughout the broad tropical regions, but few have survived the ravages of subaerial erosion or destruction by encroaching marine invasions. Fortuitous circumstances were necessary. Burial during quiet sedimentation, followed by freedom from dissection, or preserval on undissected remnants of peneplains or old erosion surfaces, or in the depressions of Karst topography, has been necessary. Many deposits occur on remnants of dissected peneplains.

Resulting Bauxite Deposits

Classes of Deposits. Bauxite deposits formed by residual concentration occur as (1) blankets at or near the surface and are approximately horizontal; (2) interstratified bedded deposits lying on erosional unconformities and consisting of beds or lenses within sedimentary formations; (3) pocket deposits or irregular masses whose bottoms are sheathed in clay, occupying solution or erosional depressions in limestone or dolomite; and (4) deposits derived from clastic deposits with sediments derived from crystalline rocks. Bauxite may also be moved from its site of formation and be redeposited in nearby sedimentary beds, or as rubble accumulations, giving rise to transported deposits.

Blanket and interstratified deposits are represented by the large deposits of Arkansas, Guyana, Surinam, and many of the important southern European fields. Pocket deposits are common in parts of southern France, Dalmatia, Hungary, Yugoslavia, and Jamaica. More than one class may occur together, as, for example, in Arkansas and in the southern European fields.

The blanket deposits generally have some soil cover but are not interstratified. They are invariably associated with erosion surfaces or older peneplains, as in Ghana (Fig. 13-5) and Alabama-Georgia. Few are deeply buried, and most have a clay base.

The interstratified deposits lie on erosion surfaces and invariably occupy unconformities (Fig. 13-5). They rest on residual clay. Their bottoms are less regular than their tops, which are generally

covered by sediments. The deposits of Guyana and Arkansas lie on erosion surfaces beneath partly eroded Tertiary sands, clays, and lignite, or nephelite syenite (Fig. 13-6). Those of France and Dalmatia lie on erosion surfaces beneath folded Upper Cretaceous or Eocene limestones.

Deposits derived from silicate rocks are, generally, fairly uniform beds that grade downward through clay into undecomposed rock. Those formed in limestone have a bottom sheath of residual clay and bottom protuberances that extend into the underlying limestone. Being residual deposits, they take the shape of an erosion surface on limestone, which, owing to solution, is generally irregular and cavernous.

The pocket deposits are restricted to limestone or dolomite, as in southern France, Hungary, Yugoslavia, Italy, Jamaica, and Georgia. They occupy solution depressions, or the cavernous openings of a karst topography (Fig. 13-4B). They also rest on clay in association with erosion surfaces and unconformities. They are shaped like huge teeth with many projecting roots. They may be covered by thick consolidated sediments, as in Yugoslavia, or by a thin mantle of surface soil, as in Alabama.

The *transported deposits* may be bedded, as in Russia and Arkansas, or they may consist of broken rubble, as in some of the Ghana deposits. Probably, most of the bauxite was washed away during formation, but some was collected and preserved.

Australian deposits are, of course, residual but probably formed on clastic or rubble rock, derived from crystalline uplifts that resulted in feldspathic fanglomerates and arkosic sediments.

Age. The optimum conditions for the formation of bauxite apparently existed between Middle Cretaceous and middle Eocene, when most of the commercial deposits of the world were formed. A few were formed earlier, and some in late Tertiary and Recent. The only Paleozoic bauxite known is at Tikhvin au Leon, Spain, in Devonian rocks; the only Triassic deposits are some in Croatia. The deposits of Arkansas are underlain by Cretaceous rocks and overlain by lower Eocene, and those of southern Europe mostly lie on Jurassic, Middle

Figure 13-5 Cross sections of Ghana bauxite deposits showing relation to erosion surfaces. A, Typical; B, Mount Ejuanema deposit. (After Cooper, Gold Coast Geol. Survey.)

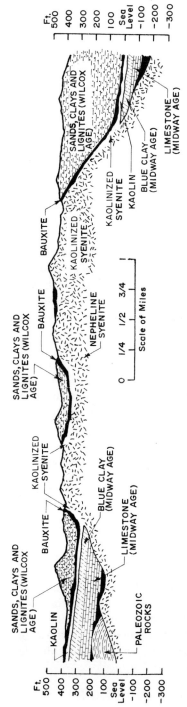

Figure 13-6 Cross section of the Arkansas bauxite deposits. (From N. M. Branlette, 1936, Inf. Circ. 8, Ark. Geol. Survey.)

Cretaceous, and Upper Cretaceous formations and are overlain by Upper Cretaceous, Lower Eocene, or Middle Eocene. The Irish and Jamaica bauxite is Oligocene, and that of the Vogelsberg Mountains, Germany, is thought to be Miocene in age. The deposits of the Guianas, Brazil, Dominican Republic, Ghana, Guinea, India, and Malaysia are latest Tertiary to Recent. Australian deposits formed during late Cretaceous to Miocene time.

It is thus evident that the conditions prevailing during the "most dramatic and puzzling event in the history of life on the earth" (Simpson)—that of the change from the age of reptiles to the age of mammals—was equally the optimum for the development of bauxite, undoubtedly the result of optimum climatic conditions at specific times at different areas as a result of continental drift.

Associations. The association of clay with bauxite is quite general. The limestone deposits rest on residual clay; the Arkansas deposits grade through clay into nepheline syenite; all the deposits of Alabama and Georgia, the Guianas, Jamaica and those of Ghana rest on clay. In most cases, the clay of the base is residual, derived from the underlying rocks.

Summary. The features of occurrence of bauxite deposits lead to the following conclusions:

1. They are all related to old peneplains or erosion surfaces. This implies the neccessity of considerable time and of flattish drainage surfaces for their formation.
2. They have mostly been formed in the late Mesozoic or Tertiary time under climatic conditions different from those that prevail in the same places today, probably the result of continental drift.
3. The bauxite deposits of Jamaica appear to be localized in karst topography not from leaching of clay, but rather from slightly impure limestone dolostone that is about 0.036 percent alumina. Considerable leaching could explain the residual concentration of alumina.

Residual Clay Formation

Clay results wherever aluminous rocks are chemically weathered, but most of it is carried off and deposited as ordinary argillaceous sediments. Under suitable conditions beds of pure sedimentary clay are laid down. With favorable climatic and topographic conditions and suitable source rocks valuable deposits of residual clays may result, and this is the most common mode of origin of all the high-grade clays. Clay is also formed by hydrothermal action of feldspathic rocks.

Types of Residual Clays. Ries classified residual clays as follows:

1. Kaolins, white in color, and usually white burning.
    (a) Veins, derived from weathering of dikes.
    (b) Blanket deposits, derived from areas of igneous or metamorphic rocks.
    (c) Replacement deposits, such as indianite.
    (d) Bedded deposits, derived from feldspathic sandstones.
2. Red-burning residuals, derived from different kinds of rocks.

Source Materials. The chief source rocks of residual clays are crystalline rocks, more especially the silicic granular rocks that are rich in feldspars and low in iron minerals, such as granite and gneiss. These yield the purest residual clays, but they have to be washed to remove quartz and limonite. Some granites high in silica yield "lean" clays of low plasticity. Basic igneous rocks yield much ferric oxide, which stains the clay, often rednering it useless. Feldspar-rich pegmatites yield dikelike masses of high-grade white kaolin that is generally very low in iron and other impurities deleterious to chinaware manufacture. Syenites yield excellent clay. Limestones, after long-continued solution erosion, leave a mantle of insoluble clayey impurities that are used for brick clays. Shale, which is largely made up of clay minerals, is used as clay material as it is, but weathering often yields a purer product. Sericitized igneous rocks also yield clay.

Mode of Formation. Clay formation results from normal weathering processes. Vegetation plus the atmosphere supply the necessary carbon dioxide, and it is noteworthy that good clays commonly underlie swamps or former swamps. Organic compounds serve to remove coloring materials and produce white clays; they change iron from the insoluble ferric to the soluble ferrous state, permitting its removal in solution and thereby bleaching the clay.

The formation of clay from silicate minerals is essentially a breaking down of the silicates to form hydrous aluminum silicates and the removal of the soluble silica and alkalies in solution. Some free quartz will remain and must be extracted to obtain pure clay. The alteration of orthoclase, for example, yields kaolinite, potassium carbonate, and silica. The last two are removed in solution and the kaolinite persists. Noll has shown that a thorough removal of the alkalies gives rise to kaolinite, whereas less complete removal favors the formation of montmorillonite. Figure 8-7 illustrates the stability field of kaolinite derived from feldspar.

Kaolinite deposits also result from hydrothermal action. Kaolinite has been produced experimentally by Schwartz, Noll, and others between 200 and 400°C although pyrophyllite will begin to form at about 350°C (Fig. 8-7). Kaolinite, dickite, and montmorillonite occur in the halo of hydrothermal rock alteration that surrounds many hydrothermal ore deposits, particularly copper deposits. Sales and Meyer show that in the alteration halo at Butte, Montana, sericite with some dickite occurs next to the veins and kaolinite and montomorillonite farther out, the last representing the earlier formed, lower-temperature alteration products. They also found that sericite dominates in the intensely altered zone and kaolinite in the less-intense zone. Lovering has noted hydrothermal kaolinite at East Tintic, Utah. These occurrences prove conclusively that kaolinite is a hydrothermal as well as a weathering alteration product. The deep kaolin of the Cornwall tin lodes has also been ascribed to hydrothermal action, as have the deep china clays of Cornwall. The clays can be traced downward into unweathered sericitized granite.

So many residual clays exhibit downward transitions through partly altered to fresh rocks, and so widespread are the mantels of clay overlying feldspathic rocks, that there can be little doubt that most residual clays result from weathering.

Occurrence and Distribution. Residual clay deposits assume roughly the form of the source rock. Dikelike deposits are derived from pegmatite dikes and are as wide as 100 meters and as deep as 40 meters. Many of the kaolin deposits of the southern states are derived from pegmatites. Those derived from crystalline rocks occur as mantles many acres in extent and tens of meters in depth. Some residual clays are covered by later formations, such as the kaolins of Monroe County, Pennsylvania, which Mansfield states were residual on an old Pennsylvanian surface and are covered by Upper Cretaceous strata. A large hydrothermal halloysite deposit was being mined recently in the Main Tintic district, Utah, but is now closed.

The color and purity vary greatly. Those derived from granitic rocks range from 10 to 50 percent white clay minerals, the remainder being mineral fragments and decomposition products of the granite that must be removed by washing. Residual clays from limestone are mostly highly colored by iron compounds.

Geologically, the residual clays are much more restricted in distribution than the sedimentary clays. Most of them are of Pleistocene age; they are nearly all post-Paleozoic, and few are of Mesozoic age.

Geographically, residual clays are widely distributed wherever suitable source rocks crop out in favorable topography and humid climates. The best kaolins come from temperate humid regions and are lacking in arid regions. The glaciated regions, although rich in sedimentary clays, contain no good residual deposits because of removal by ice and insufficient post- glacial time.

In North America, residual clays are confined mostly to the southern states and some of the western states, although some kind of clay is produced in every state in the Union. England, China, France, Germany, and Czechoslovakia contain the most important deposits of high-grade clays.

The constitution, varieties, uses, mode of origin, and examples of deposits of all kinds of clays are given under ceramic materials in Chapter 24.

Nickel Deposits Formed by Residual Concentration

Many ultrabasic igneous rocks are known to contain very small quantities of nickel in some unknown form but presumably held in the silicate lattices. Under tropical and subtropical weathering such rocks become decomposed, lose silica, and yield hydrous silicates of magnesium and nickel. Formerly, several species were thought to exist, such as genthite, pimelite, nepouite, connarite, garnierite, and noumeite. Cornwall lists more than two dozen nickel minerals and restates that garnierite, a common secondary mineral, is not a true mineral but rather a mixture of nickel serpentine, nickel talc, and possibly other silicates.

In several places, "garnierite," derived from serpentinized peridotite, has undergone sufficient residual concentration on the surface to form workable deposits of nickel ore. Such deposits have been formed in New Caledonia, Cuba, Celebes, Borneo, Australia, Brazil, Venezuela, and the Philippines. The Oregon deposits may be supergene in origin.

The deposits of New Caledonia, at one time the world's chief source of nickel, illustrate the mode of formation. Part of the island is underlain by deeply weathered serpentized peridotite of Tertiary age, and nickel is present throughout all the weathered mantle. Locally, the residual concentration has proceeded far enough to form hundreds of small deposits from 8 to 12 meters thick, mostly localized in six centers, three on each coast. The reserves of New Caledonia are 500 billion tons averaging about 1.8 percent nickel. The deposits lie on slopes rather than on plateau ridges.

At Nicaro, Cuba, the minute nickel content of serpentine has been residually concentrated to 1.4 percent nickel in a flat ferruginous surface blanket 7 to 15 meters thick. Cobalt and chromium have also undergone concentration but have not yet been commercially extracted.

The similar deposits of Celebes have been worked by the Japanese. A small residual nickel deposit was being mined at Riddle, Oregon, but recently closed. New lateritic nickel deposits are under development in Australia, especially, but also Indonesia, Philippines, Dominican Republic, and Botswanna; others in Venezuela, Guatemala, and Colombia are being investigated.

Other Products of Residual Concentration
Kyanite. High-quality kyanite is obtained from Singhbhum, Bihar, Orissa, and Ghagidih, India, in the form of massive residual boulders that have accumulated on the surface or are washed from the regolith. Their source in underlying Precambrian kyanite-quartz-granulite rocks.

In the United States at Baker Mountain, Virginia, Burnsville, North Carolina, and Habersham County, Georgia, kyanite occurs in kyanite schists formed by Precambrian metamorphism of aluminous rocks. Weathering of the schists has produced a deep clay regolith containing loosened kyanite crystals up to 7 centimeters long, and boulders of kyanite crystals, which are extracted from the clay and processed. Some residual concentration has taken place by the weathering processes, but the unweathered schist also contains abundant kyanite.

Cobalt. In Zaire, there are residual accumulations of black oxide of cobalt associated with the oxidized copper ores.

Phosphates. The "land pebble" phosphates of Bone Valley, Florida, which consist of pebbles and boulders in a matrix of sand and clay, are considered to have been weathered out of the underlying Alum Bluff phosphate formation. They are residual accumulations of weathering that were worked over by an advancing sea and included in marine sands and gravels.

Tripoli. Tripoli, a light, soft, porous, earthy substance that is nearly pure silica, is a residual product of weathering and is found in massive form, either blocky or in friable masses.

Zinc Ore. In Virginia and Tennessee, primary zinc sulfide deposits in the shaly dolomite have been oxidized to hemimorphite and smithsonite, which occur in nodules and mineable masses in residual clay overlying the dolomite.

Tin Ore. Residual accumulations of tinstone occur in Indonesia and Thailand in connection with the fluviatile and eluvial placers described under placers. Some remarkably rich pockets have been found.

Gold. Grains of gold released from their matrix during weathering have accumulated in places to form small residual deposits. Residual accumulations have been found in the Appalachian states, the western states, the Guianas, Brazil, Madagascar, Tanganyika, and Australia. The gold occurs in angular particles, showing that it has not been transported, and is generally in loose ferruginous detritus. Derby described a residual gold deposit in Brazil, which is in the lower part of a bed of laterite iron ore that represents a residual concentration in place from the weathering of underlying schist.

Selected References on Residual Concentrations
General Reference 9, Chap. 21, p. 344–378. *Residual deposition in general.*
General Reference 19. *Several chapters on residual deposits of specific minerals.*
Brown iron ores, Tennessee. 1932. E. F. Burchard. Tenn. Geol. Surv. Bull. v. 39.
Bauxite deposits of Arkansas. 1915. W. J. Mead. Econ. Geol., v. 10, p. 28–54. *Occurrence and origin by removal of silica.*
Formation of bauxite from basaltic rocks of Oregon. 1948. v. T. Allen. Econ. Geol., V. 43, p. 619–626. *Alteration first to clay minerals, then to bauxite.*
Clays and Other Ceramic Minerals. 1937. C. W. Parmelee. Edwards Bros., Ann Arbor, Mich.
Bauxite deposits of the U. S. 1965. U. S. Geol. Survey Bull. 1199. *Numerous papers by different authors on bauxite deposits of the United States.*
Geology of some kaolins of western Europe. 1932. E. R. Lilley. AIME Tech. Paper 475. *Good description and discussion of origin of residual clays.*
Zinc and lead region of southwestern Virginia. 1935. L. W. Currier. Va. Geol. Surv. Bull., 43.
Jamaica type bauxites developed in limestone. 1963. H. R. Hose. Econ. Geol., v. 58, p. 62–69. Classic paper on bauxite from limestone.
Stratigraphy and origin of bauxite deposits. 1949. E. C. Harder. Bull. Geol. Soc. Am., v. 60, p. 887–908.
Bauxite. 1970. H. F. Kurtz. U.S. Bur. Mines Min. Yearbook.
Florida kaolins and clays. 1949. J. L. Calver. Florida Geol. Surv. Inf. Cur., v. 2, 23 p., 1949.
Clays in mineral facts and problems. 1970. J. D. Cooper, U.S. Bur. Mines Bull. 650, p. 923–938.
Applied Clay Mineralogy. 1962. R. E. Grim. McGraw-Hill Book Company, New York. Also *Clay Mineralogy.* 1968. McGraw-Hill Book Company, New York, 596 p. *Leading texts on clay mineralogy.*
Nickel-silicate and associated nickel-cobalt-manganese deposits, Goiaz, Brazil. 1944. U.S. Geol. Surv. Bull., 935-E. *Residual nickel-cobalt deposits.*

MECHANICAL CONCENTRATION

Mechanical concentration is the natural gravity separation of heavy from light minerals by means of moving water or air by which the heavier minerals become concentrated into deposits called *placer deposits*. It involves two stages: (1) the freeing by weathering of the stable minerals from their matrix, and (2) their concentration. Concentration can occur only if the valuable minerals possess the three properties: high specific gravity, chemical resistance to weathering, and durability (malleability, toughness, or hardness). Placer minerals that have these properties are gold, platinum, tinstone, magnetite, chromite, ilmenite, rutile, native copper, gemstones, zircon, monazite, phosphate, and, rarely, quicksilver. Pyrite and uraninite may be placers in some Precambrian deposits.

Source Materials

The minerals that make up placer deposits may be derived from:

1. *Commercial lode deposits*, such as gold veins, for example, the Mother Lode gold veins of California.
2. *Noncommercial lodes*, such as small gold quartz stringers or veinlets of cassiterite, for example, the tin placers of Indonesia.

3. *Sparsely disseminated ore minerals,* minute grains of platinum sparsely disseminated in basic intrusives, for example, the Ural Mountains.

4. *Rock-forming minerals,* such as grains of magnetite, ilmenite, monazite, and zircon, for example, ilmenite beach sands of India and the beach sands of Australia.

Principles Involved

In the formation of placer deposits, nature has operated to produce the results achieved by man when he mines, crushes, and concentrates ores. The placer minerals are released from their matrix by weathering. The comminuted materials are washed slowly downslope to the nearest stream or to the seashore. Moving stream water sweeps away the lighter matrix, and the heavier placer minerals sink to the bottom or are moved downstream relatively shorter distances. Waves and shore currents also separate heavy minerals from light ones and coarse grains from fine ones. From thousands of tons of debris, the few heavy minerals in each ton are gradually concentrated in the stream or beach gravels until they accumulate in sufficient abundance to constitute placer deposits. Eventually, the little gold of countless thousands of tons of matrix is concentrated in relatively small volume. Rich placer gold deposits formed in this manner gave rise to the great California gold rush of 1849, to the Klondike stampede, and to the rich discoveries in Alaska, Australia, and other places that have yielded billions of dollars in gold.

The operation of mechanical concentration rests on a few basic principles involving chiefly the differences in specific gravity, size, and shape of particles, as affected by the velocity of a moving fluid. First, in a body of water, a heavier mineral sinks more rapidly than a lighter one of the same size. Moreover, the difference in specific gravity is accentuated in water as compared to air. For example, the ratio of gold (sp. gr. 19) to quartz (sp. gr. 2.6) is as follows:

$$\frac{\text{Gold in air, 19}}{\text{Quartz in air, 2.6}} = \frac{7.3}{1}$$

$$\text{whereas } \frac{\text{Gold in water, } 19-1}{\text{Quartz in water, } 2.6-1} = \frac{11.2}{1}$$

Second, the rate of settling in water is also affected by the specific surface of particles. Of two spheres of the same weight but of different size, the smaller, with its lesser surface and, therefore, lesser friction in water, sinks more rapidly. Third, the shape of a particle affects its rate of settling. A spherical pellet has less specific surface than a thin, platy disc of the same weight, and, therefore, will sink more rapidly. Thus, flaky specularite and molybdenite are difficult to concentrate by gravity despite their high specific gravity.

Now, add to these factors the effect of moving water. The ability of a body of flowing water (or air) to transport a solid depends upon the velocity and varies as the square of the velocity. A racing floodwater can carry substances that a quiet stream cannot. When the velocity doubles, the transporting power is increased about four times and stationary materials are moved. Conversely, if the velocity is halved, much of the transported load is dropped. Hence placer minerals may be dropped where the current slackens. Faster-moving water accentuates the differences in settling rate based upon specific gravity, and if particles of gold and quartz were dropped into moving water the gold might drop directly to the bottom and the quartz be swept downstream. The specific surface again enters in; of two equal-weight particles, the one with the larger specific surface is increasingly rapidly, swept away from the other with increased water velocity. Thus flaky mica is readily separated from quartz, and fine materials are separated from coarse particles.

Again, add another factor, that a particle in suspension is more readily moved by a flowing fluid than one at rest. As one stirs up sugar from the bottom of a tea cup, so eddies in streams or shore currents raise light substances from the bottom and enable currents to swish them away. Separation of light minerals from heavy ones is thus aided in bringing about concentration. Also, the swirls and eddies of stream and wave action simulate the upward pulsations of jigs and tables in ore dressing by which lighter minerals are bounced higher than heavier ones so that they can be more easily moved away by flowing water. This jigging action enables gold particles scattered through bottom gravels to become concentrated on the bottom, even though the gravels are thick—the coarser gold on the bottom, the finer above.

The various factors enumerated above operate together to separate the light and fine minerals from coarse and heavy ones, and, with long-continued action, placer minerals may eventually become sufficiently concentrated to constitute workable deposits.

For the mechanical concentration of placer minerals it can thus be seen that the water velocity must be favorable. If too low, the lighter materials will not be removed from the heavier. If too great, the placer minerals will also be swept away and perhaps dissipated. A slackening of velocity, whether it be of a stream, a shore current, or undertow, causes deposition and accumulation. In a stream, a change of gradient, meandering, spreading, or obstructions all produce reduced velocity that permit heavier minerals to drop and accumulate.

The scraping and pounding of the placer minerals during jigging and transportation eventually crushes the nondurable ones to a powder; it rounds off sharp corners of the durable ones, and compacts and flattens the malleable ones. These features are, therefore, a criterion of the amount of concentration or the distance of travel they have undergone. Sharp, angular placer gold, for example, is not far from its source, and this principle is utilized by the wily prospector in his search for the "Mother lode."

An essential for any mechanical concentration is that a continuous supply of placer minerals be made available for concentration. This means that the most favorable regions are those of deep weathering and topographic relief—the weathering to free the placer minerals, and the relief to permit the debris of weathering to move streamward or beachward. A plateau or a peneplain cannot supply much debris. Still more favorable are areas of stream rejuvenation by recent uplift where new valleys are cut into older ones, causing rewashing and reconcentration of preuplift gravels. The more such reconcentration takes place, the greater is the degree of concentration.

When weathering yields debris on a hill slope, the heavier particles move downslope more slowly than the lighter ones, giving a rough concentration into *eluvial* placers. During water transportation, concentration may occur in streams, giving *stream* placer (or *alluvial*) deposits, or on beaches, giving *beach* placers. If concentration takes place by wind, *eolian* placers result. The basic principles enumerated above apply to each of them, but special features contribute to the individual groups, and these will be considered separately.

Eluvial Placer Formation

Eluvial placers may be considered an intermediate or embryonic stage in the formation of stream or beach placers. They are formed, without stream action, upon hill slopes from materials released from weathered lodes that outcrop above them. The heavier, resistant, minerals collect below the outcrop (Fig. 13-7); the lighter nonresistant products of decay are dissolved or swept downhill by rainwash or are blown away by the wind. This brings about a partial concentration by reduction in volume, a process that continues with continued downslope creep. Fairly rich lodes are necessary to yield workable deposits by this incomplete concentration. The most important eluvial deposits are gold and tin; minor deposits include manganese, tungsten, kyanite, barite, and gemstones.

Eluvial gold deposits mined in Australia contributed much of the early placer gold production and several of the famous Australian nuggets. The first placer deposits worked in New Zealand were mainly eluvial deposits. Becker, describing early gold mining in the southern Appalachians, recorded the formation of eluvial gold deposits under secular decay 30 meters deep; at Dahlonega, the debris was hydraulicked, and the enclosed quartz boulders were crushed and amalgamated. Considerable eluvial gold was washed in California, Oregon, Nevada, and Montana; Lindgren mentions that eluvial deposits in Eldorado County, California, gave rise to litigation to determine whether they were lode or placer deposits. Eluvial gold deposits were formerly extensively worked in South America; some are now being mined in Sierra Leone, Tanganyika, Kenya, and Guiana.

Tinstone is obtained in abundance from eluvial deposits in the Malasia states; it is also being washed in Burma, Thailand, Vietnam, Zaire, Nigeria, and Australia.

Wolframite is obtained as a by-product of the eluvial tin washing in the above-mentioned states; minor amounts of eluvial manganese boulders are mined in Ghana, and also in India and Brazil. Eluvial kyanite boulders of high purity are gathered, or washed out, from the regolith in India, as are also barite masses in Missouri.

Stream or Alluvial Placer Formation

Stream placers are by far the most important type of placer deposits. They have yielded the greatest quantity of placer gold, tinstone, platinum, and precious stones. Primitive mining, undoubtedly, started on such deposits. The ease of extraction and the richness of some deposits made them as eagerly sought in early times as in recent times.

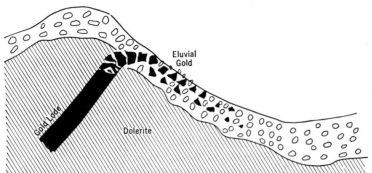

Figure 13-7 Eluvial gold ore at San Antonio vein, Chontales District, Nicaragua. (After Carter, AIME.)

They have been the cause of some of the great gold and diamond "rushes" of the world.

Concentration. Flowing water is the most effective separator of light from heavy materials. Stream water, although ever flowing downstream, does so with irregularity. It rushes through canyons sweeping everything along with it; it slackens in wide places; it swirls around the outside of bends, creating back eddies on the inside; it laps up over bottom projections, forming quiet eddies on the lee side. In these slack waters the heavy substances drop to the bottom. In streams, jigging action is particularly effective in concentrating placer minerals in the bottom gravels. Further concentration is effected by the gradual abrasion of the bottom gravels and reduction in their thickness.

During dry seasons of low stream velocity the placer minerals remain at rest, but in flood time they and the enclosing gravels may all be swept farther downstream and reconcentrated on bars, stream margins of flood plains, or other favorable places. Minute particles of gold called "colors," however, may be carried far downstream.

Places of Accumulation. The lower sluggish reaches of streams are not favorable sites for placer accu-

mulations, and neither are the upper headwaters because of the limited supply of source materials. The most favorable sites are the middle reaches. Where streams surge along polished canyon floors, placer minerals and gravels alike are rushed along with little opportunity for settling; but where they debouch into valley sections of gentler gradient, the conditions are ideal for settling and concentration into deposits.

In a rapidly flowing meandering stream the fastest water is on the outside curve of meanders, and slack water is opposite (Fig. 13-8). The junction of the two, where gravel bars form, is a favorable site for deposition of placer minerals. With lateral migration of the meander, the "pay streak" becomes covered and eventually lies distant from the present stream channel. Obviously, placer deposits do not form in the downstream meanders of sluggish old-age streams, because the stream velocity is insufficient to transport heavy minerals.

Where streams cross highly inclined or vertically layered rocks, such as slates, schists, or alternating hard and soft beds, the harder layers tend to project upward and the softer ones to be cut away. This forms natural "riffles" (Fig. 13-9A), similar to the wooden riffles nailed in the bottom of a sluice box to arrest the gold in sluicing operations. Such nat-

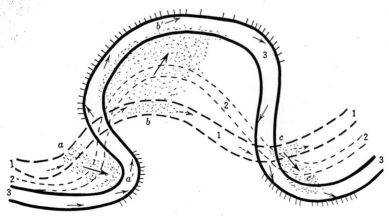

Figure 13-8 Gravel deposition and formation of pay streaks in rapidly flowing meandering stream, in which meanders migrate laterally and downstream. Stream arrows represent point of cutting. 1, Original position; 2, intermediate position; and 3, present position of stream. Deposits formed at a, b, c, or inside of meanders of stream 1, become extended downstream and laterally in direction of heavy arrow growth to a', b', c' on the present stream, and buried pay streaks result.

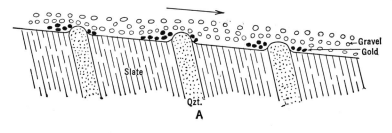

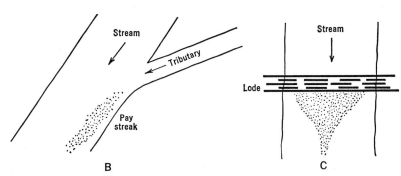

Figure 13-9 Quartzite ribs, interbedded with slate, which serve as natural riffes for the collection of placer gold (black); B, pay streak formed where fast tributary enters slow master stream; C, pay streak formed below gold lode crossed by stream.

ural riffles are excellent traps for placer minerals and may give rise to *bonanzas* or exceptionally rich streaks.

The mode of entry of the placer minerals into a stream also determines the site of accumulation. Where materials are delivered by a swift tributary into a slower master stream, they accumulate under diminished velocity, as a pay streak down the near side (Fig. 13-9*B*). If a stream crosses a mineralized lode, and through its own errosion supplies the placer minerals, the pay streak will be spread across the stream channel on the downstream side of the lode (Fig. 13-9*C*). The early placer miners followed the trail of rich pay streaks or indications upstream to where they ceased, and they then looked for the "mother lode" nearby; they found many valuable bedrock deposits in this manner. Sometimes the search led to *bench* or *terrace gravels.*

The accumulation of placer minerals necessitated a nice, long-continued adjustment between stream velocity and gravel accumulation. The gravels could not be too thick; they had to be slowly moving downstream; and they had to be water-soaked so that jigging could occur.

Gradient of Streams. Placer accumulation requires well-graded streams, where a balance has been reached between erosion, transportation, and deposition. Examples of gradients in meters per km. of some placer streams are: (1) Klondike, "white channel," about 10 meters; (2) Fairbanks, Alaska, less than 10 meters; (3) other Alaskan streams, up to 30 meters; (4) California and British Columbia,

10 to 25 meters; (5) California high-level gravels, up to 35 meters; and (6) Indonesian tin gravels, less than 10 meters. The high gradient of the high-level gravels of California has been shown by Lindgren to be in part due to a later westward tilting of the Sierra Nevada. Lindgren considers that moderate gradients of around 6 meters/kilometer probably yield the best concentration.

Gold Placer Formation

The lure of gold placers, their richness, and their ease of working caused the great gold rushes of modern times. Small placers are a "poor man's" type of deposit. The hardy miner, single-handed, requires only a shovel and a gold pan to extract the gold and in a fortnight may accumulate fabulous riches. The great California gold rush of 1849, the rush to Australia in the 1850s, and the perilous stampede to the Klondike and Alaska in 1897 initiated ephemeral placer mining. More stable lode mining followed. Behind them came settlement, agriculture, industries; great new countries were born.

The California rush of 1849 amidst hardships of laborious overland travel or around Cape Horn in inadequate ships, was, according to Rickard, but the modern counterpart of an ancient gold rush made classic by Greek legend. In the land of Colchis by the Euxine, ancient placer miners extracted alluvial gold by shoveling the gravels into sluice boxes made of hollowed trees. A lining of sheepskins entangled the gold particles; the coarse gold was shaken out, but the fine gold adhered to the wet wool, and the fleeces were hung in trees to dry so that the gold could be beaten out of them. The tale of a "gold strike" reached Greece, and Jason and his Argonauts set sail in the good ship *Argo* to seek the Golden Fleece! In Brazil today placer gold is extracted by using hollowed-out trees for sluice boxes and cowhides with the hair side up for riffles.

Today the production of placer gold is increasing again because of the increase in the price of gold. The known deposits are being reworked, and strikes and rushes are occurring. The slightly explored lands hold hope of yielding new placer discoveries of consequence. The Russian rise in gold production is due largely to the production of new placer deposits.

Source Materials. The gold in placers has come from primary workable lodes, lean lodes, or rocks sparsely traversed by gold-bearing veinlets. The source lodes need not necessarily be rich, as is popularly supposed; probably most of them are uneconomic. The rich placers of the Klondike, nor the Paleocene Pinyon formation of auriferous conglomerate in Wyoming for example, have never led to the discovery of workable lodes; some of those of Australia and California did. The richness of the placer deposit is more a result of abundant disintegration of auriferous rock and good concentration than of rich primary source material.

Pay Streaks. Gold tends to occur in concentrated pay streaks that are apt to be narrow and rich. The coarser gold is deposited in the upstream reaches of a run and the finer gold in the lower reaches. The pay streak may not lie in the present course of a stream, and, if it does, it is not necessarily in the central part. It is generally irregular in outline; it branches and splits and is absent in certain places. Although a definite pay streak may be present, some gold is also generally scattered throughout the main body of stream gravels. Thus, in many of the Alaskan gold placers the rich pay streaks were first mined by hand methods, and the low-grade gravels beyond the pay streaks are still being mined by means of large dredges.

Size, Shape, and Fineness of Gold. Most placer gold is in the form of fine specks called *dust*, but to the joy of the miner a few larger lumps called *nuggets* are found. They range from submicroscopic size to a pea or bean; nuggets from 1 to 10 pounds are common. The largest nugget recorded is the "Welcome Stranger" from Ballarat, Australia, which weighed 2280 ounces; the "Blanche Barkley" from Victoria weighed 1296 ounces. A 1050-ounce nugget was found in the Ural Mountains in 1936. The largest of the Australian nuggets were found immediately below the outcrop; therefore, they were not strictly stream placer nuggets. The largest nugget reported from the Yukon was 85 ounces. Rickard records nuggets of 47 ounces from Arizona and 40 ounces from Montana. All of these, of course, are exceptional. The general run of placer gold particles is roughly about the size of sand. Recent hypogene gold deposits discovered at Gold Acres and Carlin, Nevada, contain microscopic gold that is so small that it is seen only with an electron microscope.

An astonishing feature of placer gold is its divisibility into minute scaly particles called *flour* gold. Hite estimated that the flour gold of the Snake River averages about 5000 *colors* to equal 1 cent in value, the very smallest colors figured would require 7 to 8 millions to 1 cent. Such flour gold floats readily if exposed to air for an instant and may travel hundreds of miles. Its extreme fineness explains the wide distribution of "colors" throughout the entire thickness and width of stream gravels.

The relationship of the various units of weight of gold and some gem stones is given in Table 13-1.

The shape of placer gold is generally disclike, owing to the continued pounding to which it is subjected. The nuggets have rounded but irregular outlines. The "dust" and "flour" gold consist of small flattened pellets and thin discs that may be only 2 microns in thickness. Crystallized gold is rarely present in placers, but the Latrobe nugget (23 ounces) in the British Museum is a beautiful example of a crystal that escaped mutilation.

The purity or *fineness* of gold is usually expressed in parts per 1000 (1000 is pure gold). The fineness of placer gold varies from 500 to 999; that of veins, from 500 to about 850. Most placer gold is above 800 fine. Lindgren points out that the gold of the California veins averages 850 fine, whereas the Tertiary placer gold averages 930 to 950; also that the fineness of placer gold increases with the distance transported and with decreasing size of grains. McConnell showed that Klondike nuggets have greater fineness on the outside than on the inside. These observations suggest that some silver has been dissolved from the outside of the gold grains. The Snake River flour gold, according to Hite, is 943 fine.

Associated Minerals. Gold-bearing gravels commonly consist largely of durable quartz pebbles; the

Table 13-1 Weight Conversion

	Metric			Avoirdupois (av)		Troy (or Apothecary) (t)			
Kilograms (kg)	Grams (gm)	Carat (ct)	Pounds (lb)	Ounces (oz)	Pounds (lb)	Ounces (oz)	Pennyweight (dwt)	Grain (gr)
1	1,000.0	5000.0	2.2046	35.274	2.6792	32.151	643.01	15432.
0.00100	1	5.0000	0.00220	0.03527	0.00268	0.03215	0.64301	15.432
0.00020	0.20000	1	0.00044	0.00705	0.00054	0.00643	0.12860	3.0865
0.45359	453.59	2268.0	1	16.000	1.2153	14.583	291.67	7000.0
0.02835	28.350	141.75	0.06250	1	0.07595	0.91146	18.229	437.50
0.373324	373.24	1866.1	0.82286	13.166	1	12.000	240.00	5760.0
0.03110	31.103	155.51	0.06857	1.0971	0.08333	1	20.000	480.00
0.00155	1.5552	7.7500	0.00343	0.05486	0.00417	0.05000	1	24.000
0.000064	0.06480	0.32400	0.000143	0.00229	0.000174	0.00208	0.04167	1

other pebbles are comminuted or chemically destroyed. Alluvial gold deposits are, therefore, commonly marked by "white" runs of quartz gravel, as in the "white channels" of the Yukon, the "white bars" of California, or the "white leads" of Australia. "Black sand," which consists chiefly of magnetite and some ilmenite, is an intimate associate; and garnet, zircon, and monazite (yellow sand) are common. Any of the other placer minerals may also occur.

Relation to Bedrock. The character of the bedrock is important both in localizing gold placers and in effecting their recovery. Rocks that form smooth bedrock are less desirable than steeply dipping foliates or sediments that have irregular bottoms, or "riffles," to trap the gold. Clayey and decomposed rocks, and limestone with solution cavities make good traps. Crevices in the surface of the bedrock are generally intricately penetrated by the gold grains and a top layer has to be mined to obtain all the gold.

This table is constructed so that all values on a given line are equal.

Non-SI	SI
1 gamma (1γ)	1 * μg
1 grain	64.80 mg
1 oz (avoirdupois)	28.35 g
1 oz (apothecary or troy)	31.10 g
1 lb (avoirdupois)	0.4536 kg
1 short ton (2000 lb)	0.9072 mg
1 metric ton (1 tonne)	1 * mg
1 long ton (2240 lb)	1.016 mg

* Marks an exact SI value.

Coarse gold generally rests on or within a foot or so of bedrock; rarely it may lie within the lower 2 to 5 meters of gravels. Normally, there is a decrease from the bottom up. This is shown by McConnell's oft-quoted section of the "white channel" of the Klondike; the lower 6 feet contained $4.13 per cubic yard; the next upper 6 feet $0.18; and succeeding 6-foot sections contained 0.47, 0.04, 0.34, 0.32, 0.45, and 0.24 with gold at $20.67 per troy ounces.

In many places there is "false bedrock," which is generally a compact clay bed within the gravels. It serves as a floor for placer gravels, and other gold-bearing gravels may or may not lie beneath it.

Growth of Gold in Gravels. Do the large gold nuggets indicate some growth in the gravels? The many proponents of accretion have claimed that large nuggets and coarse grains are larger than gold particles observed in veins and that crystals of gold have been seen on nuggets and pebbles. However, the prevailing belief is that placer gold is entirely of mechanical derivation; nuggets polished and etched by Uglow showed crystalline structure typical of vein gold.

Bench Gravels. Slight uplift or sudden decrease of stream load or increase of stream volume often causes stream incision and the stranding of *terrace* or *bench* gravels on one or both sides of the valley. Repetitions may form more than one set of terraces, such as "low" and "high" benches. Their tops represent the former valley floors. They may consist entirely of alluvium or of alluvium-covered bedrock. Commonly, the gravels from such terraces are side-washed again into the stream and undergo further reconcentration. In the absence of sidewash the placer gold may persist in the benches, and much gold has been won from them. Gros Ventre Canyon, Wyoming, is typical of terrace deposits formed as the result of dams formed by landslides.

High-Level Gravels. Ordinarily, regional uplift causes rapid erosion of unconsolidated gravels. However, in California, Victoria, and New South Wales, gold-bearing gravels have been preserved by later rock covers, uplifted, and re-exposed by cross-cutting valleys. These are called variously "high-level gravels," "buried leads," and "high Tertiary gravels."

Those of California have been made classic by Lindgren. He has shown that in the Sierra Nevada region during early Tertiary time gold-bearing gravels accumulated in stream channels that traversed a gently sloping country. These were buried by thick deposits of rapidly accumulated lean gravels, and these in turn by tuffs and breccias that attained thicknesses of 500 meters. The range was elevated; new consequent streams eroded canyons to depths of 600 to 1000 meters, and these now form superimposed streams, many of which are out of adjustment with and transverse to the early Tertiary streams. The new stream bottoms now lie on an average about 800 meters below the early Tertiary stream bottoms; the present valley walls disclose elevated cross sections of the former gold-bearing stream deposits, and these have been extensively mined. Their eroded parts have supplied much of the recent stream gold.

Submerged Placers. In Alaska, California, Australia, New Zealand, and Siberia regional subsidence, stream overloading, or lava flows have buried placer gravels to depths of tens or hundreds of meters. These are reached by dredging or through shafts, but they are difficult to mine because of excessive water. In Alaska and Siberia, however, where the ground is frozen, these gravels are extracted by "drift-mining," or are excavated by dredges to depths of 7 to 10 meters; Lincoln records gold gravels at depths of 100 meters near Fairbanks, Alaska, and much drift-mining there had been carried on at depths of 15 to 50 meters. The thickness of the gold-bearing gravels ranges from 5 centimeters to 3 meters, and the gold content is from 1 to 8 dwt per yard (Table 13-2). The most notable submerged placers are some of the "deep-leads" of Australia and New Zealand. At Ballarat many rich "gutters" or channels have been found beneath 100 meters of cover.

Cement Gravels. Consolidation of gold-bearing gravels into conglomerate gives "cement gravels," and these range in age from Cambrian to Recent. A Cambrian conglomerate overlying schist in the Black Hills carries detrital gold, presumably derived from the Homestake Lode, and Cretaceous conglomerates in Oregon and California yield a little detrital gold. Similar conglomerates of Permian age in Bohemia are described by Pošepńy; of Miocene age in New Zealand, by Park; and of Permo-Carboniferous age in Australia, by Wilkinson. The celebrated gold-bearing cements of Blue Spur, New Zealand, are reported by Park to have been profitably mined for over 30 years and were the source

Table 13-2 Production Values of Some Dry-Land Placer Deposits

Metal or mineral	Place	Approximate value produced in millions of 1967 U.S. dollars	Average value per yd³ in 1967 U.S. dollars	Average value per yd³ in 1978 U.S. dollars[c]
Platinum	Goodnews Bay, Alaska	70	2.00	10.28
(at $100/troy oz)	Choco, Colombia	90	0.65	3.34
Gold	Nome, Alaska[a]	125	0.50	2.57
(at $35/troy oz)	Calif. placers	1000	?	?
	Pato, Colombia	100	0.25	1.28
	Bulolo, New Guinea	75	0.30	1.54
Tin	Bangka, Indonesia[b]	1500	0.75	3.86
	Kinta Valley, Malaysia	2800	0.50	2.57
Rutile and ilmenite	Trailridge, Florida[a]	75	0.35	1.80
	Queensland, Australia[a]	40	0.50	2.57

Source: Supplied by F. J. Lampietti, 1968, with permission of Ocean Mining, A.G.
[a] Originally concentrated in the littoral zone, although now emerged.
[b] About 10 percent of Bangka tin has come from offshore areas.
[c] With gold at $180/troy oz.

of the remarkably rich alluvial gold of Gabriel's Gully.

Distribution and Value. Stream placer deposits have been mined in most parts of the world. They were probably the chief source of the gold of the ancients and also account for a considerable part of the world's total gold supply. The most notable placer gold deposits are those of California (Fig. 13-10),

Nevada, Idaho, Utah and other western states, Yukon-Alaska, Australia, New Zealand, Siberia, British Columbia, Peru, Bolivia, and Chile. Many others have been worked throughout Europe, Asia, and Africa. Gold placers are being mined at present on a large scale in Alaska, California, Colombia, Siberia, Central Africa, and New Guinea.

An extensive ancient placer gold deposit is located in and near the Gros Ventre canyon, Jackson,

Figure 13-10 Maps of northern and central California showing locations of major streams and gold-bearing regions (shaded areas) (From: W. Clark, Calif. Geol., 1972.)

Wyoming, where the gold occurs throughout the Pinyon formation of Paleocene age. Of course, weathering and erosion of this formation have formed younger Tertiary and Quaternary placer deposits. The average tenor of gold in this formation that covers hundreds of square miles varies considerably, but portions may average higher than $1 per cubic yard at $180 per ounce.

The tenor of gold placer deposits is referred to by the yield per cubic yard or cubic meter, by the running foot of channel, or by the square foot of surface. The first is generally used in connection with large-scale operations. Bonanza yields are in dollars per pan. Individual small pockets have yielded several thousand dollars. One area 40 by 40 feet at North Zachlan, New South Wales, yielded $38,000, or nearly $24 per square foot at $20 per troy ounce.

The yield of some of the better-known placer deposits is shown in Table 13-3.

Price of Gold. In 1933, the value of gold was $20.67/troy ounce but was raised by the U.S. Congress to a value of $35/troy ounce with the U.S. government being the sole buyer within the United States. During the Depression years, the United States acquired over 25×10^9 worth of gold. Two men could recover gold from several cubic yards per day by simple methods and possibly recover sufficient gold to pay a few tens of dollars per day—a high wage during the Depression.

With the economic woes of the 1970s, the U.S. government reluctantly ceased to support the price of gold at $35/troy ounce and allowed the world market to determine the price, which rose briefly to almost $200/troy ounce dropped to about $100/troy ounce and rose to more than $230/troy ounce in 1978. Since January 1, 1975, any American citizen is allowed to own, purchase, and sell gold. Gold prices for 1968 to 1979 are shown on Fig. 13-11. In 1980, gold prices approached $700/oz.

Platinum Placer Formation

Most of the world's production of platinum and platinum metals formerly came from stream and beach placers.

The source of placer platinum is ultramafic rocks, such as dunite or peridotite, in which the metal, along with chromite, has undergone a preliminary concentration by magmatic differentiation although there is evidence of some hydrothermal processes. The enriched portions, although highly concentrated above the original magmatic content, are still too lean to constitute workable lodes (except in Russia and the Bushveld of South Africa), but through erosion the metal has been sufficiently concentrated to form commercial placer deposits.

Platinum is concentrated into placer deposits in the same manner as gold and has similar occurrence. In fact, it is commonly associated with gold placers but few gold placers contain platinum in economic amounts. It occurs in the form of dust and little flattened pellets; nuggets are extremely rare. Associated minerals are chromite, magnetite, and gold. The platinum ranges in fineness from 700

Table 13-3 Better-Known Placer Deposits

Locality	Width of deposit (ft)	Yield	
		Per linear ft[a]	Per yd³[a]
Madame Berry, Victoria	450	$7500	
Victoria, No. 2	1000	2571	
Victoria, general			$11.60–87.50
White Channel, Yukon		2200	
Klondike			52.00–290.00
Nome Creeks, Alaska	50	580	
Seward, Alaska			11.60–35.00
Red Point, Calif.	120	417	
Nevada County, California	1000	2400	
New South Wales			1.40
Fairbanks, Alaska			1.00
California, general			145–2.00
North Fork, California			0.24
Columbia			0.87

[a] Gold at $120 per ounce.

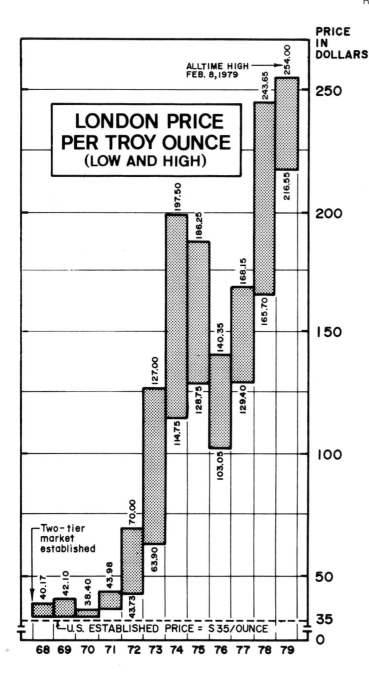

PRICE IN DOLLARS

Figure 13–11 Dual gold prices from 1968 to Feb.,1979.

to 850. Invariably, one or more of the platinum metals—palladium, osmium, iridium, rhodium, ruthenium (and the minerals iridosmine and osmiridium)—are alloyed with the platinum, as well as small quantities of copper, iron, and other metals.

Most of the placer platinum of the world has come from the Ural Mountains, where streams transect Paleozoic intrusions of enriched peridotitic rocks. Russian production is followed by South Africa and Canada. Colombia has long yielded platinum metals from Recent gravels and Tertiary conglomerates and ranks fourth in world production. Peridotite in Tasmania and New South Wales yields platinum and osmiridium placers. Small amounts of placer platinum are also obtained in Ethiopia, Japan, Panama, Papua, Sierra Leone, South Africa, and Goodnews Bay, Alaska. About one-third of the world production of platinum metals comes from placers.

The production, uses, occurrence, and deposits of platinum are given in Chapter 18.

Tin Placer Formation

Cassiterite (SnO_2), because of its weight and resistance to chemical and mechanical disintegration, ranks second in value as a placer mineral; most of the tin production now comes from placer deposits. The "stream tin," which is derived from the weathering of stockworks and lodes in granite and other rocks, has accumulated in alluvial and eluvial deposits in a manner similar to gold.

The stream tin is in the form of rounded pellets of shot, bean, or nut size. Associated minerals are similar pellets of limonite, hematite, magnetite, garnet, wolframite, tourmaline, and other placer minerals.

The greatest tin placer deposits are those of Malaysia (Table 13-2), where extensive alluvial deposits containing only half a pound of tinstone to the cubic yard are worked by some of the largest dredges in the world.

The Indonesian tin deposits resulted from deep trpoical weathering of granite containing lean stockworks and sparse veinlets of quartz and cassiterite giving rise to eluvial concentrations of cassiterite (koelits) and of alluvial concentrations (kaksas) during Quaternary time. An eluvial origin is indicated by unworn cassiterite, in high concentration, along with angular fragments of other resistant Indonesian minerals. These deposits contribute a considerable part of the tin production (Table 13-2). They are really a transient stage in the formation of the alluvial tin placers because eventually the eluvial material reaches the stream channels and there undergoes rigorous stream concentration to form the productive stream placers. During a slight subsidence of the region in Quaternary time, when the islands became separated from each other, the pay streaks of tinstone became covered by subsequent gravels and sands and the downstream parts became submerged beneath the sea, so that former stream gravels are now dredged from the sea bottom.

Small amounts of stream tin are also found in Nigeria, China, Zaire, Bolivia, Thailand, Burma, Victoria, New South Wales, and Tasmania. A trivial amount was obtained as a by-product from gold gravels in Alaska. About two-thirds of the world supply of tin is obtained from placer deposits.

See Chapter 21 for the history, production, uses, and deposits of tin.

Precious Stones Placer Formation

Diamond. The diamond is the chief precious stone found in placer deposits. Prior to 1871, when the first diamond in primary rock was found at Kimberley, South Africa, the entire world production came from placers in India and Brazil. About 95 percent of the world production has been obtained from placers in Zaire, Ghana, Angola, Sierra Leone, Guinea, Brazil, and South Africa. Small amounts come from Tanzania, southwest Africa, Guyana, Borneo, Australia, Nigeria, and Venezuela.

The original home of the diamond, so far as known, is in pipelike intrusions of ultrabasic igneous rocks, such as kimberlite or peridotite. These occur in south and central Africa and Russia, and there is one occurrence in Arkansas; the original source of the South American diamonds is not definitely known. Since the diamonds are very sparsely scattered in the pipes, an extremely high ratio of concentration must have taken place to yield the workable placer deposits.

The most remarkable placer deposits of South Africa were those of the Lichtenburg district, which witnessed the great "stampede" by 25,000 "runners" in 1926. This district is a striking peneplain developed upon an ancient dolomite. The stream gravels are sinuous embankments that project above the peneplain and stretch discontinuously across the country, marking the meandering course of a former stream. The stream was formerly entrenched in dolomite, and its valley contained diamondiferous gravels. The region developed a karst topography, the surface drainage disappeared, and the soluble dolomite plateau was slowly lowered by solution-weathering; but the protected dolomite beneath the gravels escaped solution and was left as a ridge capped by diamondiferous gravels. The diamonds now lie in the bottom gravels of the ridge. Even more unusual are sinkholes developed beneath the former valley bottom into which the diamond gravels have slumped; some of these pockets proved fabulously rich. The Lichtenburg gravels in the year of their greatest production, yielded over 2,100,000 carats valued at over $23,000,000 as against approximately 2,300,000 carats from the diamond pipes in the same year.

The diamondiferous gravels of Zaire and Liberia which yield about two-thirds of the world production, are former stream gravels, and the diamonds probably came originally from weathered kimberlite pipes, several of which are known within Zaire. The same is probably true of the adjacent Angola gravels. The diamonds of Minas Geraes, Brazil, are found in gravels derived from a conglomerate, which in turn represents a former lightly concentrated placer deposit whose stones came from an unknown source. The prized *carbonado* of Bahia, Brazil, desired for its use in diamond drilling, is obtained from present stream gravels. Extensive diamond placers are also worked in Siberia.

Diamonds are now produced synthetically by man. Even though the PT diagram (Fig. 3-4) for carbon indicates modest temperatures and high pressures are needed to synthesize diamonds, catalysts are also needed. Small tetragonal presses to large two-story pressure instruments are used. Tracy Hall, the first man to synthesize diamonds that have been used industrially for several years, is now forming diamonds of gem stone quality in Provo, Utah.

Other Gemstones. Most rubies, sapphires, chrysoberyls, aquamarines, and spinels are washed from stream gravels. Their original site was in pegmatite dikes or in contact- or regional-metamorphosed rocks from which they have been released by weathering. The sapphires, aquamarines, and zircons of Ceylon and Kashmir also come from stream gravels. Formerly, stream gravels from peridotite areas in North Carolina and Georgia supplied some fine gemstones of sapphire and ruby; sapphires also occur along the Missouri River near Helena, Montana.

Other Fluviatile Deposits. *Monazite* (see Beach Placers), an accessory mineral of igneous rocks, accumulates with other placer minerals. Stream deposits occur in the Carolinas and Idaho. Alluvial *quartz sands* are common. *Ilmenite,* which is found mainly in beach gravels, also occurs in stream gravels and is produced by a by-product of placer tin in Malaysia. *"River pebble" phosphate rock* of alluvial origin occurs in streams in Florida. Other minerals obtained in minor quantities from stream placers include zircon, garnet, native copper, chromite, quicksilver, wolframite, thorianite, thorite, and kyanite.

Beach Placer Formation

Beach placers are formed along seashores by the concentrating effects of wave and shore action. Shore currents shift materials alongshore, and the lighter materials are moved faster and farther than the heavy, thereby concentrating the heavy minerals. Wave action operates at the same time. Pounding waves throw up materials on the beaches; the backwash and the undertow carry out the lighter and finer material, which in turn is moved alongshore; and the larger and heavier materials are concentrated on the exposed beaches. A familiar man-

ifestation of this is gravel or shingle beaches from which the sand has been removed. If placer minerals are available, they are similarly concentrated.

Placer minerals may be made available for beach concentration by (1) streams that debouch upon the coast, (2) wave erosion upon sea terraces or gravel plains, (3) wave encroachment upon former near-shore stream terraces, and (4) wave erosion of rocky shores.

Placer minerals of beach deposits consist of gold, ilmenite, magnetite, rutile, diamonds, zircon, monazite, garnet, and quartz.

The beach gold is mostly very finely divided, ranging from 70 to 600 colors to the cent, giving evidence thereby of the travel and beating it has undergone. The deposits of Nome, Alaska (Table 13-3), illustrate beach placer gold. Ilmenite, rutile, and monazite have undergone unusual beach concentration at Travancore and Quilon, India, that has resulted in commercial sands containing 50 to 70 percent ilmenite and 2 to 5 percent (in places up to 25 percent) monazite, along with zircon, rutile, garnet, and other minerals. Richards Bay on the Natal Coast of South Africa is a major new beach sand operation. Somewhat lower concentration has occurred in beach sands of Florida, (Table 13-3); Redondo Beach, California; North Carolina; Senegal; Ceylon; Argentina; Brazil; Evanshead and Ballina in Australia; and in New Zealand. Zircon has reached high beach concentration in Brazil where, according to H. C. Meyer, the sands will carry 51 percent zirconium oxide, along with titanium and cerium and yttrium earths. The beach sands in Australia contain up to 75 percent zircon. Magnetite, or black sands, containing also ilmenite, chromite, and other heavy minerals have been formed extensively on Oregon, California, Brazil, Japan, India, Egypt, and New Zealand coasts.

One of the most remarkable beach concentrations was that of the diamantiferous marine gravels of Namaqualand, South Africa, which in one year yielded 50 million dollars. The diamonds occur in marine gravels that occupy wave-cut terraces 6 to 65 meters above sea level, which extend intermittently for a distance of 320 kilometers or more along the desert coast south of the mouth of the Orange River at a maximum distance of 5 kilometers inland. The best deposit was near Alexander Bay in gravels 3 to 30 meters wide and 35 meters above sea level. After the finding of the first few stones, the mode of occurrence was not realized until Merensky and Reuning discovered that the diamonds were associated with certain diagnostic oyster shells that marked definite beach lines. These beach lines could then be traced along the coast. The diamonds in places occur in exposed coarse beach shingle from which the sand has been blown away. A cupful of excellent diamonds was said to have been picked up by hand from among the pebbles over a very small area. Other diamondiferous gravels are buried under surface deposits. The diamonds are thought to have been carried down by the Orange River and to have been distributed down the coast by the prevailing southerly shore currents, and by waves. Offshore gravels are also being dredged.

Eolian Placer Formation

Wind instead of water may act as the agent of concentration and give rise to placer deposits. Obviously, this can occur only in arid regions. It is reported that some eolian gold placers have been formed in the Australian deserts from the disintegration of gold quartz lodes. The light decomposed materials have been blown away; the heavy gold particles, freed from their matrix, remained behind. The continuation of this process finally resulted in patches of surface accumulations of placer gold in a debris of sand and wind-worn pebbles. Some of the gold was recovered by "dry washing," utilizing wind instead of water to separate the debris from the gold after screening off the pebbles. Similar concentration has taken place at El Arco, in Lower California, Mexico.

Selected References on Placer Deposits

Witwatersrand Gold—The Challenge of Inflation. 1967. World Mining, London, v. 3, p. 12–17.

Gold-Recovery, Properties, and Applications. 1964. E. M. Wise. Van Nostrand Reinhold Co., New York, 367 p.

Gold resources in the tertiary gravels of California. R. W. Merwin. U.S. Bur. Mines Tech. Rept. No. 968.

The World of Gold. 1968. T. Green. Walker and Co., Salt Lake City, Utah, 242 p.

General References 19 and 20.

General Reference 9, Chap. 17. *General discussion.*

Tertiary gravels of the Sierra Nevada. 1911. W. Lindgren. U.S. Geol. Surv. Prof. Paper 73. *A fascinating correlation of Tertiary gravel formation and physiographic development of the region resulting in remnants of elevated buried gravels.*

Diamond-bearing gravels of South Africa. 1930. A. F. Williams. 3rd Empire Min. and Met. Cong., S. Africa. *Descriptions of various types of placer gravels.*

14

OXIDATION AND SUPERGENE ENRICHMENT

All "limonites," including the cellular pseudomorphs and fringing "limonites," eventually are transformed into pulverulent products.

R. Blanchard

There are those who question the significance of the study of leached outcrops nowadays when most outcrops of mineralized bodies have not only been found but many, if economic, have been mined out. On the other hand, gossaniferous zones are not limited to present-day surfaces. Evidence of these zones derived from drill cores is of primary significance, which can be better qualified by a knowledge of oxidation and supergene enrichment.

When ore deposits become exposed to the oxidation zone, they are weathered and altered along with the enclosing rocks. The surface waters oxidize many ore minerals and yield solvents that dissolve other minerals. An ore deposit thus becomes oxidized and generally leached of many of its valuable materials down to the groundwater table, or to a depth where oxidation cannot take place. The oxidized part is called the *zone of oxidation*. The effects of oxidation, however, may extend far below the zone of oxidation. As the cold, dilute, leaching solutions trickle downward they may lose a part or all of their metallic content within the zone of oxidation and give rise to oxidized ore deposits—a familiar group of deposits readily accessible to mining and made colorful in the glamorous beginnings of many mining districts. If the down-trickling solutions penetrate the water table, their metallic content may be precipitated in the form of secondary sulfides to give rise to a zone of *secondary* or *supergene sulfide enrichment*. The lower, unaffected part of the deposit is called the *primary* or *hypogene zone*. This zonal arrangement (Fig. 14-1) is characteristic of many mineral deposits that have undergone long-continued weathering. In places the supergene sulfide zone may be absent, and in rare cases the oxidized zone is shallow or lacking, as in some glaciated areas or regions undergoing rapid

erosion. Special conditions of time, climate, physiographic development, and amenable ores are necessary to yield the results shown in Fig. 14-1, but they are sufficiently common for oxidized and enriched supergene ores to occur in most of the nonglaciated land areas of the world.

The effect of oxidation has been either to render barren the upper parts of many deposits or to change ore minerals into more usable or less usable forms, or even to make rich bonanzas. The effect of supergene enrichment has been more far reaching, for it has added much where there was previously little; leaner parts of veins have been made rich; unworkable protore has been enriched to ore grade. Indeed, many of the great copper districts would not have come into existence except for the process of supergene sulfide enrichment.

OXIDATION AND SOLUTION IN ZONE OF OXIDATION

Supergene oxidation and reduction enrichment go hand in hand. Without oxidation there can be no supply of solvents from which minerals may later be precipitated in the zones of oxidation or of supergene sulfides. The process resolves itself, therefore, into three stages: (1) oxidation and solution in the zone of oxidation, (2) deposition in the zone of oxidation, and (3) supergene sulfide deposition. Each is considered separately.

The effects of oxidation on mineral deposits are profound. The minerals are altered and the structure is obliterated. The metallic substances are leached or are altered to new compounds that require metallurgical treatment for their extraction quite unlike that employed for the unoxidized ma-

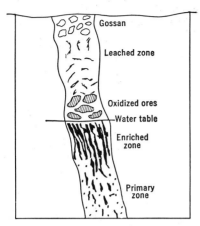

Figure 14-1 Diagram of zones of a weathered vein, with oxidized, supergene enrichment and primary zones.

terials. The texture and the type of deposit are obscured. Compact ores are made cavernous. Ubiquitous limonite[1] obscures everything and imparts to the gossan that familiar rusty color which from earliest times has attracted the curiosity of the miner. One can only infer what lies beneath, but in this inference one can be guided by the character of the oxidation products themselves.

Water with dissolved and entangled oxygen is the most powerful oxidizing reagent, but carbon dioxide also plays an important role. (Locally chlorides, iodides, and bromides play a part.) These substances react with certain minerals to yield strong solvents, such as ferric sulfate and sulfuric acid. Sulfuric acid, in turn, reacting with sodium chloride yields hydrochloric acid, which with iron yields the strongly oxidizing ferric chloride. Bacteria also promote oxidation (Fig. 1-24); they oxidize ferrous iron at low pH to ferric sulfate.

The classical approach to the study of oxidation and supergene enrichment has been based on observations and supposed reasonable chemical reactions. The effectiveness, limitations, and corrections of the past studies are now better understood through the study of solution chemistry and the adjunct comprehension of the subjects of mineral

[1] No mineral corresponds to the formula for *"limonite"* that is commonly given as $2Fe_2O_3 \cdot 3H_2O$. Actually, two minerals make up what was and is referred to as "limonite." They are goethite and lepidocrocite, both with the composition of $Fe_2O_3 \cdot H_2O$. More recently, structural compositions have been indicated for these two minerals, that is, goethite is $HFeO_2$ and lepidocrocite is $FeO(OH)$. Kelley found that poorly crystallized goethite could not be distinguished from lepidocrocite by differential thermal analyses. Only by means of X-ray analysis can the two minerals be distinguished. Lepidocrocite is extremely rare in leached outcrops.

equilibria, thermodynamics, and physical chemistry. Garrels has presented information on chemical relations in aqueous solutions at low temperatures and pressures of systems with direct application to oxidation and reduction of many mineral deposits.

Fortunately, the geologist is generally limited to the study of dilute solutions in oxidation and supergene enrichment studies. As the interrelationship of molality and activity, or thermodynamic concentration, show significant deviations in moderately concentrated solutions, it is fortunate that geologists are dealing with dilute solutions in oxidation processes where the activity of the solute is numerically equal to its molality. It is readily possible to determine total molality, M_0, which is defined as

moles per 1000 grams of water =

weight of element obtained by analysis × 1000

---

atomic weight of element
× weight of containing water

The concentration of the hydrogen ion is expressed as pH, whereby $pH = -\log a[H^+]$, which is sometimes involved in oxidation processes and can, therefore, be used as a common reference activity for the oxidation-reduction reactions. With pH as a variable, a number of reactions can be expressed, but more significant is that the same reaction will differ in its oxidation-reduction potential (Eh) with variations of pH.

Professor Willard Gibbs, of Yale University, first described the relationship between absolute temperature (T), the heat of formation (H), and entropy (S) in regard to Gibb's free energy (G) as

$$G = H - TS$$

Differences in free energy are

$$\Delta G = \Delta H - T \cdot \Delta S \qquad \text{for constant } T$$

The difference in electrical potential of an electrolytic reaction from equilibrium is, of course, related to the free-energy change, or

$$\Delta G = nfE$$

where n is the number of electrons that are involved, f the Faraday constant, and E the redox potential in volts.

The relationship between ΔG and K (the standard equilibrium constant) is

$$\Delta G = RT \ln K - \Delta G° \qquad \text{(the standard free energy change)}$$

In order to find the potential differences for a reaction under different conditions the two equations

$$\Delta G = nfE \quad \text{and}$$

$$\Delta G = -RT \ln K + RT \ln \frac{[Y]^y [Z]^z \cdots}{[B]^b [D]^d \cdots}$$

are combined to give the Nernst equation or

$$E = \frac{\Delta G}{nf} = \frac{\Delta F°}{nf} + \frac{RT}{nf} \ln \frac{[Y]^y [Z]^z}{[B]^b [D]^d}$$

$$= E° + \frac{2.303\, RT}{nf} \log \frac{[Y]^y [Z]^z}{[B]^b [D]^d}$$

For reactions at 25°C,

$$E = E° + \frac{0.059}{n} \log \frac{[Y]^y [Z]^z}{[B]^b [D]^d}$$

As an example, determining E for the Fe-Mn reaction is as follows:

$$MnO_2 + 4H^+ + 2Fe^{2+} \rightleftarrows Mn^{2+} + 2H_2O + 2Fe^{3+}$$

at 25°C, pH = 3, and $E° = 0.46$ v.

$$E = 0.46 + \frac{0.059}{2} \log \frac{[Mn^{2+}][Fe^{3+}]^2}{[H^+][Fe^{2+}]^2}$$

$$= -0.46 + 0.03 \log \frac{1}{[10^{-3}]^4} = -0.46 + 0.03 \times 12$$

$$= -0.10 \text{ volts}$$

It is evident, therefore, that Eh-pH diagrams are not only most helpful in understanding their relationships at the low temperature-low pressure environments that permit specific reactions to occur but indicate the stability fields of mineral species.

Figure 14-2 is a prime example of a pH-Eh diagram that applies to reduction-oxidation for common metal sulfides and sulfates. Note that the three sulfides involved in the diagram are stable at A where the environment is reducing. If oxidizing solutions of acid pH react with the sulfides, secondary sulfate carbonates and oxide minerals form in this

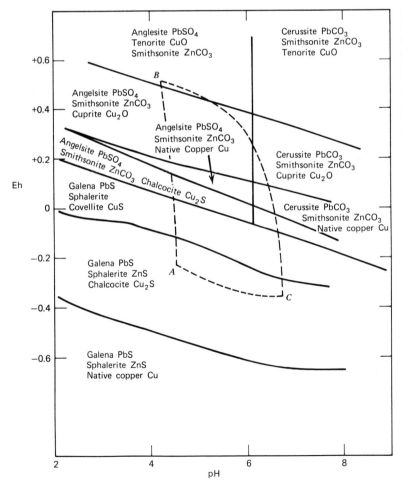

Figure 14-2 Composite diagram of stability of metal sulfides and oxidation products at 25°C and 1 atmosphere total pressure in the presence of total dissolved carbonate = $10^{-1.5}$, total dissolved sulfur = 10^{-1}. (From R. M. Garrels, 1953, Mineral species as functions of pH and oxidation potentials. Geochim. et Cosmochim. Acta, 5, 165.)

environment of weathering listed as *B*. If these secondary minerals are transported in solution to the reducing zone *C*, the pH approaches neutrality because of the consumption of H_2SO_4 and HCO_3 in forming the secondary minerals.

Secondary sulfides will form by the replacement of preexisting sulfide, generally pyrite, in zone *C*. It is apparent, moreover, that the secondary minerals are also stable in the environment of oxidation of pH value in excess of 6.

Chemical Changes Involved

There are two main chemical changes within the zone of oxidation: (1) the oxidation, solution, and removal of the valuable minerals, and (2) the transformation in situ of metallic minerals into oxidized compounds.

Most metallic mineral deposits contain pyrite. This mineral under attack readily yields sulfur to form iron sulfate and sulfuric acid; pyrrhotite does the same. The following reactions are suggested to indicate, without intermediate steps, their general trend:

[1] $FeS_2 + 7O + H_2O \rightleftarrows FeSO_4 + H_2SO_4$

[2] $2FeSO_4 + H_2SO_4 + O \rightleftarrows Fe_2(SO_4)_3 + H_2O$

Reaction 2 passes through intermediate stages during which S, SO_2, and $FeSO_4$ may form. The sulfur may oxidize to sulfuric acid. The ferrous sulfate readily oxidizes to ferric sulfate and ferric hydroxide:

[3] $6FeSO_4 + 3O$
            $+ 3H_2O \rightleftarrows 2Fe_2(SO_4)_3 + 2Fe(OH)_3$

The ferric sulfate hydrolizes to ferric hydroxide and sulfuric acid:

[4] $Fe_2(SO_4)_3 + 6H_2O \rightleftarrows 2Fe(OH)_3 + 3H_2SO_4$

Ferric sulfate is also a strong oxidizing agent and attacks pyrite and other sulfides to yield more ferrous sulfate.

[5] $Fe_2(SO_4)_3 + FeS_2 \rightleftarrows 3FeSO_4 + 2S$

Ferric sulfate, in addition, changes to various basic sulfates.

The above reactions indicate the importance of pyrite, which yields the chief solvents, ferric sulfate and sulfuric acid, and also ferric hydroxide and basic ferric sulfates. Moreover, ferric sulfate is continuously being regenerated not only from pyrite, but also from chalcopyrite and other sulfides. The

ferric hydroxide changes over to hematite and goethite and forms the ever-present "limonite" that characterizes all oxidized zones. The basic ferric sulfates, of which there are several, may be deposited as such, but generally limonite is the end product.

The part played by ferric sulfate as a solvent may be seen in the following equations. Although the end products are obtained, it is not established in all cases that the following reactions are those that actually take place.

[6] *Pyrite*—$FeS_2 + Fe_2(SO_4)_3 \rightleftarrows 3FeSO_4 + 2S$

[7] *Chalcopyrite*—$CuFeS_2 + 2Fe_2(SO_4)_3 \rightleftarrows CuSO_4 + 5FeSO_4 + 2S$

[8] *Chalcocite*—$Cu_2S + Fe_2(SO_4)_3 \rightleftarrows CuSO_4 + 2FeSO_4 + CuS$

[9] *Covellite*—$CuS + Fe_2(SO_4)_3 \quad 2FeSO_4 + CuSO_4 + S^0$

[10] *Sphalerite*—$ZnS + 4Fe_2(SO_4)_3 + 4H_2O \rightleftarrows ZnSO_4 + 8FeSO_4 + 4H_2SO_4$

[11] *Galena*—$PbS + Fe_2(SO_4)_3 + H_2O + 3O \rightleftarrows PbSO_4 + 2FeSO_4 + H_2SO_4$

[12] *Silver*—$2Ag + Fe_2(SO_4)_3 \rightleftarrows Ag_2SO_4 + 2FeSO_4$

Similarly, other minerals are dissolved yielding, except for lead, soluble sulfates of the metals. The sulfuric acid also attacks various sulfides, yielding sulfates of their metals. Chlorides, bromides, and iodides, chiefly of silver, are also formed.

Most of the sulfates formed are readily soluble, and these cold, dilute solutions slowly trickle downward through the deposit until the proper Eh-pH conditions are met to cause deposition of their metallic content.

If pyrite is absent from deposits undergoing oxidation, only minor amounts of the solvents are formed; little solution occurs, the sulfides tend to be converted in situ into oxidized compounds, and the hypogene sulfides are not enriched. This is illustrated in the New Cornelia mine at Ajo, Arizona, where a deficiency of pyrite has resulted in chalcopyrite being coverted to copper carbonate and supergene sulfides are negligible. This also happens where a supergene chalcocite zone lacking pyrite is oxidized; the chalcocite is not dissolved but is converted into copper carbonates, cuprite, or native copper.

A country rock of limestone tends to inhibit migration of some sulfate solutions; it immediately reacts with copper sulfate, for example, to form copper carbonates, thus precluding any supergene sulfide enrichment.

During the oxidation processes, alumina-silicate minerals are leached of silica and the oxidized material becomes clay by hydrogen ion metasomatism. The leached silica may exist as a gel or a crypto-crystalline material incorporated with various amounts of iron oxide dispersed through the silica. This material is jasper; *jasperoid* is a prime prospecting tool used in junction with the study of gossans and alteration.

Oxidation Separation of Metals

Oxidation of mixed ores commonly results in a separation of the contained metals, as in the lead-zinc-pyrite "manto" limestone deposits of Mexico (illustrated in Fig. 14-3). The pyrite is largely removed; the galena becomes oxidized to anglesite and cerussite, which remain in place; the sphalerite is dissolved as zinc sulfate, which migrates, encounters limestone, and the zinc is deposited as ore bodies of zinc carbonate. Thus complex metallurgy is simplified by nature, rendering the deposits of greater value than the original. At Sierra Mojada, Mexico (Fig. 14-3), for example, a body of rich lead cerussite ore was underlain by large bodies of zinc carbonate and locally by areas rich in silver that had also undergone migration. Lindgren has described similar examples from Tintic, Utah.

In similar manner, gold is separated from sulfide matrix and rendered "free milling."

Gossans and Cappings

Gossans are signboards that point to what lies beneath the surface. They arrest attention and incite interest as to what they may mask. Most ore deposits, save in glaciated regions, are capped by gossans; hence the finding of one may herald the discovery of buried wealth. Noncommercial mineral bodies, however, also yield gossans. To distinguish between them is of vital importance; but it requires experience, knowledge, and careful obser-

vation. The distinctions involve delicate differences of color and form that are difficult to transmit by written word; they must be seen by experienced eyes.

Gossan is a Cornish word used to designate the oxidized outcropping cellular mass of "limonite" and gangue overlying aggregated sulfide deposits. The Spanish refer to these oxidized outcroppings as *Colorados* because of their reddish color, and the Germans refer to them as *eiserner Huts* because of their shape, iron content, and their relative resistance to erosion.

In summary, gossan is a heavy concentration of "limonitic" material, derived from massive sulfide minerals or from their iron-yielding gossan, which has been leached in place and transported downward. *Capping* is the leached upper part of a body or rock that still contains disseminated sulfide minerals.

Gossans are not limited to the surface but may extend some distance below the surface. In some cases the gossan has contained sufficient economic mineralization to warrant mining. Figure 14-4 is a cross section of an oxidized and secondarily enriched portion typical of eight massive sulfide ore deposits discovered by M. Magee that occur in the Ducktown district, Tennessee. The gossan is predominantly "limonite" with an iron content of about 45 percent and highly porous, about 40 percent porosity. These gossans were mined for their copper content, which varied from 0.1 to 0.8 percent copper. The so-called "black copper" of the supergene zone varied in copper content from 10 to 50 percent and averaged about 25 percent copper. This supergene zone consisted of sooty chalcocite except where it had replaced primary sulfides.

Materials of Gossans and Their Formation. The limonite universally formed during oxidation of iron-bearing sulfides persists in the oxidized zone and

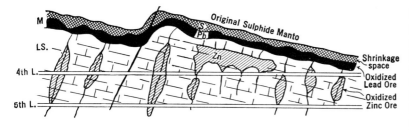

Figure 14-3 Sketch of longitudinal section of part of ore manto of Encantada mine Sierra Mojada, Mexico, illustrating the separation of metals by oxidation. Original body (M) composed of pyrite, galena, and zincblende. Oxidation yielded body of oxidized lead-silver ore (Pb), overlain by open spaces (S) due to oxidation shrinkage, and underlain by large bodies of transported oxidized ore (Zn) in iron-stained dolomite.

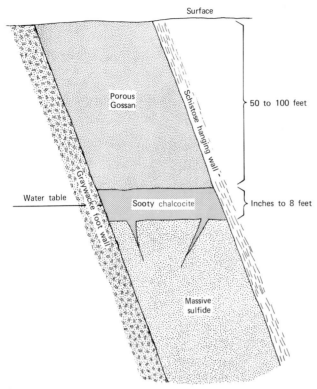

Figure 14-4 Diagrammatic cross section, showing secondary ore development and supergene-enriched copper zone. (From: Magee, Ore deposits of the U.S., AIME.)

imparts to the gossan and capping its diagnostic color. Its many forms and colors are considered separately.

Other persistent oxidized minerals in the gossan also indicate former sulfides, such as the sulfate and carbonate of lead, the carbonate and silicate of zinc, native silver, and horn silver. Gold generally persists in the croppings as native gold.

The Role of Iron in Gossans

The importance of iron in croppings cannot be overemphasized. Not only is "limonite" universally present, but also it occurs in a variety of colors, structures, and positions, each of which is significant (Tables 14-1 and 14-2). Much information has been gained in recent years regarding the composition and interpretation of limonite.

Indigenous and Transported Limonite. In croppings, iron of sulfide derivation may either (1) become fixed as oxide at the site of the former sulfide, forming *indigenous limonite*, or (2) be dissolved, transported, and precipitated elsewhere, forming *transported limonite*. The indigenous is thrown down in the insoluble ferric state; the transported is in the soluble ferrous state. Posnjak and Merwin

Table 14-1 Types of Limonite Boxworks

Boxwork	Fig.	Character	Color	Derivation
Coarse cellular	14-6	Coarse, angular; thin, wide, rigid walls; blebs, masses	Ocherous	Chalcopyrite
Coarse cellular	14-7	Siliceous; thin, rigid, angular walls	Light brown	Sphalerite
Fine cellular	—	Thin, small, friable walls; specks, blebs	Yellow orange	Bornite-chalcopyrite
Fine cellular	—	Stronger cells; "shriveled" limonite jasper	Light brown	Sphalerite
Cellular sponge	—	Rounded, thick, rigid, empty cells; crisp; much silica	Brownish	Sphalerite
Triangular	—	Triangular cells, thick, fragile, crusted	Ocherous orange	Bornite
Triangular	14-8	Triangular, curved	Ocherous orange	Bornite
Contour	14-10	Long, narrow, angular rigid cells	Chocolate	Tetrahedrite
Relief	—	No boxwork; arrangement of sulfide grains; weak, porous, relief	Maroon	Chalcocite, covellite, bornite
Limonite pitch	—	Pitchlike; varnish; no cells	Dark brown	Chalcopyrite, bornite
Limonite crusts	—	Thin, fragile, flaky concentric foils	Dark brown to black	Chalcocite
Cleavage	—	Thin parallel cubic plates of limonite jasper	Ocherous orange	Galena
Diamond mesh	14-11	Diamond-shaped meshes	Ocherous orange	Galena
Pyramidal	—	Steplike arrangement	Ocherous orange	Galena
Foliated	14-9	Smooth, thin, rounded cells	Tan to maroon	Molybdenite

Table 14-2 Oxidation Residuals of Some Common Sulfides

Mineral	Voids	Limonite	Boxwork	Color	Composition
Pyrite	Empty	Transported; halos or flooding	None	Brick red	Hematite +, jarosite +
Pyrrhotite	Empty	Transported; halos or flooding	Coarse spongy masses	Brick red	Hematite +, jarosite +
Chalcopyrite	Occupied	Indigenous	Coarse cellular; fine cellular; limonite pitch	Ocherous	Goethite +, jarosite −
Bornite	Occupied	Indigenous	Triangular; sponge crusts; relief	Ocherous to orange	Goethite +, hematite
Chalcocite	Occupied	Indigenous	Relief; crusts	Maroon to seal brown	Goethite +
Tetrahedrite	Occupied	Indigenous	Contour; coagulated	Chocolate	Goethite
Sphalerite (inert gangue)	Empty	Transported	"Moss" limonite	Yellow to brown	Goethite +, silica +
Sphalerite (reacting gangue)	Occupied	Indigenous	Coarse cellular; fine cellular; cellular sponge	Yellow to brown	Goethite +, silica +
Galena	Occupied	Indigenous	Clevage; crusts; diamond mesh; cellular sponge	Ocherous to orange, seal brown, dark chocolate	Goethite, hematite
Molybdenite	Occupied	Indigenous	Foliated; granular	Maroon	Goethite
Siderite	Occupied	Indigenous	Cleavage; mica plate	Yellow, deep brown	Goethite

Note: + means important constituent; − means minor constituent.

have shown that the oxidation of ferrous to ferric iron is retarded by free sulfuric acid and is accelerated by copper; also that the deposition of ferric oxides is retarded by free acid and aided by copper. In plain words, this means that when pyrite undergoes oxidation, the free acid generated tends to keep the iron in the soluble ferrous state and enables it to be removed in solution. The free acid, in addition, retards deposition as limonite. Also, it means that, if chalcopyrite undergoes oxidation which yields little or no acid, ferrous iron readily changes to the insoluble ferric and limonite is precipitated in situ. Thus, as pointed out by Locke and Morse, indigenous limonite indicates the former presence of copper, and transported limonite indicates a former high ratio of iron to copper or else lack of copper.

Indigenous limonite of sulfides derivation occupies the voids left by the former sulfides (Fig. 14-5A). It does not occur outside the voids. Its characteristic structure denotes the kind of predecessor sulfide. It is generally compact and fairly hard, and it has subdued colors.

Fringing limonite may have been moved no far-

ther than beyond the rim of the void, or it may have been transported tens or hundreds of feet from the site of its originating sulfides. The distance of transportation depends largely upon the precipitating power of the gangue or rock through which the solutions pass. Fringing or transported limonite thus may form halos around the empty voids (Fig. 14-5B), or it may thoroughly permeate the gangue

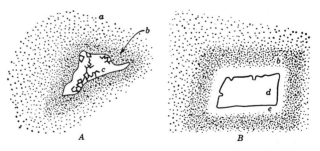

Figure 14-5 A, Sketch of indigenous limonite, (c), derived from the oxidation of chalcopyrite and remaining within the cavity; a, b, limonite fog surrounding the cavity. Cactus mine, Utah, ×10. B, Transported or fringing limonite from the oxidation of pyrite, with empty cavity d, surrounded by bleached zone e and by limonite halo b. Silverbell, Ariz., ×10. (From Blanchard-Boswell, Econ. Geol.)

or enclosing rock. A little of it, like a drop of ink in a glass of water, goes a long way; it makes, therefore, a conspicuous and exaggerated showing of iron oxide. It occurs in varied structural forms, colors, and positions. Paints, crusts, impregnations, and earthy mixtures are common.

Limonite Boxworks. The structure assumed by indigenous limonite is the most diagnostic feature of the predecessor minerals. Color and structure of the limonite combined are, in many cases, specifically diagnostic of certain minerals; some structural patterns that resemble each other closely may have distinguishing color differences.

When a grain of sulfide is oxidized and residual limonite remains in the cavity, the limonite assumes a honeycomb pattern, called *boxwork*, that persists in the cropping. Neither the goethite nor the hematite suffers alteration although the jarosite may do so. Thus the original patterns are preserved. The boxwork patterns, like colors, have been correlated in place with underlying sulfides, thus establishing their sulfide derivation. Several varieties have been described, chiefly by Locke, Morse, Blanchard, and Boswell. Blanchard in 1968 greatly expanded the descriptions, interpretations, and use of boxworks and color of cappings.

From Limonite Residues. Limonite colors, boxworks, and position, properly interpreted, can be resolved into many of the original sulfides. The diagnostic features of some common ones are listed in Table 14-2. Characteristics of some common boxwork patterns are shown by Figs. 14-6 to 14-11.

Rate of Erosion. The rate of erosion and the ensuing lowering of the surface and the water table may be so rapid that oxidation cannot keep pace with it. Any former oxidized zone would then eventually

Figure 14-7 Coarse cellular boxwork form sphalerite, Spruce Mountain, Nevada, ×2. *a*, Quartz veinlet; *b*, cell walls of limonitic jasper; and *c*, interstitial limonite. (From Boswell-Blanchard, Econ. Geol.)

be removed by erosion, and little or no oxidation would be evident on the surface. A very low rate of erosion would cause an almost stagnant position of the water level and little oxidation.

Generalizations Regarding Oxidized Ores

When it is established that given ores are oxidized, it follows that (1) such ores will change in nature in depth; (2) there is likely to be a pronounced change in tenor in depth; (3) in most cases, only shallow depth may be expected; (4) different metallurgical treatment will be required for underlying ores; (5) extraction plants should not be erected until the volume of the oxidized ore is delimited; and (6) more adequate transportation was generally necessary for oxidized than for unoxidized ores, because most ores were shipped directly to smelters. In situ leaching or heap leaching, however, is now a more common method of treating oxidizing ore because it avoids transportation, milling and reduces smelting costs.

It should be determined if the oxidized ores are indigenous or transported. If indigenous, sulfide ores may be expected to underlie them; if transported, they may be remote from their source and so may have no sulfide roots—their source is more likely to be upward and lateral, as has been found

Figure 14-6 Coarse cellular boxworks from chalcopyrite. A, Bagdad, Arizona, ×2; *a*, quartz veinlets; *b*, cell walls of limonitic jasper; *c*, *e*, interstitial limonite. B, quadrangular pattern, Duquesne, Arizona, ×5, with parallel cell walls; *b*, cross structure; *c*, *d*, thin angular web-work. C, average type, Creston Verde, Sinaloa, Mexico, ×5. (From Blanchard-Boswell, Econ. Geol.)

Figure 14-8 Triangular, curved boxwork from bornite with eye-shaped cells, ×5. A, Engel mine, California; B, Black Mountains, New Mexico (From Boswell-Blanchard, Econ. Geol.)

Figure 14-9 Foliated boxwork from molybdenite, containing limonite flakes and quartz veinlets. A, Nogales, Arizona; B, Hodgkinson, Queensland, ×20 (From Blanchard-Boswell, Econ. Geol.)

Figure 14-11 Galena boxworks. A, Regular pattern, Lawn Hill, Queensland, ×3. B, Diamond mes type from steely galena, Eureka, Nevada, ×3. (From Boswell-Blanchard, Econ. Geol.)

in many zinc and copper deposits. If transported, the size of the ore body may have no relation to the size of the source body.

Supergene sulfides rarely form by simple precipitation but only by replacement of other metallic minerals, mostly sulfides. One exception is a deposit of supergene chalcocite upon quartz in Morenci, Arizona, apparently brought about by escaping hydrogen sulfide. Instances have been recorded of the replacement of wood cells by supergene chalcocite in the red-beds type of copper deposits. The chalcocite has clearly inherited the wood cell structure, but some specimens show clear evidence that the woody material was first replaced by hypogene bornite and later by chalcocite, which inherited the cell structure from the bornite.

Supergene sulfide replacement is generally volume for volume, which is not, therefore, a molecular interchange between replacing and replaced substances. If the replacement of pyrite by chalcocite were a molecular interchange, a grain of denser chalcocite would occupy only a portion of the volume of a grain of pyrite. This is not the obvious case in nature. Hence customary balanced chemical equations depicting such changes do not correctly represent what has taken place. Such equations are usually given to indicate the trend of the change rather than the exact stochiometric chemical reaction.

Climate. The important factors of climate that affect oxidation are temperature and rainfall, but often they cannot be disentangled from the effect of time,

Figure 14-10 Typical contour boxwork (like topographic contours) from tetrahedrite. A, Hachita, New Mexico, ×5. B, Patagonia, Arizona, ×5. (Boswell-Blanchard, Econ. Geol.)

especially as many regions were closer to equatorial latitudes during earlier Tertiary time because of continental drift. Is the paucity of post-Glacial oxidation in glaciated countries due to the cool climates that prevail there, or to the short time since glaciation? Is the deeper oxidation of southwestern United States or Zaire attributable to the warm climate of the regions or to the length of time oxidation has been going on? In the Leonard mine at Butte, Montana, one end of a drift along a deep vein terminated in a very hot "dead end," the other in a cool airway; after 2 years, the hot end oxidized to a depth of 90 centimeters and the cool end to a depth of only 5 centimeters. The effect of temperature is fully expected and predicted by the van't Hoff equation. Oxidation, therefore, is favored by warm much more than by cool temperature.

Titley has described "modern leaching" and secondary enrichment in a wet climate at the Plesyumi copper prospect in New Britain, Papua, New Guinea, where he suggests that the leaching and enrichment could be the product of climatic cycles rather than any unique climatic type. Stranded lenses of tenorite exist above the chalcocite zone that is undergoing renewed leaching along with underlying primary sulfides to greater depth where a new supergene zone is forming. This is probably the result of recent uplift and the present high rainfall in the area of the southwestern Pacific.

Figure 14-12 is an Eh-pH diagram indicating the high oxidation state of tenorite which in the Plesyami prospect is undergoing oxidation while chalcocite forms in a broad pH reducing environment. Actually, most supergene chalcocite replaces preexisting sulfides, predominantly pyrite, but sooty chalcocite may form by simple reduction.

Depth of Oxidation. Several instances have been observed of oxidation to depths greater than 600 meters, which generally occurs along near vertical fracture zones and is highly limited and localized. At Kennecott, Alaska, the deepest oxidation is over 600 meters; at Tintic, Utah, it is 670 meters; at

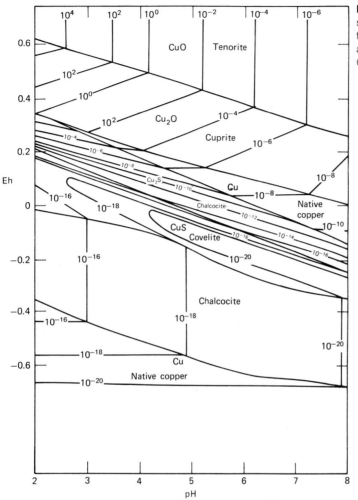

Figure 14-12 Fields of stability of some copper compounds as functions of pH and Eh. Contours are for activity Cu^{2+} + activity Cu^+ (After Garrels, 1954, fig. 2.)

Bisbee, Arizona, it is 500 meters; in the Northern Rhodesia copper mines it is known to a depth of more than 600 meters; and at the Lonely mine, Southern Rhodesia, it reaches a depth of 900 meters. In general, however, more than 90 percent of oxidation ranges in depth from a few tens to less than one hundred meters. Oxidation to the deep depths indicated here are generally limited to deep fracture zones.

Where there is no evidence of a postoxidation rise of water level, broad generalizations can be made regarding the depth of oxidation: (1) In humid regions of low relief it is generally shallow, but with high relief it will probably be shallow to medium. (2) In glaciated regions, postglacial oxidation is negligible or lacking; in areas sheltered from deep glacial erosion, preglacial oxidation may be present and is likely to be shallow in flattish areas and deep in areas of high relief. (3) In arid regions it is apt to be very deep and is generally rather deep under old

or mature topography and shallow under youthful topography.

ORE DEPOSITION IN THE ZONE OF OXIDATION

Oxidized ore deposits may result within the zone of oxidation if the down-trickling oxidizing solutions encounter precipitants above the water table.

There is abundant evidence that such solutions are present in the zone of oxidation. In copper mines, trickles and pools of copper sulfate eat into the steel rails and the nails of miners' boots. Several years ago at Bingham Canyon, it was common practice to anchor tin cans in the stream behind the homes and recover copper cans. Kennecott Copper Corporation finally covered the stream, purchased all the homes in Bingham Canyon and constructed a large copper precipitation plant at the mouth of

$$4Cu\,FeS_2 + 6H_2O + 17O_2 \rightarrow 4FeO(OH) + 8SO_4^{2-} + 8H^+ + 4Cu^{2+}$$

$$Cu\,FeS_2 + Cu^{2+} \rightarrow 2\,CuS + Fe^{2+}$$

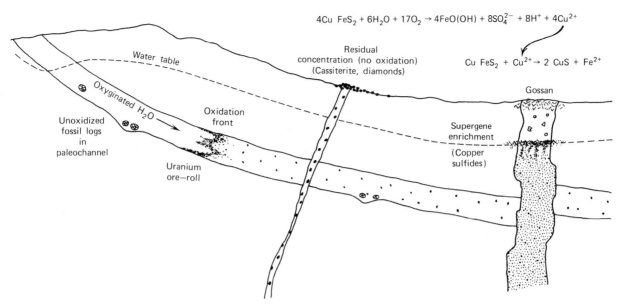

Figure 14-13 The importance and role of the water table on three quite different types of mineral deposits. (D. Garlick.)

the canyon. Since then, copper has been commercially precipitated as "cement" copper or by replacement of iron from the copper sulfate mine water at many properties. At Rio Tinto, Spain; Ruth, Nevada; Chino, New Mexico; Inspiration and Blue Bell, Arizona; Butte dumps, Montana; and Bingham, Utah, commercial heap leaching yields copper sulfate solutions from which the copper is precipitated by iron. In addition, efflorescences of sulfates of copper, iron, zinc, magnesium, manganese, cobalt, and other metals are commonly found on the walls of underground workings, indicating seeping sulfate solutions.

The economic minerals redeposited within the zone of oxidation are chiefly native metals, carbonates, silicates, and oxides. The metals of greatest commercial importance are copper, zinc, lead, silver, vanadium, and uranium, in place below the surface. Uranium is being leached in the Gas Hills region, Wyoming, by using sulfuric acid. Less valuable deposits of cobalt, manganese, iron, and other substances are similarly formed. For the most part, the ores are deposited in the lower part of the zone of oxidation.

Relation of Enrichment to Water Level

Sulfide enrichment starts at the redox interface, which may be the water table and extends below it. If the position of the top of the zone is controlled by the water table, then it is similarly related to the topography. The upper surface may be sharply sep-

arated from the oxidized zone, but generally there is an interpenetration of the two; the surface may be gently curved or highly irregular and deeply penetrated by long roots of oxidized material. The bottom of the enrichment zone is highly irregular and is a gradual transition to the primary ore. In places it terminates abruptly downward against impervious faults, or it may send roots downward along faults or other structural features.

Sulfide Enrichment above the Water Table. In masses of sulfides stranded above the water table, the supply of oxygen may be consumed by the exteriors, and supergene sulfides may be deposited in the oxygen-free interiors. This has occurred at Bingham, Utah, and at Bisbee, Arizona, although mining operations have long since removed such zones.

Similarities and differences between three quite different mineral deposits are shown in Fig. 14-13 in relation to the water table and oxidation and reduction processes.

Metal Ratios. In copper deposits, supergene enrichment can be determined from drill-hole records by plotting curves of the copper content against depth. In Fig. 14-14 are shown typical curves for hypogene and well-enriched ore. If iron and sulfur curves are also plotted, the zone of supergene enrichment is even more strikingly shown. Figure 14-14A shows the relation between copper and iron in hypogene ore, and Fig. 14-14B, enriched ore. Note that the

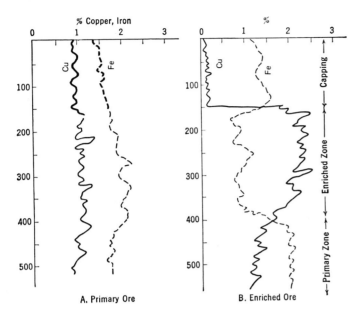

A. Primary Ore

B. Enriched Ore

Figure 14-14 Drill-hole assays of copper and iron plotted in percentage against depth. In A, unenriched ore, the curves of copper and iron are approximately parallel. In B, enriched ore, the copper content jumps up at the water table and gradually decreases toward the bottom of the enriched zone; the iron content, being replaced by copper in the enriched zone, drops down, but its curve crosses the copper curve again in the primary zone.

curve for iron crosses that of copper at the top and the bottom of the enriched zone; in the middle zone, copper has largely replaced the iron sulfide. Sulfur is negligible in the oxidized zone, less than copper in the enriched zone, and its curve crosses the copper curve at the bottom of the enriched zone.

Selected References on Oxidation and Supergene Enrichment

The enrichment of ore deposits. 1917. W. H. Emmons. U.S. Geol. Surv. Bull. 625. *Older comprehensive treatment of supergene sulfide enrichment.*

General Reference 9, Chap. 32. *Résumé.*

General Reference 10, Chap. 9. Supergene enrichment. W. H. Emmons.

Solutions, Minerals, and Equilibria. 1965. R. M. Garrels and C. L. Christ, Harper & Row, Publishers. *The standard advanced text on the measurement and utilization of Eh-pH diagrams in geologic applications.*

Oxidation Potentials. 1952. Latimer. Prentice-Hall, Inc., Engelwood Cliffs, N.J. *One of the earliest sources of many oxidation-reduction potentials.*

Leached Outcrops as Guides to Copper Ore. 1926. A. Locke. The Williams & Wilkins Company, Baltimore.

Outcrop characteristics of oxidized and enriched copper deposits, Interpretation of leached outcrops. 1968. R. Blanchard, Nevada Bureau of Mines Bull. 66. *Outcrop characteristics of sulfides not necessarily confined to the copper family.*

Mineral Equilibria at Low Temperature and Pressure. 1960. R. Garrels. Harper & Row, Publishers. *The geochemical, more quantitative approach to solution chemistry with direct application to oxidation-supergene processes and deposits.*

Oxidation at Chuquicamata, Chile. 1944. O. W. Jarrell. Econ. Geol., v. 39, p. 251–286. *Excellent discussion of processes of oxidation of large copper deposits.*

Oxidation and Enrichment in the San Manuel Copper Deposit. 1949. G. M. Schwartz. Econ. Geol., v. 44, p. 253–277.

Jasperoid in the United States—Its Characteristics, Origin, and Economic Significance. 1972. T. G. Lovering. U.S. Geol. Surv. Prof. Paper 710. *A lengthy and excellent treatise on Jasperoid.*

15

METAMORPHISM

*The economic geologist should tell
his story in plain English, then
because of the transparency of his
statement his clients or the public
can see things as they are and will
learn to refuse the highly colored
substitute offered by his quack
imitators.*

Geórge Otis Smith, in
Plain Geology, 1921

Metamorphic processes profoundly alter preexisting mineral deposits and form new ones. The chief agencies involved are heat, pressure, time, and various solutions. The substances operated upon are either earlier formed mineral deposits or rocks. Valuable nonmetallic mineral deposits are formed from rocks chiefly by recrystallization and recombination of the rock-making minerals.

TEMPERATURE AND PRESSURE

Metamorphic processes occur in adjustment to the chemical potential of any system to changes in temperature and pressure. A specific chemical reaction that cannot occur in one environment on account of negative affinity may react readily under different temperature and pressure conditions, where the chemical potentials of the reactants change in such a way that the affinity becomes positive.

An increase in pressure will cause a reaction to move in a direction in which the total volume of the system decreases. For example, increasing pressure results in the following changes from left to right with a reduction in the total molar volume:

Olivine + anorthite → garnet

Augite + anorthite → garnet + quartz

Ilmenite + anorthite → sphene + hornblende

Nephelene + albite → glaucophane or jadeite

Anorthite + gehlenite + wollastonite → grossularite

Andalusite → sillimanite → kyanite

An increase in temperature may result in endothermic reactions unless the system is not in equilibrium conditions, which then results in exothermic reactions. A possible example is the conversion of pyroxene to hornblende during the metamorphism of diabase to amphibolite. The adjustment of minerals to metamorphic rank with fairly accurate temperature measurements is shown in Fig. 3-9.

In summary, metamorphic reactions result from the tendency of mineral systems to adjust to their physiochemical environment of high temperatures and pressures in contrast to the low temperatures of weathering processes, both of which processes generally occur in the presence of water.

METAMORPHISM OF EARLIER DEPOSITS

When rocks are metamorphosed, enclosed mineral deposits may also be metamorphosed. However, unlike rocks that undergo both textural and mineralogical changes, ores undergo less mineral recombinations. Textural changes, however, are pronounced. Schistose or gneissic textures are induced, particularly with sectile minerals, and flow structure is not uncommon. Galena, for example, becomes gneissic, as in the ores of Coeur de'Alene, Idaho. It may also be rendered so fine grained that individual cleavage surfaces cannot be discerned with a hand lens. It "flows" around hard minerals, such as pyrite. Other minerals, such as chalcopyrite, bornite, covellite, or stibnite, behave similarly. The result is that ores may exhibit streaked, banded, smeared appearances with indistinct

251

boundaries between minerals of different color. The original texture and structure may be so obscured that it is difficult to determine to which class the original deposits belonged. Such deposits are then classified as "metamorphosed."

FORMATION OF MINERAL DEPOSITS BY METAMORPHISM

Several kinds of nonmetallic mineral deposits are formed as a result of regional metamorphism. The source materials are rock constituents that have undergone recrystallization or recombination, or both. Rarely, water or carbon dioxide has been added, but other new constituents are not introduced as they are in contact-metasomatic deposits. The enclosing rocks are wholly or in part metamorphosed; it is the rock metamorphism that has given rise to the deposits. The chief deposits thus formed are asbestos, graphite, talc, soapstone, andalusite-sillimanite-kyanite, dumortierite, garnet, and possibly some emery.

Asbestos Formation

There are two main groups of asbestos minerals—serpentine and amphilbole. The serpentines are hydrous magnesium silicates, chrysotile and picrolite, and are of the same composition as serpentine; the fine, silky chrysotile is the most valuable. The amphiboles are silicates of calcium, magnesium, iron, sodium, and aluminum. They comprise the minerals amosite, crocidolite, tremolite, actinolite, and anthophyllite.

Serpentine Asbestos. Chrysotile asbestos occurs in serpentine that has been altered from ultrabasic igneous rocks, such as peridotite or dunite, or magnesian limestones or dolomite; the first yields 93 percent of the world's asbestos supply.

In the ultrabasic occurrences, the fiber is in lenslike veinlets enclosed in serpentine and has three modes of occurrence: (1) *cross-fiber,* with fibers normal to the walls, their length being the width of the veinlet, or less if they contain "partings"; (2) *slip-fiber,* parallel or obliquie to the walls, and long but of poor quality; (3) *mass-fiber,* composed of a mass aggregate of interlaced, unoriented, or radiating fibers. The three modes of occurrence are found in a single deposit. Chrysotile fibers range up to 10 to 12 centimeters in length, rarely 20 centimeters; most of them are less than 2 centimeters. Chrysotile may make up from 2 to 20 percent of the rock.

The fiber veinlets are commonly short and discontinuous and crisscross in all directions, forming a network. Where numerous and closely spaced, they constitute a workable deposit. Less commonly, they occur in parallel veinlets, as in the "ribbon rock" in the Transvaal. The deposits generally constitute only a part of the ultrabasic igneous masses; the deposits in Quebec, the most valuable in the world, are as much as 250 by 75 meters in dimension.

The Cassiar orebody of crysotile asbestos in northern British Columbia has reserves of more than 16,000,000 tons amenable to open pit mining. The asbestos occurs in a sill-like serpentinite body enclosed in Paleozoic sedimentary rocks. Intrusion of the Cassair batholith produced presumably the crysotile asbestos.

In the deposits in magnesian limestone, cross-fibers in discontinuous bands of serpentine develop within the limestone beds, parallel to the bedding. Several parallel bands may alternate with unserpentinized limestone. The chrysotile veinlets are discontinuous and lie en echelon within a serpentine band. This type of asbestos is very pure, and its freedom from included magnetite makes it desirable for electrical insulation.

Amphibole Varieties. The amphibole varieties, of which crocidolite and amosite are the most important, are inferior in quality to chrysotile. These two minerals are found in slates, schists, and banded ironstones over an extensive belt in the Transvaal and Cape Province of South Africa. They occur as cross-fiber in greater lengths than chrysotile; some of the amosite attains lengths up to 30 centimeters and averages around 15 centimeters. The crocidolite deposits are said to be the most extensive asbestos deposits in the world but only makeup 3.5 percent of the world's asbestos market. They are in part associated with dolerite sills.

The other asbestiform minerals occur largely as mass fiber with some slip-fiber. The most important is anthophyllite, which is mined in the United States. It occurs as lenses and pockets in peridotite and pyroxenite, and the best quality is from weathered portions. The fibers are harsh and break to lengths of less than one centimeter. Fibrous material may make up 90 percent of the rocks. Tremolite and actinolite varieties, except for the Italian tremolite, are commercially unimportant.

Origin of Chrysotile. Chrysotile asbestos is confined entirely to serpentine and, strictly speaking, is a fibrous variety of serpentine. Serpentinization is an autometamorphic process, and in the ultrabasic rocks, such as dunite, serpentinization has proceeded along fractures; in Quebec, Dresser noted a rather constant ratio of width of serpentine bands to chrysotile of 6:1:6. Chrysotile is not formed except where there is serpentinization, but serpentine may occur without chrysotile. In some occurrences the entire rock mass may be serpentinized. Cooke recognizes two stages of serpentinization, a first general stage whereby 40 to 60 percent of the rock mass is converted to serpentine, and a second stage during which portions of the partly altered rock along fractures are completely altered to serpentine. This alteration was probably accomplished by hot residual solutions that emanated from within the intrusives. Since granitic intrusives may have come from the same reservoir, the solutions might possibly have emanated from them. In the alteration, the magnesium silicate olivine is converted to the hydrous magnesium silicate serpentine, only water being added.

The puzzling problem of origin, however, is how chrysotile, having the same composition as serpentine, was formed and how it became emplaced. Various theories have been advanced:

1. Fissure fillings in openings of hydration expansion from serpentine solutions of short distance transportation (Cirkel), or in fractures produced by dynamic stresses, by means of hydrothermal solutions of remote source (Keith and Bain).
2. Replacement and recrystallization of serpentine walls outward from tight cracks (Dresser, Graham).
3. Serpentine extracted from rock and deposited as asbestos in tight fractures, the walls of which are pushed apart by the force of growing crystals (Taber, Cooke).

Cooke, concludes that for the Quebec deposits replacement is untenable and orindary fissure filling is impossible and he, therefore, favors the theory that the fiber commenced to crystallize in tight fractures, the walls of which dilated, aided by tension set up by deformation movements.

Ultramafic rocks of periodotite, dunite, and pyroxenite are altered to form *serpentites*. When juxtaposed with basaltic volcanics, diabase, and gabbro to make up complexes they are referred to as *ophiolites*. Their origin has been considered to be magmatic, but with the recent emphasis on plate tectonics, Lockwood, Coleman, Shride, and others suggested that serpentinites may be slabs of ultra-

mafic mantle ripped from the bottom of the thin (5-km) ocean crust and rafted to continental margins where it is structurally emplaced. Note the depths in the earth, even at elevated temperatures of the equilibrium curves for the system $MgO-SiO_2-H_2O$, especially for serpentine (Fig. 15-1).

It is commonly stated that serpentinization involves considerable volume expansion because of the addition of water. Serpentine replaces olivine, and it is the experience of economic geologists that replacement is a volume-for-volume interchange. Therefore, the alteration to serpentine may not be attended by volume increase. Actually, there may possibly be a decrease in volume, thus accounting for the numerous fractures.

Amphibole. The crocidolite is thought by Peacock to have originated by molecular reorganization, without essential transfer of materials or constituents of the enclosing banded ironstones. Deep burial is thought to have supplied heat and pressure that resulted in the metamorphism of the rock constituents into the blue asbestos. This is strongly suggested by its wide distribution in similar rocks generally unassociated with igneous intrusions. The amosite is chemically dissimilar to the enclosing rocks, and its occurrence around the contact aureole of the Bushveld Complex suggests contributions from solutions of Bushveld origin in addition to static metamorphism.

Graphite Formation

Graphite or "black lead" is a form of carbon that occurs in two varieties, *crystalline*, consisting of thin, nearly pure black flakes; and *amorphous*, a noncrystalline, impure variety. It is soft and black, has a greasy feel, and marks paper; hence the term *graphite* (to write). It is debatable that the material of graphitic slate, which yields "amorphous graphite," is really graphite or amorphous carbon. True graphite yields graphitic acid when treated with nitric acid; amorphous carbon does not.

Occurrence. Graphite occurs chiefly in metamorphic rocks produced by regional or contact metamorphism. It is found in marble, gneiss, schist, quartzite, and altered coal beds; it also occurs in igneous rocks, veins, and pegmatite dikes. Most of the crystalline variety occurs in minute flakes disseminated through metamorphic rocks. The amorphous variety is in dustlike form. The deposits may be of large size, and the graphite content may be as much as 7 percent. Associated minerals are quartz,

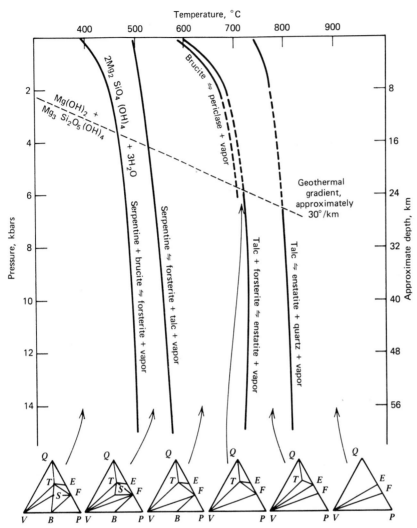

Figure 15-1 Equilibrium curves for reaction in the system MgO-SiO₂-H₂O. Each univariant curve shows equilibrium for the reaction specified. Triangles show stable mineral assemblages in the divariant areas between the curves. The triangles are similar to ACF diagrams, except that water vapor is substituted for CaO at the lower left. The light dashed line shows temperatures within the earth, on the assumption of an average gradient of 30°C per kilometer. Abbreviations: Q, quartz; P, periclase (MgO); B, brucite [Mg(OH)₂]; V, vapor (chiefly H₂O); E, enstalite (or pyroxene); F, forsterite (or olivine); S, serpentine; and T, talc. (Sources: Solid lines at low pressures (below about 5 kilobars) from Bowen and Tuttle (1949). Solid lines at high pressures from Ditahara, et al. (1966). Curve for brucite dehydration from Weber and Roy (1965). Dashed lines show interpolations and extrapolations of experimental data.)

chlorite, rutile, titanite, and sillimanite. Disseminations and fissure veins are the most important types of deposits. Synthetic graphite, produced from oil and anthracite coal, now accounts for 90 percent of the graphite production.

Origin. Graphite originates by (1) regional metamorphism; (2) original crystallization from igneous rocks as shown by its occurrence in granite, syenite, and basalt; (3) contact metamorphism, as at Calabogie, Ontario, where it occurs with contact metamorphic silicates in limestone adjacent to an igneous intrusion; and (4) introduction by hydrothermal solutions, which accounts for vein deposits and, as Beverly considers, for deposits in pegmatites and shear zones in schist in the San Gabriel Mountains, California. The graphite of 2, 3, and 4 is considered of magmatic derivation; that of 3 and

4 may have resulted from carbon bearing compound solutions given off by the magma, as believed by Weinschenk. The coal beds that have been altered to graphite in Sonora, Mexico; Raton, New Mexico; and India are clearly the result of igneous metamorphism. The volatile materials of the coal have been driven off and the residual carbon converted into the amorphous condition.

Two views exist for the deposits resulting from regional metamorphism: one, the graphite is altered organic matter formerly present in the sediments, and the other that it results from the breakdown of calcium carbonate. Black, carbonaceous limestones, when metamorphosed, yield white marbles with disseminated graphite. Either the original hydrocarbons have been broken up, causing direct precipitation of the carbon, or they have been converted into carbon monoxide and carbon dioxide, which in turn were reduced and the carbon deposited. In either case the distributed carbon has been moved into concentration centers.

The other idea, advanced by Winchell and others, is that carbonates are broken down, yielding their calcium, magnesium, or iron to form silicates and releasing CO and CO_2, which in turn become reduced to form graphite. A. N. Winchell suggests two possible reversible reactions:

$$C + 2H_2O \rightleftarrows CO_2 + 2H_2$$
$$C + CO_2 \rightleftarrows 2CO$$

Either could account for free carbon.

The occurrence of graphite in Precambrian rocks may suggest an inorganic rather than an organic origin for the carbon. Under either hypothesis, the carbon has come from the sediments. It is also possible that the carbon of the graphite found in igneous rocks, dikes, and veins was picked up from underlying carbonate rocks.

The uses, distribution, occurrence, related references, and examples of graphite deposits are given in Chapter 26.

Talc, Soapstone, and Pyrophyllite

Talc, a product of metamorphism, is a hydrous magnesium silicate $[H_2Mg_3 (SiO_3)_4]$, which, when finely ground, forms the familiar talcum powder. The pure, soft mineral is known in trade as *talc; steatite* describes a massive, compact variety; *agalite* is a special name applied to fibrous talc from New York State. Soapstone is a soft rock composed essentially of talc but also containing chlorite, serpentine, magnesite, antigorite, and enstatite, and perhaps some quartz, magnetite, or pyrite. It is a massive, impure talcky rock that can be quarried and sawed into large blocks. *Pyrophyllite,* sometimes included among soapstones, is a hydrous aluminum silicate that serves some of the same uses as soapstone.

Occurrence. Commercial talc and soapstone deposits occur in metamorphosed ultrabasic intrusives or dolomitic limestones. They are thus restricted to metamorphic areas and are largely confined to the Precambrian. The best quality talc comes from metamorphosed dolomitic limestones and is generally associated with tremolite, actinolite, and related minerals. These deposits are generally lens-shaped, in beds, and reach widths up to 40 meters. The important deposits of Ontario, New York, North Carolina, Georgia, California, Bavaria, and Austria are of this type. Talc also occurs in Europe inter-

calated in schists and gneisses, of which they are supposed to be replacements. However, Gillson suggests they may be replacements of included magnesian limestone beds.

The deposits in, and associated with, ultrabasic masses are more numerous but smaller than those in altered limestones. They occur with serpentine, whose formation preceded that of the talc, as in the case of the soapstone deposits of Virginia. The pyrophyllite deposits of North Carolina occur in slates and tuffs with interbedded volcanic breccias and flows, all of which have been metamorphosed. The pyrophyllite occurs chiefly in acid tuff.

Origin. Talc is an alteration product of original or secondary magnesian minerals of rocks. It results from mild hydrothermal metamorphism, perhaps aided by simple dynamic metamorphism but never from weathering. It is rare in ore deposits. It is pseudomorphic after minerals such as tremolite, actinolite, enstatite, diopside, olivine, serpentine, chlorite, amphibole, epidote, and mica. Lindgren states that it may be formed from any magnesian amphibole or pyroxene acted on by CO_2 and H_2O according to the reaction:

$$4MgSiO_3 + CO_2 + H_2O \rightleftarrows H_2Mg_3Si_4O_{12} + MgCO_3$$

It thus originates in (1) regionally metamorphosed limestones, (2) altered ultrabasic igneous rocks, and (3) contact-metamorphic zones adjacent to basic igneous rocks. Talc is always late in the mineral sequence. It is formed largely from other minerals that in turn represent alteration products of original minerals. Where present in serpentine, it was not formed as a result of the serpentinization but, according to Hess, by subsequent, unrelated processes by means of which the serpentine was replaced by talc. Gillson states that talc in serpentine rocks is pseudomorphic after actinolite, or after chlorite that replaced biotite. Its random orientation suggests that it cannot have formed from dynamic metamorphism alone as is believed by Harper. The magnesia is largely, if not entirely, derived from the rocks in which the talc occurs. According to Stuckey, the pyrophyllite deposits are hydrothermal replacements of silicic volcanics.

Sillimanite Group—Andalusite, Kyanite, Sillimanite

The four interesting minerals andalusite, kyanite, sillimanite, and dumortierite withstand high temperature, but change over to mullite. They are sought for high-grade refractories, and are used for similar ceramic purposes. The first three have identical composition ($Al_2O_3 \cdot SiO_2$) but differ in cyrstallization, andalusite and sillimanite being orthorhombic and kyanite triclinic. Dumortierite is a basic aluminum borosilicate (orthorhombic). At high temperatures (1100 to 1650°C) these minerals change over to mullite ($3Al_2O_3 \cdot 2SiO_2$) and vitreous silica, which is considered to be cristobalite. This material, rare in nature, remains stable up to 1810°C, therefore is heat resistant, is a good high-temperature insulator, and is particularly resistant to shock. A tentative PT diagram for the $Al_2O_3 \cdot SiO_2$ composition is shown in Fig. 15-2.

Occurrence and Origin

Kyanite. Although kyanite is a common mineral of metamorphic rocks, commerical deposits are few. The commercial deposits consist of disseminated crystals or small masses in gneiss or schist. Kyanite also occurs as lenses in pegmatite dikes and as bunches in quartz veins. It is considered to have been formed from mica schists or other aluminous silicate rocks by dynamothermal metamorphism, perhaps accompanied by magmatic emanations.

Andalusite. The mineral andalusite occurs in argillaceous crystalline rocks and also in pegmatites. Its common associates are tourmaline, garnet, corundum, topaz, quartz, and mica. The largest deposit is at White Mountain, Calif., where it occurs in irregular segregations in a quartz mass enclosed by sericite schist. The minable portions, averaging 70 to 80 percent andalusite, occur in a zone 3 to 20 meters thick. According to Kerr, the deposit was formed by a sequence of metamorphic processes during which aluminous rock (volcanic or sediment) was converted to andalusite segregations as a result of pneumatolytic action by a nearby intrusive. Another occurrence near Hawthorne, Nevada, is a veinlike deposit 0.5 to 1.5 meters thick, about 1000 meters long, and explored to a depth of 30 meters.

Dumortierite. Dumortierite occurs at Oreana, Nevada, in pegmatites or quartz veins that cut aluminous rocks, where a quartz mass in a sericite schist contains irregular lenses of andalusite altered, or partly altered, to dumortierite. Kerr recognizes three generations of dumortierite. He thinks that igneous emanations caused ''crystallization of the andalusite and later dumortierite by metamorphism along the boundaries of the quartz mass adjoining the schist. The earlier phase was probably a high temperature one when andalusite and earlier dumortierite were formed. Later, presumabbly hydrothermal metamorphism resulted in the formation

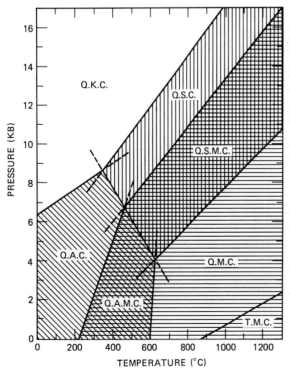

Figure 15-2 Stability relations of crystalline phases in the system SiO_2-Al_2O_3 below 17 kb and 1300° C as shown in Fig. 1. The blank area outlines the kyanite stability field; horizontal ruling—the mullite field; vertical ruling—the sillimanite field; diagonal ruling—the andalusite field. Quartz and corundum are incompatible over the entire P.T. range. (From: D.R. Waldbaum, Am. Miner., v. 50, p. 193)

of late dumortierite." These deposits are no longer commerical.

Sillimanite. Occurring as slender prisms in aluminous crystalline rocks, sillimanite results from high temperature metamorphism.

Distribution and Examples of Deposits

The United States, South Africa, Canada, Mozambique, Keyna, and India are the chief producers of these refractory minerals.

Kyanite occurs in commercial deposits in North Carolina, Virginia, Georgia, Florida, India, and Kenya. The Indian occurrences, mostly in the Singhbhum district, and those of Kenya are the largest. In India kyanite occurs in kyanite schist or quartz-kyanite schist and is in disseminated crystals or in bunches. Most of the material shipped consists of residual boulders of kyanite collected from the regolith that have been weathered out of underlying kyanite rock. The deposits are extensive, and the unweathered material has hardly been touched. In the Bhandara district both kyanite and sillimanite are found exclusive of each other. It is thought that the two minerals have been formed from chlorite-muscovite schists by gaseous and hydrothermal metamorphism resulting from granitic intrusions. The Appalachian occurrences, and those near Ogilby, California, and Kenya have been formed from aluminous schists.

Andalusite occurs at White Mountain, California; Oreana and Hawthorne, Nevada; the Transvaal; Russia; Sweden; and Korea.

Dumortierite is known in a number of localities, but the deposit at Oreana, Nevada, is no longer commercial; large deposits occur in India.

Sillimanite comes from India, where it is recovered in boulders from the weathered parts of sillimanite schist. Large inaccessible deposits are reported by Dunn in India at Khasi Hills, Assam; and Pipra, Rewa. The mineral occurs in sillimanite schist enveloped in granite and associated with corundum.

Possible sources of commerical kyanite are reported from Russia, western Transvaal; Nyasaland, Western Australia, and Black Mountain, North Carolina; recent sources of sillimanite are reported in South Africa.

Other Metamorphic Products

Garnet. There are seven varieties of garnet, of which two are of commerical importance: alamandite and rhodolite. Garnets occur as accessory minerals in many rocks, but their common home is in gneisses and schists. They are formed by regional metamorphism, contact metamorphism, and as original constituents of igneous rocks; but metamorphic processes are responsible for all the commerical deposits.

Miscellaneous Materials. *Emery,* which is a mixture of corundum and magnetite with hematite or spinel, is formed by metamorphic processes, mostly contact metamorphism. The Peekskill deposits, New York, occur in the Courtland igneous series where it contains inclusions of mica schist. The Virginia deposits occur in schist bands within quartzite and granite. At Chester, Massachusetts, pod-shaped pockets occur in amphibolite. The Grecian deposits are pockets in crystalline limestone cut by granite as are the Turkish deposits of Asia Minor.

Selected References on Metamorphic Minerals

Industrial Minerals and Rocks, 3rd ed. 1960. Joseph L. Gillson, ed. Sillimanite Group. AIME, New York, Chap. 44. *Complete bibliography up to 1960.*

Crecidolite from the Koegas-Westerberg area, S. Africa. 1961. J. J. LeR. Gilliers. Econ. Geol., v. 56, p. 1421–1437.

Kyanite, sillimanite and andalusite deposits of the S.E. states. 1960. G. H. Espenshade and D. B. Potter U.S. Geol. Survey Professional Paper 336.

Dumortierite-Andalusite, Oreana, Nev. 1935. P. F. Kerr and C. P. Tenney. Econ. Geol., 30, p. 287–300. *Occurrence, origin, production.*

General Reference 12, Chap. 44 W. R. Foster. *Brief summary of mineralogy and geology, distribution and uses. Complete bibliography.*

General Reference 19. Chapters on *Kyanite and related minerals, asbestos, graphite, and talc.*

16

SUMMARY OF ORIGIN OF MINERAL DEPOSITS

Men love to wonder and that is the seed of our science.
Emerson

In the prior chapters an attempt has been made to follow the processes of origin of mineral deposits from the source of the materials to their final resting place as mineral deposits. Magmas are an important source of many of the ingredients of mineral deposits. Some constituents, such as oxygen, carbon dioxide, or water, are derived from the atmosphere or the oceans, but even these were originally of magmatic derivation. Some of the metals are presumably derived by lateral leaching of premagma rocks the metals of which are incorporated in the magma but remain in solution until the last stages of crystallization of the intrusive.

The initial stages of magma crystallization are attended by separation by certain metallic oxides, chlorides, sulfides, and native metals. Some of these crystallize and become segregated by crystal separation into mineral deposits early in the magmatic stage; others interreact and solidify later than the rock crystals and either become segregated at the original site of accumulation or are injected into the cooled intrusive or the surrounding rocks, forming late magmatic mineral deposits.

During the progressive crystallization of the magma, the higher vapor pressure constituents become more abundant. These *mineralizing fluids* vary in composition but consist predominantly of hydrous brine solutions containing other fluids such as CO_2, H_2S, HCl, SO_2, and many other constituents. These fluids are capable of altering parts of the host intrusive rock. The resulting hydrothermal alteration may or may not be followed by later brine fluids carrying complex metal chlorides which, when encountering H_2S, result in the deposition of sulfides. The released Cl^- ion reacts with Na^+ or K^+, some of which is trapped in fluid inclusions of the minerals of the deposit.

Other processes exist that also result in mineral deposits. The bacteriogenic process does not provide the metal-bearing solution but may be quite efficacious in causing the precipitation of metal sulfide minerals, possibly in such localized abundance to provide an economic deposit.

Recent studies of many mineral deposits, some of which are stratiform in shape, and the study of the brine electrolyte solutions of the Red Sea have increased support for the formation of deposits by a submarine exhalative or volcanic process. This process is an aid in explaining the genesis of deposits that have no known association with an intrusive body as a source of the mineralization. In addition, the ore mineralogy is generally complex, as in the Kidd Creek, Ontario, deposit. The ore bodies are generally stratiform although often modified by later metamorphism such as the deposits of eastern Honshu Island, Japan; possibly the Rhodesian (Zaire-Zambia) copperbelt; and Balmat, New York.

Toward the close of the solidification of specific magmas some of the accumulated, highly silicic, mother liquors may be withdrawn to form pegmatites, accompanied or followed by the brine electrolyte solutions from which mineral deposits are deposited. Shortly before complete crystallization of the magma, high vapor pressure fluids are concentrated to such an extent that they may shatter the cupola of the intrusive and thereby provide the "plumbing system" for the remaining solutions. The residual solutions, in the form of gases, liquids, or both, are therefore drawn toward sites of lower pressure. These constitute the mineralizing solutions from which most epigenetic mineral deposits are formed. In many cases, the *hydrothermal solutions* that escape from the intrusive body may undergo chemical change by reaction with the wall rocks and thus become alkaline solutions. On the other hand, they may still remain sufficiently hot so that HCl remains undissociated; therefore, the solution is only slightly acid, allowing the solutions to transit carbonate beds before final deposition. The

258

mineral substances in these solutions may replace the rock substances, giving rise to *replacement* deposits, or they may be precipitated from solution and fill the rock openings, forming *cavity-filling* deposits. The deposition may occur at high, medium, and low temperatures and pressures, forming, respectively, the hypothermal, mesothermal, and epithermal groups of Lindgren, each somewhat characterized by distinguishing minerals and textures rather than by the *PT* conditions originally postulated by Lindgren. The magma is thus the parent of these kinds of mineral deposits. The deposits may be deposited in sufficient concentrations to constitute economic mineral deposits or they may be sparsely deposited, requiring other means of concentration to render them valuable.

Secondary processes may act upon many of these previously formed deposits or upon enclosing rocks to form yet other types of mineral deposits. Weathering releases many valuable mineral substances that are transported in solution or mechanically to sedimentary basins and they are then deposited as sediments giving rise not only to the common sedimentary rocks, but also to economic deposits of metals and many industrial nonmetallic minerals and products. Organic processes also take part in the growth of plants and animals from which coal and oil are formed. The inorganic substances were derived originally from igneous rocks and magma although they may have passed through previous sedimentary cycles. Other soluble substances released during erosion are concentrated in bodies of water of the oceans, of enclosed basins, or of the ground water, from which they are deposited by evaporation, giving rise to numerous valuable saline deposits. The circulating ground water is considered by some to be effective in dissolving, transporting, and redepositing mineral substances in more concentrated form.

Weathering, combined with the sorting action of water and air, effectively garners heavy, insoluble, and durable minerals from their enclosing rock matrix and concentrates them into valuable placer deposits of both metallic and nonmetallic minerals. Weathering alone, in its relentless attack upon the rocks, deliberately sorts out valuable and nonvaluable materials. Soluble waste products are removed from desired insoluble substances, which persist and accumulate in situ as important residual mineral deposits. Other products, such as clays, nickel minerals, or bauxite, are created during weathering and persist, while associated undesira-ble substances are removed in solution leaving accumulated masses as residual deposits of economic importance. Surficial oxidation profoundly modifies most ore deposits, rendering barren the upper parts of many deposits or changing the ore minerals into more usable or less usable forms. Metals are dissolved within the zone of oxidation, and are then carried down below the water table, where they are reprecipitated or replace existing minerals. The metals removed from above are thus added to those existing below, thereby bringing about a supergene enrichment of the upper part of the sulfide zone. Leaner parts of veins have been made richer, and worthless protore has been made workable. Thus many large and rich ore deposits have been created.

Metamorphism not only drastically changes the form and texture of pre-existing mineral deposits but it also creates new ones. Under high pressure and temperature, aided in some cases by hot waters, metamorphic minerals that are stable under the new environment are produced. The change may consist only of recrystallization, or of recombinations of materials to form the new minerals. Generally nothing, except perhaps water, is added during the metamorphism.

It is evident that the types of economic mineral deposits are many and varied and that numerous and unrelated geologic processes must be invoked to explain their origin. In certain cases more than one process is involved, and these may overlap or operate at different periods of time. In deciphering the genesis of mineral deposits, and in the application of the deductions therefrom, it is imperative that multiple working hypotheses be utilized. The student of economic geology must not only have at his command a knowledge of the other geologic sciences, but he must apply them to the problem of ore genesis. Should ore be found, then it may be assumed to exhibit mineralogy, distribution, and continuity similar in general to the initial deposit. It was such realization of the controls of mineralization that led, for example, to the remarkably rapid development of great tonnages of copper ore in the Zambian copper belt, where the geographic situation would ordinarily have been a determent. Past experience with deposits formed by one process can be applied quickly to similar new deposits under consideration.

Different processes may operate to produce distinct types of deposits of the same metal. Deposits of iron ore, for example, may be formed by magmatic, contact-metasomatic, replacement, sedimentary, and supergene processes. The distinction between them is vital, particularly from an economic standpoint. An iron deposit formed by sedimentary processes may be expected to exhibit the charactersitics of a sedimentary rock—lateral continuity, thinness, general uniformity of composition— whereas one formed by hydrothermal replacement would be laterally restricted, of smaller tonnage, and probably of irregular shape. Early lack of such distinction greatly retarded the development of the Clinton sedimentary iron ores in the eastern United States.

17

CLASSIFICATION OF MINERAL DEPOSITS

Yet do thou act as one new born.
What each day needs, that shalt thou ask,
Each day will set its proper task,
Give others' work just share of praise.
Goethe

Classifications attempt to arrange related subjects in logical order or sequence and thus help clarify a diverse assemblage. Classifications of natural phenomena, however, are seldom so comprehensive that all phenomena fit exactly into their own pigeonholes. This is the case with classifications of mineral deposits, which consist of substances that vary greatly in metal and mineral content, and in form, size, mode of origin, and value.

A classification should be logical, orderly, and permit clean-cut separation, as far as possible. It should not allow one group to fit equally well into two or more pigeonholes. In the early classifications of mineral deposits, material, form, texture, position, and origin were hopelessly given equal rank. Ever since Agricola first classified ore deposits, successive writers have attempted classifications of mineral deposits, none of which has attained unanimous endorsement. In any classification, uncertainty always arises as to which division certain deposits belong. Some investigators tend to consider classification as an ultimate objective; others refuse to worry much over it. In this book no formal classification of mineral deposits has been used other than listing deposits by genetic processes. However, a review of classifications that have been used is desirable.

EARLY CLASSIFICATIONS

Early schemes of classification appeared in the middle of the nineteenth century and involved veins alone. Among them are those of von Wissenbach, von Cotta, and LeConte; but each illogically used

for equal major divisions, form, origin, and position or materials; obviously, veins could not be so classified. The latter half of the century saw a more logical development of classifications. First, a group appeared based partly on form and partly on origin without subdivisions. Next came a more logical group based upon form alone, by von Cotta, Prime, Koehler, Callon, and Lottner-Serlo. Von Cotta divided deposits into I, Regular, with *A*, Beds, *B*, Veins; and II, Irregular, with *C*, Segregations, and *D*, Impregnations. This group although logical was too simple and not sufficiently inclusive. As to be expected, all of these classifications were based on shape, form, and other physical properties, rather than on processes or genesis. Finally, a more logical group based chiefly upon origin, by Grim, Von Groddeck, Pumpelly, and Phillips was proposed. Some of these classifications interspersed form and origin but none of them were sufficiently inclusive for the then-existing knowledge of ore deposits.

The end of the century saw the first group of classifications based entirely upon origin, such as those by Monroe, Wadsworth, Pošepńy, and Kemp. They were the forerunners of later genetic schemes, but they lacked sufficient subdivision to permit the desired separation of deposits.

LATER CLASSIFICATIONS

Around the beginning of the twentieth century much attention was directed to schemes of classification, several of which were based upon origin. Most of them utilized terms that are still widely used today. Examples of them are

Beck, 1904	Bergeat—Stelzner, 1904	Irving, 1908
I. Primary.	*I. Protogene.*	*I. Bedrock deposits.*
A. Syngenetic.	A. Syngenetic.	A. Syngenetic.
1. Magmatic segregations.	1. With eruptive rocks.	1. Igneous.
2. Sedimentary ores.	2. With sedimentary rocks.	2. Sedimentary.
B. Epigenetic.	B. Epigenetic.	B. Epigenetic.
1. Veins.	1. Cavity fillings.	1. Cavity fillings.
2. Epigenetic deposits not veins.	2. Replacements.	2. Replacements.
II. Secondary.	*II. Secondary.*	3. Contact-metamorphic deposits.
A. Residual.	A. Residual.	*II. Disintegration deposits.*
B. Placers.	B. Placers.	A. Mechanical.
		B. Chemical.

These schemes, like the accompanying ones, separated primary from secondary deposits and divided the primary into *syngenetic,* or those formed at the same time as the enclosing rock, and *epigenetic*, or those formed later than the enclosing rock. This is an advantageous distinction that permits broad scientific and practical conclusions to be applied to each group. The epigenetic groups are those formed by gases or liquids dominantly of igneous derivation and are divided into subgroups based upon processes of origin. The cavity fillings in turn are subdivided according to form into fissure veins, saddle-reefs, amygdaloidal fillings, and so forth. The above schemes are genetic and simple; the terms are everyday words and are readily usable by the geologist in the field and by the mining profession.

Beck in 1909 changed his classification as follows:

1, Magmatic segregations; 2, contact-metamorphic deposits; 3, fissure veins; 4, bedded deposits; 5, stocks; 6, secondary alterations; 7, sedimentary ore deposits; and 8, detrital deposits. This is inconsistent in that process and form are ranked equally; genetically different types must go in the same group and the same deposit could be placed in more than one group.

A genetic scheme suggested in 1908 by J. D. Irving, is as follows:

I. Bedrock deposits.
A. Syngenetic deposits: (1) igneous; (2) sedimentary.
B. Epigenetic deposits.
  1. Cavity fillings: (a) fissure veins, (b) shear zones, (c) ladder veins, (d) stockworks, (e) saddle-reefs, (f) tension-crack fillings, (g) solution cavity fillings (caves, channels, gash veins), (h) breccia fillings, (i) pore-space fillings, (j) vesicular fillings.

  2. Replacement deposits: (a) massive, (b) lode, (c) disseminated.
  3. Contact-metamorphic deposits.
II. Disintegration deposits.
  A. Mechanical. B. Residual. C. Chemical.

Although the above classification has many advantages, it has some disadvantages, among which is the difficulty common to all classifications, that of determining into which group certain deposits should be placed. The distinction between replacement and cavity filling is at times difficult, although generally it can be determined if the deposit has been formed dominantly by one or the other process. The advantages of such a classification are that it is genetic; it is readily usable in the field by the geologist or mining man; and it employs familiar terms that are themselves descriptive of a deposit. The assignment of a deposit to a division is not always dependent upon subsequent laboratory work. Further, in order to classify a deposit, it directs attention in the field to the cause of ore localization, such as structural control, and the subdivisions take cognizance of the form of the deposit, which is one of the most striking observable features of an individual deposit. In this classification, once a deposit is fitted into its proper niche, broad geological conclusions affecting origin, occurrence, and search for other deposits immediately apply. It is, therefore, a valuable classification for the working geologist. For example, if an epigenetic deposit is determined to be a cavity filling of tectonic breccia, it indicates that mineralization by circulating solutions was later than the rock shattering and that other similar deposits might be found in the region where similar brecciation is known to occur or might be expected. Also, the distribution, shape. and continuity of the deposit may be inferred.

Lindgren's Classification

The desirable emphasis on genetic schemes of classification culminated in Lindgren's, which first appeared in 1911. This classification was and is still considered by many to be a new inclusive and consistent genetic classification of ore deposits. With more detailed studies in the field and laboratory of mineral assemblages, it is evident that Lindgren's PT conditions must be extended and modified. Furthermore, newly suggested processes such as bacteriogenic and exhalative volcanic processes now include deposits Lindgren included in "deposits derived from waters of uncertain origin." He considered that most deposits have been formed by physicochemical reactions in solutions, whether liquid, igneous, or gaseous, which constitute one large class as distinct from those formed by mechanical concentration. These two subdivisions, therefore, constitute the main divisions of his classification, and practically all deposits fall under those produced by chemical concentrations. The main outlines of his classification are condensed as follows.

	Temperature °C	Pressure
I. Deposits by Mechanical Processes.		
II. Deposits by Chemical Processes.		
A. In surface waters.		
1. By reactions	0–70	Medium to high
2. Evaporation.		
B. In bodies of rocks.		
1. Concentrations of substances contained within rocks.		
a. By weathering.	0–100	Medium
b. By ground water.	0–100	Medium
c. By metamorphism.	0–400	High
2. By introduced substances.		
a. Without igneous activity.	0–100	Medium
b. Related to igneous activity.		
(a) By ascending waters.		
(1) Epithermal deposits.	50–200	Medium
(2) Mesothermal deposits.	200–500 +	High
(3) Hypothermal deposits.	500–600 +	High +
(b) By direct igneous emanations.		
(1) Pyrometasomatic deposits.	500–800	High +
(2) Sublimates.	100–600	Low to medium
C. In magmas by differentiation.		
1. Magmatic deposits.	700–1500	High +
2. Pegmatites.	575 ±	High +

This is one of the most advanced genetic schemes that has yet been produced, and it has now become the most widely adopted classification although there are increasing criticisms of different portions of the classification. Even so, its terminology is extensively used and parts of it are incorporated in many books. It is genetic, logical, but not sufficiently complete so that all deposits can be assigned to their proper pigeonholes. Moreover, as in all other classifications, it is not always possible to determine into which group certain deposits should be placed.

Some deposits find no place in any of the classification groups but are treated by Lindgren under separate headings, such as "deposits of native copper," "regionally metamorphosed sulfide deposits," and "deposits resulting from oxidation and supergene sulfide enrichment." The last certainly deserve a place in any classification, since many of the great copper deposits, for example, owed their initial operation to the process of supergene enrichment.

The chief basis of distinction between groups in the classification is the temperature and pressure of formation of deposits. Unfortunately, this is often determinable only after laboratory work or not at all, and many anomalous features appear.

Separation of deposits on the basis of temperature and pressure of formation, if determinable, should offer fairly precise lines of demarcation, but it is also obvious that it may be quite impossible to draw a distinction between deposits at the lower part of the hypothermal group, limited by a lower temperature of 500° C, which temperature has been exceeded in many deposits, from those of the upper part of the mesothermal group limited from above by a temperature of 500° C.

All good classifications have some defects, and this also applies to Lindgren's. Some objections that may be pointed out are (1) the classifications, divisions, and subdivisions do not themselves constitute appropriate names by which the deposits within them may be designated; for example, deposits of class IIA1, or "Deposits produced by chemical processes of concentration in bodies of surface waters by interaction of solutions," includes deposits as different in genesis as sandstone-type uranium deposits, Kidd Creek massive sulfide deposits, the Kuroko deposits of Japan, and Mt. Isa, to name just a few, which, except for bog iron ores, are all that are described under this grouping. The term *sedimentary deposits* at once connotes the process of formation, geologic occurrence, shape, distribution, and other features. (2) The terminology of class II*B*2*b(a)*, or "deposits produced by hot ascending waters of uncertain origin but charged with igneous emanations," which are more generally known as hydrothermal deposits and constitute most of the deposits of the American Cordillera, is insufficiently descriptive of the deposits themselves. Lindgren's classification offers a better understanding of the changes in, for example, of PT conditions in different portions of a single deposit. Thus to state that a given deposit is mesothermal does imply the approximate temperature and pressure of formation and indicates the mineral assemblage even though it gives no hint as to the exact mode of formation or the localization, form, size, or continuity of the deposit, which are desirable features to convey.

In general, most geologists unconsciously use the terminology of other classifications along with Lindgren's terminology to avoid this difficulty. Thus, instead of stating a deposit is mesothermal, they tend to say it is a mesothermal disseminated replacement deposit, using the term mesothermal as a qualifying adjective and the term *disseminated replacement deposit* to give a word picture of the deposit. The Lindgren classification does aid in studying many zoned deposits. The Butte, Montana, deposits, for example, are classed as meso-

thermal, yet the Gagnon vein extends from the central or higher-temperature zone, through the intermediate and into the outer or lower-temperature zone. The central, with its characteristic enargite, might be classed as mesothermal, but the outer part characterized by the absence of enargite and copper minerals and by the presence of sphalerite, galena, manganese, and rhodochrosite gangue would be epithermal. Thus the same vein would be both mesothermal and epithermal. Further, the mesothermal section contains abundant chalcocite and bornite and considerable covellite, minerals that are late in the paragenetic sequence and are more typical of the epithermal than the mesothermal zone. Thus, even the central part of the deposit is a composite of epithermal superimposed upon mesothermal. Similarly, Loughlin and Behre have pointed out that the Leadville, Colorado, deposits indicate transitions from hypothermal to mesothermal to epithermal, and even to the telethermal class of Graton. Some of the deposits have been built up by additions of minerals representing three of Lindgren's stages. One would, therefore, have to state that in a single deposit the magnetite is hypothermal; the quartz-pyrite, sphalerite, and chalcopyrite are mesothermal; and dolomite, silver, lead, and gold are epithermal. With the changes in pressure and temperature, the classification is valid in providing varying genetic differences for the mineral assemblages that are indicative of specific physiochemical environments.

On the other hand, some difficulties with the Lindgren classification are cited by certain deposits, such as the Cerro de Pasco in Peru where the mineralogy is mesothermal, but according to Graton and Bowditch these minerals formed, not at great depth and high pressure, but at shallow depth, equivalent to that of epithermal deposits. High-temperature minerals actually are found at shallow depths and low-temperature minerals at great depths. In general, most hydrothermal deposits are formed in successive stages with later minerals replacing earlier ones; and earlier minerals of epithermal deposits commonly appear to have formed at higher temperatures than later minerals of mesothermal deposits, so temperature alone, as to be expected, is not the controlling factor. The mineral sequence of earlier to later minerals is generally also that of higher- to lower-temperature forms, and commonly minerals characteristic of the epithermal zone replace or are later than those characteristic of the mesothermal zone.

It seems clear, therefore, that depositional factors other than temperature and pressure play a part in the formation of mineral deposits. Thus structural control, the physical and chemical effects of wall rocks, the relative ratios of concentration of different ions in solution, and chemical complexes all play a part in determining the position and mineralogic content of mineral deposits. A generalized summary of the chemical distribution of epi-meso-hypothermal zones is shown in Fig. 17-1.

Graton proposed two additions to Lindgren's classification of deposits formed by hot ascending waters, namely, the *leptothermal* group, lying above the mesothermal; and the *telethermal* group, above the epithermal. Buddington called attention to the fact that high temperature and high pressures do not necessarily exist together. Some near-surface deposits may form at high temperatures and some high pressures may exist at reduced temperatures. He proposed, therefore, the term *xenother-*

mal to apply to high-temperature deposits that formed close to the surface. Because of some difficulties in the classification of deposits that may contain sulfur of *bacteriogenic* origin and metal cations of *submarine volcanic or exhalative* origin, Lovering suggested that the term *diplogenetic* might apply to such deposits.

Lindgren in 1922, recognizing the disadvantage of concise terminology in his classification, proposed additional terminology as follows:

Deposits of Origin Dependent upon the Eruption of Igneous Rocks

A. Hydrothermal deposits.
 a. Epithermal.
 b. Mesothermal.
 c. Hypothermal.
B. Emanation deposits.
 a. Sublimates.
 b. Exudation veins, surface type.

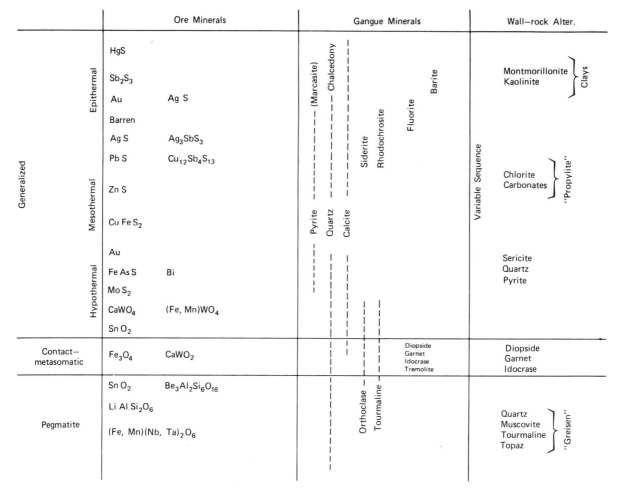

Figure 17-1 Generalized chemical and mineralogical associations with the epi-meso-hypothermal zones of the ore minerals, gangue minerals, and the wall-rock alteration. (D. Garlick.)

c. Pyrometasomatic deposits.
d. Exudation veins, deep-seated type.
C. Magmatic deposits.
   a. Orthotectic.
      1. Differentiation in situ.
      2. Injected.
   b. Pneumotectic.
      1. Differentiation in situ.
      2. Injected.

This arrangement met with objection because it distinguished emanation deposits formed from magmatic vapors from deposits formed from liquid solutions. Unfortunately, conclusive criteria for recognition of deposits formed by vapors are not available. This revision was not adopted by Lindgren in the later editions of his *Mineral Deposits.*

Other Classifications

Beck and Berg (1922) presented a classification somewhat similar to that of Beck (1909). It is (1) magmatic ore segregations, (2) contact deposits, (3) ore veins, (4) epigenetic ore stockworks, (5) epigenetic sulfide deposits, and (6) gossan formations. This classification groups incongruous deposits together, contains insufficient groupings (as for massive replacement deposits in limestones), and permits the same deposit to be classed in more than one group.

Niggli in 1925 introduced a new major separation on the basis of "plutonic" and "volcanic," similar to that made for igneous rocks. His classification is: I. Plutonic: (A) hydrothermal, (B) pegmatitic-pneumatolytic, and (C) ortho-magmatic. II. Volcanic: (A) exhalative to hydrothermal, (B) pneumatolytic, and (C) ortho-magmatic. It appears logical to state that if rocks from a magma can be separated into plutonic and volcanic, ore deposits could similarly be separated; unfortunately, mineralizing solutions do not congeal in situ; they are mostly mobile, and those in and associated with volcanic rocks may have come from a deep underlying magma reservoir that at depth produced hidden plutonic rocks.

A more extended genetic classification was introduced by Schneiderhöhn in 1932, as follows:

A. Magmatic rocks and ore deposits.
  (a) Intrusive magmatic.
    I. Intrusive rocks and liquid magmatic deposits.
    I-II. Liquid magmatic-pneumatolytic.
    II. Pneumatolytic.
      1. Pegmatite veins.
      2. Pneumatolytic veins and impregnations.
      3. Contact pneumatolytic.
    II-III. Pneumatolytic-hydrothermal.
    III. Hydrothermal.
  (b) Extrusive magmatic.
    I. Extrusive-hydrothermal.
    II. Exhalation.
B. Sedimentary deposits.
  1, Weathered zone (oxidation and enrichment); 2, placers; 3, residual; 4, biochemical-inorganic; 5, salts; 6, fuels; 7, descending groundwater deposits.
C. Metamorphic deposits.
  1, Thermal contact metamorphism; 2, metamorphic rocks; 3, metamorphosed ore deposits; 4, rarely formed metamorphic deposits.

The classification has many good points and is fairly complete although it is too subjective to find much favor. Most geologists will balk at the usage of pneumatolytic and intrusive and extrusive magmatic. Many deposits would be difficult to classify under the intrusive magmatic group, and the sedimentary grouping is not entirely logical.

Proposed Classification

The following classification is a simple genetic classification that is usable for the field or laboratory and carries familiar, well-established terminology. It is proposed primarily as a working classification for the use of the beginning student in economic geology and for the mining geologist and mine operator.

Proposed Classification

Process	Deposits	Examples
1. Magmatic concentration. High T and P.	I. Early magmatic:	
	A. Disseminated crystallization.	Diamond pipes.
	B. Segregation.	Chromite deposits.
	C. Injection.	Kiruna magnetite?
	II. Late magmatic:	
	A. Residual liquid segregation.	Taberg magnetite.
	B. Residual liquid injection.	Adirondack magnetite, pegmatites.
	C. Immiscible liquid segregation.	Insizwa sulfides.
	D. Immiscible liquid injection.	Vlackfontein, S. Africa.

2. Sublimation.
   Low T and P.

3. Contact metasomatism.
   Int. low high T and P.

4. Hydrothermal processes.
   A. T and P conditions from low to high
      1. Telethermal.
      2. Epithermal.
      3. Leptothermal.
      4. Mesothermal. } 6. Xenothermal.
      5. Hypothermal.
      (1) Cavity filling.

      (2) Replacement.
          Lo. to Hi. T & P.

5. Sedimentation (exclusive of evaporation)
   Low T and P

6. Bacteriogenic

7. Submarine Exhalative volcanism
   Low to high T and P
8. Evaporation
   Low T and P.

9. Residual and mechanical concentration
   Low T and P.
   A. Residual concentration.

   B. Mechanical concentration.

10. Surficial oxidation and supergene enrichment.
    Low T and P.
11. Metamorphism.
    Int. to high T and P.

Sublimates.

Contact-metasomatic:
   Iron, copper, gold, etc.

Cavity filling (open space deposits):
   A. Fissure veins.
   B. Shear-zone deposits.
   C. Stockworks.
   D. Ladder veins.
   E. Saddle-reefs.
   F. Tension-crack fillings (pitches and flats).
   G. Breccia fillings:
      a. Volcanic.
      b. Tectonic.
      c. Collapse.
   H. Solution-cavity fillings.
      a. Caves and channels.
      b. Gash veins.
   I. Pore-space fillings.
   J. Vesicular fillings.
Replacement:
   A. Massive.
   B. Lode fissure.
   C. Disseminated.
Sedimentary: Iron, Manganese, phosphate, etc.

Bacterial products or reduction

Sumarine volcanic

Evaporites:
   A. Marine.
   B. Lake.
   C. Ground water.

Residual deposits:
   Iron, manganese, bauxite, etc.
Placers:
   A. Stream.
   B. Beach.
   C. Eluvial.
   D. Eolian.
Oxidized, supergene sulfide.

A. Metamorphosed deposits.
B. Metamorphic deposits.

Sulfur.

Cornwall magnetite,
   Morenci (old), Iron Springs, Utah, etc.

Pachuca, Mexico.
Otago, New Zealand.
Quartz Hill, Colorado
Morning Star, Australia.
Bendigo, Australia.
Mississippi Valley types
Porphyry Copper deposits.

Bassick pipe, Colorado.
Mascot, Tennessee, Zn.
Bisbee, Arizona.

Wisconsin-Illinois Pb and Zn.
Upper Mississippi Valley Pb and Zn.

Lake Superior copper.

Bisbee copper.
Kirkland Lake gold.
"Porphyry" coppers.
Clinton iron ores

Sulfur in salt domes, Mount Isa
Red-beds copper.
Manganese nodules, Kidd Creek
Kuroka deposits.

Gypsum, salt, potash.
Sodium carbonate, borates.
Chile nitrates.

Lake Superior iron ores, Gold Coast manganese, Arkansas bauxite.

California placers.
Nome, Alaska, gold.
Dutch East Indies tin.
Australian gold.
Chuquicamata, Chile.
Ray, Arizona, copper.

Rammelsberg, Germany.
   Graphite, asbestos, talc, soapstone, sillimanite group, garnet.

There still are objections to Lindgren's modified classification. Some deposits classified as mesothermal exhibit thermal records exceeding 500°C, such as the porphyry copper deposit at Bingham, Utah. No single deposit can be classified as epithermal through mesothermal to hypothermal from its top to its bottom. As a result, these three terms are more often used as adjectives to describe, for example, an epigenetic assemblage of rocks and minerals.

It has always been both pleasing and disturbing to genetic classifiers that RST Mine Services Ltd. has used a syngenetic hypothesis in exploration for deposits in the Zambian copper belt. Their competitor, Anglo-American, has been more receptive to the hydrothermal origin. And yet, both have had rather equal success in their exploration results. This is a misleading paradox, however, as the success of both concerns are stratigraphically related and exploration is based on a stratigraphical, guided genetic observation.

Because of the difficulties of all genetic classifications, there is a tendency to return to nongenetic classifications and refer to the deposits by the type of rock in which the deposits exist. This is possibly an extreme retreat as, for example, gypsum in marine rocks has obviously an evaporite genetic origin.

This text has tried to retain a genetic process classification. The determination of the processes by which mineral deposits form is the initial thought preceding the listing of the deposits; this method is followed, but by no means dogmatically, in this text.

Selected References on Mineral Classification

High temperature mineral associations at shallow to moderate depths. 1935. A. F. Buddington. Econ. Geol., v. 30, p. 205–222.

Thermochemical data, mineral associations, and the Lindgren's classification of ore deposits. 1957. H. D. Holland. Bull. Geol. Soc. Am., v. 68, p. 1745. *Needed emphasis on thermochemical data about mineral stability fields.*

The classification of ore deposits. 1955. J. A. Noble. Econ. Geol. 50th Anniv. Vol., p. 155–169. *Review article with needed suggestions.*

Epigenetic, diplogenetic, syngenetic, and lithogene deposits. 1963. T. S. Lovering. Econ. Geol., v. 58, p. 315–331. *Suggestions on the more precise use of epigenetic and syngenetic in light of controversies about the origin of certain deposits.*

General Reference 11, Chap. 81: Lindgren's ore classification after 50 years; and Chap. 82: Changes in concepts of ore genesis, 1933–1967. *Two excellent chapters summarized by J. D. Ridge.*

General Reference 21. *Stanton discusses the pros and cons of different genetic suggestions.*

METALLIC MINERAL DEPOSITS

It is better for a man's worldly prospects to be a drunkard, than to be smitten with the divine dipsomania of the original investigator.

T. H. Huxley, 1874

In Part III the discussions of the various metals are necessarily brief. However, much of the related theoretical matter has been covered in Parts I and II, and for details of form, occurrence, and origin reference should be made to the corresponding sections in Parts I and II. A knowledge of mineralogy is presupposed.

18

THE PRECIOUS METALS

And you young men who want to leave Zion and go to California for gold—go and be damned.

Brigham Young, Salt Lake City, 1849

GOLD

Since antiquity gold has been prized as an ornament, as a concentrated form of wealth, and for monetary use. Man's urgent desire for it in ancient days led to barter, invasion, conquest, colonization, and exploration in India, Asia, Africa, and Iberia. The first gold rush started under Jason in the *Argo*. Gold was a strong incentive in the discovery of America, and its greedy acquisition by the Conquistadores was accompanied by treachery, robbery, and murder. Man has roamed the world in search of it, exploring far-off parts of the earth and undergoing almost unbelievable hardships and privations. It has little use other than decorative or monetary, but no other substance has been the cause of so much horror and misery, or has done so much good. For over 2000 years its value has generally increased, rarely decreased. It is still eagerly sought and its discovery is attended by the springing up of new communities, the pushing back of frontiers, and the rising of accompanying agriculture and industry; it has been the forerunner of civilization in many distant lands.

History, Production, and Distribution

History. Gold was first mined more than 6000 years ago, but the first noteworthy rise in gold production followed the discovery of America, when Mexico, Peru, Bolivia, and Chile poured forth streams of yellow and white metal to enrich the capitals of Europe. The United States, however, suffered a lag of 300 years, until payable gold was discovered in North Carolina in 1801 and in Georgia in 1829. Then came the sensational discovery of placer gold in California in 1848 that culminated in the great gold rush of '49. Production increased tenfold, and by 1853 the United States had become the leading gold producer of the world, a position maintained for 50 years. A similar discovery in Australia in 1851 boosted the world output to over 6 million ounces in 1850 to 1860. This flash production quickly declined, but it was offset by other discoveries in the western United States, culminating in Cripple Creek, Colorado, in 1891, which started another sensational gain in United States output, which was surpassed by the even more sensational rise in South Africa, following the discovery of the great "Rand" in 1886.

With the romantic and tragic Klondike rush of 1896 the world production in 1890 to 1900 exceeded 15 million ounces annually and by 1915 had reached a peak production of nearly 23 million ounces—a peak that stood until the change in the price of gold.

Since 1905 South Africa has assumed dominant world leadership, seconded by the United States until 1931, when Russia and Canada pushed ahead.

Prior to 1930 it was generally considered that the world would soon be facing a serious shortage of gold. Peak production had apparently been reached, and it was estimated that the great Witwatersrand would begin to decline about 1933 and have a production of only $2^{1}/_{3}$ million ounces by 1947. All this was suddenly changed, however, with the increase in the price of gold. Former marginal ores became included in reserves, low-grade deposits became commercially important, and production was greatly stimulated. World production again decreased, owing to the stable price and increased cost of production during the 1950s and more so during the 1960s. But there has been a resurgence of exploration and production of gold with the increased price even in the face of inflation that now approaches $700/ounce.

At the end of World War II, the United States held more than half of the gold mined during the history of the world, more than 25 billions dollars worth at $35/troy ounce.[1] Increased balance of payments deficits brought about by U.S. spending in other nations, resulted in payment to these nations with gold at their desire. This procedure was done at the price of $35 per ounce in order to prevent devaluation of the dollar. When the value of the gold held by the United States was reduced to less than 10 billion dollars, the United States made the decision that it could no longer support the $35 per ounce value of gold nor could the 90.5¢ per ounce price of silver be supported. Both metals now sell by supply and demand resulting in silver rising to prices in excess of $50 per ounce at its highest price and gold approaching $700 per ounce in September, 1980. (Fig. 13-9). These comparatively high values are the result of many factors. They are chiefly inflation, energy crises especially in petroleum whereby the dollars acquired by Arabian countries are converted to gold, the desire of individuals to hold gold and silver rather than currency, and the fact that gold and especially silver have become industrial metals. Almost half of the United States demand for silver is for the photography industry, and imports of silver each year are needed for more than half of the domestic consumption of silver in the United States.

Again, therefore, the production of gold is expanding throughout the world; low-grade and marginal deposits are being redeveloped with gold now worth more than 19 times what it was worth in the 1930s to the '60's. Of course, costs are also higher, which results in the use of a fine pencil to determine if the inflated higher prices result in a viable economic project on any given gold property.

Post-World War II Production. The world production has averaged about 40 million troy ounces per year during the past score of years. The total world production since gold was first known is estimated to be 80,000 tons. The chief producing countries and their average percent of world production over a number of years are as follows:

		Percent of World
1.	South Africa	65.
2.	Russia	12.9
3.	Canada	5.8
4.	United States	3.2

5.	Australia, New Zealand	1.7
6.	Ghana	1.5
7.	South America	1.4
8.	Philippines	1.1
9.	Rhodesia	1.1
10.	Dominican Republic	0.9
11.	Columbia	0.6
12.	Japan and Chosen	0.5
13.	Mexico	0.4
14.	Zaire	0.3
15.	Others	3.6

The United States average production in 1979 of 875,651 million ounces ranks the states in percents as follows:

1.	South Dakota (28)	5.	Montana (3)
2.	Utah (27)	6.	Idaho (3)
3.	Nevada (20)	7.	Colorado (2)
4.	Arizona (11)	8.	New Mexico (2)
		9.	Others (4)

The Homestake mine ranks South Dakota as first. This is the largest gold-producing mine in the Western Hemisphere. The Carlin mine ranks Nevada as second; this mine may have been worked out in a few years, but reserves have increased with changing prices.

Interestingly enough, Utah's production, second in the nation, is as a by-product from the Bingham porphyry copper deposit.

Distribution. Gold deposits range in age from Precambrian to late Tertiary and are found throughout the world, mostly where igneous activity has occurred, except for South Africa. They show a preference for intrusives of intermediate to felsic composition.

In North America the greatest gold belt lies within the Precambrian Canadian Shield. Another is associated with the multiple stocks of the Sierra Nevada and the mother lode and deposits in the stocks of the Pacific Coast region extending, with interruptions, from California to Alaska. A central Rocky Mountain belt contains numerous gold-bearing lodes in association with smaller intrusives and is flanked by two eastern outliers—an important one in the Black Hills and an insignificant one in the Appalachians. An extensive Tertiary volcanic belt containing shallow lodes projects into Arizona, Nevada, and southern California and extends through central Mexico toward Nicaragua. New large hydrothermal deposits have been found in Nevada beneath the Roberts Mountain thrust area.

In South America, small, scattered gold veins

---

[1] 1 troy ounce = 31.1 g or 20 dwt
  1 metric ton = 10^6 g or 32,150 ounces or 1.1 short ton
  1 ppm = 1 g per ton or 0.032 ounces per ton
  Gold at $100/oz = $3.22 per gram

and replacements occur throughout the Andean chain in association with Tertiary and older intrusives. Outliers occur in Brazil and the Guiana Highlands.

In Africa is the great Witwatersrand of the Transvaal and its counterpart in the Orange Free State, which dominates the world. Also, numerous smaller deposits occur in northern Africa, Rhodesia, and central and West Africa.

In Australasia many important lodes and placer deposits occur in Australia, New Guinea, Fiji, Indonesia, and New Zealand.

In Europe meager occurrences lie in Scandinavia and the mountainous parts of central and southern Europe, but numerous and important deposts lie in the Urals.

In Asia, Siberia boasts many important placer and lode deposits, and a prominent belt of small deposits extends from the mainland through Japan, the Philippines, Indochina, and Burma. The Kolar district of India is one of the large gold fields.

Mineralogy, Tenor, Treatment, and Uses

Ore Minerals. The economic gold minerals consist only of native gold and minor amounts of gold tellurides, electrum, and amalgam. The tellurides include calaverite, sylvanite, krennerite, and petzite. Nearly all gold contains some silver, but if silver is present in considerable amount the result is pale yellow to white electrum. Natural amalgam occurs in a few deposits.

Gangue Minerals. The common gangue mineral of gold is quartz, but carbonates, tourmaline, fluorspar, and a few other nonmetallics may also be present. Gold is commonly intimately contained in base-metal sulfides and related minerals, or in their oxidation products.

Association and Treatment. Gold is so commonly associated with silver that gold and silver deposits are generally considered together. Where the gold is in the native form and easily amalgamated or cyanided, it is said to be *free milling*; where it occurs in minerals difficult to treat, it is called *refractory*. Its common associate, pyrite, generally offers no trouble in treatment except for the necessity of fine grinding or roasting. Associated copper minerals make cyaniding difficult, often necessitating roasting or concentration and smelting. Associated arsenical minerals make treatment difficult and expensive.

Most gold ores are treated by cyaniding, or amalgamation, or both together, with or without flotation and roasting. Refractory ores are mostly concentrated and smelted. Massive sulfide ores are generally smelted directly; if copper is absent, they may be ground and cyanided. Leaching with NaCN or KCN is a rapidly increasing recovery technique whereby gold or silver ore is leached in place or crushed and dumped on a leaching pad to depths of 10 to 20 feet. A few feet of coarse porous material underlies the leachable ore to allow the downward percolating soluble gold or silver cyanide solution that migrates down the slight dip of the pad to a "pregnant" pad. The precious metals are recovered by treatment with zinc powder, carbon or other collectors and the cyanide solution returns to the pad where it is sprayed on the ore once again. Operations of this sort are located at Round Mountain, Goldfield, Comestock, Reveille, Keystone, Eureka, to name a few in Nevada.

Tenor. The tenor of gold ores varies widely. Formerly, the lowest-grade ores treated were those of Alaska Juneau, which averaged about 0.04 ounce of gold per ton. The South African mines range from 0.20 to 0.55 ounces per ton. The Canadian ores at $180 per ounce range from about $40 to $110 per ton. The value of gold ore changes with its price and the cost of production; rising costs have necessitated utilizing higher-grade ores, but the increasing value of gold allows the development and redevelopment of lower-grade deposits.

Uses of Gold. The foremost use of gold has been for monetary purposes, much of it being kept as bullion in reserve for notes issued, but industrial uses of gold are becoming important.

Another important use for gold is for ornamentation, chiefly jewelry, using "white," "green," and "yellow" gold. Because of its softness, jewelry gold is generally alloyed with copper, silver, nickel, or palladium. Its purity is designated in carats; one carat means 1 part gold in 24. It is also used for gold plating, glass and china inlays, gold lace, gilding, book binding, lettering, and interior decoration. For these purposes gold leaf is generally employed, which is made by pounding out gold to less than a micron in thickness; 1 ounce of gold will cover about 10 square meters of surface. Gold is also used for dentistry, glass making, and in the chemical industry.

Kinds of Deposits and Origin

The principal classes of gold deposits, and some examples of each, are

1. Magmatic deposits (?): Waarkraal, South Africa (quartz gabbro); Golden Curry, Montana.
2. Contact-metasomatic: *Nickel Plate Mine*, British Columbia; Cable mine, Montana; Ouray, Colorado; and Gold Hill, Utah.
3. Replacement deposits:
   a. Massive: *Noranda*, Quebec; *Rossland*, British Columbia; *Morro Vehlo*, Brazil; northern Black Hills, South Dakota.
   b. Lode: *Kirkland Lake, Porcupine,* and Little Long Lac, Ontario; *Homestake*, South Dakota; and *Kolar*, India.
   c. *Disseminated: Cortez* and *Carlin*, Nevada; Beattie Mine, Quebec., and *porphyry copper* mines.
4. Cavity filling:
   a. Fissure veins: *Mother Lode*, Grass Valley, California; *Cripple Creek* and Camp Bird, Colorado; *Porcupine*, Ontario; *El Oro*, Mexico; Kalgoorlie, Australia; and *Benguet*, Philippines.
   b. Stockworks: Quartz Hill, Colorado; and Victoria, Australia.
   c. Saddle reefs: *Bendigo*, Australia; and Nova Scotia, Canada.
   d. Breccia deposits: Bull Domingo and Bassick, and Cresson, Colorado; Cactus, Utah; and *Mt. Morgan*, Australia.
5. Mechanical concentrations: Placer deposits of *California*, Teton County, Wyoming; Yukon, *Alaska, Russia, Australia, Witwatersrand, South Africa*, and other places.
6. Residual concentrations: Brazil, Madagascar, Australia, western United States, and Mexico.

Origin.

Most gold deposits originate through igenous emanations or surficial concentrations. A few gold deposits have been formed by contact metasomatism, but most lodes have been formed by hydrothermal solutions. Emmons has shown a worldwide association between gold lodes and intrusive rocks that clearly indicates kinship between the two.

Mechanical concentration is responsible for the vast wealth of placer gold throughout the nonglaciated regions of the world. Supergene enrichment of gold is negligible.

Examples of Gold Deposits in the United States

The United States has produced about 15 percent of the world's gold, mainly from the following provinces: (1) Sierra Nevada, (2) Great Basin, (3) Rocky Mountain, (4) Black Hills, (5) the Southwest, (6) the Appalachians, (7) southeastern Alaska, and (8) central Alaska. The bedrock deposits are dominantly fissure veins and replacements. Placer deposits account for about 5 to 10 percent of the total production. About 39 percent of the gold produced in the U.S. is a by-product from base-metal ores. The Homestake, S.D., and the Carlin, Nev. mines are the chief U.S. gold mines. Jerritt Canyon, Nev. and the Homestake's new discovery in Calif., will exceed some of the above.

Homestake Mine, Black Hills

The Homestake mine at Lead, South Dakota, is the ranking gold producer of the Western Hemisphere. Since 1879 it has produced over 1 billion dollars in gold, from ores that ranged in tenor from $4.05 per ton in 1905, $9.00 per ton in 1933, then up to $20 per ton prior to 1972. With the price of gold reaching $650.00 per troy ounce, Homestake ore is worth more than $216.00 per ton. In 1965, Homestake production was 2,031,500 tons milled containing 128,971 ounces of silver, and 628,259 ounces of gold. It can produce about 6000 tons of ore daily and carries greater ore reserves than any other known American gold mine. In 1979, production was 260,000 on.

The ore deposits occur in altered Precambrian sediments, with included basic sills. The beds have been folded into a major anticline and syncline about 2000 meters from crest to trough, and each contains prominent minor folds and crenulations; the folds plunge 35 to 40° to the southeast. Large dikes and sills of Tertiary rhyolite intersect the beds. The ore bodies follow closely the crests of the major anticline and minor folds and plunge with the structure. (Fig. 18-1)

The ores are localized in the Homestake formation. Originally, this was a bed about 70 to 100 meters thick consisting of iron-magnesium carbonate with included thin bands of shale and quartz sand, but the folding metamorphosed it to cummingtonite and chlorite schists, and squeezed the 20 meter stratum into thicknesses ranging from a knife-blade to 100 meters. (Fig. 18-1). The Homestake formation is overlain by arches of impervious Ellison slate, which may have channelized the rising mineralizing solutions.

McLaughlin states that the major loci of ore are (1) the Homestake formation; (2) the plunging folds into which the Homestake formation is crumpled, thickened, and sheared beneath the impervious El-

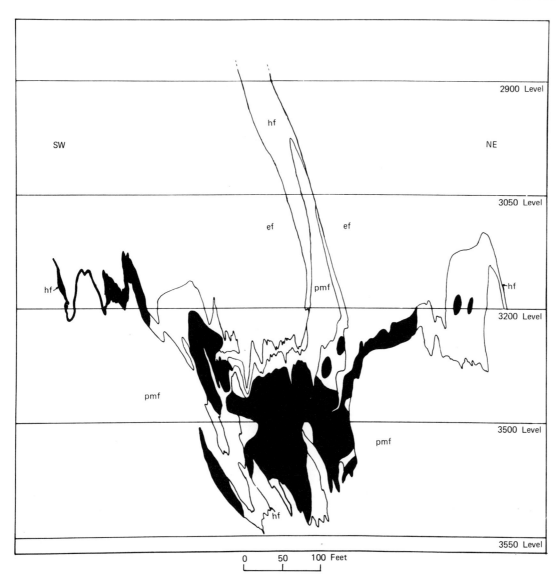

Figure 18-1 Cross section of main ore body. Homestake formation is outlined; the ore is solid black. The Precambrian host rocks surround the dilatant ore zones and are generally conformable with the ore deposit boundaries. Poorman formation, pmf; Ellison formation, ef. (From Slaughter, Ore deposits of the U.S.)

lison slate; and (3) zones of fracturing and shearing within the Homestake formation.

The ore bodies are inclined pods, veins, saddles, and lenses that lie within the Homestake formation along the crests of the plunging folds, each following its own minor fold. In plan the main deposit is 100 to 250 meters long and from 15 to 45 meters wide, but it attains widths up to 100 meters in areas of compressed minor folds. Some bodies extend down the plunge over 1400 meters. Another lower-grade ore body lies about 300 meters away from the main deposit. The ore bodies are cut and chiefly displaced by the Tertiary intrusives and faults.

The ore consists chiefly of coarse white quartz, cummingtonite, chlorite, ankerite, and minor amounts of garnet, mica, sericite, with pyrrhotite, fluorite, and pyrite and arsenopyrite, and other minor sulfides. The ratio of silver to gold is exceptionally low (0.20). Gold value averages 0.345 ounces per ton and silver tenors average 0.070 ounces per ton. The ore is generally uniform in character from the top to the bottom of the mine. Sericitization and silicification are lacking, and the ore abuts sharply against barren walls.

The deposit was thought to have been formed by hydrothermal replacement of the cummingtonite schist. Rye and Rye, however, have concluded that the origin was by fluids migrating into zones of dilation during metamorphism about 1700 million years ago. Their key evidence of the process is the alteration of the garnet isograd sideroplesite to cummingtonite. Furthermore, trace elements, including gold, are widespread in the iron formations of the Black Hills and the mobilized hot-brine solutions could have transported the ore-bearing fluid to the dilatant zones. This suggested genesis leads to the needed search for similar exploration targets. Iron formations are not noted for their gold content with the exception of the Homestake formation. With the widespread presence of gold in the district, similar dilatant zones formed during Precambrian metamorphism could have formed gold deposits not yet discovered.

Pacific Coast Gold Deposits

The Pacific Coast gold deposits were all uneconomic and dormant or closed until the 1970s, but the increased price of gold has caused a reevaluation of these deposits. Brief descriptions of the chief deposits are given because of their present increasing potential, their modes of occurrence, ore localization, and processes of formation of ore deposits.

Mother Lode, California. This famous area was not a mine but two-score mines in a belt 200 kilometers long and one and a half kilometers wide on the western foothills of the Sierra Nevada. It has yielded more than 370 million dollars during more than 100 years from gold quartz veins averaging 0.34 ounces of gold per ton, reaching a depth of almost 1800 meters.

Long, en echelon quartz veins, with intervening bodies of mineralized rock, occupied reverse faults traversing slates, greenstones, and schists of Jurassic and Carboniferous age that have undergone intense alteration. These were intruded by plutonic rocks ranging from peridotite to granodiorite, which ended with the formation of the gold veins. The veins had pinches and swells and the ore was localized in short shoots that persisted steeply in depth. A few percent of sulfides accompanied native gold. In the rock alteration much silica was drawn to the dilatant veins, which Knopf suggests was more than enough to supply the quartz of the veins. The veins are hydrothermal fillings with enlargements due to dilatant openings by renewed fault movement.

Grass Valley, Nevada City, California. This gold district has produced about 150 million dollars from ores that have averaged about $11 per ton. The gold occurs in conjugated fissure veins within granodiorite plutons that intrude Paleozoic to Jurassic rocks. The quartz veins are only 1 to 3 meters wide but are up to 2500 meters long and extend down a 35° dip for 3300 meters. The ore is in shoots up to 1500 meters long. The quartz is ribboned by repeated opening and refilling of the fractures (Fig. 18-2), which took place in three stages, with quartz, pyrite and arsenopyrite, chalcopyrite and sphalerite, and later quartz, gold, tetrahedrite, galena, and carbonates. The veins illustrate widening by repeated dilatant openings. Cavity filling by hydrothermal solutions illustrate one type of gold occurrence.

Sierra Nevada Placer Deposits. About two-thirds of the Pacific Coast gold production has come from "high-level" Tertiary and Quaternary gravels. The Sierra Nevada belt has yielded, since 1848, around 1.5 billion dollars in placer gold from an area about 250 kilometers long and 80 kilometers wide in the middle and lower reaches of the western slope of the Sierras. This is a region of gentle undulation passing upward into rugged slopes with canyonlike valleys, 1000 to 1200 meters deep.

The *Tertiary placers* represent the accumulations from erosion of hundreds or thousands of feet of quartz veins. The buried Tertiary gravels that escaped erosion now lie in old valley remnants high up on the Quaternary valley slopes. Their gold was won by drift mining and hydraulicking from a lower layer of "deep gravel" and an upper layer of "bench gravel" within about one meter of the bottom. Drift mines, taking only the richer parts, yielded up to $40 per cubic meter; hydraulic mines averaged about 10 cents at a value of $20 per troy ounce. Platinum metals have been found with the gold. Remnants of gravel deposits in Tertiary channels are estimated to aggregate 800 kilometers in length and to contain 30 million ounces of gold.

The *Quaternary gravels* contain much gold reconcentrated from the Tertiary gravels from the Feather, Yuba, and American Rivers. The richest gold, which was in the channels, has now been removed, but smaller channel deposits and larger dredging areas remain. These gravels are 6 to 20 meters in depth and yielded up to 30 cents per cubic meter prior to the present time—values now exceed $1 per cubic meter. Some gold is also won from the residual mantle.

MOVEMENT

DEPOSITION

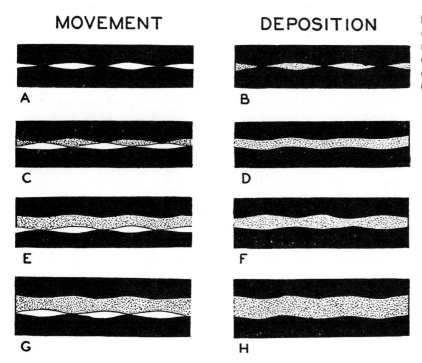

Figure 18-2 Development of thick quartz veins by alternate movement of vein walls (with continuous support), and deposition of quartz. Nevada City, California (Johnston, Geol. Soc. Amer.)

Klamath Mountain Placers. In the Klamath Mountains from Redding, California, to Coos Bay, Oregon, is a placer belt 320 kilometers long and 80 to 130 kilometers wide that has yielded around 80 million dollars since 1852. This region underwent several oscillating movements recorded in erosion surfaces and elevated marine terraces. A Pliocene uplift, which caused valley incision, was followed by a Pleistocene submergence and reelevation of 450 meters. The placer deposits consist of Cretaceous conglomerates, Tertiary and Quaternary gravels, and residual mantle deposits.

The *Creatceous gold-bearing conglomerates* are considered by Pardee to be marine shore deposits, 50 to 250 meters thick, which lie on or near the Klamath peneplain. Weathered material has been hydraulicked, yielding $2 to $3 in gold per ton at $35 per ounce. *Uncapped Miocene gravels*, cemented, occur up to 300 meters thick in the Trinity River basin. Weathered portions have been mined, and unweathered material has been hydraulicked, yielding 3 to 4 cents per cubic meter at $35 per ounce. The Quaternary gravels are in flats, terraces, and beach deposits. The flats yield 15 to 17 cents per cubic meter at $35 per ounce by dredging. The beach deposits are ancient and modern. Ancient ones are worked at elevations of 45 to 250 meters above sea level, where "black sands" with fine gold, buried beneath 6 to 30 meters of barren sand, are exposed by streams. The present beach

sands contain find gold and platinum in "black sand."

The Nevada Gold Province

This province embraces Nevada and parts of adjacent Arizona, California, Utah, and Idaho. It contains a great many gold and silver deposits that may belong to the metallogenetic province that extends into Central America. In general the formerly worked deposits are relatively shallow veins, of epithermal to mesothermal type, mostly of Tertiary age, and some of them are very rich in gold and silver. Recently discovered deposits in Nevada are disseminations of microscopic gold.

In north central Nevada, complex stratigraphic and structural studies indicate that eugeosynclinal rocks have been thrust eastward over transitional and miogeosynclinal formations. This mid Paleozoic, probably late Devonian, thrust, named by Ralph Roberts, is the Roberts Mountain thrust fault. It is significant as the underlying miogeosynclinal carbonate sediments are preferentially mineralized rather than the overlying, chemically inert, siliceous, volcanic, and clastic rocks.

Where windows in the thrust area exist, mineralization is sometimes exposed in the lower plate carbonate sediments. In some of these windows, fine colloidal (submicroscopic) gold has been found that has resulted in the development of several gold producing properties, viz., the Carlin, Bootstrap, Cortez, Gold Acres, and the Getchell deposit. The Carlin mine in the Lynn window is by far the largest operation, but will be exceeded by the Jerritt Canyon deposit.

Exploration for these properties was initiated by a U.S. Geological Survey report by Roberts who noted the alignment of mining districts in north central Nevada. The Carlin Gold Mining Co., wholly owned by Newmont Mining Corporation, began a staking, mapping, drilling, and assay program and study of the Lynn window in 1962. Three years later, gold reserves amounted to 11 million tons of 0.32 ounces per ton of gold, but now producing 131,000 oz./year (1980). (Fig. 18-3) of this deposit shows the Roberts Mountain thrust dipping northward with the western facies Vinini formation overlying and covering most of the area.

The deposit lies in tilted dolomitic siltstones of the lower Silurian Roberts Mountain formation about one hundred meters below the thrust plate. The host rock is a decalcified, decarbonated, hydrothermally altered rock now consisting largely of chalcedonic quartz, clay minerals, and dolomite. The host rock is cut by nearly vertical post-thrust faults, by dikes of quartz porphyry, and siliceous chimneys. A small pluton of quartz diorite lies about 5 kilometers to the north.

Two ore bodies follow the inclined highly altered Roberts Mountain beds. An earlier mineralization consists of sparse galena and sphalerite in barite veins and replacements with a little copper and nickel. The gold mineralization is a disseminated replacement of the altered rock and consists of microscopic to submicroscopic specks of gold, presumably colloidal. Accompanying minerals are specks of native arsenic, realgar, orpiment, cinnabar, pyrite, arsenolite, stibnite, jordonite, and tennantite. The gangue consists of preserved dolomite, chert, illite, and montmorillonite.

Hausen and Kerr consider that the highly inclined fault zones may have served as conduits for the mineralizing solutions and that the depositions of the gold occurred in the capillaries of the argillaceous siltstones, where carbonate was removed along the bedding planes. They also note a mutual association of gold and organic matter. They conclude that the deposit is a low-temperature, epigenetic, epithermal, replacement of possible hot-spring origin, and that the mineralogy has some similarity with that of the Getchell, Mercur, and Manhattan gold deposits in Nevada and Utah. Radke and Scheiner think that the organic carbon controlled the deposition of the gold, the activated carbon present absorbed gold complexes, and sub-

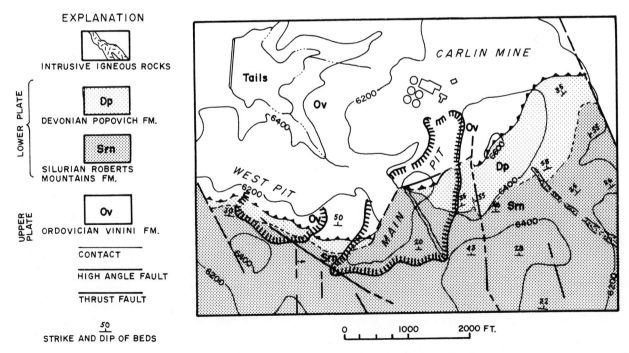

Figure 18-3 Generalized geologic map of the Carlin mine area (modified from detailed mapping by Newmont Mining Corp. and U.S. Geological Survey).

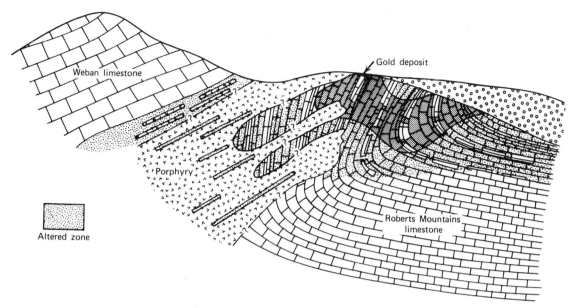

Figure 18-4 Idealized cross section through the Cortez gold deposit window from west to east showing rock units, gold deposit, alteration zone (the lower extent of which is unknown), and overlying gravels. (From Econ. Geol.)

sequent oxidation of the organic components of the gold organic compounds led to the formation of the native gold.

Hausen and Kerr, state that "based upon the evidence of argillic replacement of dolomitic host rock, major deposition of colloidal-sized gold in clay aggregates, chemical changes from rock to ore, nearby igenous activity, and elliptical chimneys of recrystallized silification," indicate that the deposit is of a low-temperature epithermal origin. Recent $\delta^{18}O$ and δD studies corroborate this genetic model with the additional recognition that the ore fluid was composed predominantly of connate or meteoric water.

Other Deposits. Few gold deposits of the rich past history have survived, but several are now being resurrected. The *Getchell* mine, with 5 million tons, has some similarities to Carlin. The Getchell vein is 2100 meters long and up to 60 meters in width but extended to only 250 meters in depth. The richest part was a hydrothermal replacement of wall rock by a porous mass consisting of minute quartz crystals, carbon and gold, along with sulfides, realgar, and orpiment. The *Manhattan* district and also *Round Mountain,* both in Nevada, have generally similar mineralogy. Round Mountain, however, has only a placer potential remaining, the gold of which is now being recovered by NaCN leaching. Man-

hattan was placer mined years ago but some underground mining is beginning again.

Of the former areas, outstanding was the unique and famous *Goldfield district,* which produced over $100 million in gold and deserves mention because it not only illustrates one type of gold deposit, but also it is undergoing a U.S. geological survey and private capital reevaluation. Here flat dipping veins, that extended only to 300 meters in depth, occur in late Tertiary volcanics, particularly in a thick dacite sill, and did not penetrate the pre-Tertiary basement. The exceptionally rich bonanza ores (up to $400 per ton at $20.67 troy ounces per ton) consisted of quartz, gold, pyrite, and primary and secondary alunite with kaolinite, along with tellurides (including goldfieldite) marcasite, wurtzite, bismuthinite, famatinite, and tennanite. Deposition was by successive precipitation in open spaces with some minor replacement. As alunite suggests acid solutions, Ramsome considered that magmatic alkaline solutions with H_2S were oxidized near the surface, supplying sulfuric acid that caused the alunitic alteration; the mingling of surficial acid waters and ascending waters caused near-surface precipitation of the gold. Actually, some of the primary solutions were acid, as primary alunite based on $\delta^{34}S$ studies is associated with the gold.

Other gold deposits of past interest and noteworthy production were Aurora, Tuscarora, Jarbidge, and Golden Arrow Nevada; and Katherine and Oatman, Arizona, all now defunct, although the Tuscarora and Golden Arrow areas are undergoing renewed development.

Rocky Mountain Region

The Rocky Mountain Region of Colorado formerly boasted scores of gold mines that contributed appreciably to the gold output of the United States. These are now largely depleted and most of the gold production of this area is presently obtained from base-metal ores. With increasing metal prices, not completely offset by costs and newer exploration techniques and development costs, some rejuvenation is taking place. Some of the famed districts are Cripple Creek, Ouray, Telluride, Silverton, Camp Bird, Smuggler, and Creede. The Idarado is a recently revived operation.

The Colorado Mineral Belt is a zone of enormous deposits extending from Boulder to Durango (Fig. 18-5 and Table 18-1). The host rocks are predominantly granite and Precambrian gneiss in the northern portion of the zone although some of the deposits were formed in Paleozoic and Mesozoic host rocks and ash flow tuffs and volcanics. The later Tertiary volcanics form the San Juan volcanic pile. Many of the deposits are associated with intrusive stocks.

The northern mountain province of the belt has produced over $3,446,000,000 of metals although Tweto points out that 76 percent of this value came from only five districts: Climax (28 percent), Leadville (15 percent), the Ouray-Telluride-Silverton triangle (14 percent), Cripple Creek (12 percent), and Gilman (7 percent).

Steven describes the locale of ore districts in the *San Juan Mountains,* Colorado, as mostly existing "within a large area of complexly overlapping volcanic subsidence structures that subsided repeatedly in response to voluminous ash-flow eruptions in middle Tertiary time." Except for the nearly circular calderas, these structures are referred to as cauldrons.

The episodic eruption of great quantities of ash that followed periods of volcanism resulted in collapse or subsidence of at least three structures or calderas into the partly evacuated magma chambers.

Episodic stages of mineralization gave rise to numerous fissure veins, some of which were long, deep, and beautifully crustified. The famous *Camp Bird* vein is a long, continuous sheeted zone about 2 meters wide and 900 meters deep. The *Smuggler*

Union vein was mined for 2700 meters in length and 1200 meters deep. Some veins in the Silverton caldera, on the western site of the San Juan Mountains, have a length of 8 kilometers and a depth of 1500 meters and are only 1 to 2 meters wide. The ores consist of gold, silver, minor sulfides, and sparse tellurides with a gangue of crustified quartz, carbonates, and fluorite.

Some geologic features of a few of the deposits

Table 18-1 Mining Districts, Colorado Mineral Belt, and Chief Products (from Ore Deposits of the U.S., AIME)

District or center	Chief products (In order of value)
1. Northgate	CaF$_2$
2. Jamestown	Au, Ag, CaF$_2$
3. Ward	Au, Ag
4. Gold Hill	Au, Ag
5. Caribou	Ag, Pb, U
6. Nederland	W
7. Magnolia	Au, Ag
8. Urad	Mo
9. Empire	Au, Ag
10. Lawson-Dumont	Au, Ag
11. Central City-Blackhawk	Au, Ag, Cu, Pb, U
12. Ralston	U
13. Idaho Springs	Au, Ag, Zn
14. Georgetown-Silver Plume	Ag, Pb
15. Montezuma	Ag, Pb, Cu
16. Gilman	Zn, Ag, Cu, Pb, Au
17. Kokomo	Zn, Pb, Ag, Au
18. Breckenridge	Au, Ag, Pb, Zn
19. Climax	Mo, W
20. Aspen	Ag, Pb
21. Sugar Loaf-St. Kevin	Ag, Au
22. Leadville	Ag, Zn, Pb, Au, Cu
23. Alma	Au, Ag, Pb
24. Granite-Twin Lakes	Au, Ag
25. Crested Butte	Ag, Pb, Zn
26. Gold Brick-Pitkin	Au, Ag, Pb
27. Tincup	Ag, Au, Pb
28. St. Elmo	Au, Ag
29. Cripple Creek	Au
30. Monarch-Tomichi	Zn, Ag, Pb, Cu
31. Browns Canyon-Poncha	CaF$_2$
32. Tallahassee	U
33. Cochetopa-Marshall Pass	U
34. Bonanza	Ag, Pb, Cu
35. Westcliff-Silver Cliff	Ag, Au, Pb
36. Ouray	Au, Ag, Pb, Cu
37. Lake City	Ag, Pb, Au
38. Telluride	Au, Ag, Pb, Cu, Zn
39. Rico	Ag, Zn, Pb, Au, Cu
40. Silverton	Ag, Au, Pb, Zn, Cu
41. Creede	Ag, Pb, Zn
42. Wagonwheel Gap	CaF$_2$
43. La Plata	Au, Ag
44. Summitville	Au, Ag, Cu

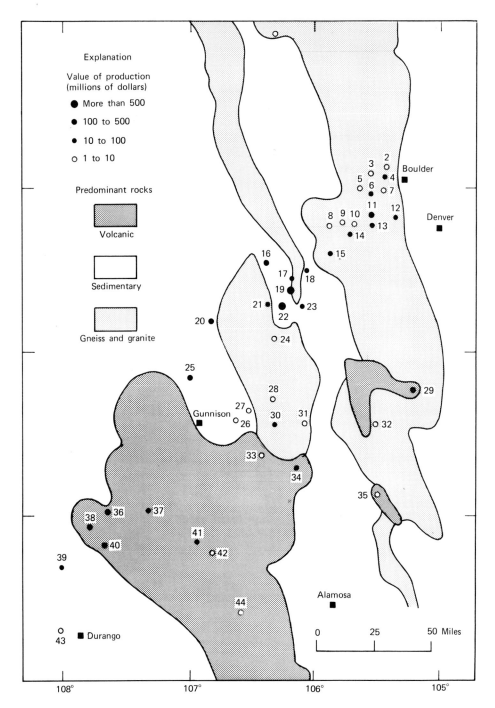

Figure 18-5 Principal mining districts in the Mountain Province of Colorado. (From: Tweto, Ore Deposits of the U.S., AIME.)

will be mentioned here because they illustrate modes of occurrence, controls, and origin of the types of ore deposits of this region, which may be applicable to deposits elsewhere or to future discoveries in this region.

The famous and historically interesting *Cripple Creek* district has yielded more than 300 million dollars from 64 mines that lie at an elevation of 3000 meters in a complex breccia plug of Tertiary volcanics about 3 to 6 kilometers across, called the Cripple Creek volcano. According to Loughlin, the plug tapers downward into mine roots or subcraters through which latitc-phonolite, syenite, phonolite, and basaltic rocks were successively intruded.

Water-laid sediments containing leaves lie within the plug. A late explosive eruption in the center formed the rich Cresson pipe 200 meters in diameter. The ore occurred as veins, sheeted zones, irregular bodies, and breccia fillings. The lodes are cavity fillings with minor replacement of wall rocks. The veins converge downward toward the roots of the subcraters; the maximum depth reached was 950 meters but few were profitable below 450 meters. The ores were noted for their paucity of native gold and abundance of numerous tellurides and fluorite. Phenomenally rich veins and cavities were found; one vug in the Cresson Mine, lined with calaverite yielded $1.2 million. The rich telluride ores are thought to have been formed from hot alkaline solutions that rose from the subcrater vents and nearer the surface spread into numerous branch fissures.

Appalachian Region

The Appalachian region of the eastern United States is of historic interest only since it was the site of the first gold mining of the country, yielding about $30 million, since 1800. Lenticular gold quartz veins in ancient crystalline rocks contained gold, common sulfides, and molybdenite, and some tellurides and enargite in a gangue of carbonates, apatite, chlorite, tourmaline, and garnet. The gold content was low. Only a few veins were mined and most of the gold production came from placers.

Alaska

Although Alaska has produced over $600 million in gold, production is now increasing over the mere 20,000 ounces produced only a few years ago, entirely from placer dredging. Its heyday followed the famous Klondike rush of 1898 in the *Fairbanks* region, where rich shallow gravels gave way to thick, low-grade gravels dredged from frozen ground up to 60 meters beneath the surface.

In the productive *Nome* district in the *Seward Peninsula*, gold has been and is being won from six beaches, two submarine and four elevated ones, the highest being 20 meters above sealevel lying under 7 to 35 meters of frozen overburden. The gold content ranged from 50 cents to $10 per cubic yard at $35 per ounce.

The lode deposits are now abandoned. The old *Treadwell* mine on Douglas Island was extinguished by a submarine cave-in. The *Alaska Juneau* Mine was noted for being the largest disseminated gold mine (but now surpassed by the Carlin mine) and the lowest-grade ore produced at the lowest cost in the world, $70 million were produced from a little over 70 million tons of ore; the pre-World War II content was 0.04 ounces of gold per ton, or a recovery of $1.15 at a total cost of 72 cents per ton. The deposit was a zone 5 kilometers long and 300 to 600 meters wide. It is a gigantic stockwork made up of stringers and bunches of quartz containing gold with minor silver, base-metal sulfides and ankerite, which were deposited in open spaces and by replacement. Most of the gold was a late introduction into fractures in the quartz. The deposit is considered to have been formed by hydrothermal solutions of Coast Range Batholith derivation into a zone of structural weakness.

A small production came from a few mines in the Willow Creek district and Chichagoff Island.

Even with the general high costs of Alaskan economy plus high inflation, the value of gold at $240 or more per troy ounce is resulting in reactivation of several Alaska gold properties.

Canadian Gold Deposits

Canada now contributes about 6 percent of the world's gold supply, or 2½ to 3 million ounces annually. Its former 140 gold mines have shrunk to 35: Ontario leads in production with 51 percent (Fig. 18-6); Quebec, 28 percent; Northwest Territories, 12.7 percent; British Columbia, 4.3 percent; and Prairie Provinces, 3 percent. Lode mines yield about 81 percent of the gold, the remainder being chiefly by-product gold. Placer mining now yields only 1 percent.

The chief area is the Canadian Shield, where gold quartz lodes occur from western Quebec to the Northwest Territories. Most of the lodes yield only gold and by-product silver is minor, but considerable gold is a by-product of massive sulfide deposits. The deposits are mostly hydrothermal veins and replacements.

Porcupine District, Ontario. Porcupine has been the premier gold district of Canada; the Hollinger Mine alone has yielded over $580 million, but it was closed in 1970. There were formerly 36 mines in the district, notably the Hollinger, McIntyre, and Dome, but most of the smaller mines are now closed. Some mines were almost 2 kilometers deep. The average tenor was $9 to $10 per ton.

Keewatin volcanics overlain by Temiskaming sediments were intricately folded into the Porcu-

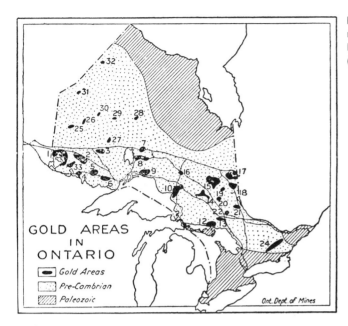

Figure 18-6 Geologic sketch map of Ontario, showing distribution of 33 known gold areas. (Ontario Dept. Mines.)

GOLD AREAS
IN
ONTARIO
- ⬭ *Gold Areas*
- ⸬ *Pre-Cambrian*
- ▨ *Paleozoic*

Ont. Dept. of Mines

pine syncline and were intruded by bodies of Algoman quartz porphyry, of which the principal one, the Pearl Lake porphyry, extended with continuity of dimensions and shape for 2 kilometers down dip. There were later intrusions of albitite, quartz diabase, and late olivine diabase dikes. Shearing and fracturing formed loci for the veins on both sides of the syncline. The deposits are mostly in the volcanics and avoid the porphyry.

The deposits consist of (a) quartz veins, (b) irregular replacement lodes, (c) pipelike bodies, and (d) intervening mineralized country rock. The veins are long and deep or form a linked system of short, closely spaced veins (Fig. 8-37B). The average width of the veins is around 10 feet, and of the irregular lodes around 30 meters; the length is up to 600 meters.

The ores consist of quartz, mineralized wall rocks (Fig. 18-7) and some gold-bearing sulfide bodies. They consist of quartz, gold, pyrite, sericite, tour-

Figure 18-7 Banded quartz lode of No. 7 vein, McIntyre mine, Porcupine district, Ontario. (From Dougherty, Econ. Geol.)

maline, and carbonates with sporadic sulfides, tellurides, scheelite, siderite, dolomite, selenite, and anhydrite (considerable carbon occurs in the McIntyre mine). The gold is largely free and in places is in spectacular masses. Hurst and Langford recognize three periods of mineralization: (1) barren quartz-tourmaline veins, (2) main quartz-gold deposition, and (3) barren quartz-calcite. Graton and McKinstry stated that the quartz of the Hollinger ore replaced ankerite, and Hurst states that open space filling dominated. The ore solutions are considered to have been derived from an underlying magma, of which the altered porphyry bodies are offshoots.

Kirkland Lake District, Ontario. This famed district formerly yielded annually about a million ounces of gold. There were seven contiguous producing mines in a length of 5 kilometers; two reached depths of over 2500 meters. The average gold tenor was $15.17 a ton at $35 per ounce of gold. This district and nearby Larder Lake have produced over $1 billion in gold and silver.

The veins are in metamorphosed Temiskaming sediments underlain by Keewatin volcanics and intruded by syenites and basic dikes. The veins are localized along two chief parallel reverse faults (throw 600 meters), which suffered post-ore movement creating strike faults that displaced the veins; diagonal veins connect the north and south breaks. The great north vein is 4 kilometers long and within the Lake Shore mine is continuously mineralized. The fracturing is narrow within the porphyry but widens to 30 meters west of the porphyry; mining widths reach 30 meters or more. The quartz in part is open space filling and in part replacement. The ore is localized in shoots up to 600 meters in length and about 2 kilometers deep. The ore consists of quartz, native gold, tellurides, pyrite, calcite, altered rock, sparse base-metal sulfides, and a little tourmaline. The ore zone is intensely silicified and carbonitized. The ore solutions are considered to be genetically related to a deep-seated magma, which also supplied the intrusives. Porcupine and Kirkland Lake both probably belong to the same metallogenetic epoch and province.

Other Areas. The Kirkland Lake gold belt extends eastward through *Larder Lake,* where gold occurs in quartz veins in carbonitized and volcanic rocks and in the hard wall rocks near prominent shear zones.

Farther east is the large *Noranda* district (Fig. 11-9), where many mines yield copper-gold ore from massive copper deposits in volcanic rocks. Still farther east is the *Val d'Or* area with several mines such as the Sigma, Lamaque, Malartic, and Manitou that yield much of the Canadian gold production. These mines lie near a large granodiorite batholith in volcanic rocks invaded by smaller bodies of porphyritic rocks. The ores are mainly quartz-tourmaline bodies in fissure veins and shear zones, containing gold, minor sulfides, some tellurides, and carbonates. In the Manitou, gold and silver are associated with massive zinc and copper ore bodies along a shear zone in sericite schist.

In the *Malartic* area some deposits occur along the Cadillac-Malartic "Break," which extends eastward for 95 to 100 kilometers. In the *East Malartic* zone, the ore occurs along a subsidiary fault in altered greenstone, porphyry, and sediments. In the interesting *Camflo* group, the gold ore is associated with a syenite stock that intrudes volcanics and graywackes; most of the ore, consisting of quartz, gold, some scheelite and tellurides, carbonate, and fluorite, occurs in and along numerous stringers in the porphyry (as in porphyry coppers) giving rise to ore zones up to 40 meters wide and 120 meters long, containing up to $\frac{1}{2}$ ounce of gold.

All of these deposits are considered to be hydrothermal fillings of fissure and shear zones of magmatic derivation.

Witwatersrand, South Africa

The Rand since 1886 has produced over 50 percent of the world's gold. Its scores of mines, centering around Johannesburg, hoist annually around 80 million tons of ore, and to date they have hoisted nearly 2 billion tons. The annual production is about 30 million troy ounces. The average recovery of gold is about 7.65 dwt[1] per ton. The greatest depth penetrated is over 3300 meters, vertically, and there are about 1100 kilometers of underground workings. The main district is about 150 kilometers long and in places 40 kilometers wide. Another Rand-type district occurs in the Orange Free State, where large volumes of hot saline waters were encountered.

Geologic Setting. The gold of the Rand occurs in thin conglomerate beds, called reefs, that lie within the Precambrian Witwatersrand system, an 8 kilometer thickness of conformable sediments that rest

---

[1] 1 dwt = 0.05 Troy ounce.

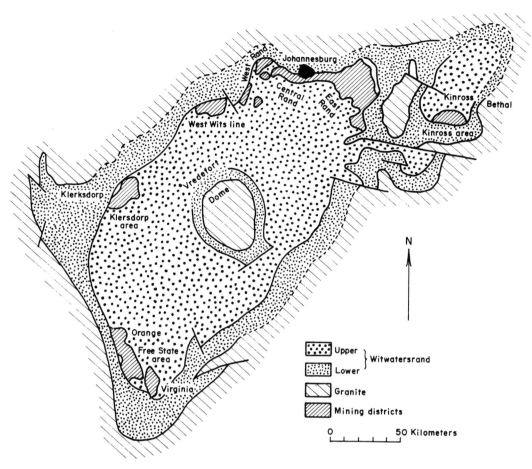

Figure 18-8 Tectonic scheme of Witwatersrand Basin with positions of mining areas. (From Bentz and Martini, 1968; Bachmann, 1976, *Introduction to Ore Deposits*, Halsted Press.)

unconformably upon ancient granite and schist. The sediments have been folded into a broad syncline; they have been traced 290 kilometers east and west and are 150 kilometers across. Most of the mines are concentrated within an 80 kilometer stretch on the northern limb (Fig. 18-8), but later explorations have extended the field eastward and westward, where payable ore has also been discovered at depth beneath younger formations, In the central part of the field the beds dip steeply; toward the margins the dip is gentle. Considerable faulting has displaced the beds.

The Lower Witwatersrand (5 kilometers thick) consists of argillaceous zones alternating with quartzites and including some iron formations and volcanics; the Upper is dominantly quartzites with conglomerate layers. The system has been invaded by dikes and sills of basic to intermediate igneous rocks.

The Reefs. There are eight principal conglomerate

zones within the Witwatersrand system. All of them contain some gold, but the payable deposits are confined mainly to the Main Reef group; the Kimberley and Bird Reefs yield ore in the West and Far East Rand, and the Bird and Livingstone Reefs in the Central Rand.

The Main Reef group, lowermost member of the Upper Witwatersrand, is a zone from 6 to 50 meters thick including several conglomerate layers within quartzites and grits. There are three payable members—the Main Reef, Main Reef Leader, and South Reef—each from less than one to more than 3 meters thick. Each in places is split by quartzite intercalations. The Leader is the most persistent and has been the chief gold producer.

In the Main Reef the pebbles are small and uniform, and the distribution of the gold is uniform. Eastward it merges into the Leader and the South Reef to form the "Composite Reef." The Main Reef Leader, which contains larger pebbles, has been mined continuously for 35 kilometers but it thins eastward. Its base is well marked and cuts across underlying filled channels; in places it lies almost on top of the Main Reef. The South Reef is separated from the Leader by 20 to 33 meters of quartzites in the Central Rand; eastward the two are in contact. The Bird Reef lies about 500 meters above the Leader. The Kimberley Reefs are composed of a dozen or so conglomerate beds characterized by high pyrite content lying about 1500 meters above the Main Reef. The individual reefs are not continuous throughout the Rand, and the spacing between them varies from place to place.

Ore Shoots. The gold occurs in elongated ore shoots or pay streaks as much as 1500 meters long and 300 meters wide; in the Central Rand they range from 80 to 140 meters in width. The longer axes, parallel to the longer axes of the pebbles, trend roughly in the same direction, or they are braided. Reincke found that the gold distribution coincided with well-graded pebbles, large pebbles, aligned pebbles, and wide sections of the reef. These features he interpreted as originating in stream channels and concluded therefrom that the gold is of placer origin. More recent studies under Pretorius' direction at the Economic Geology Research Unit, Witwatersrand University, have provided diagnostic evidence of a placer origin for the gold, and presumably the pyrite and uraninite.

Nature of the Ore. The ore is a mineralized conglomerate, and the matrix carries recrystallized quartz grains; sericite; and minor chlorite, chloritoid, rutile, tourmaline, carbon, zircon, and calcite. The gold is mostly in minute grains (0.07 to 0.1 millimeter). Veinlets of gold pass from the matrix into and across quartz pebbles, which they replace. The gold contains 8 to 9 percent silver and 3 percent base metals, mostly copper; the fineness varies from mine to mine. The gold is probably in two or more generations, but pyrite is definitely so (Figs. 18-9, 18-10, 18-11, and 18-12), probably the effect of metamorphism. Pyrite is rather abundant, in places making up 10 to 20 percent of the ore; it averages 3 percent of the Leader.

Pretorius describes pyrite and a few other sulfides occurring as (1) disseminated grains and small irregular masses in numerous stratigraphic horizons; (2) as parallel and subparallel stringers, which generally fill old erosional channels in the Witwatersrand system; (3) as individual grains and small aggregates; (4) as disseminations and blebs in

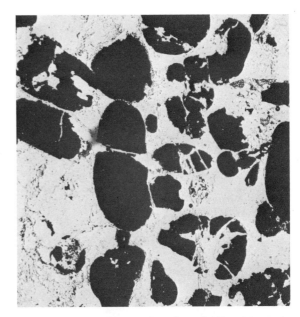

Figure 18-9 Main reef leader from Grootvlei Proprietary mines Lts., East Rand, ×62.5. Rounded grains of pyrite (black) belonging to the first generation (detrital) being replaced along fractures (possibly formed during the second generation) by sericitic material (probably third generation in age). (From Viljoen, 1963, provided by D. Pretorius, EGRU.)

quartzite and shales; (5) as masses of varying size and shape; (6) as smears; and (7) as specks and large irregular masses in dikes, silts, and lava flows of different compositions.

Heavy minerals of primary placer origin are abundant in locales of gold and pyrite. Some of these minerals are zircon, chromite, cassiterite, monazite, diamond, garnet, corundum, rutile, tourmaline, apatite, ilmenite, magnetite, and others.

Obviously, many of these minerals including pyrite, and especially uraninite, are readily oxidized when transported and deposited as detrital material, but it has been suggested that the lack of these minerals is the result of mechanical rather than chemical weathering of the source area; a cold climate, transportation, and deposition occurred under water. The environment was nonoxidizing, in fact, a controversial suggestion is that the atmosphere may have been reducing, lacking sufficient oxygen, when these blanket deposits formed about 2500 million years ago. About 40 minerals have been identified in the conglomerates, including pyrrhotite, marcasite, galena, sphalerite, chalcopyrite, bornite, covellite, cubanite, bravoite, pentlandite, niccolite, linnaeite, cobaltite, proustite, molybdenite, and others. Chromite, corundum, magnetite, diamonds, tourmaline, and many other

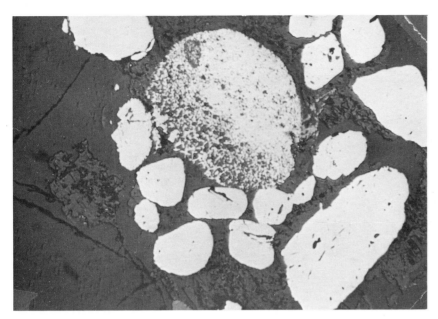

Figure 18-10 Main reef leader from East Daggafontein Mines Lts., East Rand, ×62.5. Rounded grains of pyrite (dense white) belonging to the first generation (detrital), and rounded nodule of porous pyrite (probably second generation) possibly resulting from replacement in situ of fragment of friable chert, arkosic material or iron-silica gel. Note the larger size of the second-generation pyrite grain, an indication that the original material must have had lower opecific gravity than that of the detrital pyrite. (From Viljoen, 1963, provided by D. Pretorius, EGRU.)

minerals are clearly detrital. Some were introduced after consolidation of the sediments. Both isotropic and anisotropic pyrite have been noted.

Origin. The question of the origin of the gold of the Banket has been the cause of a lengthy controversy between the placerists and the infiltrationists. The early thought was that the gold was of direct placer origin, but later microscopic examinations revealed features thought to be inconsistent with a simple placer origin. These features consisted of cross-cutting relationships, especially cross-cutting quartz veins, apparent replacement, and masses of sulfides in igneous rocks. The abundant and excellent studies done essentially by the South African geologists strongly support a placer origin not only for the pyrite, but also for the gold- and uranium-bearing minerals.

There are similar gold- and uranium-bearing Precambrian conglomerate deposits in Canada and Brazil. The well-known *Blind River, Canada,* occurrence of low-grade but huge reserves of uranium may have had an origin similar to the Witwatersrand. The similar petrologic deposit at *Serra de Jacobina, Bahia, Brazil,* is probably also of similar petrogenesis although a hydrothermal origin has

Figure 18-11 Main reef leader from Crown Mines Ltd., Central Rand, ×250. Rounded grains of first-generation, detrital pyrite (lighter gray) with aligned inclusions, partially surrounded by subhedral grains of second generation, porous pyrite (darker gray). (From: Viljoen, 1963, provided by D. Pretorius, EGRU.)

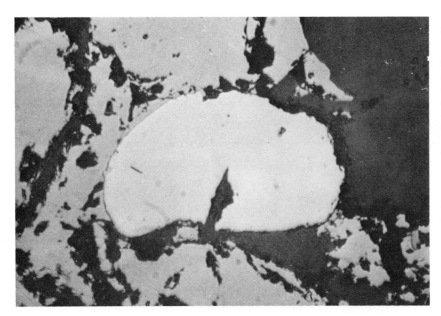

Figure 18-12 Main reef leader from Vlakfontein Gold Mining Co. Ltd., East Rand, ×250. Rounded grain of first-generation, detrital arsenopyrite (lighter gray) surrounded by third-generation quartz (darkest gray). Note the absence of second-generation overgrowths on arsenopyrite, an indication that the stability of this first-generation mineral in metamorphic environments prevailed during dissociation, mobilization, and reconstitution of detrital sulfides. (From Viljoen, 1963, provided by D. Pretorius, EGRU.)

been suggested for this deposit that contains 0.008 percent equivalent U_3O_8 and 10.0 grams of gold per metric ton of rock.

The narrow range of $\delta^{34}S$ analyses of sulfides, predominatly pyrite, of all three areas of Precambrian auriferous and uranium-bearing conglomerates is indicated in Fig. 18-13. The slight shift toward ^{34}S is typical of Precambrian sulfides but does not negate the evidence that the sulfides are of primary magmatic or hydrothermal origin whether the source be from existing high lands prior to deposition of the conglomerates or, very likely, derived from hydrothermal solutions that permeated the conglomerates. Figure 18-14 indicates little variation in $\delta^{34}S$ values possibly as the result of remobilization in various types of host rocks.

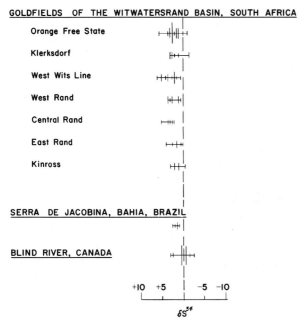

Figure 18-13 $\delta^{34}S$ values of Precambrian conglomerates containing gold and uranium.

Japan

Gold and silver mines are scattered throughout Japan. The most important production is from epithermal veins in Tertiary volcanics and sediments of Tertiary or older age. The Au:Ag ratio of these deposits is about 1:10–100.

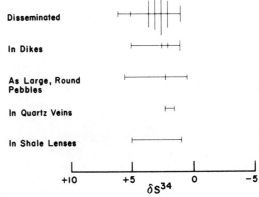

Figure 18-14 $\delta^{34}S$ values of Witwatersrand sulfides in different host rocks.

The deposits that belong to the gold-silver deposits proper with an Au : Ag ratio of about 1 : 1 are fewer in number. The deposits in the Yamagano-Oguchi district and Kagoshima Prefecture belong to this type.

A few mesothermal or hypothermal vein deposits exist but they are of little importance as producers.

Almost all of the placer deposits have been worked out with only nine being operated during the 1950s. The increase in the price of gold may activate some of these low-grade deposits.

In 1952, there were 94 working mines producing gold in Japan. All were lode mining operations. More than 79 percent of the gold production of Japan is derived from 10 mines. One is the Konomai, Hokkaido, gold-silver mine that produced 2583 kilograms of gold and 47,937 kilograms of silver in 1952. Other major producers are the Kushikino, Kagoshima; Chitose, Hokkaido; Oyo Miyagi; and the Mochikoshi, Shizuoka, mines.

The Kolar Gold Field, Mysore, India

The production of this great field is derived from a single reef—the Champion Lode—whose thickness averages between 1 and 1.5 meters. This deposit has been mined in 5 mines for 7 kilometers in length and to a depth of over 3000 meters. The ore averages 7 to 10 dwt per ton and it has yielded over $500 million in gold. The lode lies in a schist belt 80 kilometers long composed of Precambrian Dharwar conglomerates, hornblende schists, ferruginous quartzites, and older porphyritic granite, intruded by younger granite. The walls are well defined, and the lode dips about 55° conformable with the enclosing schists but is vertical at 2440 meters in depth. It lies on the limb of a major syncline. The ore is concentrated in shoots that pitch about 45° within the vein. The rich shoot of the Mysore mine had a stoping length of 250 meters, a maximum width of 10 meters, an average width of one meter, and a pitch length of 1200 meters.

The ore consists of quartz with minor base-metal sulfides; some tourmaline, pyroxene, albite, actinolite, and biotite also occur in the altered walls. The magmatic source that gave rise to the later granite is also considered to have supplied the mineralizing solutions but meteoric waters were most likely added to the mineral-bearing fluid.

Russian Gold Deposits

Although Russia ranks second in world gold production, amounting to 7.5 million troy ounces in 1975, detailed information regarding the deposits worked at present is meager. Most of the gold is won from placer deposits.

Lode Deposits. The most important lode deposits are in the Trans-Baikal region, including such districts as Beleï, Darasun Titagara, and Minusinsk. Other important lodes occur in the Altai and Ural mountains. These are mostly narrow quartz veins containing base-metal sulfides, arsenopyrite, some tourmaline, and 0.2 to 0.5 ounce of gold per ton. They are mostly associated with granitic rocks intruded at the end of the Paleozoic. The Altai lodes are pyritic quartz veins and polymetal veins associated with Paleozoic quartz keratophyres. They contain base-metal sulfides, many rare metals, and rarely tellurides, arsenopyrite, and pyrrhotite. The gangue consists of quartz, barite, and carbonates.

The Trans-Baikal lodes consist of (1) pyritic quartz veins, (2) quartz veins with base-metal sulfides and sulfantimonides, (3) quartz-tourmaline veins, and (4) quartz veins with tellurides and bismuth. These lodes occur in metamorphic rocks in association either with Paleozoic granodiorites or post-Jurassic granitic rocks.

The Beli mine is the most productive. Other smaller lodes are widely scattered throughout Siberia in the drainage basins of the Amur, Yenisei, and Lena Rivers, and in the Okhotsh regions.

Placer Deposits. About 100 or more gold dredges have been reported to be in operation in 16 districts. The pay gravels are perpetually frozen and lie at depths of 6 to 90 meters under glacial drift, recent gravels, and muck. The *Ural* gravels are mostly of low grade, are small, and lie at depths of 10 to 35 meters; they occur all along the eastern slope of the range. The *Yenisei* district has long yielded the richest gravels of Siberia, derived from small Precambrian gold quartz veins. The *Amur* and *Yakut* districts have yielded much placer gold.

Australasia Gold Deposits

Australasia with numerous mineral deposits and more being found (Fig. 18-15) occupies fifth place in world gold production. West Australia is the main source, with considerably less production from Queensland, Victoria, and New South Wales.

West Australia. West Australia gold is obtained almost entirely from lodes that average 13 dwt per ton. The region is underlain by Precambrian meta-

Figure 18-15 Locality map of the principal Australian ore deposits. (From Econ Geol., v. 67, 1972.)

morphic and sedimentary rocks intruded by gneissic granitic rocks. Most of the deposits lie in minor islands of invaded rocks. The lode deposits consist of (1) fissure lodes, (2) auriferous sediments, (3) quartz reefs, and (4) auriferous dike rocks.

The lodes are mostly replacements along shear zones in basic metamorphic rocks; they attain a few thousand meters in length and 1200 meters in depth. The associated minerals are pyrite, base-metal sulfides, and minor tellurides, scheelite, vanadinite, and bismuthite. The auriferous sediments are replacements by silica with considerable pyrite, carrying low gold values. The quartz reefs are fissure fillings of quartz, base-metal sulfides, and minor rare minerals, with the gold localized in short, shallow, but rich ore sheets. The auriferous dike rocks are alaskites and aplites that consist dominantly of quartz with low gold content, and minor iron sulfides, molybdenite, wulfenite, and crocoite in the alaskites; and of silicates, pyrrhotite, and gold in the aplites.

Kalgoorlie became famous as the "Golden Mile," which for a time made this district one of the richest gold camps of the world, with a yield of over 26 million ounces. The deposits, according to Gustafson and Miller, consist of gold-pyrite-telluride replacement lodes, mainly in a quartz dolerite sill of the regionally folded "younger greenstone." A cross fold helped localize the controlling fractures. Some of the rich bodies are pipes at fracture intersections. There are two lode systems. The Western, confined to the Golden Mile proper, consists of few but large, persistent lodes with regular ore bodies and even distribution of gold. The Eastern System consists dominantly of swarms of smaller veins and pipes with irregular distribution of gold and locally rich pockets. The famous Oroya shoot, which yielded 2 million ounces, is a freak flat pipe of ore 1400 meters long, localized by two nearly flat shears connected by a thin, nearly flat shear in a minor fold on the flank of the main Kalgoorlie syncline.

The gold occurs in structurally controlled ore shoots. The tenor with gold at $120 per ounce has declined from $140 per ton near the surface to $25 to $35 per ton at depth. The shoots bottom in the underlying, unfavorable "calc schists," and are lenticular in shape range from one to 25 meters in width, and still persist at a depth of 1200 meters. The ore minerals include yellow sulfides, native gold, and tellurides (calaverite, kremerite, petzite, hessite, and coloradoite). Accessory minerals include galena, sphalerite, enargite, pyrargyrite, löllingite, specularite, magnetite, fluorspar, tour-

maline, carbonates, and roscoelite. Spectacular seams a few centimeters across are filled by gold and tellurides.

Other Lodes. The Leonora Center district contains the Sons of Gwalia mine, which is a replacement lode in schistose dolerite, containing quartz, gold, and base-metal sulfides. The Murchison field contains the Wiluna and Great Fingall quartz lodes, and these as well as the lodes of the Meekatharra gold field are of generally similar type. Quartz pematite schists occur in the Yilgam district and auriferous sediments in the Yilgam and Peak Hill gold fields. Similar lodes, but containing arsenopyrite, occur at Coolgardie. During 1938 the Big Bell mine at Cue became an important gold producer.

Victoria. Victoria leads in gold production, being credited with about 92 million ounces. It contains the celebrated districts of Bendigo and Ballarat. *Bendigo* is noted for its unique saddle-reef deposits.

At Bendigo, Ordovician slates and sandstones have been acutely compressed into sharp anticlines and synclines, intruded by granitic batholiths, and deeply eroded. The axes of the folds are gently undulating, and the axial planes, or "center country," are steeply inclined or vertical. Fifteen parallel anticlinal lines, or "lines of reef," spaced 210 to 370 meters apart have been mapped.

The ore deposits are saddle reefs localized mainly at the crests of the anticlines between slate and sandstone beds (Fig. 18-16). The individual saddles lie mostly within 30 meters of "center country" and seldom exceed 6 meters across. The leg depth may reach 90 meters; most of them are less than 30 meters. Most legs are of unequal length or one may be lacking. The crests have many "spurs" and protuberances and are connected by steeply inclined angling faults, which Stone thinks have been the main solution channelways. One saddle has been followed for 2800 meters along the strike. Saddles underlie each other, and mining has reached a depth of 1400 meters. Five anticlinal lines have yielded most of the $300-million production, the principal ones being New Clum, Garden Gully, and Hustlers. "Inverted saddles" and "leg reefs" also occur.

The ore consists chiefly of gold-bearing quartz with small amounts of pyrite, arsenopyrite, and pyrrhotite. Minor galena, sphalerite, stibnite, molybdenite, and bournonite are present, but tellurides are absent. Dolomite and ankerite are generally present.

Evidently, the small saddlelike openings created by the folding have been the important loci of deposition, the metallizing solutions being fed through

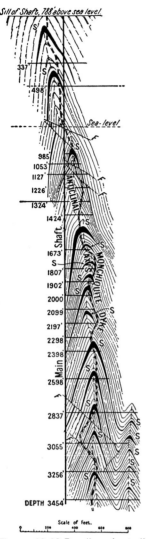

Sill of Shaft, 788' above sea level.

Figure 18-16 Bendigo, Australia. Saddle reefs shown in Great Extended Hustlers shaft. (Baragwanath, Gold Res. World.)

the connecting faults. Some open-space deposition is clearly indicated by crusts of quartz and ankerite; but the chief process of deposition, according to Stilwell, has been by replacement from hydrothermal solutions, which were probably yielded by the magmatic reservoirs that supplied the intrusive granitic rocks.

Similar saddle reefs are found at Castlemaine, also at Ballarat, Clunes, Beringa, Scarsdale, Blackwood, and other Victoria localities.

Ballarat ranks about equally with Bendigo in gold production, more than half of which, however, has come from rich placer deposits. The lodes consist of saddle reefs, and rich quartz bodies or "leather jackets" associated with the Indicator, a thin band of black slate replaced by quartz and pyrite, lying

on the east limb of an anticline. Where intersected by west-dipping strike faults, rich concentrations of gold occur; masses of gold weighing 444 ounces have been extracted from the Indicator.

The Placer deposits of Victoria rank with those of Calfiornia and have yielded over $700 million. The earlier worked placers were surficial gravels in sterams, benches, and hill cappings. Later, the rich "deep lead" gravels became very productive. These consist of

1. _Preearly basalt gravels_ found at elevations over 600 meters above present stream bottoms and overlain by sediments, and early basalt.
2. _High-level gravels_, also overlain by early basalt and capping or flanking hills above present stream bottoms. They have been regarded as marine but are probably lacustrine or floodplain deposits.
3. _Prelater basalt gravels_. These consist of three groups:
   (a) Stream gravels underlying the younger basalt and including the highly productive deep leads of Ballarat and Loddon Valley. These lie mostly below the present stream levels and contain coarse gold and many nuggets.
   (b) Stream gravels underlying sand and clay beds to depths of 100 meters below the present surface. Such deep leads of Ballarat, Tarnagalla, and Dunolly were famous for the number and size of nuggets found in them.
   (c) Buried coastsl gravels underlying basalt or sediments.
4. _Postlater basalt gravels_, formed by a late Tertiary uplift that resulted in the rapid erosion of rich reefs and former placers and their concentration in present-day streams.

Queensland. This state has produced around $600 million in gold, mainly from lodes of gold quartz or auriferous sulfides, and includes such well-known camps as Mount Morgan, Mount Isa, Charters Towers, Tympie, and Croydon. At _Charters Towers,_ productive fissure veins in Permian granodiorite have been mined to depths of 900 meters and yielded $132 million. The gold, averaging about 0.43 oz per ton, is associated with quartz and sulfides in crustified fillings. Rich ore shoots lie at intersections with basic dikes. _Mount Morgan_, a great copper mine, yielded $130 million in gold from ore averaging 4.7 dwt per ton. The ore body is a huge beet-shaped mass, 350 meters long, 230 meters wide, and 290 meters deep, and is a pyritic stock-work in tuffs carrying copper and gold.

New South Wales. has produced over $300 million from many productive gold districts. It has gold quartz fissure veins associated with granitic intrusives and carrying base-metal sulfides, as in the *Wyalong* and *Adelong fields*; rich lenticular lenses, as at *Hillend*, where $3,300,000 was recovered from 10 tons of quartz; fissure veins in sediments carrying stibnite, scheelite, and arsenopyrite; saddle reefs at *Mount Boppy*; large replacement lodes along shear zones carrying gold in massive sulfides or iron, copper, lead, and zinc at *Cobar*; and impregnations of auriferous arsenopyrite, pyrite, and pyrite in sediments.

New Zealand. has produced about $600 million in gold, but its productivity is largely past. The *Hauraki* district contains gold quartz veins in andesite, carrying arsenic, antimony, mercury, copper, lead, zinc, and iron. Rich bonanzas at *Thames* contained 1 to 6 ounces of electrum per pound of ore. This remarkable concentration has been attributed to supergene enrichment, but the evidence does not appear convincing. The *Otago* field has yielded about one-third of the production of New Zealand, most of it from Tertiary placers. The lodes are stockworks, rich at the surface, and carrying gold, pyrite, arsenopyrite, and scheelite.

New Guinea. has become an important gold producer from placers of the *Bulolo River*. Dredges have been operated, and the recovery has been about 160,000 ounces annually.

Selected References on Gold

Gold Deposits of the World. 1937. W. H. Emmons. McGraw-Hill Book Company, New York. *Gives location of many gold deposits.*

General Reference 9. *Descriptions of many gold deposits.*

General Reference 20. *Gold deposits of Canada.*

General Reference 10. *Thumbnail sketches of many western gold districts.*

General Reference 19. *Chapter on gold.*

Symposium on gold deposits of the Rand. 1931. Geol. Soc. South Africa, v. 34, p. 1–93. *Group of papers on occurrence and origin; chiefly arguments against Graton's advocacy of a hydrothermal origin.*

General Reference 16. 132 papers. *Geologic details of Canadian gold mines.*

Geology of the Homestake Mine, Black Hills, South Dakota. 1931. D. H. McLaughlin. Eng. and Min. Jour.,

v. 132, p. 324–329. *Occurrence, structure, and Precambrian origin.*

The mother lode system of California. 1929. A. Knopf. U.S. Geol. Surv. Prof. Paper 157.

Hydrothermal transport of gold. 1968. H. C. Helgesen and R. M. Garrels. Econ. Geol., v. 63, p. 622–635.

The geology, geochemistry, and origin of the gold deposits of the Yellowknife district. 1961. R. W. Boyle. Canada Geol. Surv. Mem. 310.

The solubility of gold. 1951. K. B. Krauskopf. Econ. Geol., v. 46, p. 858–870.

Mechanism and environment of gold deposition in veins. 1943. W. H. White. Econ. Geol., v. 38, p. 512–532.

Structural principles controlling occurrence of ore in Kolar gold field, India. 1947. J. W. Bichan. Econ. Geol., v. 42, p. 93–136. *Localization of gold ore by drag folds.*

Structure of a part of the Northern Black Hills and the Homestake Mine, Lead, South Dakota. 1949. J. A. Noble, J. A. Harder, and A. L. Slaughter. Bull. Geol. Soc. Amer., v. 60, p. 321–352.

Origin of the Bendigo saddle reefs and formation of ribbon quartz. 1949. F. M. Chace. Econ. Geol., v. 44, p. 561–597. *Excellent discussion of origin; fine illustrations.*

Homestake Gold Mine, South Dakota—lead isotopes, mineralization, age, and sources of lead in ores of the northern Black Hills. 1974. D. M. Rye, B. R. Rie, and M.H. Deleraux. Econ. Geol., v. 69, p. 814–822.

SILVER

Silver, along with gold, has been prized and sought since the time of the ancients. It has found use continuously as a monetary metal, although in diminishing quantities. Its call for adornment once widespread, is now more restricted because it is now an important industrial metal. Silver fluctuated in price from a high of $2.56 to a low of 24 cents in 1932 but was supported by the U.S. from 1933 until 1972 at 90.5 cents per ounce after which the price increased to more than $50, before dropping to a fluctuating value between $4 to $6, but rising to $20 in 1980.

Production. The production of silver has paralleled gold. It was mined early in Mediterranean Europe, but great production followed the discovery of America when streams of silver metal, taken from the Incas and Aztecs, flowed to Europe from Peru, Bolivia, and Mexico. This increase from 1520 to 1620 reached a peak in 1800 but fell off again until the rich discoveries of the western United States between 1860 and 1901. Since then there has been an erratic production. The chief producing countries and their normal ratio to world production are

Russia (13.8 percent)

Canada (13.5 percent)

Mexico (13 percent)

Peru (13 percent)

United States (11 percent)

This is the realignment that moved the United States from first place to fifth place primarily as a result of the disastrous Sunshine Mine fire in Idaho. In 1979, the United States production compared with Russia's production of about 40 million troy ounces with a world production of about 300 million troy ounces.

The chief producers of the United States in 1979 (with their percent of total silver production in parentheses) are

1. Idaho	(46)	5. Utah	(7)
2. Arizona	(20)	6. Missouri	(5)
3. Montana	(9)	7. New Mexico	(3)
4. Colorado	(8)	8. Others	(2)

Of the total silver produced in the United States, the relative percentage proportions supplied by the different classes of ore are

Silver ore	38	Lead ore	8
Copper ore	29	Zinc ore	2
Zinc-lead-copper ore	21	Gold-silver ore	2

Mineralogy, Tenor, Treatment, Uses
Much of the silver won is a by-product of gold, lead, copper, and zinc. It is the silver that makes many such deposits profitable.

Ore Minerals. The chief ore minerals of silver are

Mineral	Composition	Percent silver
Native silver	Ag	100
Argentite	Ag_2S	87.1
Cerargyrite	AgCl	75.3
Polybasite	$Ag_{16}Sb_2O_{11}$	80.3
Proustite	Ag_3AsS_3	65.4
Pyrargyrite	Ag_3SbS_3	59.9

Other economically important ore minerals of silver are the tellurides, argentiferous tetrahedrite, stromeyerite, stephanite, and pearcite. There are 55 well-known silver minerals. Silver also occurs in solid solution with gold and base-metal sulfides.

The common gangue minerals of silver deposits are quartz, calcite, dolomite, and rhodochrosite, and oxidation products.

Silver is most commonly associated with lead; gold is rarely free from it; most copper and zinc ores carry some silver; and cobalt is also an associate.

Tenor and Treatment. The tenor of silver ores varies with the price of silver and upon other economic considerations. In general, straight silver ores need to contain about between 4 to 10 ounces to the ton. Mexican ores are of lower tenor.

The recovery of silver from its ores depends entirely upon its association. Straight silver ores, if free-milling, may be cyanided, amalgamated, or concentrated and cyanided and the silver recovered by electroplating or precipitated from the cyanide solution by zinc or carbon. If not free-milling, the ores are concentrated and smelted to bullion. Silver-lead ores are concentrated and smelted, and silver is carried down with the lead, and the alloy is desilverized. Where silver is present in copper ores, the concentrates are smelted; the silver (and gold) is carried down with the molten copper and is then separated out by electrolytic refining, In former years about 99.5 percent of all American silver was recovered by smelting. Now cyanide leaching is increasing in importance.

Uses of Silver. Of the total world supplies of silver, about 15 percent is used for monetary purposes. This includes new production and demonetized silver. Much silver is hoarded; some is held for monetary exchange, but the largest amount is consumed industrially. Examples are sterling and plated silverware, the photographic and chemical industries, and electrolytic uses for printed circuits. Silver is the world's best electrical conductor.

Kinds of Deposits, Origin and Distribution
The silver that is a by-product of gold, copper, or other deposits is not considered separately. The principal classes of silver deposits and silver-lead deposits and some examples of them are:

Contact-metasomatic deposits: Zimapan, and *Velardeña*, Mexico; Magdalena, New Mexico.
Cavity fillings:
   a. Fissure veins—Cobalt, Ontario; Mayo, Yukon; *Pachuca*, Guanajuato, *San Francisco*, and *Fresnillo*, Mexico; *Rosario*, Honduras; San Juan, Colorado; *Sunshine*, Idaho; Tintic, Utah.

b. Stockworks—Quartz Hill, Colorado; *Fresnillo* and Guanajuato, Mexico.

c. Breccia fillings—Emma, Utah.

d. Pore-space fillings—Silver Reef, Utah.

Replacement deposits:

a. Massive:

1. Silver deposits—Shafter, Texas; Hamilton, Nevada.

2. Silver-lead deposits—*Bingham, Tintic, Park City,* Utah; *Sullivan* Mine, British Columbia; *Leadville,* San Juan, Colorado; *Santa Eulalia,* Mexico.

b. Lode—*Comstock Lode,* Tonopah, Nevada; *Potosi,* Bolivia.

c. Disseminated—Coeur d'Alene, Idaho.

Supergene Sulfide Enrichments: *Parral,* Mexico; Chañarcillo, Chile.

Origin. Most silver deposits have been formed as replacements or cavity fillings by hydrothermal solutions. Massive and lode replacements of silver-lead ores are numerous, but most of the world's silver is won from fissure veins of the mesothermal and epithermal type.

World Distribution. Most of the silver of the world comes or came from the *North American* Cordillera, where it is associated with Tertiary intrusive volcanic rocks. A belt prolific in silver extends from Utah-Nevada, through Mexico, and down to the Rosario mine in Honduras and includes such famous districts as the Comstock, Tonopah, Hamilton, Tintic, Bingham, Park City, Pachuca, Guanajuato, Real del Monte, San Luis Potosi, Zacatecas, Fresnillo, Mapimi, Parral, Sierra Mojada, and Santa Eulalia. A minor silver belt embraces the Coeur d'Alene and projects into southern British Columbia. Another silver-lead belt lies in Colorado. In eastern North America the unique Cobalt district was formerly a large producer of native silver.

In *South America* the Andean region of Peru, Bolivia, Chile, and Argentina is a silver belt which, together with Mexico, long supplied most of the world's silver. These epithermal deposits are associated mainly with Tertiary andesites. The famed mountain of Potosi contained the greatest concentration of silver in the world.

Examples of United States Silver Deposits

The United States is the fifth largest producer of silver, but it contains few large silver mines and few straight silver deposits. Most of its silver is won from silver-lead deposits and from gold and base-metal ores. The eight ranking silver producers of the United States, according to the Minerals Yearbook, are listed in Table 18-2.

Sunshine Mine, Coeur d'Alene District, Idaho. This, the premier silver mine of the United States and one of the largest silver producers in the world vies with the Galena mine and Lucky Friday as the only large, straight silver deposits north of Mexico.

The deposit consists of quartz veins in a broad zone in quartzite, which has been mined for a length of about 450 meters and a depth exceeding 1500 meters. In the upper part of the mine the ore occurs only in bunches and splits that approach each other on the 500 level and coalesce on the 1500 level, below which the vein becomes wider and richer and has yielded most of the silver production.

The ore consists of massive siderite and quartz with abundant argentiferous tetrahedrite and minor galena. The annual production has ranged between 3 and 12 million ounces of silver, from ore that has averaged 28 to 48 ounces of silver to the ton. The

Table 18-2 Major Silver Producers of the United States

	Mine or company	Location	Nature of ore
1.	Galena mine (Callahan Mining Co.) (Operated by ASARCO)	Coeur d'Alene, Idaho	Silver ore
2.	Sunshine	Coeur d'Alene, Idaho	Silver ore
3.	Kennecott Copper Corp.	Bingham, Utah	Copper ore
4.	Lucky Friday	Coeur d'Alene, Idaho	Silver ore
5.	Bunker Hill	Idaho	lead, zinc-Lead ore
6.	Anaconda	Butte, Montana	Copper, silver ore
7.	Idarado	Colorado	Copper, lead ore

adjacent Polaris mine is similar to the Sunshine mine.

Silver was also won from lead and zinc ores of the adjacent disseminated replacements in the Bunker Hill and Sullivan, Federal, and Hecla mines.

Comstock Lode, Nevada. The famous Comstock Lode was the scene of early western mining and since 1859 has procued over $700 million in silver and gold (1:40 ratio) from a group of mines scattered along its 5 kilometers of length. The bonanza period was in the 1870s; since then production has declined steadily although the change in the price of gold has brought some rejuvenation with cyanide leaching of old dumps.

The Comstock Lode (Fig. 18-17) is a fault of 900 meters throw, separating hanging wall Tertiary and older volcanic rocks from foot wall Mesozoic rocks. It branches upward and toward the bottom. The dip is about 40°; the width attains a few hundred meters, and it has been followed 900 meters vertically although the bonanza ore ceased at 600 meters. The wall rocks have been intensely propylitized and sericitized during mineralization probably occurring during Miocene time.

Most of the ore occurred in small bonanza ore shoots and in hanging wall branches or chambers and consisted of crushed quartz, some calcite and minor sulfides, with gold, electrum, argentite, polybasite, and other silver sulfo-salts, which have replaced early quartz, pyrite, sphalerite, and galena. A fine-grained later quartz is present. Bastin recognized supergene silver, argentite, and polybasite above the 500 level.

At 900 meters, the lode was flooded with hot calcium sulfate waters of 76°C temperature. The lower-level rock temperatures reached 45°C, suggesting an underlying mass of uncooled igneous rock.

In 1865, the Nevada legislature granted Adolph Sutro the right to drive a tunnel from an elevation of 1,366 meters, with the portal located about 6 kilometers east of Virginia City, for the purpose of draining the hot water in the Comstock mines and for providing cool air. Sutro, after enormous financial and operational problems, completed the tunnel in 1878 when several shafts were as much as 450 meters deeper than the tunnel level. Of course, Sutro's tunnel was a financial disaster because it was completed as the bonanza Comstock lode was nearing exhaustion.

High in the Washoe Mountains, however, Virginia City still exists with some of the original mansions, bars, and stores, all for the summer tourist trade.

Tonopah, Nevada. The Tonopah district, the last large epithermal silver district discovered in Nevada and barely in the twentieth century, has since 1900 produced over $150 million from ores with a gold-silver ratio of 1:100. Nolan, in his noteworthy study, recognizes a Tertiary sequence as follows: older rhyolitic flows, breccias, tuffs, and water-laid sediments of volcanic materials, called the Tonopah formation, followed by the Mizpah trachyte, andesitic flows, both of which interfinger. Next came an intrusion of the Extension breccia, in the western part of the district, between the Tonopah and Mizpah formations. The same contact was followed by a sill of West End rhyolite. These formations were broken first along the Halifax fault zone and then by the remarkable mineralized compound Tonopah fault, convex eastward in plan and convex upward in section, and with a total throw of 150 to 450 meters.

The mineralization was probably contemporaneous with the faulting. The first stage was wholesale albitization of the andesite, followed by quartz-adularia-sericite. The metallic minerals are electrum, pyrite, argentite, pyrargyrite, polybasite, and minor chalcopyrite, galena, blende, selenium minerals, scheelite and huebnerite with "chalcedonic" quartz, barite, and rhodochrosite. Supergene cerargyrite, iodyrite, and embolite occurred in the upper levels. The ore bodies are replacement veins along faults.

The chief control of mineralization was the Tonopah fault and its branches; locally other faults served as ore channels. The ore deposition was limited to a relatively thin shell that domes upward in the central part of the district, and the upper ore boundaries transgress both faults and formation boundaries. Nolan believes that the coincidence of the high point of the shell with the zone of intense rock alteration may indicate isogeotherms and that the shell represents the temperature range within

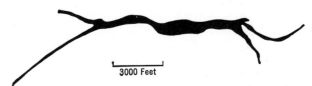

Figure 18-17 Plan of the Comstock Lode, Nevada. (After Becker, U.S. Geol. Survey.)

3000 Feet

which deposition could take place; this is further suggested by the gold-silver ratio.

The principal mines are the Tonopah, Tonopah Belmont, Tonopah Extension, and West End, whose ores averaged more than $100 per ton prior to 1904, around $15 between 1911 and 1930, and around $10 per ton or less thereafter. Recent deep hole exploration occurred in the district with financing by the late Howard Hughes.

Tintic Mining District, Utah. The total value of the entire Tintic district output is over $425 million, of which one-half has been silver and the remainder copper, gold, lead, and zinc.

The rocks of the district consist of about 4,000 meters of miogeosynclinal Paleozoic sediments, unconformably overlain by volcanics and intruded by monzonites and quartz porphyry of Tertiary age. The sediments are folded into a broad, asymmetric, pitching syncline. The folding was accompanied by numerous transverse faults, some of which have horizontal displacements of 60 meters, and by many bedding faults. Major thrust faults have been mapped that have led to the discovery of unknown ore deposits. The faults have been important localizers of ore (Fig. 8-43). Some of the ore bodies are massive replacements of the crumpled and faulted middle limestone member of the Bluebill limestone Ophir limestone, Fish Haven dolomite, Deseret formation, and others, with production as shown in Fig. 18-18. Post-crumpling, northeast fissures have yielded successively monzonite, porphyry, pebble dikes, and ore solutions. The ore solutions ascended through trough conduits at critical intersections and formed massive replacement of the limestone (Fig. 18-19) that Billingsley states extends up through "vein roots, pipe, expanding funnel, and finger-tip extensions." Ore was mined down to the conduit on the 1400 level of the Tintic Standard mine, now defunct.

The Tintic district included several other important mines, such as the Chief Consolidated, Eureka, Centennial, Mammoth, Gemini, Iron Blossom, and North Lily. The Burgin and Trixic mines are the only operating mines at the present time. These deposits are replacements in limestone and are silver-lead and argentiferous copper, zinc, or gold ores.

During World War II, a major geological study of hydrothermal wall-rock alteration and ore deposits of the East Tintic mining district was initiated by the U.S. Geological Survey under the direction of T. S. Lovering. In 1954, the second project under the direction of H. T. Morris was initiated for the purpose of a geological study of the East Tintic Mountains and included mining districts. Even with the need for metal production during World War II, the Tintic district production of known ore bodies declined and, by 1952, all mining activity had ceased. At that time the total production of the Main East, and North, Tintic mining districts had reached $425 million with mineralization and production contained stratigraphically as shown in Fig. 18-18. More than 22 percent of this sum was paid in dividends.

Through the results and stimulus of Lovering, Morris, and others, the East Tintic district underwent a revival in 1958. The development of new insights provided by the geologic study of the Paleozoic sedimentary formations, covered by pre- and postore volcanics at the surface, and the extensive development and use of geochemical prospecting and hydrothermal alteration mapping led to a bold exploration project, a project that entailed the sinking of a 1040-foot shaft. The shaft sinking and lateral drifting were done by the Bear Creek Mining Company under the direction of the late William Burgin. The shaft was located a mile east of any known ore deposits. A major thrust fault was inferred (by Lovering and Morris) to dip under the East Tintic district and was intersected during the exploration drifting, which ultimately led to the discovery of a major lead zinc-silver ore deposit and development of the Burgin Mine (Fig. 18-20). This discovery has stimulated the search for other concealed ore bodies in the area resulting in some revitalization of the district.

Canada

Canada, primarily because of Kidd Creek, became the largest silver producing country in the world. Canada obtains its metal almost entirely as by-product silver. Some of the most important producers of silver are listed in Table 18-3.

Cobalt, Ontario. This unique silver camp of northern Ontario, now declining, has produced around 400 million ounces of silver since 1904. The ores were fabulously rich, the shipments of the first few years averaging over 1000 ounces per ton; low-grade ores ran 200 ounces per ton. The famous "silver sidewalk" of the La Rose was almost solid native silver for a length of 30 meters and to a depth of 18 meters yielded 658,000 ounces of silver. Great

AGE		UNIT	Thick–ness (Feet)	Approximate Gross PRODUCTION (Dollars)
Middle or Upper Eocene	Intrusive	Silver City Monzonite		
		Swansea Quartz Manzonite		12,000,000
Middle Eocene	Extrusive	Laguna Latrite Series		
		Packard Rhyolite Series		
Mississippian		Humbug	650	
		Deseret	875	46,500,000
		Madison upper member	500	
		Madison lower member	250	
Miss.–Dev.		Piñon Peak	160	43,000,000
Devonian		Victoria	350	
Dev.–Sil.		Bluebell	500	110,000,000
Ordovician		Fish Haven	280	10,000,000
		Opohonga	875	9,000,000
U. Cambrian		Ajax	640	60,500,000
		Opex	500	7,500,000
		Cole Canyon	600	7,000,000
		Blue Bird	180	
M. Cambrian		Herkimer	375	10,000,000
		Dagmar	90	
		Teutonic	390	
M.L. Cambrian		Ophir	440	84,500,000
L. Cambrian		Tintic	3000	25,000,000
TOTAL GROSS PRODUCTION – 425,000,000				

8155 feet

Figure 18-18 Distribution of the total production of the Main and East Tintic mining districts from various rock units.

slabs of almost pure silver were extracted; one famous specimen 2 meters long weighs 742 grams and contains 9715 ounces of silver. The deposits illustrate low-temperature hydrothermal fissure filling.

The rock formations consist of Keewatin greenstones overlain by Cobalt conglomerate (Huronian) and intruded by a tilted Keeweenawan diabase sill, 300 meters thick (Fig. 18-21). The productive veins occur in the conglomerate beneath the sill.

The deposits were steep joint or fault fissure

Table 18-3 Important Canadian Silver Producers.

Mine or company	Location	Nature of ore
1. Kidd Creek mine, Ecstall	Timmons district, Ontario	Zinc-copper-silver-tin
2. United Keno mine	Galena Hill, Yukon Territory	Silver-lead
3. Consolidated Mining and Smelting Co.	British Columbia	Lead-zinc
4. International Nickel Co.	Sudbury, Ontario	Nickel-copper
5. Hudson Bay Mining and Smelting Co.	Flin Flon, Manitoba	Zinc-copper
6. Noranda mine	Noranda, Quebec	Copper-gold

veins 10 to 13 centimeters wide, one hundred meters long, and mostly less than 90 meters deep. About 100 veins are known.

The ore consists of an unusual assemblage of minerals. The gangue is calcite or dolomite and may be absent. The chief ore mineral is native silver. Dyscrasite, argentite, tetrahedrite, stromeyerite, polybasite, stephanite, and ruby silver also occur. Bismuth, galena, pyrite, chalcopyrite, pyrrhotite, and sphalerite are known; and arsenopyrite is abundant. The striking feature of the mineralogy is the unique group of nickel and cobalt arsenides of which the following have been checked by Thomson:

1. Smaltite	$CoAs_2$	5. Rammelsbergite	$NiAs_2$	9. Cobaltite	$CoAsS$
2. Chloanthite	$NiAs_2$	6. Skutterudite	$CoAs_3$	10. Glaucodot (CoFe)	AsS
3. Löllingite	$FeAs_2$	7. Arsenopyrite	$FeAsS$	11. Niccolite	$NiAs$
4. Safflorite	$CoAs_2$	8. Gersdorffite	$NiAsS$	12. Breithauptite	$NiSb$

Of these, numbers 9, 4, 7, and 11 are the most abundant. Later R. J. Holmes discarded 1 and 2 as mineral species. Some of the minerals are isomorphous mixtures. Dendritic forms are common, the centers of which are commonly occupied by native silver or calcite. The general mineral succession, according to Thomson, was first the arsenides and sulfarsenides of nickel and cobalt, followed by the iron compounds, and then by the monoarsenides; later some silver, argentite, and bismuth; and lastly the rare sulfides and silver sulfarsenides. Surface oxidation products of ''cobalt bloom'' (erythrite) and ''nickel bloom'' (annabergite) lent pink and green colors to the outcrops.

The ores are cavity fillings with minor replacement. They have been formed by hydrothermal solutions, and the presence of bismuth shows that the last minerals were deposited below a temperature of 271°C. The solutions are considered to have sprung from the same reservoir that gave rise to the diabase intrusion. The Cobalt deposits have counterparts at Annaberg and Schneeberg in Saxony and the deposits of Gowganda, Ontario, are generally similar to those of Cobalt.

Great Bear Lake. Rich silver ores associated with uranium-radium ores occur in veins in Precambrian sediments in the Great Bear Lake district. The ores contain native silver, pitchblende, cobalt and nickel arsenides, copper sulfides, and bismuth, along with quartz and rhodochrosite.

Mexico

Mexico had long been the leading silver-producing country of the world, and camps may still be seen where silver mining has been almost continuous since the early 1500s. With the increasing production of other countries, notably, the United States, Canada, Peru, and USSR, Mexico competes with Peru for third place in silver production. Its annual production ranges from 30 to 40 million ounces at the present time. About 10 companies exceed 1 million ounces annually. The most important producers are listed in Table 18-4.

Fresnillo, Zacatecas. This famous mine, discovered in 1570 and worked almost continuously since then, is stated to have produced more than 200 million ounces of silver, chiefly from oxidized ores. Some lead, zinc, and copper are also obtained from deeper sulfide ores. Gold is present in both. The oxide ores average about 6 ounces of silver and 0.3

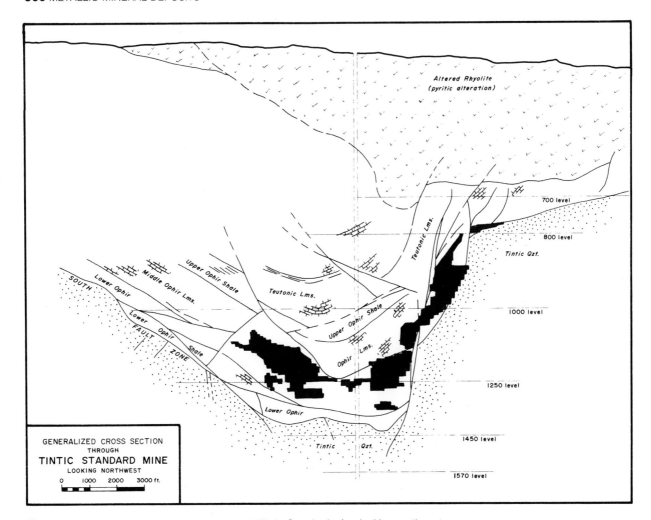

Figure 18-19 Generalized cross section through Tintic Standard mine looking northwest.

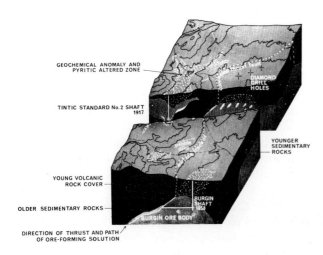

Figure 18-20 Sketch showing geology, ore deposits and Burgin mine of the East Tintic mining district. (From Minerals Day Collected Papers, Int. Min. and Chem. Logs, 1965.)

dwt gold. The oxide ore is cyanided; the sulfide ore is concentrated.

The deposits occur in slate, graywacke, and altered volcanics in two forms—fissure veins and stockworks. Main veins from below spray out into minor veins and veinlets above, forming a stockwork several hundred feet across (Fig. 18-22), that has yielded about 15 million tons of ore from an open cut. Seven main veins occur below and numerous small ones extend outward from the stockwork. The ores within a shallow depth display a vertical zoning with 2 to 6 percent copper ores beneath, grading upward into lead-zinc ores poor in copper, and into silver ores, free from base metals, above. The primary ore minerals are base-metal sulfides, quartz, and calcite.

El Potosi Mine, Santa Eulalia, Chihuahua. The El

Table 18-4 Leading Mexican Silver Producers

Mine or district	Location	Nature of ore
1. Fresnillo Co.	Fresnillo, Zacatecas	Oxidized and sulfide silver ore
2. El Potosi	Santa Eulalia, Chihuahua	Silver-lead-zinc
3. San Francisco	San Francisco del Oro, Chihuahua	Silver-lead-zinc
4. Inversiones Co.	El Oro, Zacatecas	Silver ore
5. Pachuca	Pachuca, Hidalgo	Silver ore

Potosi mine, famed since 1703 as a rich silver-lead producer, but has had no significant production for 30 years. Its 500 kilometers of workings disclose mantos, chimneys, irregular replacements, and fissure veins, all in gently dipping Cretaceous limestone. The Santa Eulalia district is credited with a metal production of over $600 million from 800 kilometers of underground workings.

The limestones have been folded into a gentle anticline along which the ore bodies are localized. Rhyolite flows and tuffs of premineral age cap the limestones, which are cut by numerous dikes and sills and by fractures that have helped localize the ores.

The deposits are massive replacements of limestone along fracture zones in several favorable beds (Fig. 18-23). The ore bodies may be mantos following along single beds, or chimneys that cut across several beds and connect with mantos. The almost horizontal mantos attain lengths of 1000 or more meters and cross sections from 100 to 5000 square meters. The great "P" chimney is 310 meters high and 1100 square meters in cross section. It throws off mantos into upper and lower beds. Other bodies are fissure replacements that extend as flanges into favorable beds. Limestone is metamorphosed by selective replacement to dolomite. The mantos ore bodies end abruptly against adjacent carbonate beds.

The ore is of three classes: (1) high-grade sulfide mill ore; (2) low-grade sulfide silicate ore; and (3) oxidized ores, now largely exhausted. The massive sulfide ore consists of carbonate and base-metal sulfide with minor silver minerals, carrying 9 to 10 ounces of silver and 9 to 10 percent each of lead and zinc. The oxidized counterpart of this carries 15 ounces of silver, 13 percent lead, and 2.5 percent zinc. The unusual silicate ore consists of quartz, silicified limestone, ilvaite, actinolite, hedenbergite, fayalite, magnetite, and hematite, with minor base-metal sulfides and a silver content of 20 ounces in the unoxidized and 33 ounces in the oxidized portions. Zinc and lead are low. Oxidation is complete to the tenth level, partial to the thirteenth, and absent below that. The reserves are running low.

These interesting ores indicate both intermediate and fairly high-temperature conditions of formation combined in one mine. The silicate ores overlap, grade into, and lie above the sulfide ores, and there are no indications of separate periods of formation.

San Francisco del Oro, Chihuahua. These deposits

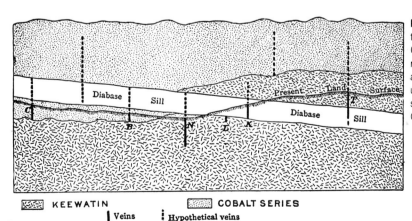

Figure 18-21 Generalized section through productive part of Cobalt area, Ontario, showing the relation of veins to diabase sill and overlying Cobalt series and underlying Keewatin greenstones. (From Knight and Miller, Ont. Dept. Mines.)

KEEWATIN   COBALT SERIES

Veins   Hypothetical veins

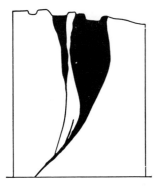

Figure 18-22 Cross section of South Branch and Catillas stockwork ore bodies, Fresnillo, Zacatecas, Mexico. (Livingston, mine records.)

lie northwest of the Parral district and have been highly productive in precious and base metals from strong fissure veins. There are nine veins, of which the San Francisco is the most important. The ore averages about 6 to 7 ounces of silver, 7 percent lead, 9 percent zinc, and a little gold.

Pachuca, Hidalgo. This famous district, including the Real del Monte, since its discovery in 1534 has been an almost continuous producer of silver, being credited with over one billion troy ounces. Filled fissure veins, crustified and brecciated, occupy faults in Lower Cretaceous sediments and Tertiary extrusives and intrusives. Wisser states that after intrusion, warping, gentle folding, and intrusion, a great series of northwest fault-fracturing followed

These northwest fault fractures were cut by a late north-south system closely associated with mineralization. Hulin states the first deposition was barren quartz, then later, rhodonite, bustamite, base-metal sulfides, and lastly silver minerals, chiefly argentite and minor polybasite and stephanite; gold occurs in traces. The andesite country rock is widely propylitized and the vein walls are chloritized and silicified. The veins are 2 to 5 meters wide and mostly less than 600 meters in depth. The ore is in structurally controlled ore shoots affected by fault movements contemporaneous with metallization. Silver deposition took place during reopening. Zoning is indicated; the silver zone is overlain by a barren zone up to 400 meters thick, which in places is entirely removed by erosion. Oxidation extends to 700 meters, and supergene sulfide enrichment is absent. The veins are hydrothermal fissure fillings of epithermal type.

Guanajuato. This great district since 1548 has yielded more than a half billion dollars in silver. According to Wandke and Martinez, pre-Cretaceous shales, sandstones, and conglomerate are cut by rhyolite, andesite, granite porphyry, and monzonite and are overlain by later volcanics. Faulting then occurred on a large scale. One mineralized fault, the Veta Madre vein, extends 25 kilometers. The faults became metallized and the resulting veins have been dislocated by postore faulting.

The ore bodies are of three types: (1) the great

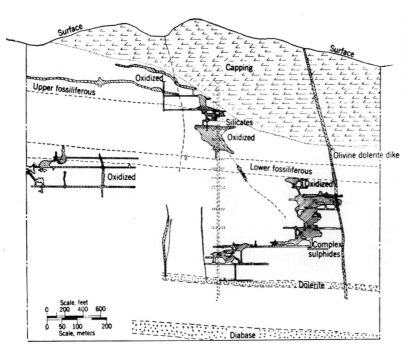

Figure 18-23 Cross section of part of the ore bodies of the Potosi mine, Santa Eulalia, Mexico, showing mantos and pipes in limestone (white). The ore is shaded. (From Walker, U.S. Bur. Mines.)

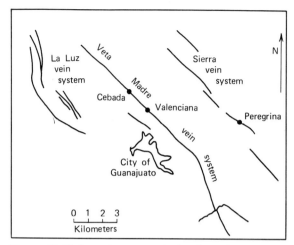

Figure 18-24 Ore vein systems in the Guanajuato mining district (modified from Wandke and Martinez, 1928; Detruck and Owens, Econ. Geol.)

Veta Madre vein, (2) smaller fault-fissure veins, and (3) stockworks.

The famed Veta Madre (mother lode) vein is a rotational fault with an average displacement of 800 meters, a maximum width of 20 meters, and an average width of 6 to 8 meters. About 5 kilometers of its 25-kilometer length contain most of the ore bodies. The ore is concentrated in shoots 200 to 400 meters long, where enlargements have resulted from fault motion (Fig. 18-24). Other large fault fissures have been traced for 1½ to 7 kilometers. The vein matter is in part open space deposition and in part replacement of brecciated rock that filled the fissure.

The stockworks lie in the hanging walls of fault fissures where their downward movement over a hump in the dip induced shattering. They are most common along the Veta Madre, where they form large ore bodies 75 meters wide, 200 meters long, and 500 meters deep (Fig. 18-25).

The ores carry silver and gold in the ratio of 100:1. The ore minerals of silver sulfides, selenides, and electrum are disseminated or in bands in crustified quartz-carbonate gangue and consist of numerous silver sulfides and antimonides, base-metal sulfides, along with quartz, carbonates, and adularia, with some rare fluorite, barite, and zeolites. Oxidation has extended only 20 to 30 meters and the ore minerals are characteristically hypogene and typical of epithermal fissure veins.

Parral-Santa Barbara, Chihuahua. These two camps, 20 kilometers apart, are of interest because of silver ores in similar geological environment but

in contrasting mineralogy. The deposits are filled fissures, and the ores are dominantly silver with minor gold and low copper; lead and zinc are present in both areas. At Parral, Schmitt recognizes quartz, chalcedony, fluorite, garnet, and calcite containing blende, galena, argentite, proustite, chalcopyrite, pyrite, and specularite. These are classed by Lindgren as epithermal deposits (note specularite and garnet). Considerable silver enrichment has occurred. Santa Barbara, in addition, has closely associated high-temperature ores classed by Lindgren as hypothermal. They are both siliceous and massive sulfides. Accompanying the silver, lead, zinc, and copper minerals, however, are surprising gangue minerals, including, according to Schmitt, quartz, fluorite, garnet, pyroxene, orthoclase, epidote, and zoisite and also ilvaite and fayalite. These minerals denote high temperature. According to Barry and Schmitt, the high-temperature phase cuts the low-temperature phase, and vice versa; they are both phases of the same mineralization in the same mineral district.

Other Deposits. Rich silver ores have been mined from the many veins (Fig. 18-26) of Zacatecas since 1548, and numerous other camps have produced tens to hundreds of millions of dollars in silver.

South America

The former glory of South America as a silver region has diminished until now South America produces only about 25 percent of the world's silver. More than half of this amount is derived from the copper ores of Cerro de Pasco (Chapter 19) in Peru which produced 15 percent of the world's silver in 1972 making this mine one of the world's largest producers of silver. Of the total South American production, Peru now yields 75, Bolivia 10, and Chile 5 percent. Many of the famous districts of the past are now largely exhausted.

Potosi, Bolivia. This district, famed as the richest silver hill on earth, is said to have produced well over 2.5 billion ounces of silver since its discovery in 1544. It is located in a tin-silver belt that extends for several hundred miles through the south western part of Bolivia. Potosi's conical volcanic peak rising 760 meters above its surroundings to an elevation of 4900 meters consists of a core of a granitic intrusion surrounded by Pliocene sediments and tuffs that gently dip outward (Fig. 18-27). Numerous steep veins cut the mountain both in rhyolite and

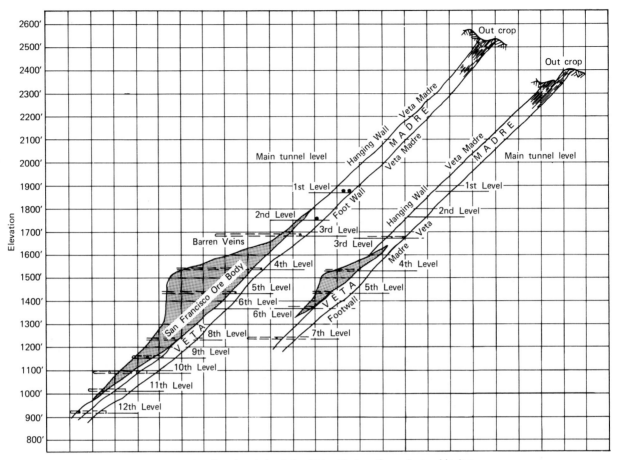

Figure 18-25 Hanging wall ore bodies (stockworks) of the Veta Madre, Guanajuato, Mexico. (Wandke, Econ. Geol.)

sediments. The veins are sharp, well-defined, and frozen, and the vein filling is banded and drusy. The walls are bleached and altered and in places shattered and impregnatged by ore. The veins (Fig. 18-28) branch upward, are closely spaced, and intersecting, so that most of the mountain is mineralized. They range in thickness up to 4 meters. The veins are open-space fillings and replacements of sheeted zones. Oxidation extended to 300 meters, and to this depth the ores were very rich, many above 100 ounces silver per ton. These ores, now largely exhausted, contained cerargyrite, native silver, and some argentite and ruby silver. No enrichment in tin had been noted.

The primary ores average 10 ounces silver and 1 to 4 percent tin. They are fine-grained aggregates consisting of quartz, alunite, cassiterite, stannite, base-metal sulfides, and silver minerals. The silver minerals consist of argentiferous tetrahedrite, andorite, ruby silver, matildite, and jamesonite. Chalcocite and covellite are supergene; and oxidation

products described by Lindgren and Creveling are jarosite, halloysite, hematite, goethite, sideronatronite, and voltaite.

The ores deposited at shallow depths at temperatures ranging from 200 to in excess of 400°C suggest a classical xenothermal deposit. In northern Bolivia, in contrast, the deposits are classified as mesothermal and hypothermal.

Huanchaca, Bolivia. This district produces 4 to 5 million ounces of silver annually. The Pulacayo for some time has been the largest silver-producing company of Bolivia. The deposits are veins occurring in a mountain composed of red shale and conglomerate cut by a central mass of trachytic porphyry surrounded by granite porphyry. The veins are known for a length of 2350 meters and a depth of 750 meters. They range in width from 1 meter to 4 meters and branch upward.

The veins are crustified fissures with marginal crusts of quartz and pyrite, followed by bands of

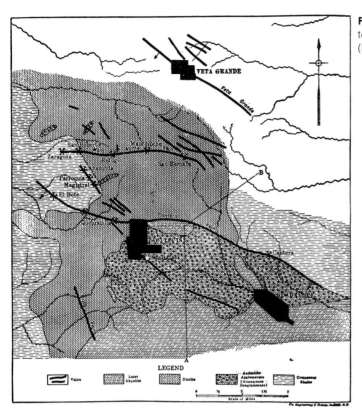

Figure 18-26 Veins of the Zacatecas silver district, Mexico. (Bastin-Botsford, Econ. Geol.)

pure pyrite, sphalerite and tetrahedrite, galena-chalcopyrite-quartz, and in the middle by blende, tetrahedrite, bournonite, and boulangerite. Rare ruby silver, stibnite, wurtzite, jamesonite, and bismuth occur. The galena and tetrahedrite are high in silver. Galena decreases, and blende and pyrite increase, with depth. The rich tetrahedrite shoots carry 180 to 600 ounces of silver per ton. The average ore runs 50 ounces of silver, 5 percent copper, and 5 percent zinc.

Chocaya, Bolivia. This district contains complex silver-tin ores and is now the second largest silver producer of Bolivia. The deposits are filled fissures

up to 1 meter wide carrying base-metal sulfides, tetrahedrite, ruby silver, and cassiterite in quartz, barite, and calcite gangue. They are shallow-seated cavity fillings. Similar deposits occur in the old Oruro and Colquiri districts in Bolivia.

Colquijirca, Peru. This mine yields ores averaging 60, 40, and 20 ounces of silver, respectively, from three mantos in folded limestone beds. Lindgren states that three beds have been replaced to form three mantos, separated by a few meters of shale. The upper manto (5 meters) contains rich silver-copper ores. Underneath this is a manto higher in lead, and a lower pyritic copper manto with enargite

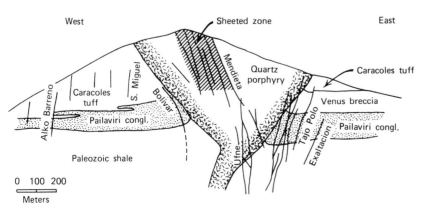

Figure 18-27 Cross section of Cerro Rico de Potosi showing rock types and main vein systems. (Evans, Econ. Geol.)

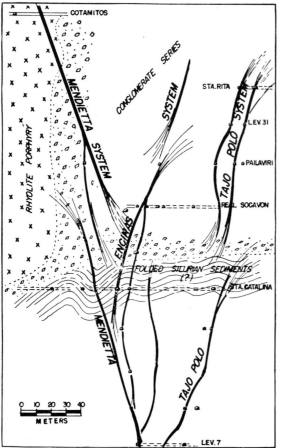

Figure 18-28 Typical section of Potosi facing north, showing echelon character, vein relationships, and upward branching. (Evans, Econ. Geol.)

lies nearer a monzonite intrusion. In places the three merge. The different minerals in them are noteworthy.

The silver manto consists of chert, barite, dolomite, tennantite, stromeyerite, pyrite and some wittichenite, galena, pearcite, sternbergite, galena, and sphalerite; enargite is rare. Magnificent specimens of supergene native silver occurred in the upper zone. The mineralization closed with a late specularite-marcasite phase. The richness of the ores has made this single mine one of the most productive of the world.

McKinstry considers that cross fracturing helped localize the ores and that the mineralogical difference of the mantos is small. The general mineral sequence was (1) rock alteration; (2) early sulfides: pyrite, marcasite, blende, tennantite, enargite, and galena; (3) later botryoidal sulfides: blende, galena, pyrite; (4) stromeyerite; (5) specularite-marcasite; and (6) supergene native silver. Here is a peculiar intermingling of mesothermal and epithermal min-

eralization. Most of the deposition is distinctly low-temperature, and telescoping is absent. McKinstry suggested that there is a zonal "gradation northward from the intrusive from copper to silver-lead mineralization."

Other Districts. Other smaller silver mines are scattered throughout Peru and Bolivia and also in northern Chile, such as the camps of Oruro, Colquiri, Colquechaca, and Negrillos in Bolivia; those of Morococha in Peru; and Chañarcillo, Caracoles, Huantajaya, and Tres Puntos in Chile. In the Argentine, which ranks third in South American production, most of the silver is a by-product of lead-zinc and copper deposits, such as those of Famatina, Iglesia, Hoyada, Los Cobres, and Aguilar.

Australia

Australia is the sixth ranking country of the world in silver production, New South Wales accounting for 65 percent of it, Queensland 25 percent, and Tasmania 8 percent. Most of the silver is by-product metal from lead, zinc, and copper ores. In the following ranking properties the ore is the lead-zinc-silver type; all the properties are located near Broken Hill, New South Wales, except Mount Isa, located in Queensland: (1) Mount Isa, (2) North Broken Hill, (3) Broken Hill South, (4) Sulfide Corporation, and (5) Zinc Corporation. Since these deposits are dominantly lead-zinc, and silver is secondary, they will be described under "lead" and "zinc."

Selected References on Silver
General Reference 10. *Brief references to silver deposits.*
General Reference 19. *Chapter on Silver.*
Alluvial Prospecting and Mining. 1960. S. V. Griffith. Pergamon Press, Elmsford, N.Y., p. 245. *Recent updating in placer operations.*
The Main Tintic mining district, Utah. 1967. H. T. Morris. General Reference 11. *Results of district study, especially since 1942.*
Geology and ore deposits of the East Tintic mining district, Utah. 1967. W. M. Shepard, H. T. Morris and D. K. Cook. General Reference 11. *Detailed paper on ore deposits, and successful exploration.*
Tonopah district, Nevada. 1930. T. B. Nolan. Nevada Univ. Bull. 24, pt. 4, p. 35, *A fine illustration of careful field work, and structural and geothermal ore localization.*
Pachuca Silver district, Mexico. 1937. E. Wisser. AIME Tech. Publ. 753. *A good study of detailed structure.*

Guanajuato mining district, Mexico. 1928. A. Wandke and J. Martinez. Econ. Geol., v. 23, p. 1–44. *Details about the great Veta Madre vein.*

PLATINUM GROUP

Platinum may be considered one of the precious metals even though about 80 percent of that consumed in the United States is for industrial purposes. From 1828 to 1841 it was used for coinage in Russia until its value exceeded that of the coins. Its industrial use is increasing—in electrodes, thermometry, chemical catalysts, the manufacture of nitric acid, dental and medical devices, and in many other areas. If the platinum catalysts are used to control nitrogen oxide fumes from automobile exhausts, the consumption of platinum will "skyrocket." Because of its qualities, it is used as the prototype international standard of the meter length.

Platinum is only one of a group of related metals consisting of osmium, iridium, palladium, rhodium, and ruthenium. They are not only associated together but also are generally alloyed, and are called, therefore, the "platinum metals." They are very heavy, resistant to corrosion and most acids, melt at temperatures of 1549 to 2700° C, and range in hardness on Moh's scale from 4.8 to over 7. Iridium is the heaviest and osmium the hardest.

Platinum is a modern metal. During the past its production was small. At the present time 45 percent of it comes from Russia, 45 percent from South Africa, and 7 percent from Canada. World production is about 6,000,000 ounces. The United States produces less than one percent of the platinum metals that it consumes, but recovers secondary platinum.

Mineralogy, Treatment, and Uses
Most platinum used today is a by-product in the refining of other metals, chiefly nickel.

Ore Minerals. The chief mineral is the native metal, but platinum also occurs as *sperrylite* ($PtAs_2$) *cooperite* ($PtAsS$), *stibio-palladinite* (Pd_3Sb), and *braggite* ($PtPdNiS$). There are also arsenides of the other platinum metals. The platinum metals form natural alloys with one another, such as *osmiridum* or *platiniridium*, also with iron to form *ferroplatinum* (16 to 21 percent Fe) and *polyxene* (6 to 11 percent Fe), and with copper to form *curopla-tinum* (8 to 13 percent Cu). Native platinum is never pure platinum.

Treatment. By-product platinum is separated from the containing metal during electrolytic refining. Placer platinum is concentrated by delicate gravity concentration. Lode platinum is recovered by combined gravity and flotation concentration. The individual metals of the platinum group are separated by complex refining methods.

Associations. Platinum is almost invariably associated with ultramafic and mafic rocks and with the ore minerals characteristic of those rocks. Most of the platinum of the world is intimately associated either with chromite or nickel sulfide. Even platinum placers are derived from mafic rocks rich in chromite. The platiniferous nickel ores also contain copper and appreciable quantities of gold and silver.

Uses. The chief uses for platinum are in the electrical, chemical, and petroleum industries. Its white color and hardness make it a desirable setting for diamonds. In the electrical industry it is used for resistors and contacts in the more delicate instruments, such as telephones, TV, electrodes, and radios. Platinum finds wide uses in the chemical industry, such as for containers, wire, electrodes, coils, acid making, X-ray equipment, and as a catalyst. Large acid stills utilize considerable quantities.

The major primary platinum producers are Russia and South Africa.

Kinds of Deposits, Origin and Distribution
The classes of deposits that yield platinum metals, and some examples, are

Magmatic concentrations:
1. Early magmatic:
   (a) Disseminations—sparse disseminations with chromite in dunites, the erosion of which yields placers—Urals, Alaska, Colombia, and Stillwater Complex, Montana.
   (b) Segregations, by fractional crystallization—*Merensky Reef,* Rustenburg, South Africa.
2. Late magmatic:
   (a) Immiscible liquid segregations—Vlackfontein, South Africa.
   (b) Immiscible liquid injections—possibly Frood mine, Sudbury, Ontario
Contact-metasomatic deposits—Tweetfontein and Potgietersrust, South Africa.
Hydrothermal—Waterburg, South Africa; Medicine Bow Mountains, Wyoming.
Placer deposits—Urals; Colombia; Goodnews Bay Alaska.

Origin. The home of platinum is in ultramafic igneous rocks, where it has been concentrated by magmatic processes. The erosion of disseminated magmatic concentrates has yielded the placers of the Urals, Colombia, and Alaska. Richer magmatic concentrations have formed lode deposits in South Africa. The by-product platinum from the nickel and copper ores of Sudbury, Ontario, may have been formed by hydrothermal processes.

Distribution. Large placer deposits were found in South America by the Spanish conquest, but little mining was done. There are three main centers of platinum in the world, namely, the Soviet Union; Sudbury, Ontario; and the Bushveld of South Africa. Less important localities are the Choco district of Colombia and Goodnews Bay, Alaska. More than 30 ultramafic complexes have been found mainly in the Alaskan panhandle. Similar deposits occur in the Urals, South Central British Columbia, and Venezuela. Platinum was obtained as a by-product from Alaska, Oregon, and California gold placers. Minor production, less than 1 percent, comes from Ethiopia, Sierra Leone, Shaba in Zaire, New South Wales, Victoria, and Tasmania. Small amounts have also been found in Papua, India, Indonesia, New Zealand, and the Philippines. A potentially significant deposit is being studied once again in the Stillwater Complex, Montana.

Examples of Deposits

Sudbury, Ontario. The main source of platinum in Canada is the Sudbury district, where it is concentrated to an unusual degree in the massive copper-nickel sulfide bodies. Although small amounts of by-product platinum had formerly been obtained from the other nickel-copper mines, it was not until the rich Frood ores were discovered that Canada became an important producer of platinum. The platinum content is stated to average 0.66 to 0.9 ppm.

With the advent of the electron probe, platinum metals are being found in numerous Sudbury mines by L. Cabri and J. Gilles LaFamure. By this technique they can detect as little as 0.6 ppm Pt. A large number of platinum group mineals have been identified occurring in numerous different rocks and minerals and in many different deposits.

The South Range and offset deposits are characterized by differences in their Pd/Pt ratios as the result of the different mineral assemblages and the different geochemistry of the deposits. Those deposits containing platinum metal group minerals are Strathcona; Copper Cliff, South and North; Cream Hill, Creighton, Frood, Vermillion, Coleman, Levack West, and others. Many of the deposits contain platinum group metals in solid solution in other minerals.

In the Strathcona deposit, platinum ore occurs in two environments, namely, in the main ore zone, which is a leucocratic sublayer breccia below the norite; and in a group of subparallel sulfide lenses intruded into basement gneisses up to 360 meters below the mafic norite of the eruptive (Fig. 18-29). Similar offset deposits exist in the footwall below the iruptive in the Coleman ore body (Fig. 18-30).

Russia. For 100 years after the discovery of platinum in the Ural Mountains in 1819, Russia was the leading and almost the only source of platinum. The metal was almost entirely mined from placers, but primary deposits in mafic rocks are now being worked.

At *Kola Peninsula*, in European Russia, where the *Monchegorski* property was developed prior to World War II, and at *Norilsk*, in Siberia, in the *Talnakhski* and *Octiabrski* ore bodies, developed in 1942, large reserves of complex nickel sulfide ores occur containing platinum group metals. Oxide nickel ores have also been found in the Southern Urals near *Orsk* and in northwestern *Kazakhstan* and *Akhtiubinsk* containing commercial quantities of platinum group metals and cobalt. These deposits average about 0.7 percent nickel, 1 percent copper, and about 11 grams per ton of platinum group metals, principally platinum and pladium.

Another nickel-platinum producer is the *Pechenga*-nickel deposit that was formerly known as *Petsamo* before being annexed from Finland at the end of World War II. When Inco operated Petsamo, the ore was rich containing 3.8 percent nickel, 1.8 percent copper, and 800 grams per metric ton platinum metal. The *Hituri* nickel-copper deposit in the Kotalahti nickel belt contains platinum metals that average 0.027 grams per ton Pt and 0.048 grams per ton Pd in the ultramafic body.

Most of the Russian placer deposits of platinum occur in the northerly 500 kilometers of the Ural Mountains, where small oval-shaped masses of dunite contain magmatic concentrations of platinum in irregular pockets, veins, and bands of chromite. The dunite is richer than the pyroxenite.

South Africa. There are two main groups in South Africa; namely: magmatic concentrations and con-

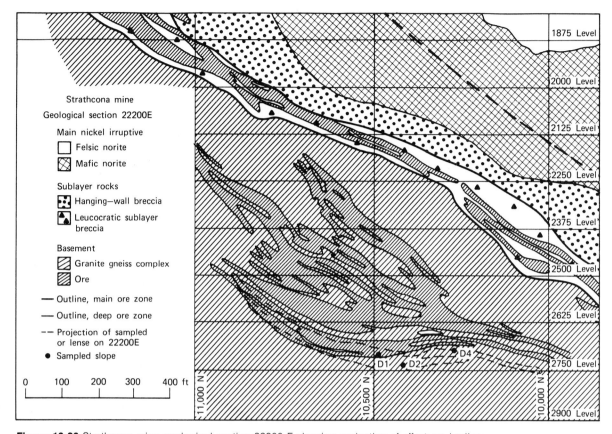

Strathcona mine
Geological section 22200E

Main nickel irruptive
☐ Felsic norite
▨ Mafic norite

Sublayer rocks
▨ Hanging—wall breccia
▲ Leucocratic sublayer breccia

Basement
▨ Granite gneiss complex
▨ Ore

— Outline, main ore zone
— Outline, deep ore zone
-- Projection of sampled or lense on 22200E
● Sampled slope

0   100   200   300   400 ft

Figure 18-29 Strathcona mine geological section 22200 E showing projection of offset ore bodies in basement rock. (Section after Cowan, 1968, with minor modifications in Econ. Geol., v. 71, 1976.)

tact-metasomatic deposits. The former consist of magmatic concentrations with (1) hortonolite dunite, in pipes, (2) chromitite, and (3) nickel-copper sulfides.

The *hortonolite-dunite pipe deposits* are unique. The rock itself is the heaviest silicate rock known and the richest in ferrous iron. It consist of hortonolite (olivine) with subordinate phlogopite, hornblende, diallage, ilmenite, chromite, and magnetite, and is a differentiate of the Bushveld igneous complex. Some 60 occurrences are known in the lower part of the norite zone; only 3 are of commercial importance, and these are all pipes. They are carrot-shaped bodies, up to 18 meters in diameter and as much as 300 meters deep. The bodies, which transgress the pseudostratification of the norite, consist of a central zone of platinum-bearing hortonolite dunite with local segregations of chromite, surrounded by envelopes of olivine dunite, and pyroxenite, grading outward into norite (Fig. 18-31). The central zone only is mined. The platinum content varies considerably, assays up to 1200 dwt being obtained, the mining averages ranging be-

tween 4 and 20. At the Onverwacht pipe, Wagner states that the upper levels averaged 20 dwt, the 250-foot level 18.4 dwt, and the 750 level 9.5 dwt; the Mooihoek pipe averaged between 6 and 7 dwt. These peculiar ultrabasic pipes, crossing the stratification, with the most basic part in the center, and intrusive into the norite, suggest that the most basic portions remained fluid longest and are an enigma; possibly they may be pegmatitic.

The *chromitite segregations* are in the lower part of the differentiated norite zone and occur within the *Merensky Horizon* or "Reef"—a zone that has been traced scores of kilometers. In places the platinum is sufficiently concentrated to be workable, as at Rustenberg, Potgietersrust, and Lydenburg. At Rustenburg the zone dips gently, conforming with the pseudostratification of the differentiation zones in the B.I.C. Bushveld igneous complex. The platinum layer is about 30 centimeters thick and averages 10 to 12 dwt platinum. This lies directly and sharply on anorthositic norite and is overlain by bands of pyroxenite, spotted norite, mottled anorthosite, and again by anorthositic norite, which in

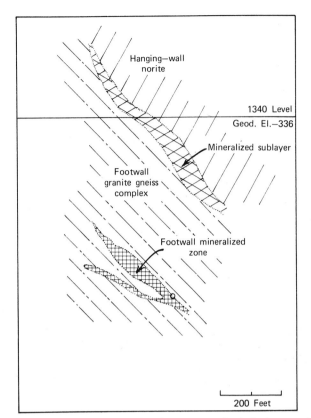

Figure 18-30 Cross section, Coleman ore body in basement footwall. (From Econ. Geol., v. 71, 1972.)

turn underlies the Bastard Reef 12 meters above the main reef. The ore consists of unusual pyroxenite or harzburgite with chromite, native platinum, magnetite, and specks of pyrrhotite, pentlandite, chalcopyrite, cubanite, millerite, and nickelferous pyrite. Considerable palladium accompanies the platinum, and minor quantities of rhodium, osmiridium, gold, and silver.

The *magmatic nickel-copper sulfide deposits* of platinum ore occur at Vlackfontein, Rustenburg. They are isolated masses of irregular shape composed of disseminated or massive sulfides in bronzitite. The ore minerals are pyrrhotite, pentlandite, and chalcopyrite. Only the oxidized portions are worked for platinum.

The *contact-metasomatic* deposits occur northwest of Potgietersrust, where Bushveld rocks rest on dolomite and ironstones. The dolomite has been silicated to a diopside-grossularite rock impregnated by platinum-bearing yellow sulfides, with unusually large crystals of sperrylite. Cooperite is also present, and the ores carry nickel and copper. The platinum tenor is 7 dwt.

Other platinum occurrences in South Africa include some pegmatitic bodies, minor placers, osmiridium in the Rand conglomerates, and the unusual Waterbug deposits, where platinum occurs in a vuggy quartz vein carrying quartz, chalcedony, specularite, pyrite, and chrome mica. It is presumably a shallow-seated, high-temperature hydrothermal deposit.

United States. The former small platinum production of the United States came from by-product refining of gold and copper ores, and mainly from Goodnews Bay, Alaska, where high-grade placers were worked in nonglaciated stream deposits near two periodotite intrusions. The best gravels lie within a meter of bedrock, and chromitic magnetite and ilmenite occur with the platinum. All six platinum metals are present, platinum constituting 68 to 75 percent, iridium 6 to 13 percent, and palladium is low.

Johns-Manville Corp. is developing promising platinum-palladium deposits in the west fork of the Stillwater River. In 23 sample locations, assays of

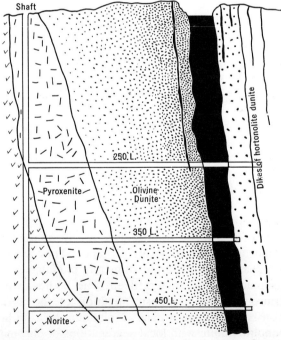

Figure 18-31 Mooihoek platinum pipe, South Africa. Center, hortonolite dunite with chrome, carrying platinum, surrounded by olivine dunite, then by pyroxenite, all enclosed in country rock of norite, repeated on other side. (Based upon Wagner, modified by sketches made by A. Bateman at the mine.)

platinum and palladium average 0.5 ounces per ton. The mineralized surface zone is 1000 meters long and over 300 meters in a vertical distance.

Australia. With the discovery of nickel sulfide ores in 1966 in Western Australia, the occurrence of platinum group metals has been detected in numerous locations in the Yilgarn Block (Fig. 20-16) of Archean greenstone belts in the Precambrian shield of Western Australia. The precious metal deposits range from low-grade disseminated sulfides (typically pentlandite, pyrite, millerite, and heazlewoodite) in intrusive ultramafic bodies in the high-grade massive or matrix-type pyrrhotite and pentlandite deposits located at the base of volcanic-ultramafic units.

Keays and Davidson report that the "typical" nickel sulfide ore samples from Western Australia with 3.3 percent nickel contain 360 ppb palladium, 110 ppb gold, 220 ppb ruthenium, 110 ppb osmium, and 60 ppb iridium. Its platinum content is probably less than 200 ppb.

Selected References on Platinum

Platinum Deposits and Mines of South Africa. P. A. Wagner. Oliver and Boyd, Edinburgh, 1929. *The pipes and "reefs" in the Bushveld igneous complex.*

General Reference 1. 1975. *Vol. 71, No. 7 issue is devoted entirely to platinum-group elements and deposits.*

General Reference 19. *Excellent chapter on platinum-group metals.*

Economic Geology of the Platinum Metals. 1969. J. B. Mertic, Jr. U.S. Geol. Surv. Prof. Papers 630. *A comprehensive survey of all platinum deposits. Extensive bibliography.*

Strategic Mineral Supplies. 1939. G. A. Roush. McGraw-Hill Book Company, New York. *Occurrence, distribution, and technology of platinum.*

Platinum at work in 1942. E. M. Wise. Min. and Met. v. 23, p. 421–425. *Chiefly properties and uses.*

Some aspects of the genesis of platinum deposits. 1962. E. F. Stumpfl. Econ. Geol., v. 57, p. 619.

Geochemistry of the platinum metals. 1965. Wright and Fleischer, U.S. Geol. Surv. Bull. 1214A.

The Soviet Challenge in Base Metals. 1971. A. Sutulov. The University of Utah Printing Services, Salt Lake City, 195 p.

19

THE NONFERROUS METALS

When we practiced gunboat diplomacy, we really didn't need it. Now when we need it, we can't use it.

Forbes Raw Materials,
August 1971

COPPER

Production

The tremendous growth in the use of copper is indicated by the fact that of the total world production of copper during the last 100 years, about 80 percent was mined in the last 25 years and more than one half of it in the last dozen years (Fig. 19-1).

Annual world production ranges around 7.5 million metric tons of metallic copper. The chief producing countries and their approximate percentage of current world production follows: United States 18, Chile 14, Zambia 11, Canada 10, USSR 10, Zaire 6, Peru 4, Australia 3, South Africa 2, and Japan 2. The remaining production comes chiefly from the Philippines, China, Mexico, Yugoslavia, Southwest Africa, Bulgaria, and Finland.

From early times until 1800 copper was widely produced in small quantities. From 1801 to 1810 the annual world production was only 18,200 tons, equivalent to less than 1 month's production of many present-day mines. England was the world's leading producer until 1850, when Chile assumed first place, held until 1883. Since then the United States has been the world leader.

In the Western Hemisphere copper was mined in Chile before the arrival of the Conquistadores, but the first Spanish mining was in 1601 and the first modern smelting in 1842. Although copper was discovered in the United States in 1632, none was worked until 1705, when the Simsbury mine in Connecticut was opened. Similar ores were worked in the Schuyler mine, New Jersey, 1719, and at Gap, Pennsylvania, in 1732. Considerable copper was mined in 1800 by the Spaniards from what is now the Chino mine, at Santa Rita, New Mexico. Records show that the Lake Superior copper was known in 1771, but the first real mining did not take place until 1830. The chalcocite ores of Bristol, Connecticut, where large chalcocite crystals occur, were discovered in 1836, and workings reached a depth of 80 meters by 1853. Mining started at Ducktown, Tennessee, in 1843. By 1874 the mines of Butte and Bisbee became prominent and by 1904 the "porphyry copper" era began in the United States and Chile. Subsequently, large production has come from Canada, Russia, and central Africa, especially the Zambian copper belt formerly known as The Rhodesian copper belt. Numerous new large deposits are being developed in Panama, South America, Canada (especially New Brunswick and British Columbia), the United States, and Australia including the west and south Pacific areas.

The order of production in the United States is Arizona, Utah, New Mexico, and Montana. Production in 1979 in the U.S. was 1,943,000 metric tons.

Distribution

Although copper is widely distributed, 80 percent of the world supply emanates from six regions, the southwestern United States, the USSR, the Andean belt, the central African belt, and the Canadian Shield. A sixth area of the Phillipines and the southwest Pacific area is increasing production rapidly.

In *North America* one of the greatest concentrations of copper in the world centers in Arizona and the Cordilleran parts of the United States, Canada, and Mexico, and includes all the well-known North American "porphyry coppers" and a host of other famous districts. All the ores are associated with felsic types of intrusions. A compact, productive Montana province centers around a granodiorite massif at Butte. Other provinces include the Appalachian, the fruitful Lake Superior district, and

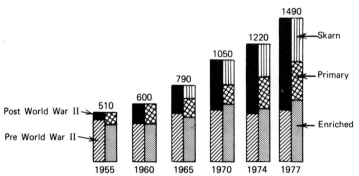

Figure 19-1 Southwestern U.S. porphyry copper production by start up and type of mineralization shown in thousands of short tons of metal. (From Skilling, Mg. Rev., Oct. 11, 1975.)

the sprawling Cascadian-Coast Range belt extending from the Yukon Territories through northern British Columbia to the state of Washington. A lone but rich center was at Kennecott, Alaska, which is now defunct. Figure 19-2 indicates the location of many North American and South American porphyry copper deposits.

The *Canadian Shield* from Manitoba to New Brunswick includes the Hudson Bay, Sudbury, Noranda, Heath Steele, Kidd Creek, and other deposits, with a large number of recently discovered deposits in New Brunswick.

The *Andean belt* (Fig. 19-3) from Chile to Panama includes the large Chuquicamata, Braden, Potrerillos, El Salvador, Cerro Colorado, Rio Blanca, Toquepala, and Cerro de Pasco deposits, and many more, also associated with monzonitic intrusives.

The *central African* province constitutes the most concentrated copper belt in the world and includes the prolific mines of Zambia (Fig. 19-4) and adjacent Zaire. They are bedded stratigraphically controlled deposits.

Exploration in the *islands adjacent to Australia,* namely, New Zealand, Fiji, New Hebrides, Buganville, British Solomon Island protectorate, the Territory of Papua, New Guinea, and West Irian has resulted in the discovery of several new porphyry copper deposits that form a new copper province. Some of the deposits that are just beginning production are Penguna, Ok Tedi, Frieda, and the high-grade deposit of Carstenz.

Other belts are the *Uralian province* of Russia, the outer Japanese Island arc, and isolated European centers in Spain-Portugal (Rio Tinto), Bor in Yugoslavia, Mansfeld in Germany, Outokumpu in Finland, and Boliden in Sweden. *Australia* includes several copper centers such as Mount Lyell, Mount Morgan, Mount Isa, Cobar, Tennant Creek, and Mount Oxide.

The exploration for copper deposits has resulted recently in the discovery of new large deposits in

known copper province areas and in somewhat unsuspected areas such as British Columbia, Panama, New Guinea, Fiji, New Idria, Brazil, Puerto Rico, New Brunswick, Afghanistan, Phillipines, Solomon Islands, and other areas. Exploration in Brazil, especially in the northwest, has been accelerated by the Brazilian government road building to improve access to this remote areas (Fig. 20-21). Extensive exploration in British Columbia has resulted in many new discoveries such as in the Highland Valley area exemplified by the large Lornex open pit copper-molybdenum deposit. Another large copper district in central British Columbia is under development after a 10-year search by Falconbridge. Unfortunately, both the federal and provincial governments of Canada have adopted or are considering more restrictive mineral land rights, taxation, and securities regulations, which has resulted in a curtailment of mineral exploration, especially in British Columbia.

Three of the largest discoveries made during the past decade include the Cerro Colorado porphyry copper deposit in Panama, where 2,200,000,000 tons of 0.8 percent copper has been recognized with reserves still being developed.

The Ecstall deposit known as the Kidd Creek mine is one of the fabulous discoveries of this century. It was located under a thin layer of glacial moraine in the northern portion of the well-known Porcupine-Timmons mining district in Canada. Even though the deposit was discovered as recently as 1959, the deposit now provides more than two-thirds of the income to the parent company, Texasgulf Inc.

Near the center of the Arizona porphyry copper province, Anaconda is producing copper from the Twin Buttes deposit that required the removal of at least 625 feet of Quaternary alluvium before the mineralized bedrock was reached. The discovery of this deposit resulted from geophysical studies of an extensive pediment that led to the finding of the

Figure 19-2 Map showing the distribution of porphyry deposits in North and South America. (After Paul Eimon.)

Pima, Mission, and Esperanza deposits. The Twin Buttes deposit is also located on the east flank pediment of the Sierrita Mountains.

Mineralogy, Tenor, Treatment, and Uses

The copper of commerce occurs in several forms and in a variety of minerals. Mineralogically, copper ores are divided into four large groups, namely, native, sulfide, oxidized, and complex. Each requires a different metallurgical treatment to obtain the copper, and different ore tenor to form economic deposits.

The sulfide ores are the most valuable; the complex ores contain copper admixed with lead, zinc, gold, and silver minerals.

Mineralogy. About 165 copper minerals are known, but the chief economic ones are

Native: Mineral	Composition	Percent of Copper
Native copper	Cu	100.0
Sulfides:		
Chalcopyrite	$CuFeS_2$	34.5
Bornite	Cu_5FeS_4	63.3
Chalcocite	Cu_2S	79.8
Covellite	CuS	66.4
Enargite	Cu_3AsS_4	48.3
Tetrahedrite	$Cu_8Sb_2S_7$	52.1
Tennantite	$Cu_8As_2S_7$	57.0
Oxidized:		
Cuprite	Cu_2O	88.8
Tenorite	CuO	79.8
Malachite	$CuCO_3Cu(OH)_2$	57.3
Azurite	$2CuCO_3Cu(OH)_2$	55.1
Chrysocolla	$CuSiO_3 \cdot 2H_2O \pm$	$36.0\pm$
Antlerite	$Cu_3SO_4(OH)_4$	54.0
Brochantite	$Cu_4SO_4(OH)_6$	56.2
Atacamite	$CuCl_2 \cdot 3Cu(OH)_2$	59.4

The chief gangue minerals of copper ores are rock matrix, quartz, calcite, dolomite, siderite, rhodochrosite, barite, and zeolite. Contact-metasomatic silicates may also be present in ores of that type. Most copper ores are free from other associations, except for gold, silver, and molybdenum. A few zinc, lead, and silver ores may carry copper as a by-product, and in complex ores copper may also be associated with zinc and lead. In general, copper sulfide ores are associated with intrusions of quartz monzonite and related rocks, less commonly with mafic intrusives.

Tenor and Treatment. The type of copper ore and its tenor generally determine the method of treatment. The lowest-grade ores are the simple and easily treated native copper deposits, which may run as low as 0.4 percent. Sulfide ores run as low as 0.55 percent or less for the lowest-grade ores; high-grade ores may range from 5 to 30 percent. Most oxidized ores range from 0.4 to 10 percent, the lowest mined by acid leach processes. Ores carrying 4 percent or more copper are generally smelted directly to avoid concentration losses; low-grade ores of about 0.6 percent copper are either leached or concentrated and the concentrates are smelted. Such ores are smelted with siliceous or base ores, flux, and coke, and the matte is refined in a converter to blister copper. This may be sold as "blister," or fire-refined, but generally it is electrolytically refined to remove impurities and to recover the precious metals.

Low-grade sulfide ores are floated, and the concentrates are smelted. Selective froth flotation permits the separation of undesired pyrite or pyrrhotite and of zinc and lead minerals. Concentrates range from 10 to 40 percent copper. About 95 percent of all copper ores mined in the United States are concentrated.

Oxidized ores are smelted directly or are leached. Specific solvent solutions or acid leaching is used for carbonate ores in nonreacting gangue. Water is used for leaching sulfate and chloride ores and for "heap leaching" sulfide ore dumps, sometimes with the aid of bacteria (Fig. 1-24) and ore in place. The copper is precipitated from such liquors as "cement" copper or recovered by electrolysis or by replacement of scrap iron.

Uses of Copper. Copper is one of the very essential minerals in modern industry. It is normally a prosperity metal, that is, used when electrical expansion takes place, but it is also an essential metal of war. The United States, the world's largest consumer, uses between 2 and 2.5 million tons annually, as listed in Table 19-1.

Most wires and electrical equipment are made of pure copper, and considerable alloy copper is used, chiefly as brass and bronze. The brasses are copper-zinc alloys (55 to 99 percent Cu), and the bronzes are copper-tin-zinc (88 percent Cu, 10 percent Sn, 2 percent Zn). There are also nickel, aluminum, and steel alloys of copper; minor special alloys utilize arsenic, beryllium, cadmium, chromium, cobalt, iron, lead, magnesium, manganese, and silicon.

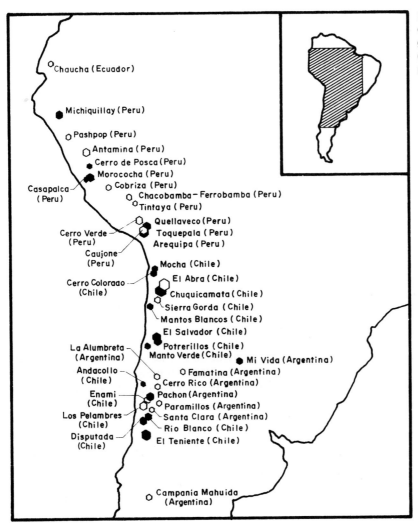

Figure 19-3 The main porphyry copper deposits of the Andes region, following Eimon's map (1974). (From J. Kulina, Lithospheric Plate Motions)

O < 1,000,000 tons Cu   ⬡ 1,000,000 to 10,000,000 tons Cu   ⬡ > 10,000,000 tons Cu

● IN OPERATION OR EXHUSTED RESERVES   ⬡ ESTIMATED RESERVES

Electrolytic copper is the form normally distributed, but considerable fire-refined copper is used. Low-grade blister is commonly utilized for making copper sulfate. Considerable scrap copper, or copper and brass cuttings and reclaimed metal, is used along with primary copper; the proportion may reach 0.8 pound scrap to 1 pound of primary copper.

The world reserves of copper are large, and as economies in production are achieved, the reserves will increase even with the rapid increase of glass fiber bundles substituting for copper in transmission use.

Kinds of Deposits and Origin. The principal classes of copper deposits and some examples of each are

1. Magmatic: Copper-nickel deposits— *Insizwa*, South Africa; *Merensky Reef*, Transvaal; *Sudbury*, Ontario; *Thompson*, Manitoba; and the Duluth gabbro, Minnesota.
2. Contact-metasomatic: Older deposits of *Morenci*, Arizona; *Bingham*, Utah; and *Cananea*, Mexico.
3. Hydrothermal:
   A. Cavity Filling:
      (1) Fissure veins—*Butte*, Montana; and *Walker*, California.
      (2) Breccia filling—*Nacozari*, Mexico; *Bisbee*, Arizona; and *Braden*, Chile.

Table 19-1 Uses of Copper in the United States

Use	Percent of total	Use	Percent of total
Electrical manufacture	24.6	TV and radios	5.7
Automobiles	13.2	Refrigerators and air conditioning	2.3
Miscellaneous wires	11.8	Ammunition	1.7
Light and power lines	9.6	Manufacture for export	4.2
Buildings	8.2	Other uses	14.1
Telephone and telegraph	4.6		

(3) Cave fillings—*Bisbee*, Arizona.
(4) Pore-space fillings—Urals, Russia.
(5) Vesicular fillings—Keweenawan, *Lake Superior* district.

B. Replacement:
   (1) Massive—*Bisbee, United Verde*, Arizona; *Bingham, Tintic*, Utah; *Ducktown*, Tennessee; *Noranda*, Quebec; *Flin Flon*, Manitoba; *Granby*, British Columbia; *Rio Tinto*, Spain; *Boliden*, Sweden; *Cerro de Pasco*, Peru; *Outokumpu*, Finland.
   (2) Lode—*Kennecott*, Alaska; *Magma*, Globe, *Bisbee*, Arizona; *Butte*, Montana; *Britannia*, British Columbia.

(3) Disseminated—The "porphyry coppers": *Bingham*, Utah; *Ely*, Nevada; *Ray, Miami, Twin Buttes, Inspiration, Ajo*, and *Clay* ore body, *Morenci*, Arizona; *Santa Rita*, New Mexico; *Branden, Chuquicamata, El Salvador*, and *Potrerillos*, Chile.

4. Sedimentary: *Kupferschiefer*, Mansfeld, Germany; *"red-beds"* copper deposits.

5. Bacteriogenic: *"red-beds"* copper deposits; *Mount Isa*, Queensland; *Zambian copper belt*; *Sullivan* British Columbia; and Kuperschiefer.

6. Submarine exhalative: *Kidd Creek*, Ontario; *Sullivan*, British Columbia; *Kuroko* deposits, Japan; and *Red Sea* hot brines.

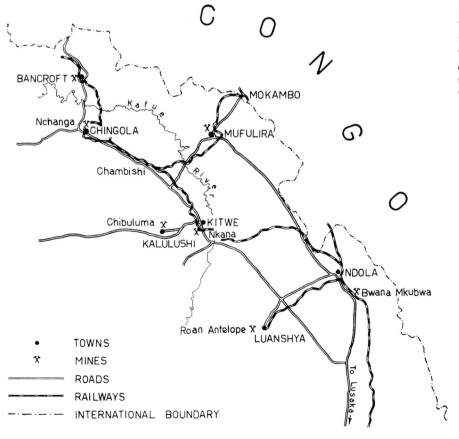

Figure 19-4 The Copper belt: Towns, mines, and some important physical features. (Modified from *Geology of the Northern Rhodesian Copperbelt*, Mendelsohn, ed., 1961, Macdonald & Co., London.)

7. Surficial oxidation enrichment: *Bisbee* and *Globe,* Arizona; *Chiquicamata,* Chile; and *Shaba,* Zaire.
8. Supergene sulfide enrichment: Some "porphyry coppers"; Cananea porphyry, Mexico; *United Verde Extension,* Arizona.

Origin. Most copper deposits have been formed by hydrothermal solutions, with replacement dominant over cavity filling. Contact metasomatism accounts for a few. Some of the sedimentary bacteriogenic and submarine volcanic copper deposits are controversial. Most copper deposits in unglaciated regions have undergone oxidation and some supergene enrichment. Many deposits have been almost completely converted to oxidized compounds, and there are some deposits whose economic importance is attributable solely to supergene enrichment. It is thus evident that copper deposits have originated by diverse processes, but most of them are either the direct result of hydrothermal activity, submarine exhalative, bacteriogenic, or of weathering processes.

The "Porphyry Coppers"

Mining of "porphyry copper" deposits of the western-southwestern province of the United States sprang into prominence between 1905 and 1910 and now constitutes the backbone of American copper production. One mine, the Utah Copper mine, at Bingham, Utah, is the greatest copper-producing mine of the world with a gross production in excess of $15 billion. Some others or the group are shown in Table 19-2.

All these deposits have similar characteristics: They are of low grade and are operated on a large scale by comparatively low-cost methods; they are associated with stocklike intrusions of monzonitic porphyries; and they are disseminated replacements and veinlets in porphyry, volcanics, or intruded schists. Some are of blanket shape, with greater horizontal than vertical dimensions while others are pipelike in shape. Their primary mineralogy is generally similar and is accompanied by hydrothermal alteration of the host rocks; many were overlain by leached cappings and have been subjected to less supergene sulfide enrichment than originally thought; and all have similar modes of origin. The differences between them are in details of host rock, shape, size, tenor, oxidation, and degree of supergene enrichment. They are mined either by large-scale, open-cut methods (Figs. 1-18 and 1-19) or underground caving, and the ores are readily amenable to flotation, with high ratios of concentration. Some of their common features are summarized in Table 19-3. Jerome has succinctly summarized many generalizations of porphyry copper deposits in Fig. 19-5.

Intrusives. The porphyry coppers are closely associated with intrusions of monzonite, quartz monzonite, or diorite porphyry of Mesozoic or Tertiary age. The New Guinea Highland deposits are mid to late Miocene with one deposit believed to be Pleistocene in age. The Ruth, Nevada, intrusive is Mesozoic in age. These are stocks or chonoliths of about a thousand meters across and in irregular shape, which have entered along major tectonic

Table 19-2 Some Porphyry Copper Deposits of the United States

Mine	Copper ore tenor average percent	Copper hypogene tenor only average percent	Total ore tonnage × 10⁶	Copper produced (pounds × 10⁶)	Host. rock composition		Average depth of capping (meters)
					Gr Rhy	Qm Grd, etc.	
Ajo	0.75	0.75	245	4,500		X	15
Bingham	0.65	0.65	>1,000	17,000		X	30
Esperanza	0.51	0.30	<100	600			
Inspiration	0.90	0.1–1.2	<500	2,027	X		80
Morenci	0.88	0.1–0.15	>500	1,760	X		35
Ray	0.8	0.1–0.8	130	3,000		X	80
Ruth	1.1	—	225	5,200		X	35
San Manuel	0.75	0.75	20	—		X	200
Santa Rita	0.97	0.1–0.2	230	—		X	35
Silver Bell	0.75	0.3–0.4	55	865	X	X	

Table 19-3 Composite Model of Commercial Porphyry-Copper Deposits of the Cordilleran Belt

Geometry of intrusive complex: total area of intrusive: 336.4 acres
length of perimeter of intrusive: 2.8 miles
area of porphyritic phase: 61.6 acres
area of breccia and diatrems: 28.3 acres
Contact alteration zone: width: 1300 feet
Geometry of ore body:[a] area of horizontal section: 122.4 acres
length: 3240 feet
width: 1740 feet
thickness; 678 feet
tonnage: 275.3 million short tons
Portion of tonnage of ore body in wallrock: 46.2%
Portion of tonnage of ore body in intrusive: 53.8%
Economic characteristics of ore body:[b] overall grade: $9.08 per short ton
copper grade: 0.67%
molybdenite grade: 0.029%
Pb, Zn, Ag, Au values: $1.38 per ton
total gross value: 2,451 million dollars
average mining rate: 21,000 tpd
average stripping ratio: 0.8/1
total capital investment required to
reach the production stage: $123 million
Geometric of alteration zone: total area: 484.5 acres
area of K-biotite alterations: 142.3 acres
area of pyritization: 413.2 acres

[a] Based on 0.35% Cu cutoff or its MoS_2 equivalent.
[b] Based on $0.50/lb price for copper and $1.70/lb price for MoS_2.

breaks. All are bared by erosion with the exception of the Sierrita deposits, Arizona, and El Salvador. The ore occurs in their cupola or outer crackled parts or in the intruded rocks. The host rocks have been metasomized and hydrothermally altered during the early prephase of mineralization.

The outlines of the abundant feldspars in the intrusive rock are commonly indistinguishable, and the rock is bleached to a creamy color. Superimposed upon this hypogene alteration is a supergene kaolinization, which extends as deep as the supergene sulfides.

With the resurgence of the subject of plate tectonics, the origin of porphyry copper intrusions is better understood. These intrusions are believed to be triggered in the upper mantle by underriding plates along continental or island arc margins as illustrated in Table 19-4.

Character of the Deposits. The deposits are huge blankets roughly parallel to the topography and extend downward in irregular protuberances. Their tops are generally smoothly undulating and are sharply demarked from the overlying leached capping, marking present or former positions of the water level. The capping ranges in depth from 15 to over 130 meters, and the thickness of the deposits is measured in scores of meters, more than a hundred meters. Areally, they range from a few hundred to 3000 meters across.

The bodies are a combination of disseminations and stockworks in shattered porphyry and invaded rocks. Closely spaced veinlets of quartz and sulfides ramify in all directions; and discrete grains of sulfides, in places hardly visible to the eye, give a pepper-and-salt effect to the altered host rock. The lower boundary of the ore is gradational, merging into primary yellow sulfides. The total sulfide content amounts to 5 to 18 percent of the rock, and the copper minerals around 1 or 2 percent.

Mineralogy. The mineralogy is extremely simple. The primary sulfides consist of pyrite, chalcopyrite, and bornite with minor sphalerite and molybdenite. In the capping these sulfides are largely or wholly removed, leaving voids occupied by limonite of diagnostic colors and patterns. Below the capping the yellow sulfides are coated, or partially or wholly replaced by chalcocite and covellite. The molybdenite is unaffected and is obtained in considerable quantity. At Utah Copper the percentage distribution of ore minerals is chalcopyrite, 80; chalcocite,

Geophysical Expression

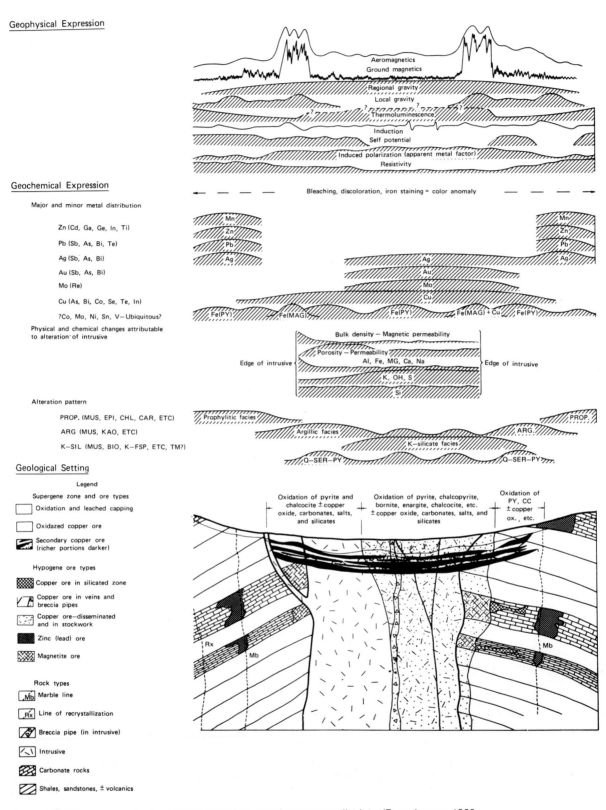

Geochemical Expression

Major and minor metal distribution

Zn (Cd, Ga, Ge, In, Ti)

Pb (Sb, As, Bi, Te)

Ag (Sb, As, Bi)

Au (Sb, As, Bi)

Mo (Re)

Cu (As, Bi, Co, Se, Te, In)

?Co, Mo, Ni, Sn, V—Ubiquitous?

Physical and chemical changes attributable
to alteration of intrusive

Alteration pattern

PROP. (MUS, EPI, CHL, CAR, ETC)

ARG (MUS, KAO, ETC)

K–SIL (MUS, BIO, K–FSP, ETC, TM?)

Geological Setting

Legend

Supergene zone and ore types

☐ Oxidation and leached capping

☐ Oxidazed copper ore

▨ Secondary copper ore
(richer portions darker)

Hypogene ore types

▨ Copper ore in silicated zone

▱ Copper ore in veins and
breccia pipes

▱ Copper ore—disseminated
and in stockwork

■ Zinc (lead) ore

▨ Magnetite ore

Rock types

Mb Marble line

Rx Line of recrystallization

▱ Breccia pipe (in intrusive)

▱ Intrusive

▨ Carbonate rocks

▱ Shales, sandstones, ± volcanics

Figure 19-5 Some generalizations and speculations, porphyry copper districts. (From Jerome, 1966 *Geology of the Porphyry Copper Deposits*, Titley and Spencer, eds.)

Table 19-4 Characteristics of Large, Disseminated Copper Deposits in Various Environments

Island Arcs e.g., Phillippines, Solomons, New Britain	Continental Margins e.g., British Columbia, E. Australia	Complex Continental Areas e.g., Southwestern United States
Country rocks Andesites and sediments	Marginal basin and eugeosynclinal assemblages	Sediments and (locally) volcaniclastics (Bisbee Basin)
Age Upper Tertiary	Ordovician-Triassic	Pre-Cambrian-Cretaceous
Intrusions Diorite and quartz diorite	Quartz-diorite-quartz monzonite	Quartz monzonite
Age Miocene-Pliestocene	Silurian-Cretaceous	Mid Tertiary (excepting Bisbee—170 million years) Ruth—130 million years
Size Large stocks (4-5 km across) Minor intrusive phases Andesite-Dacite dykes	Large stocks and batholiths Granodiorite porphyry and dacite-quartz-feldspas porphyry dikes	Small stocks (1-2 km across) Quartz porphyry dikes and plugs
Alteration Potassic → propylitic → phyllic Superimposed Chlorite and biotite with ore	Potassic → propylitic → phyllic Superimposed All types with ore	Potassic → phyllic → propylitic Well-zoned ± argillic Potassic and phyllic types in ore zone

Source: By G. Ballantyne. (unpublished).

9; covellite, 7; and bornite, 4. Pyrite is in excess, especially in the ore base intrusive core.

Oxidation and Enrichment. All the porphyry coppers have undergone various amounts of oxidation and all except Ajo and Copper Mountain have been secondarily enriched. The capping is leached of most of its copper, except at Ajo, where the sulfide has been converted in situ to carbonate. At Miami-Inspiration, Bagdad, Ray, and San Manuel, the capping also carried some carbonate and oxide copper.

It was formerly thought that all the porphyry coppers were made commercial by supergene enrichment, but this is true of very few of them; the primary ore is of commercial grade in many places, for example, Ajo and Ely; even the shallow ore at Utah Copper was primary. Secondary enrichment is rather complete at Miami-Inspiration, Ray, and Chino but is lacking at Ajo.

Mode of Formation. These huge deposits result from the intrusion of plugs of monzonite porphyry along old faults or other lines of weakness. The upper and outer margins became shattered, probably due to adjustments or to shrinkage either from cooling or from the high vapor pressure of the late mineralization fluids. This shattered area and enclosing rock permitted penetration by pulsations of uprising hydrothermal solutions given off from the magma chamber that supplied the porphyry. First, widespread rock alteration occurred, which further increased the rock permeability. During the alteration some iron, magnesia, and soda were extracted, and potash and silica were added. The cracks became filled by quartz and sulfides, and little grains of sulfides replaced silicates in intercrack areas and penetrated in places into the adjacent invaded rocks, mineralizing them also. Thus a great volume of rock became metallized with sufficient copper to constitute ore. Subsequent deep erosion and weathering of the metallized portions released copper, giving rise to a zone of supergene sulfide enrichment.

Lowell and Guilbert have illustrated the concentric alteration and mineralization of the San Manuel and Kalamazoo intrusion (Fig. 19-6). The Kalamazoo deposit was dropped to depths exceeding 1 kilometer by the San Manuel normal fault and only recently discovered by 20 deep drill holes.

Individual Properties. The porphyry coppers together produce more than 80 percent of the total United States copper. Their similarities are greater than their differences even though they form in different environments (Fig. 19-4). A composite model of porphyry copper is given in Fig. 19-5.

Utah Copper is located southwest of Salt Lake City in the Oquirrh Mountains and is composed of a porphyry mass that intrudes Pennsylvanian and Permian rocks. Replacement deposits of copper, zinc, and silver-lead are zonally arranged in limestones outward from the intrusive. The disseminated ore is confined to the outer zone of altered porphyry and is an oval body 2,500 meters long, 1,700 meters wide, and 1,000 or more meters deep (Fig. 19-7A and B). The shape of the intrusive is unknown in depth. The present mining tenor is 0.65 percent copper. Molybdenite is separated from the concentrates. The ore is shot down in benches, loaded by electric shovels into electric trains, and is transported to the Garfield and Magna concentrating mills of 130,000 tons daily capacity. About 120,000 to 150,000 tons of waste are removed daily. Over 10 million tons of metallic copper have been mined; 212,300 tons of copper were produced in 1979.

At *Ruth, Nevada*, are numerous funnel-shaped monzonite porphyry intrusions, aligned east-west along a major fault system (Fig. 1-18). Several of these intrusions are metallized and four have been mined, namely, in the Ruth underground mine, the Liberty Pit mine, and the Kimberly Pit, which was the last property to be operated. The Ruth porphyry copper district is now closed down. The Ruth ore body is 500 by 400 meters across and 100 meters thick; the Liberty Pit is 1500 by 800 meters across and has provided 70 percent of the production. These two mines have produced about 225,000,000

tons of ore averaging 1.09 percent copper. The upper parts were well enriched; the lower parts were primary chalcopyrite ore. The mill handled 24,000 tons per day but had to reduce production because of high SO_2 emissions.

The *Chino mine* at Santa Rita, New Mexico, is an elliptical ore body 1,800 by 2,500 meters divided by a barren island in the northern part. It is worked as an open pit. The ore is predominantly chalcopyrite, but parts of the body consist of oxidized ores containing native copper, cuprite, and malachite.

At *Ray, Arizona*, Precambrian Pinal schist, porphyry, and diabase have been metallized over an area of 3 by 2 kilometers to form a blanket. Some lean protore has been built up to commercial grade by secondary enrichment, which is fairly complete.

At *Ajo, Arizona*, the ore in a body 1,500 by 2,500 meters in area is in monzonite porphyry and some intruded andesite-rhyolite flows. Deficiency in pyrite has prevented sulfide enrichment by pyrite replacement. The upper 55 feet are oxidized to carbonate without the loss of copper because of the lack of pyrite. The ore from this part of the deposit was treated by sulfuric acid leaching. The underlying primary ore, consisting of chalcopyrite, minor bornite, and trivial pyrite, extends to a maximum depth of 350 meters and an average depth of 140 meters. It is worked by shovel in an open pit.

The *Miami-Inspiration* deposits occur both in porphyry and in Pinal schist. The highly enriched ores occur as an inclined blanket whose upper end has been converted to carbonate at the surface; this blanket dips beneath a later cover of Gila conglomerate. The ore bodies are about 3 kilometers long and 500 meters across.

The *Globe-Miami* mining district has been mined since 1881, when the railroad was completed to Casa Grande, even though mineral locations were made as early as 1873. The Inspiration Consolidated Copper Co. has been the major mine and milling developer in the district and a pioneer in the development of copper leaching techniques, processes developed as early as 1926 when their better grade sulfide ore was almost depleted, but large reserves of oxide copper remained. Leaching by sulfuric acid-ferric sulfate with the copper from the solution electrolytically precipitated was the initial method, which has now been changed to the "Dual Process." Only a partial leach is now being made, as chalcopyrite occurs in varying quantities and is not leached.

Underground block caving was the mining

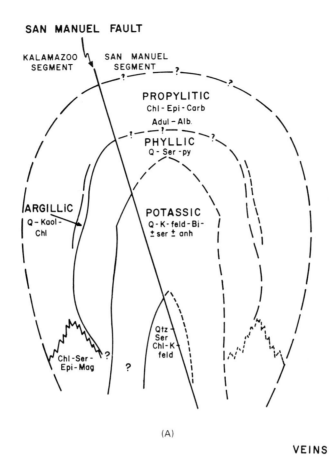

(A)

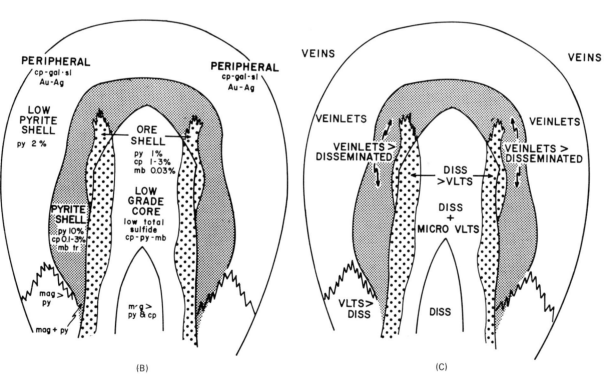

(B)   (C)

Figure 19-6 Concentric alteration-mineralization zones at San Manuel-Kalamazoo. (a) Schematic drawing of alteration zones. Broken lines on Kalamazoo side indicate uncertain continuity or location, and on San Manuel side extrapolation from Kalamazoo. (b) Schematic drawing of mineralization zones. (c) Schematic drawing of the occurrence of sulfides. (From Lowell and Guilbert, 1970, Econ. Geol., v. 65.)

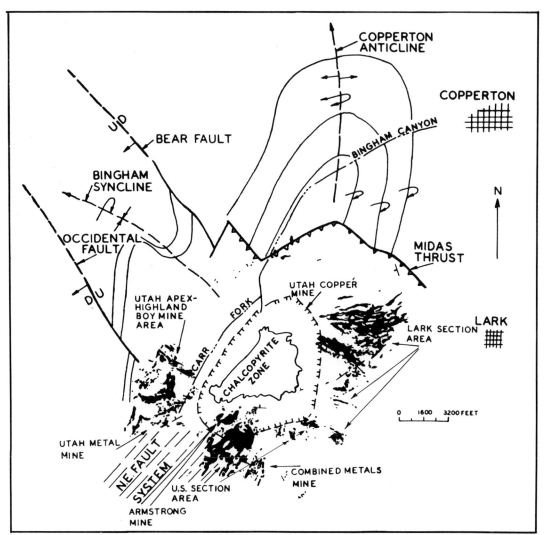

Figure 19-7a Principal structures, metal zones, and composite stoping of the Bingham mining district. Modified from 1961 Kennecott map. (From Smith, 1969, Graton-Sales Vol., p. 888.)

method used until 1961 when open-pit mining began and continues to the present day.

The *Miami Inspiration* deposits occur in the Pinal Schist, much of which is ore where it borders the mineralized Schultz granite (more correctly a porphyritic quartz monzonite) on the north side and it also constitutes a large part of the ore on the south side of the Liberty pit. The ore consists of a mixture of supergene and hypogene copper-bearing minerals.

In 1950, development of the Copper Cities Mine began. Equipment from the depleted, inverted, open-pit Castle Dome mine was used in the Copper Cities operation beginning in 1953. In fact, the savings in capital costs made possible by utilization of the Castle Dome plant and equipment was an im-

portant factor in the decision to mine the Copper Cities deposit.

More recently, based on extensive drilling, Miami Copper Company has reactivated the Castle Dome mine, now referred to as the Pinto Valley area. An open pit operation is now active in the area producing 40,000 tons per day of about 0.49 percent copper. The sulfide ore is milled at the mine and the concentrates are pumped over several miles as a slurry to the Inspiration Consolidated Copper Company smelter.

As the copper grade of Copper Cities is low, induced leaching operations were started in 1962. Owing to the presence of pyrite in the dumps, it is not necessary to add acid to the leach solutions.

A resurgence of discovery and the development

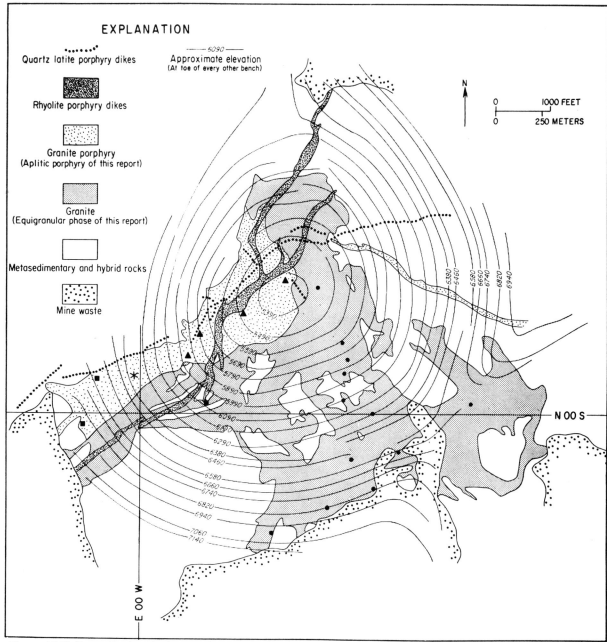

Figure 19-7b Generalized geologic map of composite Bingham stock. (Adapted from Smith, 1969; Moore and Nash, 1974, Econ. Geol., v. 69, p. 663.)

of new deposits in the Globe-Miami district has occurred during the past score of years. The Copper Cities deep pit, Diamond H pit, Inspiration oxide deposit and the Bluebird deposit are developments in which much of the copper is recovered by leaching. Some of these ore bodies have been located at greater depth than the overlying depleted deposits; others have been known of since the early history

of the district but were not thought to be economic until World War II time and since then.

The new Miami East copper mine was to be in production in 1977 but this has been postponed. This deposit was discovered by surface drilling in 1970 and contains ore reserves estimated at 50 million tons of 1.95 percent copper. To provide access to the ore body, the old Miami No. 5 shaft was extended from a depth of 1150 ft to 3500 ft although the initial ore contact is approximately 830 meters northeast of the No. 5 shaft. Dimensions of the sulfide deposit are about 1000 meters long, 500 meters wide, and averages about 75 meters thick. At its widest point, it measures 130 meters. The maximum dip is 34°.

The *Clay mine* at Morenci, Arizona, is an open-cut mine that contains more than 600,000,000 tons of 0.88 percent chalcocite ore lying beneath a capping of variable thickness. It yields 60,000 tons of ore per day, and 120,000 tons of copper a year.

The *San Manuel mine* contains some 460,000,000 tons of 0.75 percent copper ore with some molybdenum and minor gold and silver in quartz monzonite. The ore contains pyrite, chalcopyrite, and bornite with a chalcocite layer, and occurs as a blanket that is mostly buried by Gila conglomerate up to depths of 600 meters. The water table is 200 to 250 meters deep. The Kalamazoo ore body is in fault contact with the San Manuel deposit with portions of the ore body more than 2 kilometers deep.

Butte, Montana. Butte has changed a lot since the late John Gunther[1] described Montana's mineral capital in this way: "This is the toughest, bawdiest town in America . . . By night it has a certain inferno-like magnificence. By day it is one of the ugliest places I have ever seen." The gambling joints and bordellos that once lined Venus Alley have now disappeared. The Berkeley pit, 2,400 meters long, almost 2 kilometers wide, and 600 meters deep is eating into the town's business district. By 1985, Butte at its present site is a city that could simply cease to exist.

The famed district, hardly more than 3 by 6.5 kilometers in area and almost 2 kilometers deep, and with over 5000 kilometers of underground workings, has produced since 1879 over 3 billion dollars in copper with considerable silver, gold, zinc, lead, and some manganese. Only a few mines are now operated because the Berkeley pit, begun in 1955, has cut into many of the older underground mines and part of the town. Butte is called the richest hill on earth but it is greatly exceeded by Bingham, Potosi, and the Rand gold deposits in metallic wealth extracted.

The district lies near the exposed margin of the quartz monzonite Boulder batholith, which intrudes Cretaceous andesites. Barren aplite and quartz por-

phyry dikes cut the mineralized granodiorite, and are in turn cut by seven systems of fissures, and by later barren rhyolite dikes.

The Ore Deposits. The deposits are steeply dipping penecontemporaneous fissure veins of the first three systems, the oldest being largely lode replacements, and the later ones filled faults, all with generally similar fillings. The deposits range in widths from a few meters to a few tens of meters and are as much as 2,500 meters long. The recurrent and sequential faulting has formed a complex parquetry of vein or fault segments, cleverly resolved by C. Meyer and Reno Sales and their staff.

The older, or Anaconda, are east-west, nonfault tension fissures of southerly dip. They extend for 3,500 meters, are wide (up to 40 meters), deep (over 2,000 meters), rich, and fairly continuously mineralized and have been the great producers of the district. Eastward, where faults are numerous, they exhibit unique "horsetail" structure (Fig. 19-8). The individual "hairs" of the tails are also accompanied by disconnected en echelon segments. The result is an intensely fractured and metallized area. Well-known veins of the Anaconda system are the Anaconda-St. Lawrence-Original-Steward, Rainbow-Black Rock, State, Badger, and Mountain Con.

The Blue or Northwest veins are strike-slip faults of steep southwest dip that fault to the left. The movement is mainly horizontal and reaches 100 meters. The veins contain thick fault gouge; their walls are crushed; and they include drag ore of the Anaconda veins. Their ore is identical with that of the Anaconda system, except that it occurs in shoots. They are long and productive.

The Steward or Northeast fault veins dip steeply south and displace veins of the two earlier systems. They carry similar ores in minor quantity. They commonly strike-fault the East-West veins and are themselves faulted by other northeast faults.

These three systems contain identical ores and apparently all were formed during the period of metallization, which was dying out while the Steward system was being formed. Although each system displaces its predecessor veins, all may have been formed essentially contemporaneously or recurrent as a result of rotational stresses.

The subsequent systems are postmineral and contain only drag ore, which causes trouble by cutting and displacing members of all the preceding systems, forming disconnected vein segments.

[1] *Inside U.S.A.*, Harper & Row, Publishers, New York, 1951.

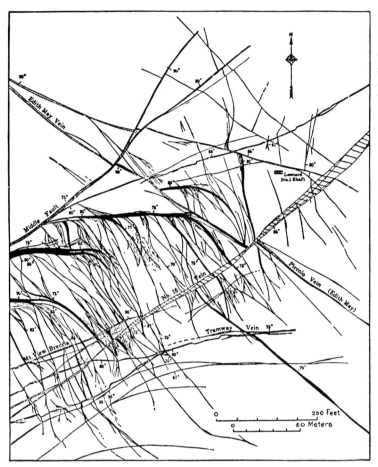

Figure 19-8 Part of 1200-level, Leonard mine (now defunct) Butte, Montana, showing "horsetail" (now open pit) structure in the Colusa-Leonard vein. Shows also, E-W, NW, and NE veins and fault relations. (Sales, AIME.)

Mineralization. The chief ore minerals are chalcocite, enargite, bornite, chalcopyrite, tennantite, tetrahedrite, and covellite. Pyrite and quartz dominate. Sphalerite is locally dominant; deep molybdenite occurs; galena is rare; huebnerite occurs as specimens; and "colusite" (tin-bearing tetrahedrite) is known. Rhodochrosite, calcite, rhodonite, barite, and fluorite also occur. Molybdenite and both steely and "sooty chalcocite" are present, and digenite and djureleite are locally abundant in the lower levels. About 70 minerals are known.

The metals and minerals are zonally arranged (Fig. 19-9 and Fig. 19-10) outward from the central zone. A single, long vein at Butte may contain minerals characteristic of each zone. The ores average 0.6 to 4 percent copper and 1 to 3 ounces silver; the central zone is the richest in copper. The mineralization of each vein system began with quartz-pyrite and ended with chalcocite-chalcopyrite.

The accompanying rock alteration is intense in the central zone, where little fresh granodiorite can be seen and diminishes in the intermediate zone to

bands paralleling the walls, and in the peripheral zone is much less. It is also less in the veins of the later systems than in the earlier. Sales and Meyer have shown that the rock alteration sequence is, from the veins outward, sericitic, argillic, (kaolinite

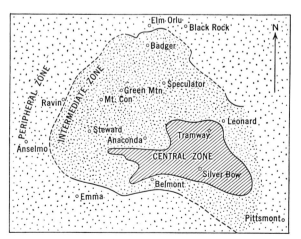

Figure 19-9 Zonal arrangement of ore minerals at Butte, Montana, at 4600-foot elevation. (After Sales, AIME.)

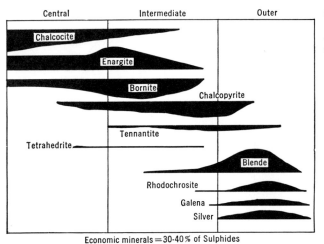

Economic minerals = 30-40 % of Sulphides

Figure 19-10 Zonal range of economic minerals at Butte, Montana. (After Hart, Copper Res. of World.)

next to sericite-quartz and montmorillonite farthest out), and fresh quartz monzonite (Fig. 19-11). This sequence is significantly the same for all veins, regardless of size or age, and indicates a continuous mineralization by the same solutions. In contrast, Lovering and Morris stress sequential stages, one after the other, of alteration fluids for the East Tintic district, Utah.

Oxidation and Enrichment. The depth of the oxidized zone averages about 100 meters. The copper and most of the silver have been completely removed from it, the galena is oxidized in situ, and the manganese carbonate has been altered to oxides, forming some residual deposits. A shallow zone of supergene enrichment, characterized by "sooty chalcocite," in places reaches a depth in fractures of 300 meters. Within this zone the pyrite of the wall rocks has been enriched to form wide

stopes. Some supergene silver enrichment has occurred in the peripheral zone.

Origin. The Butte deposits are replacements and fissure fillings formed by hydrothermal solutions emitted from the reservoir of the intrusive. These solutions must have continued through the first three periods of faulting. It is striking that the first minerals in veins of *each system* are quartz and pyrite, slightly preceded by wall-rock alteration, followed by copper minerals ending with "chalcocite," and not the earlier minerals in the first fissures and later minerals in the later fissures, as one might expect. This repetition of sequence in each of the later fissures suggests that the solutions underwent changes in composition while traversing the rocks. This is also shown by the sequence of rock alteration outward from each vein.

The central and intermediate zone mineralization

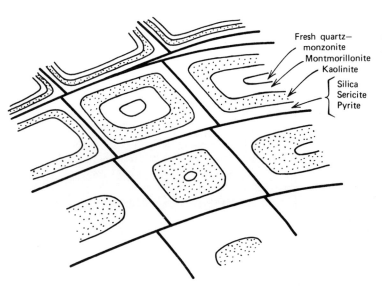

Fresh quartz—
monzonite
Montmorillonite
Kaolinite
{ Silica
Sericite
Pyrite

Figure 19-11 Idealized diagram of hydrothermal alteration minerals at Butte, Montana. (D. Garlick.)

would be classed as mesothermal; the peripheral as epithermal; thus one vein may contain mesothermal and epithermal ores.

Ducktown, Tennessee. The Ducktown, Tennessee, district produces metallic copper, iron oxide sinter, sulfuric acid, sulfur dioxide, and other related chemicals. Annual production is more than 1.5 million tons of sulfide ore.

The deposits consist of several sulfide replacement lenses in folded metamorphosed Precambrian schists and graywackes. The lenses are conformable with the host Ocoee metasediments.

The primary ore consists of pyrrhotite, pyrite, and chalcopyrite, with minor zincblende, bornite, specularite, magnetite, quartz, chlorite, and iron-magnesium-lime silicates; the copper content averages about 1.5 percent. The deposits outcrop as strong gossans, which were shipped for iron ore. A thin, rich, supergene sulfide zone from 3 to 8 feet thick and carrying 5 to 2 percent copper underlay the gossan.

The original ore formed as a volcanogenic deposit in Precambrian time but underwent moderate to high later metamorphism. Temperatures as high as 550 to 650° C may have existed based on the presence of kyanite, staurolite, the absence of cordierite, and iron content in sphalerite. Mauger has suggested an original hot spring origin based on a sulfur isotope study.

Lake Superior Copper District, Michigan. The Lake Superior district is one of the greatest copper districts of the world and the outstanding deposit of native copper. It forms a belt in the Keweenaw Peninsula 3 to 6 kilometers wide and 150 kilometers long, of which 40 kilometers has been highly productive. First mined by a prehistoric race, it was discovered by the Jesuitts in the seventeenth century, and modern mining started in 1845. It was the premier district of North America until 1887, when Butte surpassed it. New deposits have been found in the district but are closed because of labor management discord. It has supported 100 mining companies, yielded ores that averaged 1.27 percent, and

paid dividends approaching $350 million. At the present time, only the Copper Range Company deposit of different origin is in production.

The district is underlain by about 400 basaltic lava flows with 20 to 30 intercalated zones of felsitic conglomerates and sandstones of Keweenawan age, totaling 25,000 feet. The copper belt beds lie in the south limb of a huge syncline that dips 40° northwest beneath Lake Superior and reappears in Ontario. To the south, the beds are faulted against Cambrian sediments. Many of the flows are amygdaloidal. The copper lodes occur near the middle of this Keweenawan series.

The Lodes. The copper lodes (Fig. 19-12) are of three types: (1) conglomerate lodes, (2) amygdaloidal lodes, and (3) fissure veins. The chief mineral is native copper. More than 90 percent of the production has come from six lodes, namely, the Calumet conglomerate, and the Baltic, Isle Royale, Kearsarge, Osceola, and Pewabic amygdaloids.

The *conglomerate lode* at Calumet is a reddish conglomerate lens from 3 to 7 meters thick composed of pebbles in a sandy matrix. Native copper fills the interstices and replaces matrix and pebbles; the ore portions are bleached. There is a little adularia, epidote, calcite, and quartz. The ore is localized in an ore shoot shaped like an upright funnel. It has been mined for 6,000 meters in length and for 3,000 meters down the 38° dip. It was opened by 10 shafts and 300 kilometers of workings. The tenor gradually decreased with depth.

The *amygdaloidal lodes* occur in the upper, permeable parts of flows classed as coalescing, or fragmental. These tops are reddish owing to included hematite formed by oxidation during solidification; where mineralized, they are bleached. Native copper occupies the vesicles along with quartz, calcite, epidote, chlorite, adularia, sericite, pumpellyite, ankerite, and zeolites. The lodes averaged 5 meters in thickness and contained from 0.6 to 1.5 percent copper but averaged about 0.8 percent. The Quincy lode had been followed 3,500 meters down the dip. The amygdaloids have furnished about half the total copper.

The few veins were noted for their "mass" cop-

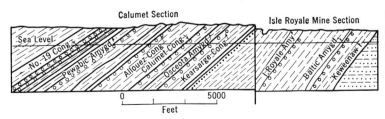

Figure 19-12 Section through Michigan copper range showing conglomerate and amygdaloid lodes. (Broderick, 16th Internat. Geol. Cong.)

per—one mass weighed 500 tons; they also yielded beautiful specimens of native copper and native silver within clear calcite, interesting sulfides, and arsenides.

The district was thought to be almost exhausted until new deposits were located and developed after World War II. Unfortunately, in this area of historically bitter labor-management discord and strikes, there is little production because developed properties have been closed by long strikes.

Magma Mine, Superior, Arizona. Since 1914 the Magma mine, halfway between Ray and Miami has produced both copper and silver from ore that averages about 5 percent copper and 4 ounces silver. Its daily production is 61,000 tons of ore per day and about 140,000 tons per year of metals.

The Magma "vein" occupies a fault zone 2 to 14 meters wide that cuts through inclined Paleozoic sediments and into two underlying diabase sills at least 800 meters thick. Quartzite and limestone constitute the walls in the upper levels and diabase in the lower levels.

The "vein" had an insignificant leached outcrop. The main ore body apexed at a depth of 150 meters, and to 275 meters it consisted of rich supergene ore, below which is rich primary copper sulfide. In the upper levels the "vein" was about 180 meters long and 2 meters wide, but in the underlying diabase it is longer, wider, and richer. An eastern ore shoot extends from the 150 to the 1900 level; a western one from the 1200 to the lowest level. Its apex is sphalerite-galena ore which gives way shortly to bornite ore. Both shoots are sheathed by zinc-lead ore and sericitized wall rocks. The horizontal workings are 2,800 meters long, and the vein has been explored below 1,600 meters in depth. The deposit is a replacement lode with no open-space filling.

Kennecott, Alaska. The unique deposits of Kennecott, noted for their large bodies of rich, almost pure chalcocite ore, yielded around 1.2 billion pounds of copper and much silver from four mines, now closed and completely defunct.

The ores are confined to the lower dolomite belts of the Triassic Chitistone limestone, which overlies 1,800 meters of altered basaltic flows or greenstone that contains disseminated sulfides. All beds dip gently northeastward, being part of a major anticline whose center is excavated by the Kennecott glacier.

There were four classes of deposits: (1) Wide, steeply dipping replacement veins strike normal to

the bedding. They bottom on limestone bedding planes, are widest at the base, and pinch stratigraphically upward from 80 to 180 meters above their base. Their vertical and horizontal dimensions are thus small, but they extend down dip for 1,300 meters. Few veins outcrop, and some do not reach within 300 meters of the surface. (2) "Flat" ore or tabular bedding replacements are localized by fissures restricted within certain beds. (3) Glacier ore, consisting of pieces of chalcocite eroded from the Bonanza vein outcrop, is incorporated as lateral moraine within a small glacier during its growth. The country rock is ice. (4) Slide ore, consisting of fragments of outcrop chalcocite, is incorporated within large talus slopes.

The ores consist of hypogene digenite and orthorhombic chalcocite, covellite, very minor bornite, rare traces of enargite, luzonite, chalcopyrite, and tennantite, and an occasional speck of sphalerite and galena. Pyrite, quartz, and other introduced gangue minerals are absent. The mineral assemblage is thus unusual. The sulfides constitute from 10 to 100 percent of the ore; pure chalcocite in masses of tens of thousands of tons were common. The Jumbo Main vein contained one chalcocite mass 30 meters wide, 50 meters long, and 140 meters down dip that averaged over 60 percent copper. The average tenor of all ores mined was 12.4 percent copper. Note (Fig. 19-13) that it had the highest copper grade of any major mine in history. Supergene sulfide enrichment is absent, but preglacial partial oxidation to copper carbonate extends to a depth of 800 meters along vertical fissures.

Lindgren correctly had difficulty in determining the genesis of this deposit and he, therefore, included it in a chapter (Reference 9) entitled, "Deposits formed by concentration of substances contained in the surrounding rocks, by means of circulating waters."

Lindgren refers to the monzonite porphyry, which has been suggested as the source of mineralization as "1½ miles distant and apparently not directly related to the deposit." He further suggests the possibility that the "meteoric warm waters searched the underlying Nikolai greenstone, whose copper content antedates the main deposits, have carried copper solutions up into the limestone where it was precipitated as chalcocite."

Canadian Copper Deposits
Although Canada ranks fourth in world copper production, most of it comes from ores in which other

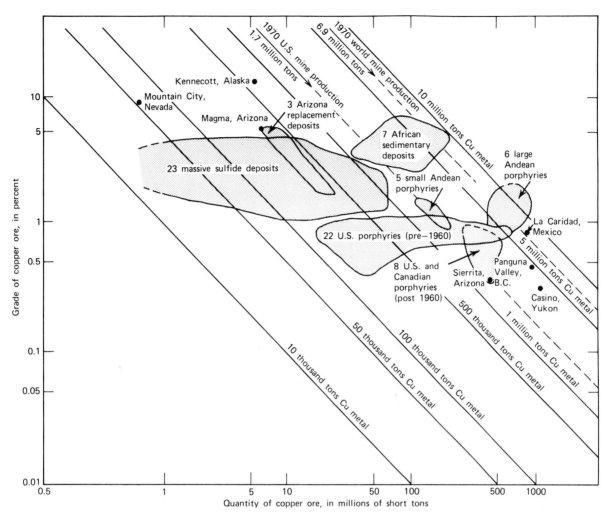

Figure 19-13 Size and grade characteristics of selected copper ore bodies. Diagonal lines show tons of copper metal per deposit; world and U.S. mine production of copper metal are shown for comparison by diagonal dashed lines. (Data from unpublished compilation by A. R. Kinkel, Jr., and others from Peters, 1970.)

metals dominate, as for example, at Kidd Creek. The Canadian Shield has been the main source but the more recent development of porphyry copper deposits in British Columbia are adding to Canada's copper production. There are no supergene enrichment ores. The important districts or mines in order or rank are listed in Table 19-5.

Massive sulfide deposits have some common characteristics that may be of aid in understanding not only their origin, but also prospecting for such deposits. Figure 19-14 illustrates the premetamorphic generalizations of the Noranda-type, massive sulfide deposits. Of prime consideration is the sandwich nature of these deposits with fractured felsic volcanics below and andesite above, but with metamorphic modifications. The localization of massive sulfide bodies in volcanics is obvious in

many of these deposits and has been used as a prospecting tool.

The Quemont and Horne massive sulfide and magnetite deposits also illustrate this association with volcanics, specifically again, with rhyolite below and andesite above as shown in Fig. 19-15. Generally, similar but smaller deposits occur in the nearby *Waite-Amulet, Normetal,* and *Aldermac* mines.

The geology of the various properties in the Timmins or Porcupine districts differ substantially. In the Kamiskotia area, located north of the Kidd Creek Mine, the *Kam-Kotia, Jameland, Con Jamieson,* and *Genea* mines are located (Fig. 19-16) in a thick, largely fragmented rhyolitic pile. The deposits occur predominantly in felsic pyroclastic rocks or a brecciated zone confined to a particular

Table 19-5. Important Copper Districts or Mines in the Canadian Shield

District or mine	Location	Chief associated metals
1. Kidd Creek	Timmins, Ontario	Zinc-silver-tin
2. International Nickel Co.	Sudbury, Ontario	Nickel-copper-platinum-gold
3. Bathurst	New Brunswick	Copper-lead-zinc-silver
4. Noranda	Rouyn, Quebec	Gold-copper
5. Porphyry copper deposits	British Columbia	Copper-molybdenite
6. Churchill District Flin Flon Mine	Manitoba-Saskatchewan	Copper-zinc-silver-gold
7. Kimberly	Sullivan district	Lead-zinc-copper

stratigraphic unit; this unit is marked by an abundance of native volcanic flows. It is suggested that a high level, possibly subvolcanic intrusion within specific portions of the volcanic pile may be the source of the mineralizing fluids. Pyke and Middleton express caution about this source as they suggest that the intrusive and volcanics appear to have had the same structural and metamorphic history; it is not yet proved that the intrusion belongs to the volcanic cycle.

The Timmons or Porcupine area is well known for its gold production, which was nearing exhaustion in 1959 when McIntyre Porcupines Mines, Ltd., discovered a porphyry copper deposit in a felsic intrusive that appeared to have genetic cor-

relation with the gold mineralization. This had been suggested by Horst as early as 1936.

The *McIntyre Pearl Lake* porphyry copper deposit is located in a leucocratic quartz-feldspar porphyry. The ore zone is about 200 meters long, 500 meters wide and plunges 50° east parallel to the pronounced lineation so common in the Timmons area. The deposit has reserves of more than 80 million tons of ore containing disseminated chalcopyrite-bornite mineralization with minor tetrahedrite, tenorite, molybdenite, and native silver. The average grade is not fully established, but a cutoff grade of 0.5 percent was established several years ago and is probably less at the present time.

An interesting aspect of this porphyry deposit

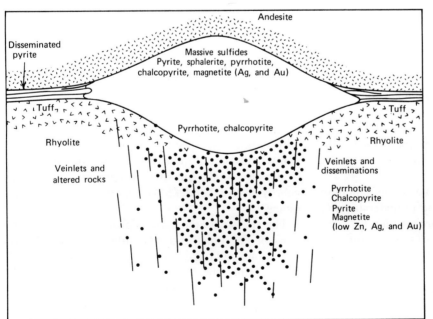

Figure 19-14 Submarine-volcanic-exhalative features common to many Noranda-area sulfide deposits. (S. M. Roscoe, 1956.)

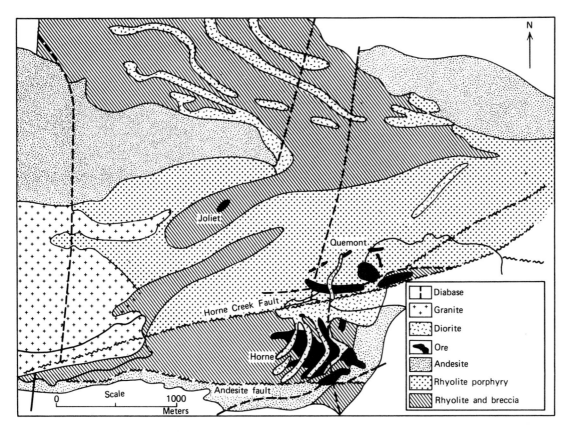

Figure 19-15 Generalized geology adjacent to the Quemont and Horne deposits (After Campbell, 1963; from J. Lusk, et al., 1975, Econ. Geol., v. 70, p. 1970–1083.)

has been indicated by Pyke and Middleton that even though it formed during Early Precambrian time, the level of erosion has not removed the deposit.

The nickel deposits (*McWatters, Horst, Inco,* and *Texamont)* that are located south of Timmons form in alpine-type ultramafic intrusions that are confined almost exclusively to a felsic volcanic sequence.

Although the Timmons area had been studied geologically and by numerous drill holes, it is about 97 percent covered by postore blankets of glacial-lacustrine deposits. The open pit area itself was overlaid with 4 to 30 meters of muskeg, and glacially unstratified and varved clay. Figure 19-16 is a geological map of the region completed by Pyke and Middleton, of the Ontario Department of Mines. The map indicates additional deposits discovered before and since 1964. Kidd Creek is included in Chapter 11.

Churchill District, Manitoba-Saskatchewan. This Cu-Zn-Pb-Ag-Au-Cd-Te Churchill district in the Canadian hinterland includes the *Flin Flon* mine of the Hudson Bay Mining and Smelting Co. Ltd., the

Sherritt-Gordon, the exhausted *Mandy* mine, and a dozen other properties only half of which are still in production. The Flin Flon deposit reached production in 1930 and was one of the largest copper and largest zinc mines in Canada in 1966 when it produced 4630 tons per day of 2.23 percent Cu and 4.57 percent Zn. It lies in Amist volcanic strata of "quartz porphyry" (rhyolite and dacite) layer that is overlain by andesite flows, coarser fragmental beds of andesite. Breccia containing disseminated sulfides that underlie the massive ore are in a somewhat typical arrangement as shown in Fig. 19-17.

The ore body is a large, solid, massive, sulfide lens, bifurcated in places, 900 meters long on the surface and up to 150 meters wide. At a depth of 300 meters it is 330 meters long and 12 meters wide, but at greater depths (1,200 meters) it is still longer and wider. The core of the body is massive sulfide, and this is flanked by disseminated sulfides (Fig. 19-17). The ore consists of an intimate mixture of iron, zinc, and copper sulfides, with some quartz, calcite, and schist residuals and rare arsenopyrite, galena, and magnetite. It averages 2.4 percent Cu, 5.2 percent Zn, 0.09 ounce Au, and 1.3 ounces Ag.

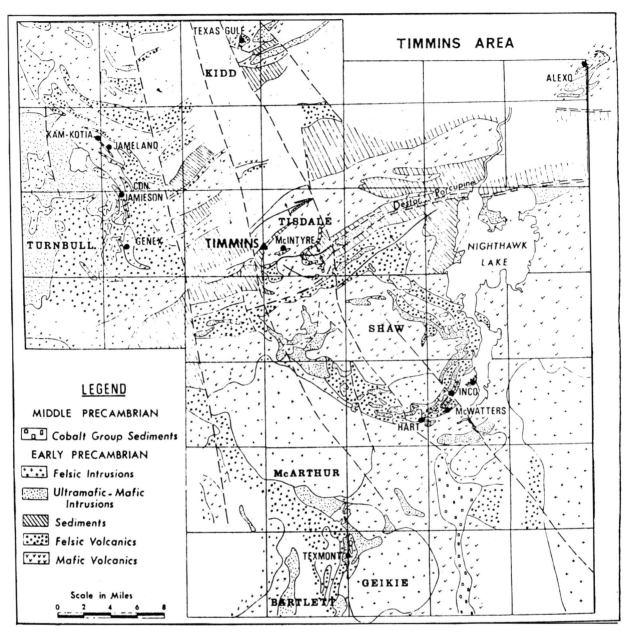

Figure 19-16 The Timmons area (From Bull Can. Int. Mining and Metall.)

In addition, cadmium, selenium, and tellurium are recovered. The deposit is bounded by volcanics, a characteristic of massive sulfide deposits and may, therefore, have an exhalative volcanic origin. Presumably, the other ore deposits in the area are of a similar origin. Intense sericitization of the walls accompanied ore deposition.

The *Sherridon* deposit consisted of two tabular sulfide bodies in Precambrian sediments invaded by granite and pegmatite. The two ore bodies were 1,400 and 1,800 meters long and 5 meters wide, consisting of base-metal sulfides that carried 2.6 percent Cu and 2 percent Zn, with low gold and silver. The gap between the two ore bodies coincides with a cross-anticline, and it is presumed that the two lenses were originally one body, the ore in the gap having been removed by erosion. During its life from 1931 to 1932 and 1937 to 1951 when it closed, it produced 133,122 tons copper, about 74,500 tons Zn, 101,026 ounces Au, and 3,218,364 ounces Ag.

Noranda, Rouyn district, Quebec. The Rouyn district of northern Quebec is dominated by the large

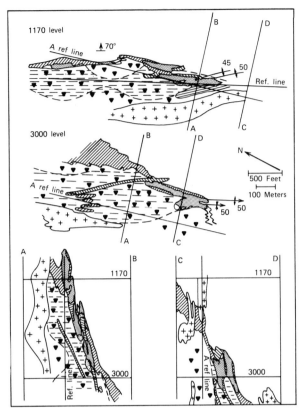

Figure 19-17 Flin Flon mine, Manitoba. (After the staff, Hudson Bay Mining and Smelting Company, Ltd.).

Noranda mine, which illustrates massive sulfide deposits. The enclosing rocks consist of altered Keewatin rhyolite and andesite extrusives cut by dikes of quartz diorite, syenite porphyry, and later gabbro. These rocks have been dragfolded over a width of 650 meters. The deposits are massive sulfide bodies in andesite breccias or tuffs, formed by hydrothermal replacement, but the ore-bearing solutions may have an exhalative volcanic source indicated by the association of rhyolite porphyry overlain by andesite not only for the Horne deposit but also for the Quemont and Joliet deposits (Fig. 19-15). There are also rich, siliceous gold-quartz replacement veins. About 24 sulfide lenses are known, dominated by the large H and lower H bodies. The large H is 50 meters wide and extends to 450 meters; the lower H extends from the 1600 level to the 2600 level and is 200 meters across. The ores are solid pyrrhotite-pyrite bodies with cores and fringes of chalcopyrite and minor zinc blende and magnetite. The massive pyrrhotite of the upper levels has given way to pyrite in the lower levels. The ores average around 1.5 percent Cu and 0.14 ounces Au per ton, and the reserves amount to

about 30 million tons. Profound alteration, consisting of silicification and chloritization, accompanied ore deposition.

Highland Valley Mining district, British Columbia. The majority of the porphyry copper deposits of the southwestern United States are associated with felsic stocks of limited size. About 125 miles northeast of Vancouver, British Columbia, however, a large batholith more than 40 meters long, Jurassic in age, and known as the Guichan Batholith contains at least six porphyry copper deposits. They are the *Alwin, Valley Copper, Bethelehem, Lornex,* (Front Plate), *Highmont,* and *Craigmont* deposits. All are clustered together in an area of more than 300 square miles near the center of the batholith in the Highland Valley mining district.

The batholith apparently formed in sequential phases with older quartz diorites toward the margin, younger quartz diorite as an intermediate ring, and a still younger granodiorite forming the central cone. All three exist within the Highland Valley district. These phases are cut by a swarm of clastic and rhyolite porphyry dikes. Most of the dikes are steeply dipping and have northerly strikes while others swell and branch to form irregular intrusions and associated intrusive breccias. When the latter intersect west-northwesterly contacts of the major phases, they have formed the Highmont, Bethlehem, Troyan (South Seas), and Krain (North Pacific) deposits.

The Highland operations operate at more than 25,000 tons per day from two pits, one of which will be 30 meters deep yielding 123.5 million tons of about 0.29 percent Cu and 0.042 percent MoS_2; the other will be 200 meters deep yielding 26.5 million tons of ore averaging 0.27 percent Cu and 0.093 percent MoS_2.

Britannia Mine, British Columbia. The Britannia mine, near Vancouver, is in a band of steeply dipping metamorphosed sediments and igneous rocks that form a roof pendant 11 kilometers long by 3 kilometers wide in the Coast Range batholith. The deposits are in a shear zone up to 700 meters wide that occurs mainly in quartz-sericite schist. It includes five large lenticular replacement deposits consisting of schist impregnated with and replaced by ore and localized by structures of the shear zone. They have been followed to 830 meters in depth.

The main minerals are base-metal sulfides with minor hematite, magnetite, and gold. Wide lenses of anhydrite, replaced by gypsum, occur in the

Table 19-6 Characteristics of Some British Columbian and East Australian Copper Deposits

Deposit and location	Country rocks	Intrusion Age-Composition-Size	Minor intrusions	Veins (alteration selvedges) (earliest → latest)
Bethlehem, B.C.	Triassic volcanics	Lower Jurassic quartz diorite batholith	Grandiorite porphyry	
Highmont, B.C.	Triassic volcanics	Lower Jurassic quartz diorite batholith	Granodiorite porphyry	(1) Qtz − cpy − born − mo (seric − tourm − Ksp) (2) Qtz − cpy ± pyr + mo (no alteration) (3) Qtz − mo − clay (argillic selvedge) (4) Barren qtz (no selvedge)
Gibralter, B.C.	Permian eugeo-synclinal assemblage	Cretaceous (foliated) quartz diorite batholith	Quartz diorite porphyry	(1) Qtz − pyr − cpy (qtz − seric − clay − pyr) (2) Qtz − chl − pyr − cpy − inte − calcite (3) Qtz ± chl ± cpy ± born ± epid ± calcite + pyr (4) Qtz − mo − cpy − pyr − inte − calc. (5) Qtz ± chl + cpy
Brenda, B.C.	Upper Triassic eugeo-synclina assemblage	Lower Jurassic quartz diorite "Stock" (actually a narrow attenuation of a batholith)	Dacite dikes	(1) Qtz − K − spar − cpy − mo (2) Biot − cpy (3) Qtz − mo − pyr (4) Epid − inte − mo
Cadia, N.S.W.	Ordovician-Silurian andesite and siltst.	Silurian-Devonian? quartz diorite-quartz monzonite large stock (3-4 km across)	Quartz-feldspar porphyries	Weak potassium silicate alteration Moderately-strong propylitization with better mineralization
Copper Hill, N.S.W.	Ordovic-Silurian andesite and siltst.	Silurian-Devonian quartz diorite stock (3-4 km across)		Pervasive propylitization in better fracture areas
"Typical[a] porphyry copper"	Precambrian paleozoic to Cretaceous sediments, and local volcanics	Small (−1 to 2 km in diameter) Mid-Tertiary (laramide) quartz monzonite-stock	Quartz plugs and dikes	Well-zoned concentric alteration pattern Potassic → phyliic → (argillic) → propyl. Ore with potassic and phyllic types

Source: By G. Ballantyne.
[a] Of the southwestern United States.

upper levels. There is a little supergene chalcocite, covellite, and marcasite. The copper content averaged about 1 percent, but gold and silver added to the value. The mine had produced 48 million tons of ore as of 1966 but closed in 1974.

A summary of some of the characteristics of a few of the British Columbia copper deposits is given in Table 19-6.

Mexican Copper Deposits

Mexican copper production comes mainly from the extension of the Arizona belt into Sonora. Lower California accounts for about one-fifth, and the remaining production is by-product copper obtained from base-metal ores carrying silver and gold. Cananea and Nacozari in Sonora are the outstanding copper deposits.

Cananea. This notable district has yielded over 2.5 billion pounds of copper and 35 million ounces of silver from porphyry copper deposits, breccia pipes, and minor limestone replacements and contact-metasomatic deposits. The region has a complex geologic history involving Paleozoic sediments, later extrusives, profound deformation, and porphyry in-

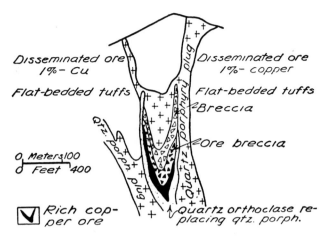

Disseminated ore 1% – Cu

Flat-bedded tuffs

Qtz. porph. plug

Quartz porphyry

Disseminated ore 1% – copper

Flat-bedded tuffs

Breccia

Ore breccia

0 Meters 100
0 Feet 400

▼ Rich copper ore

Quartz orthoclase replacing qtz. porph.

Figure 19-18 Sections of Colorada breccia pipe, Cananea, Mexico. (Emmons-Billingsley, AIME.)

trusions, accompanied by widespread mineralization.

The *pipelike* bodies, of which the rich *Colorada* (Fig. 19-18) is outstanding, are the most important. They may have been the feeder funnels to overlying deposits. The unique Colorada ore pipe, according to Sales and Perry, is goblet-shaped, apexing at the 500 level as a quartz-sulfide ring 200 meters by 170 meters across. The ore ring converges into the goblet stem of solid ore pipe at the tenth level, and the base of the goblet is pegmatitic quartz and phlogopite with sulfides. Also, stringers of quartz and sulfides penetrate the walls. The ore minerals are bornite, chalcopyrite, and molybdenite with minor pyrite, tennantite, covellite, and chalcocite. The sequence of events was a preporphyry breccia pipe intruded by quartz porphyry, which divided into upper and lower elements with the ore fluids trapped in the lower one and later injected along the sides and into the fractured walls of the upper element; subsequently, the sulfides were fractured by later slumping. The unusual features suggest to Sales that the deposit may exemplify an ore magma injection. The possibility of a replacement origin, characteristic of the overlying deposits, is not eliminated, however.

Other ore pipes are the *Capote,* extending from the 400 to the 1600 level and underlying the Capote ''porphyry copper'' deposit, and the *Duluth* breccia pipe, an oval-shaped pipe 400 by 70 meters extending from the surface to 470 meters in depth. A minor pipe at the *Veta* mine extends upward into enriched ''porphyry copper'' ore. Other deposits are Nacozari-La Caridad.

South American Copper Deposits

An extensive summary (Table 19-7) of the regional characteristics of some of the more important South American porphyry copper deposits has been provided by V. F. Hollister, who includes data on 29 deposits located in Argentina, Chile, Ecuador, and Peru.

South American deposits are prime examples of the results of an oceanic plate plunging (Table 19-4) under western South America and resulting in the generation of the upper mantle of rising intrusive bodies, many of which are mineralized. The effects of this mechanism of magma generation are still occurring, as suggested by the youthful age of some of the deposits, some thought to be as recent as Pliocene. Similar mineralization, hydrothermal alteration, and composition of host bodies are evident even though considerable age differences exist, as much as 50 million years. Even though the deposits occur at variable distances from the leading edge of the subduction plate, the composition of the host porphyries varies slightly. The correlation of plate tectonics with mineral deposits is still a complex problem even though it has been helpful in providing thought-provoking hypotheses.

Breccia pipes and stockworks, the former occurring with associated tourmaline that is contemporaneous with the ore mineralization, provide two different South American porphyry-type deposits.

Chile and Peru are the two chief copper countries of South America, with minor production from Ecuador, Bolivia, and Argentina. Chile's copper belt includes some of the largest deposits (Fig. 19-19), some of which are listed in Table 19-7, all lying on the western flank of the Andes. The Peruvian deposits are also on the western flank; those of Bolivia are in the heart of the Andes; and the Argentina occurrences are on the eastern slopes. All of these deposits in Chile and Peru have been nationalized by the respective governments.

Chuquicamata mine, Chile. This deposit, one of the largest of all ''porphyry coppers,'' lies in Antogafasta, 90 miles from the Pacific, at an elevation of 3,100 meters. It is a large, open-pit, deposit of low-grade oxide ores noted for its huge ore reserves and the diversity of its ore minerals.

Granodiorite porphyry intrusions accompanied the intense folding and faulting of Mesozoic sediments and volcanics at the end of the Cretaceous period. This was followed by andesitic eruptions, prolonged erosion, Pliocene uplift and volcanism,

Table 19-7 Summary of Major Known Porphyry Copper Deposits in South America

Deposits	Age	Intrusive	Intruded	Structural type	Hypogene grade	Primary regional fault	Secondary reg. fault	Alteration zones present	Pyritic halo dimen.
Argentina									
Campana Mahuida, Neugnen	Upper Tert.	Qtz. Mon. Por.	Cret. Sed.	Stockwork	N.A.	N35°E		Phy. Arg. Prop.	3 × 2 km
Co. Rico (MiVida) Catamarca	5.9	Qtz. Mon. Por.	Paleo (?)	Stockwork	N.A.	N.A.		Phy. Arg. Prop.	2 × 1 km
La Alumbrera Catamarca	8.0	Qtz. Mon. Por.	Tert. Volc.	Stockwork	0.4 Cu 0.04 Mo	N4SW		Phy. Arg. Prop.	2 × 1 km
Paramillos Sur. Mendoza	Trias?	Dio. Qtz. Mon. Por.	Per.-Tria. Sed. & Vol.	Stockwork	0.38 Cu 0.02 Mo	N-S		Pot. Phy. Arg. Prop.	2 × 1 km
Pachon, San Juan	Tert.	Qtz. Mon. Por.	Tert. Volc.	Tour. Bx.	0.65 Cu	None		Phy. Arg. Prop.	8 × 6 km
Chile									
Chuquicamata, Anto	29.2	Qtz. Mon. Por.	Granodiorite	Stockwork	1.3 Cu 0.04 Mo	N10E		Pot. Phy. Arg. Prop.	8 × 3 km
Mocha, Tarapaca	56.4	Qtz. Dio. Por.	Tert. Volc.	Stockwork	N.A.	N40W		Phy. Arg. Prop.	4 × 3 km
El Salvador, Atacama	39.1	Grdr. Por.	Tert. Volc.	Stockwork	0.9 Cu 0.04 Mo			Pot. Phy. Arg. Prop.	3 × 3 km
Potrerillos, Atacama	34.1	Tonalite Por.	Jurassic Sed.	Stockwork	N.A.			Pot. Phy. Arg.	6 × 4 km
Los Pelambres, Coquimbo	9.96	Grdr. Por.	Grdr. Sed. and volc.	Stockwork				Pot. Phy. Arg. Prop.	4.5 × 1.5 km
Co Blanco (Cerro) Santiago	4.59	Qtz. Mon. Por.	Tert. Volc.	Tour. Bx.	1.38 Cu 0.03 Mo	N15W	N45E	Phy. Arg. Prop.	11 NS × 5 km EW
Los Bronces, Santiago		Dacite	Grdr. Volc.	Tour. Bx.	1.7 Cu	None	None	Phy. Arg. Prop.	11 NS × 5 km EW
El Teniente, O'Higgins	4.32	Dacite	Tert. Volc.	Tour. Bx.	1.7 Cu 0.05 Mo	None	None	Phy. Arg. Prop.	
El Abra, Anto	33.2	Qtz. Mon. Por.	Grdr.	Stockwork	N.A.	N10E		Phy. Arg. Prop.	2.5 × 1 km
Mantos Blancos, Anto		Dac. Por.	And. Volc.	Stockwork	0.7 Cu	N.A.		Phy. Arg. Prop.	3 × 1 km
Ecuador									
Chaucha, Azuay	9.9	Qtz. Mon. Por.	Grdr.	Stockwork	0.7 Cu 0.03 Mo	E-W	N10E	Phy. Arg. Prop.	4 × 3 km
Peru									
Toquepala	58.7	Dacite	Cret. Volc.	Tour. Bx.	0.7 Cu 0.04 Mo	None	None	Phy. Arg. Prop.	3 × 3 km
Michiquillay	20.6	Qtz. Mon. Por.	Cret. Qtzte.	Stockwork	0.6 Cu 0.02 Mo	N45W	N50E	Phy. Arg. Prop.	2 × 2 km
Cuajone	Lower Tert.	Qtz. Mon. Por.	Cret. Volc.	Tour Bx.	0.7 Cu 0.03 Mo	None	None	Phy. Arg. Prop.	4 × 3 km
Quelloveco	Lower Tert.	Qtz. Mon. Por.	Cret. Volc.	Stockwork	0.6 Cu 0.03 Mo	E-W		Phy. Arg. Prop.	2 × 1.5 km
Morococha	7.0	Qtz. Mon. Por.	Perm. Cret.	Stockwork	0.7 Cu 0.02 Mo	N25W	E-W	Phy. Arg. Prop.	5 × 3 km
Cerro Verde	58.8	Qtz. Mon. Por.	Grdr.	Tour. Bx.	N.A.	None	None	Phy. Arg. Prop.	3 × 3 km

Source: From Hollistor V. F., 1973. Mining Eng. v. 25, p. 52.

Abbreviations:

N.A.—Not available	Grdr.—Granodiorite	Arg.—Argillic
Tert.—Tertiary	Perm.—Permian	Phy.—Phyllic
Cret.—Cretaceous	Dac.—Dacite	Pot.—Potassic
Qtz.—Quartz	Volc.—Volcanic	Prop.—Porpylitic
Mon.—Monzonite	And.—Andesite	

erosion, gravel formation, and recent uplift and aridity. Mineralization, localized by strong fissuring, accompanied the latter quartz monzonite porphyry intrusions, and the Pliocene erosion exposed the deposits and oxidized them to soluble minerals that persisted in the extreme arid climate. The age of the mineralization is about 29.2 million years ago.

The deposit is a pear-shaped porphyry over 500 meters deep. The host rock is intensely sericitized and silicified, five types of alteration being recognized. The ore minerals occur in stringers and specks throughout the altered rock. Operations were initially confined to the oxide ores, but the underlying sulfide ores are now being mined.

The primary mineralization apparently was by hydrothermal replacement of the porphyry. Mineralization resulted in the deposition of quartz, sericite, hematite, pyrite, and enargite, with rare tetrahedrite, chalcopyrite, zinc blende, bornite, and molybdenite. The considerable amount of covellite and the late alunite that are present are probably supergene.

Weathering yielded an upper leached zone, not everywhere present, an underlying oxide zone that now constitutes the ore body, and an underlying zone of mixed oxide-sulfide ore, and below this are sulfide ores (Fig. 19-20). Ore reserves are 1400 million metric tons of 1.2 percent Cu and 0.04 percent Mo.

The oxide ores include unusual copper minerals which are generally absent from humid surfaces but which form in the arid Atacama Desert. They are listed below in order of importance.

Common		Less Common
Antlerite $Cu_3(SO_4)(OH)_4$	Brochantite $Cu_4(SO_4)(OH)_6$	Native copper
Atacamite $Cu_4Cl_2(OH)_6$	Natrochalcite $Na_2Cu_4(SO_4)_4(OH)_2 \cdot 2H_2O$	Cuprocopiapite $CuFe_4(SO_4)_5(OH)_4 \cdot 7H_2O$
Chalcanthite $CuSO_4 \cdot 5H_3O$	Chrysocolla	Lindgrenite $Cu_3(MoO_4)_2(OH)_2$
Krohnkite $Na_2Cu(SO_4)_2 \cdot 2H_2O$	Turquois	Pisanite $(CuFe)SO_4 \cdot 7H_2O$
	Cuprite	

In addition, there are many uncommon oxidation products, including several hydrous sulfates of iron, combinations of magnesia, soda, and potash, also halite, and gypsum. The copper from these ores is extracted by leaching and electrolytic precipitation.

The supergene sulfide ores, known from drilling, consist of chalcocite-covellite ore with unreplaced remnants of the original sulfides. Between the oxide zone and the mixed zone there may be an intervening leached zone caused by a rising of the water level, which resulted in the leaching of the soluble oxide minerals.

The primary mineralization attended the porphyry intrusion. Oxidation and leaching during the late Tertiary climate produced a leached capping and a deep supergene sulfide zone. The Pliocene uplift initiated erosion, removed most of the capping, and during the arid climate converted the upper part of the sulfide zone to the oxide ores.

Potrerillos mine, Chile. This former subsidiary of the Anaconda Copper Co. lies in the Atacama Desert at an elevation of more than 3500 meters, 150 kilometers from the Pacific port of Barquito, and is a smaller replica of the Chuquicamata mine. It is near the newly developed El Salvador deposit.

The ore deposit consists of stockworks and disseminations in shattered tonalite porphyry. The veinlets of the stockworks have been filled, the intervening rock has been partly replaced, and the host porphyry has been intensely sericitized and chloritized. There are three separate bodies: the central body (Fig. 19-21) is 700 by 370 meters in area and tapers downward to a depth of 500 meters; the south body is 550 by 100 meters in area and also tapers down to a depth of 650 meters. The north body is known only by drilling. The pyritic halo measures 4.0 by 6.0 kilometers.

The oxide ore zone is 87 to 120 meters deep and averages 1.08 percent copper. The chief ore minerals are brochantite and antlerite with some copper oxides, carbonates, and silicates. The tenor is higher than that of the sulfide ore because it represents the oxidation of the richer upper part of the supergene sulfide zone. Below the oxide ore is another leached zone, up to 55 meters thick, formed by a fluctuating water table during its depression in response to oncoming aridity. The supergene sul-

Figure 19-19 Mine location map of northern Chile.

fide zone averages 1.65 percent copper and is from 100 to 110 meters thick. The upper part is highly enriched to chalcocite ore with a little covellite and residuals of chalcopyrite and pyrite.

Braden copper deposit, Rancagua, Chile. This large, Kennecott Copper former subsidiary lies east of Santiago at an elevation of about 3000 meters. The deposit was nationalized by Chile and is now part of CODELCO, the largest copper concern in the world. CODELCO grossed 1.15×10^9 in 1974 and makes up 25 percent of Chile's GNP and 85 percent of Chile's exports.

Lindgren and Bastin considered that the Braden deposit (Fig. 19-22), lies in and around an explosive volcanic vent through igneous masses and that ore fluids followed the same paths. Their interpretation involves three periods of intrusion, three periods of hypogene metallization, uplift, erosion, and supergene enrichment. Tertiary volcanics intruded by andesite, followed by high-temperature mineralization, was interrupted by an explosion that produced a vent, later filled by fragmental material forming a breccia pipe that some geologists consider to be the most favored ore host deposit. Solutions rose around the periphery, producing copper protore with 0.5 to 1.5 percent copper. Then dacite and latite porphyry and breccia (Teniente breccia) penetrated the plug and walls, accompanied by tourmalinization of the breccia. A third upwelling of ore fluids produced the "bornite body," augmented the pyrite and chalcopyrite, and added tennantite, molybdenite, rare galena, and zincblende, huebnerite, and enargite, by replacement. Carbonates, barite, and anhydrite also came in, and the final deposition was minor sulfides, crystals of barite and quartz, and mammoth crystals of gypsum.

Uplift, tilting, and erosion initiated supergene enrichment that raised the ore grade of the upper levels by the addition of chalcocite and covellite, but in the deeper ores only a trace of enrichment is visible under the microscope.

El Salvador, Chile. This deposit is located near the Potrerillos deposit area of iron-stained slopes of Cerro Indio. Muerto has been of interest since 1922. It was not until the last stages of World War II that the area was mapped with the conclusion that the Camp area 3 kilometers from the unsuspected El Salvador deposit was of economically low tonnage. Several years later, Perry urged Anaconda to undertake a major exploration effort to test for a su-

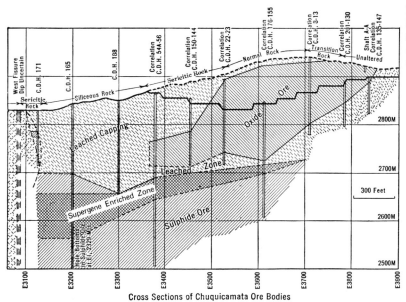

Figure 19-20 Cross section of ore body at Chuquicamata, Chile, showing ore zones, open-cut levels, and drill holes. (After Taylor, Copper Res. of the World.)

Cross Sections of Chuquicamata Ore Bodies

pergene secondarily enriched zone beneath Turquoise Gulch by more mapping and drill hole exploration. As reported in a detailed, excellent paper, by Gustafson and Hunt, with some of the first colored maps and figures ever printed in the Journal of Economic Geology, the fifth drill hole in the area "intercepted high-grade secondarily enriched ore beneath the barren outcrops of Turquoise Gulch, and it was evident that a major discovery had been achieved." The El Salvador deposit was placed in production in 1959 and was operated for 12 years during which 80 million tons

of sulfide ore, predominantly chalcocite, averaging 1.5 percent copper was recovered. Actual ore reserves prior to 1957 were about 300 million tons averaging 1.6 percent copper. Gustafson and Hunt report that this is roughly one-third to one-half of the total amount of copper deposited in the district. The *finale* of this successful geological project is that the property was nationalized by the Chilian government under President Allende, but with his overthrow and death, the new government still retains the property.

The Turquoise Gulch area of mineralization con-

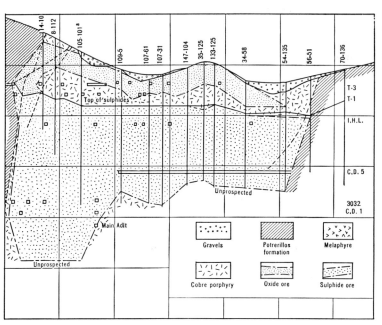

Figure 19-21 Cross section of central ore body, Potrerillos mine, Chile, showing oxide and sulfide ore bodies. (After March, Copper Res. of the World.)

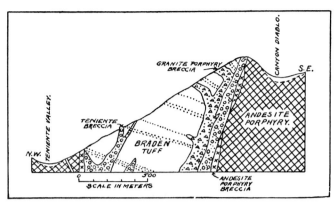

Figure 19-22 Section across so-called Braden crater, Chile. (Lingren-Bastin, Econ. Geol.)

tains a complex variety of siliceous to intermediate felsic intrusions. Gustafson and Hunt have described the mineralogy, form, and origin of these bodies including their geochemistry, isotopic factors, and the details of hydrothermal alteration of the area followed by the late alteration and final mineralization of the area (Fig. 19-23).

The sulfide zone at El Salvador shows the extensive lateral and overlying pyritization of the area in which most of the secondarily enriched zone has formed by covellite replacing preexisting sulfides although there is evidence that some of the covellite may be primary.

An interesting mineralogical aspect of some of the intrusive rock types at El Salvador is the relative abundance of primary anhydrite. More than half of the sulfur of the area is contained in anhydrite than is contained in all of the sulfides. The suggestion is that the partial pressure of oxygen must have been comparatively high for the stability of the sulfate. Also, the $\delta^{34}S$ analyses of the deposit exhibit near-equal depletion of ^{34}S in the sulfides and the enrichment of ^{34}S in the sulfates compared to the near-zero permil $\delta^{34}S$ compositions of sulfides in almost all other porphyry copper deposits. Possibly, the temperature conditions, along with other chemical factors, could have resulted in near-equal action and reaction of the following equation:

$$H_2S + 2O_2 \rightleftarrows SO_4^{2-} + 2H^+$$

with the result that essentially equal amounts of SO_4^{2-} and S^{2-} were produced. Certainly, much more study will be required to determine the validity of this hypothesis.

Gustafson and Hunt conclude that the deposit displays many similar features to porphyry copper deposits. "Very similar rock textures, types of veining, patterns of alteration-mineralization assemblage, and the same evolutionary trend can be seen in many deposits. *Yet each deposit is unique in detail!*"

Cerro de Pasco, Peru. The Cerro de Pasco mines are at Cerro de Pasco, Morococha, Yauricocha, and Casapalca, the first being the main center of copper production. All are reached from Lima, Peru, Morococha being 120 kilometers distant and at an elevation of 4930 meters and Cerro being 290 kilometers and at an elevation of 4730 meters. The deposits are heavy sulfides that also yield lead, zinc, silver, gold, bismuth, cadmium, and arsenic.

At *Cerro de Pasco,* for 200 years after 1630, there was active production of silver and prior to that by the Incas. Copper mining was desultory from 1890 to 1915, when large-scale operations commenced.

McLaughlin, Petersen, and others describe the geologic setting as that of metamorphosed sediments overlain by Carboniferous and Mesozoic sediments, dominantly limestones, and buried by thick early Tertiary extrusives.

Intermittent folding and faulting was followed by felsic intrusions that appear to have existed as a complex explosive vent. Mineralizing fluids rose along the eastern and southern margins of the vent forming a huge pyrite-silica body, lead-zinc ore bodies, copper-silver veins, and the so-called silver-pyrite ore bodies. A cross section of these bodies is shown by Fig. 19-24. It is presumed that the deposit underwent a comparatively long period of mineralization during which temperatures dropped from about 500 to 100°C. Sphalerite, for example, is well zoned varying from dark brown and black colored cores to a light almost colorless rim suggesting the wide range in temperatures.

The copper deposits are of three classes: (1) copper bodies in a massive pyrite deposit, (2) transverse copper-silver veins, and (3) copper mantos in limestone. The first is a massive pyritic body 1700

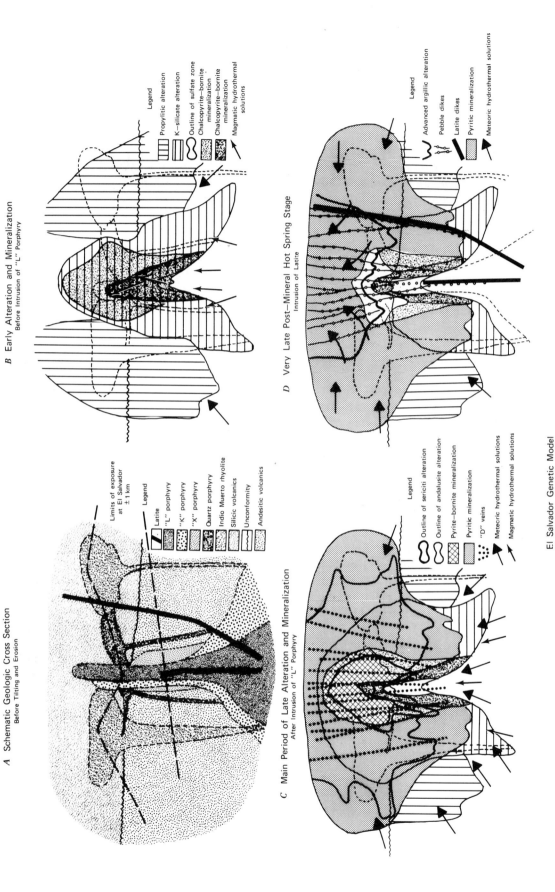

A Schematic Geologic Cross Section
Before Tilting and Erosion

Limits of exposure
at El Salvador
± 1 km

Legend

Latite
"L" porphyry
"K" porphyry
"X" porphyry
Quartz porphyry
Indio Muerto rhyolite
Silicic volcanics
Unconformity
Andesitic volcanics

B Early Alteration and Mineralization
Before Intrusion of "L" Porphyry

Legend

Propylitic alteration
K–silicate alteration
Outline of sulfate zone
Chalcopyrite–bornite
mineralization
Chalcopyrite–bornite
mineralization
Magmatic hydrothermal
solutions

C Main Period of Late Alteration and Mineralization
After Intrusion of "L" Porphyry

Legend

Outline of sericitic alteration
Outline of andalusite alteration
Pyrite–bornite mineralization
Pyritic mineralization
"D" veins
Meteoric hydrothermal solutions
Magmatic hydrothermal solutions

D Very Late Post–Mineral Hot Spring Stage
Intrusion of Latite

Legend

Advanced argillic alteration
Pebble dikes
Latite dikes
Pyritic mineralization
Meteoric hydrothermal solutions

El Salvador Genetic Model

Figure 19-23 El Salvador genetic model. A, Schematic geologic cross section before tilting and erosion; B, early alteration and mineralization before intrusion of "L" porphyry; C, main period of late alteration and mineralization after intrusion of "L" porphyry; and D, very late post mineral hot spring stage, intrusion of latite. (From Gustafson and Hunt, 1975, Econ. Geol., v. 70, p. 857.)

343

by 135 meters, and 480 meters deep, that contains irregular bodies and transverse veins of copper ores carrying abundant enargite and some bismuth, silver, and gold. Other portions of the pyrite mass carry silver-lead-zinc ores and pyritic-silver ores. The transverse veins are curving fault-fissure, copper-silver veins near the pyrite body that are up to 1000 meters long and 3 to 5 meters wide. The replacement mantos in limestone are all oxidized and carry mainly lead-zinc-silver ores with subordinate copper.

All the ore bodies formerly yielded rich supergene chalcocite and covellite ore blankets that extended to a maximum depth of 180 meters beneath the oxidized zone. Oxidation also extends several hundred feet below the water table. Only trivial supergene silver enrichment is discernible. The silver remained in the oxidized zone, forming a thick mantle of siliceous silver-bearing ground called "pacos," of which parts are mined as low-grade silver ore.

The mineralization solutions were part of the early Tertiary igneous activity that gave rise to the volcanics and to the Cerro quartz monzonite porphyry. They were localized along major fractures from which replacement proceeded. A rough zoning is evident, with the pyrite body on the vent margin, copper-gold ores being close, lead-zinc and pyritic-silver ores lying outside, and copper-free, lead-zinc-silver ores lying most remote from the igneous center.

U. Petersen estimates that the Cerro de Pasco deposit probably contained in excess of 2 million tons of lead, 4 million tons of zinc, and 100 million tons of pyrite. Petersen's figures as published by the Cerro de Pasco Copper Corp. are given in Table 19-8.

It is obvious that the zinc grade diminishes upward and lead increases upward. The silver grade remains about constant, but Petersen indicates that its distribution is somewhat erratic.

The *Morococha* deposits were mined for silver in

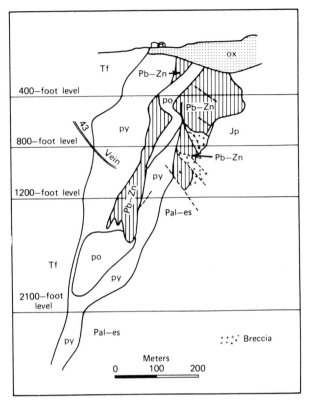

Figure 19-24 Generalized east-west section, Cerro de Pasco, looking north. (Rivera unpublished, 1960). py = pyrite-silica body; Pb-Zn = lead-zinc ore; po = pyrrhotite masses; 43 vein = copper vein; ox = oxides; Tf = fragmental and Rumiallana agglomerate; Jp = Pucára formation; and Pal = celsior formation. (From Petersen, 1965, Econ. Geol., V. 60.)

pre-Spanish and colonial times, but large-scale copper production started only in 1905. The ores occur in altered Permian-Cretaceous limestone cut by Tertiary quartz monzonite porphyry intrusives and consist of sulfides of iron, copper, zinc, enargite, magnetite, and lead, sulfo-salts, tennantite-tetrahedrite, iron oxides, and rare aikenite, bournonite, cubanite, jamesonite, luzonite, and matildite. Quartz, carbonates, and contact-metasomatic silicates are associated. The tenor of hypogene Cu is 0.7 percent and 0.02 percent Mo.

Table 19-8 Production Distribution of the Cerro de Pasco Copper Corp.

Between levels (ft from surface)	Average grade			Metal ratios	
	% Zn	oz Ag	% Pb	An:Ag:Pb	Ag:Pb
300–375	16.9	5.1	12.0	1:0.3:0.7	1:2.4
500–600	18.2	4.9	8.0	1:0.3:0.4	1:1.6
800–1000	18.1	5.4	7.8	1:0.3:0.4	1:1.4
1200–1400	21.0	3.1	6.5	1:0.2:0.3	1:2.1

Ecuador. Geologists for the United Nations discovered a porphyry copper deposit at Chaucha, Ecuador, using a chemical stream sediment sampling program. The exploration and drilling of the property by a Japanese concern developed 74 million tons of ore with a grade, including supergene ore, of 0.7 percent Cu and 0.04 percent Mo, or 100 million tons of ore grading 0.5 percent Cu have been indicated. The supergene enriched zone, up to 100 meters thick in the central zone, varies between 0.6 to 1.5 percent Cu. Hypogene mineralization averages about 1.4 percent Cu.

The intersection of two major strike slip faults (Fig. 19-25), which Goossens and Hollister believe are due to ENE displacement and plunging of the Nazca plate, has been effective in localizing the deposit. They further state that this plate tectonic model deposit appears to be controlled by two plates in movement relative to each other. The fault intersection with adjacent shattering provides enough deep conduits for the rising magma, with the associated shattering providing the conduits for disseminated mineralization.

Brazil. The location of the *Caraiba* district in Bahia State, Brazil, (Fig. 19-26) is the country's largest known copper deposit with reserves estimated at 90 million tons containing between 1.4 to 1.5 percent copper. The mining of this deposit both by open-pit and underground methods will reduce Brazil's current reliance upon imports of copper, which reached 130,000 tons, as domestic production was only 2000 tons by primary recovery in 1975. Caraiba produced 60,000 tons per year by December, 1976.

The ore deposits (Fig. 19-27) exist in a mineralized belt that extends about 100 kilometers. The mines are located in the enriched portions of this belt. The ore bodies are associated with basic intrusions, with the ore occurring near contacts with the underlying Archean granite–gneiss batholith.

Other Deposits. At *Corocoro, Bolivia,* are interesting native copper deposits, without known igneous association, which are thought to have been deposited by the reducing action of ferric oxide in red Pliocene sediments. These grade upward into chalcocite-covellite-domeykite ores. At *Famatina, Argentina,* and near by, shallow fissure veins carry rich copper-silver ores of enargite, famatinite, and sulfides of iron. copper, zinc, and lead. Some po-

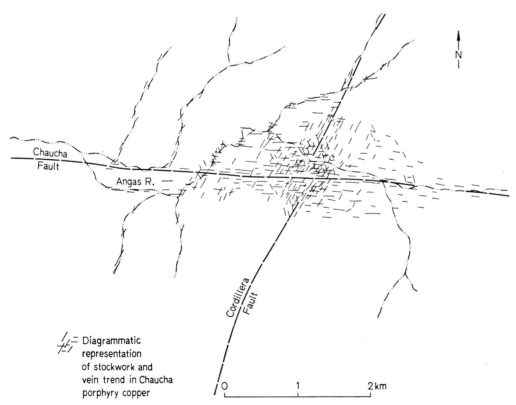

Figure 19-25 Mineralized area in the Chaucha porphyry copper showing the general trending of stockworks and veins. (From Goossons and Hollister, 1973, Mineral Deposits, v. 8.)

Figure 19-26 Location of Caraiba, Brazil.

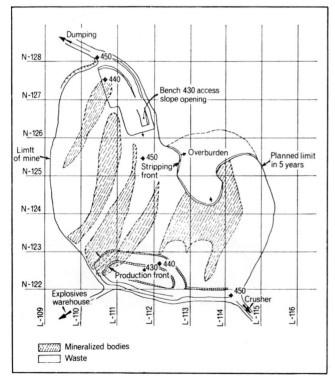

Figure 19-27 Plan of Caraiba open pit area showing surface outcrops of mineralized zones and the planed pit limits after 5 years' work. (From: Min. Magazine, p. 529, 1976.)

tential porphyry copper deposits are being studied and developed in Argentina.

Russia. In 1963, Russia surpassed the copper output of both Chile and Zambia. In 1976, Chile regained its second place and became the second largest producer of copper, being surpassed only by the United States. By 1978, the United States produced 18 percent, Chile 14 percent, Canada and Russia 10 percent. The United States is first in smelter production followed by the USSR, Japan, Chile, and Zambia.

The Russian copper resources and production, according to A. Sutulov, are concentrated in four principal regions (Fig. 19-28).

1. *Urals Mountains.* This region is still playing a leading role in copper production as the result of new large contact-metasomatic deposits discovered since 1953 prior to which the older deposits were approaching exhaustion. The discovery of the Uchaly, Gal, and Sibai deposits have greatly increased the production capacity of this region.

2. *Central Asia.* The large copper porphyry deposits of this area, principally Kazakhstan, have made this area one of increasing potential for the development and discovery of other porphyry copper deposits. The largest at present is Dzhezkazgan, which may develop into a 300,000 metric tons of copper per year operation.

3. The only large copper producer presently known to operate in Siberia is the Severonickel combine at *Norilsk.* The discovery, however, of the great porphyry copper deposit at *Udokan* and its attended exploration may make Siberia one of the largest copper producing areas of the USSR.

4. *Massive pyritic replacement* bodies of the Rio Tinto type yield some of the Russian copper. These bodies range from 50 to 600 meters in length, are about 50 meters wide, and carry 1.4 to 2.7 percent copper. Some deposits are secondarily enriched to depths of 100 to 250 meters. The reserves are large.

The *Dzhezkazgan deposits* in Kazakhstan Atlai, one of the two largest copper reserves of Russia, are impregnations in Carboniferous sandstone, of the Rhodesian type, forming gently dipping lodes extending outward from feeder faults; 620 square kilometers of these beds have been surveyed and 1.5 kilometers have been explored by drilling, disclosing 37 ore bodies. The tenor of the workable ore is stated to be 2.17 percent copper and 3.25 million tons of copper reserves are claimed. One of the largest individual deposits that has been found in

Russia is the huge deposit of *Udokan,* which is located 320 kilometers northwest of Lake Balkai in Siberia. It is similar to the Dzhezkazgan deposits.

In the Kola peninsula, northern Russia, and at Eriser River in Siberia (Norilsk), chalcopyrite is associated with pyrrhotite and pentlandite in erratic low-grade disseminations of 1 to 3 percent copper grade. The host rock is peridotite and carries 0.5 percent nickel.

Red-beds copper deposits of small production apparently make up the hundreds of small copper deposits in the Ural Mountains.

Other *Porphyry-copper-type* deposits have been developed in Asia, particularly around Kounrad. They consist of disseminated yellow sulfides and supergene chalcocite in small porphyry masses and invaded rocks. Only shallow, enriched zones, averaging around 1.1 percent copper, are workable. The protore carries 0.5 to 0.7 percent copper. Several zones have been explored and are claimed to aggregate 7.9 million tons of metallic copper, but noncommercial protore is included in these estimates. The Kounrad deposits are stated to contain 2.6 million tons of copper in an area of one-quarter square mile, which must mean that a vast amount of unprofitable protore is included. Similarly, 3.5 million tons of copper is claimed to be in the Tashkent deposits. The Transcaucasian porphyry deposits of Agarak contain copper-molybdenum ore.

The *Permian red-beds* of the Urals and Don Basin contain small sandstone layers 0.1 to 0.4 meter thick, carrying disseminated copper minerals with an average tenor of 1.5 to 1.9 percent copper. Similar deposits occur in the Silurian red-beds of the Lena region.

Sar Cheshmeh, Iran. This large porphyry copper deposit is located in southeast Iran (Fig. 19-29). Other porphyry copper deposits have been located farther east in Afghanistan. The intrusives of this large area can be explained by the localization of plate tectonics.

The Sar Cheshmeh deposit contains about 430 million tons of 1.13 percent copper using a cutoff grade of 0.4 percent. The overall tonnage, however, is far greater than this, as the deposit has not been fully tested at depth, and there is a considerable tonnage of oxide material that could be processed. A supergene zone of 90 million tons grades at about 2 percent copper; this tonnage is included in the 430 million tons.

Preproduction stripping was initiated in 1975 and

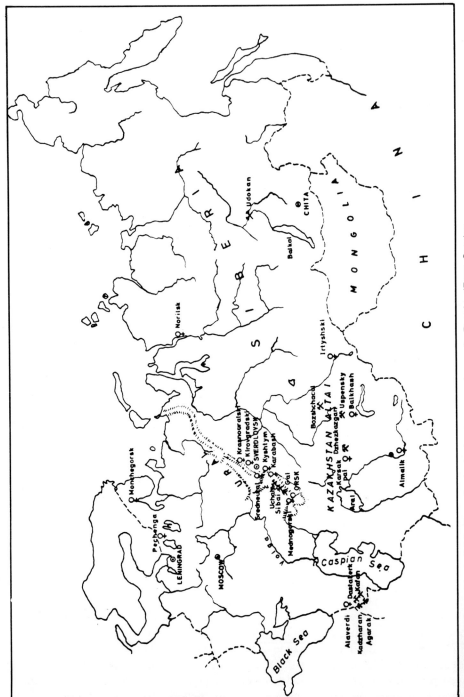

Figure 19-28 Geographical distribution of principal copper producers in Russia. (From Sutulov, *Copper Production in Russia,* University of Concécion, Chile.)

Figure 19-29 Map of Iran showing location of the Sar Cheshmeh copper project. (From Min. Mag., August 1975.)

was completed in 1977. The exploration and development management on the deposit is being done by an Iranian subsidiary of Anaconda, which has a 14-year management contract that involves the generation of the mine beneficiation plant and metallurgical facilities, as well as the training of Iranian nationals in the relevant technologies. The Sar Cheshmeh Copper Mining Co. is changing its name to the National Copper Co.

Yugoslavia. The important deposits of Bor consist of three massive bediike bodies in hydrothermally altered andesite porphyry. The largest consists of pyrite and enargite, with minor luzonite and famatinite, and supergene chalcocite and covellite, carrying 5 to 6 percent copper and 1 to 4 grams gold. The deposit has yielded over 4 million tons of ore from an open cut 100 meters deep. The second ore bed yields 2000 tons of 5 to 6 percent ore daily, and a third body is pyrite-chalcopyrite ore, carrying 1 to 2 percent copper. The prewar annual production was about 45,000 tons of copper metal.

Germany. Since A.D. 1150 most of Germany's copper has come from the Kupferschiefer of Mansfeld, where a thin but widespread cupriferous bed of black shale 40 centimeters thick has been mined for

its 2 to 3 percent copper content. It occurs in folded Permian beds.

The origin of the ore bed is controversial. Originally, the deposit was thought to be epigenetic. Advocates of a syngenetic sedimentary origin hold that the ore was deposited in a shallow sea where sulfate-reducing bacteria provided the sulfur as H_2S that reduced and precipitated the sulfides of predominantly copper but also lead, zinc, and some silver and pyrite. The source of the metals has recently been suggested as an exhalative source such as submarine hot-spring solutions that might be widespread enough to provide a thin cuperiferous formation covering 6,000 square kilometers.

Oceania. Porphyry copper deposits in the Philippines are numerous. Figure 19-30 and Tables 19-9 and 19-10 indicate the location of these deposits, the reserves, copper tenors, and gold tenors of the deposits. The characteristics of hydrothermal alteration and veining are also indicated in Table 19-9.

The occurrence of the numerous porphyry copper deposits in the Philippines, South Pacific Island arcs, and New Guinea is not at all surprising based upon the derivation of magmas by plate tectonics. Table 19-4 illustrates the derivation of porphyry copper deposits east of the Pacific rise with the descending lithospheric plate giving rise to calc-alkaline magmas.

Although copper has been mined in the Philippines as early as the Ming dynasty, A.D. 1368 to 1644, the recent spectacular discoveries of porphyry copper deposits began about 1955. During the last 20 years some porphyry copper deposits were in production and 16 other occurrences have been demonstrated to contain important proven or probable reserves. Since 1955, the annual production of copper has increased from 12,000 tons to 250,000 tons. The largest reserves are at the Atlas mine with 898 million metric tons averaging 0.46 percent Cu. The Sipalag mine has reserves of 662 million metric tons of 0.50 percent Cu. Total copper ore reserves in the Philippines are about 3230 million metric tons of ore with 0.46 percent Cu and 0.30 grams of gold per ton.

The most notable difference between Philippine porphyry deposits and their counterparts in North America is that the deposits are generally associated with fine-grained diorites rather than the coarser-grained monzonites common to continental deposits. The difference may possibly be ascribed to passage of the fluids through a thinner section of the oceanic crust existing beneath the island arcs.

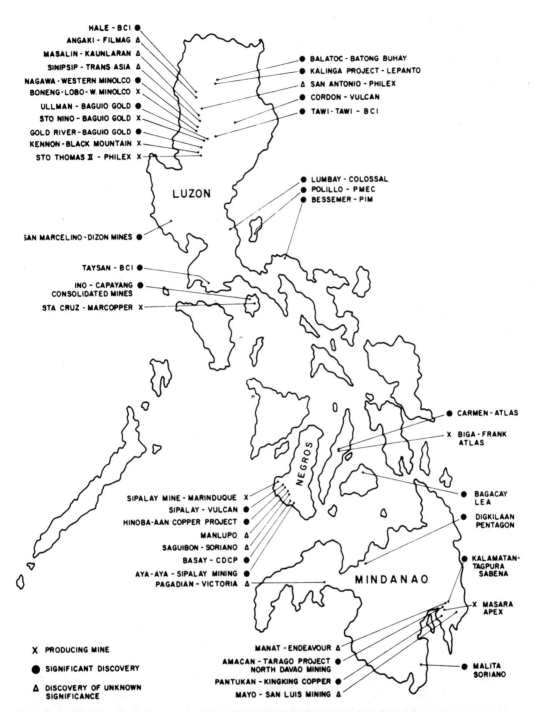

Figure 19-30 Major mineral deposits of The Philippine Islands, predominately porphyry copper deposits.

Also notable about the Philippine porphyries is that they usually carry higher gold values and lower molybdenum values than the North American deposits.

In 1977 development of the Carmen porphyry copper deposit was started by the Atlas Consolidated Mining and Development Co., which produces 45 percent of Philippine copper output and ranks second only to Bouganville in the Far East. Atlas properties include two open pit operations and an underground mine, which are the Frank and Greater Biga open pits and the Lutopan mine, respectively. The latter is a block caving operation. Reserves at Carmen are about 320 million tons

Table 19-9 Characteristics of Some Island Arc Disseminated Copper Deposits

Deposit, location	Country rocks	Intrusion (age, composition, and size)	Minor intrusions	Alteration (dominant types underlined) and veining
Marcopper, Philippines	Eocene-Oligoc. Andesites and metasediments	Miocene diorite stock (4 km × 5 km)	Andesite dikes	*Biot.*, *Chlorite*, Ksp, clay, all present. Rock is up to 50% chlorite in ore zone. Copper mostly in quartz veins.
Atlas Deposits, Philippines	Oligoc-Miocene sediments	Late Pliocene diorite porphyries—a cluster of small irregular bodies	Dacite dikes	Strong *biotite* and *chlorite*, weak sericite in ore zone. Also pervasive silicification.
Sipalay Deposits, Philippines	Lower-Tertiary metavolcanics	Upper Tertiary diorite stock (3 km × 5 km)	?	*Silification, sericitization,* kaolinization, and chloritization in that order of importance
Bougainville	Upper Olig-L-Mioc Andesites	Pliocene quartz diorite and quartz diorite porphyry stocks	None reported	Two overlapped spatial sequences: *Biot → chl/epid →* fresh rock *Biot → chl/seric →* Chlorite and Biotite interleaved.
Plesyumi and You You, New Britain	Miocene-Pliocene ?	Miocene-Pliocene quartz diorite →	Granodiorite	Propylitic alteration with weak phyllic alterations superimposed
Frieda Riv., New Guinea	Miocene metasediments	No major intrusive phase exposed	Miocene, andesite-dacite dike swarm (dated at 7 million years)	Alunite, sericite, chlorite, and calcite reported
Yanderra, New Guinea	Miocene metasediments	Miocene granodiorite batholith (12.5 m.y. old)	Andecite-dacite dike swarm (dated at 7 million years)	*Mod-strong propylitic alterations* superimposed as weak potassic alterations
OK Tedi,[a] Papua	Miocene shales	3,000 foot diameter Plio-Pliestocene Quartz-latite-stock	None (Biotite age is 1.3 million years)	Potassium silicate alteration decreases outward, weak clay-sericite alteration superimposed

Source: By G. Ballantyne. (unpublished).
[a] The OK Tedi deposit occurs within the continental eastern part of New Guinea.

averaging 0.43 percent Cu. Even with this low grade, the Carmen ore body is reported to have a favorable ore to waste ratio of 3.43:1. The planned mining rate is 32,000 tons per day of ore. Drilling is still being done on the *OK Tedi*, Papua, porphyry copper deposit.

The objectives of the present drilling is to increase the potentially mineable ore reserves with the ultimate target of defining reserves in the range of 200 to 250 million tons averaging better than 0.8 percent Cu. The gold and molybdenum content of the deposit are appreciated but not presently accurately known with the exception of one hole, No.

83, which from 0 to 43 meters contains 3.16 grams per ton of gold.

The host intrusion of quartz latite is about 1000 meters in diameter and of a comparatively young age of Plio-Pleistocene with biotite dated at 1.3 million years. It intrudes Miocene shales and limestone and has formed extensive skarn contact rocks that apparently are well mineralized with intervals of lower-grade ore. Tests, however, have proved that the skarn rock will be difficult to concentrate.

Potassium-silicate alteration decreases outward with associated superimposed weak propylitic-sericitic alteration.

Table 19-10 Porphyry Copper Deposits of the Philippines

Name	Owner	Reserves	Cu %	Au oz/t	Ag oz/t	Mining rate	Disc decade	Production	Remarks
Philippines									
Biga and Lutopan (Cebu)	Atlas Consolidated	753 MT	0.47	Rec.	Rec.	70,000 T/day	1950s	75 MT 0.5% (1956–1971)	
Marcopper (Marinduque)	Marcopper	108 MT	0.64	0.007	0.03	28,000 T/day	1960s		
Santo Tomas II (Luzon)	Philex	48 MT	0.50	Rec.	Rec.	19,000 T/day	1950s		
Sipalay (Negros)	Marinduque Mining	80 MT	0.79	Rec.	Rec.	14,500 T/day	1950s	21 MT 0.76 Cu (1957–1969)	
Kennon (Luzon)	Black Mountain	27 MT	0.47	0.003	0.02	6,000 T/day	1950s		
Santo Nino (Luzon)	Baguio Gold	50 MT	0.50	Rec.	Rec.	3,300 T/day	1950s		
Boneng-Lobo (Luzon)	Western Minolco	70 MT	0.45	0.013		10,000 T/day	1960s	Starts spring 1974	Construction underway
(Luzon)	Dizon Copper	75 MT	0.50	0.007		12,000 T/day (planned)	1960s		Mitsubishi drilling
Ino-Capayang (Marinduque)	Consolidated Mines	70 MT	0.625	Rec.	Rec.	15,000 T/day (planned)	1960s		$30 million capitalization
Lubuagan (Luzon)	Batong-Buhay	32 MT	0.62			6,000 T/day (planned)	1970s		French loan of $18 million
Bessemer (Luzon)	Philippine Iron Mines	20 MT	0.22				1960s		Plus 0.03% Mo and 11.2% Fe
Tawi Tawi (Luzon)	Benguet Consolidated	100 MT	0.50				1960s		Under exploration
Hinobaan (Negros)	Lepanto Consolidated	80 MT	0.50				1970s		Under active exploration
Upper Masara (Mindanao)	Samar Mining	50 MT	0.55	0.01			1960s		Under active exploration
Basay (Negros)	CDCP Mining	80 MT	0.51				1960s		Under active exploration
Taysan (Luzon)	Tayson Copper Mines						1960s		Under evaluation
Cordon (Luzon)	Vulcan Mining						1970s		Tie-up with International Nickel
Lumbay (Luzon)	Colossal Mining						1970s		Starting drilling

A summary of some drill hole depths and assays are given in Table 19-11.

Australia. Several porphyry copper deposits are under development or operation in Australia (Table 19-6). The *Copper Hill mine,* NSW, is an old mine that lay dormant for more than a century before deep drilling in 1966 to 1967 disclosed a porphyry copper deposit but of very low grade, with subsidiary gold more prominent than molybdenum. The deposit has the porphyritic characteristics of multiple intrusions and hydrothermal alteration, inter-mineralization intrusive activity, and the existence of a secondary enrichment blanket of chalcocite (digenite).

In the *Cobar-Mount Hope* district, NSW, Australia, iron-rich copper-ore deposits occur according to Raynor as epigenetic, high-temperature hydrothermal deposits derived from a widespread homogenized ore fluid. These deposits are emplaced in sedimentary rocks, and the hydrothermal solutions are believed to have been derived from deep-seated magmas even though the only igneous rocks in the district are two small pipes of orthoclase porphyry located 18 kilometers southeast of Cobar. Granite occurs along the orogenic belt and is evident north and south of the district but whether or not it exists at depth below the district is not known. This district is large and highly mineralized.

The *Cadia NSW* deposit consists compositionally

Table 19-11 Summary of Ok Tedi Drill Assay Results

Hole	Meters	Copper (%)	Hole	Meters	Copper (%)
83	0–68	0.08[b]	92	18–35	0.11
	0–123			35–48	3.87
84	10–47	1.10		48–103	0.12
	47–92			103–222	0.76
85	92–183	1.92			
		0.13	93	0–90	0.13
		0.61		90–136	<0.32
				136–184	<0.30[a]
86	0–170	0.09			
	170–185	0.72	94	0–38	0.06
	185–210	0.28		38–90	0.10
	210–242	0.78		90–137	0.23
	242–300	0.31			
	300–335	0.71	95	0.85	0.05
	335–352	0.23		85–164	0.13
				164–396	1.09
87	0–116	0.07			
	116–308	0.91	96	0–62	0.89
	308–367	0.21		62–137	0.35
88	0–102	0.06	97	88–280	0.78[a]
	102–136	0.47			
	136–275	0.13	98	0–107	0.05
				107–140	0.47
89	0–107	0.05		140–143	—[a]
				143–152	0.60
90	0–133	0.05			
	133–360	0.81	99	[a]	[a]
91	0–47	0.07	100	0–3	0.13
	47–214	1.23		3–21	0.88
				21–152	0.33

Source: Mining Magazine, July 1976.
[a] = Assays incomplete.
[b] Includes 0—43m 3.16 g/ton gold.
check assays and gold, nonsulfide copper and molybdenum assays in progress.

of a quartz diorite to quartz monzonite intrusion that measures about 3 to 4 kilometers in diameter. The host rocks are Ordovician-Silurian andesites and siltstones. Weak potassic-silicic alteration exists with moderate to strong propylization occurring with better metallization.

African Deposits. Information about the huge copper deposits of the *Zambian copper belt* and those deposits in Zaire are included in Chapter 11. In *Namaqualand,* the *O'okiep* and *East O'okiep mines* contain low-grade copper sulfide ores disseminated in norite, for which both a magmatic and replacement origin have been claimed. At *Messina, Transvaal,* hypogene chalcocite, bornite, and chalcopyrite, with minor pyrite and specularite, and quartz, feldspar, prehnite, delessite, and zeolites, occur in

granite gneiss altered to zoisite quartz. The ores extend to 700 meters in depth. At *Tsumeb, Southwest Africa,* are two hydrothermal replacement lenses in dolomite adjacent to aplite, which consist of massive aggregates of chalcocite, galena, and sphalerite with some enargite, germaninite (?), stibioluzonite, and vanadium compounds. Limited oxidation has yielded an astonishing variety of oxidized minerals. At *Insizwa, South Africa,* are small magmatic copper sulfide deposits carrying cubanite; similar low-grade nickel-copper deposits occur in the Bushveld rocks at *Pilansberg.* Massive gold-copper lenses carrying pyrrhotite, chalcopyrite, and pyrite have recently been disclosed at *Macalder, Kenya.*

The *Prince Leopold mine,* Zaire, is a sulfide pipe-like replacement deposit in limestone, extending up

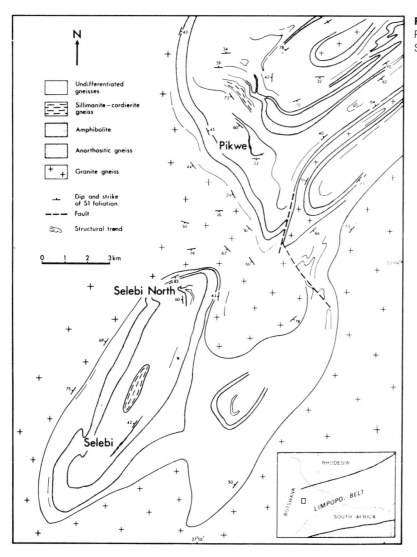

Figure 19-31 Location of Selebi-Pit copper-nickel project. (From Skillings Mag. Rev., 1977.)

into Kundelungu beds. Rich ores consist of bornite, chalcopyrite, galena, and considerable chalcocite and silver, but no cobalt or gold. Some disseminated sulfide ores, similar to the Rhodesian ores, have been developed near Tshinsenda.

Selebi-Pikwe, Botswana. The *Selebi-Pikwe* ore deposits are situated in eastern Botswana, 55 kilometers east of the main rail link connecting South Africa with Rhodesia, Zambia, and Zaire. The mining complex includes underground and open pit mining operations, concentrator, and a smelter at the larger Pikwe deposit. The copper-nickel deposits comprise two stratabound sulfide deposits 15 kilometers apart of Archean age situated in the highly deformed and metamorphosed gneissic successions of the Limpopo field belt. Both disseminated and massive sulfides occur in silllike bodies of amphibolites that are conformable within quartz-

feldspar-biotite-hornblende gneiss. It is suggested that the deposits have a syn- or pre-tectonic intrusive origin. Post ore folding has resulted in a variable thickness of the ore. Widths range from about 40 meters to less than 1 meter (Fig. 19-31).

According to Valentine, the Pikwe ore body forms an arcuate shape at the surface and has a strike length of 3300 meters, which reduces below to 400 meters. The deepest point of intersection of the ore zone is 370 meters. The deposit can be divided roughly into three zones: a southern massive plus 70 percent sulfide zone with grades up to 2.5 percent Ni and 1.5 percent Cu, enclosed in grey gneisses; a central semimassive 40 to 70 percent sulfide zone underlaid with poorly mineralized host amphibolite; and a northern low-grade 0.8 percent nickel and 0.8 percent copper zone of generally disseminated ore with localized bands of massive sulfide distributed throughout the host amphibolite.

Other Important Deposits

Rio Tinto (Huelva) Spain. The massive pyrite deposits of the famed Rio Tinto or Huelva district (Fig. 19-32) of Spain-Portugal have been mined for 3000 years, first for gold and then for copper and sulfur. They are the largest pyritic copper deposits of the world.

The ores are associated with sheared quartz porphyry intrusives in Paleozoic slates, and both are cut by diabase dikes. The deposits lie in porphyry, in slate, or at contacts. Some 50 ore bodies have been worked, mostly in huge open cuts. Eight large ones are concentrated near Rio Tinto, and a similar group at Tharsus.

The deposits are collossal lenses of solid sulfide flanked by disseminated sulfides and sericitized wall rock, merging into fresher rock with diminution of sulfides. The San Dionisio Eduardo, and South Lode are connected bodies 3,000 meters long, up to 2,500 meters wide, and 500 meters deep. As the deposits were formed at different levels, erosion has removed some, disclosed others, and not reached others.

The ore is massive pyrite with 1 to 5 percent silicates, carrying chalcopyrite, minor sphalerite and galena, traces of tetrahedrite, enargite, and pyrite, and uncommon arsenides, selenides, and antimonides. A little chalcocite and covellite occur in a shallow supergene zone. The ores yield sulfur, copper, a little lead, zinc, gold, and silver, and small amounts of nickel, cobalt, and other metals. They consist of smelting ores, leaching ore, copper-sulfur ore, and sulfur ore.

Their origin is controversial. A magmatic origin has been maintained with vigor, and a hydrothermal origin has been strongly advocated. The associated hydrothermal rock alteration and the nature of the ore, however, indicate that these interesting deposits do not differ from other massive sulfide deposits of suggested exhalative volcanic origin.

Selected References on Copper

Preliminary study of vein formation, Butte, Mont. 1949. R. H. Sales and C. Meyer. Econ. Geol., v. 44, p. 465–484. *Theories of formation.*

The Porphyry Coppers. 1957. A. B. Parsons. AIME, New York.

XVI International Geological Congress, Washington, 1933. Guidebooks: 2, *Ducktown;* 14, *Chino;* 23,

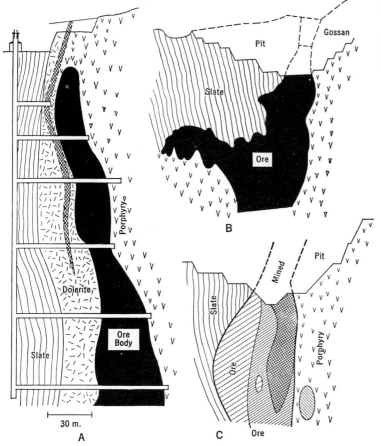

Figure 19-32 Cross section of Rio Tinto ore bodies, Huelva, Spain. A. Perrunal deposit; B. San Dionisio lode, central section; C. Smith Lode. (After D. Williams, Inst. Min. and Met.)

Butte; 17, *Bingham* ; 14, *Bisbee*; 27, *Lake Superior*; 19, *Colorado. Good summaries.*

The Ajo mining district, Arizona. 1946. J. Gilluly. U.S. Geol. Survey Prof. Paper 309. *Detailed geology of a "porphyry copper."*

Grade and tonnage relationships among copper deposits. 1975. D. A. Singer, D. P. Cas, and L. J. Drew. U.S. Geol. Survey Prof. Paper 907A. *Grade and tonnage data on 267 deposits and descriptions of porphyry, stratabound, and massive copper deposits.*

Geology and mineral resources of Japan. 1960. Geol. Survey Japan, p. 176–180.

General Reference 1. 1975. Special issue devoted to Japanese mineral deposits, v. 70.

Characteristics of Philippine porphyry copper deposits and summary of current production and reserves. 1976. W. Saegart and D. E. Lewis. AIME Preprint 76-1-79. *Excellent summary and details about Philippine porphyry copper deposits.*

The Soviet Challenge in Base Metals. 1971. A. Sutulov. University of Utah Printing Service, Salt Lake City. *Succinct summary of Russia's base metal deposits, production, and smelters, in comparison with production of other nations.*

Zoning of the White Pine copper deposit. 1966. A. C. Brown and J. W. Trammel. Econ. Geol., v. 61.

The copper mineralization in the Corocoro Basin, Bolivia. 1964. P. Ljanggren. Econ. Geol., v. 59, p. 110–125.

The porphyry copper deposit at El Salvador, Chile. 1975. L. B. Gustafson and J. P. Hunt. Econ. Geol., v. 70, p. 857–912.

Hot Brines and Recent Heavy Metal Deposits in the Red Sea. 1965. E. T. Dagens and D. A. Ross, eds. Spring- er-Verlag, New York, Inc., p. 600. *Numerous papers on the important source of metals at the present day.*

General Reference 19. Chapter on copper.

Geology of the Porphyry Copper Deposits, Southwestern United States. 1966. S. Titley, and C. L. Hicks, eds. University of Arizona Press, Tucson. *Twenty-three papers on deposits and other topics pertaining to prophyry copper deposits.*

Geology and Economic Minerals of Canada. 1968. R. J. W. Douglas, Sci. ed. Economic Geology Rept. No. 1. *Comprehensive succinct summary of Canadian geology and mineral deposits.*

Copper Production in Russia. 1967. A. Sutulov. University de Concepcion, Chile. *A review of Russian copper deposits on metallurgical plants and smelters.*

Metamorphism and volume losses in carbonate rocks. 1957. J. R. Cooper. Bull. Geol. Sci. Am., v. 18, p. 577–610.

LEAD AND ZINC

Despite chemical dissimilarity, geological conditions favor the formation of lead and zinc together. The world over, the companion sulfides galena and zinc blende, or their oxidation products, yield most of these important metals of modern industry. They are also associated with copper and other base sulfides; less commonly, lead and zinc occur separately. The association is so general, however, that the two are considered together.

History

Lead has been known since the beginning of history. Lead water pipes found in Pompeii show that its present-day use for plumbing was known to the Romans. In fact, it has been suggested that the Fall and Decline of the Roman Empire may have resulted from lead poisoning of the aristrocratic Romans as lead affects the fecundity of women. Those Romans of lower class drank or sipped their wine from clay containers, not lead cast decanters. Lead was even added to the wine supposedly to improve its taste. The Chinese used it for money and debasing coins before 2000 B.C. despite Pliny's statement that it was discovered by Midas, King of Phrygia, about 1000 B.C. Ancient silver-lead deposits were worked in the Mediterranean countries, India, China, Persia, and Arabia. The famed Laurium deposits of Greece were worked in 1200 B.C. Lead was used by the ancients for ornaments, coins, solder, bronzes, vases, and pipes. Later it was extensively mined in Spain, Greece, the Pyrenees, the Harz, and Silesia.

Zinc was discovered as a metal in 1520, but bracelets filled with zinc were found in the ruins of Cameros, burnt in 500 B.C. The Greeks and Romans unknowingly used it to make brass, as they found that copper melted with smithsonite resulted in a metal more yellow than bronze. In the sixteenth century zinc was imported into Europe from India and China, and mining of zinc in Europe began in 1740. In the Americas, lead was mined by the pre-Spaniards, and then by the Spaniards in Bolivia and Peru. The first real mining in the United States was in Missouri in 1720. The first production of zinc was at the Washington arsenal in 1838 from New Jersey ores.

Production

The large industrial demand for lead and zinc has caused exhaustion of the earlier mined small deposits throughout the world, and large steady production now comes from relatively few regions. Five countries, for example, produce more than two-thirds of the World production of lead.

Table 19-12 Chief Producers of Lead in Order of Rank

Lead	Percent of world	Zinc	Percent of world
1. United States	16	1. Russia	14
2. Russia	12	2. Canada	9
3. Australia	12	3. Australia	9
4. Canada	12	4. United States	8 (Smelter production = 15)
5. Mexico	10	5. Poland	4
6. Peru	3	6. Mexico	3
7. Germany	4 (Prod = 8)	7. Germany	2
8. Yugoslavia	4	8. Japan	1.5 (Smelter production = 13)
		9. Zambia	1
		10. Zaire	1

The annual production in the 1970's has ranged between 3.5 to 4.0 million tons per year for lead and about 6.0 million tons of zinc. The chief producing countries in order of rank and their average ratio to world production are listed in Table 19-12.

The countries with the greatest lead reserves are the United States (39,185,000 tons), Australia (18,500,000 tons), Russia (18,000,000 tons), and Canada (15,850,000 tons). About two-thirds of the world's zinc and lead are each derived from the four countries.

In the United States, the chief ranking states for lead are Missouri (90 percent), Idaho, Utah, Colorado, Montana. For zinc: Tri-State Missouri, Oklahoma, Kansas, Idaho, Arizona, Utah, and Tennessee.

Missouri has been the leading state in production of lead since about 1905, producing 519,000 tons in 1979.

Uses

Lead and zinc rank next to copper as essential nonferrous metals in modern industry. Normally, the chief uses are for storage batteries, ammunition, electrical cables, caulking lead, pipe, solder, and red and white lead. Lead is used over other materials because of its demonstrated corrosion resistance over a wide range of conditions, and after a long and useful life, it has a high salvage value.

Lead pigment in paints and its use in tetraethyl gasoline are declining because of lead toxic affect and environmental restrictions. Still, lead as a gasoline antiknock additive consumes 20 percent of total lead consumption; batteries consume 42 percent.

Zinc is used in galvanizing, die castings, alloyed with copper to form brass, wire, tubes and pipe, and other uses. Diecasting alloy (40 percent) is by far the greatest use of zinc followed by galvanized sheets (18 percent).

Zinc galvanizing, the process of coating steel with zinc to prevent rusting, is done by dipping sheets into molten zinc. "Sheradizing" is the coating of sheets by zinc vapor. Zinc constitutes 30 to 40 percent of brass and is also a component of certain bronzes and other alloys. Zinc die castings are used for automobile carburetors, pumps, hub caps, and drills, and for many other purposes. Rolled zinc is used for glass jar tops, batteries, cans, and similar purposes. Zinc also finds application in precipitating gold, and in medicines and chemicals. Little secondary zinc is recovered.

Mineralogy, Tenor, and Treatment

The mineralogy of lead and zinc ores is simple. Only three lead and six zinc minerals are commercial sources of these metals; two are sulfides and four are oxidation compounds.

Lead Minerals		Percent Pb	Zinc Minerals—Cont.		Percent Zn
Galena	PbS	86.6	Hemimorphite	$Zn_4Si_2O_7$	
Cerussite	$PbCO_3$	77.5	(Calamine)	$(OH)_2 \cdot H_2O$	54.2
Anglesite	$PbSO_4$	68.3	Zincite	ZnO	80.3
			Willemite	Zn_2SiO_4	58.5
Zinc Minerals		**Percent Zn**	Franklinite	(Fe, Zn, Mn)	
Zinc blende (sphalerite)	ZnS	67.0		$(Fe, Mn)_2O_4$	15–20
Smithsonite	$ZnCO_3$	52.0			

Associations. Galena and zinc blende may occur separately but are generally associated, and zinc blende rarely occurs without galena. In oxidized ores, however, the two become separated. Galena is rarely free from contained silver (argentiferous galena); without silver it is called "soft lead." Both minerals may carry gold. Cadmium is a common associate of zinc, and bismuth and antimony of lead. Pyrite and chalcopyrite are common associates, and silver minerals are likely to be present.

Tenor. The tenor of lead ores depends upon the associated metals. In Missouri, silver-free ores average 2 to 4 percent lead. Many 2 to 3 percent lead ores containing silver are profitably mined; commonly they average 6 to 8 percent, and rich ores of 20 percent or more are rather common. Zinc ores average 2 to 3 percent zinc in the Tri-State district and range from 4 to 12 percent elsewhere.

Treatment. Simple galena and sphalerite ores of low grade undergo gravity concentration by jig or table, or they may be concentrated by froth flotation. Complex ores undergo differential flotation, largely eliminating pyrite and yielding lead concentrates low in zinc, zinc concentrates with a little lead, and copper concentrates.

The lead concentrates are smelted, and the molten lead carries down any silver and gold present. The lead is then disilverized and purified. Sphalerite concentrates are roasted, and the zinc is volatilized, condensed to liquid zinc, and cast into slabs. In electrolytic treatment, the zinc is dissolved by sulfuric acid from roasted concentrates, copper and cadmium being removed from the solution, and then the zinc is electrolytically deposited.

Kinds of Deposits and Origin

The principal classes of predominantly lead and some zinc deposits and some examples of each according to Morris, Heyl, and Hall are as follows;

1. *Stratabound deposits of syngenetic origin.* Kuperschiefer, Germany; Dzhazkazgan, Kazakhstan, USSR.
2. *Stratabound deposits of epigenetic origin.* Southeast Missouri; Tri-state; Upper Mississippi Valley; Pine Point, Northwest Territory; Canada; Alpine lead deposits; Kuperschiefer; Laisvall, Sweden; Silesia-Cracow, Poland; and many others.

3. *Volcano-sedimentary deposits.* Kuroko, Japan; Atasu and Achisai, Kazakhstan, USSR; Bathurst, New Brunswick, Canada; Broken Hill, NSW, and Mount Isa, Australia; Kidd Creek, Canada; Sullivan, British Columbia, Canada; etc.
4. *Replacement deposits.* Cerro de Pasco, Peru; Tintic, Utah; Bingham, Utah; Gilman and Leadville, Colorado; Central Mexico; Sardinia, Italy; Trepca, Yugoslavia; Tsumeb, Southwest Africa; Turlan, USSR.
5. *Veins.* Coeur d'Alene, Idaho; Harz Mountains, Germany; Silverton area, Colorado; North Pennine, England; Santa Barbara, Fresnillo, and Taxco, Mexico; Clausthal and Freiberg, Germany, East and West.
6. *Contact pyrometasomatic deposits.* Kamioka, Obori, Chichibu, and Nakatatsu, Japan; Tetyukhi and Siberia, USSR; Trepca (?), Yugoslavia; and Darwin (?), California.

Origin. Most lead and zinc occurs as cavity fillings and replacements formed by low-temperature hydrothermal solutions. Prevailingly, they occur in limestones or dolomites. Considerable difference of opinion has existed in the past as to the origin of many zinc-lead deposits in limestone, such as those of the Tri-State district. Three views are current. They were formed by (1) connate solutions, (2) ascending artesian meteoric waters, and (3) hydrothermal solutions of igneous derivation. Support has steadily been decreasing in favor of the last theory of origin with greater acceptance of connate solutions, especially in Mississippi Valley type deposits.

In many regions lead and zinc deposits have been oxidized; much of the Mexican production, for example, comes from oxidized ores. Even the pyritic Rio Tinto deposits of Spain carry some lead and zinc (Fig. 19-32).

World Distribution of Lead and Zinc

Despite wide distribution, most of the world production comes from relatively few regions. The belts, however, are less well defined than those of gold and copper.

North America. The world's greatest concentration of zinc-lead ores occurs in the Mississippi Valley region, centering around the Tri-State district of Missouri, Oklahoma, and Kansas and diminishing

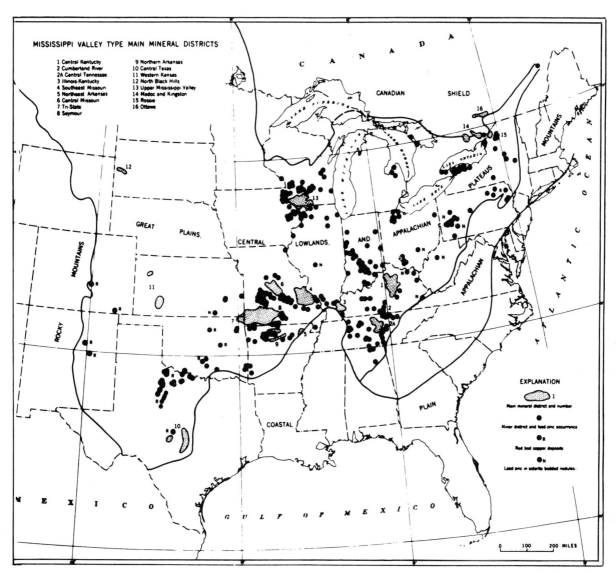

Figure 19-33 Map showing the central and eastern United States and the Mississippi valley-type districts. 1, Central Kentucky; 2, Cumberland River; 2A, Central Tennessee; 3, Illinois; Kentucky; 4, Southwest Missouri; 5, Northwest Arkansas; 6, Central Missouri; 7, Tri-State; 8, Seymour; 9, Northern Arkansas; 10, Central Texas; 11, Western Kansas; 12, North Black Hills; 13, Upper Mississippi Valley; 14, Mades and Kingston; 15, Rosse; and 16, Ottawa. (Heyl, Econ. Geol. Monograph 3, 1967.)

northward into Wisconsin (Fig. 19-33). A parallel zinc belt lies in the lower Appalachians, embracing the Mascot-Jefferson City region of Tennessee and Austinville, Virginia. A great isolated lead district and the Viburnum Trend (Fig. 19-34) lie between these belts in southeastern Missouri. Two other small isolated areas occur at Franklin, New Jersey, and Edwards, New York. A Rocky Mountain province of lead-zinc ores centers in Colorado and extends into New Mexico and Utah. The Coeur d'Alene district of Idaho is a rich lead; zinc prov-

ince that extends northward into Canada and embraces the greatest lead-zinc mine of the continent. In Manitoba, zinc-copper ores center around Flin Flon, and similar ores lie around Noranda, Quebec. An isolated lead-zinc occurrence, not unrelated to those of Gaspé, lies at Buchans, Newfoundland.

In Mexico, a rich province of lead-silver-zinc ores lies in the center of the country, expanding in Chihuahua and Coahuila.

About half of the world's lead and zinc comes from the North American Continent.

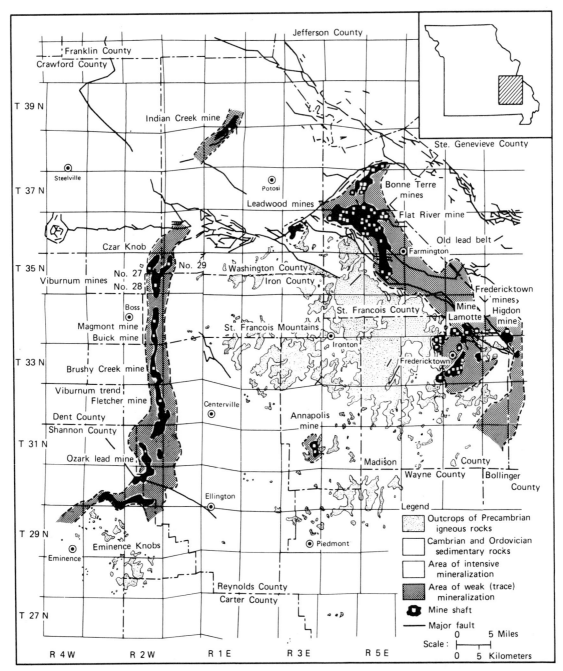

Figure 19-34 The southeast Missouri lead-zinc-copper mining district. (Kisvarsanyi, 1977, Econ. Geol., v. 72.) Production from the old lead belt was roughly 8 million tons of lead during the last 100 years. Potential of the Viburnum trend is an estimated 30 million tons.

Southeastern Missouri. This district, centers around Flat River (Fig. 19-34) and is called the "lead belt." The deposits occur chiefly as disseminated replacements in the flat-lying Bonneterre dolomite of Cambrian age. They lie at shallow depth, within a vertical range of 70 meters, and average 3 to 5 meters in thickness, 70 to 100 meters in width, and 300 meters or more in length.

The Southeast Missouri lead district (Fig. 19-34) was one of the greatest lead districts in the world having produced more than 9 million tons of pig lead. Discovered in 1720 by French miners and slaves, the "Old Lead Belt," from which more than 90 percent of the previously mentioned production has been derived, is now near its economic death.

A new district, however, forming a north-south narrow band is located about 80 kilometers west and has revitalized Southeast Missouri's mining industry (Fig. 19-34). The band is known as the *Viburnum lead belt* and contains deposits that are classical examples of the "Mississippi Valley type."

The ore deposits are stratiform in character and are localized in a narrow carbonate bar and algal reef environment (Fig. 19-35) in the flanks of exposed Precambrian rocks of the St. Francis Mountains. Snyder and Gerdemann describe ore structures as primary depositional features such as pinchout zones, disconformities, ridge structures, reefs, and submarine gravity slides. The ore is predominantly lead with minor zinc. Copper, nickel, cobalt, cadmium, and silver occur in small amounts.

Lead-zinc-barite mineralization exists in every formation from upper Cambrian Lamotte sandstone to lower Ordovician Jefferson City, the youngest formation in the district. Nevertheless, the Southeast Missouri lead district is arbitrarily defined as the area in which stratiform deposits of lead-zinc-copper occur in the Bonneterre formations.

Theories have been abundant to explain the origin of these deposits, but even the lack of association with known intrusions has not dismissed entirely a magmatic hydrothermal origin. The preponderence of evidence favors the source of the low-temperature solutions and their contained metals from deep, adjacent sedimentary basins. These brine solutions were capable of leaching the trace metals from sediments and transporting them as metal chloride complex ions.

The abundance of sulfate-rich, presumably connate water in the vicinity of Mississippi Valley-type deposits in the United States is well known as it is often tapped in drilling and mining operations. It is possible, therefore, that the sulfate in this source could have undergone bacterial reduction to H_2S as the solutions passed through bioherm or reeflike, more porous zones that contained enough residual organic matter to provide an energy source sufficient to support the sulfate-reducing bacteria contained therein. The H_2S there formed would be a most effective reductant of even trace amounts of lead or zinc added to, or contained in, the same connate fluids. Recent studies by Doe and Delevaux however, on lead isotopes suggest a juxtaposed acquifer formation source for the Southeastern Missouri deposits.

The slow, prolonged passage of such solutions through a trap such as a porous bioherm structure would permit continued increase in the concentration of lead and zinc sulfides to the slowly growing but large crystals.

Tri-State district. This broad district located in Kansas, Missouri, and Oklahoma was not only the greatest zinc district of the world, but it yielded the lowest-grade zinc ores. Some lead was recovered but no other metals. Its remarkable waste dumps, dotting the flat surface like sand dunes on a desert, attest to the extensiveness of the underground workings. Its area is about 5,000 square kilometers, and it has yielded over $2 billion in metals.

These deposits differ from other U.S. Mississippi Valley-type of deposits that occur in Cambrian and Ordovician sediments, in that Tri-State ores occur in Mississippian Limestones containing abundant chert.

The geology is relatively simple. Nearly flat-lying sediments include the Boone formation, which is separable into 16 beds and is the host rock of the ores. It is unconformably overlain here and there by uneroded patches of Pennsylvania shale. Beneath the pre-Pennsylvanian erosion surface solution breccias developed, with sinkholes, underground drainage channels, and caves, in the underlying limestone and particularly along the Grand Falls chert member. Chert and jasper beds are numerous in the Boone. Two prominent faults are known, one of which is mineralized. Also, there are prominent shear zones that brecciated the cherts and localized the ores. These structures appear to be related to major structural features of the Mississippi Valley.

The deposits consist of (1) runs and circles of ore near the surface, and (2) sheet or blanket ore along the Grand Falls chert. The first are the most productive. The runs fill and follow shallow solution channels along cherty horizons. They range from 3 to 100 meters in width and are as much as 25 meters thick and 3 kilometers long. The circles encircle sinkholes and attain dimensions of 250 by 200 meters; the ore is deposited in thicknesses up to 10 meters, around fragments of slumped chert. The sheet ground, which may be 5 meters or more thick, is in the Grand Falls chert where the limestone has been removed from around the chert nodules.

The minerals consist chiefly of zinc blende and subordinate galena with minor amounts of wurtzite, marcasite, pyrite, and chalcopyrite. Enargite and millerite have also been identified. The gangue minerals are quartz, chert, carbonates, and a little barite. There is also a large group of oxidation products. The minerals occur in definite sequence.

Geologists of the Eagle Pitcher Industries, Inc., namely, Brockie, Hare, and Dingess summarize the enigmatic genesis of these deposits by concluding that "warm, saline, ore-bearing solutions evidently

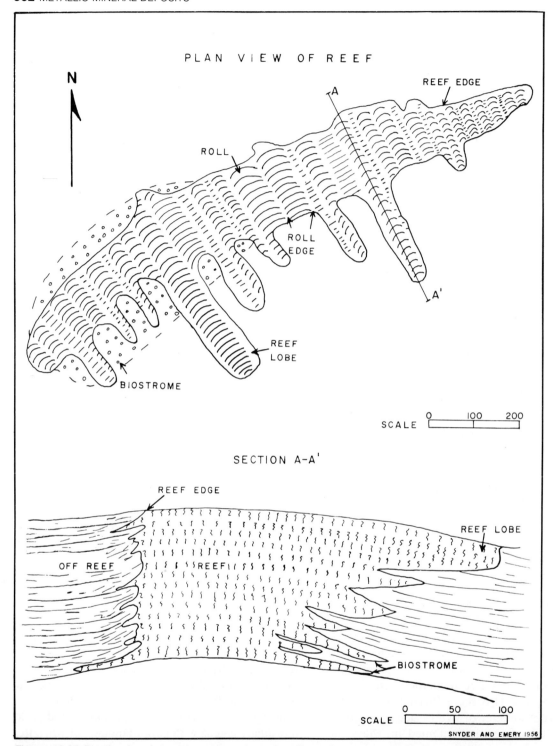

PLAN VIEW OF REEF

REEF EDGE

ROLL

ROLL EDGE

REEF LOBE

BIOSTROME

SCALE 0 100 200

SECTION A-A'

REEF EDGE

REEF LOBE

OFF REEF

REEF

BIOSTROME

SCALE 0 50 100

SNYDER AND EMERY 1956

Figure 19-35 Details of reef structure, plan and section. (From: Snyder and Gerdemann, 1968, Graton-Sales Vol., p. 342.)

were derived from some distant source, migrating through the Cambro-Ordovician sediments until a zone of structural weakness, such as the Miami Trough, combined with a window in the Chattanooga shale (and a certain thinness of the North-view shale) permitted concentration and access of the ascending solutions into the Mississippian formations. The solutions migrated laterally outward from the Trough and deposition occurred in ground already prepared by dolomitization and brecciation.'' The time of mineralization may have been diagetic or epigenetic.

Uplifts associated with the Ozark uplift during post-Ordovician, pre-Mississippian time resulted in truncation and omission with unconformities of some pre-Mississippian formations. By Mississippian time, carbonate sediments were being deposited over the entire area with some minor folding and doming. The Miami trough and Pitcher anticline presumably formed during this time and played significant roles in localization of the ore-bearing fluid.

Generally, similar deposits are scattered over the upper Mississippi Valley, particularly in Wisconsin. These ores, now largely exhausted, occur mainly as fillings of caves, pitches, and flats, and solution joints in the Galena limestone. They also probably have a low-temperature origin.

The Appalachian Zinc Provinces. This important belt of straight zinc ores includes Mascot, Jefferson City, and Embree, Tennessee; and Austinville, Virginia; and is the third most important zinc district in the United States. It illustrates deposits formed in breccias.

The Mascot-Jefferson City deposits were formed by the filling and replacement of tectonic and collapse breccias, localized by faults and fault intersections, in dolomitized limestone beds within the gently folded Knox dolomite. Light yellow sphalerite, dolomite, and calcite fill the breccia interstices, and the fragments are partly replaced. Traces of galena, marcasite, and barite are present. The ore bodies follow dipping beds and average 5 to 30 meters in thickness, 40 to 50 meters in width, and have been followed to a depth of 300 meters. At Austinville, galena, chalcopyrite, and pyrite are also present, and oxidized ores are abundant.

The origin of these ores is similar to the Mississippi Valley region. The view that they are low-temperature deposits is well accepted.

Franklin Furnace, New Jersey. The unique zinc deposits of Franklin and Sterling Hill have no exact counterpart in the world. Zinc mining began about 1840. The deposits illustrate high-temperature replacement or metamorphism of an earlier oxidized hydrothermal deposit. The region is made up of Precambrian granitic gneisses and crystalline marble cut by pegmatite dikes.

Both deposits resemble a half-canoe in shape with one side higher than the other and the keel pitching gently underground. In plan and section they are hooklike. They are enclosed within the marble. The thickness of the curved layer ranges from 4 to 30 meters. The inclined keel has been followed over 500 meters and to a depth exceeding 300 meters. The ore is interrupted by trap and pegmatite dikes.

The mineralogy is unique. The ore minerals are zincite. willemite, and franklinite occurring as abundant grains and bunches in calcite. But the deposit is a mineralogists' paradise, since more than 100 minerals have been identified, many occurring nowhere else. Most of them, such as garnets, amphiboles, pyroxenes, and complex silicates, resulted from metamorphism. Some sulfides and arsenides occur in postore veins. The franklinite is used for making zinc oxide for paints, and the residual manganese is recovered. The zincite is also used for zinc white, and the willemite for metallic zinc.

The origin is most puzzling. The chief problem is to account for abundant zinc in the form of oxides and silicates instead of the customary sulfides found elsewhere. Wolff thought they were metamorphosed sediments. Spencer thought they resulted by injection of magmatic emanations. Lindgren considered them as pyrometasomatic. Spurr and Lewis thought them to be an ore magma of sulfides subsequently metamorphosed. Tarr considers them to have been originally deposited in limestone as sulfides and carbonates, then weathered to oxides and hydroxides, and later matamorphosed. Palache suggests that hydrated oxides and silicates replaced limestone and then were metamorphosed when the limestone was recrystallized, which is one of the better theories. The mineralogy indicates high temperature and pressure, and the last theory has much to recommend it.

Edwards, New York. At Edwards and Balmat are two unusual zinc deposits. They occur in Precambrian marble, cut by silicic intrusives, that J. S. Brown believes is part of a large plunging anticline with Balmat mine on one limb and Edwards on the other (Fig. 19-36). Both mines are over 600 meters deep.

The deposits may be hydrothermal replacements along folded limestone bedding resulting in V-shaped lodes. The primary ore consists dominantly of zinc blend and pyrite with minor galena and traces of chalcopyrite and pyrrhotite accompanied by carbonate, diopside, talc, quartz, tremolite, hematite, magnetite, phlogopite, chlorite, barite, ilvaite, and garnet, in order of abundance. The Balmat ore averages 16 percent each of sphalerite and

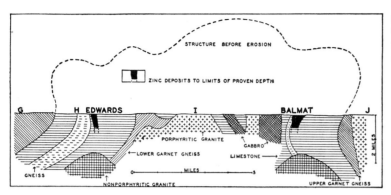

Figure 19-36 Longitudinal section, Edwards-Balmat zinc mine, New York. Black is ore. (Brown, Econ, Geol.)

pyrite, 1 percent galena, one-third carbonate, and one-third silicates. Some willemite also occurs.

Brown shows that considerable weathering has taken place, resulting in surficial masses of red hematite. He also proposes the bold conclusion, not accepted by all, that the supergene hematite is accompanied by supergene magnetite, specularite, chlorite, and willemite and that about one-fourth of the Balmat zinc blende, galena, and chalcopyrite are also supergene sulfides.

Wisconsin massive sulfide deposits. Massive sulfide deposits have recently been found in northern Wisconsin where they occur in Precambrian volcanic rocks of the Rhinelander-Ladysmith greenstone belt that strikes easterly across the northern part of the state. This belt is about 100 kilometers wide and 240 kilometers long. Metamorphism of the volcanics has produced low greenschist to granulite facies similar to the Precambrian geology of the 2.5 to 2.7 billion year old Superior Province geology in Canada. Some galena, however, from the Flambeau and Pelican River deposits, indicates 1.8 billion years for these deposits located also in Northern Wisconsin.

The Crandon deposit located in northern Wisconsin east of the above mentioned deposits was discovered under 30 to 60 meters of glacial drift in July, 1975 by Exxon geologists. Drill hole exploration has developed in-place reserves of 62 million metric tons of ore grading about 5 percent zinc and 1 percent copper with substantial lesser values of gold, silver, and lead.

The Crandon ore deposit is a near vertical tabular body striking N 85° W and consists of a stratabound massive zone and a stratigraphically underlying stringer zone. The body is underlain by felsic volcanic pyroclastic rocks and overlain by intermediate volcanic rocks and their derived sedimentary rocks. As is typical of massive sulfide deposits,

these conditions and other geological observations indicate a submarine-volcanic-exhalative origin.

Other known Wisconsin massive sulfide deposits are the Flambeau and Thornapple deposits of Kennecott Copper Corporation and the Pelican River deposit of Noranda Exploration, Inc., located immediately east of Rhinelander. These deposits occur in an east-west zone extending for more than 180 kilometers in the Precambrian Greenstone belt of northern Wisconsin.

Coeur d'Alene, Idaho. This district is a major producer of lead, zinc and the first silver producer in the United States.

Mining has progressed since 1885 and through 1961 its mines have yielded 754,300,000 ounces of silver; 7,010,000 tons of lead; 2,555,000 tons of zinc; and 125,000 tons of copper, in addition to substantial amounts of cadmium and antimony. The total value of the metals produced exceeds $2.3 billion.

Folded and faulted Precambrian quartzites, argillites, and siltstones of the Belt Supergroup are intruded by Cretaceous monzonite. Great faults are barren, but small faults and shear zones in quartzite have been filled and replaced to form replacement lodes or fissure veins. They attain 2000 meters in length, average about a meter in width, and have been followed downward 2500 meters, but other veins continue below present mine workings. The ore occurs in definite shoots.

The ore consists of disseminated grains and masses of siderite and galena with minor pyrite, sphalerite, and argentiferous tetrahedrite. Boulangerite and pyrrhotite are present as are also quartz, calcite, dolomite, and barite. The ores carry 3 to 4 percent Pb and Zn and 1 ounce Ag per ton. Oxidation is shallow and irregular.

The deposits are related to the monzonite intrusions; age dating, however, is inconclusive even though contact-metasomatic deposits occur close to the monzonite; nearby veins also carry considerable pyrrhotite and magnetite with garnet and tourmaline, but the siderite replacement lodes lie well away from the contact zones.

Bingham, Utah. The Bingham district, if it were not overshadowed by the colossal Utah Copper mine, would be better known for its lead-zinc-silver deposits. It once ranked fourth in the country in lead production, fourteenth in zinc, and third in silver. Since 1865 about $650 million has come from these ores in such mines as the Highland Boy, United States, Utah Apex, Yampa, Jordan, Lark, and U.S. mines.

All of these replacement deposits have been worked out and abandoned. Anaconda Copper Corporation has recently discovered another contact deposit west of the Bingham pit as the result of deep drilling since acquiring the land in 1948. It is known as the Carr Fork mine. Discovery of ore was made in 1969 and development work is now underway. The project will consist of a deep underground mine and a 10,000-ton-per-day concentrator plant located in Pine Canyon on the west side of the Oquirrh Mountains. Reserves are 61.2 million tons of 1.84 percent Cu.

The ore occurs as replacement bodies in Pennsylvanian limestones that are located 600 to 1500 meters below the collar of the production shaft.

Around the porphyry copper deposit of the Utah Copper stock are synclinal folded beds of Pennsylvanian quartzites and intercalated limestones. The limestones near the stock are metamorphosed and contain replacement deposits of pyritic copper ores. Figure 19-37 illustrates the geological reasons for the notorious Highland Boy versus Utah Apex legal battle pertaining to the extralateral and apex lode

claim laws of the United States. As illustrated, both concerns were mining ore from the limestone beds from the 1000-foot level of the Utah Apex and from the 700-foot level of the Highland Boy concern. After months of testimony from dozens of expert geological witnesses, the judge spent more than a year in studying the case before making a decision in favor of Utah Apex. Highland Boy then declared bankruptcy. Both deposits were worked out scores of years ago.

Zonally distributed outward in the same but unaltered limestone beds are the lead-zinc ores, and still farther out are siliceous silver ores. The lead-zinc deposits are fissure veins, replacement lodes, and irregular blanket replacement deposits. The ore minerals are argentiferous galena and sphalerite with tetrahedrite, pyrite, and chalcopyrite. Ore body after ore body were found in these beds with remarkable persistency.

The deposits display a close connection with the monzonite intrusives from whose reservoir emanated the waves of hydrothermal solutions that produced the metallization: first, the core of pyrite surrounded by the disseminated copper metallization in the Utah Copper porphyry, then adjacent contact metasomatism and pyritic copper replacement in altered limestone that displays outward a gradual diminution in alteration coincident with a change from copper, to copper-lead-zinc, to lead-zinc-silver, and then to siliceous silver ores. With the exception of Carr Fork deposit, all of the replacement deposits are defunct.

Other Districts. Many other districts contribute appreciable quantities of lead and zinc. *Park City, Utah,* is an example of prolific lode fissures and bedded replacements in limestone that yield sulfide

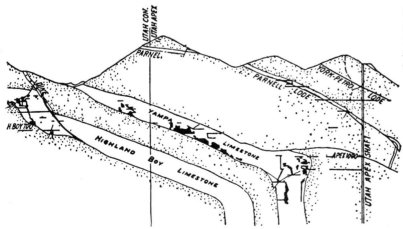

Figure 19-37 Section through part of Bingham, Utah, showing limestone beds that contain lead-zinc-silver ores. (Hunt, 16th Int. Geol. Cong.)

and oxidized ores. The last mine active in the Park City district, the Ontario mine, which had constructed a new concentration plant and was hoped to revitalize the district has recently closed down. But Park City is a "booming town" with former city lots of 25 × 75 feet that were abandoned less than a decade ago, now selling for $6000! The reason is that Park City is a major ski resort in the winter and a city of condominiums, golf courses, and so forth, during the summer.

Pyritic lead-zinc-silver limestone ores from Leadville, Colorado, illustrate gradations from mesothermal replacements on a grand scale, into epithermal, hypothermal, and even contact-metasomatic deposits. Many veins in the *San Juan* of Colorado contain lead and zinc. Mixed ores are numerous in the *Tintic* and *Park City* districts of Utah and the *Pecos* mine of New Mexico, and contact-metasomatic zinc deposits are found at *Hanover*, New Mexico.

Canadian Lead and Zinc Deposits

Much of Canada's lead and zinc is derived from mixed ores. The ranking mines and districts are listed in Table 19-13.

Sullivan mine, Kimberly, British Columbia. This great mine once yielded 98 percent of Canada's lead and three-quarters of the country's zinc and was the largest lead-zinc mine in the world, but Kidd Creek now surpasses Sullivan. The deposit is a lens of massive sulfide about 2000 meters in length and 90 meters in maximum width, along a zone in tilted Precambrian quartzites and argillites. The body parallels the bedding and has been formed partially by replacement, but the deposit may have a syngenetic origin, preserving the original bedding in the ore (Fig. 19-38). The lode contains two ore shoots separated by 200 to 330 meters barren massive pyrite. The ore consists of argentiferous galena and zinc blende, with pyrite dominant in the upper shoot and pyrrhotite dominant in the lower. There are small amounts of garnet, actinolite, tourmaline, and cassiterite. The margins of the body contain less lead and zinc. The ore averages 6.6 percent Pb and 5.7 percent Zn, is high in silver, and contains some tin and cadmium. The deposit is now thought by some to be illustrative of a volcanic, exhalative syngenetic deposit.

Other Canadian Deposits. The *Flin Flon mine*, Manitoba, described under copper deposits, is the fourth largest producer of zinc in Canada.

Canada has added to her lead and zinc production with the discovery of Mississippi-type ore deposits in the *Pine Point district*, Northwest Territory. These ore bodies occur within the Middle Devonian Presqu'ile and Pine Point Formations near the margin of the eastern flank of the Western Canada sedimentary basin (Fig. 19-39).

Those deposits that are exposed or exist at shallow depth are amenable to open pit mining (Fig. 19-40). The ore averages about 7 percent Zn and 2.6 percent Pb in a 30 to 6 meter belt of barrier reef. Known reserves exceed 50 million tons.

Reef structures are by far the prime host for the mineralization. Dolomitization of the carbonate probably provided the porosity that allowed the solutions to leach lead and zinc from sediments, migrate up dip (Fig. 19-41) to the reefs, and slowly deposit the metals at temperatures that rarely exceeded 100° C as determined by $\delta^{18}O$, $\delta^{13}C$, and $\delta^{34}S$ isotopic studies.

Australian Lead and Zinc Deposits

Mount Isa. Although the Mount Isa deposit was discovered in 1923, controversy continues as to the

Table 19-13 Canadian Lead and Zinc Districts or Mines

District or mine	Location	Chief metals
1. Ecstall (Kidd Creek)	Kidd Creek, Ontario	Lead-zinc-copper-silver-tin
2. Consolidated Mining and Smelting Co. (Sullivan mine)	Kimberly, British Columbia	Lead-zinc-silver-copper-gold
3. Bathurst-Newcastle Area	New Brunswick	Copper-lead-zinc-silver
4. Flin Flon district	Flin Flon, Manitoba	Copper-zinc-silver-gold
5. Normetal	Noranda, Quebec	Copper-zinc-silver-gold

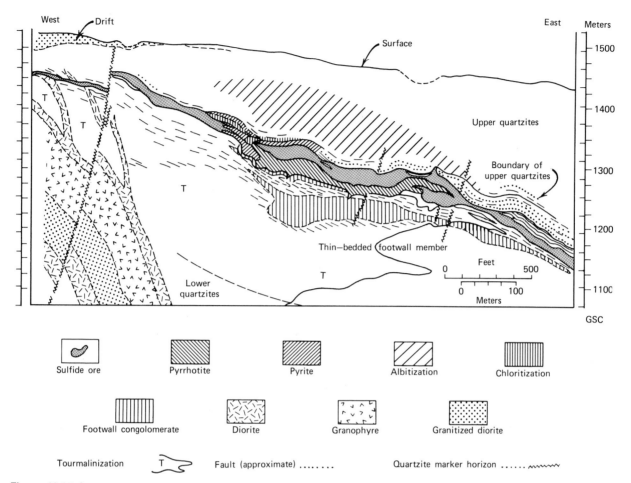

Figure 19-38 Cross section showing conformable nature of Sullivan ore body, British Columbia. (After A. C. Freeze.)

genesis of the ore deposits. This copper-silver-lead-zinc deposit was the largest lead-zinc producer in the world until the recent discovery of Kidd Creek. Nevertheless, it remains as one of the best examples of a concordant deposit.

The Urgwhart shale formation of the Mount Isa Group sequence of Precambrian formations contains the stratabound ore bodies of the area that are confined to a 1,000 kilometers stratigraphic zone. The lead-zinc sulfides in the formation range in thickness from 0.25 to 5 cm with even the thinnest beds showing remarkable continuity over many tens of meters (Fig. 10-2). The copper zones of mineralization are spatially distinct from beds of lead-zinc mineralization (Fig. 10-2) but occur in a variety of horizons.

The Precambrian rocks of this area are structurally deformed and metamorphosed. There are no nearby intrusives. The grade of metamorphism is

of relatively low rank but the rocks are intricately crenulated (Fig. 10-3).

Mineralogically, zinc, lead, and iron sulfide are the chief metallic sulfides along with minor chalcopyrite, arsenopyrite, tetrahedrite, silver minerals, valloriite, pentlandite, marcasite, and appreciable graphite.

Even though it was suggested that the deposits are epigenetic and formed by hydrothermal solutions, the lack of nearby intrusions, the extremely fine layers of sulfides over appreciable distances, and the suggestive evidence that folding is postore has led to a gaining suspicion that the mineralization is syngenetic, probably submarine volcanic and the sulfides may have been deposited by bacteriogenic H_2S during sedimentation. Sulfur isotopic analyses with $\delta^{34}S$ values ranging from about +4 to +30 per mil suggest that bacterial reduction of SO_4^{2-} in lagoonal areas provided the H_2S reduction that

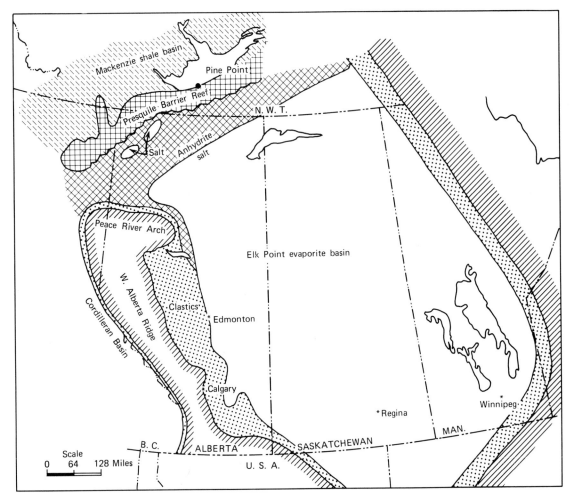

Figure 19-39 Location map illustrating the position of Pine Point and the Paleogeography of western Canada during deposition of the Middle Devonian Presqu'ile barrier reef complex (modified from Grayston, Sherwin, and Allan, 1964). (From Jackson and Follinsbee, Econ. Geol., v. 64.)

brought about this metal concentration. Such a hypothesis raises several problems as to the difficulty of including the copper deposits in the same genesis, and the complexities of metamorphism and structural effects upon the ore deposition.

The nearby MacArthur River deposit is quite similar to Mount Isa but has the added factor of lacking metamorphism, which suggests that Mount Isa may have been formed before metamorphism.

Broken Hill. Controversies still exist on basic aspects of the origin of the highly mineralized Broken Hill area (Fig. 19-42) and the Broken Hill ore body as to whether or not it has experienced high-grade metamorphism followed by several retrograde metamorphic events, as would be expected if the ore deposit formed syngenetically. According to Hobbs, Ransom, Vernon, and Williams, "There is no evidence at the moment that unambiguously es-

tablishes that the sulfides were present prior to the Willyama metamorphism. If they were present prior to or during the Willyama metamorphism and there is no period of deformation prior to the first recognizable folds at Broken Hill, then the sulfide mass was initially grossly discordant with bedding." Carruthers states that as the field evidence indicates that the ore occurs only in the highly metamorphosed rocks and is everywhere conformable with these rocks, he concludes that the Broken Hill type of mineralization as a whole has been metamorphosed.

In conclusion, either the mineralization is prior to the Willyama metamorphism, which supports the syngenetic origin, or the mineralization occurred during the Willyama metamorphism resulting in an epigenetic magmatic hydrothermal origin. The former theory is gaining credence.

The lead-zinc ore bodies of the *Zeeman* mineral

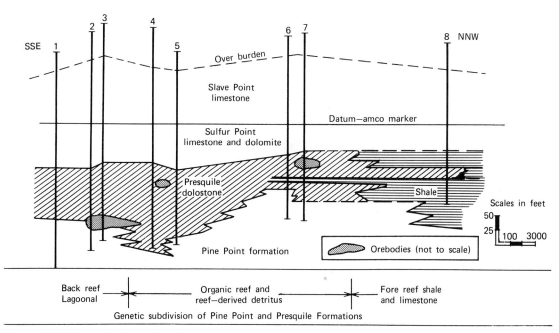

Figure 19-40 The ore bodies at Pine Point occur in the Presqu'ile dolostone and at its interface with the Pine Point formation and are controlled in part by the paleoecology of the barrier complex. (From: Jackson and Follinsbee, 1969, Econ. Geol., v. 64.)

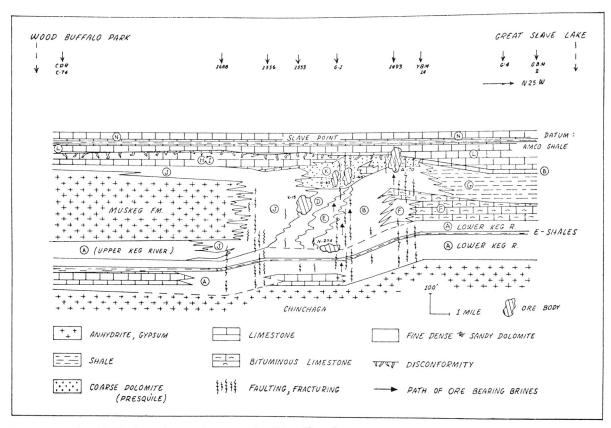

Figure 19-41 Post-Middle Devonian emplacement of sulfides. (From Skall, 1975, Econ. Geol., v. 70, p. 44.)

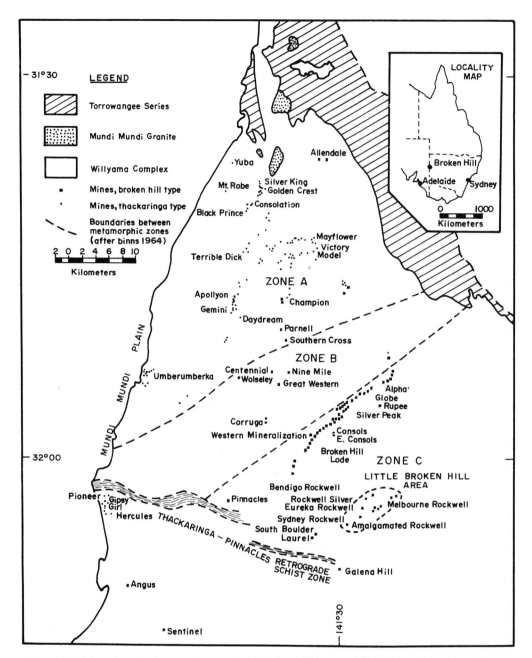

Figure 19-42 Location of mines and prospects of Broken Hill-type and Thackaringa-type mineralization in the Willyama complex. (From Howard, 1972, Econ. Geol., v. 67.)

district, *Tasmania*, Australia, are well zoned around the Heerskirk granitic intrusive. As the intrusion is highly enriched in radiogenic strontium and contains much xenolithic material, it is regarded as a high-level, possibly cupola-type, body as both factors suggest a crustal origin of the magma. The high enrichment of ^{34}S in sulfides within the intrusion also supports the crustal origin of the magma.

The host rocks are Ordovician to Silurian ande-

site and siltstone. The intrusion measures 3 to 4 kilometers in diameter and consists of quartz diorite. It is dated as mid Tertiary. The hydrothermal alteration consists of pervasive propylization localized in fracture areas.

Other Important Lead-Zinc Deposits

Europe is fairly rich in lead and zinc deposits, but, with the exception of Russia no major country produces its own requirements of these metals. Southern Asia has only one large representative—Burma.

Poland-Germany, Silesia. The chief zinc-lead district of Europe lies at the former Polish-German boundary. Bedded zinc-lead deposits in slightly warped Triassic dolomites overlie Carboniferous coal seams—a fortunate combination. There are three ore zones, one at 80 to 90 meters deep at the base of the dolomite, another 15 meters above it, with another minor bed above this. The lowest deposit is about 4 meters thick, is regular, extensive, and nearly horizontal; rare pipes and irregular cavities occur. The ore consists of sulfides of iron, zinc, and lead, with rare jordanite and meneghinite along with clay and dolomite, and carries 15 to 23 percent Zn and 3 percent Pb. The sulfides exhibit colloform structure. The upper ore beds are largely oxidized. The deposits are mainly replacements with minor cavity filling, formed by low-temperature, hydrothermal solutions.

Italy. The most important deposits of Italy are in Sardinia. Others occur at Raibl and in Trentino and Trieste.

Those of Sardinia were worked by the Phoenicians and Romans. In the Arbus district, according to Wright, granitic intrusives cut Paleozoic schists. Two systems of veins are found, one parallel and another normal to the contact. Of the former, the Main Lode, worked in the Montevecchio and Ingurtoso mines, supplies most of the lead and zinc of Italy. It is 10 kilometers long and 20 meters wide. In the eastern section lead dominates, and westward in the Brassy section it splits into three lodes and zinc dominates. The ore shoots are from 400 to 600 meters long, with equal barren intervals. The ore consists of blende and galena, in a gangue of barite, carbonates, and quartz. The ore body is a wide brecciated mass with the ore forming the matrix. It carried 10 to 12 percent of combined zinc and lead. Oxidation extends 5 to 10 meters in depth.

Near *Iglesias, Sardinia,* are large replacement Mississippi Valley-type deposits in limestone, and important veins of zinc-lead ores occur in northern Italy at *Raibl* and *Auronzo.*

Spain. Spain was once the leading European producer of lead. The chief district is the Linares-Carolina, in the Sierra Morena, where old Roman workings and dumps are still in evidence. Here, rich silver-lead fissure veins are numerous, the outstanding one being the *Los Guindos* vein, which

has been followed underground continuously for 11 kilometers, except for a distance of $1\frac{1}{2}$ kilometers. The vein is enclosed in quartzite and slate and is filled by irregular blocks and fragments of these rocks, surrounded by ore. The inclusions suggest a premineral rubble of a fracture zone around which vuggy ore was deposited, rather than replacement residuals. The width varies a few meters, in places containing 2 meters of pure galena. Zinc blende is absent, and there are traces of pyrite and chalcopyrite in a gangue of quartz, carbonates, and barite. The ore is localized in shoots, one of which is 300 meters long and yields 10 to 35 percent lead with 100 to 300 grams of silver. Similar veins are found at *La Rosa.* Near Linares, the great *Arrayanes Lode,* over 6 kilometers long and only a meter wide, occurs in granite. Similar ore is in shoots up to 300 meters long and 500 meters deep. All these veins are part of a silver-lead province of cavity-filling deposits.

Yugoslavia. This country is both an ancient and a recent producer of lead and zinc. Ancient workings around *Trepca* were revived when modern production commenced in 1930 at the *Stantrg* mine (Fig. 19-43) after the development of over 3 million tons of ore averaging 8.7 percent Pb and 8 percent Zn. In an area explored by 1000 ancient shafts up to 200 meters deep, an inclined pipelike mass of early Tertiary volcanics has been disclosed between underlying limestone and overlying Paleozoic schist. Hydrothermal solutions followed the contact, and ore replaced the limestone to form an inclined troughlike body, which has been followed on a 40° dip for 600 meters vertically. The ore is made up of base metal sulfides, jamesonite, and magnetite with a gangue of quartz, four carbonates, actinolite, garnet, and amphibole, and represents a combination of mesothermal and hypothermal metallization.

Burma. The *Bawdwin* mine (Fig. 19-44) in Burma has been one of the world ranking lead mines. Here an ore zone 2400 meters long and 150 meters wide has a hanging wall of feldspathic grits and a foot wall of rhyolite lavas and tuffs, all of early Paleozoic age. The ore bodies are enormous replacements within a shear zone. Within the ore zone are three lode systems, the principal one of which has been broken by faulting into three segments: the *Shan, Chinaman,* and the *Meingtha.* The ore shoots are up to 400 meters long; in places there are solid sulfides 15 meters wide composed of galena and

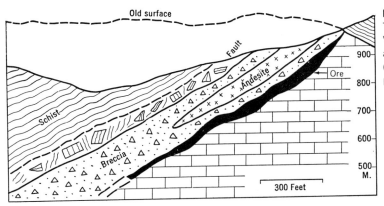

Figure 19-43 Stantrg mine, Trepca Mines, Ltd., Yugoslavia. Vertical section showing ore along pipelike mass of breccia. (From Forgan, and private company report.)

zinc blende, with minor pyrite and chalcopyrite in altered country rock and quartz.

Finland. The Outokumpu mine in southeastern Finland began production in 1913. Many years passed before at least half-a-dozen additional deposits entered production in Finland. One of the larger deposits is the *Pyhasalmi* mine with 1,100,000 metric tons of ore per year from which copper, zin, and pyrite concentrates are recovered. The *Kotalahti* mine produces about 450,000 metric tons of ore from which nickel and copper are concentrated. In

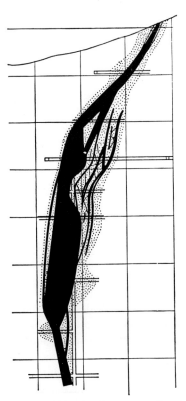

Figure 19-44 Section across Bawdwin lode, Burma. Black is solid sulfide; stippled, disseminated ore. (Loveman, AIME.)

1950, the *Aijala* mine and the near-by *Metszamonttu* mine ceased operations until further exploration located the continuation of the Metzamonttu deposit from which 80,000 metric tons of ore per year is now being produced containing lead, zinc, and copper.

Another deposit, the *Koranos* mine, ceased operations in 1961 but was restarted in Dec. 1964. Lead and lanthanide concentrates are recovered from 100,000 metric tons of ore produced per year.

Chromium is recovered from the *Kemi* mine with a capacity to treat 400,000 metric tons of ore per year.

The *Vihanti* mine is located in the parish of Vihanti in Oulu province. Boulders containing mineralization were noted between 1936 to 1941 by the local people in glacial moraine. The Geological Survey of Finland geologists located the ore-bearing formation using geological and geophysical techniques. Outokumpu Oy took over the exploration and development in 1951; by 1957, over 400,000 metric tons of zinc ores were being produced. Since then, production has increased to 750,000 metric tons of ore which includes 250,000 metric tons of pyritic ore.

The mine complex of *Vihanti* is associated with a high-rank metamorphic schist formation composed of mica-gneisses, mica-schists cordierite gneisses, quartzites, black schists, dolomites, and skarn rocks. The structural nature of the region of the deposit is extremely complex.

The known length of the complex is about 2 kilometers and it reaches about 500 meters at its thickest point. The strike is east-west and the dip is about 45 degrees south. To the east, it rises to the surface and to the west it has been located by drilling below the 600 meter level.

There are both zinc and pyritic ore bodies with grades varying over a wide range. The host rocks from the ores are mostly dolomites, skarn rocks, and quartzitic rocks. The ore bodies follow the general trend of the country rocks and the dips vary from 90 degrees to 35 degrees south. There are structural similarities of these bodies to the dilatant ore deposits of the Homestake deposits in South Dakota.

The most important and abundant ore mineral is sphalerite accompanied by some chalcopyrite and galena. The largest zinc ore bodies are located at the northern edge of the complex near the footwall contact of quartzitic and dolomite-skarn rocks. The structure of the deposits is complicatred due to metamorphism, faulting, folding, and replacement changes.

The Ristonaho-Valisaari zinc ore body at Vihanti is about 400 meters in length and 10–40 meters in width. The composition of the zinc ore bodies is 7–14 percent zinc, 0.5–0.8 percent copper, and 0.4–0.7 percent lead with small amounts of silver and gold.

The pyritic ores that are mined in the Outokumpu deposits consist of pyrite, of course, but with varying amounts of pyrrhotite. Sphalerite and chalcopyrite are present only in small quantities. Because of the complex nature of the ore mineralogy of these deposits, Finland is noted for her development of large, efficient, automated smelters and refineries. Their plants are sold, and constructed in many different nations including Russia.

Outokumpu Oy is a government-led company whose capital stock amounts to 96 million Fmk. The biggest shareholder is the Finnish state with 51 percent of the shares. Kansanelokelaitos (Old Age Pensions Scheme) has about 49 percent, and 0.001 percent of the shares are owned by five other shareholders.

Each of the mines has its own distinctly unique Sauna where civilized men enjoy the brisk beating of their bodies with birch bundles collected even for their need in the winters. This is followed by cooling-off with sausage and excellent beer, and stimulating, intellectual, chit-chat.

Other Provinces. South America contains important local lead-zinc deposits in Peru, Bolivia, and Argentina. Europe is fairly bountifully supplied with lead and zinc. There is a Mediterranean province with deposits in Spain, Roumania, Turkey, and Tunisia; a central European province includes parts of Czechoslovakia, Germany, Poland, and England. A minor northern province has representatives in Sweden, Norway, and Finland, and another one lies in the Urals. Belts of abundant lead-zinc mi-

neralization occur in parts of Australia and in Burma, and minor localized areas occur in Indochina, Japan, Siberia, southern Rhodesia, and southwest Africa.

Selected References on Lead and Zinc
Zinc and lead deposits of Canada. 1930. F. J. Alcock. Can. Geol. Surv. Econ. Geol. Ser., No. 8, Ottawa. *Résumé of world deposits.*
General Reference 19. *Chapters on lead and zinc.*
Geology and ore deposits of the Darwin Quadrangle, Calif. 1963. W. E. Hall and E. M. MacKevett. U.S. Geol. Survey Prof. Paper 368, p. 87. *Area of sulfide mineralization now undergoing redevelopment.*
Southeastern Missouri lead deposits. 1936. W. A. Tarr. Econ. Geol. v. 31, p. 712–754, 832–866. *This article proposes a hydrothermal origin.*
Broken Hill, Australia. 1926. E. C. Andrews. Econ. Geol. v. 21, p. 81–89. Also N.S. Wales Geol. Survey Mem. 8, 1922. *One of the great lead-zinc districts of the world.*
Sullivan mine, British Columbia. 1948. C. O. Swanson and H. C. Gunning. General Reference 15, p. 219–230. *Description of one of the world's greatest lead-zinc mine.*
Mount Isa, Australia, rock deformation, and mineralization. 1942. R. Blanchard and G. Hall. Aust. Inst. Min. Met. v. 125, p. 1–60. *Details of geology, occurrence, and origin.*
Massive sulfide deposits and volcanism. 1969. C. B. Anderson. Econ. Geol., v. 64, p. 129–146.
Genesis of stratiform lead-zinc-barite-fluorite deposits. 1967. J. S. Brown, ed. Econ. Geol. Mem 3., p. 443. *Results of international meeting on lead-zinc-barite-fluorite deposits with numerous genetic processes described.*
Mineral Industry of the British Empire and Foreign Countries: *Lead,* 2nd ed., 1933, and *Zinc,* 2nd ed., 1930. Imperial Inst., London. *Brief summary of occurrences.*
General References 8, 10, 11, 12, 15, 16, and 10.
Genesis of stratiform lead-zinc-barite-fluorite deposits. 1967. J. S. Brown, ed. Mem. 3, Econ. Geol. Publ. Co.

TIN

Tin is believed to have been one of the first metals employed by man. So far as is known, it was first used as a constituent of bronze. The Phoenicians regularly shipped tin from Cornwall, England, between 1500 and 1200 B.C. where it was probably acquired in large part from placer deposits. Tin was also acquired from the Orient and some possibly from western Asia. It has been a standard metal of commerce at least since the late Bronze Age (3500 to 3200 B.C.).

A bronze rod with 9.1 percent tin found in Egypt dates back to 3700 B.C., and many bronze objects with 10 to 14 percent tin have been found in excavations of different ancient civilizations. Tin free from lead and silver was used for an Egyptian mummy wrapping in 600 B.C.. The earliest Mediterranean bronzes are presumed to have been made from Asiatic tin. By 500 B.C., 100 tons of tin were exported annually from Britain. Ancient uses for tin were chiefly for making bronze for implements, weapons, and ornaments, but medieval times found greater use for bell casting, bronze armor, and pewter. This supply came mainly from Britain, and later from Saxony and Bohemia, and then from Asia. Malay tin was in high demand from 1400 to 1800, and by that time, tin also came from Malaysia and Indonesia, and China. By the end of the eighteenth century, tin production was exceeded only by three metals: iron, lead, and copper. About the middle of the nineteenth century, tin production started in Australia, followed by that in South America and finally in Africa in 1910. North America is singularly lacking in tin but consumes more than 30 percent of the world's production of tin.

Uses

Many of the early uses of tin, such as for pewter, the lining of cooking vessels, and tin roofing, have been displaced by the newer uses of bronze bearings and tin-plated food containers. The United States annually consumes almost 65,000 metric tons of tin.

The six principal use categories of tin, in decreasing order of importance, are tinplate (43 percent), solders, bearing alloys, bronze, chemicals, and coatings other than tin plate. An alloy of tin and niobium has been found to be a superior superconductor. Such a superconductor, consisting of more than 20 kilometers of niobium-tin ribbon wound on a wheel of 5-foot diameter, when activated in liquid helium, will produce a magnetic field so strong that the entire assembly will float in space in its own magnetic field.

Production

The annual world production of primary tin has remained slightly below 220,000 metric tons of metal during the past several years. About 90 per-

cent of the world's tin is produced by six countries. Malaysia, which once was almost the sole source of tin, now produces 32 percent; Bolivia produces 15 percent; China produces 11 percent, and Indonesia produces 12 percent as also does Thailand. Australia produces about 6 percent and minor production is also derived from Russia, Brazil, Nigeria, and the United Kingdom. More than 10,000 tons of tin has been produced as a by-product from Sullivan, British Columbia, Canada. Kidd Creek, Ontario, Canada, is also a tin producer. Tin production in the United States as a by-product of molybdenum production at Climax, Colorado, is negligible. Minor tin reserves do exist in Alaska.

Until new discoveries of tin added to the world production, the tin producers of Malaysia held a monopoly on tin production. With new discoveries a virtual cartel existed, which even today is rather unique of all the mineral commodities in world trade. Tin is regulated by international agreements, imposed by the International Tin Council. Almost all major producers of tin contribute funds to the Council in order to create a buffer stock of tin which is disposed of, or acquired by, the Buffer Stock Manager who must buy or sell tin as the world market price goes below or above predetermined prices. At the present time, tin has reached an all time high cost exceeding 876¢/pound. As the Tin Council resembles an international cartel, the United States is not a member but still the largest consumer of tin, about 62,000 metric tons/year.

Mineralogy, Tenor and Treatment

Mineralogy and Associates. Tin metal exists in two forms: "white" tin of the tetragonal crystal system and specific gravity of 7.31, and "gray" tin of the cubic system and specific gravity of 5.75. Below a temperature of 13.2°C., white tin changes to gray tin and crumbles to a powdery mass. Alloying tin with any other metal presents this change; hence tin is rarely used in its pure elemental form although some Mexican tin can be hammered into ornamental tin leaves.

Practically all the tin of commerce is obtained from the mineral cassiterite or tinstone; stannite, the tin-copper-iron sulfide; and teallite, which supplies a minor amount in Bolivia. The accompanying gangue minerals are chiefly altered granite, quartz, and white mica. A common associate is tungsten; molybdenum or silver may be present; and gold is generally absent.

Tenor and Treatment. Placer deposits range in tin content from 0.4 to 5 pounds per cubic yard, and lode deposits carry from 1 to 8 percent tin.

Pure cassiterite is smelted with only oxygen to be eliminated. This is done by adding coal, coke, or charcoal. However, iron and silica cause tin to go into the slag, necessitating careful manipulation of fluxes and resmelting of the slag. The molten tin is drawn off and purified by agitation with air or by electrolytic refining.

Kinds of Deposits and Origin

The principal classes of tin deposits are placer deposits, stockworks, fissure veins, and disseminated replacements.

Origin and Distribution. The lode deposits contain high-temperature minerals and, except for Bolivia, are always in close association with silicic granites, from whose magmatic reservoirs they are generally considered to have been derived by hydrothermal action. Since Daubree's famed synthesis of cassiterite, which was one of the earliest minerals to be synthesized, it is commonly assumed that the tin was transported from the magma chamber as gaseous tin fluoride or tin chloride, which by reaction with water formed cassiterite, releasing HF or HCl. The granite wall rocks are generally altered to muscovite, quartz, and topaz (greisen), presumably by attack of the acid gases. The associated fluorine-bearing minerals, topaz and fluorite, suggest this mode of origin, which, however, is only a hypothesis. A hydrothermal origin applies to the Bolivian tin deposits and perhaps also to the others.

The greatest tin province of the world is the belt of placers found in a strip of country 1,600 kilometers long and 200 kilometers wide along the Malaysia peninsula, including Indonesia to the south, and Burma and Thailand to the north. The province also projects into Yunnan, China. Separate provinces occur in Bolivia; Cornwall, England; Erzgebirge, Germany; Nigeria, Zaire; Australia; and South Africa.

Examples of Tin Deposits

Malaysia. The chief occurrences of tin are in the hundreds of mines in the Malaysian States of Perak, Selangor, Negri Sembilan, and Pahang. Perak produces over 60 percent of the output from the rich and accessible deposits of the Kinta Valley. The ore is mostly alluvial, but some lodes are also worked. The bedrock is limestone, granite, and schist, overlain by a thick mantle of tropical residual soil. The tinstone contained in stockworks and veins resisted weathering and so accumulated in the residual soils. The soils were washed into the valley, leaving the heavy resistant tinstone to accumulate on the pinnacled bedrock, particularly near valley sides. As these sites contain valuable agricultural lands, the placer miner is not welcomed because of his absorption of water, destruction of agricultural land, and undesired tailings disposal. Probably much placer ground still remains. In places the rich eluvial deposits in the residual soils of the valley slopes can be recovered by sluicing or open pitting. The reserves of placer tin are large, and little is known as to whether the source lodes may prove productive beneath the soil mantle.

Generally, similar conditions exist in other parts of the Federated States except that granite is generally the bedrock. In many places the deeply decomposed granite is directly mined for its tin content.

The Malayan tinstone is treated at smelters located in Penang, Singapore, and Selangor, which also treat the ore from Thailand, Burma, and other neighboring provinces.

Indonesia. The three Indonesian islands of Banka, Billiton, and Singkep, lying south of Malaysia, have long been producers of tinstone, about two-thirds of it coming from Banka, one-third from Billiton, and small amounts from Singkep and adjacent islands. The Banka deposits are all alluvial, but some lodes are worked on the other islands. The ore goes to smelters at Banka, Singapore, and Arnhem, Holland.

Geologically, the tin occurrences are generally similar to those of the Malayan States. According to Westerveld, humid, tropical weathering of post-Triassic granite containing stockworks of cassiterite gave rise to residual, hill-slope, eluvial placers (koelits). Part of these in turn were swept into the adjacent valleys, adding tinstone to the residual accumulations already present in the valley bottoms (kaksas). These were covered by alluvial top beds of sands, which in places contain upper "hanging" placer beds. Quaternary submergence drowned the lower parts of the river valleys with their bedrock tinstone to depths of 30 meters below sea level. These have been traced out to sea and recovered on a large scale by sea dredges.

The primary deposits consist of (1) hydrothermal stockworks in greisen containing cassiterite, along with wolframite, tourmaline, topaz, and minor base-metal sulfides, and (2) contact-metasomatic deposits in the intruded sandstone-shale formation containing cassiterite, magnetite, abundant base-metal sulfides, contact silicates, siderite, and fluorite.

Bolivia. Bolivia, the second-ranking tin country, derives its ores entirely from unique lode deposits, chiefly from *Llallagua-Uncia, Huanuni,* and *Potosi,* but there are 10 tin centers. The ores carry about 2 to 4 percent tin with silver, and normally have been smelted in Europe but were also treated briefly at Texas City, Texas. Formerly, the upper-level ores carried 8 to 12 percent tin.

The *Llallagua-Uncia* group of the Patiño Company produces about 60 percent of Bolivian tin. Turneaure describes the ore as occurring in a great network of veins within and contiguous to a small quartz porphyry stock of late Tertiary age and near-surface origin, a network that contracts in depth between walls of folded Cretaceous sediments. The veins are of two distinct intersection structural types (Fig. 19-45). One group is composed of strong, continuous, parallel fissures one to 2 meters wide, and the other is a maze of discontinuous small cracks. Many do not outcrop. Their close spacing may give rise to stopes 5 to 12 meters wide. Sillitoe has illustrated an idealized porphyry tin deposit of the Bolivian tin province (Fig. 19-46) and the associated hydrothermal alteration. This suggestion correlates with Turneaure's description of the Llallagua deposits.

The veins are fracture fillings. The walls are well defined, and the filling is crustified and drusy. In the major veins the ore occurs in distinct shoots, parts of which are phenomenally rich, containing a meter or so of solid cassiterite. Cassiterite, quartz, pyrite, and marcasite make up 90 percent of the filling. Silver is absent. The episodes of vein formation are: (1) vein fillings of quartz, bismuthinite, cassiterite, and franckeite; (2) replacement of franckeite by pyrrhotite, and (3) replacement of pyrrhotite by wolframite, stannite-sphalerite, marcasite, arsenopyrite, pyrite, and siderite; (4) crustification and some replacement by later sphalerite and pyrite, chalcopyrite and gangue; (5) wavellite; and (6) supergene minerals.

The ores are pronouncedly zoned (Fig. 19-47). Vertically, a rich hypogene zone lies under the zone of oxidation but extends up into it (Fig. 19-48); horizontally there is a central cassiterite zone flanked by sulfides. Turneaure believes that the rich cassiterite zones are clearly hypogene and that supergene tin enrichment is absent. Hydrothermal alteration is pronounced (Fig. 19-46).

The deposits were formed under shallow cover by medium- to high-temperature liquid solutions that changed to low-temperature in the late stages, when rapid release of pressure permitted concentrated deposition of a cassiterite core. Telescoping is indicated.

The other tin deposits of Bolivia are mostly Tertiary fissure veins that carry stannite and teallite as well as cassiterite and silver. At *Potosi* the silver-tin veins are richer in tin at depth. At *Oruro,* Chace describes three stages of vein mineralization: 1, quartz-pyrite-cassiterite; 2, silver-sulfo-salt; and 3, oxidized ores.

The *Potosi* stock, located in the Andean ranges

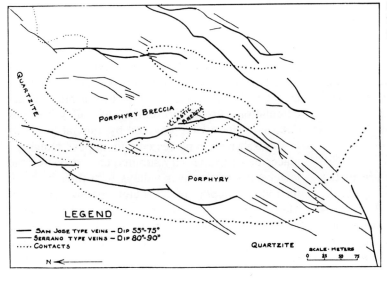

Figure 19-45 Fissure systems of the Llallagua tin mine, Bolivia. (Turneaure, Econ. Geol.)

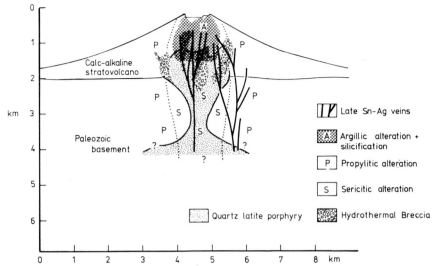

Figure 19-46 An idealized reconstruction of a porphyry tin deposit, using observations from the Bolivian tin province. (From Sillitoe, Halls, and Grant, 1975, Porphyry tin deposits in Bolivia, Econ. Geol., v. 70, p. 813.)

Legend:
- Late Sn-Ag veins
- [A] Argillic alteration + silicification
- [P] Propylitic alteration
- [S] Sericitic alteration
- Quartz latite porphyry
- Hydrothermal Breccia

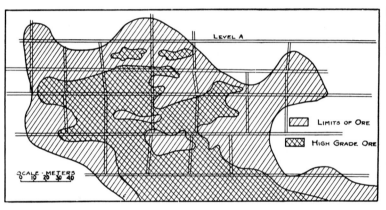

Figure 19-47 Zonal arrangement of tin ores with high-grade ores in the central part, Llallagua, Bolivia. (Truneaure, Econ. Geol.)

Legend:
- Limits of Ore
- High Grade Ore

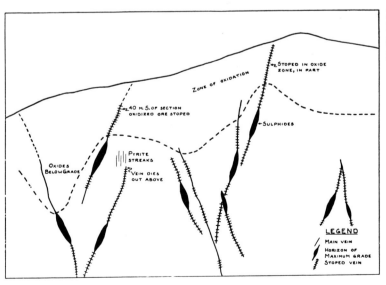

Figure 19-48 Vertical section of Llallagua tin veins showing the position of rich ore shoots. (Turneaure, Econ. Geol.)

of central and southern Bolivia, is in a region that includes major tin and tin-silver deposits as well as tin-zinc, silver, and tungsten deposits. Fluid inclusion data by Kelly and Turneaure indicate that the ore-forming solutions were high-temperature (to 560° C), highly saline, and of low CO_2 content that probably boiled at times, as often believed by pneumatalists.

Other Deposits. *China* has long been an important producer of tin from lodes in the Kochiu district, Yunnan, where one mine, operated for 500 years, is over 1,000 meters deep. *Nigerian* tin comes mainly from the Bauchi plateau, where the weathering of veins and stockworks in granite has yielded alluvial tin. *Zaire and Thailand* are important producers of alluvial tinstone.

Cornwall, England, was long the famous tin district of the world. The total production since 500 B.C. is estimated at 3.3 million tons. The deposits are now largely exhausted. The early tin came from placers and outcrops. Lode mining on a large scale began in 1600 and culminated in the middle of the nineteenth century. The lodes were high-temperature fissure veins, carrying cassiterite, occurring in, or close to, granite intrusives. They offer the earliest, best-known examples of mineral zoning; some silver-lead veins at the surface passed downward into copper veins, and copper veins in depth yielded to tin veins. In later years the veins averaged about $1\frac{1}{4}$ percent tin.

Selected References on Tin
Geologic Features of Tin Deposits. 1912. H. G. Ferguson and Alan M. Bateman. Econ. Geol., v. 7, p. 209–262. *General features of occurrence and origin.*
Geology of Malayan Ore Deposits. 1928. J. B. Scrivenor, London.
General Reference 19. 1973. Tin. C. G. Samsbury and B. G. Reed. *Recent summary on tin geochemistry, geology, and reserves.*
Mineralogy, paragenesis, and geothermometry of the tin and tungsten deposits of the Eastern Andes, Bolivia. 1970. W. C. Kelly and F. S. Turneaure. Econ. Geol., v. 65, p. 609–680. *Detailed study of geothermometry by different techniques of different zones and minerals of these deposits.*
Tin in mineral facts and problems. 1970. R. A. Heindl. U.S. Bur. Mines Bull. 650, p. 759–771.
Porphyry tin deposits in Bolivia. 1975. R. H. Sillitoe, C. Halls, and J. N. Grant. Econ. Geol., v. 70, p. 913–927. *Similarities and differences between this porphyry and porphyry copper deposits.*

The search for tin. 1965. K. F. G. Hosking. Ming. Mag., v. 113, p. 261–273.
Tin resources of the world. 1969. C. L. Sainsbury. U.S. Geol. Survey Bull. 1301, 55 p.
Tin ores of the Dutch East Indies. 1937. J. Westerveld. Econ. Geol., v. 32, 1019–1041.
Tin deposits of Llallagua, Bolivia. 1935. F. S. Turneaure. Econ. Geol., v. 30, p. 14–16, 170–190.
Tin-silver veins of Oruro, Bolivia. 1948. F. M. Chace. Econ. Geol., v. 43, p. 333–383, 435–470. *Detailed geology of bonanza tin-silver veins of shallow depth of formation.*
Geology of Colquiri tin mine, Bolivia. 1947. D. F. Campbell. Econ. Geol., v. 42, p. 1–21. *High- to low-temperature cassiterite veins.*
Ore deposits of the Eastern Andes of Bolivia. 1947. F. S. Turneaure and K. K. Welker. Econ. Geol., v. 42, p. 595–625. *Laterally zoned high-temperature tin-tungsten veins and sulfides.*
A survey of deeper tin zones, Carn Brea area, Cornwall. 1946. B. Llewellyn. Bull. Inst. Min. Met., v. 477, p. 1–19; Disc. v. 479, p. 1–28. *Geologic data of the classical tin deposits.*
Tin-bearing placers near Guadalcazar, State of San Luis Potosi, Mexico. 1949. Carl Fries, Jr., and Eduardo Schmitter. U.S. Geol. Survey Bull. 960-D. *Finely divided cassiterite and other metals in huge alluvial fan.*

ALUMINUM

Aluminum is the latest metal to find general large-scale use in modern industry. Although the most abundant metal in the earth's crust, it long defied commercial extraction in the metallic state. It is an important constituent of such common substances as clays, soils, and rock silicates, but it is obtained commercially almost entirely from a mixture of minerals commonly referred to as "bauxite"; a little andalusite with 30 to 40 percent Al_2O_3 has been utilized at Boliden, Sweden; alunite is now becoming a source; and experimental runs have been made on anorthosite and clays. Although aluminum was isolated in 1827, it was not until 1886 that Hall and Heroult discovered the cheap process for its extraction by its electrolysis in fused cryolite. It is largely a twentieth-century metal.

Uses

The lightness of aluminum, its high strength compared with its weight, its resistance to atmospheric corrosion, and its electrical conductivity make it a popular modern metal. It enters into almost every conceivable use. It is a desired metal for airplane construction, motor vehicles, electrical equipment and supplies, fabricated metal products, beverage cans, containers and lately, in the home as aluminum foil. Plastics, wood, and other metals including copper, steel, lead, zinc, magnesium, and titanium could be substituted for aluminum although none of these materials has all of the properties of aluminum or the market potential. Both alumina (Al_2O_3) and "bauxite" are used in the manufacture of refractories, abrasives, and chemicals.

The chief use of aluminum is in the form of alloys, whose light weight and strength make them particularly desirable for airplanes, motor cars, and trains.

Of the "bauxite" ore mined, about 88 percent is used for the production of aluminum, about 6 percent for chemicals, and about 6 percent for abrasives, refractories, and alumina cements. Only 2 percent of U.S. "bauxite" production is used to manufacture aluminum oxide abrasives, yet this represents about half of the total production of all artificial abrasives and more than five times the total production of natural abrasives.

Production

The production of aluminum, unlike that of most other metals, takes place largely in countries distant from the source of the ore. Because of the large amount of electrical power necessary for its reduction, the ore moves to available cheap power. Thus countries such as the United States, Germany, and Canada are large producers of the metal but not of the ore. About 4 kilowatts of electrical power is required to produce one kilogram of aluminum.

Annual production of "bauxite" did not reach 1 million metric tons until 1917, during World War I. Production reached 11 million metric dry tons in 1951 at which time the Paley Commission report estimated that aluminum consumption would increase by 300 percent during the 1970 to 1980 decade. This percentage increase was reached by 1959! In 1978, world production of "bauxite" reached 84.5 million metric tons, more than double what the Paley Commission predicted. The percentage provided by the different countries ore as follows.

Australia, 31 percent.

Jamaica, 14 percent.

Guinea, 15 percent.

Surinam, 7 percent.

All others, 33 percent.

Prior to 1970, Jamaica was the largest producer of "bauxite" accounting for 59%, followed by Surinam (24%), and other countries (17%). Australia obviously has become a large producer during the past few years.

The chief producing countries of aluminum in order of their output are the United States, Japan, Russia, Canada, Germany, France, Norway, Italy, and Switzerland.

In the United States the chief production of "bauxite" is from Arkansas, with minor amounts from Alabama, Georgia, and Mississippi, where large low-grade deposits have been developed.

Mineralogy and Associations. "Bauxite" is not a mineral species but a colloidal mixture of hydrous oxides of aluminum and iron and water. Commercial "bauxite" occurs in three forms: pisolite or öolite, sponge ore, and amorphous or clay ore. Specifically, "bauxite" is a complex impure mixture of aluminous minerals. The principal minerals are gibbsite, $Al_2O_3 \cdot 3H_2O$ and diaspore, which has the same composition as boehimite but is more dense and harder. With varying mixtures of these minerals "bauxite" varies in the water of hydration but is usually assigned an approximate composition of $Al_2O_3 \cdot 2H_2O$.

Tenor and Treatment. The tenor of "bauxite" varies with the use to which it is put and the nature and amounts of the impurities present (see Table 19-14).

High-grade "bauxite" should contain about 50 percent Al_2O_3 and have a SiO_2 content of less than 5 percent. Low-grade "bauxite" may have the same Al_2O_3 content, but the SiO_2 content may vary between 5 to 10 percent or slightly more depending upon the amount of leaching (Fig. 19-49) as controlled by pH.

Terra rossa is a red soil typically higher in SiO_2 and lower in Al_2O_3 than "bauxite." Both terra rossa and "bauxite" contain a few to tens of percent of Fe_2O_3.

For aluminum production, "bauxite" is washed, purified, ground, and leached under pressure, by the Bayer process with hot, strong, sodium hydroxide, which dissolves the alumina as sodium aluminate, leaving the impurities. The solution is agitated with aluminum hydroxide, causing precipitation of the metal as hydrated aluminum oxide, which is

Table 19-14 Variations in the Tenor of Bauxite (in Percent)

Purpose	Alumina	Allowable SiO₂	Allowable Fe₂O₃	Allowable TiO₂
Chemicals	55–58	max 5–12[a]	max 2	—
Abrasives	min 55	max 5	max 6	min 2.5
Refractories	59–61	max 1.5 to 5.5	max 2	2.5–4
	85–90	3 to 6	1.5–7.0	3–4

Source: From Patterson and Dyni, U.S. Geol. Survey Prof. Paper 820, p. 36.

[a] Some U.S. "Bauxite" exceeds 15% SiO₂ and the Russian nephlene syenite used for alumina may exceed 3% SiO₂.

then filtered, dried, and calcined. This product is converted to metallic aluminum by electrolysis, in a bath of fused cryolite at 950° C, contained in a cell of baked carbon. Four tons of bauxite yield about 2 tons of pure alumina and about 1 ton of metal.

Kinds of Deposits

All "bauxite" deposits result from residual weathering and occur as (1) blankets at or near the surface, (2) interstratified deposits on unconformities, (3) pocket deposits in limestone, and (4) transported deposits.

As summarized by Patterson and Dyni,[1]

"Bauxite" minerals are formed mostly by weathering of aluminous rocks. Various deposits of bauxite in different parts of the world are known to have formed from virtually every type of rock that contains aluminum. Conditions favorable for the formation of bauxite are (1) warm tropical climate, (2) abundant rainfall, (3) aluminous parent rocks having high permeability and good subsurface drainage, and (4) long periods of tectonic stability that permit deep weathering and preservation of land surfaces. During weathering, the bauxite becomes enriched in aluminum by removal of most of the other elements in the parent rocks mainly by solution by subsurface water. Because of chemical weathering and removal of essential plant nutrients, soils developed on bauxite deposits commonly support a cover of undernourished, dwarfed vegetation.

Useful guides in exploration for surficial bauxite are the recognition of old land surfaces and dwarfed vegetation; such guides led to the discovery of the bauxite deposits in Oregon and Hawaii, as well as several very large deposits in other parts of the world.[1]

World Distribution of "Bauxite"

Commercial "bauxite" deposits are widely distributed over the tropical and temperate zones of the continents.

[1] Patterson and Dyni, U.S. Geol. Prof. Paper 820.

In *North America* the important commercial deposits are confined to Arkansas, with minor ones in Georgia and Alabama, and rather small deposits in Tennessee, Virginia, Mississippi, and New Mexico. *South America* contains an important belt in Guyana and Surinam and minor deposits in Minas Geraes, Brazil. Haiti, Jamaica, and the Dominion Republic are important producers of bauxite.

The most extensive deposits of "bauxite," however, occur in southern *Europe* parallel to the Mediterranean, reaching their maximum development in France (near Baux from where the name is derived) and extending through parts of Italy, Yugoslavia, Hungary, Greece, and Rumania. This group, at one time, supplied about 60 percent of the "bauxite" of the world. Minor deposits occur in Germany, Spain, and Ireland, and more extensive ones in Russia. *Asia* contains relatively few deposits. Some occur in India, and deposits in Indonesia, Celebes, and Palau are being worked.

Africa, a tropical continent, contains commercial "bauxite" only in the Gold Coast. Large deposits are reported from the Atlas Mountains, Morocco, and from the Tichenya Plateau, Nyasaland. Small deposits are worked in Portuguese East Africa, and other deposits are reported in Tanganyika, Sierra Leone, Madagascar, and Ethiopia. *Australia* yielded a small tonnage from Victoria and New South Wales but now has massive production from the Northern Territory. Australia and Jamaica are now the leading producers of "bauxite." Both have reserves of more than 5 million tons of high-grade "bauxite."

Arkansas. These deposits occur in an area that was slowly covered by advancing Eocene sediments that were subsequently stripped off in places to disclose small areas of syenite and "bauxite." Some beds of lignite occur in the overlying sediments, and the operators say that wherever they see lignite they expect good "bauxite" beneath it (Fig. 19-50). According to Branner, there are five

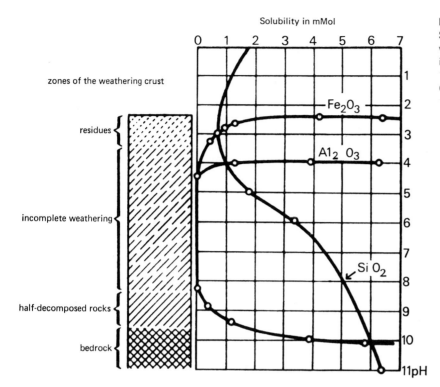

Solubility in mMol

zones of the weathering crust

residues

incomplete weathering

half-decomposed rocks

bedrock

Fe_2O_3

Al_2O_3

SiO_2

pH

Figure 19-49 Solubility of Fe^{3+}, Si, and Al in relation to the pH-value of solution in the weathering crust. (From V. I. Smirnov, 1970, in Baumann, Introduction to Ore Deposits, Halsted Press, 1976.)

stages readily discernible to the eye, in the formatijon of "bauxite": (1) unaltered nepheline syenite; (2) partly kaolinized syenite, nepheline partly removed; (3) completely kaolinized syenite, nepheline more completely removed; (4) "bauxite" high in silica; and (5) merchantable "bauxite," low in silica (Fig. 13-3). The "bauxite" is pisolitic, "granitic," or amorphous.

The ore beds reach 12 meters in thickness and average 4 meters. The maximum overburden is 25 meters, and the "bauxite" is extracted by power shovel from large open cuts. The annual production

has ranged from a prewar rate of 400,000 tons to a wartime rate of 6 million tons, to about 2 million tons in 1974. Three grades of ore are produced—aluminum, abrasive, and chemical ore. The bauxite runs from 56 yo 59 percent Al_2O_3, and some is mined with less than 5 percent SiO_2 and 2 to 6 percent Fe. The known reserves of high-grade ore are being rapidly depleted.

Alabama-Georgia. The deposits occur along a 100 kilometer belt and consist of pockets and irregular masses of residual bauxite lying in and on a mantle

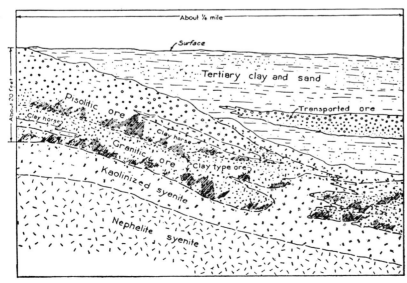

About ¼ mile

Surface

Tertiary clay and sand

Pisolitic ore

Transported ore

Clay horse

Clay horse

Granitic ore

Clay type ore

Kaolinized syenite

Nephelite syenite

About 20 feet

Figure 19-50 Section of the bauxite deposits of Arkansas showing the relation between syenite, bauxite, and Tertiary beds. (Branner, 16th Internat. Geol. Cong.)

of residual clay that marks an Eocene erosion surface on the Knox dolomite.

France. The French deposits occur mainly in Var and Hérault. They consist of pockets and blankets on karsted erosion surfaces developed on Jurassic and Cretaceous limestones. Consequently, their lower surfaces are highly irregular and discontinuous. They are mostly covered by Upper Cretaceous sediments and are, therefore, of Cretaceous age. Some of the deposits have slumped into collapsed solution cavities; others have been worked over by the transgressing sea (Fig. 13-4). They have been formed by the alteration of argillaceous material, some of which may have been residual clay and some shale beds let down by erosion from a higher horizon.

There are three main varieties of "bauxite": white, white banded, and dark red. They carry 57 to 60 percent Al_2O_3, about 20 percent Fe_2O_3, and 3 to 5 percent SiO_2, and they are rather low in water. Some of the bauxite is hard and splintery. It is chiefly used for making aluminum and chemicals. Much of it has to be extracted by underground methods of mining.

Reserves have been estimated at 16 million tons with a maximum content of 7 to 8 percent silica.

Guyana. The Guyanan deposits, and also the deposits of Surinam, according to Harder, were formed on an old base-level surface that truncates schists and gneisses and basic and silicic igneous rocks. First, residual clays were formed, which were then changed to "bauxite." A Tertiary coastal depression caused inundation, and sand and clay covered some of the bauxite accumulations; others were washed into the advancing sea. Elevation, still proceeding today, permitted erosional cutting through the unconsolidated sediments, exposing the "bauxite" deposits as beds or flat lenses up to 15 meters thick (Fig. 19-53). Several deposits have been discovered under a cover of white sand. Small pipes and veins extend down into the underlying clay; the top is siliceous; and the overburden is up to 20 meters thick. The deposits are worked by power shovel in large open cuts. The ore contains 58 to 63 percent Al_2O_3, 2 to 5 percent SiO_2, and 3 to 6 percent Fe_2O_3.

Ghana. The deposits of Ghana, studied by Cooper and others, constitute large resources. They lie on flat erosion remnants of a peneplaned surfaced developed across folded sediments, volcanics, and shales. They mostly rest on a thin band of "lithomarge," a cream-colored, colloidal clay-like material that represents a residual clay from which the "bauxite" was derived. The best "bauxite" occurs on the most dissected peneplain remnants, and Cooper thinks it underwent enrichment during dissection. It is overlain by a meter of red soil and in places by swampy materials. The bauxite beds are as much as 20 meters thick. One 5 meter bed in Yenahin averages 64 percent Al_2O_3, 1.9 percent Fe_2O_3, 2.4 percent SiO_2, and 4.1 percent TiO_2. About 60 percent of the Yenahin deposits contain 55 percent alumina. The average thickness is 10 meters. Reserves are estimated in tens of millions of tons.

Hungary, Yugoslavia, Greece, and Italy. These countries yield more than twice as much bauxite as the United States. The deposits here, as in France, occur in limestones, are really a part of the same belt of deposits, and are of similar origin and occurrence. They lie in pipes, pockets, and lenses, in residual clay that mantles the old surface and lines solution holes in Jurassic, Cretaceous, and lower Eocene limestones. Many lie in karst depressions. They are unconformably overlain by Upper Cretaceous or Eocene sediments. They were formed under different climatic conditions than exist today and during erosion intervals that preceded the Upper Cretaceous, lower Eocene, or middle Eocene. Those of Dalmatia (Fig. 19-52) are of Upper Cretaceous and lower Eocene age; those of Gánt, Hungary, are of middle Eocene age. Invariably, the deposits lie on unconformities.

These bauxites range from 50 to 65 percent alumina and are low in silica and iron.

Jamaica. It had been suggested that the *Jamaican* deposits formed from carbonate sediments as residual high alumina surficial or blanket deposits. Recent studies by Comer, however, including mineralogical and chemical evidence suggest that Miocene volcanic ash is a more likely "bauxite" parent source. Bentonitic clay occurs in the middle and upper Miocene limestones both as interbeds and dispersed material. The high alumina and low silica content of the ash are ideal sources of the comparatively rich and extensive "bauxite" deposits of Jamaica. The chronological events and geomorphic changes leading to the formation of these deposits are shown diagrammatically in Fig. 19-51.

The Jamaican deposits occur in karst topography and fault controlled lowlands that provided optimum drainage conditions and extensive subaerial leaching; the karst topography also led to the earlier

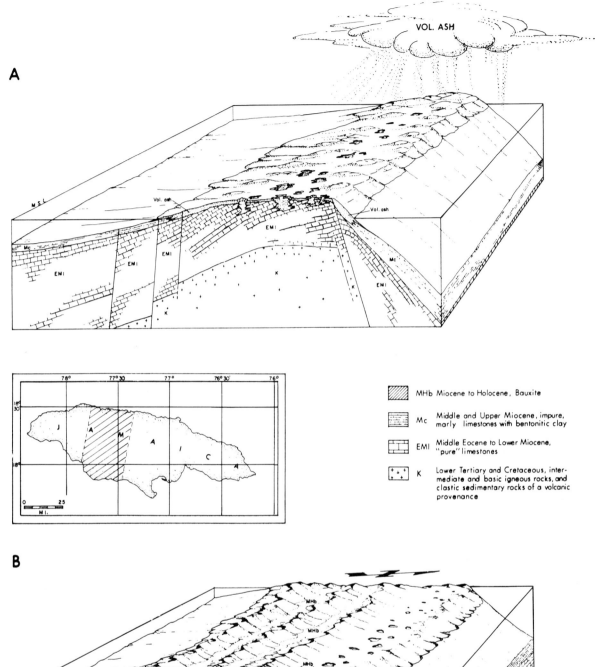

A

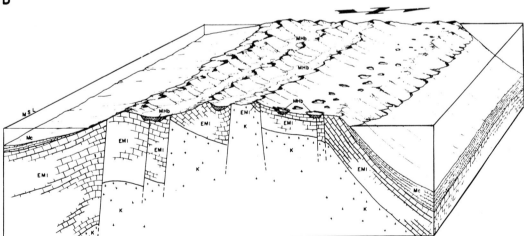

	MHb	Miocene to Holocene, Bauxite
	Mc	Middle and Upper Miocene, impure, marly limestones with bentonitic clay
	EMl	Middle Eocene to Lower Miocene, "pure" limestones
	K	Lower Tertiary and Cretaceous, inter-mediate and basic igneous rocks, and clastic sedimentary rocks of a volcanic provenance

B

Figure 19-51 Schematic block diagrams illustrating the geologic history of central Jamaica and the formation of bauxite. A. Volcanic ash eruption during middle and upper Miocene time. B. Uplift, faulting and the formation and concentration of bauxite in karst depression and valleys from Pliocene to Holocene time.

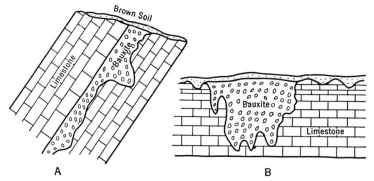

Figure 19-52 Section of typical bauxite. A, bauxite bed in Dalmatia; B, bauxite pocket in Istria. (After Harder, Aluminum Industry, V. 1.)

genetic theories of leaching of limestone. Comer's studies, however, indicate that the limestone thickness calculations based on the abundance of alumina are inconclusive by themselves, but less inconclusive when the effects of host limestone erosion, "bauxite" erosion, and solution losses are included.

Other Deposits. *Russia* contains extensive deposits of bauxite in the Leningrad district near Tikhvin, in the Urals, and in the Belovo region, Siberia. The Ural Devonian bauxites lie between marine strata and are thought by Arkhangelskii to be marine sedimentary deposits derived from Paleozoic weathering of lavas and tuffs, a conclusion with which others differ. The Triassic deposits of the Urals and the Carboniferous deposits of the Leningrad district are interpreted to be transported bauxites deposited in lakes. The Ural deposits contain up to 62 percent Al_2O_3, from $3\frac{1}{2}$ to $5\frac{1}{2}$ percent SiO_2, and up to 27 percent Fe_2O_3. The Tikhvin bauxites are much higher in silica. Diaspore and chlorite have devel-

oped in the altered bauxite of Ivdel.

Indonesia has large producers of bauxite on the islands of Batam and Bintang. The beds are about 12 feet thick and contain 53 percent Al_2O_3, $2\frac{1}{2}$ percent SiO_2, and 13 percent Fe_2O_3. In *India* bauxites in the Jabalpur and Khaira districts are thought to have been formed from underlying original kaolin beds after the covers had been removed. Fermor states that some of the deposits resulted from the surface weathering of an amygdaloidal trap. In *Germany* bauxite is also reported to be derived from basalt. The bauxite of *Minas Geraes, Brazil*, occurs in lenticular lenses, between itabirite and slate, associated with residual clay. The Australian Weipa bauxite deposits occur in laterite that extends for over 160 kilometers. The bauxite ranges in thickness from a meter to 10 meters with a solid overburden at only a meter.

Selected References on Bauxite

General References are given under "Bauxite," Chapter 13.

Bauxite in British Guiana. 1936. E. C. Harder. Can. Inst.

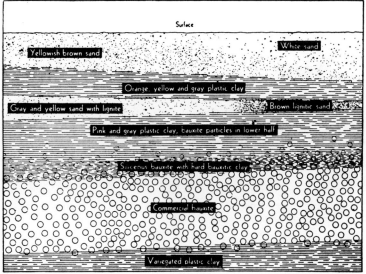

Figure 19-53 Vertical section through bauxite bed and underlying clay, Trevern mine, Guyana. (Harder, Bull. Can. Min. Inst.)

VERTICAL SECTION THROUGH TYPICAL BAUXITE BED, BRITISH GUIANA

Min. and Met. Bull., November. 1936. *Excellent description of occurrence and origin.*

Bauxite deposits of the Gold Coast. W. G. G. Cooper. Gold Coast Geol. Survey Bull. 7, 1936. *Occurrence and origin.*

Relations of bauxite and kaolin in Arkansas bauxite deposits. M. Goldman and J. I. Tracy, Jr. Econ. Geol. v. 41, p. 567–575, 1946. *Thesis that bauxite is derived from syenite rather than intermediate kaolin.*

Stratigraphy and origin of bauxite deposits. 1949. E. C. Harder. Geol. Soc. Am. Bull., v. 60, p. 887–908. *Broad discussion of origin of bauxites; good bibliography.*

General References 8, 11, 13, and 19.

The origin of bauxite in the coastal plain of Surinam and Guyana. 1969. L. Krook. Surinam Govt. Geol. and Mineral. Service Med 20. Rotterdam. *Contributions to the Geology of Surinam.*

Professional Paper 1199. 1965. U.S. Geol. Surv. *Numerous papers by different authors on U.S. Bauxite deposits.*

Bauxite. 1971. H. F. Kurtz. U.S. Bur. Mines Yearbook, 13 p.

Alcoa jungle lab in Brazil for aluminum exploration. 1971. Mining Eng., v. 23. p. 6.

Bauxite reserves and potential aluminum resources of the world. 1967. S. H. Patterson. U.S. Geol. Survey Bull. 1228. 178 p.

20

IRON AND FERROALLOY METALS

Iron is the principal metal employed in modern industrial civilization.

Klemic, James, and, Eberlein

IRON

Although iron is the second most abundant metal in the earth, the character of its natural compounds prevented its use as early as some other metals. It was known by 4000 B.C., and the Egyptian Pharaohs regarded it more highly than gold, but this probably was the rare meteoritic iron. Apparently by 1200 B.C., iron was manufactured but was still rare, and its industrial use did not commence before 800 B.C., which marks the start of the Age of Iron. Steel came into use about 800 years later, and the blast furnace in the fourteenth century. During the sixteenth century the forests of Great Britain, including Robin Hoods Sherwood Forest were denuded to supply charcoal to smelt iron ore, but this waste became unnecessary when the great discovery was made about 1710 that coal could be used to reduce iron ore. This was the beginning of the great industrial age of iron that culminated in the steel age made possible by Bessemer's discovery in 1856. In the nineteenth century Great Britain, with her resources of iron and coal, became the first of the modern industrial nations.

In America, iron was discovered in Virginia in 1608, and the first large-scale smelting began in Massachusetts in 1664 and in Pennsylvania in 1730. With the use of anthracite for iron making in the early nineteenth century the industry centered around the anthracite region of eastern Pennsylvania, but with the advent of coke as fuel it jumped across the mountains to what is now Pittsburgh. The great Lake Superior iron-ore deposits, discovered in 1844, ushered in the industrial age of the United States. Where coal and iron met, in Pennsylvania and along the Great Lakes, great industrial centers arose, railroads were pushed afar, and a new era of United States development began.

Uses

Iron is the backbone of modern civilization. Few are aware to what extent we have become dependent upon it in homes, farms, cities, machines, automobiles, trains, and ships, Without it we should have to spin our clothes by hand and travel in wooden carts over dusty roads. When iron, or steel, is not suitable for certain uses, it is alloyed with other substances to make it suitable. To enumerate the various uses of iron would be to compile a history of the innumerable creations of modern civilization and industry. Each of the main types of iron—steel, cast iron, wrought iron, and iron alloys—has its particular sphere of use; steel, of course, exceeds all others. Some of the uses are shown in Fig. 20-1.

Production

The world production of iron ore responding to economic conditions, ranges between 800 and 1000 million metric tons annually about half of which is exported or imported (Fig. 20-2). Steel production ranges from 700 to 800 million tons. The chief producing countries of iron ore and their approximate normal percentages and changes from 1966 to 1975 of world production are in Fig. 20-3.

The United State's, declining production is 85,000,000 long tons of iron ore per year. This is crude ore of which only a few percent is direct shipping ore. The remainder is beneficiated, pelletized, and results in about 85,000,000 tons of useable ore that averages about 60 percent iron. In 1978, 55,824,000 tons of crude ore was produced in the United States, 90 percent of which came from the Lake Superior region (Table 20-1). There are sufficient reserves and potential resources of crude iron ore in the United States to meet about 75 percent of domestic needs for many decades.

The sources of steel are shown in Fig. 20-4. The imports are predominantly from Canada and Japan. The sources of U.S. iron ore and chemical compositions of the ore are given in Tables 20-1, 20-2, and 20-3.

Iron ore reserves in the United States are about 9000 million tons. The total identified U.S. iron ore

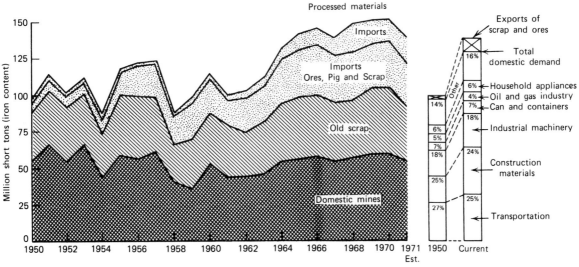

Figure 20-1 United States supplies and uses of iron, 1950–1971. (From Mineral position of the U.S., SEC, E. Cameron, ed., 1972, p. 41.)

resources, including reserves, are 108,000 million tons with consumption since World War II averaging less than 100 million tons per year (Table 20-1), but down to 85 million tons in 1979.)

World production of iron ore is approaching a billion tons each year. Australia and especially Rus-

sia exhibit the greatest growth. The raw material base for the steel industries in Russia is 128 working mines; 58 of them are open pit, which supply more than 230 million tons per year of high-grade ore and iron ore concentrates. Russia's mineral production of pig iron and steel from 1960 to 1975 is shown in

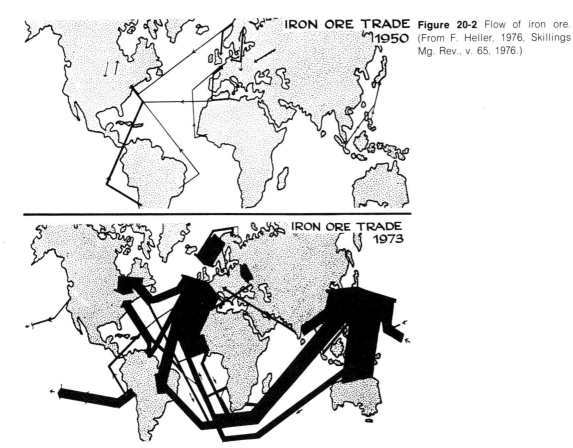

Figure 20-2 Flow of iron ore. (From F. Heller, 1976, Skillings Mg. Rev., v. 65, 1976.)

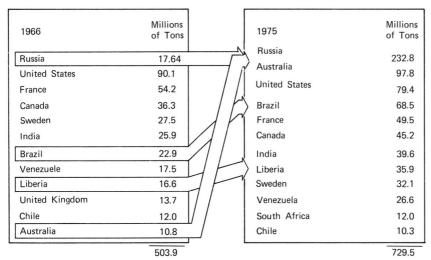

Figure 20-3 Iron ore production. (From Skillings Mg. Rev., June 23, 1977.)

1966	Millions of Tons
Russia	17.64
United States	90.1
France	54.2
Canada	36.3
Sweden	27.5
India	25.9
Brazil	22.9
Venezuele	17.5
Liberia	16.6
United Kingdom	13.7
Chile	12.0
Australia	10.8
	503.9

1975	Millions of Tons
Russia	232.8
Australia	97.8
United States	79.4
Brazil	68.5
France	49.5
Canada	45.2
India	39.6
Liberia	35.9
Sweden	32.1
Venezuela	26.6
South Africa	12.0
Chile	10.3
	729.5

Table 20-4. It is the greatest producer of iron ore in the world, but it has not joined the "Association of Iron-Ore Exporting Countries," which came into being in 1970. The Association consists presently of the countries of Peru, Mauritania, Algeria, Venezuela, Chile, India, and Australia,

The Great Lakes Region had a combined annual production of iron ore pellets of 92 million tons in 1975 and 95.7 million tons in 1976. Pellet production capacity is projected to reach 115 million tons by 1985. Pellets are the most desirable feed for blast furnaces today. This is because they have a high

Steel Imports Into The U.S. (1974)

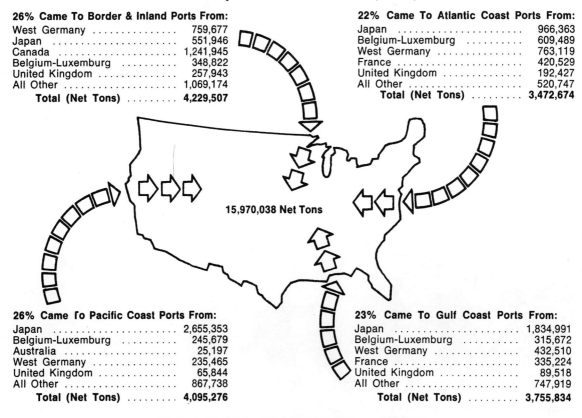

26% Came To Border & Inland Ports From:

West Germany	759,677
Japan	551,946
Canada	1,241,945
Belgium-Luxemburg	348,822
United Kingdom	257,943
All Other	1,069,174
Total (Net Tons)	**4,229,507**

22% Came To Atlantic Coast Ports From:

Japan	966,363
Belgium-Luxemburg	609,489
West Germany	763,119
France	420,529
United Kingdom	192,427
All Other	520,747
Total (Net Tons)	**3,472,674**

15,970,038 Net Tons

26% Came To Pacific Coast Ports From:

Japan	2,655,353
Belgium-Luxemburg	245,679
Australia	25,197
West Germany	235,465
United Kingdom	65,844
All Other	867,738
Total (Net Tons)	**4,095,276**

23% Came To Gulf Coast Ports From:

Japan	1,834,991
Belgium-Luxemburg	315,672
West Germany	432,510
France	335,224
United Kingdom	89,518
All Other	747,919
Total (Net Tons)	**3,755,834**

3% Came To Offshore Ports (Net Tons) 416,747

Figure 20-4 Steel imports into the U.S., 1974. (From *Steel Facts*, No. 1, 1975.)

World Pellet Production

Major pellet producers (millions of tons)	Present capacity	Under construction	Total
United States	68.2	27.5	95.7
Canada	31.7	6.0	37.7
Sweden/Norway	13.2	—	13.2
Australia	12.5	—	12.5
Japan	9.6	—	9.6
Europe/U.K.	5.6	3.0	8.6
Brazil	5.0	19.5	24.5
Liberia	4.4	2.4	6.8
Other So. America	4.2	5.5	9.7
Other	7.3	4.5	11.8
	161.7	68.4	230.1

Table 20-1 The Identified Iron-Ore Resources of the United States, in Millions of Metric Tons (United Nations, 1970) (Figures rounded. Include estimates for some deposits for which precise data are not available)

Region	Reserves[a]	Identified resources[b] (reserves + potential ores)
Lake Superior	6,500	78,000
Northeastern	200	1,400
Southeastern	100	9,400
Central-Gulf	600	700
Central-western	1,100	2,300
Western	600	1,200
Alaska	—	6,600
Hawaii	—	1,000
Total	9,100	100,600

Source: U. S. Geol. Survey Prof. Paper 820, p. 34.
[a] Reserves: Identified deposits from which minerals can be extracted profitably with existing technology and under present economic conditions.
[b] Identified resources: Specific, identified mineral deposits that may or may not be evaluated as to extent and grade, and whose contained minerals may or may not be profitably recoverable with existing technology and economic conditions.

iron content (65 percent), increase blast furnace output, conserve valuable coke, reduce the cost of hot metal, and are environmentally attractive.

Magnetic taconite ores range from about 25 to 45 percent iron. Much of this is not ore unless upgraded. The widespread world occurrences of low-grade iron formations have resulted in other terminology for taconite such as *itabirite* in Brazil,

Table 20-2 Summary of Elements Found in Iron Ores Used by the American Iron and Steel Industry

Source	Fe	P	SiO_2	Al_2O_3	Mn	S	CaO	MgO	Ti	V	Cr	Zn	Mo	As	Pb	Sn	Co	Ni	Cu
Cuyuna	55.83	0.367	10.61	4.60	12.92	0.093	0.84	0.35	0.090	0.009	0.019	0.006	0.005	0.006	—	—	—	0.002	0.011
	32.75	0.104	24.82	0.94	0.47	0.008	0.15	0.07	0.005	*	*	*	*	*	—	—	—	*	*
Marquette	61.57	0.180	5.93	4.66	2.46	0.210	0.93	1.53	0.325	0.020	0.010	0.005	0.005	0.006	*	*	*	0.006	0.012
	37.13	0.014	44.98	0.59	0.06	0.008	0.28	0.16	0.03	0.004	0.001	0.002	0.0018	0.002	*	*	*	0.002	0.006
Menominee	59.60	0.465	4.33	3.97	5.69	0.259	1.49	4.20	0.235	0.035	0.015	0.041	0.004	0.014	*	*	*	0.020	0.025
	37.11	0.015	44.57	0.89	0.05	0.006	0.24	0.18	0.065	0.016	0.002	0.012	0.0012	0.005	*	*	*	0.009	0.007
Mesabi	63.20	0.118	3.18	11.48	5.48	0.073	0.72	0.21	0.21	0.010	0.018	0.007	0.005	0.008	*	—	*	0.005	0.019
	49.07	0.028	21.68	0.45	0.07	0.005	0.09	0.07	0.008	*	*	*	*	*	—	—	—	*	*
Steep Rock	60.69	0.033	2.97	2.36	0.28	3.60	0.23	0.22	0.30	0.012	0.072	0.007	0.004	0.03	—	—	0.10	0.02	0.10
	50.80	0.019	10.35	0.98	0.10	0.016	0.08	0.04	0.01	0.006	0.013	*	*	*	—	—	—	*	*
Michipicoten	51.54	0.027	9.81	2.91	2.90	0.098	3.02	8.40	0.075	0.009	0.004	0.010	0.003	0.024	*.	0.075	*	0.004	0.025
	49.31	0.016	11.79	1.59	2.79		2.98	7.46	0.056	0.005	0.001	0.005	*	0.020	*	*	*	0.002	0.005
Eastern Penna. (Cornwall)	62.83	0.005	5.10	1.52	0.072	0.031	1.40	1.60	0.08	0.011	0.013							0.013	0.021
Texas (Lone Star)	47.52	0.18	15.34	6.90	0.32	0.10	0.15	0.030	0.026	0.032	0.020	—	0.004	0.013	—	—	—	0.008	0.004
California (Eagle Mt.)	55.00	0.080	11.00	1.60	0.08	0.40	1.80	3.50	0.13	0.003	0.005	—	0.001	0.005	—	0.003	0.020	0.005	0.05
Wyoming (Sunrise)	52.40	0.090	12.68	0.751	0.09	0.036	1.25	0.36	0.07	0.020	0.013	0.058	0.004	0.011			0.002	0.046	0.005
Utah	57.81	0.310	4.29	1.17	0.06	0.070				0.001				0.019				0.005	0.014
Tenn. C and I (Red Ore)	37.70	0.32	15.50	3.00	0.15	0.30	13.35	0.60	0.080	0.018	0.002	0.001	0.001		0.006	0.001	0.001	0.003	0.004
Tenn. Copper Co. (Sinter)	69.00	0.005	1.50	0.45	0.10	0.07	0.10	0.60		0.008			0.20	0.01			0.05	0.01	0.12
Venezuelan	64.85	0.13	0.98	2.45	0.10	0.043	0.017	0.084	.20	0.006	0.009	—	*	—	—	*	—		0.003
Chile (Tofo)	63.00	0.05	0.80	1.80	0.08	0.020	*	*	0.09	—	—	—	—	—	—	—	—	—	*
Labrador	60.30	0.042	7.49	1.76	0.09	0.021	1.41	2.00	0.59	0.28	0.008	—	—	—	—	—	—	0.15	0.010
Brazilian (Itabira)	58.00	0.058	8.40	0.84	1.30	0.012	0.17		0.005	0.001	0.008	—	0.001	—	—	0.001	—		0.005
Swedish (Kiruna)	68.50	0.035	0.39	0.76	0.05	0.008	0.01	0.03	0.014	—	0.008	0.005	0.004	—				0.006	0.018
Liberian (Bomi Hills)	66.35	0.485	1.97	0.50	0.11	0.006	2.24	0.94	0.040	0.074		0.010					0.022	0.009	0.004
	69.50	0.055	0.11	0.19	0.16	0.011	0.03	0.15	0.04	—	0.012	0.002	0.003	—			—	0.002	0.004

—None
*Trace

Source: Ohle, Econ. Geol., v. 67. p. 962, 1972.
[a] Modified from Jacobs, et al. (1954), Table 1.

Table 20-3 Compositions of Iron-Formation Types

Type	Algoma	Superior		Clinton	Minette
Facies	Oxide (1)	Oxide (2)	Silicate carbonate (3)	(4)	(5)
Fe	33.52	33.97	30.23	51.79	30.97
SiO$_2$	47.9	48.35	49.41	11.42	28.06
Al$_2$O$_3$	0.9	0.48	0.68	5.07	5.79
Fe$_2$O$_3$	31.7	45.98	16.34	61.83	29.81
FeO	14.6	2.33	24.19	11.00	13.08
CaO	1.45	0.1	0.1	3.32	1.92
MgO	1.8	.32	2.95	0.63	1.54
Na$_2$O	0.2	.33	0.03	ND	0.33
K$_2$O	.32	.01	.07	ND	.53
H$_2$O$^+$	.47	2.0	5.2 ⎫	1.94	13.10
H$_2$O$^-$	.1	0.04	0.38 ⎭		
TiO$_2$	.05	.01	.01	0.015	0.18
P$_2$O$_5$	.1	.04	.08	1.96	1.59
MnO	.3	.025	.65	0.17	0.16
CO$_2$	ND	.03	.22	2.15	2.89
S	ND	.013	.05	0.023	ND
C	ND	.08	.15	ND	ND
Location	Timagami	Knob Lake		Wabana	Peace River

Source: Geology and Economic Minerals of Canada, Econ. Report No. 1, 1970.

(1) Average of four analyses for 50 feet of section sampled systematically; Algoma type, mainly oxide facies, magnetite—quartz iron-formation; Timagami Lake area, Ont. (Analyses, Geol. Surv. Can. Laboratories.)

(2) Average of six analyses for 335 feet of section sampled systematically; Superior type, oxide facies, hematite—magnetite—quartz iron-formation; Knob Lake iron ranges, Quebec and Labrador. (Analyses, Geol. Surv. Can. Laboratories.)

(3) Analysis of 50 feet of section sampled systematically; Superior type, silicate—carbonate—chert facies; Knob Lake iron ranges, Quebec and Labrador. (Analysis, Geol. Surv. Can. Laboratories.)

(4) Analysis of large composite sample; Clinton-type iron-formation; Wabana, Newfoundland. (Courtesy, Wabana Mines.)

(5) Average of eleven analyses representing 15 feet of section; Minette type, oolitic siderite—chamosite—limonite iron-formation; Peace River District, Alberta. (Mellon, 1962.)

hematite-quartzite in India, *banded jaspilite* In Australia, and *quartz-banded* ore in Sweden.

The United States has an installed production capacity of about 140 million tons per year but does not use this full capacity because of foreign competition, imports, and sluggish markets.

During the past 10 years, revolutionary changes have been demanded by blast furnace operators in the quality of iron ore. This has resulted in the forming of high-grade concentrates and pellets made from low-grade iron formations, but preferably in higher-rank metamorphic zones. Certain horizons of the Lake Superior Precambrian cherty iron formations contain sufficient iron as magnetite or hematite, with a grain size and texture that will allow adequate mineral liberation, to yield an ore-quality product after grinding and concentration. The magnetite-rich deposits are termed magnetic *taconites* and are magnetically beneficiated. Hematite may be concentrated by flotation techniques or by converting it to magnetite.

The Reserve Mining Co. at Silver Bay, for example, produces more than 10 million tons of iron ore pellets that have a natural analysis of 60.8 percent iron with 8.6 percent silica. The mining and milling of this deposit is a prime example of environmental and ecological problems.

Environmental Problems

On March 31, 1975, the U.S. Supreme court refused for the third time to intervene in a long-simmering case involving the efforts of the U.S. government, three states, and several environmental groups to stop the Reserve Mining Company of Minnesota from dumping 67,000 tons per day of asbestos-bearing tailings into Lake Superior.

An estimated $4 million has been spent by the federal government alone in 6 years of litigation with Reserve Mining. It is known that inhalation of asbestos fibers is carcinogenic but it is not known if ingestion of the fibers in drinking waters poses a health threat. The key issue has been whether the potential threat to health is great enough to justify

Table 20-4 Russian Production of Iron and Steel (metric tons)

Year	Pig iron	Steel
1960	46,757,000	65,300,000
1961	50,893,000	70,755,000
1962	55,665,000	76,306,000
1963	58,691,000	80,231,000
1964	62,377,000	85,034,000
1965	66,184,000	91,020,000
1966	70,264,000	96,907,000
1967	74,812,000	102,735,000
1968	78,788,000	106,532,000
1969	81,000,000	109,000,000
1970	86,000,000	116,000,000
1974	94,000,000	—
1975	100,000,000	140,000,000

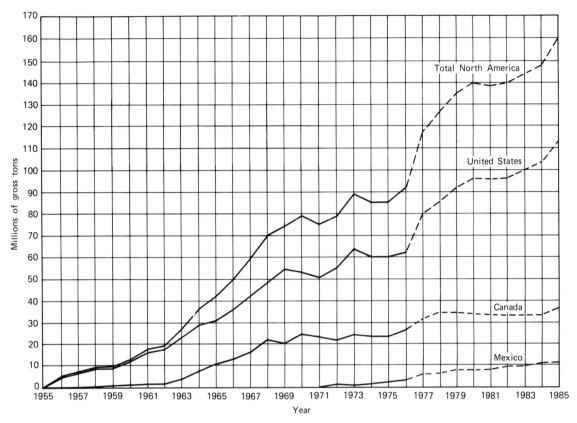

Figure 20-5 World pellet production and Great Lakes pellet facilities. (From Skilling Mg. Rev., Dec. 4, 1976; Min. Eng., v. 27, p. 20, 1975.)

the termination of operations. The environmentalist viewpoint has been that since a potential health threat has been clearly established, the burden of proof that dumping activities are not hazardous should rest with the company. But this argument is difficult to sell at a time when layoffs of workers resulting from closedown of the plant would be disastrous for the local economy. The Justice Department saw to that when in response to White House political pressure it suddenly pulled out of the legal battle to stop Reserve Mining from dumping 67,000 tons of potentially dangerous and polluting iron-mining wastes into Lake Superior each day. Because the Justice Department withdrew its support, the Supreme Court, as expected, denied a request by environmentalists and the states of Minnesota, Michigan, and Wisconsin to close the plant if it does not begin dumping the waste on land.

The court refused to close the plant because "no harm (from the asbestos fibers) has been shown to have occurred to this date." As one of the appeals court judges had said earlier: "Show me a dead body." Recently, however, Reserve Mining has initiated the development of a land dump for the tailings.

Mineralogy

The economic iron ore minerals are shown in Table 20-5.

Other iron-bearing minerals, such as pyrite FeS_2, pyrrhotite $Fe_{1-x}S$, marcasite FeS, and chamosite $Fe_2Al_2S_1O_5(OH)_4$ do not occur in significant amounts in high-grade large deposits and are not considered as potential iron reserves in the foreseeable future. Of the iron ore minerals, magnetite is the richest but of minor quantity; hematite is the mainstay of the iron industry; limonite and siderite are of minor importance in America but important in Europe.

Associations. Common impurities in iron ores are silica, calcium carbonate, phosphorus, manganese (especially in hematite), sulfur, alumina, water, and

Table 20-5 Economic Iron Ore Minerals

Mineral	Composition	Percent Fe	Commercial classification
Magnetite	$FeO \cdot Fe_2O_3$	72.4	Magnetic (or black) ores
Hematite	Fe_2O_3	70.0	Red ore
"Limonite"	$Fe_2O_3 \cdot nH_2O$	59–63	Brown ore
Siderite	$FeCO_3$	48.2	Spathic, black band, clay-ironstone

titanium. A summary of elements in iron ores used by the United States is shown in Table 20-2.

Recovery. The making of a usable product of iron involves two steps: first the reduction of the iron ore to pig iron; and second, the treatment of the pig iron to make cast iron, wrought iron, or steel.

The ore is smelted with coke and limestone. Air or oxygen blown in at the bottom, burns coke to carbon monoxide, which removes oxygen from the iron ore, reducing it to the metal. The limestone slags off the silica, alumina, and other impurities. Two tons or less of good iron ore and scrap iron yield 1 ton of pig iron. The pig iron always contains carbon, and the percentage of included constituents determines the use to which the pig is put. Various types of pig iron are: basic open hearth, foundry, Bessemer, and low phosphorus, malleable, and forge. Specialized pig, high in silicon or manganese, is also produced. With the use of iron pellets and oxygen, Bessemer production is decreasing rapidly. For certain castings the pig iron may be purified or other ingredients added to it. For *wrought iron*, which is rather pure, the pig goes to a puddling furnace, where the impurities are slagged off by stirring, and the red-hot is hammered to desired shapes. Wrought iron is malleable and ductile, and it resists corrosion.

Steel is iron with alloyed carbon, which is generally less than 1 percent but may reach 1.6. "Mild" steels with low carbon approach wrought iron. Steel is made by several processes. In the *open-hearth* methods, molten pig iron, some hematite, and limestone are melted in a tilting furnace; the excess carbon and silicon are slagged off by the oxygen of the hematite; sulfur is vaporized and also unites with manganese to form MnS. Phosphorus is removed by uniting with calcium from the refractory brick lining, to prevent the steel being "cold short" (brittle). Pig high in phosphorus is treated in a basic open hearth or one lined with basic brick. An acid open hearth is used for low-phosphorus pig. In the

Bessemer process, low-sulfur, low-phosphorus pig is rapidly converted into steel in a tilting barrel-shaped furnace through which oxygen is blown to slag off the impurities. Desired amounts of carbon and manganese are added and mixed by air blowers. For the various ferroalloys, proper amounts of alloy metals are added to yield the desired steel.

Kinds of Deposits and Origin

The various types of commercial deposits of iron ore, and some important examples of each are

1. Magmatic: Magnetite, titaniferous magnetite—Iron Mountain, Wyoming; and Adirondacks, New York.

2. Contact-metasomatic: Magnetite, specularite—Fierro, New Mexico; Cornwall, Pennsylvania; and Iron Springs, Utah.

3. Replacement: Magnetite, hematite—Lyon Mountain, New York; Dover, New Jersey; and Iron Mountain, Missouri.

4. Sedimentary: Hematite, limonite, siderite—Clinton ores of New York to Alabama; Wabana, Newfoundland; Minette ores of central Europe; Jurassic ores of England; and Brazil.

5. Residual: Hematite, magnetite, and limonite—Lake Superior; Appalachians; Western Australia; Kudremukh, India; Russia; Bilbao, Spain; Pao, Venezuela; Labrador-Quebec, and Brazil.

6. Oxidation: Limonite, hematite—Rio Tinto, Spain.

7. Volcanic exhalative—Kiruna and Taberg, Sweden.

The Lake Superior-type of deposits have been formed by a combination of two or more processes, for example, sedimentation, followed by residual enrichment and metamorphism.

Distribution of Iron Ores

As would be expected of multiorigin deposits, iron ores are widely distributed under various geologic conditions.

In *North America*, magnetite deposits occur in the deeply disected regions of plutonic intrusions, such as the northeastern states and the Cordillera. Hematite deposits outcrop around the margins of the great sedimentary basin from Alabama to New York to Wisconsin and in Newfoundland. Residual deposits occur in the eroded Appalachians and in Cuba.

In *Central Europe*, great sedimentary deposits underlie parts of Lorraine, Luxembourg, France, Belgium, and Germany. There are rich oxide deposits in Sweden. Farther east and north are the extensive deposits of the Ukraine, Siberia, and European Russia. Huge deposits of high-grade iron ore exist in the Koztamus area of Soviet Karella located about 40 kilometers east of the Finnish border. Finland is developing the area and constructing a railroad line to Kontiomake, the transportation center of central Finland. The ore will be pelletized in Finland and subsequently shipped to foreign nations.

In *Africa*, good-quality ores lie near the Mediterranean in Morocco and Algeria and large bodies of low-grade magnetite lie to the south in the Transvaal.

In *Russia* and *Australia*, many rich deposits occur that have placed these two countries as the largest producers of iron ore in the world.

In *South America* there are extensive deposits in Brazil, Chile, and Venezuela. Southeast *Asia* has stupendous resources in India and lesser ones in China.

Examples of Deposits in the United States

Lake Superior Region. The iron ranges in Minnesota, Wisconsin, and Michigan (Fig. 20-6), known as the *Cuyuna, Vermilion, Mesabi, Marquette, Menominee,* and *Gogebic,* are examples of Precambrian banded iron formations (BIF).These deposits dwarf all other types of iron ore deposits in the United States.

Perhaps the most unusual fact to emerge from recent geochronological studies of the banded iron formation deposits is that they seem to have formed within a single epoch of earth history, some time between 1900 and 2500 million years ago. The amount of iron deposited in this epoch of sedimentation, and still preserved, is enormous—at least 10^{14} tons and possibly 10^{15} tons. Even though there are other banded-iron formation deposits both younger and older than the above epoch, the total amount of iron in them is far outweighed by the iron in the limited epoch and now represented by those deposits of Labrador and the Lake Superior region of North America, of Krivoi Rog and Kursk in Russia, of South Africa, and of the Hamersely, and Pilbara group of Western Australia, all of which are at least approximately contemporaneous and possibly strictly so.

Iron formations have been defined in various ways, but the most commonly accepted definition has been given by H. James as ''a chemical sediment, typically thin-bedded or laminated, containing 15 percent or more iron of sedimentary origin, and commonly but not necessarily containing layers of chert.''

The iron formations of the United States up to 300 meters thick, are part of a thick, sedimentary series with intercalated volcanic rocks. These have been strongly folded and crumpled in most of the ranges and metamorphosed into slates, quartzites, marbles, schists, cherts, and jaspers. In some places, regional and contact metamorphism has converted the iron formations to magnetite-amphibole schists and the hematite to specularite. Some early geologists (Van Hise and Leith) attributed silica in the iron formations to a magmatic source and deposition on the sea floor, followed by leaching of silica (Fig. 20-7). Others (Grout and Gruner) believe that they came from the weathering of an adjacent Precambrian land mass and were deposited as regular sedimentary beds. The work of Moore and Maynard demonstrating the natural solution, transportation, and deposition of iron and silica supports this view. Woolnough suggests that such deposits represent epicontinental sediments formed as chemical precipitates from cold natural solutions in isolated closed basins of peneplaned surfaces.

The ore in the Cuyuna Range and the Marquette Range occurs mainly in pitching synclinal troughs (Fig. 20-8 and 20-9) in large bodies that are irregularly shaped and are banded or bedded. They generally bottom on impervious basements that guided and controlled the water circulation. In most of the ranges, the enclosing rocks are strongly folded with high dips, and in such places the ore extends to the surprising depths of 1000 to 1100 meters although the best ore is generally above 300 meters. Most of the Gogebic deep deposits are complex in structure (Fig. 20-9). In the Mesabi range, however, which has yielded about 70 percent of the Lake Superior ore, the rocks dip gently and the ore has been concentrated over a wide area to depths mostly less than 60 meters (Figs. 20-9 and 20-10), which permits mining on an enormous scale from large open cuts by power shovel; the maximum depth of ore is 300 meters.

The rhythmic banding of iron oxide and silica may be the result of several factors, such as seasonal variations, the lack of oxygen, and increased

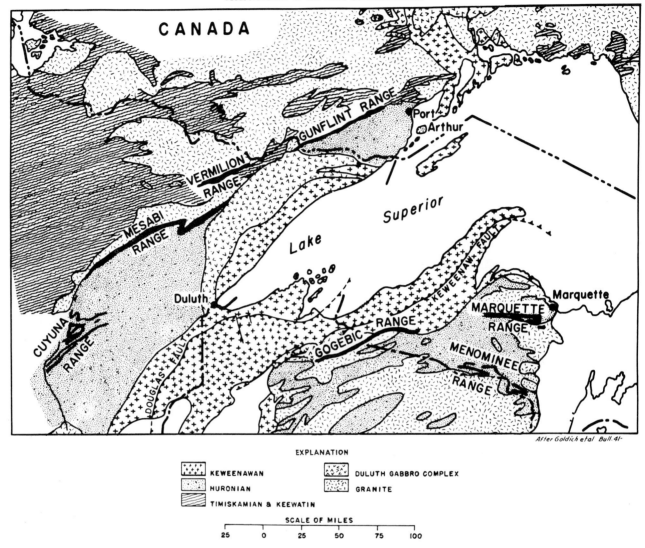

EXPLANATION

KEEWEENAWAN   DULUTH GABBRO COMPLEX

HURONIAN   GRANITE

TIMISKAMIAN & KEEWATIN

SCALE OF MILES

25   0   25   50   75   100

After Goldich et al Bull. 41-

Figure 20-6 Geological map of the Lake Superior region in the United States and adjacent Canada. (From: J. Ridge, ed., *Ore deposits of the United States*, 1968.)

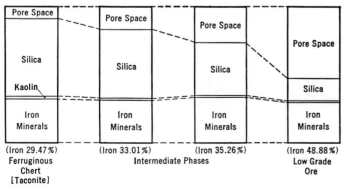

(Iron 29.47%)
Ferruginous
Chert
[Taconite]

(Iron 33.01%)   (Iron 35.26%)
Intermediate Phases

(Iron 48.88%)
Low Grade
Ore

Figure 20-7 Alteration of Lake Superior iron formation to iron ore by leaching of silica. (After Leith, U.S. Geol. Survey.)

394

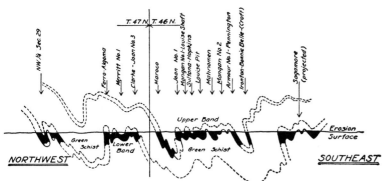

Figure 20-8 Geologic section across part of Cuyuna range, Minnesota, showing structure, the relation between upper and lower ore formations, and the relation to surface. (From Zapffe, Econ. Geol.)

carbon dioxide in the early stages of the earth's atmosphere. Under tropical conditions, the soil and streams would be acidic resulting in the leaching of iron from the red soils and transport of iron oxide to the sea. During wet seasons, the abundance of rainwater would dilute the groundwater, raising the pH, and allow the transport of silica. These repetitive bands are quite thin, sometimes averaging about 1 millimeter in thickness. At later times,

chemical weathering processes leached siliceous and carbonate minerals from the iron formations resulting in an upgrading of the iron content.

The general setting in which deposition of the Huronian iron formation occurred is summed up by H. L. James as follows:

1. During the early part of the Huronian (Fig. 20-11), the land area was one of very low relief, shelving off gradually to the open ocean. The rocks of the land area

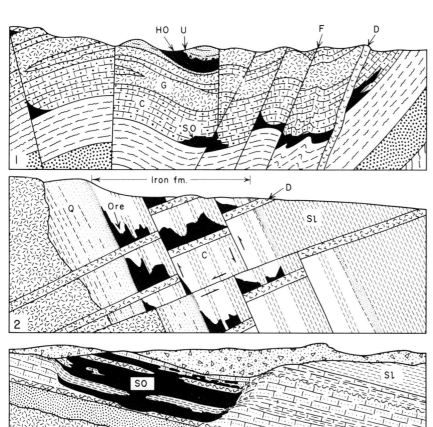

Figure 20-9 Diagrammatic sections in Lake Superior iron districts showing the occurrence of ore. (1) Marquette range, Michigan. HO, hard crystalline ore; SO, soft ore; F, fault; D, dike; G, gabbro sill; C, siliceous iron carbonate rock of iron formation from which the iron ore is formed by oxidation of the carbonate and leaching of the silica. The solutions that did this work were guided by structural troughs in the formation, created by folding, faulting, or intrusion. (2) Gogebic range, northwestern Michigan-northern Wisconsin. Sl, slate hanging wall; Q, quartzite foot wall of iron formation; and D, dike. (3) Mesabi iron range, northern Minnesota. (Reproduced by permission from *Minerals in World Affairs*, T. S. Lovering. Copyright 1943, by Prentice-Hall, Inc., Englewood Cliffs, N.J.)

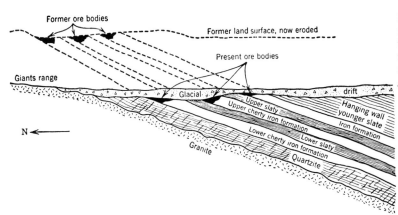

were undergoing deep chemical weathering. The sediments deposited on the broad shallow shelves consisted of clean, washed quartz sand, and of dolomite.

2. Slight structural disturbance at the end of the lower Huronian resulted in the development of broad offshore buckles in a belt peripheral to the land area. The restrictions imposed on circulation with the open ocean permitted abnormalities of concentration and bottom conditions in the marginal basins thus formed. In these marginal basins were deposited the chemical products of the deep weathering of the land area—chiefly iron and silica or derivatives from the deep seas.

3. The marginal basins may have been and probably were relatively small in size. For example, the Gogebic range is directly on the strike projection of the Marquette range and lies only 100 meters to the west, yet dissimilarities in iron-formation stratigraphy indicate deposition in separated basins. Marginal basins on the order of 800 kilometers in length and 160 kilometers in width would be adequate to account for the present distribution of the ranges. The implications of this statement are that elsewhere in the region, where the bordering sea was not restricted by offshore buckling, normal sedimentation would prevail. The absence of iron-formation in the type Huronian section north of Lake Huron might appear to support this contention, though other possibilities, such as questionable correlation of Huronian strata, may account for the fact equally well.

4. The epoch of iron deposition was ended in the region by strong structural disturbances that produced profound changes in both the land area and the sea-bottom configuration. Chemical sedimentation gave way abruptly to clastic deposition, and the broad swells that earlier acted as barriers for shallow sea developed into a chain of volcanic islands.

The concept of deposition in restricted basins here expressed in very similar to that outlined by Woolnough, who suggests that many of the unusual chemical deposits were precipitated in "barred basins." Woolnough also stresses, by analogy with conditions prevailing in Australia, the possible great importance during long periods of geologic history of deep chemical weathering of an

intensity not now observable on the present surface of the earth. This combination of conditions—namely, the barred or restricted basin of deposition coincident with deep chemical weathering of the land surface—appears necessary to explain depositon of the iron-formations.

Not all Precambrian banded iron deposits are formed by the widely accepted sedimentary processes. A magmatic hydrothermal replacement origin has been suggested for the Precambrian banded iron ore deposits in the Morrison formation, at Cleveland Gulch and Iron Mountain, Rio Arriba County, New Mexico.

The evidence cited by McLeroy to support this epigenetic origin is cross-cutting relationships proximity of the deposits to intrusive bodies; the presence of preexisting, banded metamorphic rock composed of quartz bands alternating with chlorite-rich, mica-rich, or other mineral-rich bands; and in the mineralized sequences, the presence of magnetite-rich bands with a much greater percentage of these other minerals than the quartz bands; magnetite bands cross-cutting post metamorphic minerals; and other factors.

The deposit consists of alternating magnetic-rich and quartz-rich layers averaging about 4 millimeters of thickness.

Other Magnetite Deposits. Other deposits of magnetite occur in New York, New Jersey, North Carolina, Missouri, Wyoming, and Utah. At *Lyon Mountain*, New York, Gallagher describes valuable bodies of nontitaniferous magnetite, that occur as high-temperature replacements of gneissic granite. The largest body is 1700 meters long and 7 meters wide and extends 500 meters down the dip. The ore carried 60 to 70 percent Fe, and milling ores ran about 30 to 50 percent Fe. Within the ore are miarolitic cavities, large enough to admit a man, containing hugh crystals of orthoclase, quartz, albite,

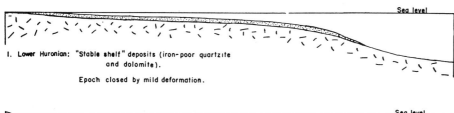

1. Lower Huronian: "Stable shelf" deposits (iron-poor quartzite and dolomite).

Epoch closed by mild deformation.

Figure 20-11 Diagrammatic outline of Huronian depositional history in Michigan. (From H. L. James, Econ. Geol., v. 49, p. 278.)

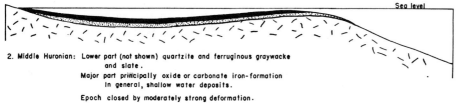

2. Middle Huronian: Lower part (not shown) quartzite and ferruginous graywacke and slate.
   Major part principally oxide or carbonate iron-formation
   In general, shallow water deposits.

Epoch closed by moderately strong deformation.

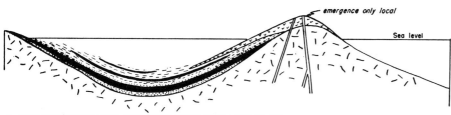

3. Upper Huronian (lower part): Conglomerate grading upward or laterally into quartzite, graywacke, and slate, with intercalated carbonate or sulfide iron-formation and explosive volcanics.

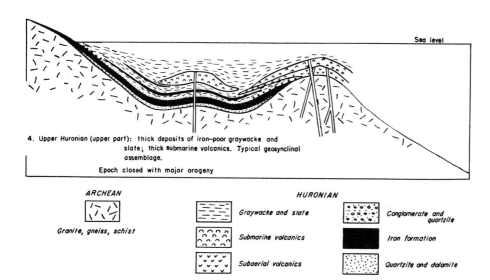

4. Upper Huronian (upper part): thick deposits of iron-poor graywacke and slate; thick submarine volcanics. Typical geosynclinal assemblage.

Epoch closed with major orogeny

ARCHEAN	HURONIAN	
Granite, gneiss, schist	Graywacke and slate	Conglomerate and quartzite
	Submarine volcanics	Iron formation
	Subaerial volcanics	Quartzite and dolomite

hastingsite, aergirinaugite, magnetite, titanite, and water under high pressure.

Similar ores are mined at *Dover, New Jersey*, and *Mineville, New York*, where magnetite and apatite occur with pegmatitic minerals. A fair-sized replacement body of hematite-magnetite is worked at *Iron Mountain, Missouri*. In the *Adirondacks* numerous deposits of titaniferous magnetites are associated with anorthosite and gabbro, and one at Lake Sanford yields concentrates of ilmenite and magnetite.

A recent development of a Lake Superior type of iron ore deposit is the *Iron Mountain mine* in the Atlantic City district, located near the south end of the Wind River Range, Fremont County, Wyoming. Ore is being mined by open pit methods where the iron formation is thickened by tight folding and faulting. About 1.3 million tons per year of iron ore pellets are shipped from this deposit to the Geneva Steel plant at Provo, Utah.

Exploration of the iron ore in this former gold-producing district began in 1954, and the first ore was shipped in August, 1962. Indicated iron formation (about 30 percent Fe) reserves are reported to be 300 million tons. This is the chief source of Rocky Mountain iron ore, as it now exceeds the production from the Iron Springs area of Southern Utah, when ore occurred as contact metasomatic deposits.

Other Hematite Deposits. Replacement bodies of hematite at *Pilot Knob* and *Iron Mountain,* Missouri, constituted important local sources of iron ore prior to 1966, when the latter deposit ceased operations.

Clinton Iron Ore Beds. The Clinton sedimentary iron ore beds rank next in importance to the Lake Superior deposits. These beds outcrop across Wisconsin and New York and south to Alabama. They have been mined at numerous places, and at Birmingham, Alabama, they have given rise to a large industrial steel center owing to the contiguous coking coal and limestone flux.

The ore occurs as sedimentary beds intercalated with shale, limestone, and sandstone of Clinton (Silurian) age. There may be from one to four beds, which thicken and thin from a few centimeters to 10 meters. Their greatest development is at Red Mountain, Birmingham, where two of four beds are workable, the production being mainly from the Big Seam.

The types of ore are (1) oölitic ore, composed of round oölites of hematite 1 to 2 millimeters across, enclosed in a matrix of hematite and calcite; (2) fossil ore, composed of fossil fragments coated and partly replaced by hematite and enclosed in amorphous and oölitic hematite, and high in calcium carbonate; (3) flaxseed ore, composed of flattened concretions of hematite surrounded by hematite mud and replaced fossil fragments. The primary ore is hard and is high in calcium carbonate, which is leached out near the outcrop, producing thereby enriched "soft" ore. The soft ore may carry 50 to 60 percent iron, but the "hard" ore runs only 35 to 38 percent. The high calcium carbonate content makes the ore largely self-fluxing; insoluble matter runs 12 to 14 percent.

At Birmingham, the Big Seam outcrops for a length of 30 kilometers and dips on an average of less than 30°. It is 5 to 10 meters thick, of which 2 to 4 meters is minable; in places it is divided by a shale parting. The ore averages about 36 percent iron and is mined from inclined shafts that extend 300 to 2000 meters down the slopes. The annual

production ranges from 5 to 6 million tons of ore, and Burchard estimates the reserves at 1.4 billion tons of good ore and 0.5 billion tons of second-grade ore.

The origin of the Clinton ores has emerged from much controversy with processes described as different as hydrothermal replacement to simple sedimentation. Such origins had important bearing upon the development of the deposit. Certainly, a sedimentary origin would suggest a larger extension of the deposit than a contact-metasomatic origin.

With the advent of Eh-pH studies, the development of the hematite-siderite-pyrite stability fields, as shown in Fig. 9-2, allowed a most reasonable hypothesis for the formation of the Clinton-type deposits. Castaño and Garrels showed by laboratory experiments that significant quantities of ferrous iron can be carried in solution in aerated river waters of pH = 7 or lower. If such waters enter a marine environment with solid carbonate in equilibrium with the seawater, the iron is essentially completely precipitated as ferric oxide or may replace calcium carbonate shells or oölites. With a gradually decreasing sea bottom from shore to the sea, clastics would be dropped near shore but the soluble ferrous iron would be carried farther seaward, where the marine pH of about 7.8 would result in an iron-rich precipitation. Such a suggested genetic model clarifies clearly the origin of the Clinton type of iron deposits.

The oölitic sedimentary hematite ores of Wabana, Newfoundland, are generally similar to the Clinton ores, except that, according to Hayes, they contain considerable chamosite and siderite. The beds are several centimeters to 10 meters thick, and one of them has been mined for a distance of more than four kilometers under Conception Bay. The ore averages 50 percent Fe, 11.8 percent, and 0.9 percent P.

Cornwall, Pa. At Cornwall, a thick quartz diabase dike and sill intrude Cambrian limestone, resulting in two large contact-metasomatic bodies of magnetite containing subordinate iron and copper sulfides, and some cobalt. The enclosing rock now consists chiefly of diopside, actinolite, phlogopite, chlorite, calcite, and dolomite. Hickok shows that the minerals are arranged zonally outward from the diabase contact, first a diopside and then an actinolite zone within which the magnetite bodies occur. The magnetite occurs in three forms: normal isometric, anisotropic, and maghemite. The exhalations from the diabase, in the order of occurrence,

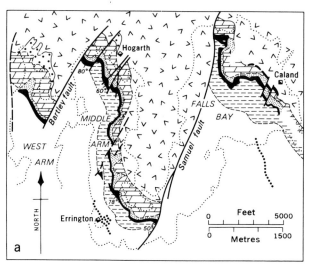

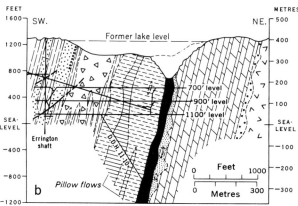

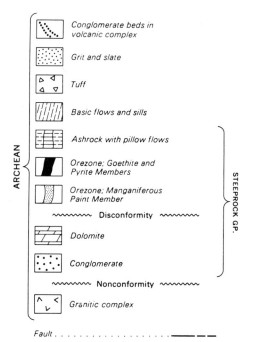

Conglomerate beds in volcanic complex

Grit and slate

Tuff

Basic flows and sills

Ashrock with pillow flows

Orezone; Goethite and Pyrite Members

Orezone; Manganiferous Paint Member

~~~~~ Disconformity ~~~~~

Dolomite

Conglomerate

~~~~~ Nonconformity ~~~~~

Granitic complex

ARCHEAN

STEEPROCK GP.

Fault . . . . ▬ ▬ ▬
Mine shaft . . . . ▪
Former shoreline . . . . ·········

GSC

contained potash, silica, iron oxides, base-metal sulfides, and hydrous silicates. Magnetite constitutes 40 to 60 percent of the ore. The western body is 1500 meters long, 70 to 200 meters wide, dips 25°S., and diminishes in depth. The eastern body is slightly smaller. The deposit, which has been mined since 1742, has yielded over 40 million tons and produces about 1 million tons annually.

Other contact-metasomatic iron deposits occur at *Iron Springs, Utah,* where monzonite porphyry intrudes limestone, forming hematite and magnetite, and at *Hanover, New Mexico,* where granodiorite porphyry has yielded a large aureole of contact-metasomatic silicates, including bodies of magnetite with hematite and sulfides of iron, copper, and zinc.

Canadian Iron Ores

Iron-ore production in Canada is growing rapidly. Iron ores exceed nickel in value and are only surpassed by copper as the most valuable metallic commodity produced in Canada.

The opening of the Helen mine in the Michipicaten district in Ontario in 1939 marked the beginning of a new period of development and rapid growth of the Canadian iron ore industry after a nonproductive stage of nearly 20 years. The Steeprock Range (Fig. 20-12) was developed during World War II to help offset increased continental market needs.

More than 94 percent of the iron produced in Canada is derived from Precambrian rocks of the Canadian Shield. In the single decade after 1950, iron ore production rose from 3.5 million tons to 22 million tons and exceeded 60 million tons by 1975 (Fig. 20-13). This is attributed to phenominal expansion in exploration and growth in the industry. Iron ore from the Labrador-Quebec Ranges (Fig. 20-14) is a prime example of the first ore shipped in 1954 from these massive deposits.

Superior-type iron formation in the Labrador Geosyncline extends more than 1100 kilometers from west of Ungava Bay to Schefferville, Quebec, thence an additional 300 kilometers within the Grenville province along the northern portion of Wabush Lake near Mt. Wright and Mt. Reed.

The iron formations are composed mainly of iron oxides and silicate-carbonate. The best ore has a

Figure 20-12 Steeprock Lake area, Ontario (Jolliffe, 1966): (a) geological map; (b) section through Errington mine. (From *Geology and Economic Minerals of Canada,* 1970.)

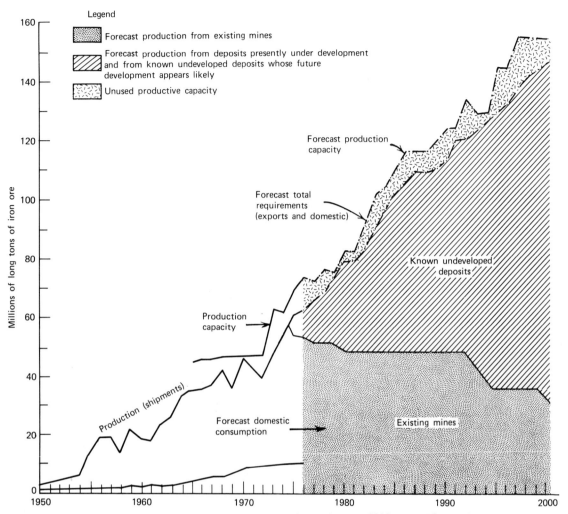

Figure 20-13 Forecast Sources of Canadian iron ore production to the year 2000 to meet forecast requirements. (From Skillings Mg. Rev., v. 64, 1975.)

fine-grained, cherty character that allowed presumably Mesozoic groundwater to penetrate permeable zones in the iron formations. Leaching of silica resulted, along with oxidation and remobilization, which concentrated some of the iron by secondary enrichment processes.

More than 45 deposits of this kind, some of which contain as much as 50 million tons of ore are present at Schefferville near the border of Quebec and Labrador. Ore of direct-shipping quality usually contains from 51.5 to 65 percent Fe, as much as 12 percent Si, 0.045 percent P, 1.5 percent A1, and small amounts of lime and magnesia. Manganese content varies, averaging less than 3 percent.

Basically, two different types of Precambrian iron deposits exist in Canada. These are the Algoman type and the Lake Superior type; other types also exist but they are not of major importance. Analyses of the altered genetic types are given in Table 20-4.

The Algoman type was deposited with volcanic and sedimentary rocks characteristic of eugeosynclinal facies. Typical beds are composed of grey or red jasper chert interlayered with magnetite-and hematite-rich layers; but other mineral suites, such as massive siderite and pyrite-pyrrhotite beds, occasionally form part of the formation. This type of iron formation ranges in thickness from a several centimeters to more than 30 meters but is usually less than a few kilometers long.

Algoman-type beds are abundant in Precambrian volcanic belts, and similar belts also occur in the Appalachian and Cordilleran geosynclines. Almost every belt of volcanic and sedimentary rock in Su-

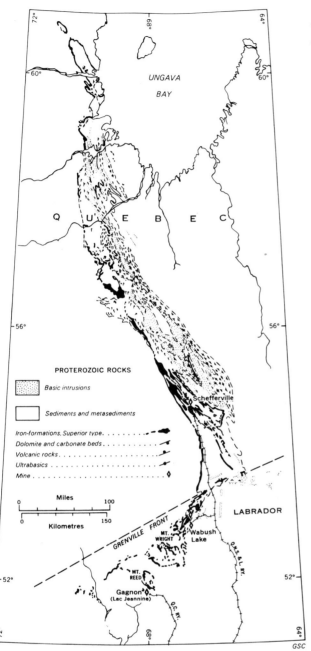

Figure 20-14 Iron formations of the Labrador geosyncline (by G. A. Gross). (From Geol. and Econ. Mineral Deposits of Canada, 1969.)

perior and Churchill Provinces contains some iron formation of the Algoman type. Most of these deposits are highly metamorphosed.

The Superior-type deposits are typical of the Lake Superior type and the iron formations are not as metamorphosed as the Algoman type but form the host rock for the hematite-goethite ore deposits.

The advent of pelletizing high-grade concentrates also began in Canada in the 1950s, and direct ship-

ping quality ore is no longer considered to be premium material.

The hematite-goethite ore deposits of the *Steeprock* iron ranges consist of tabular ore bodies (Fig. 20-12), which are faulted segments of one stratigraphic zone. Jolliffe postulates that the interlayered goethite, minor pyrite lenses with graphite, and thinly banded chert were deposited on an old erosion surface where iron, in the form of colloidal oxide and hydroxide gels, accumulated in a manner akin to modern bog-iron deposits. The iron-rich bed was then covered by basic tuff. The rock succession was later folded and deformed while oxidation and alteration took place in the ore zone during a late stage in its history; part of the tuff and banded chert was leached and transformed to a clay or paint-rock that resembles many other altered iron formations. This fact and other evidence suggest that the hematite and goethite were derived from siderite-pyrite and banded chert iron formation similar to that at Michipicoten. The sediments were later transformed to ore by supergene enrichment processes. Neither hypothesis accounts fully for all the facts known, but new information from more extensive and deeper mine workings is helping to resolve the problem of ore genesis. Ore from the mines, before processing pelletizing, contains about 55 percent iron, 5 percent silica, 1.5 percent alumina and lime combined, and 0.027 percent phosphorus. The dried ore contains about 60 percent iron.

Partially because of the majority of studies of these large deposits made during the last decade, theories of origin are only suggested. There seems little doubt, however, that the deposits are similar to Lake Superior types and are, therefore, of similar origin.

Australia. Australia is the second largest producer of iron ore in the world, having jumped from sixth place to second place within a few years. Almost all of the ore being produced is shipped to Japan, a country of furnaces, smelters, and people.

The Precambrian iron ore deposits of Australia are apparently confined to sedimentary rocks having a depositional age of 2000 million years or more. For such rocks, the banded iron ore is quantitatively more significant at about 2000 million years.

Figure 20-16 shows the six geographic areas of iron formations of Australia. Some minor occurrences are excluded because they are strongly metamorphosed and of uncertain age and relationships.

The Pilbara (Fig. 20-15) and Hamersley (Fig. 20-16) *ore provinces* of western Australia are two of the major Precambrian iron ore areas of the world and contain one of the major iron ore formations, the Brockman iron formation. The high-grade ore deposits of this formation have an age between 2000 and 2100 million years. In the central part of the outcrop area, the formation is 730 meters thick and is underlain by nearly 7000 meters of iron formation, dolomite, shale, sandstone and mixed basic

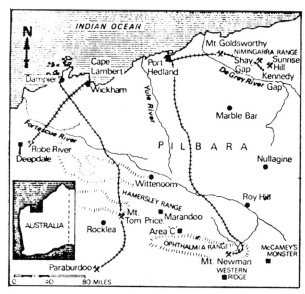

Figure 20-15 Map of Pilbara area showing main iron ore deposits.

volcanics of the Hamersley and Fortescue groups.

In the northern part of the outcrop area, dips are generally less than 4° although there is local steepening due to minor folding.

Lateral continuity of the bands of iron and silica, some less than 1 millimeter thick, have been described over a distance of 296 kilometers. On the basis of regional sampling, some microbands are continuous throughout the outcrop of the Dales George member that has an area of about 85,000 square kilometers. Possibly, therefore, stratigraphic continuity may be present throughout the 1000-meter vertical thickness of the Hamersley group iron formation over a total area of about 150,000 square kilometers.

European Iron Ores

The Alsace-Lorraine Minette Ores. The "Minette" ores or oölitic limonites of the *Lorraine-Luxembourg* region provide the bulk of the iron ore used by Western Europe. They also extend into parts of Belgium and Germany. The ores are sedimentary deposits interbedded with Jurassic shales, limestones, and sandstones; they occur in synclinal basins that pitch southwestward.

Formerly, the Lorraine outcrops were worked cheaply in open cuts by Germany, but now most of the ore comes from the down-dip portions in France, just west of Lorraine, at depths of 200 meters; but they are known by drilling to a depth of 1000 meters. Bismark moved the German border

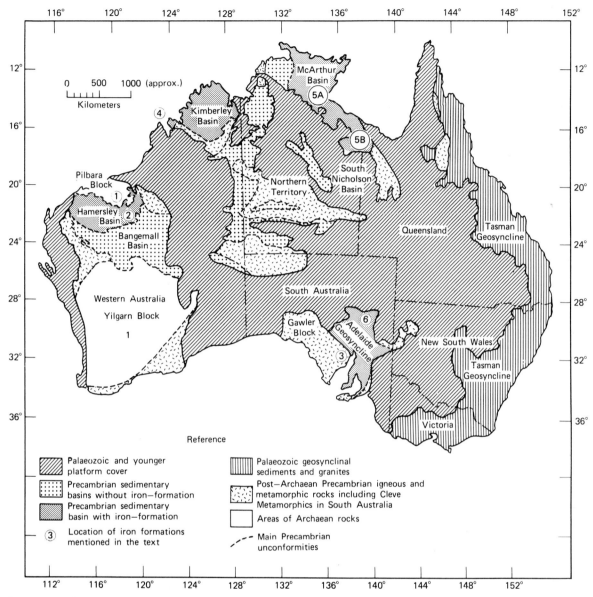

Figure 20-16 Geotectonic sketch map of the Australian continent, showing the locations of the main iron formations. (From Trendall, 1973, Precambrian iron-formations of Australia, Econ. Geol., v. 68, p. 402.)

west in 1871 in order to include the outcropping iron-bearing formations within Germany. Some of the most bitter fighting during World War I occurred in this region in attempts to rectify Bismark's abysmal lack of geological ability by not realizing that the beds extended downward into France.

The ore is soft, earthy, and contains oölitic limonite and hematite with some siderite and iron silicates. The grade is quite low, averaging 30 to 35 percent Fe with 0.5 to 1.8 percent P, 5 to 12 percent

CaO, and 7 to 20 percent SiO_2. A little magnetite in the form of oölites is present, and rarely pyrite. The ores are clearly of sedimentary origin, and, although Cayeux thought in 1909 that the iron oölites replaced calcite oölites, they probably are no different from the Clinton type of oölitic iron ores.

The deposits cover a wide area. When Lorraine was part of Germany, there were about 50 mines yielding around 20 million tons of ore annually, which supplied former German needs for iron ore. An equal number of mines yielded similar tonnage in France. Now there are fewer mines, and the French production from these ores has ranged from 32 to 50 million tons. Despite their low grade they are of great economic importance. The reserves are estimated at 5 billion tons of ore.

German Iron Ores. After 1918, Germany fell back on low-grade sedimentary limonites, chiefly in the Siegerland, Salzgitter, Lahn-Dill, and Weser areas, and in parts of Bavaria and Württemberg. These ores are mainly sedimentary limonites, but some are siderite or hematite. They are rather thin beds, and distant from coal, and carry about 28 percent iron; higher-grade ores are imported for mixing. The high-phosphorus Salzgitter ores, carrying 30 percent Fe and 25 percent SiO_2, are inclined, oölitic sedimentary beds, which are being mined by surface and underground methods at depths of as much as 700 meters. These ores yield about 5 million tons a year, and the reserves are stated to be 1 billion tons. German ores supply about 5 million tons annually.

The deposit of *Erzberg* in *Austria* is a low-grade (30 percent Fe) siderite body that can be calcined up to 50 percent iron and low phosphorus and yields 4 million tons annually.

Great Britain. Great Britain achieved greatness with its coal and iron, but newer and richer iron deposits elsewhere have provided severe competition to the low-grade English ores. British production ranges from 7 to 8 million tons annually, nine-tenths of which comes from low-grade Jurassic sedimentary beds, averaging 28 percent iron. This is a phosphorus ore of oölitic siderite and silicates in beds from 2 to 8 meters thick, which are mined by open cuts in the Northampton, Rutland, Frodingham, Cleveland, Leicester, Oxford, and intervening districts. The total reserves are estimated at 3 billion tons of ore.

Spain. The large iron ore deposits of *Bilbao*, northern Spain, are yielding 7 million tons annually. A 80 meter bed of Cretaceous limestone is cut by minor diabase and trachyte of Tertiary age. The limestone has been hydrothermally replaced by masses of siderite with minor ankerite and pyrite. Weathering has produced surficial residual concentrations of hematite and limonite averaging 50 percent iron.

Sweden. Sweden contains the largest deposits of high-grade iron ore in the world. The annual pro-

duction is 2 to 3 million tons, most of which comes from the great *Kiruna* area in Lapland.

The Kiruna deposit forms a ridge 2.8 kilometers long and 350 meters high, the backbone of which is the steeply dipping magnetite body, which is from 30 to 150 meters wide (Fig. 5-3). The enclosing rocks are Precambrian. The foot wall is syenite porphyry. It is penetrated by dikelets of magnetite which in places form "ore breccia" composed of fragments of syenite in a magnetite matrix. Similar contact relations exist in the hanging wall, quartz porphyry.

The ore consists predominantly of magnetite with intergrowths of apatite of simultaneous or later crystallization. Some pyroxene is present. The ore carries from 57 to 71 percent Fe, 0.03 to 1.8 percent P, >2.0 percent S, 0.7 percent Mn, 1.5 percent SiO_2, SiO_2, 0.7 percent Al_2O_3, 3 percent CaO, and 0.3 percent TiO_2, and is mostly of Bessemer grade. Across the lake from Kiruna is the similar but smaller *Luossavaara* body, which is 1300 to 1700 meters long and 40 meters wide.

The Kiruna ore is generally considered by some geologists to be a late magmatic differentiation in depth; others are strong advocates of an exhalative volcanic origin. According to the late magmatic classical viewpoint, it is an extreme differentiation to almost pure iron oxide, with minor titanium, phosphorus, and fluorine. Vogt considered that the differentiation took place by crystal settling and later remelting, a view not in keeping with Bowen's conclusions of the lack of sufficient heat to cause smelting. Geijer concludes that the "ores represent the last crystallizing parts." The syngeneticists stress the sedimentary nature of the iron ore, the associated volcanics above and below the ore deposits, and specific geochemical evidence.

The estimated reserves of the Kiruna deposits are 226 million tons of developed ore or 520 million tons of possible ore, and those of Gellivare are 239 million tons of possible ore. Most of the Swedish ore is normally exported to Germany and England.

Russia. Forty years after the industrialization process was initiated in Russia, the metallurgical industry has increased its overall output 25 times and today occupies second place in the world's output of metals, after the United States. Steel industry output has multiplied 27 times since the industrialization started and 13 times since the end of World War II. Russia now produces 28 percent of the world's iron ore, 4.2 times the U.S. production of iron ore, having first outproduced U.S. production in 1958.

The growth of Russian production in iron ore and steel is shown in Table 20-4.

The Russian iron ore formations are thought to be correlative, but the detailed stratigraphy, such as the number of banded ore members, differs from one locality to another. The iron formation in *Krivoy Rog* (Fig. 20-17) consists of seven members having a total thickness of 1300 meters, whereas that of *Kursk* consists of five members having a thickness of 500 meters.

Reconstruction of the ancient environments in which the iron formations were formed suggests deposition under both miogeosynclinal and eugeosynclinal conditions. The iron was concentrated in the middle formation of the sequence, corresponding to a sedimentary metallogenic epoch, which probably can be correlated throughout the gigantic metallogenic province of the Russian platform. The major iron formations in the province are about

2000 million years old, essentially of the same age as the banded iron formation deposits of the United States. Some of the smaller iron formations in other areas are apparently either older or younger than the main ones. Although five hypotheses have been proposed to explain the origin of the Precambrian iron formations and the banding in Russia, their origin is similar to that of the Lake Superior deposits. A volcanic-sedimentary origin is desired by some Russians, as was formerly favored by American geologists for the Lake Superior deposits.

Most of the Russian iron ore mines are Precambrian banded iron formation deposits. Alexandrov's summary of these deposits is as complete and detailed as possible. He states that the Precambrian iron formations in the European part of the Soviet Union occur mainly along a trend that can be followed between the Azov Sea coastal region in the south and Kola Peninsula in the north. The deposits of Krivoy Rog and Kremenchug, in the Donetz Basin (Fig. 20-17) within the Ukrainian Shield, and

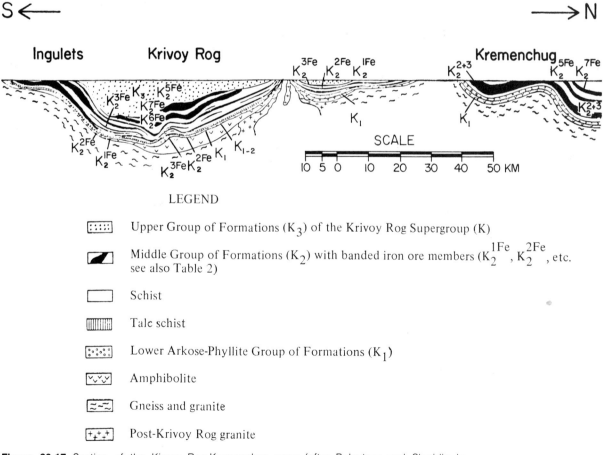

Figure 20-17 Section of the Kirvoy Rog-Kremenchug zone (after Belevtsev and Skuridin in Grigor'yev, 1969). (From Alexandrov, 1973, Precambrian banded iron formations of Russia, Econ. Geol., v. 68, p. 1044.)

those of the Kursk magnetic anomalies (Voronezh plate) form a gigantic metallogenic province. Precambrian banded ores also occur in the Urals and the Asiatic part of the Soviet Union in the cores of Caledonian and Hercynian orogenic belts.

Lithologically, the iron formations consist of intercalated, rather continuous bands of silica and iron oxides, and rarely of iron carbonates. The average iron content of the iron formations is about 35 percent. The iron formations are part of a marine geosynclinal sequence and typically are represented by magnetite and hematite facies, the latter being considered by most Russian geologists as the product of sedimentation in relatively deep water far offshore. They are part of the transgressive series.

Africa. Iron ore deposits occur in northern and southern Africa. The northern ones, located in Algeria, Tunisia, and Spanish Morocco have been exploited for many centuries. The Algerian deposits are siderite replacements of Cretaceous limestone. Metasomatic replacement processes formed the Spanish Morocco deposits. Their production is about 5 million tons per year.

In Southern Africa, the Precambrian iron formation occurs in four different tectonic environments, according to Beukes, as follows: (1) greenstone belts; (2) the Limpopo metamorphic belt; (3) the cratonic basins of the Pongola, Witwatersrand, and Transvaal supergroups, and the Shushong group; and (4) the Damara mobile belt. Numerous deposits exist in these different tectonic environments in the geographic areas of South Africa, Rho-

desia, and Botswana, and Kingdom of Lesotho and Swaziland South West Africa, and Mozambique.

The iron formations appear to have been more prominantly developed in these areas during a period prior to 2000 million years. They exhibit volcanic, clastic, and chemical stratigraphic associations. Beukes suggests that the silica in the iron formations of the volcanic association was derived from acid volcanism, while the iron and silica in most of the other associations possibly originated from the weathering of preexisting rocks.

South America. Dorr has indicated that with the exception of increasing recent studies, little effort has been devoted to study of the iron formations from which the great iron ore deposits of South America formed. An exception would be the Venezuela deposits that have been mined for some time (Fig. 20-18). The South American deposits are widespread as shown by Fig. 20-19.

Brazil. Iron formations are found in the Guayana and Brazilian Precambrian Shields as a common rock type and also occur in Chile and astride the Bolivian-Brazilian border. Only the carbonate and oxide facies are known, the former being quite rare. The dominant oxide facies occurs in major units averaging more than 100 meters in thickness and extending over hundreds of square kilometers, generally in a miogeosynclinal or intracratonic basin environment. The relation of such deposits with volcanism is tenuous and obscure, if indeed there is any direct relation. Smaller units of oxide facies

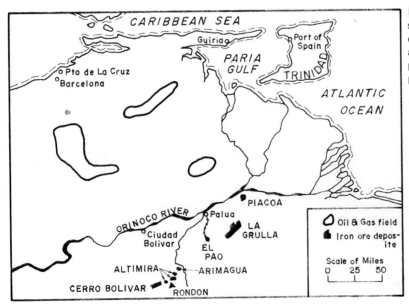

Figure 20-18 Location map showing iron ore deposits of Venezuela. (From Lake, 1950, Eng. and Min. Jour., v. 151, No. 8, Copyright McGraw-Hill Book Company, New York.)

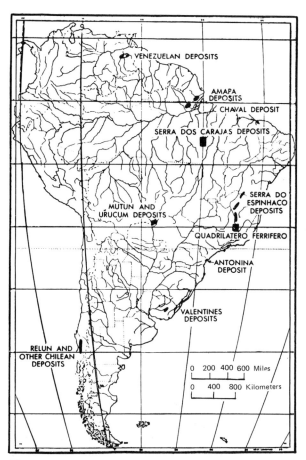

Figure 20-19 Index map showing the general location of iron formations in South America. (From J. V. N. Door, II, 1973, Iron formations in South America. Econ. Geol., v. 68, p. 1007.)

iron formation occur in many minor beds from widely varying geologic environments. The carbonate facies is found in a eugeosynclinal suite in Minas Gerais, Brazil, and is of the Algoman type.

The deposits range in age from about 3200 million years to late Precambrian or early Paleozoic; although the major epoch of deposition is debatable, it probably was about 2000 million years ago.

The South American oxide facies iron formations of early and middle Precambrian age average about 40 percent in Fe and the same in SiO_2. Younger iron formations are still richer, averaging perhaps 50 percent Fe. The iron is present as magnetite, hematite, and martite; most rocks have been metamorphosed, and accordingly it is not known how much of the magnetite is metamorphic and how much is diagenetic or depositional in origin. Hematite and martite are dominant in most iron formations.

Iron formations of South America are quite similar in lithology and occurrence to the major depos-

its in Africa and India and possibly formed when these continents were contiguous. These formations differ from those in the Northern Hemisphere in having a narrower range in lithologic facies and a generally higher iron content. In few areas can any direct relation with volcanism be demonstrated.

Miller and Singewald suggested that the deposits at Coquimbo near Cruz Grande, Chile, are magmatic differentiations of the magma that gave rise to the enclosing diorite. These deposits exceed 100 million tons in reserves.

Brazil has the largest iron ore reserves in South America (Tables 20-6 and 20-7). Most of the larger iron ore deposits of Brazil are located in the Minas Gerais area known as the "Iron Quadrilateral," or better *Quadrilatero Ferrifero* (Fig. 20-20) located 320 kilometers northwest of Rio de Janeiro. They are found in an extensive formation of *Itabirite*, which is several hundred meters thick and extends over a large area. Measured, indicated, and inferred reserves are estimated at 11,000 million tons of material grading between 40 and 69 percent iron content. The Precambrian *Itabirite* deposits contain about 35 percent silica which, where leached and replaced, form the economic deposits, especially where the silica was replaced by iron oxide that was subsequently dehydrated to form hematite.

Within this formation, several large deposits exist of *rebosaderos* or high-grade hematite. They contain not less than 0.52 percent phosphorus, 0.01 percent sulfur, 0.14 percent silica, and 0.04 percent manganeze. The more important among those deposits are *Piracicaba, Itabira, Cocais, Caue, Conceicao, Catas Altas, Aqua Quenta, Ouro Preto,* and *Itabirito* (Tables 20-6 and 20-7).

One of the world's largest iron ore districts is

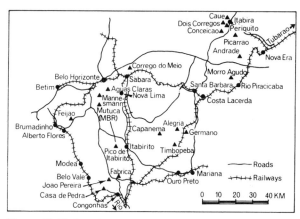

Figure 20-20. Iron Quadrilateral of Brazil.

Table 20-6 State-by-State Breakdown of Brazilian Iron Ore Reserves

State	District	Ore reserve tonnages (millions)			Grade % Fe
		Measured	Indicated	Inferred	
Amazonas	Urucará	15.98	76.71	59.42	58
Mato Grosso	Corumbá	51.86	49.33	53.27	51–62,5
	Ladário Corumbá	89.99	78.06	103.57	55–62
Minas Gerais	Brumadinho	204.22	35.53	4.46	44–67
	Congonhas do Campo	236.18	130.24	300.00	48–66,8
	Itabira	2,495.65	97.01	243.02	51–67
	Itabirito	291.61	394.66	154.48	48–66
	Mariana	600.92	—	981.05	41–67
	Nova Lima	312.41	188.04	174.03	40–68,5
	Ouro Preto	517.10	282.11	329.68	42–69
	Santa Bárbara	540.52	165.93	577.20	44–66
	Other	771.22	452.53	411.95	43–68
Pará	Marabá (Carajas)	1,657.34	2,527.98	11,302.88	>66,5
	São Félix do Xingu	36.65	28.55	197.46	>66,5
Paraná	Antonina	3.30	0.34	—	40
Pernambuco	São José do Belmonte	2.89	4.23	7.16	—
São Paulo		0.81	0.01	—	—
Total		7,837.65	4,511.30	14,899.62	

Source: DNPM (1974).

Brazil's Serro dos Carajas in the Marabá district of the State of Pará with estimated reserves of 16 billion metric tons of ore with an average grade of 66.7 percent iron, 2.2 percent combined silica and alumina, and 0.05 percent phosphorus on a dry basis. Hematite is predominant and no significant sulfur or other deleterious elements are present.

Serra dos Carajas is located in northern Brazil (Fig. 20-19) about 550 kilometers southwest of the coast town of Belém (Fig. 20-21) but the most prac-

Table 20-7 Principal Brazilian Iron Ore Mines

Company	Mine	Location	Approx. current capacity (tons)
CVRD	Caue	Itabira	46,000,000
CVRD	Conceiçao	Itabira	12,000,000
CVRD	Dois Corregos	Itabira	
CVRD	Picarro	Itabira	1,500,000
CVRD/Acesita	Periquito	Itabira	7,000,000
MBR	Aguas Claras	Belo Horizonte	12,000,000
MBR	Mutuca II		
MBR	Jangada	Belo Horizonte	3,500,000
MBR	Pico de Itabirito		
Samitri	Alegria	Algeria	6,500,000
Samitri	Morro Agudo	Rio Piracicaba	3,200,000
Samitri	Andrade	Nova Era	1,250,000
Samitri	Corrego do Meio	Belo Horizonte	200,000
Ferteco	Fabrica	Congonhas do Campo	5,500,000
Ferteco	Feijao	Brumadinho	1,500,000
CSN	Casa de Pedra	Lafaiete	4,000,000
Mannesmann	Mutuca I	Belo Horizonte	1,000,000
W. H. Muller	Muller	Brumadinho	1,000,000
W. H. Muller	Muller	Congonhas do Campo	

Source: Min. Mag., v. 136, No. 3, 1977.

Figure 20-21 New roadways in Brazil. In 1960 the roads northwest of the dark lines did not exist. Solid lines, existing roads; dashed lines, planned roads. (From C. de Meira Mattos, 1975, Brasil-Geopolitica e destino.)

tical access to the deposit is by aircraft at the present time. It will not be long, however, before the extensive network of roads (Fig. 20-21) will connect to the deposit.

At least 277 diamond drill holes ranging from 25 to 400 meters in depth, plus 14 adits and one tunnel were used to determine the grade of reserves.

The area of the deposits has about 60 elevated, sparsely vegetated, flat mesas and plateaus surrounded by precipitous slopes. Almost all of the iron mineralization is located within the plateaus or where the newly named Carajas formation outcrops. The formation is probably Precambrian. Conformable volcanic formations overlie and underlie the formation and the three units have been named the Crão Pará Group. the group overlies Precambrian metasediments and crystalline basement rocks.

The Carajás formation is a typical iron formation or *itabirite,* where unleached, consisting of alternating laminae of quartz or chert and magnetite and hematite. The *itabirite* contains about 45 percent Fe, 35 percent SiO_2, 0.5 percent Al_2O_3, 0.002 percent P, 0.09 percent Mn, and 0.2 percent ignition loss. The Carajás where completely leached is usually about 250 meters thick; however, much of the

formation has been leached of its silica content from the surface downward to depths of over 400 meters.

In addition to the 16 billion metric tons of measured, indicated, and inferred iron ore, a substantial tonnage of ore, containing 55 to 64 percent Fe is also present. The immense reserves of *itabirite* have not been estimated.

Brazil holds by far the majority of the iron ore deposits of South America (Tables 20-6, 7 and Fig. 20-3). The Brazilian government is actively promoting the development of these deposits by economic incentives and provides access to many of the deposits (Fig. 20-21).

Chile. Important deposits occur in Coquimbo, near Cruz Grande, where there are 100 million metric tons of 65 percent iron ore low in silica, phosphorus, and sulfur. Annual production is about 12 million metric tons of ore containing 63 percent Fe. The ores consist of hematite and some magnetite, with hematite replacing magnetite. They occur in diorite, and Miller and Singewald believed they are magmatic differentiations of the magma that gave rise to the enclosing diorite.

Venezuela. Near Pao, one of the larger deposits of

South America, composed of hard hematite ore, is undergoing development.

Other Iron

India. India produced about 35 million tons of iron ore in 1976 containing 62 percent Fe from reserves estimated at 10,536 million tons of which 8600 million tons are hematite. The chief sources are the Bihar and Orissa (Singhbhum) district, which Jones states contain 4 billion tons of 60 percent ore; others include Madhya Pradesh, Karnataka, and the recently developed Kudremukh deposit. The ores occur in the Archean iron ore series (Newer Dharwar) consisting of a sedimentary-volcanic series including a banded hematite-quartzite member with iron ore bodies. The bedded ores consist of massive, laminated, and powdery hematite with a little magnetite and martite, carrying 60 to 69 percent iron and low phosphorus and manganese. The banded ironstones extend for 50 kilometers, have a width up to 300 meters, and dip 70°.

Jones considers that the banded hematite-quartzites resulted from original deposition, giving rise to a rock containing 28 to 30 percent iron and identical in nature and origin with the original iron formations of the Lake Superior region. Subsequent leaching of parts of this removed silica, leaving residually concentrated iron ores in beds up to 40 meters thick. Dunn believes that the banded ironstones were originally tuffs and sandy beds laid down under subaerial conditions and replaced in part by silica and iron oxides to yield secondary quartzites and iron formation. Part of the silification may have been contemporaneous with deposition. The subseqeuent removal of silica produced the high-grade hematites. Some low-grade lateritic iron ores also occur.

Cuba. Residual concentrations of iron ore resulting from the weathering of serpentine occur in the Mayari, Mao, and San Felipe districts. They are surficial mantles on plateaus, which, in Mayari, cover an area 6 by 16 kilometers to an average depth of 5 meters and contain several hundred million tons of workable ore. The ores consist of hematite, limonite, and some magnetite. Typical ore contains about 54 percent Fe, 11 to 13 percent H_2O, 5 to 7 percent SiO_2, 10 to 12 percent Al_2O_3, about 1.7 percent Cr, 0.01 to 0.5 percent Ni, 0.5 percent Mn, 0.01 percent P, and water. The chromium and nickel content represents concentrations of original constituents in the serpentine and is sufficient to be economic. The deposits are iron-rich laterites. Contact-metasomatic deposits with magnetite, hematite, and high sulfur occur near Santiago.

China. Fairly large reserves of iron ores are found in China. The most important are contact-metasomatic deposits of hematite and magnetite averaging 56 percent iron, with low phosphorus and moderate sulfur. The best of these occur in a belt along the lower Yangtze Valley. Extensive Precambrian sedimentary beds of oolitic hematite in Chihli contain 55 percent iron; low-grade Archean banded ironstones also occur in Chihli. Extensive reserves occur in Manchuria. China, however, is not yet a large producer of iron ore.

Mexico. Contact-metasmatic iron ore deposits are being developed in the tectonic western portion of Mexico. Two of these deposits occur in the contact

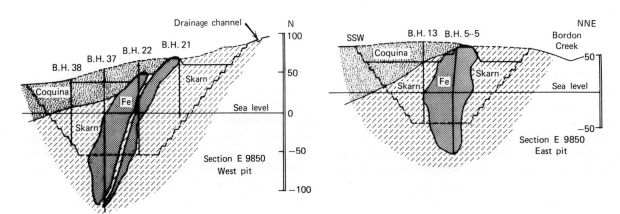

Figure 20-22 Geological cross sections of the Ferrotepec mine with ground levels as at June 1976. Left, West pit; above, East pit. The Ferrotepec deposit is made up of mineralized bodies oriented E-W up to 150 m deep. (From Mining Mag., 1976.)

zone between granodiorite intrusions of Early Tertiary age and a series of Precretaceous andesites overlain by Middle Cretaceous calcareous marine sediments. The iron ore deposits of Las Truchas and Ferrotepac (Fig. 20-22) are dispersed in an area 4 kilometers (North-South) by 8 kilometers (East-West). Estimated reserves of Ferrotepac are about 8.2 million tons of almost entirely magnetite. Measured reserves of the Las Truchas deposits are about 76 million tons.

Selected Refferences of Iron

Iron Ores, Occurrence, Valuation, and Control. 1941. E. C. Eckel. McGraw-Hill Book Company, New York. *Broad treatment of iron ores of the world.*

Geology of the Lake Superior region. 1911. C. R. Van Hise and C. K. Leith. U.S. Geol. Surv. Mon. 52. *A classical monographic survey.*

Hydrothermal leaching of iron ores of the Lake Superior type. 1937. J. W. Gruner. Econ. Geol. v. 32 p. 121-130.

Brown iron ores, Tennessee. 1934. E. F. Bruchard. Tenn. Geol. Surv. Bull. 39. *Residual ores.*

Precambrian iron formations of the world. 1973. Econ. Geol., v. 68. No. 7. *Entire volume contains numerous papers on iron deposits.*

Sedimentary facies of iron formation. 1954. H. L. James. General Reference *Classic paper on iron formations as chemical sediment.*

Evaluation of iron ore deposits. 1972. E. L. Ohle. Econ. Geol., v. 67, p. 953-964.

Survey of World Iron Ore Resources. 1970. United Nations, Dept. of Econ. and Social Affairs, United Nations, 479 p.

United States Mineral Reserves. General Reference 16.

General Reference 19. Chap. on Iron.

Canadian iron ore industry review—1972. 1973. Skilling's Mining Review, v. 62, p. 10-13.

World iron ores and future North American supplies. 1973. Skilling's Mining Review, v. 62, p. 10-11.

Birmingham District, Alabama. 1933. E. F. Bruchard. Int. Geol. Cong. Guidebook 2, Washington, D.C. *A concise summary of sedimentary Clinton hematite.*

Iron ore deposits of Bihar and Orissa. 1934. H. C. Jones. India Geol. Surv. Mem. 63, 1934. Also J. A. Dunn, Econ. Geol. v. 30, p. 643, 1935. *Geology of India iron-ore deposits.*

Solution, transportation and precipitation of iron and silica. 1919. E. S. Moore and J. E. Maynard. Econ. Geol. 44, Nos. 3, 4. *Experiments on cold water solution and deposition of iron.*

Wabana iron ores of Newfoundland. 1931. A. O. Hayes. Econ. Geol. v. 26, p. 44-64. *Description, occurrence, and origin of these sedimentary deposits.*

Solutions, Minerals, and Equilibria. 1965. R. M. Garrels and C. L. Christ. Harpers Geoscience Series. *Numerous Eh-pH diagrams of Fe stability relationships.*

Canadian Iron-Ore Deposits. General Reference 16, p. 414-429. *General descriptions.*

Mineralogy and Geology of the Mesabi Range. 1946. J. W. Gruner. Iron Range Res. Rehab. State Office, St. Paul, Minn. *Concise detailed report on geology, minerals, structure, and types.*

Mineral resources of China. 1946. V. C. Juan. Econ. Geol. v. 41, p. 424-433. *Occurrences, nature, and resources.*

Development of Lake Superior soft iron from metamorphosed iron formations. 1949. S. A. Tyler. Geol. Soc. Am. Bull., v. 60, p. 1101-1124. *Broad discussion of origin and concentration by oxidation and solution.*

FERROALLOY METALS

Under this heading are included a group of important metals whose chief use, but not their only use, is for alloying with iron to yield special steels of desired properties. They include *manganese, nickel, niobium, chromium, molybdenum, tungsten, vanadium, cobalt,* and *titanium.* Silicon and phosphorus are also used for certain steels. The addition of a few percent of these metals to steel not only improves its properties but also adds new ones, such as hardness, toughness, strength, durability, lightness, ability to retain temper at high temperatures, combat fatigue, and resistance to corrosion. Despite the comparative small volume in which they are used, they are essential to modern industry, particularly for dependable high-speed machines. The large industrial nations lack most of them, and the United States is deficient in all of them except molybdenum and titanium.

MANGANESE

Manganese is the most important of the ferroalloy metals. Not only is it necessary for the making of high-manganese steels but it is also an absolute essential, for which there is no substitute, in the making of all carbon steel, 13 to 20 pounds being required for every ton of steel produced. About 95 percent of the consumption is for metallurgical purposes, of which a minor amount is used for other alloys, such as some bronzes. Its chief purpose in steel making is as a scavenger to remove oxygen and sulfur in order to produce sound, clean metal.

It is added in the form of ferromanganese (80 percent Mn, 15 to 18 percent Fe), for which ore containing a minimum of 46 percent manganese is desired, but it is also used in the form of spiegeleisen (20 percent Mn; the remainder Fe). For chemical uses, manganese ore of high purity is required for dry batteries, the glass industry, paints, pigments, dyes, and fertilizers. Manganese steel is used where hardness and toughness are desired; it enters into armor plate, projectiles, car wheels, railway switches, safes, crushers, cutting and grinding machinery, machine tools, cogwheels, structural and bridge steel, and so forth.

Production and Distribution

The world production of manganese ore is about 24 million tons, of which Russia produces about one-

third, followed in order by South Africa, Brazil, Australia, and Gabon. Minor production comes from China, Czechoslovakia, Indonesia, Hungary, Italy, Malaysia, Philippines, Romania, Chile, Mexico, North Africa, and Egypt.

The United States production of commercial ore ceased in 1970. The annual domestic consumption of manganese ore with 35+ percent manganese is about 2 million tons and is derived from the following sources: Brazil, Australia, India, Gabon, Republic of South Africa, and France.

Mineralogy and Tenor

The commercial minerals and classes of manganese ores are:

Mineral	Composition	Percent Mn	Classes of Ore	Percent Mn
Pyrolusite	MnO_2	63	Chemical	82–87
Manganite	$Mn_2O_3 \cdot H_2O$	62.4	Manganese	35+
Psilomelane	$MnO \cdot MnO_2 \cdot 2H_2O$	45–60	Metallurgical	46
Hausmannite	Mn_3O_4	72.5	Ferruginous manganese	10–35
Rhodochrosite	$MnCO_3$	47.6	Manganiferous iron	5–10
Rhodonite	$MnSiO_3$	41.9		
Bementite	$2MnSiO_3 \cdot H_2O$	39.1		

Treatment

Manganese ore is generally hand-sorted or washed to bring it to shipping grade. It is then furnaced with iron ore to make ferromanganese. Lower-grade ores are similarly treated to make spiegeleisen. A little metallic manganese is made electrolytically.

Types of Deposits and Origin

Manganese occurs in four valent forms, namely +2, +3, +4 and +6. Divalent manganese readily substitutes for Ca^{2+} and Fe^{2+} and is, therefore, commonly associated with these elements. Under reducing or near neutral conditions, divalent manganese is readily soluble but become concentrated when the redox potential is higher, that is, in a more oxidizing environment, where the manganese acquires a higher valence and becomes relatively insoluble. The most soluble iron compounds found in nature are less soluble than divalent manganese. Iron is generally precipitated before manganese because the ferrous ion is precipitated more readily than the manganese ion. The sedimentary process is effective, therefore, in the separation of

iron and manganese unless the Fe:Mn ratio is abnormally high. Both sedimentary and supergene processes are effective in producing manganese deposits.

Primary manganese deposits may be classed in three types—sedimentary (including sea-floor nodules), volcanogene, and hydrothermal—which to some extent are intergradational. Supergene processes are important in modifying those near-surface deposits where manganese oxides being relatively insoluble, form residual deposits.

Most of the world's supply is obtained from sedimentary and residual deposits. The latter are surficial deposits that have resulted from the weathering of manganese-bearing minerals of schists, pegmatites, sedimentary beds, fissure veins, and replacement deposits.

Examples of Deposits

Russia. The annual production of manganese ore in Russia increases each year and is now about 5 million tons or about one third of the world's production. Ninety-three percent of it comes from two groups of sedimentary deposits: Chiaturi in Georgia and Nikopol in the Ukraine; the rest comes from west Siberia, the middle Volga, and the south Urals.

The *Nikopol* district consists of two areas of Oligocene beds of sandy clay more than 100 square kilometers in area. The ore bed averages 3 meters in thickness and lies at a depth of 20 to 80 meters. It contains oölitic earthy lenses and nodules of pyrolusite, with wad, polianite, manganite, and iron oxides. This deposit is mined by open pit methods. J. Van N. Dorr describes the environmental restoration of this area where the ''rich black topsoil of the Ukraine was carefully removed, the sterile rock overburden replaced after the underlying manganese bed had been extracted, and the topsoil replaced. After 3 to 4 years, orchards or grain were planted and seemingly thrive. Lakes, the shores of which had been landscaped, had formed in some of the older unrestored large pits.''

Well over 100 million tons of material are moved each year from the mines. The ore carries 28 to 33 percent manganese and is concentrated to 42 to 52 percent shipping ore. The reserves exceed 3 billion tons.

These deposits are unique in exhibiting the intertonguing of the near-shore manganese oxides that formed in oxygenated water while the carbonates formed basins in deeper water of a more reducing environment (Fig. 20-23).

The *Chiaturi* district has long been a large producer of manganese ore. The deposits occur in several horizontal lenses of the Nikopol formation of early Oligocene age consisting of marly sand from one to 8 meters thick in a zone 30 kilometers long by 8 to 10 kilometers wide. Access is by open pit

with an overburden of 80 to 200 meters or more of sands and shales. The ore bed consists of oölites of psilomelane and pyrolusite imbedded in a cement of wad and calcium carbonate with which small quantities of iron copper, nickel, phosphorus, and barium are associated. About 2 to 3 tons of ore after washing and concentration yield 1 ton of concentrates carrying 50 to 53 percent manganese.

India. Extensive and rich manganese deposits occur in the Madhya Pradesh (M.P.), Madras, Bihar, Bombay, and Mysore, but the Balaghat district of the M.P. and the Sandur State in Madras are the most important producers.

The deposits of the *Madhya Pradesh,* particularly, are believed by Fermor to have resulted from an ancient (perhaps Precambrian) oxidation, chiefly of Early Precambrian crystalline ''gondite'' schists, containing spessartite and rhodonite. Most of the silica and alumina is believed to have been removed during weathering and the resulting manganese minerals are chiefly psilomelane, pyrolusite, and wad. One body is 500 meters wide and almost 750 meters long. Fermor believes that originally the manganese was deposited as a sediment that was later metamorphosed to schists with bands of pure manganese oxides and silicates, which in turn underwent surface alteration to form residual enrichment deposits.

The *Sandur* deposits of Madras occur in Dharwar schists that have undergone surface alteration to form principally psilomelane and wad, with minor pyrolusite and manganite, which have in part replaced the country rock. The larger deposits are about 30 by 230 meters, and the district is said to

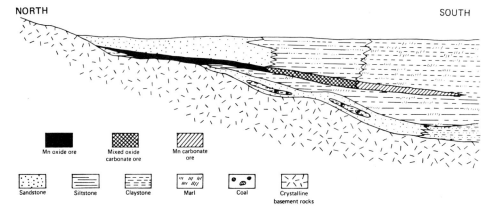

NORTH   SOUTH

Mn oxide ore | Mixed oxide carbonate ore | Mn carbonate ore

Sandstone | Siltstone | Claystone | Marl | Coal | Crystalline basement rocks

Figure 20-23 Diagrammatic cross section of the Nikopol district, USSR. (From Hewitt, 1966, Oxides and carbonates of manganeze, Econ. Geol., v. 61, p. 431–461.)

contain 10 million tons of ore. Somewhat similar deposits are near Vizagapatam, where peculiar "Kodurite" rocks containing manganese garnet and manganese pyroxene have been altered and residually enriched to manganese oxides. Many other generally similar deposits have been described by Fermor in a 1300-page memoir.

South Africa. The *Postmasburg* deposits, northwest of Kimberley, give South Africa an annual output of 600,000 to 700,000 tons of 40 percent ore; shipments average 49 percent. There are at least two belts, an eastern and a western, the latter being 60 kilometers long. The deposits lie at the contact of an eroded surface of dolomite of the Transvaal system and the unconformably overlying Gamagara shales and quartzites. It is exposed in disconnected troughs and outcrops. The ore forms an irregular bed 2 to 7 meters thick on the dolomite itself or in the overlying cherty ferruginous breccia. The ore is rather hard, carries 56 to 60 percent combined manganese and iron, and occurs in botryoidal and steely crystalline varieties. The minerals are stated by DuToit to be pyrolusite, psilomelane, polianite, braunite, sitaparite, hematite, barite, and diaspore, but under the microscope some of these specific manganese minerals are seen to be intimate mixtures of several oxides, suggesting a former colloidal deposition. The present crystalline character is presumed to be due to metamorphism. The origin is not clear. Nel thinks the deposits are caused by residual concentration and subsequent metamorphism, and DuToit advocates a hypogene replacement origin. Reserves may amount to a billion tons.

Ghana. Kitson states that the oldest deposits are in sediments, later than these are their metamorphosed representatives, and the youngest are residually concentrated surficial deposits. The manganiferous rocks occur over a distance of 1000 kilometers, but only one deposit is worked, namely the Nsuta mine, at Dagwin, 50 kilometers from the port of Sekondi. This is the largest manganese mine in the world. The ore crops out on a ridge for four kilometers, where it is mined by power shovels. It occurs in ancient altered rocks in massive and concretionary bodies. The capping of the hill 10 to 12 meters thick, contains detrital ore of about 40 percent manganese resting upon "black" ore bodies, which are massive and lenticular and contain 50 to 55 percent manganese. Kitson states that the deposits have resulted from residual accumulation through tropical weathering of manganiferous bed

rock that removed silica and alumina and redeposited the manganese near the surface. Junner, however, believes that they were formerly ancient high-grade manganese ores that were only slightly enriched by meteoric waters. One theory would indicate a shallow depth for the ores, the other that they might persist to considerable depth.

Brazil. The most important manganese deposits of Brazil are near Lafiete in Minas Gerais, although large ones occur at Urucum in Matto Grosso and Bahia. The chief deposits at Lafiete, according to Wright, represent residual concentrations from the weathering of ancient crystalline rocks containing lenses of manganese carbonate, tephroite, rhodonite, and spessartite. The ore averages 48 to 51 percent manganese. The Morro da Mina has six ore bodies; the largest (No. 3) is 4,300 meters long and 70 meters wide and is known for 220 meters vertically. Reserves are estimated at 6 million tons of ore.

The unweathered stratified manganese at *Urucum* near Bolivia have been extensively explored. The Band Alta formation consists of two horizontal beds, one 3 meters thick containing 44 to 49 percent manganese and 9 to 14 percent Fe. A larger deposit at Amapa is being developed.

Cuba. In the foothills of the Sierra Maestra range, Oriente Province, Cuba, are many layered manganese deposits. Two of the largest are the *Quinto mine* near Cristo and the *Chario Redondo*. These unusual deposits are volcanogene or submarine exhalative in origin. Most of the larger deposits are bedded; others are nonbedded. The ores contain from 15 to 55 percent manganese and according to Park and Cox contain psilomelane and pyrolusite with minor manganite, rancieite, braunite, orientite, and piedmontite along with calcite, cryptocrystalline silica, and rock remnants. The low-grade ores of the Quinto mine carrying 18 percent manganese are sintered and concentrated to 50 to 52 percent manganese.

The sources of the manganese in these deposits are presumed to be derived from fluids evolved from a magma at depth during volcanism. The manganese evidently was transported some distance and deposited in a sedimentary environment with limestone and water-laid tuff.

United States. The minor United States past production of manganese ore was obtained from small deposits that are mostly low in grade and were relatively costly to work. There are, however, extensive emergency reserves of low-grade material. None is now being mined. Minor residual deposits occur throughout the southern states, but the main production has come from vein deposits in Montana, manganiferous iron ores in Minnesota, and the Three Kids deposit in Nevada.

The southern residual deposits are derived from manganiferous sediments, and most of them underlie remnants of dissected erosion surfaces and consist of nodules in residual clay or replacements of underlying limestones.

In *Arkansas*, the rich residual ores of Batesville occur as "buttons" in the Cason shale and its residual clay and as masses of manganese oxides in residual clays occupying solution depressions in the weathered Fernvale and other limestones. The clays attain 25 meters in thickness, and the ores range from 25 to 50 percent manganese. The manganese is considered to have been deposited originally in the Cason shale and in the Fernvale limestone and subsequently to have been weathered to form accumulations, residual in the clays. More recent work suggests that the richer Batesville ores were derived from the weathering of epigenetic rhodochrosite deposits in limestone.

In *Georgia*, the important deposits of Cartersville, according to Crickmay, consist of (1) boulders of manganese oxides in gravel derived from the weathering of quartzite; (2) hard, concretionary masses; and (3) soft "chemical ore" in veins in clay and gravel. The *Virginia* Blue Ridge deposits, including the Crimora mine, consist of nodular masses of oxides in clays derived from the Shady dolomite and are considered to have grown by replacement of the clays. Other deposits are fillings and replacements of breccia zones in sandstone. The manganese accumulated during or after the early Tertiary peneplaning of this region.

Montana produced about two-thirds of the high-grade ore of the United States from rhodochrosite veins at Philipsburg and the Emma mine at Butte, which carry 35 to 37 percent manganese. This was concentrated and nodularized to a high-grade product suitable for battery ore and for ferromanganese.

In *South Dakota* the extensive low-grade deposits near Chamberlain are estimated to contain in excess of 100 million tons of manganese. Manganiferous iron carbonate nodules, which contain 18 percent Mn, 11 percent Fe, 12 percent CaO, and 11 percent SiO_2, occur in a 12 meter horizontal shale bed that caps remnants of a dissected plateau. The ore is an isomorphous mixture of the carbonates of mangenese, iron, calcium, and magnesium. About 12 to 14 cubic yards of shale yield 1 ton of nodules, and 4.5 tons of nodules are required to make 1 ton of sinter containing 68 percent Mn. The nodules could be recovered cheaply by power shovel and screening. Other low-grade emergency deposits have been partly developed in Nevada, New Mexico, Arizona, and California.

In *Nevada*, the Three Kids deposit in a Pliocene section consists of a lens about 1000 meters long with a maximum thickness of 20 meters, from which about 2 million tons of ore containing 20 to 25 percent manganese has been mined.

In *Arizona*, in the Artillery Mountains, a similar but larger deposit than the Three Kids deposit has been extensively explored. This deposit extends for several miles and attains a thickness of 20 meters, but the manganese content ranges only from 5 to 15 percent. In both of these layered deposits, the beds are part of a section that consists of layered volcanic debris and flows laid down in continental basins.

Deep Sea Nodules. Manganese nodules (Chapter 11) may be one of the world's largest resources of manganese, copper, nickel, cobalt, and other metals. The actual amount of these nodules (Chapter 11) has been estimated at billions of tons—more than enough to make the nodules an almost inexhaustible resource. Typically, the Pacific nodules contain about 20 percent manganese, about 1.5 percent of both copper and nickel, roughly 0.3 percent cobalt, and other metals.

Summary

Dorr, Crittenden, and Erl suggest that the principal hopes of finding domestic reserves or resources of conventional types may lie in (1) finding the source of the manganese of the Pierre Shale, conceivably buried under Pleistocene sedimentary rocks in central or western Minnesota or adjacent areas; (2) finding another Molango-type deposit by careful analysis of the distribution of manganese in certain miogeosynclinal carbonate rocks; or (3) finding the source of the high manganese concentrations in the Salton Sea brines. Much more promising modes of relieving dependence on foreign sources are by vigorously attempting to perfect techniques of effectively exploiting sea-floor nodules and by resolving legal impediments to large-scale investment in subsea mining.

Selected References on Manganese

Mineral Resources, Production and Trade of Brazil. July 1941. C. W. Wright. Foreign Minerals Quart., v. 4, No. 1. U.S. Bur. Mines, Washington, D.C. *Much information on manganese in Brazil.*

Manganese deposits of Cuba. Charles F. Park, Jr. *Manganese deposits in part of the Sierra Maestra, Cuba.* Also C. F. Park, Jr., and M. W. Cox. *Geology and manganese deposits of Guisa-Los Negros area, Oriente Province, Cuba*; and W. P. Woodring and S. N. Daviess. *Geology of the manganese deposits of Cuba.* S. F. Simons and J. A. Straczek. U.S. Geol., Sur. Bull. 835-B, F, and G.

Occurrence and hydrothermal replacement origin of various manganese deposits in Oriente Province, Cuba. 1942–1944 and 1958. U.S. Geol. Surv. Bulls. 935-B, F, G, and 1057.

On the ore-deposition and geochemistry of manganese. 1970. H. Borchert. Mineralium Deposita, v. 5, p. 300–317.

Manganese in the United States. 1962. M. D. Crittenden and L. Pavlides, U.S. Geol. Surv. Min. Inv. Res. Mgs. MR23.

Primary manganese ores. 1968. J. V. Dorr. Soc. Brazileira Geol. XXII Congresso, Belo Horizonte, v. 22, 12 p.

Manganese deposits at Philipsburg, Montana. 1940. E. N. Goddard. U.S. Geol. Surv. Bull. 922-G *Geology of one of the more important deposits.*

Manganese. Hearings before the Subcommittee on Mines and Mining, Feb 12–27, 1948. 80th Congress, Hearing 38, pp. 497. *Covers distribution, uses, technology, statistics, and exploration for, and distribution of, domestic and important foreign deposits.*

Manganese deposits of Serra do Navio, Amapa, Brazil. 1949. J. V. Dorr, II, C. F. Park, Jr., and G. de Paiva. U.S. Geol. Surv. Bull. 964-A. *Brazil's newest deposit.*

Manganese deposits of Mexico. 1948. P. D. Trask and J. Rodriguez Cabo. Jr. U.S. Geol. Surv. Bull. 954-F. *Broad survey of various types of deposits.*

Manganese deposits of the Soviet Union. 1970. D. Sapozhnikov, ed., in Proc. of the Conf. on principal genetic types and the geochemistry of manganese deposits in the USSR, 522 p. *Details on Soviet manganese deposits.*

The geology and mineralogy of the Kalahari manganese field north of Sishen, Cape Provima, South Africa. 1970. Geol. Survey, Dept. Mines, Pretoria, Mem. 59, 84 p.

General Reference 11. Chap. 25.
General Reference 19. Chap. on manganese.

NICKEL

Nickel ranks second to manganese among the ferroalloys, but it has also a multitude of other applications. Most of the world reserves are centered at Sudbury, Ontario, and in Russia. New Caledonia, however, has out-produced Russia during recent years. Finland is a minor producer. Australia is increasing production rapidly.

Uses of Nickel.

Nickel is preeminently an alloy metal, and its chief use is in the nickel steels and nickel cast irons, of which there are innumerable varieties. It is also widely used for many other alloys, such as nickel brasses and bronzes, and alloys with copper (Monel metal), chromium, aluminum, lead, cobalt, manganese, silver, and gold. Nickel steels contain from 0.5 to 7 percent nickel in the "low-nickel steels" and 7 to 35 percent in "high-nickel steels," nickel-irons from 0.5 to 15 percent, nickel brasses and bronzes from 0.5 to 7.5 percent, copper-nickel alloys from 2.5 to 45 percent, Monel metal 67 percent, and special alloys up to 80 percent nickel (Table 20-8).

Nickel imparts to its alloys toughness, strength, lightness, anticorrosion, and electrical and thermal qualities. Consequently, nickel steels and alloys are preferred for moving and wearing parts of innumerable machines, tools, shafts, bolts, axles, and gears, in automobiles, airplanes, and ships and in railway, power, agricultural, crushing, mining, milling, and pressing equipment. Its use for coinage and plating is widespread.

Production.

At the present time the free world requirement of nearly 800,000 metric tons of nickel per year is supplied from deposits of nickel sulfides, mostly in

Table 20-8 U.S. Consumption of Nickel in 1970

	Consumption (million pounds)
Steel	124
Superalloys	23
Nickel-copper alloys	13
Permanent magnet alloys	5
Other nickel alloys	71
Cast irons	10
Electroplating	50
Chemicals	2
Other	14
Total	312

Canada and Philippines; and of nickel laterites, mainly in New Caledonia. World resources from these types of deposits are estimated to total 90 million tons (180 billion pounds) of nickel in 7 billion tons of material averaging about 1 percent Ni. An additional 7 billion tons averaging 0.2 percent Ni, or 14 million tons of nickel, is estimated for sulfide deposits in the United States. The 0.2 to 0.4 percent Ni universally disseminated in peridotite serpentinites throughout the world amounts to a figure equal to orders of magnitude greater than 70 million tons, as does the quantity of nickel contained in deep-sea manganese nodules; but new technological developments will be required to recover nickel successfully from these two types of occurrences.

Canada (245,000 metric tons) produces less than twice as much nickel as does Russia (168,000 metric tons). Caledonia (109,000 metric tons) lags behind Russia in the mine production of nickel. The leading smelter production, however, occurs in Russia, Canada, Japan, and New Caledonia in decreasing order. The U.S. consumes about 150,000 tons/yr.

A large reduction plant at Nicaro, Cuba, yielded nickel oxide during World War II but is now being operated by Cuba. The United States is the second greatest consumer, following Western Europe and the United Kingdom.

Mineralogy and Tenor

The chief commerical, primary mineral of nickel is pentlandite (Fe, Ni_9S_8), which is always associated with pyrrhotite and chalcopyrite. Hydrous nickel silicate ores, the ore minerals of New Caledonia, carry 1 to 4 percent of nickel. The sulfide ores of Sudbury averaged about 1½ percent Ni and 2 percent Cu.

Treatment.

The nickel-copper sulfide ores are first roasted to release sulfur and then smelted to a Ni-Cu-Fe matte, which is Bessemerized to a matte of 75 to 80 percent Cu-Ni. Some matte is used directly to make Monel metal, the rest being specially smelted to separate copper and nickel sulfides, the latter being roasted, reduced with carbon, and electrolytically refined to pure nickel.

Kinds of Deposits.

There are only two kinds of commerical nickel deposits—residual concentrations of nickel silicates from the weathering of ultrabasic igneous rocks and nickel-copper sulfide deposits formed either by replacement or magmatic injection.

The core of the earth is thought to contain about 3 percent Ni which is, of course, unavailable to man. Mason estimates that the crust of the earth contains only 0.003 percent Ni. The ultramafic mantle, like similar rocks exposed at the surface, may contain 0.1 to 0.3 percent Ni. As many nickel deposits are associated with ultramafic rocks, it may be suggested that the upper mantle is the original source of most of the rocks that may contain nickel ore bodies.

Nickel-Copper Deposits of Sudbury, Ontario. The great deposits of Sudbury dominated the world nickel production. Since 1905, Sudbury has produced almost half the world's supply of nickel. The annual production is up to about 250,000 tons of nickel and almost as much copper along with about 400,000 ounces of platinum metals, 1.5 million ounces of silver, 35,000 ounces of gold and considerable amounts of selenium and tellurium. Iron and sulfuric acid are additional by-products. Production comes from about 40 mines, which have yielded over 6 million tons of nickel and about as much copper. The ore now averages about 1.1 percent Ni and 0.9 percent Cu. The ore reserves exceed 400 million tons with about 10 million tons of copper-nickel. The Sudbury district is thus one of the world's greatest producers of nickel rising with New Caledonian area production.

Geologic Setting.

The region is noted for its layered complex—the "nickel irruptive"—which crops out as an elliptical ring 11 kilometers long and 5 kilometers wide (Fig. 20-24) forming a basin. The outer part of the basin consists of norite members with overlying micropegmatite and a discontinuous sublayer of quartz diorite and breccia, which also occurs as dikes extending into the footwall. The outer contact generally dips steeply inward but in places outward. The interior of the basin, overlying the micropegmatite, consists of breccias, tuff, quartzite, slate, and arkose, called the Whitewater series. The footwall of the basin consists of steeply dipping, metamorphosed volcanics and sediments cut by basic intrusions and granite, and by genesises and recrystallized mafic rocks. These have undergone extensive brecciation that has not affected the nickel irruptive or the Whitewater series. The nickel-copper deposits occur around the periphery of the norite and in the

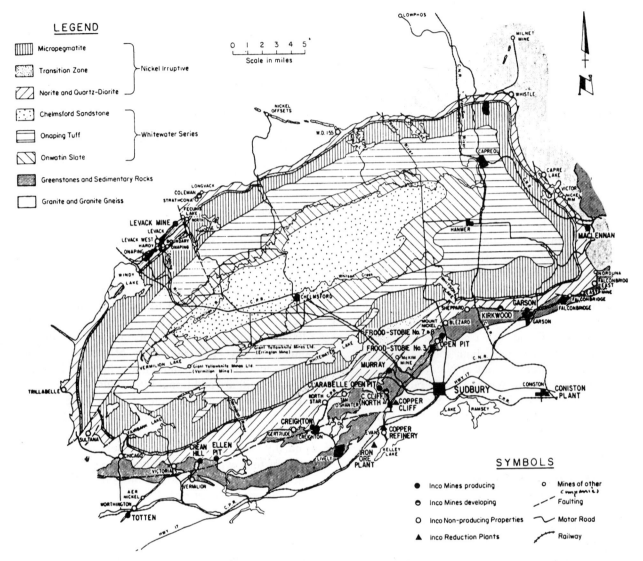

Figure 20-24 Sudbury geology and mines. (From Eng. and Mining Jour., 1965.)

offsets in quartz diorite, and basic norite and breccias.

The margins of the basin are cut by numerous faults, particularly at the ends. Those at the west end, according to Souch, et al., have exposed the basal part of the irruptive and sublayer over a prefaulting vertical range in excess of 9 kilometers. Mining has followed the ore of the outer contact continuously to a vertical depth in excess of 3 kilometers in the Creighton mine.

Isotopic analyses indicate that the norite and sublayer are essentially contemporaneous at about 1.9 billion years and may be older than the micropegmatite, but the transition zone indicates contemporaneity.

Ore Deposits

The ore deposits are of four characteristic types: (1) disseminated bodies, largely in quartz diorite, (2) massive sulfide bodies, (3) breccia ores, and (4) vein and stringer bodies, all with generally the same mineralogy.

The deposits lie mainly in the sublayer at the base of the norite and in offsets below the irruptive; very little ore occurs in the norite. The chief deposits such as the Creighton, Frood-Stobie, Murray, Copper Cliff, Garson, and Falconbridge lie along the south limb, and the Levack, Hardy, Strathcona, and Lake Fecunis are on the north limb.

The Creighton mine illustrates disseminated and

massive ore. According to Souch, the ore zone follows a troughlike depression at the base of the irruptive to a depth of 1580 kilometers. The slightly mineralized, hanging wall norite with immiscible sulfide droplets is underlain by disseminated ore occurring as a matrix surrounding inclusions of gabbro and norite. This passes downward into a zone of larger inclusion of pyroxenite, peridotite, and gabbro with sulfides in the matrix. This in turn passes downward into massive sulfide ore enclosing angular fragments of footwall rock, in part aligned along shear zones and extending out into the footwall rocks. Generally, similar conditions exist in the Murray mine.

In the Garson mine, where the contact dips steeply outward, disseminated and massive ore lies along shear zones at the contact and extends inward into underlying norite. The Frood ore body also illustrates an offset deposit. This huge deposit occupies a dikelike zone parallel to the irruptive contact but separated from it by a mile of Murray granite. The deposit is 1300 meters long, 300 meters across at its widest part, and 1000 meters deep. It consists of a wide upper part of quartz diorite breccia containing blebs of sulfides in disseminated ore. Downward this grades into massive sulfide ore containing rounded blebs of quartz diorite, which Hawley considers to be "immiscible silicate globules in a sulfide melt." The massive sulfide contains contorted schist inclusions. The quartz diorite inclusions consists of rounded pieces of gabbro, pyroxenite, peridotite, and angular wall rock.

The Falconbridge deposit consits of breccia made up of a variety of rounded to angular rocks with a matrix of disseminated and massive sulfides and silicates aligned along a contact shear that extends for 3000 meters in norite, quartz diorite, and preirruptive (or footwall) greenstone.

The north rim deposits are disseminated and massive sulfides in granite breccia lying between the norite and footwall gneiss. One of these, the Strathcona, illustrates the offset deposits below the irruptive (Fig. 20-25).

Origin.

The origin of these ores has been controversial, and the literature voluminous. For detail, the reader is referred to the more recent contributions. The older conception of Barlow and Coleman of magmatic segregation by early settling of the sulfides into embayments in the norite is not tenable because

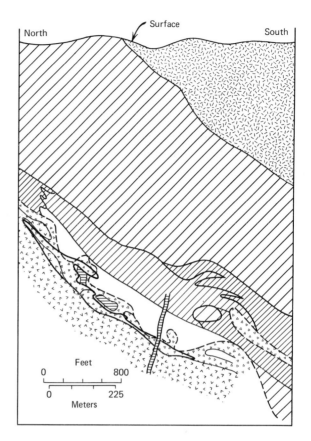

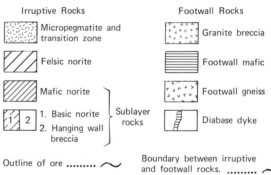

Figure 20-25 Vertical section through the Strathcona deposit, north rim of the Sudbury basin, Ontario. (Naldrett and Kullerud, 1967; from *Econ. Mineral Deposits of Canada*, 1969.)

most of the ore minerals are later than the rock silicates and invade and enclose them. The ore is now known to be later than the norite. Also, between the solidification of the norite and the introduction of the sulfides, there were injections of quartz diorite, basic norite, and others, beneath the base of the norite, and shearing and brecciation occurred. Consequently, there could have been no magmatic segregation of sulfides in situ within the norite.

A hydrothermal replacement origin was vigorously upheld by early company and other geologists, but this hypothesis now attracts few geologists.

Later investigations by Hawley, Naldrett, and Kullerud, and Souch, Podolsky, and company geologists, agree that the ores are definitely magmatic in origin. The sulfides probably accumulated as immiscible separations and some were subsequently injected along with accompanying varieties of norite, and sulfide melts, into late intrusives, breccias, shear zones, and faults around the margins beneath the norite.

Although the origin of the Sudbury Basin has been used as the classic example of a lopolith, there are few who accept this genesis. In contrast, the origin of the basin by meteorite impact with subsequent magmatism and eventual deformation by tectonism is a recent hypothesis. Brocoum and Dalziel have recently detailed such a hypothesis. The Sudbury breccia in which much ore is located is interpreted to be an impact breccia.

French found abundant evidence that the 1.5 kilometers thick Onaping tuff breccia, overlying the "micropegmatite," represents the fallback breccia subsequently floated up by the Sudbury magmas.

The Thompson District. As summarized by Cornwall, the discovery of the Thompson district, Manitoba, in 1949, and particularly the discovery of the Thompson mine in 1956, revealed a new type of nickel-sulfide deposit. Data collected by the company geologists indicated that the nickel was derived from peridotite and serpentinite where metamorphosed to a garnet-amphibolite-sillimanite assemblage, indicating a temperature of 600°C or higher. The ore contains 2.8 percent Ni and 0.18 percent Cu.

Unmetamorphosed peridotites in the area (as elswhere in the world) contain 0.25 to 0.3 percent Ni; metamorphosed peridotite (metaperidotites) in the area contain significantly less. There is evidence that sulfur was available in hydrothermal solutions during metamorphism; metamorphosed iron formation in the area has been partly converted to pyrrhotite ($Fe_{1-x}S$). It is possible that nickel was mobilized by hydrothermal solutions during metamorphism and deposited at present sites as the sulfide (pentlandite), as first proposed by Michener in 1957.

If this mechanism for the formation of the Thompson deposits is valid, peridotites throughout the world could have associated nickel-sulfide deposits, when sufficiently metamorphosed, in an environment where sulfur was available—for example, in proximity to black shales, which commonly contain sulfur in disseminated pyrite.

United States. The most recent producing nickel mine in the United States is the *Nickel Mountain mine*, Riddle, Oregon, which is now closed. The newly discovered *Minnamax mine*, near Babbitt, Minnesota, will soon, however, be in production. About 10 percent of the domestic consumption of nickel was derived from the Oregon deposit. The mine consists of two ore bodies with saprolite constituting the bulk of the ore. It is thought that the ore deposits were derived from periodotite and dunite through chemical decomposition by downward percolating meteoric groundwater. They are thus similar to the residual lateritic type of nickel deposits. The nickel content is about 1.5 percent and the deposits were mined by open pit methods.

More than 15 years of geological exploration were required before the *Minnamax* copper nickel deposit near Babbit, Minnesota, exploration work only at Minnamax. Deposit remains subeconomic with no expectation of development and production for some years. This time includes several years of developing an acceptable environmental program. This has involved extensive stream water sampling, measuring the general health of the ecosystem at the site, describing the existing physical conditions of the ecosystem recorded prior to project operations, and developing information on the environment for use for future industrial planning. The field studies include surface and groundwater hydrology, aquatic biology, soil studies, and microclimatology.
oldest to the youngest are the Biwabik iron formation, the Virginia formation, and the Duluth Gabbro complex. This layered basic intrusion is the host for the copper-nickel mineralization that occurs in the basal unit of the complex and is also present along portions of its northern and western contact. Possible economic concentrations have been identified in three discontinuous and isolated areas, one of which is included in the Minnamax project.

A 14-foot-diameter shaft has been bottomed at 1710 feet. This shaft is providing the means of acquiring a bulk sample for metallurgical studies from the mineralized zone that occurs 1400 to 1800 feet below the surface.

Australia. A new nickel area is under development in the Forrestania region of western Australia. The area is located in sand plains 400 kilometers south-

Table 20-9 Average Analysis of a Series of Drill Samples (in percent)

	Chemical analysis	Mineral analysis		Ferronickel analysis	
Ni	0.28				
Fe	5.2	Serpentine	74	Ni	37.9
Ca	0.02	Olivine-Enstatite	16	Ca	0.72
MgO	39.9	Spinel	4	Cr	0.10
		Others	6	C	0.28
Si	37.4			C	0.08
Al_2O_3	0.5			S	0.07
				P	0.017
Ignition loss	11.0			Fe	60

west of Kalgoorlie. The deposits are confined to ultramafic units associated with metamorphosed volcanic rocks and intercollated metasediments including banded iron formations. The folding is so intense that the iron formations are almost vertical.

The nickel mineralization lies in numerous pods or pockets each usually too small for individual exploitation. Recently, however, sufficient reserves have been proved to encourage exploitation. Numerous such deposits exist in the Yilgarn block, western Australia as shown in Fig. 20-26.

Dominican Republic. Extensive nickel laterite deposits have been operated since 1971 near Bonao in the Dominican Republic. It is one of the country's major industrial employers with a workforce of 2100 persons, of whom about 95 percent are Dominicans.

The deposits are very similar to those located in Cuba. The lateritic ore bodies mined contain nickel oxides enriched as the result of weathering of the ultramafic rocks under tropical climatic conditions. The rock contains between 0.15 percent to 0.35 percent Ni. The average analyses of a series of drill samples located in the mineralized area and taken from depths of more than 10 meters below the surface is given in Table 20-9.

The removal of ferruginous material by natural leaching has produced economic grade nickel deposits containing an average of 1.58 percent Ni in the project zone. The ore reserves are estimated as of January 1, 1975, at 66 million short tons with an average nickel content of 1.58 percent.

Open pit mining is done from the zone of surface enrichment that varies in thickness from 1 or 2 meters to more than 40 meters.

Some 2.5 million tons of ore were extracted in the year 1974 from which 79,643 tons of ferronickel with 30,492 tons of contained nickel were produced from the start of mining operations to the end of 1974, a total of 6.6 million tons of dried ore have been mined. More than 9 million tons of wet ore have been mined.

The ferronickel produced metallurgically from the ore has found worldwide acceptance among steel producers. Its principal markets are in Europe, the United States and Japan.

One of the most important factors, during both the construction and the operation of the plant, has been the need to protect and preserve the environment. Extensive systems for controlling the level of water contamination have been designed and constructed, and reforestation programs have been carefully planned to restore the mine and plant areas to their natural state. Alongside the Mining Department, a Forestry Unit has been set up, with agronomists supervising the environmental rehabilitation programs that are already under way at the Falconbridge Dominican project.

Other Deposits. Lower-grade deposits occur in the *Lynn Lake* area of *Manitoba*. A small deposit of the same type occurs at *Rankin Inlet* on *Hudson Bay*, where a segregation at the bottom of a pyroxenite sill carries 4.6 percent Ni and 1.2 percent Cu. Many minor deposits with a close spatial relation to ultrabasic rocks occur in Ontario, Quebec, British Columbia, and Alaska. Somewhat similar deposits have recently been developed near *Kalgoorlie, Australia,* carrying 4.6 percent Ni and 0.5 percent Cu. At *Insizwa,* southern Africa, magmatic segregation nickel-copper ores occur at the base of an ultrabasic

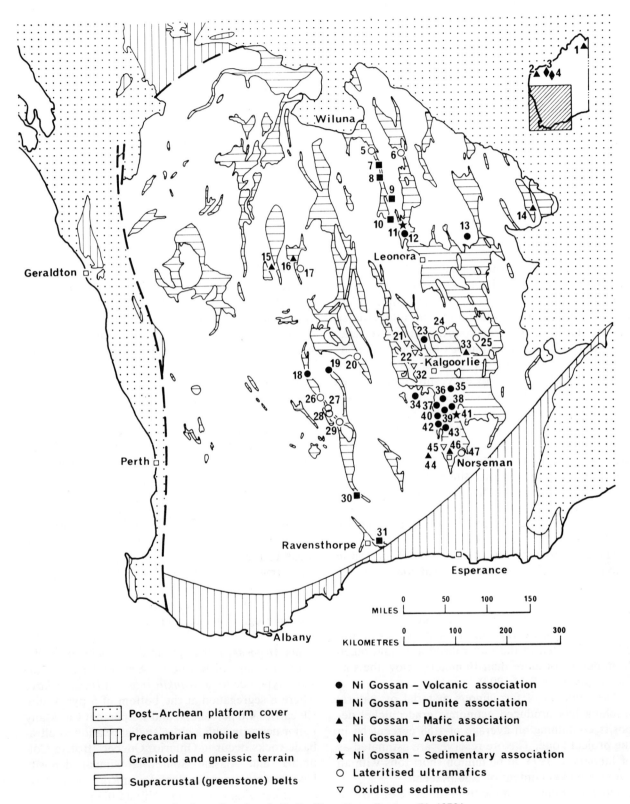

Figure 20-26 Nickel mineralization in southwestern Australia. (From Econ. Geol., v. 71, 1976.)

Ni Gossan – Volcanic association
Ni Gossan – Dunite association
Ni Gossan – Mafic association
Ni Gossan – Arsenical
Ni Gossan – Sedimentary association
Lateritised ultramafics
Oxidised sediments

Post–Archean platform cover
Precambrian mobile belts
Granitoid and gneissic terrain
Supracrustal (greenstone) belts

intrusive and with magmatic platinum deposits of the *Bushveld Complex*, South Africa. Many small, generally similar magmatic deposits occur in *Fennoscandia*, and for a while the world's supply of nickel came largely from southern *Norway*. The *Petsamo* deposit in the USSR (annexed from Finland) closely resembles Kambolda and carries 1.5 percent Ni and 1.3 percent Cu. Other sulfide deposits at *Nittis; Kumajie;* and in the *Urals*, USSR, make this country the largest nickel producing country in the world. Small deposits occur in *Greece*. Magmatic sulfide nickel deposits are lacking in the United States although the old *Gap mine* in *Pennsylvania* was a former nickel producer.

Lateritic nickel deposits developed by weathering of ultrabasic intrusives have been receiving world wide attention. The Nicaro and Moa Bay deposits of Cuba are lateritic nickel deposits. Other lateritic nickel deposits with cobalt have been developed or explored in Brazil, Venezuela, Guatemala, the Dominican Republic, Phillipines, Indonesia, and the USSR.

Nickel laterites are considered to be the major source of much of the world's future nickel production. These are surface or near-surface deposits in which the nickel has been concentrated as a residual deposit by leaching of other material by the weathering process. Primarily, they are magnesium silicates with nickel, iron, and cobalt substituted for some of the magnesium in the parent rock. For many years, they have been mined and processed for their nickel content but are becoming increasingly more important as a nickel source because of the high cost of mining sulfide deposits of declining grade.

Selected References on Nickel

The Sudbury Ores: Their Mineralogy and Origin. J. E. Hawley. University of Toronto Press, Ontario, Canada. *Comprehensive treatment of geology, occurrence, mineralogy, and origin.*

Investigations of nickel-copper ore and rocks of Strathcona mine, Sudbury, Ontario. 1965–1966. A. J. Naldrett and G. Kullerud. Annual Report, Director Geophysical Laboratory. p. 302–329; also in 1966–1967 Report p. 403–406. *Rocks, ores, emplacement, and geochemistry.*

Geology of the Strathcona ore deposit, Sudbury, 1968. J. C. Cowan Can. Min. Inst. Bull., v. 61, p. 38–54. *A north rim magmatic deposit in breccia.*

The sulfide ores of Sudbury: Their particular relationship to inclusion-bearing facies of the nickel irruptive. 1969. B. E. Souch and T. Podolsky and geological staff. Magmatic ore deposits, Econ. Geol., Mon. 4, p. 252–261. Also papers by L. C. Kilborn, et al., p. 276–293; by A. J. Naldrett, p. 359–365. *New Information of ore occurrence and origin.*

Geology of the Nickel Mt. mine, Riddle, Oregon. 1968. J. T. Cumberlidge. Graton-Sales volumes, p. 1650–1672. AIME, New York. *Occurrence and origin.*

Nickel. 1968, Min. Engr. v. 20, no. 10, p. 69–116. *Brief survey of important world deposits.*

Emplacement of ore at the Strathcona mine, Sudbury, as a sulfide-oxide magma in younger noritic intrusions. 1968. A. J. Naldrett and G. Kullerud. 23rd Int. Geol. Cong., v. 7 p. 197–213. *Sulfides are magmatic.*

La genese et l'evolution des gisements de nickel de la Nouvelle-Caldeonie. 1947. E. de Chetelat. Bull. Soc. Geol. France, v. 17 p. 106–160. *Discussion of ores and origin.* General Reference 19. Chapter on nickel.

CHROMIUM

Chromium, like manganese, occurs mainly in countries that use little of it, and most of the large consuming countries are deficient in it. It was first mined in Norway in 1820 and then in Maryland in 1827. Formerly, chromite was used chiefly as a refractory mineral, but since the rapid development of stainless steels it has become a prized steel alloy.

Uses.

The main uses for chromium are: metallurgical, 67 percent; refractories, 18 percent; and chemical, 15 percent. The metallurgical uses include a great variety of alloys, mainly with iron, nickel, and cobalt. Some of the important ones are

	Percent Cr		Percent Cr
Low chrome steels	0.5–5	Super stainless steels	12–30
Low chrome iron	0.2–4	Chrome-nickel ferrous alloys	14–30
Medium chrome steels	3–12	Electrical resistance alloys	8–20
Stainless irons	12–15	Chrome-cobalt alloys	20–35
Stainless steels	12–18		

Chromium imparts to alloys strength, toughness, hardness, and resistance to oxidation, corrosion, abrasion, chemical attack, electrical conductivity, and high-temperature breakdown. Consequently, it finds manifold uses. The great strength of chrome steels allows a reduction in the weight of metal in automobiles, airplanes, and trains, for example. The stainless steels, containing 18 percent Cr and 8 percent Ni have become popular as noncorrosive, tough, hard, strong steels. Special chrome alloys, such as stellite, using also tungsten, molybdenum, and cobalt, yield hard, high-speed tool steel. Chromium plating has become more popular than nickel plating.

As a refractory, the mineral chromite is widely used for furnace linings, and chemical compounds are used for dyeing, tanning, bleaching, pigments, and oxidizing agents.

Production and Distribution.
World leadership in chromite production has passed from the United States (1860) to Turkey, to Russia, to New Caledonia, in 1922 to Rhodesia, and now to Russia and South Africa. The annual world production is about 10 million long tons with South Africa as the leading producer of 2.4 million tons; followed by Russia (2.2 million long tons; and Albania (700,000 long tons), which recently replaced Turkey in third place. Other producers are Rhodesia, the Phillipines, India, Greece, New Caledonia, Finland, Iran, and Canada. Brazil, Cuba, and Bulgaria are minor producers.

During the past year, the United States has consumed about 1.2 million short tons of chromium ore, equal to about 9 percent of total world production but has mined none at all since 1962 (Fig. 20-27) during which time the entire supply of chromite for the United States has been imported from the Eastern Hemisphere especially South Africa and Russia. Rhodesian production has been less because of a trade embargo by the U.S.

Mineralogy, Tenor, and Treatment.
There is only one ore mineral, chromite, that theoretically carries 68 percent Cr_2O_3 and 32 percent FeO, but Al_2O_3, Fe_2O_3, MgO, CaO, and SiO_2 may displace some Cr_2O_3, reducing the Cr_2O_3 content to as little as 40 percent. Chromite chemically varies within wide limits permitted by the formula $(Mg, Fe^{+2}) (Cr, Al, Fe^{+3})_2O_4$. Commercial ores should contain 45 percent Cr_2O_3. The chrome-iron ratio should be above 2.5:1 for metallurgical chrome.

Chrome is marketed as lump chromite after hand sorting or rough concentration. Most chromite ores are not adaptable to concentration processes. The chrome ore is smelted in an electric furnace with fluxes and carbon to ferrochrome, in which form most of it is marketed.

A compilation of the chemical compositions of typical chrome ores from major producing countries is as listed in Table 20-10.

Primary chromium deposits occur only in ultramafic or closely related anorthosite rocks in two basic forms, namely, stratiform (layered), which make up 98 percent of the chromite resources of the world; and pod-shaped. The Bessveld igneous complex (BIC) and the Stillwater complex are examples of the former while those deposits in mountain belts, such as the Appalachians, Urals, and belts rimming the Pacific Ocean are examples of the latter.

All chromite deposits are believed to have formed by early crystal settling or by late gravitative liquid accumulation. These processes aid in understanding how layered deposits form, but where differential pressures exist, liquid injection is more likely to form podiform deposits.

Kinds of Deposits and Origin.
Almost all chromite deposits are magmatic segregations in ultrabasic igneous rocks, as described in Chapter 5. Chromite occurs in the host rock as masses, lenses, and disseminations. In some places, disseminated grains have undergone residual concentration.

Chromite is a rock-forming mineral that is closely associated with magnesium and nickel in ultramafic rocks. These rocks contain an average amount of about 1000 to 3000 ppm Cr. Gabbro contains about 200 ppm and granite about 5 ppm Cr.

Chromite also exists in secondary deposits. Because it is heavy and resists chemical weathering, it accumulates in black sand and placer deposits. Such deposits have been worked in the United States and Japan on a small scale but are recovered on a larger scale in Rhodesia by flotation.

According to Gross, under favorable tropical or subtropical conditions, magnesium silicates may be leached from chromite-bearing ultramafic rocks resutling in a residual lateritic soil containing as much as 50 percent Fe, 2 to 4 percent Cr_2O_3, 2 to 2.5 percent nickel, and about one-tenth as much cobalt as nickel.

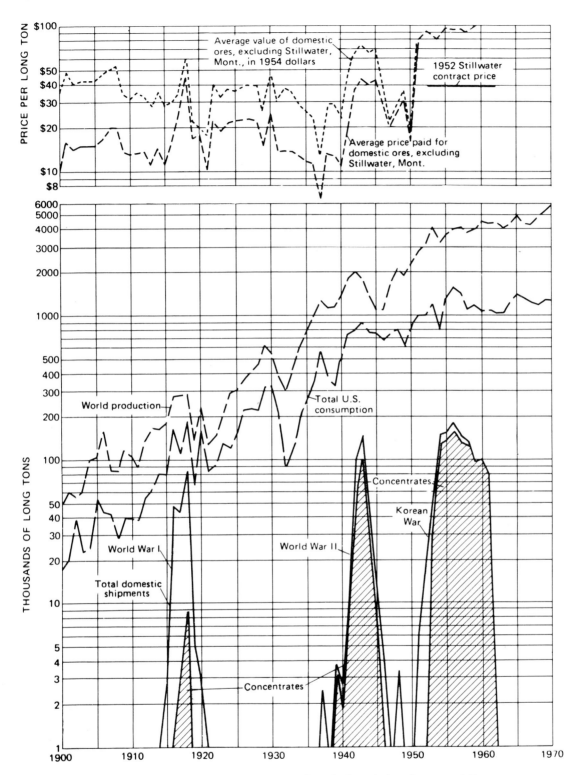

Figure 20-27 World production and U.S. consumption and production of chromite ore and the relation of price to domestic production from 1900 to 1970. Data from U.S. Geological Survey and from U.S. Bureau of Mines (1942–1970). (From T. P. Thayers, 1972, Chromium in U.S., Geol. Surv. Prof. Paper 820.)

Table 20-10 Chemical Compositions of Typical Chrome Ores from Major Producing Countries

Country: District, Type or Use	Cr_2O_3	Total Fe as FeO	Al_2O_3	MgO	CaO	SiO_2	Ign. loss	Cr/Fe
Cuba								
Camaguey	30.84	13.18	29.03	18.92	0.91	5.33	1.25	2.08
Moa Bay	36.00	14.51	27.50	16.53	0.45	3.85	0.77	2.19
Philippines								
Masinloc	32.10	12.72	30.20	18.06	0.44	5.00	0.35	2.21
Aoje concentrate	49.32	18.80	11.37	15.85	—	3.50	—	2.30
Transvaal								
Standard friable	44.50	24.71	15.03	10.05	0.31	3.86	+	1.58
Hard lump	42.41	24.80	15.95	8.35	0.36	6.27	+	1.50
High-grade	48.40	22.85	13.94	10.72	0.20	2.55	+	1.86
Rhodesia								
Selukwe	49.70	14.37	13.10	13.06	0.46	6.77	2.24	3.05
Selukwe	47.00	11.93	12.65	15.46	1.77	5.71	3.85	3.47
Selukwe refractory	42.63	15.65	13.80	15.77	0.32	8.59	1.86	2.40
Dyke	50.73	16.38	13.01	13.23	0.75	4.33	0.89	2.73
Dyke	48.50	18.25	11.51	13.37	0.07	5.62	0.85	2.35
Turkey								
Metallurgical	48.26	14.11	13.03	16.85	0.96	5.08	0.97	3.02
Kefdag	42.44	12.50	16.75	18.92	0.25	6.21	1.34	2.97
Refractory	37.10	13.94	24.35	17.74	0.22	4.34	1.02	2.36
New Caledonia								
Metallurgical	51.10	14.15	11.16	16.96	0.10	5.75	—	3.20
USSR								
Metallurgical	53.87	12.56	9.63	13.27	1.06	5.82	2.43	3.77
Refractory	39.08	13.98	17.35	16.07	9.65	9.38	2.90	2.46
Yugoslavia								
Metallurgical	49.73	15.16	10.40	15.90	0.42	5.98	1.25	2.90
India								
Orissa	49.90	13.85	10.85	15.26	0.17	6.25	2.20	3.17
Karnataka, refractory	41.53	22.41	13.40	12.34	0.06	6.91	2.08	1.63
Greece								
Tsagli no. 1	45.18	15.16	17.08	15.82	0.35	3.80	1.02	2.6
Domokos	35.80	16.30	22.12	14.50	0.54	5.96	1.97	1.92
United States								
Stillwater concentrate	38.50	23.20	15.45	13.43	—	6.00	+	1.47
Pakistan								
Baluchistan	48.80	15.20	10.80	16.37	0.30	4.98	1.87	2.82
Sierra Leone								
Kambui Hills	42.00	14.08	21.23	16.16	0.05	4.81	1.48	2.53

Source: Ind. Min. and Rocks, p. 247, 1960.

Examples of Chromite Deposits

Stillwater, Montana. The Stillwater complex in Montana is of the layered type. The layers occur in an east-west belt about 50 kilometers long by about 1 kilometer wide and dips from 50 to 90° (Fig. 20-28) to the north. Thirteen high-iron, chromite-rich layers are known that range in thickness from a few centimeters to 4 meters; some can be traced laterally for more than 25 kilometers.

The thicker minable parts of the Stillwater complex, the principal domestic sources containing 80 percent of U.S. reserves, are complexly faulted and sheared off at depth. Sporadic mining has been attempted, especially during World Wars I and II, and the Korean War when 800,000 long tons were mined but stockpiled because of its low grade (38.5 percent Cr_2O_3).

Rhodesia. The Great Dyke in Rhodesia is a layered complex where high-grade deposits occur at Gwelo, which includes Selukwe and Victoria, Lomagundi, and Hartley. The deposits occur in or near the amazing Great Dyke, which is 550 kilometers long, averages 6 kilometers in width (Fig. 20-29), and

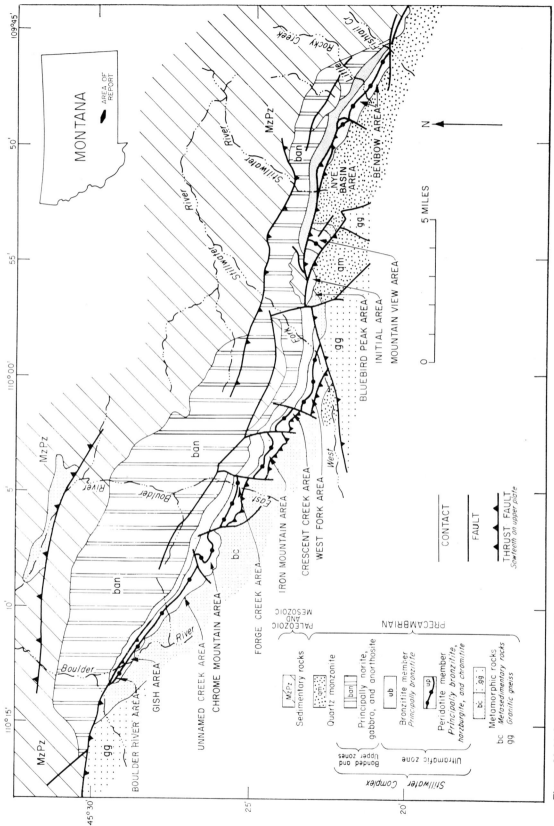

Figure 20-28 Geologic map of the Stillwater complex, southwestern Montana. (Simplified from Page and Nokleberg, 1974 and Jones, et al., 1960.)

MONTANA

AREA OF REPORT

N

5 MILES

0

MzPz

gg

MzPz

ban

bc

ban

ban

MzPz

ban

qm

gg

gg

BENBOW AREA

NYE BASIN AREA

MOUNTAIN VIEW AREA

INITIAL AREA

BLUEBIRD PEAK AREA

CRESCENT CREEK AREA

WEST FORK AREA

IRON MOUNTAIN AREA

FORGE CREEK AREA

CHROME MOUNTAIN AREA

UNNAMED CREEK AREA

GISH AREA

BOULDER RIVER AREA

Boulder

River

Boulder

River

East

West

Fork

Stillwater

River

Stillwater

River

Rocky

Creek

Fishtail Cr.

CONTACT

FAULT

THRUST FAULT
Sawteeth on upper plate

PALEOZOIC AND MESOZOIC

PRECAMBRIAN

MzPz | Sedimentary rocks

qm | Quartz monzonite

ban | Principally norite, gabbro, and anorthosite

Bonded and Upper zones

ub | Bronzitite member *Principally bronzitite*

up | Peridotite member *Principally bronzitite, harzburgite, and chromitite*

Ultramafic zone

bc | Metamorphic rocks *bc Metasedimentary rocks*

gg | *gg Granitic gneiss*

Stillwater Complex

427

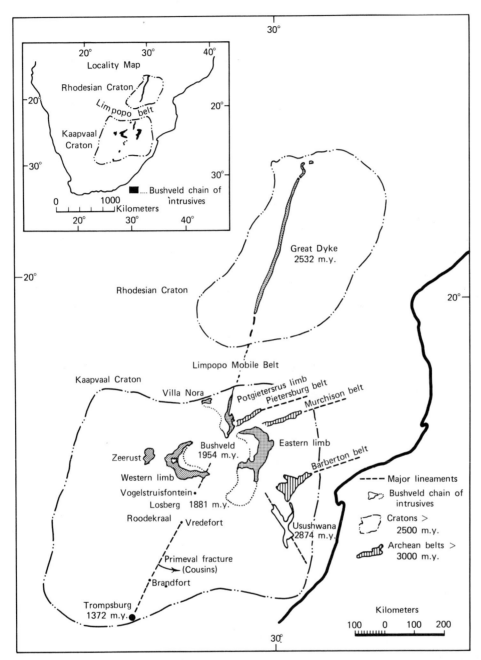

Figure 20-29 Sketch map showing principal lineaments and tectonic setting in southern Africa. (From Econ. Geol., v. 71, 1976.)

consists of layers of ultrabasic rocks now largely altered to serpentine. The large Selukwe deposit is podiform and occurs in Archean rocks that predate the Great Dyke. It appears to be only accidentally juxtaposed with the Great Dyke.

The dike deposits are bands of chromite varying in thickness from 5 to 50 centimeters that parallel the pseudostratification. The deposits outside of the dike are lenses in talc schists and serpentine from 45 to 135 meters long. Thayer considers that the

deposits are early magmatic segregations formed by crystal settling, but Sampson concludes that they are late magmatic accumulations. Nearly all are of the high-chromium variety. Only those layers 15 centimeters or more in thickness are mined below the outcrops, some of the thicker layers are mined more than 300 meters down the dip.

Busveld Igneous Complex, South Africa. The Busveld complex is the largest layered complex of the world. It is a crudely basin-shaped, layered mass 65,000 square kilometers in area. As many as 40 bands of chromite are evident that vary from less than 1 centimeter to slightly more than a meter in thickness and extend for tens of kilometers.

Because of the size of this complex, the reserves are enormous and difficult to determine with any precision. An estimate of 6 billion tons is given, but it is thought to be conservative. The grade is about 43 percent Cr_2O_3.

The total thickness of the Busveld igneous complex is based upon measurements spread out over tens of kilometers apart. The near basal ultramafic rock can be viewed in several locations with intermediate igneous rocks in other areas. One well known outcropping of the supposedly late differentiate of "red granite" is located on a knoll alongside of a small African hut site. Strontium-rubidium analyses suggest that the red granite is derived from the crust while the mafic magma is derived from the upper mantle.

The origin of the chromite in the complex is now thought to be late liquid separation and crystallization with some cross-cutting relationships formed by late liquid injection.

Rhodes recently suggested that the Busveld-Vredefort structures (Fig. 20-30) are the result of the high-velocity impacts of a tidally disrupted asteroid.

Russia. The high-chromium chromite ores of the Eastern Hemisphere are mostly podiform and, according to Thayer, are associated with three major tectonic belts: (1) the Ural mountains, (2) the Tethyan mountain chain between the Alps and Himalayas, and (3) around the western margin of the Pacific Ocean.

The older deposits of Russia center around the Sverdlovsk region on the east side of the Urals, where they occur as magmatic segregations of lenses, stringers, and disseminations in serpentine and soapstone. These deposits have been known since the ninth century.

The southern Ural Mountains appear to include the greatest known concentration of large podiform deposits of the world. The ore, however, is low grade and is used for refractories and in the chemical industry. Thayer estimates reserves at about 25 million tons but states that no accurate figures are available.

The principal sources of Russian metallurgical grade chromite ore occur in Kazakhstan near Aktiubinsk and in Khrom Tau. The ore mined was once of sufficient grade to be used directly after a minimum of preparation. Now it must be benefi-

ciated in a gravity concentration plant installed at the Donskoie open pit near Aktiubinsk. A ferroalloy plant was built in this town during World War II, which, together with the Cheliabinski ferroalloy plant, produces about 30 types of chromium alloys. Yet Russia exports 70 percent of its chromite ore principally to the United States and also, Japan, Sweden, and West Germany.

Apart from the Urals, chromite finds were reported in the Caucasus, Transcaucasia, and Transbaikalia. There are more than 31 chromite mines and 21 washeries in Russia, according to Sutulov.

Canada. The *Muskox intrusion* of recent discovery is a layered ultramafic complex of Precambrian age located in the Northwest Territories of Canada. It is continuously exposed from its feeder to its roof and its entire internal form is well preserved. It is exposed for 118 kilometers with a northward plunge of only 4°, which has provided the exposure from the feeder to the roof.

Bronzite gabbro and picrite compose the feeder with other rock types as shown in Fig. 20-31. Although Ni-Cu-Fe sulfides are distributed throughout the complex, their greatest abundance is in the sublayer. Chromite-rich horizons exist in the upper part of the central layered series.

This complex has only been studied since 1960, but as it provides a uniquely exposed dissection of a layered igneous complex, it will continue to provide a better understanding of their nature.

Chromite layers in the Muskox intrusion are not too significant in comparison to other complexes where layers exist measuring up to 3 meters thick and extending for tens of kilometers laterally. One of the chromite layers of the Muskox is about a centimeter thick and the other up to 10 centimeters thick, but their extent is about 100 square kilometers with one thin layer being traced about 20 kilometers by outcrops. It seems apparent, that as most of the chromite is disseminated in olivine to the extent of 1 to 3 percent, separation of the chromite into chromotoid layers was insignificant. If the chromite had separated from the more than 1200 meters of dunites, peridotites, and picrite, which make up about two-thirds of the crosssection of the intrusive, extensive and thick layers of chromite would be evident in the Muskox intrusion.

Turkey. Turkey contains numerous scattered deposits, the most important ones occurring near Bursa, south of the Sea of Marmora, and along the south Mediterranean coast. The most important deposits are the Guleman mines, near Ergani Maden

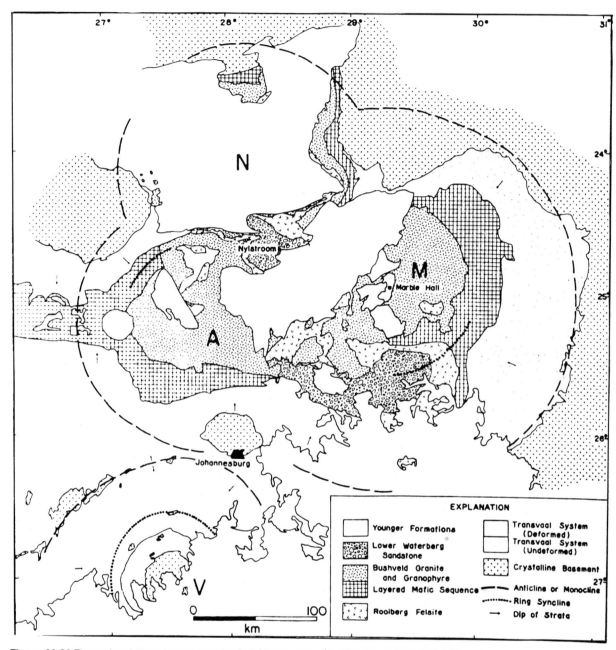

Figure 20-30 The regional tectonic setting of the Bushveld complex. Note that the gravity anomalies suggest four separate Bushveld compartments, each presumably with a separate feeder (Modified after Vermaak, 1976; from Vermaak, 1976, Econ. Geol., v. 71.)

and the Gololan ore body. All are associated with altered ultrabasic intrusives and are thought to be magmatic segregations. The ores are hard and of high quality; many mines average 50 to 52 percent Cr_2O_3, others yield 43 to 50 percent ore.

Other Deposits. In *Cuba* important refractory ores carrying 33 to 43 percent Cr_2O_3 are found in Camaguey, Oriente, and Matanzas. There are also huge tonnages of low-grade chromiferous nickel-iron ores in Mayari and Mao Bay. The *Philippines* have been an important producer of 46 to 50 percent ore. Small deposits of high-quality ore occur in Baluchistan in *Pakistan*, and Karnataka, Bihar and Orissa, in *India*. *Yugoslavia*, *Greece*, *Brazil*, and *New Caledonia* are minor producers. In the *United States* small deposits have been mined mainly in California, Oregon, Maryland, North Carolina, and

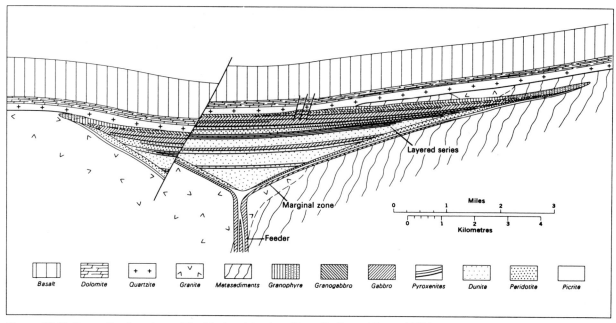

Figure 20-31 Generalized geology of the Muskox intrusion. (From Smith and Kapp, 1963.)

Alaska. A deposit is also known in Newfoundland, *Canada.*

The chromite deposits near *Campo Formosa, Bahia,* were formerly thought to be great blocks or possibly podiform deposits. Recent studies suggest that the deposits are fault blocks of a layered chromite deposit that is 16 to 18 kilometers long. Quartzite debris hid much of this evidence but the continuance of the layers was proved by drilling through the quartzite overburden.

Selected References on Chromite

Chromite deposits of the eastern part of the Stillwater complex, Montana. 1940. J. W. Peoples and A. L. Howland. U.S. Geol. Surv. Bull. 922-N. *Montana low-grade deposits.*

Chromite Deposits of the Busveld igneous complex. 1937. S. Africa Geol. Surv. Geol. Ser. Bull. 10. *Low-grade magmatic deposits.*

Chromium, United States mineral resources. General Reference 19.

The Soviet Challenge in Base Metals. 1971. A. Sutulov. University of Utah Press, Salt Lake City, p. 195. *Late reference on Russian mineral production.*

Occurrence and characteristics of chromite deposits— eastern Bushveld complex. 1969. E. N. Cameron and G. A. Desborough, in Magmatic Ore Deposits, Econ. Geol Mon. v. 4, p. 23–40.

Geology and chromite deposits of the Camaguey district, Camguey Province, Cuba. 1948. D. E. Flint, P. N. Guild, and J. F. de Alber. U.S. Geol. Surv. Bull. 954-B, p. 39–63.

Chromite in the United States. 1962. T. P. Thayer and M. H. Miller. U.S. Geol. Surv. Mineral Inv. Resources Map MR-26. General Reference 19. Chapter on chromite.

MOLYBDENUM

Metallic molybdenum was isolated in the 1770s and produced commercially in 1893, but only 10 tons of the metal had been made before 1900. Large scale use began in 1913, and it attained commerical importance in 1927.

Uses.

Practically all molybdenum produced is used as an alloy in iron and steel, where it is displacing other alloy metals. In steel it acts like tungsten, but more powerfully, requiring only one-half to one-third as much to produce similar results. It is the most potent hardening alloy. It increases the strength of steel and of cast iron and also increases their ductility. It serves best in steels containing nickel, manganese, or chromium. Only small quantities are needed, usually less than 2 percent.

Molybdenum steels are used widely in aircraft, automobiles, oil machinery, shafts, valves, pumps, gears, high-speed tools, pins, bolts, guns, armor plate, dies, wire, and many other materials. It greatly improves cast iron for industrial, railroad, and automotive uses. Minor amounts are used for catalysts, dyes, lithographic inks, and lubricants.

Production and Distribution.

The world production of molybdenite (MoS_2) is about 200 million pounds per year, of which the United States produces about 61 percent (120 million pounds in 1978) and the remainder comes from Canada, Chile, Russia, and Peru. Other minor producers are Mexico, Central America, Yugoslavia, and China.

The United States currently supplies about 60 percent of the world's total molybdenum; Canada about 17 percent; Russia and Chile about 10 percent each, and all other producing countries, 3 percent. About 95 percent of the world molybdenum production comes from porphyry molybdenum or porphyry copper deposits where, in the latter, molybdenum is generally in such abundance that it is more than a by-product but sufficient to refer to such deposits as porphyry copper-molybdenum deposits.

Mineralogy, Tenor, and Treatment.

Practically all molybdenum is obtained from molybdenite. Wulfenite ($PbMoO_4$) and ferromolybdite are minor contributors.

The Climax ore averages 0.3 to 0.5 percent MoS_2; smaller deposits range from 1 to 3 percent MoS_2. The Utah Copper deposit ores carry about 0.04 percent MoS_2 as a by-product. The Henderson deposit in Colorado averages 0.49 percent molybdenite.

Molybdenite ores are concentrated by differential flotation. The product is generally marketed in the form of molybdenite concentrates. For adding to steel, this is roasted with lime to make calcium molybdate (40 to 50 percent Mo), is converted to ferro-molybdenum (50 to 65 percent Mo), or is made into briquettes of molybdenum oxide.

Kinds of Deposits and Origin.

The types of commercial deposits are

1. Porphyry or disseminated ore including stock works and breccia pipes: Climax, Urad-Henderson, Questa, Bingham, and numerous porphyry copper-molybdenum deposits.
2. Contact-metasomatic: Pine Creek, California; Helvetia, Arizona; Knaben, Norway; Northeastern Caucasus, Russia; China; and Azegour, Morocco.
3. Quartz veins: Questa, New Mexico.
4. Pegmatites and aplite dikes: Most mine, Quebec; Val d'Or and Preissac, Quebec.
5. Bedded deposits in sedimentary rock: Sandstone-type uranium deposits, Arizona, Utah, Colorado, New Mexico, Wyoming, South Dakota, and Texas.

Disseminated Deposits. Three-fourths of the world's reserves of molybdenum are in the Western Cordilleran of North and South America. The remaining one-fourth of the world's reserves occur in the porphyry deposits now being developed in locales extending from the Philippines to New Guinea; in Russia; in parts of Europe; in southwestern Asia, including China; and in Australia.

These moly-porphyry deposits have an average MoS_2 content varying from 0.1 to 0.5 percent, while the copper porphyries average from 0.015 to 0.1 percent MoS_2. The Bingham porphyry copper deposit, which produces more than 120 thousand tons of ore per day, has at times actually exceeded the production of molybdenite of any other deposit in the world. Actually, one-fourth of the world's molybdenum production is derived from porphyry copper deposits.

Climax and Urad-Henderson, Colorado; Questa, New Mexico; and Lornex, British Columbia (frontispiece) are classical examples of the porphyry molybdenum ore deposits.

Examples Of Deposits

Climax, Colorado. One of the greatest molybdenum deposits of the world, at Climax, Colorado, illustrates large-scale disseminated mineralization. It lies at an altitude of 3,390 meters, and ore is mined in excess of 43,000 tons per day of less than 0.4 percent MoS_2.

The deposit lies in the Colorado mineral belt in Precambrian rocks that have been metasomatized to a granitic rock. During Mid Tertiary time, a wet silicialkalic magma penetrated the Precambrian rocks and formed a composite stock that consists of at least four intrusions. Hydrothermal activity in the first three formed a complex stockwork of alteration and ore bodies that are circular in plan and arcuate, or convergent upward, in cross section (Fig. 20-32). Each ore body is related spatially and genetically to that magmatic phase from which it is derived. The fourth phase contained no economic mineralization. Finally, the Mosquito fault, with at

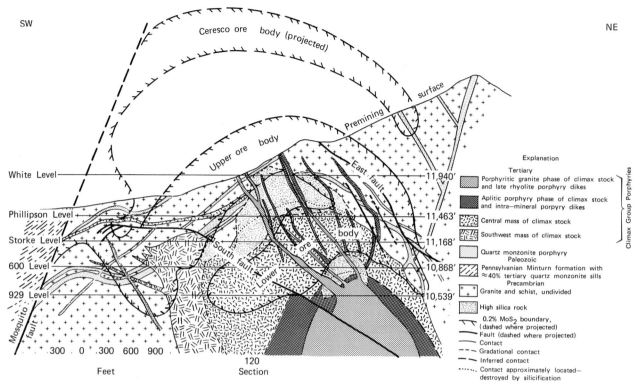

SW   NE

Ceresco ore body (projected)

Upper ore body

Premining surface

White Level — 11,940'

Phillipson Level — 11,463'

Storke Level — 11,168'

600 Level — 10,868'

929 Level — 10,539'

−300 0 300 600 900

Feet

120 Section

Explanation

Tertiary

Porphyritic granite phase of climax stock and late rhyolite porphyry dikes

Aplitic porphyry phase of climax stock and intra—mineral porpyry dikes

Central mass of climax stock

Southwest mass of climax stock

Quartz monzonite porphyry

Paleozoic

Pennsylvanian Minturn formation with ≈40% tertiary quartz monzonite sills

Precambrian

Granite and schist, undivided

High silica rock

0.2% MoS₂ boundary, (dashed where projected)

Fault (dashed where projected)

Contact

Gradational contact

Inferred contact

Contact approximately located—destroyed by silicification

Climax Group Porphyries

Figure 20-32 Section showing generalized geology and ore zones of Climax, Colorado. (From Wallace, et al., 1968, Ridges, ed.) in *Ore Deposits of the U.S.*)

least 2700 meters of displacement dropped part of one ore body to below the Paleozoic sedimentary strata on the hanging wall. Pleistocene glaciation removed a large portion of one ore body and uncovered the top of a second. The ore bodies are mined by shrinkage-stopping rather than by open pit methods even though the deposit is similar in overall aspects to a porphyry copper deposit; it would require a prohibitive waste/ore ratio for open pit mining.

The three ore bodies are referred to as the Upper ore body, the Ceresco ore body, and the lower ore body (Fig. 20-32). The first two are similar, but the Lower ore body differs in being smaller and lower grade than the other two ore bodies. Each successive ore body is smaller and formed closer to the upper contact of its intrusive source.

Hydrothermal alteration is vast, extending over an area of more than 2 kilometers in diameter. Potassic alteration and silicification are prominent. The latter is so prevalent that it has formed high silica masses or bodies by almost complete replacement of preexisting minerals.

The deposit produces by-products of huebnerite (tungsten mineral), cassiterite (tin mineral), and monazite (thorium mineral).

Figure 20-33 is a plot of δD values versus total salinity in ppm (sodium, potassium, calcium, magnesium, and chlorine), contained in fluid inclusions of quartz and one sample of fluorite. The postmolybdenite quartz-pyrite mineralization and hydrothermal fluids that formed the high silica ore bodies appear to have been diluted with more meteoric water than the fluids that produced the upper molybdenite ore bodies.

Urad-Henderson, Colorado. The Urad ore body has been known for scores of years. The ore is localized in a zone of fractured and hydrothermally altered granite in a portion of a composite granite porphyry stock. The Henderson molybdenite ore deposit lies 600 meters below the Urad deposit and is separated by altered rocks containing traces of molybdenite.

Reserves in the Henderson deposit exceed 300 million tons of 0.49 percent molybdenite. The deposit has just recently been brought into production.

The Colorado mineral belt contains clusters of mineralized areas including the stockwork molybdenite deposits. These deposits are almost entirely enclosed within a −300 milligal gravity contour (Fig. 20-34) that probably closely outlines the hy-

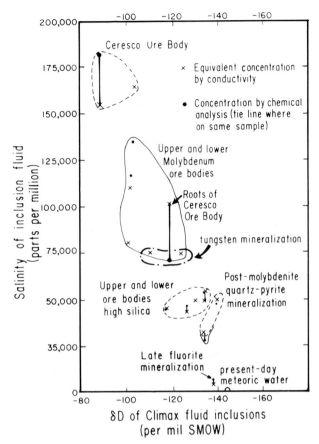

Figure 20-33 Plot of δD versus salinity of fluid inclusions in quartz, Climax, Colorado. (From Econ. Geol., v. 69, 1974.)

understood, but it is probably the result of molybdenum becoming concentrated in shales commonly associated with carbon. Leaching of the metal ion along with the uranyl ion results in the occurrence of both together.

Selected References on Molybdenum

Climax molybdenum property of Colorado. 1933. B. S. Butler and J. W. Vanderwilt. U.S. Geol. Surv. Bull., v. 846, p. 191–237. *A great low-grade, disseminated deposit in granite.*

The occurrence and production of molybdenum. 1940. J. W. Vanderwilt. Colo. School of Mines Quart., v. 3 p. 1.

Questa molybdenite deposit, New Mexico. 1938. J. W. Vanderwilt. Colo. Sci. Soc. Proc., v. 13, No. 11. *A fissure vein of molybdenite.*

Molybdenum in the United States. 1970. R. V. King. U.S. Geol. Surv. Bull. 1182-E, p. 90.

The geochemistry of molybdenum. 1954. P. K. Kuroda and S. B. Sandell. Geochim et Cosmochim Acta, v. 18 p. 35–63.

Characteristics of disseminated molybdenum deposits in the western Cordillera of North America. 1970. K. P. Clark, AIMR Preprints 70-1-90, 16 p.

General References 11 and 19. *Papers on Climax by S. R. Wallace, et al., and recent review of molybdenum deposits by R. V. King, et al., respectively.*

General Reference 20. *Details of recent molybdenum porphyry discoveries in British Columbia.*

pothesized mineral-belt batholith and coincides with an element of structural weakness of Precambrian ancestry expressed as zones of recurrent shearing.

Questa, New Mexico. Molybdenite in this deposit occurs in massive quartz veins in a stockwork of discontinuous veinlets and disseminated flakes in a fractured zone several thousand meters wide. The zone is juxtaposed with a composite granite stock and the overlying andesite volcanic rocks.

The deposit is huge even though the average grade is about 0.1 percent MoS_2 and will be a major producer of molybdenum.

Bedded Sandstone-type Uranium Deposits. Significant amounts of molybdenum occur in these deposits. The minerals jordisite and ilsemannite are present in amounts ranging from a few ppm to more than 0.5 percent. In large uranium deposits, the molybdenum is recovered as a by-product. The origin of molybdenum in these deposits is not fully

TUNGSTEN

Discovered in 1781 and isolated in 1783, tungsten was first used in 1847, but, with the discovery in 1898 that steel containing it could be used to cut other steels at high speed, it came into general use as a steel alloy. Today it is an essential alloy metal.

Uses.

The chief use for tungsten is for making high-speed cutting steels that retain their hardness even at red heat. These contain 15 to 20 percent tungsten with some chromium and vanadium. Molybdenum may replace some of the tungsten. Self-hardening steels contain 2 to 12 percent tungsten. The tungsten steels are used for various cutting tools and materials subjected to hard wear. It is widely used in stellite, an alloy of tungsten, chromium, and cobalt, for hard facing materials. A newer use is in making tungsten carbide and boron nitride, which, next to the diamond, are the hardest known cutting agents.

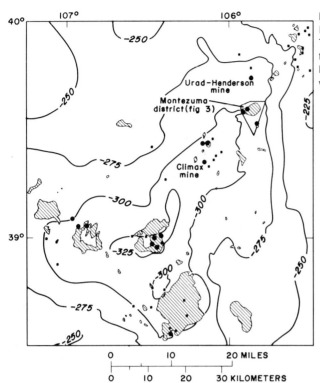

Figure 20-34 Map showing molybdenum deposits in relation to Tertiary stocks and gravity contours, Colorado. (From: G. J. Neuerburg, et al., U.S. Geol. Survey Circular 704.)

Tungsten filaments for electric light bulbs are familiar to everyone. Tungsten is also used for electrical contact parts, electrical apparatus, radio, X-ray, pigments, textiles, armor plate, guns, and projectiles.

The basic forms in which tungsten is used, according to Hobbs and Elliott are tungsten carbide (45 percent), alloy steels (25 percent), nonferrous alloys (11 percent), tungsten and tungsten-base alloys (18 percent), and chemicals and others (1 to 2 percent).

Production and Distribution.

The production of tungsten in the United States is unpredictable, primarily because of the variable impact of large reserves in southeastern China (50 percent) and North Korea and Russia (30 percent). The remaining 20 percent of the tungsten reserves are scattered throughout the rest of the world.

In 1977, world output of tungsten (Wolfram) ore concentrates was about 43,000 metric tons. The leading producers were China (9,400 tons) and Russia (7,800 tons). The United States produces about 3,700 tons, which increases sharply when the price of WO_3 increases. South Korea and Bolivia produced 2500 and 2000 tons, respectively.

The U.S. production of tungsten has varied widely as shown in Fig. 20-35. Production correlates obviously with price, but properties must be in an almost standby condition to correlate with the start periods of short duration of higher prices. During the past 20 years, the United States has had to import about 25 percent of its demand from Canada, Australia, Bolivia, Peru, Portugal, and South Korea. The U.S. production comes from Nevada, California, Idaho, Colorado, Montana, and North Carolina. The price and grade of tungsten is expressed in units, where one unit equals 1 percent per ton or 20 pounds WO_3 containing 15.9 pounds of tungsten metal.

Mineralogy, Tenor, and Treatment.

More than 20 tungsten minerals are known, but the only economic ones are scheelite, wolframite, ferberite, and huebnerite. These are all compounds of calcium, iron, or manganese, with WO_3. The tenor of tungsten ores is mostly less than 1 percent, and the standard grade of concentrates is 60 percent WO_3.

Tungsten ores are mined by underground methods; the ores are readily concentrated, because of the high specific gravity of tungsten minerals, by means of jigs and tables. The concentrates are treated to yield metal or ferrotungsten. For metal,

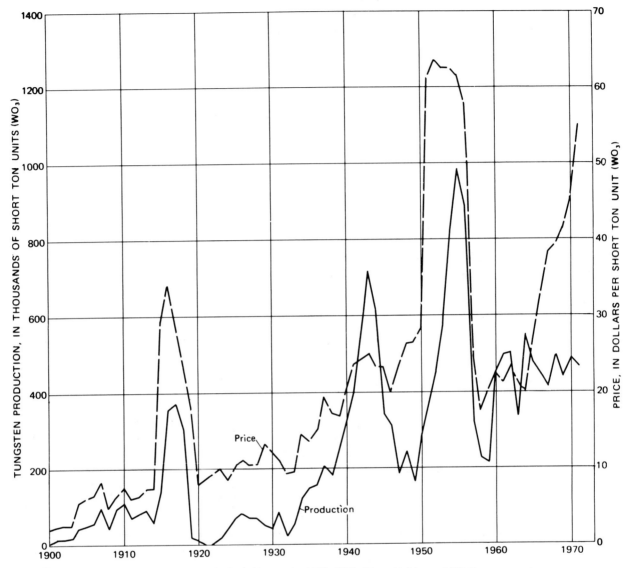

Figure 20-35 United States production and price of tungsten, 1900–1971. (From Hobbs and Elliott, Tungsten, U.S. Geol. Survey Prof. Paper 820, 1973.)

they are converted chemically to tungstic oxide and reduced to metal by heating in hydrogen or carbon. Ferrotungsten (75 to 80 percent W) is produced in the electric furnace by heating with carbon and is added to steel in this form.

Types of Deposits and Origin.

The types of commerical deposits and examples are

Contact metasomatic: Mill City, Long Mine, Tempiute, and Osgood Range, Nevada; and Pine Creek and Darwin, California

Quartz veins and pegmatites: Oreana, Nevada; Atolia, California; Boulder district, Colorado;

Ima, Idaho; Boriana, Arizona; Burma; and especially southeastern China.

Stockworks: Gulch, and Great Creek, Montana; Henderson and Climax, Colorado; Hamme, North Carolina; and Bolivia.

Placers: Kiangsi, China; Burma; and Atiola, California

Tungsten exists in a wide variety of environments not only those listed here, but presumably also in solution, for example, in the brines of Searles Lake, hot springs, in soils, and from other unsuspected sources. Tungsten is apparently carried in ore-forming solutions as tungstic acid, sodium tungstate, or a heteropolyacid, the relative amounts being controlled by pH, P and T conditions, and the silica content, according to Krauskopf. Precipitation is affected by the relative amounts and chemical potential of iron, manganese, calcium, carbon dioxide, and fluorine.

The tungsten-bearing mineral huebnerite is removed as a by-product at Climax, Colorado, even though the grade is a few hundredths of 1 percent. Tungsten as a by-product metal is rare, but in deposits similar to Climax it has become a potentially important source.

Tactite is a dark-colored, contact-metasomatic rock formed by metasomatism of impure limestone or dolomite. Various minerals are formed in these zones including garnet; epidote; hornblende; tremolite; and very commonly, scheelite, the calcium tungstate mineral. The content of scheelite in such zones is low, generally a few tenths of 1 percent WO_3. One of the Mill City deposits contained 10 percent although the total deposit averaged about 1 percent.

EXAMPLES OF DEPOSITS

United States. At *Mill City, Nevada,* are four important mines, the Stank, Humboldt, Sutton, and Springer. The deposits are all similar and, according to Kerr, are "veinlike" contact-metasomatic deposits produced in steeply dipping, thin limestone members of a siliceous sedimentary series by a post-Jurassic granodiorite intrusive. The ore bodies occur within 600 meters of the intrusive and are 1.5 to 2 meters thick, swelling in places to 60 meters where faulted. The strike length ranges from 150 to 300 meters and the present depth is 400 meters. The ore beds consist mainly of garnet, epidote, and quartz, in which are scattered small grains of scheelite and a few rich concentrations. Other contact-metasomatic silicates and common sulfides are present. The scheelite is white to pale brown in color and shows up strikingly underground with ultraviolet light. The tenor averages 1 to 1.5 percent WO_3.

At *Oreana, Nevada,* a unique and important scheelite deposit is described by Kerr as of pegmatitic mineralization. There are two modes of occurrence. In one the tungsten is in vertical pegmatite dikes that terminate at shallow depth. In the other it occurs in a string of pegmatitic lenses (Fig. 20-36), within intrusive metadiorite parallel to its inclined contact with limestone. The pegmatites are associated with post-Jurassic granitic intrusives. The pegmatite dikes are irregularly mineralized, mainly with beryl, oligoclase, and masses of fluorite and quartz; scheelite masses are several feet across. The lenses are mainly scheelite, pholgopite, and oligoclase. Other minerals of this unusual ore are orthoclase, albite, rutile, tourmaline, and some pyr-

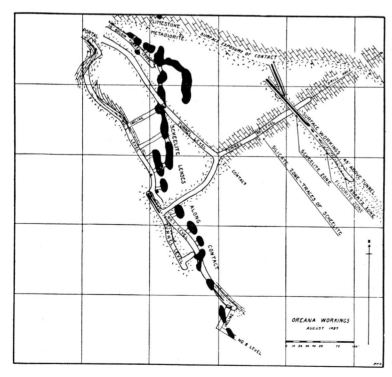

Figure 20-36 Scheelite ore bodies along limestone-diorite contact disclosed underground, Oreana, Nevada. (Kerr, Econ. Geol.)

ite and pyrrhotite. The ore is exceptionally rich. The first shipments ran 30 percent scheelite; the subsequent average is about 5 percent.

At *Silver Dyke, Nevada,* near Mina, are scheelite fissure veins filled by scheelite and comb quartz, with albite and minor sulfides. The veins occur along the contact of Triassic volcanics and intrusive diorite that has been hydrothermally altered and kaolinized and cut by albitized dikes. The steeply inclined veins are 5 to 60 centimeters wide and occur in a zone 8 to 60 meters wide; intervening plates of rock have been replaced by quartz, scheelite, and albite up to widths of 5 meters. Workings extend to 250 meters. The deposits are epithermal to mesothermal fissure fillings.

Before, during, and after World War II, the U.S. Geological Survey carried out a tungsten Strategic Minerals program with the object of increasing tungsten production and reserves within the United States. The program concentrated in the western part of the country, especially in Nevada and California. During low prices of tungsten, few of the tungsten mines of Nevada are in operation.

Near *Bishop, Calif.,* the Pine Creek mine is an extensive vertical pipelike deposit of scheelite that was developed as pods over a vertical distance of more than 1800 meters in a deep septum of metamorphosed limestone. The septum consists of skarn or tactite rock with quartzite and slates although the scheelite is confined to the metamorphosed limestone. One of the hundreds of Jurassic stocks that form the Sierra Nevada batholith is the source of the mineralization. A tactite zone of scheelite mineralization forms the Brownstone mine (Fig. 20-37), where the ore is associated with the intrusive and not the late alaskite veins.

Union Carbide, which owns the Pine Creek mine, began developing the *Tempiute, Nevada,* tungsten property after World War II and initiated production in 1976. This mine and mill will increase U.S. tungsten production to a major extent, especially if the present price of about $300/unit persists. Two small Laramide stocks, each about one-half mile in diameter are associated with the deposit. A tactite zone has formed on the western border of the southernmost intrusion, where the scheelite ore is concentrated.

Brazil. The discovery of rich scheelite ores in northeastern Brazil in 1942 has made Brazil the second largest tungsten-producing country in South America. About 60 scheelite localities are known, in most

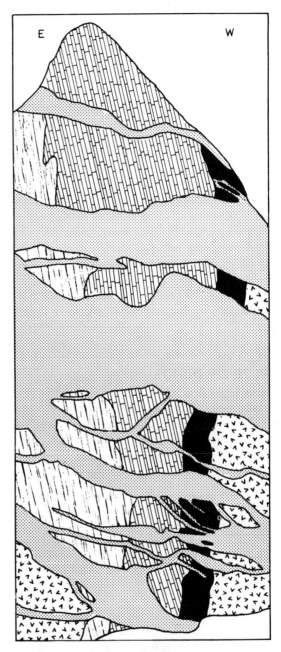

EXPLANATION

Figure 20-37 Vertical cross section through the Brownstone mine, California. (From J. Ridge, ed., *Ore Deposits of the Western U.S.,* 1968.)

of which the deposits are of contact-metasomatic origin, where granitic Precambrian rocks intrude schists and limestone. Tactites and contact-metasomatic minerals enclose scheelite, molybdenite, and bismuth minerals. Some scheelite also occurs in quartz veins. Cassiterite is rare.

Other Deposits. Economic tungsten deposits occur at *Yellow Pine, Idaho* (now closed), along with gold and antimony in a shear breccia; in highly productive scheelite-quartz veins at *Atolia, California;* in ferberite-wolframite veins at *Boulder, Colorado,* along with chalcedony and wurtzite with structures suggesting gel deposition; in tactites at *Tungsten City, California,* and *Nightingale, Nevada;* in an unusual blanket of tungsten-bearing limonite and manganese of hot spring origin at *Golconda, Nevada;* in quartz veins carrying huebnerite and scheelite in the *Hamme District, North Carolina;* and in the *Paradise Range, Nevada,* where a one meter vein carries scheelite and leuchtenbergite.

Bolivia yields considerable tungsten ore from the fissure veins in the tin belt carrying quartz, wolframite, some scheelite, common sulfides, and rarely cassiterite. The occurrence is similar to the tin veins already described. In *Argentina* are many shallow tungsten quartz veins carrying wolframite, with minor scheelite and ferberite, and with bismuthinite, molybdenite, common sulfides, and notable amounts of niobium and tantalum minerals. The *Australian* production comes from contact-metasomatic, pegmatite, and vein deposits in which are wolframite and scheelite along with bismuth and molybdenum. The *Portuguese* veins yield chiefly wolframite, generally accompanied by minor bismuthinite, cassiterite, and columbite-tantalite. In the *Nelson area, British Columbia,* scheelite is won from contact-metasomatic tactites. In *Nigeria* tungsten associated with granite occurs in quartz veins with abundant columbite related to tin mineralization.

China. Tungsten production started in China in 1915 and by 1919 led all other countries. The deposits lie mainly in the Nanling region, including parts of Kiangsi, Human, and Kwangtung; in Kwangsi; and in Yunan, a major tungsten province. The important Kiangsi area contains 17 districts and 80 mining localities grouped in 5 structural zones localized in major anticlines. The deposits are steeply dipping fissure veins either in granite or in nearby invaded sediments and range up to 3 meters wide, 1000 meters long, to depths of more than 550 meters. Many pegmatite veins are also present. The ore is mainly wolframite with minor scheelite accompanied by bismuthinite, molybdenite, quartz, muscovite, topaz, fluorite, and common sulfides. Cassiterite is sufficiently abundant in many veins to yield tin concentrates. The wall rocks have been intensely altered to tactite. Tungsten ore reserves in China are about 130×10^6 short tons of WO_3.

Selected References on Tungsten
Tungsten in the United States. 1962. D. M. Lemmon and O. L. Tweto, U.S. Geol. Survey Mineral Inv. Res. Map MR-25.
Burma-Mawchi-Tavoy. 1938. J. A. Dunn. India Geol. Survey, Rec. v. 73, p. 209–245. *Detailed geology of Burmese occurrences.*
Geology of the tungsten deposits, Mill City, Nevada. 1934. P. F. Kerr. Univ. of Nevada Bull., v. 28, p. 2. *Tungsten mineralization at Silver Dyke, Nevada.* P. F. Kerr. Univ. of Nev. Bull, v. 30, p. 5, 1936. Tungsten mineralization at Oreana, Nevada. P. F. Kerr. Econ. Geol., v. 33, p. 390–425, 1938. *These three papers give details of geology and origin.*
Geology and tungsten mineralization of the Bishop district, California. 1965. U.S. Geol. Survey, Prof. Paper 470, 208 p.
General Reference 19. Chapter on tungsten.
Bolivia tungsten. 1939. C. W. Wright. Foreign Minerals Quart. v. 2, no. 4, p. 30–38, *Cassiterite-stannite veins, with silver.*
Tungsten mineralization in the United States. 1946. Paul F. Kerr. Geol. Soc. Am. Mem. 15. *Occurrences and origin of deposits and summation of all U.S. occurrences; extensive bibliography.*
Tungsten, 3rd ed. 1955. L. C. Li and C. Y. Wang. Am. Chem. Soc. Mon. 94., 506 p., Van Nostrand Reinhold Company, New York. *Comprehensive treatment of geology, uses, and recovery of tungsten. Covers world occurrences, notably those of China. Good bibliography.*
Tungsten deposits of Vance County, North Carolina. 1947. G. H. Espenshade. U.S. Geol. Survey Bull. 948-A. *Brief descriptions.*
Tungsten Deposits in the Sierra Nevada, Near Bishop, California. 1941. D. N. Lemmon. U.S. Geol. Survey Bull. 931-E. *Large contact-metasomatic deposits.*
Tungsten deposits in the Republic of Argentina. 1947. Ward C. Smith and E. M. Gonzales. U.S. Geol. Survey Bull. 954-a. *Geology of vein deposits.*

VANADIUM

The domestic consumption of vanadium increased rapidly during the 1960s, and a growth rate of demand higher than that of any other ferrous mineral has been predicted to the year 2000. Foreign uses, consumption, and predicted requirements have a similar pattern.

Vanadium toughens steel desired for axles, pistons, crankshafts, and pins, and has various other uses where strain, shock, and fatigue are involved. Such steels use 0.1 to 1.25 percent vanadium. A little added to any steel helps remove oxygen and nitrogen, and gives a uniform grain size. For high-speed steels, 4 to 5 percent vanadium is used. It is also added to chromium, molybdenum, and tungsten steels; and to cast iron, brass, and bronze. A newer and rapidly increasing use is as a catalyst, replacing platinum, a little is employed in the electrical, chemical, ceramic, paint, dye, and printing industries.

Production and Distribution.

The United States had been the major producer of vanadium until about 1960 when Russia and South Africa each produced more vanadium than the United States. Domestic production and consumption in the United States has dropped to 6000 tons yearly, because of the decrease in steel production.

Figure 20-38 indicates the production sources by countries from 1910 to 1970. Vanadium metal, with several other metals added as alloys, is a candidate

for a fuel-cladding material for liquid-metal. Because cooled vanadium is used in fast-breeding nuclear reactors, its demand could be vastly increased. Fischer has summarized the estimated demand to increase from the present 6000 tons to between 25,500 to 37,500 tons by the year 2000. Domestic reserves now are only about 115,000 tons of recoverable vanadium.

Mineralogy and Treatment.

The chief economic vanadium minerals are carnotite, patronite, roscoelite, vanadinite, descloizite, coulsonite, and montroeseite. Actually, there are more than 60 valid vanadium minerals, but most of the vanadium being produced is recovered from ores or materials in which no specific vanadium ore mineral is recognized.

Types of Deposits and Origin.

Although vanadium is considered to be a minor element, it does occur in uncommon abundance, but in only a few specific types of occurrences. It is rare in hydrothermal deposits, as it exists in the relatively insoluble trivalent state. Vanadium is concentrated in magmatic deposits, especially those

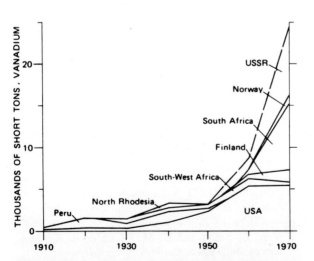

Figure 20-38 Generalized graph showing the principal geographic sources of vanadium, 1910–1970. Data from published figures of the U.S. Geological Survey (1910 and 1920) and the U.S. Bureau of Mines, except USSR data for 1960 and 1970, which are estimates. (From U.S. Geol. Survey Prof. Paper 820, 1973.)

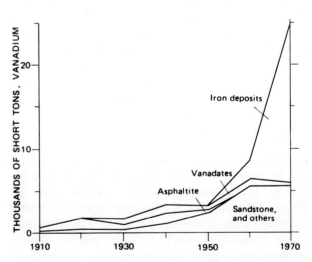

Figure 20-39 Generalized graph showing the principal geologic types of deposits from which vanadium was recovered, 1910–1970. Compilation by author, based on data in figure 75; (from U.S. Geol. Survey Prof. Paper 820, 1973.)

that are titaniferous. Figure 20-39 illustrates the principal geologic types of deposits and production from which vanadium is recovered.

During the weathering of vanadium-bearing rocks, oxidation forms the relatively soluble pentavalent form of vanadium whereby it undergoes transport and ultimate reduction and precipitation. Its occurrence in sandstone-type uranium deposits is probably the result of leaching, transport, and ultimate reduction in the more permeable and porous sandstone beds, which, of course, is similar for the process by which uranium is concentrated; hence their common occurrence.

Examples of Deposits.
By far the largest producer of vanadium at the present time is *Russia,* where all of the production is derived from the large titaniferrous *Mount Kachkanar* deposit. Possibly, some is also derived from similar ores at *Kusinkoe* and *Pervouralakoe.* Reserves at Mount Kochkanar exceed 13,580,000 metric tons of 0.10 to 0.12 percent vanadium.

Similar titaniferrous magnetite deposits containing recoverable vanadium occur at Otammaki, Finland; Rodsand, Norway; and at El Romeral, Chile. The latter is a nontitaniferrous magnetite deposit.

One of the world's greatest producers is at *Mina Ragra, Peru,* where vanadium occurs in a mass 10 meters wide and 120 meters long enclosed in shales and limestone intruded by porphyry dikes. The ore was mined from 1907 to 1955 and consisted of a red calcium vanadate (50 percent V_2O_5), and the low-grade material was calcined and burned to an ash containing 22 percent V_2O_5.

The *United States* production is obtained from (1) vanadium-bearing carnotite deposits in Colorado and Utah, (2) the so-called roscoelite ores of Colorado and Utah, (3) oxidized silver-lead-molybdenum ores in Arizona, and (4) as by-product vanadium from the treatment of sedimentary phosphate rock in Idaho, Utah, and Wyoming.

The uranium "boom" of the 1950s allowed the recovery of vanadium in such large amounts that the United States was an exporter of this metal. Typically, the vanadium occurs in Mesozoic continental stream deposits of sandstones several tens of meters thick and several kilometers long. It is most likely that oxidation of the vanadium has allowed its transport epigenetically to reduction sites where it became concentrated.

Selected References on Vanadium
Broken Hill, Rhodesia. 1929. XV Int. Geol. Cong. Guidebook C22, p. 13–16, Pretoria, South Africa. *Vanadium is a by-product of oxidized silver-lead ores.*
Tsumeb, Southwest Africa. 1929. XV Int. Geol. Cong. Guidebook C21, p. 38–45, Pretoria, South Africa. *An unusual copper deposit with vanadium.*
Lower Paleozoic vanadiferous basin of the Northern Tyan-Shan and the western part of Central Kazakhstan Pt. L. 1961. S. G. Ankkemovich. Alma-Ata, Kazakh, SSR, Akoud. Nauk Kazakh. Izadatel, p. 99–123. *Russian deposits.*
Bibliography on the geology and resources of vanadium to 1968. 1970. R. P. Fischer and J. O. Ohl. U.S. Geol. Survey Bull., v. 1316, p. 168.
Elranadio en el Peru. 1957. S. H. Aguije. Reviewed by Hugh McKinstry. Econ. Geol., v. 53, p. 324–325.
Vanadium deposits of Colorado and Utah. 1942. R. P. Fischer. U.S. Geol. Survey Bull. 936-P. *Tabular bedded deposits in sandstones.*
Vanadium deposits near Placerville, Colorado. 1947. R. P. Fischer, J. C. Haff, and J. F. Rominger. Colo. Sci. Soc. Proc., v. 15, p. 115–132.
Vanadium. H. D. Taylor. In Minerals Yearbook, U.S. Bur. Mines, 7 p., 1973.
General Reference 19. R. P. Fischer. *Vanadium.*

COBALT

Cobalt is another minor element used for carbides and special steels and desired particularly for magnet steel (35 percent of use), stellite steels for metal cutting, and temperature-resisting alloys. It is desired also for high-speed steels, dies and valve steel, welding rods, carbide-type alloys, and corrosion-resisting steels. Its ability to maintain strength at high temperatures makes it sought for jet airplane engines. It is also used for making the blue color in glasses and enamels, as a catalyst, and in paint driers.

The total production amounts to about 39 million pounds annually, of which Zaire is the largest producer followed by Zambia; and copper, nickel, and iron ore deposits associated with mafic rocks in Canada and Finland. The United States has been a small intermittent producer for more than 100 years. The Gap nickel mine, Lancaster County, Pennsylvania, was an earlier producer along with production from the southeast Missouri, Mine LaMotte, and Fredericktown areas. Domestic production was once derived almost solely as a by-product of magnetite production at Cornwall, Pennsylvania, and the Grace mine near Morgantown, Pennsylvania.

Selected References on Cobalt

Cobalt. 1948. R. S. Young, Amer. Chem. Soc. Mon. 108, p. 181. Van Nostrand Reinhold Company, New York. *Occurrence, uses, properties, and metallurgy.*

General Reference 19. Chapter on cobalt.

Cobalt. In minerals Yearbook, v. 1-2, U.S. Bur. Mines, p. 397–402, 1971.

Cobalt. 1970. In Mineral Facts and Problems, U.S. Bur. Mines Bull. 650, p. 263–274.

OTHER MINOR FERROALLOYS

Titanium is also used as a minor alloy for steel, but it is dominantly a pigment mineral and will be described elsewhere. *Ferrosilicon*, with 7 percent silicon, is used for electrical machinery. *Ferrophosphorus, silicomanganese zirconium,* and *calcium-molybdenum* are other minor alloys.

21

MINOR METALS AND RELATED NONMETALS

One sees the past better than it was;
One finds the present worse than it
is; One hopes for a future happier
than it will be.

A. A. Brant

The minor metals are used in relatively small quantities, yet several of them play an indispensable role in modern industry, such, for example, as antimony for type metal or mercury for electrical apparatus. Several are by-products of other ores and are not mined separately.

ANTIMONY

For a minor metal, antimony has many diversified and indispensable uses both for normal industry and for military use. Its striking property of expanding, rather than contracting, upon cooling makes it desirable for making type metal alloy, which when cast does not change size. Its main use is to impart stiffness and hardness to various lead alloys, thus permitting lead to be used for purposes not otherwise possible. Antimonial lead or hard lead contains 1 to 25 percent antimony, produced by alloying the two metals or by smelting mixed antimony and lead ores.

The chief uses for antimony-lead alloys are storage battery plates, sheets and pipe, sheathings for electrical cables, collapsible tubes (toothpaste), foil, bullets, type metals (10 to 15% Sb), solder, antifriction bearings, and as an intermetallic compound for semi conductors.

"White metal" alloys include Britannia metal (Pb-Sb-Cu), pewter (Pb-Sn-Sb), Queen's metal (Sn-Sb-Cu-Zn), and Sterline (Cu-Sb-Zn-Fe).

Metallic antimony is little used as such except for ornamental castings and bric-a-brac, but compounds are used for flame-proofing, pigments, enamels, safety matches, glass, vulcanizing, and medicines. Military uses include shrapnel balls and bullet cores, detonating caps, and bursting charges of shrapnel to yield the dense white marker smoke.

Production and Distribution.

Antimony production in the United States ranges from about 450 to 900 metric tons during peace time but reached a peak of more than 4,500 metric tons during World War II, predominantly from the Yellow Pine mine, Idaho, as shown in Fig. 21-1.

Most of the antimony used in the United States is imported from South Africa, Mexico, and Bolivia. Although the United States consumes about 40 percent of the primary production of the world's supply of antimony, it produces only 15 percent of the antimony consumed and most of this is recovered as a by-product from silver, copper, and lead-zinc ores mainly from the Sunshine silver mine, the Yellow Pine mine, and the Coeur d'Alene district, all in Idaho. Additional amounts are recovered from antimonial lead from lead smelters.

Identified resources in the United States are estimated at 90,000 metric tons according to Miller with world resources at 5,000,000 metric tons located predominantly in China, Bolivia, Russia, South Africa, and Mexico. China has close to 70 percent of the world reserves. Minor amounts occur in Hungary, Austria, Yugoslovia, Italy, Turkey, and Morocco. The United States production comes from Idaho, Alaska, California, and Nevada. Antimony is found with lead ores, and the hope of future production is from the Mississippi Valley type of lead deposits.

Mineralogy, Tenor, and Treatment.

There are many minerals of antimony, but stibnite and lead ores yield most of the commerical metal. Native antimony and oxidation products, such as cervantite and sinarmonite, contribute a little. Ores range in tenor from 3 to 8 percent. Ores are hand-

443

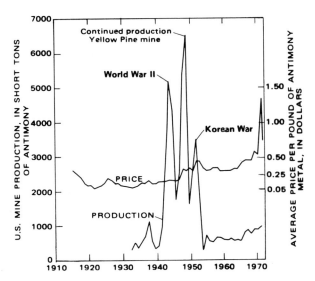

Figure 21-1 Price and production of antimony in the United States, 1916–1970. (From Antimony, 1973, U.S. Geol. Survey Prof. Paper 820, p. 96.)

sorted or concentrated; the sulfide product is enriched by liquidation, whereby the low-temperature melting point of stibnite permits the melt to drain off from unfused enclosing substances. This produces crude stibnite. Low-grade ores are roasted to the oxide. Antimony metal is produced by smelting the oxide or the crude or high-grade sulfides in a blast furnace with iron, which unites with the oxygen or sulfur, freeing the antimony. Oxide and electrolytic antimony are recovered from tetrahedrite by the ''Lee-Muir'' electrolysis process, which separates the components.

Types of Deposits and Origin.

Most antimony deposits are formed by hydrothermal solutions at low temperatures and shallow depth, giving rise to filled fissures, joints, and rock pores and to irregular replacement deposits. Some primary deposits have been enriched in oxidized products through residual weathering. Antimony occurs in epithermal veins, pegmatites, and replacement and hot spring deposits.

Antimony deposits occur principally as either stibnite or native antimony in a siliceous gangue commonly with some pyrite. Complex antimony deposits consist of stibnite associated with pyrite, arsenopyrite, cinnabar, or scheelite with varying amounts of copper, lead, silver, as well as the common sulfides of these metals and zinc.

Cinnabar and stibnite are often deposited together. This is the result of the epithermal or hot spring origin of mercury and antimony deposits. According to Dickson and Tunnel, stibnite is soluble enough for transport in H_2O, in boron, and sulfide solutions.

China. Even though China has by far the largest antimony reserves of the world. South Africa (23 percent) and Bolivia (21 percent) have placed China (17 percent) in third place in production followed in order by Russia (10 percent). The U.S. production at the present time is less than 1 percent of the world total.

China yields annually about 10,000 metric tons of antimony metal and for years yielded two-thirds of the world's supply. The principal deposits are in Hunan, near Changsha. Other deposits occur in Yunan, Kwenchow, Kwangtung, Szechwan, and Kwangsi. The largest field is the Hsi-Kuang-Shan, Hunan, where reserves are in excess of 1.5 million metric tons of metal. The Panchi is the largest mine. Here, in a belt 5.5 × 3 kilometers, stibnite deposits occur in deformed Silurian sandstone in veins, seams, pockets, and lenses, with the richest deposits, according to Schrader, beneath anticlinal structures. The deposits are of the simple type ore, with tenors of about 5 percent, but is hand-sorted to 55 or 60 percent.

Mexico. The chief deposits are in San Luis Potosi, Oaxaca, and Queretaro, with minor ones in Sonora and other states. These deposits, which consist of stibnite with some oxide minerals, occur in veins in limestone intruded by porphyry. In Hidalgo, jamesonite replacement deposits occur in Cretaceous limestone cut by monzonite. A unique deposit of livingstonite ($HgSb_2S_7$) carrying 1 to 2 percent Sb and 0.3 percent Hg is mined at Huitzuco, Guerrero. Considerable antimonial lead is produced at local lead smelters.

Bolivia. In Bolivia large reserves occur in a 250 kilometers belt from Lake Titicaca to Atocha, and the chief centers are Uncia and Porco. The high-grade deposits are shallow quartz veins in Paleozoic shale containing stibnite and base-metal sulfides.

Most of the deposits of Mexico, Bolivia, Peru, China, South Africa, Yugoslavia, Algeria, Hungary, Czechoslavakia, Italy, and Japan are of the simple type of antimony deposits.

Algeria. A mineralized belt across Algeria from Morocco to Tunis contains, near Constantine, antimony veins in limestone with primary stibnite and the oxidized minerals cervantite, senarmontite, and valentinite.

Other Deposits. Stibnite veins occur at Loznica, Yugoslavia, and very rich, narrow Tertiary veins of stibnite and gudmundite are formed with nickel minerals at Turhal, Turkey. Low-grade ores occur in Slovakia, and minor deposits are in Peru, Japan, Russia, Argentina, Germany, Morocco, and other places.

Selected References on Antimony
Antimony deposits of Tejocotes, Oaxaca, Mexico. 1947. D. E. White and R. Guiza, Jr. U.S. Geol. Survey Bull. 953-A. *Occurrence of rich oxide and stibnite deposits.*

San José antimony mines, Wadley, San Luis Potosí, Mexico. 1946. D. E. White and Jenaro González R. U.S. Geol. Survey Bull. 946-E. *Most important antimony district of Mexico.*

Antimony deposits of Yellow Pine district, Idaho. 1940. D. E. White. U.S. Geol. Survey Bull. 922-I. *Geology of a large shear-zone deposit with gold and tungsten.*

General Reference 19. Chapter on antimony.

General Reference 11. *Mercury and antimony deposits associated with active hot springs in the Western United States.*

ARSENIC

Arsenic metal is little used except for a few alloys, such as with lead for making shot. However, in the form of calcium and lead arsenate and Paris green it is widely used as an insecticide. White arsenic (As_2O_3) is a common weed killer and extensively employed. It is, of course, toxic to man. Arsenic compounds are also used to help fuse glass; as a wood and leather preservative; and in dyes, pigments, medicines, fireworks, and chemicals.

Production and Distribution.
The United States uses about 20,000 to 30,000 tons per year of arsenic but produces only about one-tenth of this amount. Imports chiefly from Sweden, France, and Mexico provide the needed imports. Other producers are Belgium, Australia, and Japan. Sweden could produce enough for the whole world.

Types of Deposits and Origin.
No ore deposits of arsenic are being mined solely for arsenic. The metal is recovered entirely as a by-

product. Some arsenical gold deposits were the first producers of arsenic, especially in the United States, but were replaced by other deposits, which Gualtieri has listed in the order of importance as follows:

1. Enargite-bearing copper-zinc-lead deposits.
2. Arsenical pyritic copper deposits.
3. Native silver and nickel-cobalt arsenide deposits.
4. Arsenical gold deposits.
5. Arsenic sulfide and arsenic sulfide gold deposits.
6. Arsenical tin deposits.
7. Other deposits.

The arsenic-rich Boliden, Sweden, copper deposit is the largest source and producer of arsenic in the world.

There are more than 25 different arsenic minerals, most of which are either arsenides or sulfides. The more common primary minerals are arsenopyrite (FeAsS), löllingite ($FeAs_2$), smaltite ($CoAs_2$), niccolite (NiAs), tennantite ($Cu_8As_3S_7$), enargite (Cu_3AsS_4), and proustite (Ag_3AsS_3).

Selected References on Arsenic
General Reference 19. Chapter on arsenic.

The Boliden sulfide deposit—A review of geoinvestigations carried out during the lifetime of the Boliden mine, Sweden, 1924–1967. 1970. E. Graip and A. Wirstam. Sreviges Geol. Undersoknig Arsbok, v. 64, p. 68.

Arsenic, in mineral facts and problems. 1970. J. Paone, U.S. Bur. Mines Bull. 650, p. 479–487.

General Reference 3. 1955. Geochemistry of arsenic. H. Onishi and E. B. Sandell.

BERYLLIUM

Beryllium is a light metal (specific gravity = 1.85) that imparts high strength and fatigue resistance to copper; also to cobalt, nickel, and aluminum. It is so light that an airplane engine constructed of beryllium could be lifted by one man. The addition of $1\frac{1}{2}$ to 3 percent to copper increases the tensile strength from 33,000 to 200,000 pounds/in² and gives it fatigue resistance greater than that of spring steel; vibrator springs stressed by 2 billion vibrations show no deterioration. The alloy finds use for various types of instrument springs, control parts, valves, and airplane carburetors and instruments.

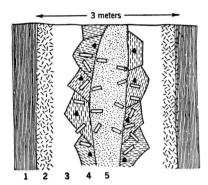

Figure 21-2 Morada Nova beryl mine, Brazil. 1, schist; 2, muscovite zone; 3, pegmatite; 4, microline with beryl crystals; and 5, quartz core. (After W. D. Johnston, Jr., Geol. Soc. Amer.)

The alloy is nonsparking, making it desirable for use in explosive and petroleum industries. The oxide is used as a refractory. The metal finds application in X-ray tubes, fluorescent lamps, neon signs, and cyclotrons. Alloys with nickel, cobalt, and aluminum are commanding attention. It is an important fatigue and heat resisting metal in space ships and the nuclear reactor industry.

Some of the metal is obtained from beryl, which occurs chiefly in pegmatite dikes (Fig. 21-2) and rarely as a gangue mineral of tungsten and tin ores. The output is about 1000 tons of beryllium annually, and the supply is limited. The metal is extracted from beryl with difficulty by electrolysis in a fluoride bath containing sodium and barium.

Even though the United States once imported more than 90 percent of its required beryllium, three-fourths of which came from Brazil as beryl, exploration for nonpegmatitic deposits has resulted in some success in the western United States. The development of the disseminated type of deposit adjacent to the Thomas Range, Utah, topaz-bearing rhyolite tuff has resulted in commerical low-grade deposits of the beryllium-bearing minerals bertrandite and phenakite. This is the largest source of beryllium in the western world, containing about 0.5 percent BeO.

The success of the discovery has led to exploration in other areas (Fig. 21-3) especially where highly siliceous, alkalic igneous rocks occur. A neutron activation field instrument is most helpful in locating beryllium mineralization.

Utah is the world's largest producer of beryllium and is now receiving beryl from Africa and South America for production of BeO at the Utah plant.

Selected Reference on Beryllium

Beryl-tantalite pegmatities of northeastern Brazil. 1945. W. D. Johnston, Jr. Geol. Soc. Am. Bull. v. 56, p. 1015–1069. *The largest beryl district of the world; in pegmatites.*

General Reference 19. Chapter on beryllium.

Geochemistry of Beryllium. 1966. A. A. Baus. W. H. Freeman and Company, Publishers, San Francisco, p. 401.

Beryllium resources of the tin-Spodumene belt, North Carolina. 1964. W. R. Griffiths. U.S. Geol. Survey Circ. 309.

Beryllium content of volcanic rock. 1966. D. R. Shawe

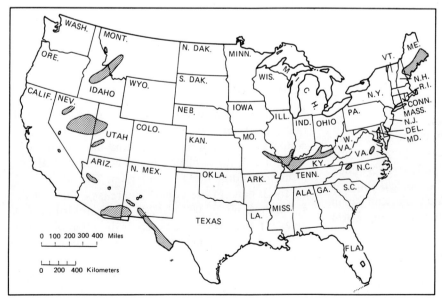

Figure 21-3 Areas of the coterminous United States in which beryllium deposits are most likely to be found. (From U.S. Mineral Resources, 1973, U.S. Geol. Survey Prof. Paper 820, p. 92.)

and S. Bernold. U.S. Geol. Survey Bull, v. 1214C, p. 11.

Geology and mineral deposits of the Thomas and Dugway Ranges, Juab and Tooele Counties, Utah. 1964. M. H. Staatz and W. J. Carr. U.S. Geol. Survey Prof. Papers 415, p. 188.

BISMUTH

Bismuth is a metal consumed in small quantity, but certain peculiar properties create important uses for it. The metal possesses curative medicinal properties, has a low melting point, and expands upon solidifying. The greatest use of bismuth is for medicinal and cosmetic preparations. Its salts cure wounds and soothe digestive disorders and also form insoluble compounds for internal X-ray examinations. Their smoothness makes them desirable for cosmetics. Bismuth salts are also used to give a glaze to porcelain, for enameling, printing fabrics, and making optical glass.

Bismuth forms various alloys with lead, tin, cadmium, and antimony, and its alloys will melt far below the melting points of the separate metals. For example, bismuth melts at 271°C, but Wood's metal (Bi-Sn-Cd) melts at 60°C and other alloys at even lower temperatures. Such a metal, therefore, will melt in hot water or in a very hot room. Consequently, the alloys are used for automatic water sprinklers for fire protection. When a room heats up, the plug melts and water is released. They are similarly used for electric fuses, boiler safety plugs, and hand grenades. The expansive qualities of bismuth cause it to be used for molds, for making casts of delicate objects, and for electrotypes. Combined with brass and bronze, it gives antifriction metals. Bismuth wire is used in electrical apparatus and delicate measuring instruments and in the production of some selenium rectifiers. There are also innumerable other minor applications.

Production and Distribution.

The annual domestic output is only from about 500,000 to 600,000 pounds of metal. Consumption varies greatly, having reached a peak of 2,227,000 pounds, 1,500 pounds in 1971. World production (excluding China and Russia) is about 4,000,000 pounds of bismuth metal. The major source of imports for the United States are Mexico, Peru, and Japan with minor production from Germany, Spain, Australia, and Bolivia.

Mineralogy, Ores, and Treatment.

The commercial ore minerals of bismuth are native bismuth, bismuthinite, and bismuth ochre. Few deposits are mined for bismuth alone. It is obtained (1) from deposits of other metals such as tin, copper, or silver, and (2) as a refinery by-product, chiefly from lead ores.

Refinery production of bismuth in the United States is mainly from the American Smelting and Refinery Co. in Omaha, Nebraska, and the Southern California Chemical Co., Los Angeles, California.

Types of Deposits and Origin.

Lead deposits are the major source of bismuth. Such deposits from North Carolina to Maine contain 0.05 ppm to as much as 1 percent according to Hasler, Milten, and Chapman. Some of these deposits are red-beds copper deposits. Not all deposits are related spatially to intrusive rocks and those that are may be related to Precambrian to Devonian granitic bodies in the eastern United States. Some bismuth has been reported in gold placer and black sand deposits in Alaska and the southwestern United States. Bismuth is also related to sandstone-type uranium deposits but not in commerical amounts.

Lead-zinc replacement deposits in limestone have probably been formerly the most important sources of bismuth in the United States. Such deposits are those at Leadville and Colman, Colorado; Tintic and Park City, Utah; Darwin, California; and Patagonia, Arizona. Certain deposits in the Coeur d'Alene, Idaho district; San Juan area, Colorado; and other hypogene deposits of the western United States contain some bismuth.

The large resources of *Bolivia* consists of bismuth minerals in the tin deposits of Potosi and La Paz, and the Aramayo mines at Tasna, Ukeni, and Chorolgue. The bismuth averages about 1 percent. Richer bismuth ores are obtained from the Tasna mine, where a hand-sorted product contains 25 percent Bi.

In *Peru*, concentrates from the San Gregorio mine carry 20 percent Bi and the Cerro de Pasco ore yields bismuth in smelter flue dust. The bismuth of *Mexico* is contained in the lead ores, as in Saxony, *Germany*. In *Australia* and *Japan* bismuth is obtained from tin, tungsten, and molybdenum ores. In Cordoba, *Spain*, rich bismuth ores are mined in small quantity from fissure veins; in one near Torrecampos the ore carries 25 percent Bi; those at

Azuel carry 1 to 7 percent Bi along with nickel, cobalt, and silver.

Selected References on Bismuth

Bismuth in Bolivia and Peru. 1940. C. W. Wright. Foreign Minerals Quart., v. 2, No. 4, and v. 3, No. 1. *Information from first-hand observations.*

General Refernces 13 and 19.

Bismuth. 1971. D. A. Bàncel. Eng. and Mining Jour., v. 172, p. 115–116.

Bismuth. 1972. D. Cook. Eng. and Mining Jour., v. 173, p. 99–1015.

Bismuth occurrences in Alaska. 1970. E. H. Cobb. U.S. Geol. Survey Mineral Res. Map. MR-53.

CADMIUM

Cadmium is entirely a by-product metal that substitutes for zinc in zinc minerals. It has a low melting point, is ductile, is softer than zinc, and emits a peculiar sound when bent. It alloys readily with other metals, and this is its chief use.

Alloyed with nickel, or silver and copper, it forms a high-pressure, antifriction-bearing metal for automobile bearings. It is a constituent of stereotype plates and of many low-melting alloys described under bismuth. Cadmium hardens copper; it makes silver resistant to tarnish; with silver it makes tableware metal; and it is used for making green gold.

An important application is in cadmium plating, particularly on iron, where it forms a thin, rustless surface alloy. Thus automobile bolts, nuts, locks, and other parts are made nonrusting. It also serves as an adhering bond between iron and other plating metals.

Cadmium compounds find use in chemicals, photographic materials, phosphorus for TV picture tubes, paint, rubber, soaps, fireworks, textile printing, and as pigment for glass and enamels. The sulfide forms an enduring yellow paint.

Production and Distribution.

World production of cadmium is about 17,000,000 tons with the United States producing about Nil percent. Other major producers are Canada, Mexico, and Australia with lesser amounts from Germany, Norway, and Southwest Africa. Cadmium comes from zinc regions, since its abode is with zinc ores, and minor production comes from other zinc regions.

Minerals, Ores, and Treatment.

The ore minerals of cadmium are the sulfides greenockite (=xanthochroite), the carbonate otavite, and the oxide cadmium oxide. They occur in association with sphalerite ores and their oxidized equivalents. The cadmium is volatilized from them, or is recovered from dust chambers and in electrolytic slimes.

Examples of Deposits.

Since cadmium is a by-product, there are no deposits worked for cadmium alone. The zinc ores of the Tri-State district yield most of the United States production. Some also is recovered from the zinc ores of Butte, Montana, and of Utah.

Selected References on Cadmium

General References 13 and 19.

Cadmium in mineral facts and problems. 1970. R. A. Heindl. U.S. Bur. Mines Bull. 650, p. 655–570.

LITHIUM

There are several different types of lithium deposits such as spodumene pegmatites, greisan bodies, low temperature veins, and evaporite deposits. The evaporite deposits have taken most of the market away from spodumene deposits. The Great Salt Lake brines contain 34 to 58 ppm of lithium, which is not yet being recovered from treatment of the bitterns remaining from operations of the Great Salt Lake Minerals and Chemical Corporation at Little Mountain, Utah. Searles Lake, California, brine contains 0.015 percent Li_2O by weight whereas the Silver Peak, Nevada, subplaya deposits average about 0.03 percent Li_2O.

The need for lithium is increasing rapidly as a result of the needs of the ceramic and glass industries, but the major potential is for the lithium battery.

World-proved and probable reserves of lithium, most of which are in the United States are 1,200,000 tons or about 400 times the 1972 consumption.

Selected Reference on Lithium

Lithium in mineral facts and problems. 1970. A. M. Cummings. U.S. Bur. Mines Bull. 650, p. 1073–1081.

General Reference 19.
Lithium resources of North America. 1955. J. J. Norton and D. M. Schlegel, U.S. Geol. Survey Bull. 1027-G, p. 325–350.

MAGNESIUM

Because magnesium is the lightest metal known (specific gravity 1.74) and is fairly strong, it is in demand for the light alloys that are used extensively in airplanes and automobiles and many other materials requiring lightness. Its alloys are noncorrosive in air but not in seawater. The chief alloying element is aluminum, generally with some zinc and manganese. Such alloys have low weight for unit bulk, high rigidity, and greater strength than most aluminum alloys. They are also used for microscope mountings, field glasses, cameras, surveying instruments, artificial limbs, shuttles, bobbins, musical instruments, and so forth. A little magnesium hardens lead for cable sheaths. A remarkably light alloy consists of magnesium and beryllium.

Other important uses of magnesium are for structural shapes and sheets, and for deoxidizing and desulfurizing nickel, Monel metal, brass, and bronze. Because it burns at low temperatures with a strong actinic light, it is used for flashbulbs, fireworks, and signal flares. Incendiary bombs are filled with 93 percent Mg and 7 percent Al.

Production and Distribution.
World production of metallic magnesium is currently about 130,000 tons, about 50 percent by the United States.

Sources and Treatment.
Magnesium is manufactured from (1) natural brines, (2) seawater, (3) magnesite, (4) dolomite, and (5) brucite. Consequently, supplies are available almost everywhere. Considerable amounts are obtained from brines; $Mg(OH)_2$ is precipitated from seawater by the addition of calcined dolomite (as at Moss Landing, California) or calcined oyster shells (Texas).

Brines and potash salts contain $MgCl_2$. Metallic magnesium is obtained from melted $MgCl_2$ by electrolysis. With magnesite, the metal is produced by (1) electrolysis of $MgCl_2$ obtained by chlorination

and (2) reducing calcined magnesite with carbon in an electric furnace and chilling the metallic vapor. The impure metal is redistilled in a second furnace and condensed.

The major source of magnesium is from the evaporation of brine waters, especially from large solar evaporation ponds such as around the edge of the Great Salt Lake, Utah, (Fig. 21-4). Its brines with a dissolved mineral content have a gross market value of about $75 million. The lake contains one of the highest magnesium contents in North America; is rich in potassium, sodium, and sulfates; and contains significant and economically recoverable amounts of boron, bromine, and lithium. These brine deposit operations are still in the development stage but may become the primary source of domestic magnesium production in the next few years.

The *Gabbs, Nevada*, deposit has been the primary domestic producer of magnesite. The deposit was discovered in 1927, drilled in the 1930s, and was expanded as the result of the demand for magnesium metal for World War II, as it makes up 93 percent of incendiary bombs.

The United States War Production Board authorized construction of the world's largest magnesium plant at Henderson, Nevada, located about 15 miles from the Hoover Dam's electrical power source. Much of the magnesite ore that was reduced to magnesium metal at this plant was derived from the Gabbs deposit.

This deposit is large, having proved or positive reserves of 27 million tons of magnesite containing less than 5 percent CaO, and about 18 million tons containing between 5 to 26 percent CaO. Production has amounted to more than 85,000 tons of ore.

The magnesite bodies are scattered over an area of about a square mile in juxtaposition to a prong of a granodiorite stock. Mineralization occurs in the Triassic Luning formation that is a schistose, dark, fine-grained, regionally metamorphosed dolomite that has been marmorized to a light-colored, coarse-grained, massive dolomite that is intergrown with magnesite.

It has been suggested that the magnesite is sedimentary and formed contemporaneously with the dolomite in an enbayment of the Triassic sea. With the later intrusion of the granodiorite stock, there is evidence that the magnesite bodies were modified by the intrusion. Others believe that the deposits formed epigenetically by hypogene solutions derived from a deeper portion of the granodiorite intrusion.

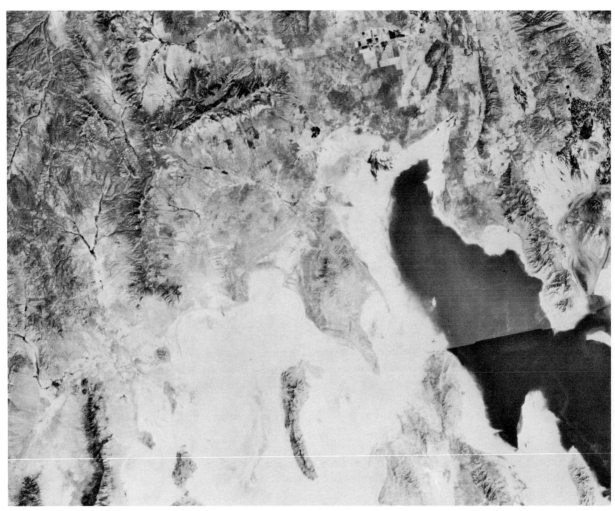

Figure 21-4 Landsat 1 image of Great Salt Lake and salt flats west of the lake, Utah. Evaporation plants exist around Great Salt Lake and near Wendover, Utah, for the recovery of various salts of metals magnesium, potassium, sodium, and lithium. The Southern Pacific railroad has an earth fill crossing over the lake with only two small culverts that effectively produce two lakes of slightly different elevations and decidedly different compositions as much more fresh water flows into the southern lake. The different intensity of light reflectors is the result of different bacteria in the lakes.

Selected References on Magnesium

General References 19.
Magnesium Compounds. 1971. E. Chin. U.S. Bureau of Mines Minerals Yearbook, Preprint, p. 8.
The Gabbs magnesite-brucite deposit, Nye County, Nevada. J. H. Schilling in General Reference 11.
Magnesium. 1971. E. Chin, U.S. Bur. Mines Minerals Yearbook, Preprint, p. 6.

MERCURY

Mercury, or quicksilver (living silver), was known to Aristotle and Theophrastus (315 B.C.). The Chinese apparently knew of it early, since a relief map of China constructed in 210 B.C. depicted the ocean and rivers by liquid quicksilver. Pliny wrote that 10,000 pounds a year were brought to Rome from Almaden, Spain. He also records that later it was used in the recovery of gold. The alchemists delighted in such a unique and mysterious material as liquid metal, and it has found constant use since that time. After the discovery of America, quicksilver was used in large quantities for the recovery of gold and silver by amalgamation.

Quicksilver is unique in being the only metal that is liquid under ordinary temperatures. For this reason and for other physical and chemical properties it has become an essential mineral today. Its chief uses in order of consumption are (1) electrical apparatus, such as switches, clutches, rectifiers, and radio equipment; (2) pharmaceuticals; (3) dry cell batteries; (4) as a catalyst; (5) industrial and control instruments; (6) agricultural insecticides and fungicides; (7) fulminate for munitions and blasting caps; (8) antifouling paint; (9) electrolytic preparations of chlorine and caustic soda (major ore); (10) electrical and scientific instruments, thermometers, dental preparations, recovery of gold and silver, vapor lamps, and many other uses.

Production and Distribution.

Years ago, the United States produced more than enough mercury to supply its needs. Now it consumes one-third of the world's production while producing one-tenth of the world's supply.

World production of mercury (Fig. 21-5) reached a peak in 1965 when the price of mercury reached an all-time high of close to $600 per flask (one flask = 76 pounds). At this high price, many dormant mines entered production only to face disaster as the price plummeted in 1970. During this time mercury had a value of close to $8 per pound. As some deposits could produce mercury at a cost of about $9 per ton, ore grade could be less than 0.1 percent mercury! Over 290,000 flasks were produced in 1969, a record world production. More than a thousand mines in the world have produced at least 100 flasks of mercury, but three-fourths of the world production has come from only six mines or districts.

Those deposits and their approximate cumulative production of flasks are

Almaden, Spain,	7,500,000.
Idria, Trieste,	3,000,000.
Monte Amiata, Italy,	2,000,000.
Santa Barbara, Peru,	1,500,000.
New Almaden, California,	1,100,000.
New Idria, California,	700,000.

During the past decade, mercury has been recognized as a highly toxic substance that has resulted in the death of thousands of human beings who ate mercury-treated grain seed or animals that fed upon such grain, or fish caught near mercury-polluted water sites, especially in Japan. Apparently, fish accumulate mercury by feeding on planktonic organisms that convert mercury to mercurial ions. At the present time, fish, or any food, containing more

than 0.5 ppm mercury is prohibited from marketing in the United States.

Mineralogy, Tenor, and Treatment.

The commercial minerals of mercury are cinnabar, metacinnabarite, calomel, and a little native mercury; the ores range in tenor from almost nil to 8 percent. Mercury is readily extracted by volatilization of its ores, and the vapor is condensed to a liquid in cooling tubes and drained to collecting tanks. It is then bottled for market in steel flasks holding 76 pounds. Because of this simplicity, low-grade ores can be worked. It is interesting that the consumption of mercury in the United States is almost independent of variations in the price as is shown by Fig. 21-6.

Types of Deposits and Origin.

All mercury deposits are formed from hydrothermal solutions at relatively low temperatures. The chief types of deposits (Fig. 21-7) are replacement deposits, *Almaden,* Spain; fissure veins, *Monte Amiata,* Italy; breccia fillings, *Idria;* stockworks, *New Idria* and *New Almaden,* California; and pore-space fillings, *Idria, Trieste.* The mercury deposits may occur in any kind of rock that has been fractured, thus permitting ingress of solutions. They are most typically associated with late Tertiary volcanism.

Spain. The region in which the famous Almaden mine occurs consists of folded and faulted Ordovician to Devonian age quartzites and slates, intruded by quartz porphyry and diabase. The ore deposits are parallel, steeply dipping, replacement lodes in quartzite, of which there are three, the San Pedro, San Nicolas, and San Francisco, the last two joining together at one end and all three terminating against a fault. The veins have been worked to a depth of 750 meters. The widest vein is 10 meters with a central core 3 meters wide that in some places contains 10 percent mercury, although the average grade of the deposit is about 2 percent. The length of the minable veins is about 300 meters.

Italy. The chief districts are Monte Amiata in Tuscany and Idria in Trieste.

The *Monte Amiata* includes six zones of which the Abbadia San Salvador is the third largest mercury mine of the world. The deposits were worked by the Greeks and Romans. They lie at intersec-

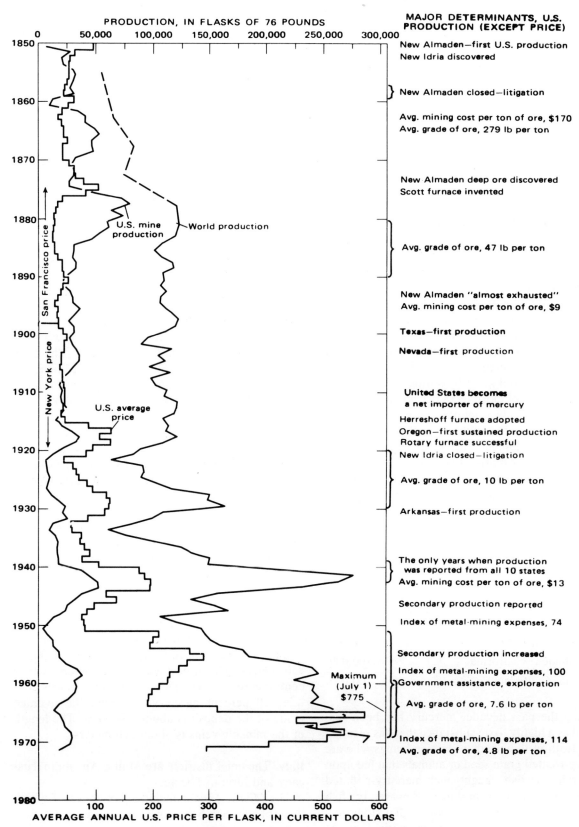

Figure 21-5 Mercury production and prices, 1850–1971. (From U.S. Bureau of Mines Minerals Yearbooks; United States Mineral Resources, U.S. Geol. Survey Prof. Paper 820, p. 404.)

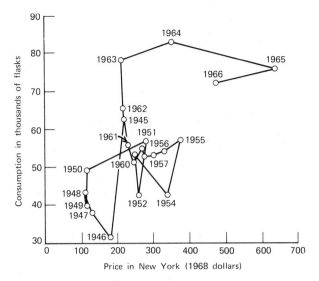

Figure 21-6 Consumption of mercury in the United States is not related to price. The jointed line traces the U.S. consumption of mercury and corresponding price from 1945 to 1966. Figure shows that price and production display the same lack of correlation. (From U.S. Bur. of Mines *Minerals Yearbook*.)

tions of minor fractures with a major fracture that cuts Eocene sediments capped by a trachyte flow from the Monte Amiata volcano. The deposits, of late Pliocene age, occur in limestone, sandstone, clay marls, trachyte conglomerate, and trachyte. They are mainly low-temperature, hydrothermal fillings of solution cavities, crushed zones, sandstone pores, and particularly of trachyte conglomerate. Cinnabar is the chief mineral, accompanied by some stibnite. The ores average about 0.4 percent mercury.

At *Idria*, cinnabar impregnates brecciated blocks of Triassic dolomite overthrust on Carboniferous sediments. Impregnations in sandstones and minor fissure fillings also occur. Some metacinnabar and native mercury are present. The ore is richest in open-textured sandstones, rich in the dolomite breccia, and lean in the bedding planes and small fractures of the bedded dolomite. It has been mined to a depth of 400 meters.

United States. The United States deposits occur in a Pacific Coast belt about 120 kilometers long in California, projecting into Oregon and Nevada, and in single districts in Texas and Arkansas.

The *California* deposits exist in veins, stockworks, and disseminated deposits in serpentine, and also in sandstone and cherts of the Franciscan group. The chief mines are the New Idria and New Almaden. The *New Idria* was in operation in 1853. Here, two fault-fissure veins lying between serpentine and sandstone carry cinnabar and metacinnabar, and much pyrite, quartz, opal, and carbonates, deposited from uprising hydrothermal solutions. The *New Almaden* in its 140 years of operation has yielded over $75 million worth of quicksilver and reached a depth of 735 meters. Its production is now small. The deposits are impregnations in a shatter zone of Franciscan sandstone and shales lying above and about a sloping peridotite intrusive contact. The mineralization is similar to that of New Idria.

Generally similar deposits occur in *Oregon* where the Opalite mine and the rich Brety mine are intermittent producers. In *Nevada*, also, many small deposits belong to the same period of metallization as the Opalite district.

Near *Terlingua, Texas*, low-grade deposits occur in folded and shattered Cretaceous limestone. The

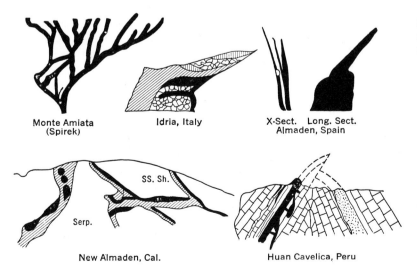

Figure 21-7 Cross section of some quicksilver deposits. (Mostly from Schuette, AIME.)

deposits, of which the Chisos mine is an example, are impregnations in cavities and caves that occur along faults or near the crackled crests of anticlines and are occupied by boulders and clay. The overlying Del Rio clay acted as an impervious barrier to uprising hydrothermal solutions that deposited cinnabar. Surface alterations yielded native mercury, terlinguaite, eglestonite, montroydite, and calomel. The ore averages 0.5 percent mercury, but some attains 2 percent.

Other Deposits. In *Mexico* are numerous small deposits, of which those around Guadalcázar, San Luis Potosi, Nuevo Mercurio and Sain Alto, Zacatecas, Canoas, and Huitzuco, Guerrero, have been the most productive. At Huan cavelica, *Peru*, cinnabar ores were deposited in openings in folded Cretaceous sandstone and limestone intruded by andesite. Nearby hot springs attest to recent igneous activity. Other minor deposits are worked in Japan, China, Czechoslovakia, Algeria, and Russia.

Mercury is also used as a prospecting aid in the detection of hidden mineral deposits and locating geothermal sites. As mercury occurs as a trace element in most mineral deposits, it constantly yields mercury vapor that ultimately reaches the surface of the earth. Various instruments are used to collect this mercury vapor, the amount of which is measured in nanograms (10^{-9} gr.) with good precision. Some measure the amount escaping to the surface at the collection site; others draw the soil-gas to the surface with a suction fan, pass the gas through a silver or gold screen of about 80 mesh, where the mercury is collected as an amalgam with the silver or gold screen. The mercury is released by heating and the amount may be measured with an atomic absorption instrument. Some instruments measure the changing conductivity of gold as it absorbs mercury. Backround mercury measurements vary from a few to 100 ppb, but anomalies associated with geothermal areas have exceeded 1000 ppb.

Selected Reference on Mercury

Mercury industry in Italy. 1948. Edwin B. Eckel. AIME Tech. Pub. 2292. *Geology of the most productive mercury mines in the world.*

Quicksilver-Antimony deposits of Huitzuco, Mexico. 1945. James F. McAllister and David Hernandez Ortiz. U.S. Geol. Survey Bull. 946-B. *An unusual deposit with livingstonite.* Geology of the Cuarenta Mercury district, State of Durango, Mexico. 1946.

Quicksilver deposits in the Steens and Pueblo Mountains, Southern Oregon. 1942. Clyde P. Ross. U.S. Geol. Survey Bull. 931-J.

The Wild Horse Quicksilver district, Lander County, Nevada. 1942. Carle H. Dane and Clyde P. Ross. U.S. Geol. Survey Bull. 931-K.

Quicksilver deposits of the Opalite district, Malheur County, Oregon, and Humboldt County, Nevada. 1942. Robert G. Yates. U.S. Geol. Survey Bull. 931-N.

Quicksilver and antimony deposits of the Stayton district, California. Edgar H. Bailery and W. Bradley Myers, 1942. U.S. Geol. Survey Bull. 931-Q. The last four references give various modes of occurrence of mercury.

General Reference 19. *Chapter on mercury.*

Mercury and base metal deposits with associated thermal and mineral waters. D. E. White. In General Reference 11, v. 2

Mercury—Its occurrence and economic trends. 1964. E. H. Bailey and R. M. Smith. U.S. Geol. Survey Circ. 496, 11 p.

TANTALUM AND NIOBIUM

Tantalum and niobium (formerly columbium) are modern technology rare elements of increasing importance because of their metallurical, electronic, chemical, and nuclear use. World production of niobium-tantalum concentrates increased seven times from 1960 to 1970.

The minerals of niobium and tantalum have strong geochemical coherence because of their similar ionic radius and valence states, but in certain rock types, one element is generally more abundant than the other. For example, nepheline syenite and carbonatites contain more niobium than tantalum. The chief rock types in which mineralization occurs are alkalic-rich complexes and associated carbonatites. Placer deposits have been significant producers. Small concentrations occur in some complex pegmatites. Many minerals contain these two elements, which are chiefly oxides, multiple oxides, hydroxides, as well as a few silicates and one borate. Example are Columbite-tantalite, euxenite, fergusonite, loparite (variety of perovskite), microlite, and pyroclore.

Production and Reserves.

During the Korean war, niobium-bearing steel alloys were in critical need for jet engines. Because supplies of this element were low, the U.S. government promoted successful exploration through a guaranteed purchase program. In fact, so successful was the program that it ended in 1959 when the purchase contracts were filled. The discovery of several new large volume deposits of niobium through this program has allowed production to increase rapidly throughout the world.

Brazil has the largest niobium world reserves, 300 million tons of ore containing 9 million tons of Nb_2O_5 in the *Axara* deposit; followed by Africa with 2,744,000 tons in Uganda, Nigeria, Kenya, and Zaire in decreasing order of reserves. North American deposits have reserves of about 2 million tons, more than 90 percent of which is in Canada with 70 percent occurring in two deposits. *Oka,* and *St. Honore,* Quebec, both with ore reserves of 325 million tons with no tantalum content. One of the largest potential deposits in the United States is in the *Iron Hill* carbonatite near Powderhorn Colorado, which has reserves of 40 million tons of 0.25 percent Nb_2O_5 mostly as pyroclore. *Magnet cove* contains local concentrations of niobium perovskite and niobium rutile. Some of the rare-earth deposits, many of which are carbonatite deposits containing Nb, are listed on Table 21-1.

Types of Deposits.
Formerly, the sole source of niobium- and tantalum-bearing minerals was by hand cobbing from complex pegmatites. Additional higher tonnage and vastly larger reserves have been found in alkalic rock complexes; placer deposits; and most important of all, carbonatite bodies.

Some of the more representative carbonatites throughout the world are Iron Hill and Gem Park, Colorado; Oka, Quebec; Axara, Brazil, Fen, Norway; Lueshe, Zaire, Sukulu and Uganda; Kaiserstuhl, Germany; Alno, Sweden; Mbaya; and Tanganyiki.

A successful placer deposit exists at Bear Valley, Valley County, Idaho. In 4 years, 1955–1959, two floating dredges produced more than 1 million tons of combined niobium and tantalum oxide. There are sufficient reserves to allow production for 30 years at the rate maintained during the 4 years when 8000 cu. yds per day were produced.

Selected References
General Reference 19.
Radioactive Black Sand Is Yielding Columbite Concentrate at Idaho Mill. 1958. S. H. Dayton. Mining World, v. 20, p. 36–41. *The Geology of Carbonatites.* 1966. E. W. Heinrich. Chicago Rand McNalley & Co. p. 555.
Geochemistry of Niobium and Tantalum. 1968. R. L. Parker and M. Fleischer. U.S. Geol. Survey Prof. Paper 611, p. 43.

TITANIUM

Titanium is a common metal occurring in igneous rocks in the minerals ilmenite and rutile, but is rarely concentrated into workable lode deposits. It is a common impurity in magnetite, generally rendering it useless as an ore of iron. It holds to its bound elements with such tenacity that it is difficult to separate.

Titanium has been a latecomer into the field of usefulness, but it holds promise of becoming an outstanding new metal of this century. Demand for it has been increasing rapidly because it makes the whitest of all paints. Its opacity is twice that of zinc oxide and three times that of lead oxide. It is used in the form of titanium oxide, called titanium white, and this can be mixed with other pigments without decreasing its opacity. The titanium paints are pure titanium oxide or are combined with other paint compounds. They are used for most inside paints and many of the outside white paints. The pigment is also used in toilet articles, linoleum, artificial silk, white inks, colored glass, pottery glazes, for tinting artificial teeth, and for dyeing leather and cloth. Metallurgically, ferrocarbon titanium is alloyed with steel for high-speed steels and with chrome steels; the oxide is used for alloys and cemented carbides and also as a paper filler, as a coating for welding steel, and in electrodes for arc lamps. In the form of chloride, titanium is used for removing colors from cloths, and in the form of tetrachloride for making smoke screens and sky writing. Metallic titanium holds unusual promise for many applications as it compares favorably with some steels at low temperatures. More important, however, the strength-weight ratio of titanium is more favorable at higher temperatures up to 800°C above which it weakens and oxidizes rapidly at temperatures in excess of 1300°C. This metal does resist corrosion in seawater better than stainless steel. Titanium metal is heavier than aluminum alloys but is much stronger and more chemically inert than aluminum.

Production and Distribution.
The United States has substantial reserves of ilmenite but is dependent almost entirely on foreign sources for rutile, chiefly from the black sands of Australia that produce 88 percent of the world pro-

Table 21-1 Identified Resources[a] of the Rare Earths and some Potential Rare-Earth Deposits of the World

Country	Deposit	Type	Rare-earth oxides[2] (tons)	Remarks	References
Canada	Blind River-Elliott Lake	Conglomerate	1,200,000	Potential byproduct of U; some has been produced.	Shenk (1971); Griffith and Roscoe (1964).
	Oka	Carbonatite	325,000,000	Potential byproduct of Nb. No Ta.	Rowe (1958); U.S. Bur. Mines (1963).
	St. Honoré	do	325,000,000	do	Valee and Dubuc (1970).
United States	Mountain Pass, Calif	do	5,000,000	Reserve in bastnae-site ore.	Parker (1965).
	Music Valley, Calif	Metamorphic rocks	Not evaluated	Xenotime and mona-zite	Evans (1964).
	Idaho-Montana placers, includes Bear Valley, Idaho.	Placers	147,600	Monazite and euxenite	Eilertsen and Lamb (1956).
	Lemhi Pass, Idaho-Mont	Thorite veins	84,000	Potential coproduct Th	U.S. Atomic Energy Commission (1969); Staatz and others (1972).
	Mineral Hill district, Idaho-Mont	Carbonatite veins	Not evaluated	Monazite, thorite, an-cylite.	Anderson (1958).
	Bald Mountain, Wyo	Conglomerate	15,000	Monazite	Borrowman and Ro-senbaum (1962)
	Wet Mountains, Colo	Thorite veins	1,750	Potential coproduct Th.	Christman and others (1959); Staatz and Conklin (1966).
	Powderhorn district, Colo.	Thorite veins and car-bonatite.	4,000	40 million tons of 0.25% Nb_2O_5, Some Th.	Olson and Wallace (1956); J. C. Olson (unpub. data).
	Gallinas Mountains, N. Mex.	Veins	Not evaluated	Bastnaesite with fluor-ite.	Perhac and Heinrich (1964).
	North and South Carolinas, and so forth (Piedmont area).	Stream placers, flood plains.	600,000	Monazite, minor xeno-time in numerous scattered deposits.	Overstreet and others (1959).
	Hilton Head Island, S.C.	Beach placers	60,000	Monazite	McCauley (1960).
	Mineville, N.Y.	Magnetite deposit	Not evaluated	Rare-earths in apatite, minor bastnaesite.	McKeown and Klemic (1956).
	Dover, N.J.	do	do	Bastnaesite, doverite, xenotime.	Klemic and others (1962).
Brazil	Atlantic coast beaches and some inland areas.	Beach and fluviatile placers.	180,000	Monazite	Parker (1965).
	Morro do Ferro	Alkalic complex	300,000	Allanite, bastnaesite	Roser and others (1964).
	Araxá	Carbonatite	90,000	Potential byproduct Nb and apatite(?).	Do.
Uruguay	Atlantida	Beach placers	288	Monazite. Represents one sampled area.	Bogert (1959).
Bolivia	Llallagua area	Veins	Not evaluated	Monazite, minor xeno-time in tin veins.	Gordon (1944).
U.S.S.R.	Kola area	Alkalic rocks and car-bonatite.	6,800,000	Apatite, loparite. Tonnage based on potential byproduct from apatite processing.	British Sulphur Corporation, Ltd. (1971).

Table 21-1 (Continued)

Country	Deposit	Type	Rare-earth oxides[2] (tons)	Remarks	References
	Kyshtym area (Vishnevye Mountain).	Alkalic rocks, placers.	Not reported	Apatite, cerite, bastnaesite, allanite. Apatite contains 0.29–3.3 percent rare earths.	Zil'bermints (1929); Ganzeev and others (1966).
Nigeria	Plateau Province	Placers	do	Small production from tin placers.	Mackay and others (1949).
U.A.R. (Egypt)	Nile delta	Fluviatile placers	120,000	Monazite. Intermittent production.	Higazy and Naguib (1958).
Burundi	Karonge mine	Veins	Not reported	Bastnaesite. Intermittent production.	Thoreau and others (1958).
Malawi	Kangankunde Hill	Carbonatite and residual soils.	68,000	Monazite	Deans (1966).
Kenya	Mrima (Jombo)	Alkalic complex and carbonatite.	300,000	Monazite; pyrochlore	Do.
Tanzania	Panda Hill (Mbeya)	Carbonatite	Not reported	Potential byproduct Nb and apatite.	Fawley and James (1955); Davidson (1956); Fick and Van der Heyde (1959).
Republic of South Africa.	Glenover	Carbonatite	do	Synchisite, monazite	Verwoerd (1963).
	Steenskampskraal	Vein	9,000	Monazite. Has had large production.	Pike (1958); U.S. Bur. Mines (1964)
Zaire (formerly Republic of the Congo).	Shinkolobwe	do	Not reported	Monazite	Derriks and Vaes (1956).
Malagasy Republic	Ft. Dauphin area	Placers	60,000	Monazite	Murdock (1963).
India	Chiefly in Kerala	Beach placers	3,000,000	Monazite	Wadia (1956).
	Bihar and west	Alluvial placers		do	Parker (1965).
Sri Lanka (Ceylon)	Chiefly west and northeast coasts.	Beach placers	6,000	Monazite. Tonnage given is from one northeast coast deposit.	Davidson (1956).
Malaysia	Various	Placers	Not reported	Monazite, xenotime. Byproduct of Sn.	Do.
Indonesia	do	do	do	Monazite	Olson and Overstreet (1964).
Taiwan	Coastal areas	Beach placers	6,000	do	Shen (1965).
Korea	Various	Stream and beach placers.	54,900	Monazite. Tonnage from evaluated deposits only.	Olson and Overstreet (1964).
Australia	do	Beach placers	480,000	Monazite	Parker (1965).
	Mary Kathleen	Contact metamorphic	Not reported	Allanite, stillwellite in U deposit.	Matheson and Searl (1956).
	Radium Hill	Veins	do	Brannerite	Barrie (1965).
Total (rounded).			18,666,000		

Source: Jolley, *Minerals Yearbook*, vol. 1, 1973.

[a] Identified resources: Specific, identified mineral deposits that may or may not be evaluated as to extent and grade, and whose contained minerals may or may not be profitably recoverable with existing technology and economic conditions.

[b] Monazite assumed to contain 60 percent and euxenite 15 percent rare-earth oxides.

duction of rutile. Australia is followed in production of rutile by Sierra Leone, Sri Lanka, and India.

Canada has the largest reserves and is the leading producer of ilmenite. The United States is second in reserves.

The United States is the largest consumer of titanium, exceeding 500,000 tons per year at the present time, which is expected to more than double by the year 2000.

Types of Deposits.

As summarized by Klemic, March, and Cooper, rutile, ilmenite, and other titanium-rich minerals occur in primary deposits in mafic, igneous, and metamorphic rocks, in placer and residual deposits, and in much minor amounts in sea-floor iron and manganese nodules.

With the exception of Australia's placer deposits, primary deposits are by far the most important for large production. Canada's *Allard Lake* deposit, Quebec, with over 200 million tons of ore containing about 20 percent combined oxides; is an ilmenite-hematite-rich, anorthosite, ultramafic deposit. Even though some Fe-Ti minerals exhibit euhedral grains, the evidence at Allard Lake suggests a late-magmatic genesis for the ilmenite.

At *St. Urbain, Quebec,* a large deposit consists of dikclike masses of ilmenite-hematite with some rutile and sapphirine enclosed in anorthosite. The ores contain 35 to 40 percent TiO_2. Mawdsley and Osborne consider them to be magmatic injections, but Gillson concluded that they represent high-temperature replacement deposits. Other primary deposits occur in the Ilmen Mountains (part of the Urals) and in the Kola Peninsula, Russia; at Tahawas, Essex County, New York; Bushveld Igneous complex, South Africa; and Teelnes, Norway. Some high-grade ilmenite ores occur in amphiboles in Finland.

Richards Bay on South Africa's Natal coast is the site of a major beach sands mining and beneficiation project. The heavy mineral deposits of the Natal coast in South Africa have been known for many years but, until recently, most development interest had centered around areas immediately to the south of Durban. In the 1950s ilmenite, rutile, and zircon were produced at *Umgababa* about 40 kilometers south of Durban, and this has been the only relatively large-scale producing operation so far. However, a major mining and metallurgical project to produce rutile, zircon, titania slag, and pig iron based on heavy mineral sands deposits north

of *Richards Bay* on the northern Natal coast will commence in 1977 to 1978. It is estimated that revenues should approach about $200 million per year by 1980 based on current prices.

Selected References on Titanium

General Reference 19.

Titanium, Its Occurrence, Chemistry, and Technology. 1966. J. Birksdale. Ronald Press, New York, p. 691.

Titanium. 1972. F. E. Noes. U.S. Bureau of Mines Minerals Yearbook, v. 1, p. 1109–1121.

Geology of titanium and titaniferous deposits in Canada. 1969. E. R. Rose. Canada. Geol. Survey, Econ. Geol. Report 25, P. 177.

Vanadium-bearing magnetite-ilmenite deposits, Lake Sanford, New York. 1945. R. C. Stephenson. AIME Min. Tech. V. 9, no. 1, pp 1–25. *Petrography, mineralogy, and geology; a late magmatic residual liquid injection origin is* advocated.

Notes on the ilmenite deposit at Piney River, Virginia. 1947. D. M. Davidson, F. F. Grout, and G. M. Schwartz. Econ. Geol., V 41, p. 738–748. *Conclusions for a magmatic origin.* Reply by C. S. Ross. Econ. Geol. v. 42, p. 194–198. *Arguments for a replacement origin.*

ZIRCONIUM

Zirconium is a relatively recent entry into industry especially in the chemical and nuclear reactor industries. The mineral zircon is worn as a semiprecious stone. The oxide (zirconia) and the metal are used industrially. Zirconia is one of the most refractory oxides known and is employed for crucibles, laboratory wares, chemical-resisting furnace bricks, and high-temperature cements. Its use is rapidly expanding to give opacity to enamelware, paints, and automobile enamels and lacquers. It is also used as an insulator for heat and electricity; as an abrasive; for gas mantles and certain incandescent lamps; as a polisher; for toughening rubber; and for white inks. The metal is used for electronic tubes, flashlight bulbs, electrical condensers, X-ray filters, lamp filaments, spot-welding electrodes, rayon spinnerets, and as an alloy. Zirconium steels make good armor plate, and projectiles and, with nickel, make high-speed and sharp cutting tools. With copper it imparts properties similar to that of beryllium, and with aluminum and vanadium in ferroalloy it serves as a steel refiner in furnaces.

The commercial minerals are zircon ($ZrSiO_4$) and baddeleyite (ZrO_2), the latter being obtained only in Brazil. Zircon is an accessory mineral of igneous rocks and is readily concentrated in placer deposits. Over 900,000 tons are produced annually from Australia with minor producion from Brazil, and Thailand where it occurs along with ilmenite, rutile, and monazite in beach sands. A little is obtained from the beach sands of Florida and Georgia placers. The beach sands at Trail Ridge, Florida, yield zircon as a by-product of ilmenite production.

MISCELLANEOUS MINOR METALS

Barium metal is produced in small quantities and is used as a "getter" in electronic tubes to promote vacuum. Its source is the mineral barite of which 80 percent is used in the petroleum industry to form heavy drilling "mud." A large sedimentary deposit has been located in Lander Country, Nevada, and is now in production.

Boron is used as a hardener for steel and for atomic energy.

Cesium is in demand for photoelectric cells, which are used for talking pictures, television, traffic controls, automatic door opening, and similar purposes. Cesium is also used in radio tubes. It is obtained from the mineral pollucite, which is mined near Custer, South Dakota.

Calcium is considered a coming metal and is already being used as a scavenger in melting steel, copper, nickel, lead, as an alloy with steel and nonferrous metals, as a reducer in the production of rare metals, and in making hard bearing metal.

Cerium is alloyed with 30 percent iron to form *ferrocerium,* a brittle alloy that when abraded emits sparks and is used as "flints" or sparkers for cigarette lighters. It is also used in signaling, photography, glassware, and radio tubes, and as a metal scavenger. It occurs in monazite, from which it is extracted as a by-product in obtaining thorium. The monazite is found in beach placers along with ilmenite.A main source is Bastnasite from Mountain Pass, California.

Gallium and *rhenium* are minute by-products of the copper ores of Mansfeld, Germany, and of coal ashes. Some gallium is found in the zinc ores of Joplin, Missouri. Gallium is one metal other than mercury that is liquid at low temperatures. It is used in electrical fields, reflecting surfaces, military devices, atomic piles, and thermometry. Rhenium is used for special plating and to impart long life to tungsten filaments in bulbs.

Germanium is obtained from zinc-refinery slimes and is contained in the minerals germanite and argyrodite. Deposits occur at Tsumeb, Southwest Africa, and in Bolivia. Its chief use is medicinal, as a specific for pernicious anemia, and for sleeping sickness. It is used also in radionic devices, optical glass, and alloys with precious metals, and as a catalyst.

Indium is also obtained from zinc residues. It is used as a precious-metal alloy, as a dental alloy to resist corrosion, as a nontarnishing plating for silverware, for heavy-duty bearing metal, for low-melting-point alloys, to give amber colors to glass, and for atomic work.

Mesothorium is produced from monazite sands in small quantity. It is used in the treatment of cancer and skin diseases and as a luminous paint.

Neodymium and *praseodymium* are recovered from cerium minerals. The former gives glass a delicate violet color, and the latter a fine greenish yellow tint. Mountain Pass, California, is a main source of these elements.

Rhenium is produced from molybdenite roaster flue dust. It can be used as an alloy with platinum metals for noncorrosive purposes, electrical parts, pen-nibs, vacuum tube components, thermocouple elements, and electrical contacts.

Rubidium is used in mercury-vapor lamps.

Thallium is recovered from cadmium-containing flue dusts and is used as a rodenticide and insecticide, for signaling devices, and for certain alloys.

Other more important metals, such as uranium, are covered adequately in the processes portion of this book.

An unusual but important source of *rare earth* is the *Mountain Pass mine,* owned and operated by the Molybedenum Corporation of America (Molycorp). It is considered to be the largest rare-earth metal mine in the United States and as such is, and will be for many years, the principal supplier for many of these metals.

There are 17 of these chemically similar metallic elements. They are not earthy and some are not rare; they usually occur together in groups of four to eight in one mineral. The principal mineral at Mountain Pass is bastnasite, a fluocarbonate ($RFCO_3$ where R = rare earth elements) containing principally cerium, lanthanum, neodymium, and praseodymium, but also small amounts of samarium, gadolinium, and europium. In addition the Mountain Pass ore contains the minerals barite, calcite, silica, and others. Surprisingly enough, niobium is rare.

Even though this area was intensively prospected

during the last hundred years and many small silver, gold, lead, zinc, and copper deposits were found, the existence of rare earth metals was not recognized until 1949. All of the area containing economic rare earth deposits was acquired by Molycorp in 1950 and 1951.

The principal deposit at Mountain Pass exceeds 600 meters in length, 120 meters in width, and 100 meters in depth. The average ton of ore contains 140 pounds of rare earth metals, of which half is cerium. It is mined by open pit methods from the ridge directly north of the new plant.

Silicon. (Formerly ferrosilicon), Silicon alloys of silicon metal, provide steel with important properties such as stainless and heat resisting, high strength, and electrical qualities. The increasing use of electronic-grade polycrystalline silicon and on waters and boules of monocrystalline silicon is a mushrooming business and highly competative not only within the United States but internationally. For example, the U.S. Department of Commerse has established interim controls on exports to the USSR of electronic-grade silicon including polysilicon systems and to saving and surface finishing equipment used in the processing of semiconductor waters.

In May, 1980, the U.S. exports of ferrosilicon totaled 2,133 tons with a value of $1,712,199. Canada purchased more than half of this tonnage. Silicon is a prime example of a low-cost, plentiful mineral resource that through technology has provided a multibillion dollar industry.

IV

NONMETALLIC MINERAL DEPOSITS

"Our civilization is founded on coal, more completely than one realizes until one stops to think about it."

George Orwell
in "The Road to Wigan Pier."

NONMETALLIC MINERALS

The materials of the nonmetallic kingdom are more commonplace and familiar than metallic ores. Most of them are rather abundantly distributed throughout the world, and the value of many depends less on the material itself than on the ability to use it nearby. Their economic value is determined largely by the cost of transportation. Their specifications vary widely, and, unlike the constancy in a metal, the specifications are in general determined by the uses to which the material is put. Nonmetallic minerals are used essentially in the form in which they are extracted, and few are broken up into elemental parts. The gross value of all nonmetallic products annually greatly exceeds that of metallic ores.

Nonmetallic products defy simple classification. One substance will occupy more than one pigeonhole or be formed by more than one process. The outstanding feature of them, and indeed the one that often determines their value, is the purpose for which they are used. With ores, one of the chief features of interest is the deposit, but with nonmetallics this is often subordinate and interest centers in usefulness. Consequently, in this section they will be grouped according to their chief uses.

For scientific purposes, however, a genetic classification according to process of formation is proposed in Table IV-1, which shows also the main uses to which the minerals are put. The table also indicates the use groups followed in Part IV, and the chapters in which processes and the various nonmetallic materials are considered. No attempt has been made to include innumerable, minor nonmetallic minerals.

Many of the nonmetallic materials have already been described in Part I.

Table IV-1 Genetic and Use Classification of Important Nonmetallic Materials

Process of formation and important products	Chap.	Mineral fuels	Ceramic materials	Structural-building materials	Metallurgy, and refractory materials	Industrial and manufacturing materials	Chemical minerals	Fertilizer minerals	Abrasives	Gemstones, ornamental	Water supplies
A. Igneous processes:											
I. Rocks:	4										
Building				⊗						X	
Soapstone				⊗		X			⊗		
Pumice					⊗				⊗		
Corundum									X	X	
Diamond									⊗	⊗	
II. Pegmatites:	4										
Feldspar			⊗		X				X	X	
Quartz (silica)			X		X	X			⊗		
Mica				⊗		⊗					
Cryolite			X		⊗						
Spodumene						⊗					
Gemstones						X				⊗	
III. Magmatic emanations:	5										
Pyrite (sulfur)							⊗	X			
Fluorspar					⊗	X	X				
Barite and witherite						⊗	X				
Magnesite					⊗		X				
B. Sedimentary processes:	9										
I. Sedimentary rocks:											
Building stones				⊗							
Lime, dolomite, magnesite				⊗	X	X	X	X			
Hydraulic cements				⊗							
Clay, shales		⊗		⊗	X	X					
Phosphates								⊗			
Sand, sandstones				⊗	X				⊗		
Bentonite, fuller's earth, diatomite				X		⊗			X		
II. Chemical precipitates:	12										
Rock salt							⊗				
Gypsum				⊗				X			
Potash							X	⊗			
Nitrates							X	⊗			
Borax and borates					X	⊗	X				
Sodium compounds						⊗	⊗				
Miscellaneous chemicals							⊗				

Table IV-1 Genetic and Use Classification of Important Nonmetallic Materials

Process of formation and important products	Chap.	Mineral fuels	Ceramic materials	Structural-building materials	Metallurgy, and refractory materials	Industrial and manufacturing materials	Chemical minerals	Fertilizer minerals	Abrasives	Gemstones, ornamental	Water supplies
III. Organic deposits:	22 and 23										
Coal		⊗				⊗					
Oil, gas		⊗					X				
Sulfur							⊗	X			
Bitumens				⊗		⊗					
C. Weathering processes:											
I. Residual concentration:	13										
Bauxite			X	X	⊗				X		
Clays			⊗	⊗	X	X					
Mineral pigments						⊗					
Tripoli									⊗		
II. Mechanical concentration:	13										
Sands			X		⊗	⊗			X		
Monazite and zircon					X	⊗	⊗				
Ilmenite					X	⊗					
Phosphates							X	⊗			
Gemstones									X	⊗	
D. Metamorphic processes:	15										
Asbestos				X		⊗					
Graphite					⊗	X					
Emery, garnet									⊗	X	
Talc			X		X	⊗					
Sillimanite minerals			X		⊗	X					
Gemstones									X	⊗	
Roofing stones				⊗							
E. Groundwater processes:	12										
Water supplies						X	X				⊗
Brines (salt)							⊗				
Bromine, iodine						⊗	X				
Nitrates							X	⊗			
Sulfur							⊗	X			
Gemstones										⊗	
Salines (except salt)						X	⊗				

⊗ indicates the chief uses.

22

ENERGY AND COAL

"There is nothing more difficult to carry out nor more doubtful of success, nor more dangerous to handle, than to initiate a new order of things."

Niccolo Machiavelli

Energy is derived predominantly from fossil fuels that include coal, petroleum, natural gas, oil shales, and tar sands. Energy is also available from burning wood, running water, earth heat (geothermal), the sun's radiation, and atomic disintegration. The changes in the use of energy by man are illustrated in Fig. 22-1 and the increasing rate of growing oil imports to the United States in Fig. 22-2.

Prior to World War I, coal was by far the chief source of energy not only in the United States but in the entire world, primarily in the industrial nations. Petroleum and natural gas began making inroads on coal that resulted in a decrease in coal production until about 1960. Then the accelerating needs for energy resulted in a continual increase in coal production, which increased slowly in the mid-1970s in spite of the so-called energy crisis. Coal, therefore, only accounts for less than 25 percent of the present U.S. energy requirements. The development of coal resources, especially in near-surface deposits where strip mining can be done, is increasing slowly because of environmental restrictions. Large electric power plants such as the planned Kaiparowits plant in southern Utah, now postponed indefinitely because of many environmental problems, were to be powered entirely by previously untapped coal fields. Coal could be an almost immediate source of energy and the reserves are abundant, enough to last for hundreds of years.

Wood and water power supplies can never be considered as comparatively important energy sources. Oil and natural gas reserves (note the winter of 1976 to 1977 in the eastern United States) are decreasing at a rate unsuspected by some a few years ago. Oil shale research in Colorado and Utah was increasing with the energy crisis, but high interest rates have postponed the Colony concern de-

velopment in Colorado. The development of underground mining of oil shale in Utah began in 1974, but it is also being terminated. The planned ultimate estimate of oil to be derived from oil shale was about 10^6 barrels per day or only about 5 percent of the petroleum now consumed in the United States.

The capital and fuel life cycle costs for a uranium power plant have been holding at 18 cents per million Btu, while coal-fired plants have escalated to 70 cents and gas-fired plants to $1 per million Btu. Almost all former gas-fired electric plants have converted to coal at the present time.

Geothermal research is increasing, and such energy sources and plants in Italy and Wairaki, New Zealand, are well known. An operating geothermal plant at the Geysers, California, is operating and being expanded even though it is already the world's largest producer of electricity by geothermal energy. These plants are located at sites where surface steam was, and is, evident. The present search for geothermal sites is being made by geological-geochemical-geophysical techniques to locate heat sources at depth that sometimes require drilling to as much as 7,000 meters. Many such sites have been located, and electric power and petroleum concerns are actively developing a few of these sites. Geothermal energy is a "clean" source of energy and the terrific noise "pollution" can be controlled by mufflers. There is some evidence, however, that removing hot water or steam and reinjecting water back into the earth may trigger earthquakes in certain seismic locales.

Solar energy is an ultimate source of clean energy that can be afforded by NASA on space satellites and in experimental homes. The cost of the solar cells used in solar panels are exhorbitant at the

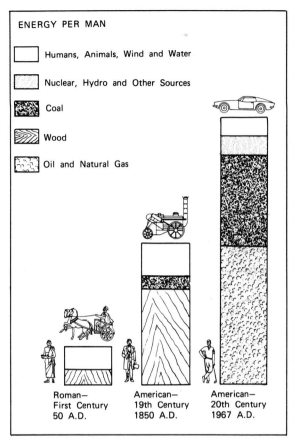

Figure 22-1 Energy available to man. (Copyright Standard Oil Company of California Bull. October 1967.)

homes to heat fluids or metal panels from which the heat is used in water or space heating, but not for electricity.

Nuclear power plants, using fissionable ^{235}U, are increasing at a slower rate than earlier estimated. There are 72 operable Nuclear power generating reactor units in the U.S. at the present time that are producing about 12 percent of U.S. electric power, 74 more are being built, and 85 are being planned of which 61 of the 85 are already ordered. The power being generated by present reactors is about 50 million KW, but if all of the above 231 reactors were operating, more than 232 million KW would be produced. Unfortunately, however, there may be insufficient readily available nuclear fuel for all of these planned reactors.

Nuclear or breeder reactors using ^{232}Th and ^{238}U, that makes up 99.3 percent of uranium, which when bombarded by neutrons produce U^{233} and plutonium, respectively. This enormous energy source is still in the development stage in the U.S., even though the first reactor operated by the A.E.C., ERDA, and now DOE at Arco, Idaho, in 1951 was a breeder process reactor. France has a breeder reactor that has been operated for the past ten years. Canada's CANDU reactors use ^{232}Th as a fuel that does not result in ^{239}Pu. The delay in the building of breeder reactors is primarily design and engineering problems, but engineering has come a long way since 1951. Breeder reactor construction is needed because at present there are not enough proven reserves of uranium in the United States to provide ^{235}U for the fission reactors to operate for the next 35 to 40 years, the expected lifetime of these nuclear power plants. Major lawsuits have arisen be-

present time but as energy costs rise and research produces lower cost solar panels, this source of energy is looked upon with some promise, at least in geographic areas of clear sunlite skies. In the mean time, solar energy is being used in many

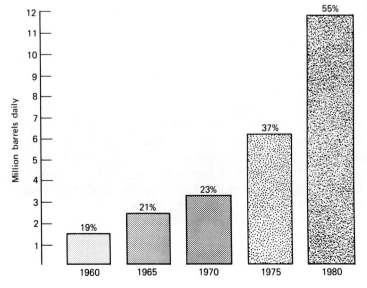

Figure 22-2 Growing oil imports. Percent of U.S. oil demand from 1960 to 1980.

tween reactor construction firms and buyers of reactors because the construction firms cannot acquire sufficient start-up uranium for the reactor construction contracts they hold. The search for uranium is a continuing endeavor, but no new uranium province has been found in the United States in the last 19 years. And the cost of U_3O_8 has skyrocketed from about $8 per pound to over $60 per pound!

In terms of cost, one pound of U_3O_8 is equal in potential energy to 10 tons of coal and 40 barrels of oil. One 25-ton truck load of U_3O_8 has the energy equivalent of 20,000 truck loads of coal. Splitting all of the atoms in a golfball-sized piece of uranium could unleash energy equivalent to that consumed by flying a fleet of Boeing 707s coast to coast.

In a fissionable reactor, the fuel is composed of about 2 to 3 percent ^{235}U (nuclear bombs contain more than 90 percent ^{235}U). The 97 to 98 percent of ^{238}U in the fissionable reactor is capable of being converted to ^{239}Pu, which is an excellent fissionable radioactive isotope. Reactors constructed of a shell of uranium surrounded by another shell of ^{239}Pu would not only provide energy but would also breed more ^{239}Pu from bombardment of the ^{238}U in the inner shell. Unfortunately, however, because of a presidential decree, no ^{239}Pu is being separated from spent uranium fuel. This altruistic desire is understandable as this isotope is used in the construction of atomic bombs, and apparently without great difficulty, even by terrorists, despots, and so forth. Even so, no other nation has joined President Carter's altruistic lead.

The Alaskan pipe line in contrast has the potential of providing 1.5 million barrels per day or more than 8 percent of the U.S. present consumption of petroleum.

On June 19, 1977, the Alaskan pipeline began the movement of oil from the North Slope to Valdez. A rundown of some of the major facts about the trans-Alaska pipeline follows.

1. Pipeline length: 800 miles, approximately half below ground. Cost—approximately $10 billion, all in private funds.
2. Capacity: 1.2 million barrels a day initial capacity, could be increased with the addition of pumping stations and terminal facilities up to 2 million barrels a day.
3. Estimated North Slope crude oil recoverable reserves: 9.6 billion barrels.
4. Valdez ship loading capacity: 80,000 to 110,000 barrels an hour.

5. Schedule: Construction started April 1974; first pipe in place, March 1975; last pipe in place, December 1976.
6. Pipe size: 48 inches in diameter, in lengths of 48 feet and 60 feet and thicknesses of 0.462 inch and 0.562 inch.
7. Right-of-way width of pipeline: 54 feet.
8. Vertical supports: Almost 80,000 vertical support members were used to elevate about one-half of the pipeline above ground.
9. Pipeline designer, builder, operator: Alyeska Pipeline Service Co., a consortium of eight major oil companies whose holdings are Sohio Pipe Line Co., 33.34 percent; BP Pipelines Inc., 15.84 percent; ARCO Pipe Line Co., 21 percent; Exxon Pipeline Co., 20 percent; Mobil Alaska Pipeline Co., 5 percent; Union Alaska Pipeline Co., 1.66 percent; Phillips Petroleum Co., 1.66 percent; and Amerada Hess Co., 1.50 percent.

Now, years later, Alaskan oil is becoming a major problem in San Diego that lacks the pipelines to ship the oil eastward. More and more tankers, therefore, are taking the Panama Canal route to the eastern seaboard.

Other sources of energy, such as synthetic fuels, wind, wood, water, biomass, and tides are often mentioned. Some have been used generally locally, but none even approach any potential for the amount of energy now being consumed without tremendous development and costs. A summary of energy sources and uses is shown in Fig. 22-3.

COAL

Coal is widely distributed, and reserves of it are sufficient to last hundreds of years. It has long been

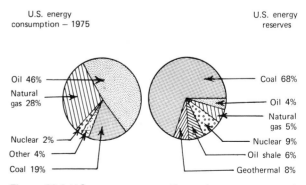

Figure 22-3 U.S. energy consumption and energy reserves in 1975.

the backbone of industrial life. Those countries endowed with it have risen commercially and politically; those lacking it have mostly become agricultural, handicraft, or aggressive nations.

Coal was known in ancient times and in the ninth century entered household use in England. By the thirteenth century, trade in coal was active. The invention of the steam engine in the eighteenth century stimulated active coal mining; the industrial age commenced in England. Coal gradually replaced man power, and eventually mechanical power held sway. When iron ore was smelted by charcoal and the forests of England were vanishing, it was discovered that anthracite was a smelting fuel. This was another stimulus to coal mining. Later, when coke made from bituminous coal was found to be a more readily controlled fuel during burning, the coal industry received a great impetus, and huge industrial expansion ensued. A further stimulus to the coal industry occurred when cities began to produce artificial gas from coal for domestic and industrial use. Its high position receded under the competition of oil and gas.

In North America, the Indians burned coal in 1660; bituminous mining began (in Virginia) in 1787, and anthracite mining in 1805.

Uses of Coal

Coal is a primary source of heat and power. It is estimated that, of the coal produced, 80 percent is used for fuel and 20 percent is used as raw coal in producing pig iron (10 percent), steel (7 percent), and gas (3 percent).

The bituminous coal used to make coke for metallurgical and other purposes yields important by-products. One ton of coal yields about 70 percent coke and by-products.

World Coal Reserves

The total world reserves of coal, to a depth of about 1,000 meters are about 7600 billion metric tons of which 56 percent belongs to Russia with its reserves located principally in Siberia, Kazakhstan, and Southern Europe. The reserves are listed in Table 22-1.

The estimated total coal resources of the world by continents are shown in Table 22-2.

As indicated, therefore, the total world coal resources are estimated to be about 16,830 billion short tons of which 9500 billion tons are classed as identified, and 7330 billion tons are classed as hy-

Table 22-1 World Reserves of Coal

	10^9 metric tons	Percent
Soviet Union	4310	56
United States	1406	20
Asia (without USSR)	681	9
Canada and Mexico	601	8
Western Europe	377	5
Africa	109	1.5
Australia and Oceania	59	—
Southern and Central America	14	—

pothetical. The United States contains 20 percent of the estimated total world resources. North Dakota has the largest identified reserves remaining in the ground, followed by Montana, Illinois, Alaska, Wyoming, and West Virginia. Thirty-seven states have coal within their borders, not all of which is economic.

Wyoming is ranked first in the total identified and hypothetical coal reserves, followed by North Dakota, Colorado, Montana, and Alaska. The first four states listed above have 58 percent of the U.S. identified and hypothetical reserves.

Figure 22-4 illustrates the amounts of coal resources by depth. Obviously, the resources are great but only about 7 percent of these resources are amenable to stripping and only 3 percent is available for use and economically recoverable. Technology will undoubtedly increase this percentage becasue of the increased value of coal, during the present and undoubtedly future energy crisis, allowing more overburden to be removed even at increased costs.

If one is confused by the discussion on reserves

Table 22-2 Estimated World Resources of Coal (in billions of short tons)

Continent	Identified resources	Hypothetical resources	Estimated total resources
Asia	7,000	4,000	11,000
North America	1,720	2,880	4,600
Europe	620	210	830
Africa	80	160	240
Oceania	60	70	130
South and Central America	20	10	30
Totals	9,500	7,330	16,830

Source: Averitt, U.S. Geol. Survey Prof. Paper 820.

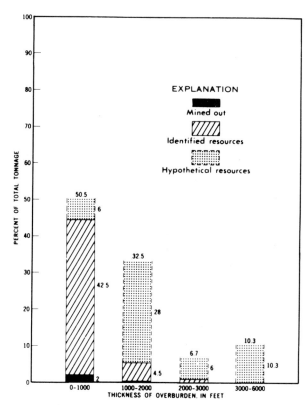

Figure 22-4 Probable distribution of total estimated U.S. coal resources according to the thickness of overburden. (From Averitt, 1973, U.S. Geol. Survey Prof. Paper 820.)

and resources of coal, the following possible dialogue of Canadian coal reserves should indicate that the term *reserve* can refer to various things as suggested by Robertson.

Q: Are our coal reserves really 120 billion tons?
A: This is the figure quoted by the experts of the Canadian Federal Government.

Q: Is this all proven reserve?
A: No, less than 10% of that is proven, say 10 billion tons.

Q: Could we presently open mines on 10 billion tons?
A: No, present economics would restrict us to about 1.5 billion tons.

Q: Is this strippable coal?
A: No, only about 700 million tons is strippable.

Q: Will this strippable coal last for some hundreds of years?
A: No, Alberta alone expects to burn 700 million tons of coal in the next 25 years.

Classification of Coal

Vitrain (or *anthraxylon*) constitutes thin bands of bright, glassy-looking, jetlike coal with conchoidal fracture. The woody structure is not visible megascopically. Its brilliance, approaching jet, varies with the rank of coal. Vitrain supplies coking qualities.

Durain is dull coal, lacking luster and having a matte or earthy appearance. It is hard, black to lead-gray in color, and consists of cuticles, spores, and so forth, formed in water less toxic than for vitrain.

Clarain forms thin bands in coal characterized by bright color and silky luster. It is composed largely of translucent attritus.

Attritus is finely divided plant residue composed of the more resistant plant products.

A new classification of coal was adopted for American usage in conformity with the provisions of the Joint Geological Survey-Bureau of Mines Resource Classification Agreement of November 21, 1973, covering all mineral resources. This classification will be used in future resource/reserve studies on coal conducted by agencies of the Department of the Interior.

This system employs a concept by which coal beds are classified in terms of their degree of geologic identification and economic and technological feasibility of recovery.

Coal is also classified by *rank* (Fig. 22-5).

Ranks and Kinds of Coal and Classification

In common usage, coals are divided into four main groups: (1) anthracite or hard coal, (2) bituminous or soft coal, (3) lignite, and (4) cannel coal. The last is a special type; the others are divided into ranks. Thus from the lowest rank upward are lignite, brown coal, subbituminous, bituminous, superbituminous, (three ranks), semianthracite, and anthracite. Peat lies below lignite, and graphite above anthracite.

Peat. Peat is not coal even though it is a fuel. It is an accumulation of partly decomposed vegetable matter that represents the first stage in the formation of all coals.

Lignite. Lignite (and brown coal) represents the second stage. It is brownish black and is composed of woody matter embedded in macerated and decomposed vegetable matter. It is banded and jointed and, because of its high moisture content, slacks or disintegrates after drying in the air. It is subject to spontaneous combustion and has low heating value. It is used for local fuels and to make

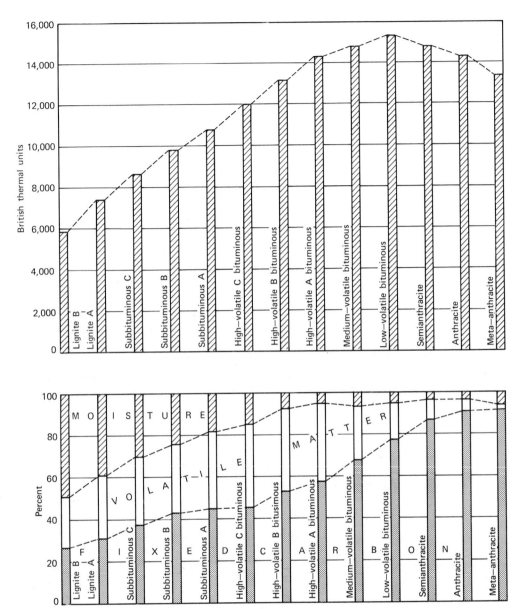

Figure 22-5 Heat value of coal of different ranks compared to proximate analysis. (From: Skillings Mining Rev., Sept. 27, 1975.)

producer gas, and in powdered form for heating and steam raising; it had supplied synthetic petroleum in Germany.

Subbituminous. This intermediate coal is often difficult to distinguish from bituminous coal. It is dull, black, and waxy. It shows little woody material, is banded, and splits parallel to the bedding but lacks the columnar cleavage of bituminous coal. Some varieties distintegrate upon exposure. It is a good clean fuel but of relatively low heating value.

Bituminous. Bituminous coal is a dense, dark, brit-

tle, banded coal that is well jointed and breaks into cubical or prismatic blocks and does not disintegrate upon exposure to air. Vegetable matter is not ordinarily visible to the eye. Dull and bright bands and smooth and hackly layers are evident. It ignites readily and burns with a smoky yellow flame. Its moisture is low, volatile matter medium, fixed carbon high, and heating value high. It is the most used and most desired coal in the world and serves for steam, heating, gas, and coking.

Cannel coal is a special variety of bituminous coal that breaks with a splintery or conchoidal fracture, is not banded, does not soil the fingers, and is lusterless. It is made up of windblown spores and pollen. It is clean, burns with a long flame, and is preferred for fireplace coal.

Bituminous coals are subdivided into several varieties as shown in Fig. 22-5.

Anthracite. Anthracite is a jet-black, hard coal that has high luster, is brittle, and breaks with a conchoidal fracture. It ignites slowly, is smokeless, burns with a short blue flame, has a low sulfur content, and has high heating value. It is restricted in distribution and was used almost exclusively for domestic heating although some is now blended with coking coal and some is used to produce carbon. It also is subdivided into varieties.

Constitution of Coal

Chemical. Chemically, coals are made up of various proportions of carbon, hydrogen, oxygen, nitrogen, and impurities. As shown in Fig. 22-6, from lignite to anthracite there is a progressive elimination of water, oxygen, and hydrogen, and an increase in carbon. The carbon is present as *fixed carbon* and in *volatile matter*, and the ratio of these to each other is an important characteristic of coal. Thus

$$\frac{\text{fixed carbon}}{\text{volatile matter}} = \text{fuel ratio}$$

cellulose	(2)	protoplasm	(1)
pentosanes	(2)	starches	(2)
lignose	(7)	cutin	(8)

The numbers in parentheses represent, according to David White, the order of assailability to bacterial action. Resin and waxes, being the least attacked, are present in most coals.

Fusain, or mineral charcoal, or "mother of coal," is carbonized wood resembling charcoal. It is high in ash. It has been thought to have come from burned wood, but White shows that the ducts full of resin argue against burning.

Origin of Coal

That coal is of vegetal or plant origin has been accepted since 1825, but there have been various forms of this theory, namely, accumulation in a sargassa sea, transported vegetable debris, fresh- or saltwater deposition, lacustrian deposition, and growth of vegetation in situ. Today it is considered

The fuel ratio is high in anthracite and low in lignite; it is the main feature determining the rank of coal. The volatile matter is that which burns in the form of a gas; it causes ready ignition, but too much gives a long, smoky flame or poor storing qualities. The fixed carbon is the steady, lasting source of heat. It produces a short, hot, smokeless flame.

Sulfur is an injurious impurity commonly present in most coals in the form of marcasite or pyrite. More than 1.5 percent S excludes coal for making gas or coke. Environmental requirements are less than 1 percent and also requires recovery of SO_2 from stacks.

Ash is the residual of noncombustible matter in coal that comes from included silt, clay, silica, or other substances. The less present, the better the quality of the coal. Too much ash may put a high-rank coal in a low grade. Too much iron in the ash makes objectionable clinkers.

There are other minor constituents of coal. In practice, however, only the above main constituents are determined, in what is called a "proximate analysis," along with the heating value, expressed in British thermal units (Btu).

Physical Constitution. Banded coals are made up of partly decomposed and macerated vegetable matter, mainly vascular land plants, of which the following parts have been recognized:

resin	(10)	pigments	(3)
gums	(6)	oils	(5)
waxes	(9)	fats	(5)

that all ordinary coals are of plant, nonmarine origin and that the common banded coals were formed in situ. There is some diversity of opinion as to whether the different kinds of coal originated because of different kinds of vegetation or by different degrees of alteration, the latter being the prevailing American opinion as well indicated in the different ranks of coal correlated with different degrees of folding and metamorphism, especially across the Appalachian mountains from the Allegany plateau to the graphite in Rhode Island. The present concept is that all coal originated in swamps and went through a peat stage.

Coals have been found in places where they would not now be formed, as in Spitzbergen, Greenland, and Antarctica which are undoubtedly the result of continental drift whereby these sites have moved from warmer climates, where swamps existed to their present colder positions.

Decomposition and Toxicity. The change from plant debris to coal involves biochemical action producing partial oxidation, preserval of this material from further oxidation, and later dynamochemical processes. The type of coal depends upon the environment, the kind of plants, and the nature and duration of bacterial action. The rank of coal depends upon the degree of metamorphism.

Source Materials. The microscope reveals that the raw material of banded coals was vascular swamp vegetation not unlike that growing in present-day, peat-forming swamps. Over 3000 plant species have been identified from Carboniferous coal beds. Roots and stumps found in underclays below coal beds attest that the vegetation grew and accumulated in situ. Luxuriant vegetation flourished, and consisted mainly of ferns, lycopods, and flowering plants, with conifers and other varieties. Ferns were treelike rushes that grew 27 meters high; lycopods (small shrubs today) attained 30 meters in height. Bulbous and arched roots attest to trees that lived in water. *Sigillaria* and *Lepidodendron* were among them. Some coals have roots in the roof also. None of the plants found are saltwater species. The same kinds of plant remains are found in coals of all ranks.

Places and Conditions of Accumulation. The extensive distribution of individual coal seams implies swamp accumulation on broad delta and coastal plain areas, on subsiding base-level continental regions, on broad interior basin lowlands that have been nearly base-leveled, and where shallow waters persist throughout the year. Most coals are underlain by carbonaceous shales of lake bottom deposition. Low-lying surrounding lands are necessary, else there would be too much inflow of silt. The freshwater swamps necessary for the growth of the plant species found in coal may be separated from the sea by only a sand bar, or barrier of vegetation, since marine beds commonly alternate with, or rest unconformably upon, coal. A subsiding shore line would supply the best conditions of accumulation. Coal fields show cycles of deposition. They are not rhythmic.

The rate of accumulation depends upon the vegetation and climate. J. Volney Lewis estimates that it would take 125 to 150 years to accumulate enough material for 1 foot of bituminous coal and 175 to 200 years for 1 foot of anthracite.

Climate. The climatic conditions that favored coal accumulation were mild temperate to subtropical, with moderate to heavy rainfall well distributed throughout the year. Severe frosts were absent, but the climates were not without dry spells, since some of the plants have water containers in their trunks and roots. The lush vegetation and the large growth means, however, a well-distributed rainfall also. The Carboniferous growth rings are poor, indicating the nonseasonal type of climate. Probably, the climate was much like that of the Carolinas or Florida.

When a tree falls upon dry land, it decays. The complex constituents become broken up into CO_2 and H_2O, with which it started, and most of it is returned to the air. No coal accumulates. When, however, vegetation falls into water or a reducing environment, a similar decay sets in, but it operates more slowly. A necessary step is that the decay be arrested before complete destruction ensues so that the residue can accumulate. This is accomplished by a reducing environment and the lack of oxygen which prevents further decay of the vegetable tissues and permits their preservation and accumulation.

The biochemical changes liberate oxygen and hydrogen and concentrate the carbon. The bacteria are most active near the surface and first attack the most easily decomposable constituents, such as the protoplasm, cellulose, and starches. The resistant substances, such as the waxes, resin, and cutin, along with woody fragments, drop to the bottom of the swamps, where the toxicity slows up, or prevents, further decay and humus accumulates. The kind of bacteria and the duration of bacterial action largely determine what constituents collect in the humus. The bacterial attack may take out certain ingredients of the wood so that it appears to have been chewed up. The parts not destroyed collect in a jellylike mass that becomes more pasty and finally forms the binder of the whole. Under normal conditions of water supply and no agitation of the surface, all but the least resistant of the vegetable matter is preserved. However, heavy rains will oxygenate the water, lessen the toxicity, and foster further decay. Floods are catastrophic in that toxicity is lessened and material is washed away. In a dry season the water will be lowered and detritus may be lost by exposure to the air, where decay proceeds. The water level thus plays an important part—a stable level gives vitrain; dilution gives thin layers of cutinous material; attritus or wash material gives durain. The accumulation of this detrital material in the swamp gives peat, the initial stage of coal.

Carbonization of Progressive Metamorphism. The progressive change from peat to anthracite involves chemical, physical, and optical changes. After the peat stage bacteria presumably play little part, and most of the changes are chemical, induced by pressure and slight increases of temperature, which result from deposition of overlying sediments. In peat

the oxygen has already been reduced about 10 percent. The further progressive elimination is caused by loss of H_2O, CO_2, and CH_4.

The chief changes, according to David White, effected by progressive metamorphism are

PHYSICAL
1. Compaction, drying, induration, and lithification.
2. Jointing, cleavage, and schistosity.
3. Reconstruction.

4. Optical changes.
5. Dehydration up to anthracite.
6. Color change—brown to black.
7. Increase of density.
8. Change of luster.
9. Fracture changes—from bedding to cleavage to conchoidal.

CHEMICAL
1. Progressive elimination of water up to anthracite.
2. Progressive loss of oxygen, lower Eh.
3. Conservation of hydrogen up to graphite stage.
4. Progressive increase of ulmins.
5. Progressive loss of bitumens.
6. Development of heavy hydrocarbons.
7. Large loss of H in anthracite.
8. Increased resistance to solvents.
9. Increased resistance to oxidation.
10. Increased resistance to heat.

The result of these changes is the successively higher-rank coals previously described. The same plant constituents are observed in the higher-rank coals as in peat.

The change in rank is largely a result of pressure and time. The older the coals, the more likely they are to be more deeply buried, which increases the pressure and accelerates the metamorphism. Lateral pressure induced by folding (accompanied by increased temperature) is also important. Folded Tertiary coals of Alaska are high-rank coals; the nonfolded ones of Dakota are lignite. In Pennsylvania the metamorphism of the coal increases with the intensity of the folding; anthracite coals are in closely folded beds; bituminous coals of the same age are in gently folded beds. The more competent the enclosing beds, the greater will be the metamorphism of the included coal. In places where local igneous intrusions occur, the rank of the coal is increased, even to natural coke.

Occurrence and Age

Coal occurs as a sedimentary rock within "coal measures," which consist of alternating beds of sandstone, shale, and clay, mostly of freshwater origin. The coal beds, or seams, in a large way are flat lenses, although some are remarkably persistent. The Pittsburgh seam, for example, underlies 38,000 square kilometers. Coal seams are generally underlain and overlain by shale, although clay or fireclay is a common "underlie," and sandstone may constitute the roof. The character of the roof affects mining operations, since a weak shale roof

may need much support, and a very strong sandstone roof may not cave uniformly during mining.

The thickness of coal seams ranges from a mere film up to 30 meters. One of the thickest beds in the United States is at Adaville, Wyoming. This is 25 meters thick. Recent drilling in the Powder River Basin has located coal seams that exceed 50 meters thickness. The famous Mammoth seam in Pennsylvania is 15 to 18 meters, but most seams range between ½ and 3 meters and are rarely 6 meters; the widespread Pittsburgh seam averages 2 meters. Most coal seams exhibit considerable variations in thickness, as would be expected from a swamp origin. Generally, there are several coal seams in a given vertical section, not all of which are workable, however. For example, in Pennsylvania there are 29 seams aggregating 32 meters of coal; in Alabama 55 seams; in Indiana 25 seams totaling 27 meters, of which 9 are minable; in England there is an aggregate thickness of 25 meters, and in Germany 36 meters. There are more than half-a-dozen coal seams in the Navajo coal mine in New Mexico, when the thinner upper seams are removed along with shale during the first pass of the overburden stripping.

Coal seams exhibit the structural features of sedimentary rocks. They commonly contain "partings," or thin bands of shale or clay, called "bone" or "slate," which interfere with mining of clean coal; thick partings are called splits. The Commentary seam of France on one side of the basin divides into six splits of sandstone, showing inroads of sediment from that side during accumulation. Seams also exhibit "cut outs" caused by erosion along a channel and subsequent filling by later sediments; "rolls" where the floor bulges up into the coal; "horsebacks," where a body of foreign matter lies in the coal; and "clay slips."

Folding produces protuberances of roof and floor into the coal and pinches and swells of the soft coal along the limbs and crests of folds; many supposedly thick seams are merely thin seams squeezed into anticlinal bulges. Faulting and slickensiding are not uncommon. In a general way the anthracite coals are in regions of close folding and the bituminous coals and lignite are in regions of gentler folding.

Age. Coal occurs in all post-Devonian periods. The Carboniferous (Mississippian and Pennsylvanian) received its name because of its world-wide inclusion of coal. Permian coals are less widespread but are found in China, Russia, India, South Africa, Australia, and perhaps in Kansas. The Triassic contains coal in Australia, central Europe and eastern Asia, and the Jurassic in Alaska, China, Australasia, and Austria. The Cretaceous ranks next to the Carboniferous in importance, containing extensive beds in western North America and central Europe. The Tertiary yields most of the lignite of the world, although high-rank Tertiary coals occur in Alaska and elsewhere. Miocene coal occurs in Antarctica.

Distribution of Coal

In *North America* the world's most bountiful supply is in the United States (Fig. 22-6) with northward extensions into eastern and western Canada and a great blank area between the Rockies and the Atlantic north of the Great Lakes. Coal is sparsely distributed in northern Mexico. In *South America* small coal basins occur in Colombia, Peru, and Chile. In *Europe* extensive coal deposits occur in England, west-central Europe, and Russia but are sparse in Scandinavia, the Mediterranean countries, Switzerland, Bulgaria, and Rumania. *Asia* contains vast fields in Russia, China, and India and moderate ones in Japan, Manchuria, Vietnam, and Iran. *Africa* is deficient in coal; moderate-sized deposits occur in South Africa, Zaire, Rhodesia, and Nigeria and small deposits in Madagascar, Nyasaland, East Africa, and Ethiopia. *Oceania* holds

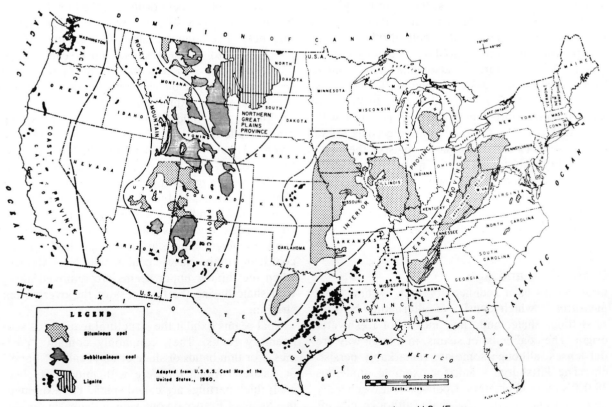

Figure 22-6 Bitummins and subbituminus coal and lignite fields of the conterminus U.S. (From: Skillings Mining Rev., April, 1975; adapted from U.S. Geol. Survey Coal Map of the United States, 1960.)

large deposits in Australia and moderate ones in New Zealand, Indonesia, North Borneo, and the Philippines.

Coal Fields of the United States

The economic exploitation of coal involves a number of interacting industries and factors: complicated mining and complex processing procedures; large-scale transport both within the United States (Fig. 22-7) and elsewhere in the world (Fig. 22-8); the meeting of safety standards and environmental restrictions; the strip-mining problems; and so forth. All of these factors, especially the high cost, have had to be considered by major coal-producing concerns.

Overburden thickness affects estimates of mineable coal as only those seams that exceed 1 meter in thickness are included in reserves.

The United States is so bountifully supplied that few states are more than one state away from coal; it is mined in 33 states. The various bituminous and lignite fields are shown in Fig. 22-6, the Appalachian, Interior, and Rocky Mountain being the most important.

Appalachian Field. This field embraces tne coal areas of Pennsylvania, Ohio, West Virginia, Virginia, Maryland, Kentucky, Tennessee, Alabama, and Georgia, or an area of around 70,000 square miles, 75 percent of which contains workable coal. The thick coal measures consist of westward-thinning, intercalated lenses of sandstone, conglomerate, shale, limestone, fireclay, and coal. They are mostly of Pennsylvanian age, but the Mississippian also contains coal. The beds were involved in the Appalachian orogeny and are strongly folded in eastern Pennsylvania where three anthracite basins

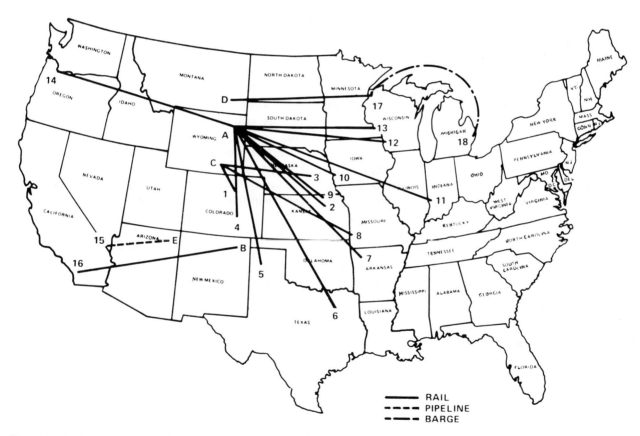

Figure 22-7 Map showing selected long distance transportation systems for coal in the U.S. 1, Denver, Colorado; 2, Linn County, Kansas; 3, Omaha, Nebraska; 4, Pueblo, Colorado; 5, Amarillo, Texas; 6, Cason, Texas; 7, Flint Creek, Arkansas; 8, Kansas City; 9, Marysville, Kansas; 10, Council Bluffs; 11, Sullivan, Indiana; 12, Lansing, Iowa; 13, Alma, Wisconsin; 14, Messner, Oregon; 15, Mohave Power Station; 16, Fontana, California; 17, Superior, Wisconsin; 18, Detroit, Michigan; A, Gillette, Wyoming; B, Kaiser Steel; C, Hanna, Wyoming; D, Decker Coal; and E, Black Mesa, Arizona. (From Skillings Mining. Rev., September 27, 1975.)

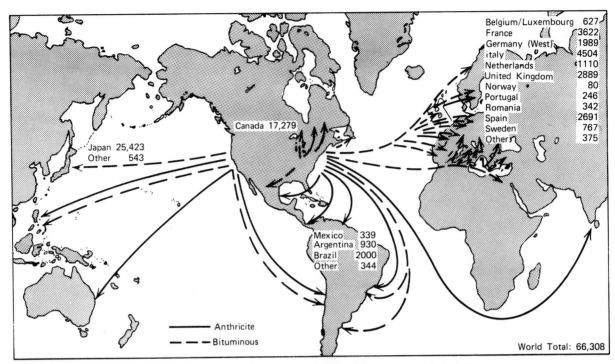

Belgium/Luxembourg 627
France 3622
Germany (West) 1989
Italy 4504
Netherlands 1110
United Kingdom 2889
Norway 80
Portugal 246
Romania 342
Spain 2691
Sweden 767
Other 375

Canada 17,279

Japan 25,423
Other 543

Mexico 339
Argentina 930
Brazil 2000
Other 344

———— Anthricite
– – – – Bituminous

World Total: 66,308

Figure 22-8 United States coal exports for 1975. Anthracite and bituminous in thousand net tons. (From Steel '76, Am. Steel Ind.)

exist, less so in the west, and steeply folded and faulted in the south. Economically, the field may be divided into three units, namely, the anthracite unit, the northern unit, and the southern unit, the southern being included within the Southern States.

The *anthracite area* occupies 1240 square kilometers in 3 basins in northeastern Pennsylvania, where erosion has left only synclinal residuals in intricately folded coal measures. These residuals represent about 5 percent of the original extent before destruction by erosion. About 6.5 billion tons have been extracted to date and about 20 billion tons remain. As bituminous coal is preferred especially for producing coke, the anthracite area is and has been destitute for some time. The several coal seams range from a few centimeters to the 18 meters of the Mammoth seam, which, however, splits in places to form several seams. The close folding has produced much crushed coal or "fines." The steep dips, the hard coal, and many thin beds have made anthracite more costly to mine than bituminous coal but since 1957, anthracite coal has been mined by open pit methods. It is separated from rock, sized, and cleaned in large "breaker" plants.

The *northern bituminous area* is the heart of the American coal industry. Here the coal measures form hill cappings or occur in gently dipping basins;

northeastward the folding is greater and the bituminous belt is flanked by semianthracite coal. Numerous seams are mined, among them the Pittsburg, in Pennsylvania and Ohio, which averages 1.9 meters in thickness over an area of 6400 square kilometers; the Pocahontas No. 3 bed, underlying 770 square kilometers in West Virginia, averages 2 meters in thickness. The coal seams are persistent, regular, lie at shallow depths, and are gently dipping. Consequently, they are cheaply mined, by massive strip mining techniques. The coals are clean, low in ash, and embrace all ranks. In general, the lower-rank coals lie to the west and the higher-rank to the east, where folding is more accentuated. The coals of western Pennsylvania, Ohio, and Kentucky are mainly noncoking. High-grade coking coals occur in the eastern and central part. Individual districts are widely known for superior coals, such as the Connellsville, Pennsylvania, for coking coals; the Kanawha, West Virginia, for gas coal; and the famed Pocahontas coal of the Virginias, for its smokeless steam coals. This field furnishes the great eastern industrial centers with steam and gas coals and much of industrial United States with coke.

The *southern bituminous area* is one of folding and faulting, with steeply dipping coal seams distributed in four basins. There are many seams, but few exceed 2 meters in thickness. There are good coking and fair steam coals, but numerous partings do not give cheap or clean coal. In proximity with Clinton iron and limestone, however, this coal forms the backbone of the large Birmingham steel industry.

Interior Fields. There are four Interior fields, the Eastern, Western, Northern, and Southwestern.

The *Eastern Interior field* is an oval-shaped basin in Illinois, Indiana, and Kentucky; mining is restricted to the margins. There are 9 minable seams out of 25, ranging from 0.6 to 2 meters in thickness. High ash and sulfur make most of the coals undesirable for coking, but they are excellent steam and domestic coals. The production ranks next to the Appalachian field because of the proximity to large industrial regions. Much of this coal is mined by shovel stripping.

The *Western Interior field* lies in the Great Plain states from Iowa to Arkansas, where the coal measures are mostly flat and shallow. The coals are all bituminous, except for some semianthracite coals in Arkansas, but they are of lower rank and grade than eastern coals. They have been used for local steam and domestic fuel because of strong competition with petroleum and natural gas; but with the present crisis of energy, the fields are undergoing increased production, and others are being studied for development, such as the Dakota deposits.

The *Southwestern field*, in Texas, is an extension of the Western field. There are three workable bituminous seams of noncoking coal of Pennsylvanian age.

The *Northern field* is a smaller, shallow basin in Michigan, containing irregular Pennsylvanian bituminous seams that yield fuel coal.

Rocky Mountain Field. This extends along the Rockies from Montana to New Mexico and contains more than 60 percent of the U.S. identified and hypothetical reserves. The coals are bituminous and lignite, with some local anthracite where altered by igneous intrusions. They are mostly Upper Cretaceous in age (Fig. 22-9). The field consists of several disconnected basins with coals of variable rank, depending upon the amount of folding. Some of the coal is excellent for coking and contains low sulfur, especially in Utah, but much of it is subbituminous; the lignites are mined only locally. The chief bituminous industry is in Colorado, particularly around Trinidad, where good steam and coking coals are mined. Utah yields coking coal for the U.S. Steel plant in Provo, Utah. The Utah reserves are large and because of the quality of Utah coal and its low sulfur content, it is being shipped to eastern states where it complies with environmental requirements.

Other Fields. The *Pacific Coast* field consists of small Tertiary basins in Washington containing lignite to bituminous coal, some of which is coking coal. In *Alaska,* in the Matanuska and Bering River areas, Tertiary coals range from lignite to high-rank bituminous, depending upon how much folding they have endured. Attempts to mine the bituminous coals failed because of their folded, faulted, and crushed character. Lignite and subbituminous coals are mined for local consumption. The vast re-

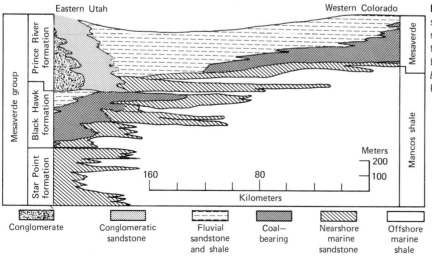

Figure 22-9 Stratigraphic cross section of Upper Cretaceous rocks of eastern Utah and western Colorado showing coal-bearing strata. (From LaPorte, *Encounter in the Earth*, Canfield Press.)

sources of good quality coal credited to Alaska are uneconomic. Carboniferous coals occur in the Arctic.

Coal Fields of Other Countries

Canada. There are three coal areas in Canada—Nova Scotia, western provinces, and the Pacific Coast islands.

Nova Scotia contains four areas of fair bituminous coal of Pennsylvanian age, two of which contain coking coals. In two areas the beds are folded and dip steeply. At Sydney, the most important area, there are several seams aggregating 0.3 to 14 meters of coal. The minable seams dip gently seaward, and a strong roof permits slope mining for 5 kilometers out from shore.

The *western provinces,* Alberta and British Columbia, contain Cretaceous coals in the eastern Rockies and the foothills. The measures are folded and faulted; the rank of the coal rises with the folding from low-rank bituminous and even to a little anthracite. Good bituminous coals are mined at Canmore, Frank, Lethbridge, and Bankhead in Alberta; and Crows Nest, British Columbia. There is considerable coking coal, but much is high in ash and low in fuel value. In places there are 15 to 20 seams aggregating 30 meters of coal; the minable seams range from 1.2 to 2 meters in thickness, although a 25 meter seam is worked at Corbin. The extensive reserves formerly credited to Alberta are now known to be greatly overestimated.

The *Pacific Coast* area, Vancouver and Queen Charlotte Islands, contains Upper Cretaceous coals around Nanaimo and Comox. The measures are folded and faulted, and individual seams pinch and swell. The rank is subbituminous to bituminous, and some coals are coking. The reserves have been greatly overestimated.

Europe. Extensive coal fields in Europe once yielded over half of the world production but now yield only 20 percent. Carboniferous coals are well distributed; there are a few areas of Mesozoic coal; and Tertiary lignites and brown coal underlie broad areas in central and eastern Europe. East-coast U.S. power plants are obtaining low sulfur coal from Poland, which is cheaper than shipping coal from the western United States by railroad. Sweden and Italy have only meager amounts of coal, and Norway, Denmark, Greece, and European Turkey are almost entirely lacking in it.

Great Britain is one of the four great coal-producing countries. All of her coals are of high rank, from bituminous to anthracite. The coals of England and Wales are Pennsylvanian, and those of Scotland are Mississippian.

Major exploration and development work is being done in the United Kingdom in order to increase its coal production (Fig. 22-10). There are three main coal areas in the United Kingdom: the southern, central, and northern. In the southern area, in Wales, there are 8 to 40 seams with a maximum of 36 meters of coal, in coal measures 3,600 meters thick. The coal is of high rank, high grade, and low in ash and sulfur. The central area, in the *Midlands,* has good bituminous coal of lower rank, of which only a minor part is coking. The seams are regular, fairly thick, and are mined to depths of more than 1,000 meters. This coal is the mainstay of the great Midland industrial area of England. The *northern* fields contain thin seams of excellent coking coal, low in ash and sulfur. Thin seams were mined by young teenaged children. The use of child labor in English coal mines during the nineteenth century was supposedly one of the reasons why Karl Marx wrote *Das Kapital.*

Germany has about equal amounts of bituminous and brown coal, the latter occurring in western Germany (Figs. 22-11 and 22-12). The best bituminous coals center around the *Ruhr*, as do the great industrial centers. Carboniferous coal measures dip northward beneath younger rocks, and blind outcrops apex below a peneplaned surface. Some 30 meters of coal are known here, divided among many thin but regular seams. The coals are of lower rank than British coal and are gassy. Excellent coking coal makes the Ruhr field more valuable than other German fields. In the *Saar* (Fig. 22-13), Carboniferous strata contain 40 meters of workable bituminous coal, but it is noncoking, high in ash, and of low fuel value. The *Silesian* field underlies the old boundaries of Germany, Poland, and Czechoslovakia and contains numerous thick seams of Carboniferous bituminous coal.

In *France* and *Belgium*, the Valenciennes-Namur basin has long supplied the coal of this industrial region. The coal is a high-volatile, noncoking bituminous variety, occurring in thin beds that are much folded and faulted and deeply buried. Some 69 seams are known but none exceed 2 meters in thickness. The *Aachen-Limburg* field has been extensively worked by the Dutch.

The USSR contains enormous quantities of Carboniferous coals in the Moscow Basin, the Donetz, and in the Urals. In fact, the USSR has coal in 24

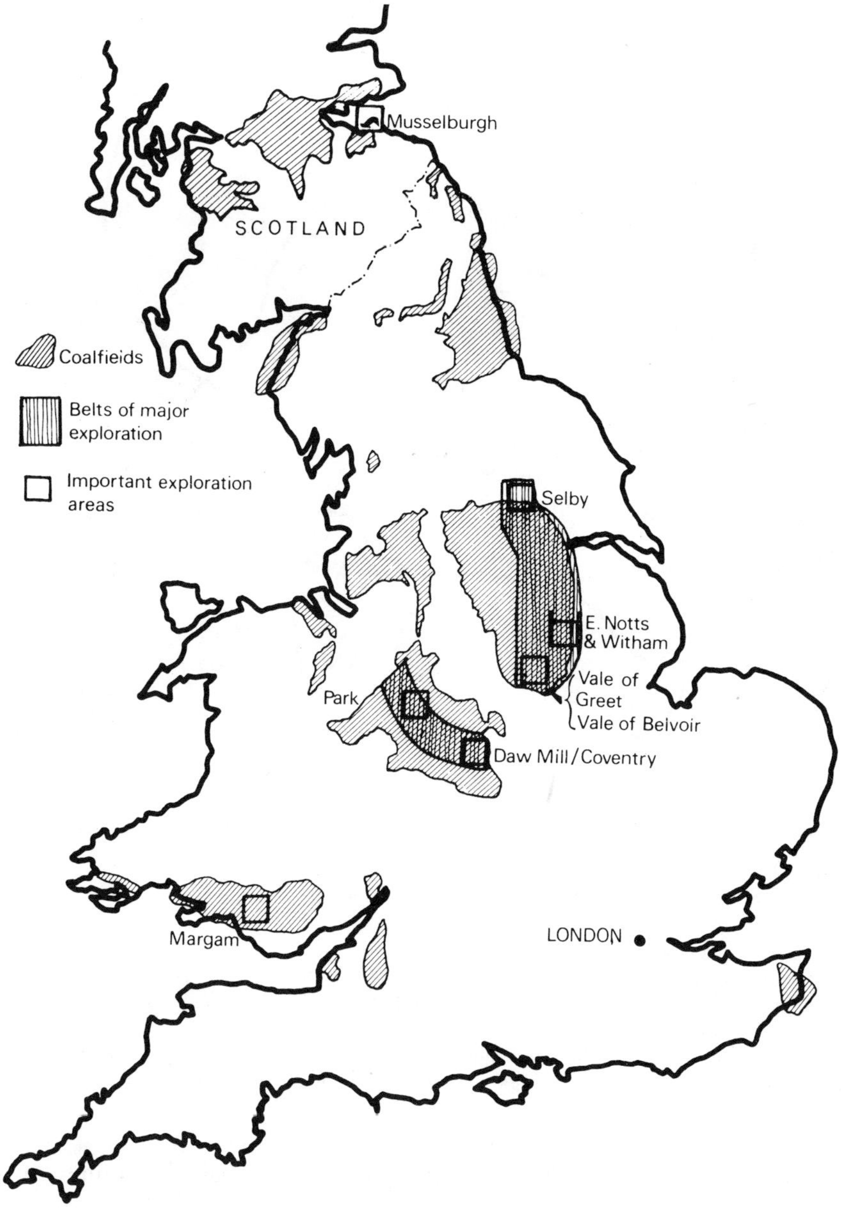

Figure 22-10 United Kingdom coal fields. (From Mining Mag. 1976.)

different basins or areas with the Lena region being the largest, but the greatest mining potential for coal exists in the Kansk-Achinsk basin in Siberia, the Kuznetski basin in Kazakhstan, and the Donetzki basin in the Ukraine. The USSR's resources of 8669 billion metric tons rank it first in the world with 51 percent of the identified and hypothetical coal resources. The Donetz basin contains excellent low-ash, noncoking anthracite, and bituminous coals. Some 135 workable seams are known in a much disturbed area. Bituminous and cannel coals are deeply buried in the Moscow Basin. Mesozoic bituminous coal and Tertiary lignites are abundant in Siberia and in the Caucasus.

Asia. Extensive bituminous and lignite coals have been actively developed throughout *Siberia*. Also, *China* has vast quantities of good Carboniferous to Tertiary coals ranging in grade from lignite to anthracite. In Hunan, one anthracite seam is 15 meters thick, and several average 4 to 5 meters. Large resources of high-rank coal have long been mined in *Manchuria*. Some of the Tertiary, low-rank, bituminous seams are of remarkable thickness; the Chien-Chin-Chai seam ranges from 40 to 60 meters in thickness, with 100 partings aggregating 7 meters. *Japan* has small Mesozoic and larger Tertiary low-rank bituminous coal, some of which is coking. *India* has large deposits of Permian, low-rank, bi-

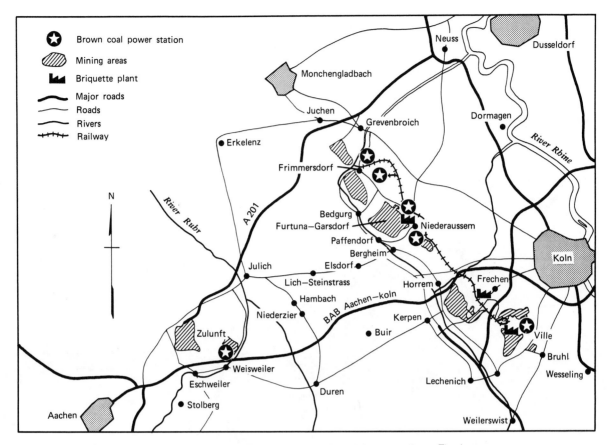

Figure 22-11 Map showing locations of major West German lignite mining operations. The large scale open pit lignite operations in Europe were among the first mining undertakings to develop continuous bulk handling machines. Although early attempts to devise bucket wheel excavators were made as long ago as the 1860s, it was not until the 1920s, after practical endless rubber belt conveyors became available that successful systems were introduced into the German lignite mines, first for removing overburden and later for mining the lignite. In the years that followed, machine sizes gradually increased as frontiers advanced in engineering designs, high strength steels and metals became available, and advances were made in electrical circuitry, motors, and gearing. Today the largest and most productive mining systems in the world are installed in the lignite mines in the Rhineland area. (From Mining Mag., November 1976.)

tuminous coal in the Gondwana series, but good coking coals are scarce. The resources of Asia are about 2540 billion metric tons.

Africa. Coals of Karoo age, in many places overlying the Permian Dwyka tillite, are mined in Rho-

desia and South Africa. These coals are bituminous but are very high in ash. The Wankie deposit of *Rhodesia* consists of seams from 0.3 to 4 meters thick of high-ash, high-volatile, bituminous coking coal that supplies steam coal and coke during friendly times, to the large copper mines of Zambia

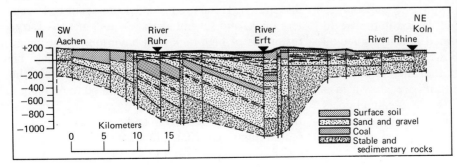

Figure 22-12 A geological section through lignite mining area of West Germany. (From Min. Magazine, November 1976.)

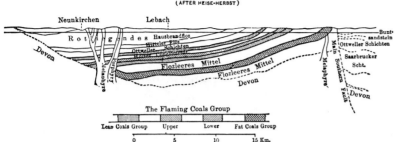

SECTION THROUGH THE SAAR COAL BASIN
(AFTER HEISE-HERBST)

Figure 22-13 Section through the Saar basin coal beds, Germany. (From More, Coal.)

and Zaire. The Karoo bituminous coal of *South Africa* carries the same flora (Permian) as the coals of India and Australia.

Australia. Australia contains the largest coal reserves south of the equator. Valuable Paleozoic coals in *New South Wales* adjoin the coast near Sydney and Newcastle and supply high-grade bituminous coal for steaming, gas, bunkering, and coking purposes. Similar and increasingly developed coals also occur in *Queensland*; one Paleozoic seam at Blair Athol contains 20 meters of clean coal at shallow depth. Most of the Queensland coals now mined are Mesozoic in age. The production at the Goonyella, Peak Downs, and Sarafi mines in Queensland were shut down in the mid-1970's as the result of labor-management nonagreement but have recently reopened.

Victoria contains the thickest seams of brown coal in the world, at Moreland, where a 300 meter bore hole cut three successive seams with thicknesses of 80, 68, and 50 meters, respectively. This coal is mined by open pit and is utilized for electric generation and briquetting. It contains only 33 percent fixed carbon.

Medical Geology of Coal

The trace elements released to the atmosphere with the increased use of coal are being studied because of their potential affect on humans and their environment. Some such elements and their concentrations in ppm as determined by Coleman, et al., in coal or coal ash are antimony (0.4 to 12.5), arsenic (2 to 120), beryllium (5 to 100), cadmium (1.0 to 2.1), zinc (35 to 352), and even some mercury. Of course, sulfur is a major, nonhydrocarbon, element in coal that oxidizes during burning to SO_2, the major pollutant provided in coal burning. It converts to H_2SO_4 at a half-life rate of production of less than 1 day, which then reacts with limestone-faced structures to form gypsum that results in spalling or exfoliation of limestone or marble. Even more important is that SO_4^{-2} is carried as an aerosol in moist atmospheres causing damage to lung tissue.

Selected References on Coal

Coal, 2nd ed. 1940. E. S. Moore. John Wiley & Sons, Inc., New York. Comprehensive textbook.

Geology of Coal. 1940. Otto Stutzer. University of Chicago Press, Chicago. *Excellent for constitution, and for European and brown coals.*

Environments of coal deposition. 1969. E. S. Dapples and M. E. Hopkins. Geol. Soc. America. Spec. Publ. 114, p. 209.

A definition of coal. 1956. S. M. Schopf. Econ. Geol., v. 43, p. 207–225.

Coal resources of the United States. 1975. P. Averitt. Bull. U.S. Geol. Survey 1412. *Latest resource estimates of the United States.*

Coal resource classification system of the U.S. Bureau of Mines and U.S. Geol. Survey. 1972. Bull. U.S. Geol. Survey 1450-B. *Gives definitions of terms used in Table 12-2.*

Some structural features of North Anthracite Region, Pennsylvania. 1940. M. R. Campbell. U.S. Geol. Survey Prof. Paper 193-D.

Bibliography on Coal (general and American). 1939. In *Economic Geology*. H. Ries. John Wiley & Sons, Inc., New York. *Earlier bibliography.*

Classification of coal. 1938. T. A. Hendricks. Econ. Geol., v. 33, p. 136–142.

Coal reserves of Canada, Report of Royal Commission on Coal and Maps. 1947. B. R. Mackay. P. 113. Ottawa, Canada. *Canadian coal reserves revised to 10 percent of 1913 estimates.*

The Nature and Origin of Coal and Coal Seams. 1939. A. Raistrick and C. E. Marshall. I-VII, 13-282. English Universities Press, Ltd., London. *The formation of coals and coal measures with considerable reference to British coals.*

Coal geology: An opportunity for research and study. 1949. G. H. Cady. Econ. Geol., v. 44, p. 1–12. *Problems of the natural history of coal.*

Appropriate research in natural history of coal beds.
1949. E. C. Dapples. Econ. Geol. v. 44, p. 598–605.
Problems of coal sedimentation.

Problems of coal geochemistry. 1949. H. P. Miller. Econ.
Geol., v. 44, p. 649–662. *Relations between chemical and physical properties of coal.*

General Reference 19. *Recent survey of coal in the
United States and extensive bibliography.*

World resources and the world middle class. 1976. Key-
fitz, N. Sci. Am., v. 235, p. 28–33. *Economic development and population problems related to energy consumption expressed in coal consumption.*

Mineral resource estimates and public policy. 1976. V.
E. McKelvey. In *Focus on Environmental Geology*,
R. W. Tank, ed. Oxford University Press, New
York.

23

PETROLEUM AND NATURAL GAS

The Industrial Revolution moves on wheels of steel, cut by industrial diamonds, and lubricated and powered by oil.

Coal production and consumption will increase as a source of energy again, at least extending into the twenty-first century, until it can be gradually supplanted by nuclear, geothermal, and the ultimate solar source. In the meantime, petroleum and natural gas will remain as the main sources of heating and fuel energy for a few more decades in the United States. They are still being used in greater amounts for the generation of electricity even though energy from coal and nuclear power are gradually becoming increasing sources.

More than 50 percent of the production cost, which totals more than $70 billion, of all natural mineral resources is now being paid for petroleum and natural gas. The domestic search for hydrocarbons continues within the United States and offshore. Excellent textbooks and articles are abundant pertaining to this subject and it would be ludicrous to dwell on the subject with a long chapter. Rather, for the sake of completeness, a short summary of petroleum geology follows.

HISTORY OF PETROLEUM

Petroleum was known in antiquity. Both Neolithic and Paleolithic man used bitumens in building, and the Egyptians used it for mummy preservation and for making boats of woven bullrushes. The Japanese used "rock oil" for light over two millenniums ago, and the Chinese are reported to have drilled for it about 221 B.C. Herodotus in 450 B.C. referred to seeps in Persia and Greece, and Pliny told of how the Romans obtained it for illuminating fluid. The oil of Baku has been exploited for over 300 years.

In America, survivors of the DeSoto expedition repaired their ships in 1543 with asphalt from the Sabine river; oil seeps were known in 1627 in southwestern New York; and in 1650 seeps were observed near Cuba. In "Oyl Creek" in Pennsylvania oil was discovered in 1755 and mapped in 1791. In the eighteenth and early nineteenth centuries it was largely a medicinal curiosity, fetching high prices. Up to about 1850 it was quite widely sold for medicinal purposes and thereafter it became used for the extraction of illuminating oil, or kerosene, which gradually displaced the coal oil distilled from coal. The oil for such purposes was obtained from oil seeps and from wells that were drilled for salt. Thus a salt well yielded some oil in West Virginia in 1806, another salt well in Kentucky in 1829 yielded much oil, and prior to 1850 a 400-foot salt well at Tarentum, Pennsylvania yielded oil.

In 1859, "Col." Edwin L. Drake an unemployed conductor on the N.Y., New Haven, and Hartford Railroad, completed the first oil well to a depth of 69 feet at Titusville, Pennsylvania. That well is still producing from a small stratigraphic trap.

Definition

Petroleum ("rock oil") and natural gas ("earth gas") are a mixture of solutions with dissolved gas that are composed predominantly of organic compounds of carbon and hydrogen.

Hydrocarbons have been identified in the 3 billion year old Fig Tree Shale of the Swaziland System, South Africa, and from unmetamorphosed, carbonaceous, middle Precambrian, undoubted microfossiliferrous chert (1.9 billion year old Gunflint iron formation, Ontario). It is possible, however, that these hydrocarbons are present because of contamination and derivation from younger sources of hydrocarbons. As Hoering states, "Vast quantities of geologically younger petroleums are a part of our present environment and can appear in unsuspected places.''

In contrast, the organic matter in the Nonesuch Shale point clearly, according to Barghoorn and others, to a biogenic origin for its organic matter. Be all this as it may, the majority of all oil produced exists in reservoir rocks of less than 100 million year age, followed by Paleozoic reservoir rocks.

The world's known oil and gas reserves are distributed according to the age of the reservoir rocks by 29 percent in Cenozoic, 57 percent in Mesozoic, 8 percent in the Paleozoic, and 6 percent in the Precambrian. The U.S. production in Cenozoic and Paleozoic rocks is much higher and that for Mesozoic is much lower. The figures indicate the impact of the Mesozoic production in the Middle East. As offshore discoveries by the United States increase, however, Cenozoic and Mesozoic production will increase at the expense of Paleozoic derived oil and gas.

Petroleum exploration within the United States has emphasized increased depth of production throughout its history with wells nearing 10,000 meters, and still going down. Even so, by far the majority of petroleum production is from reservoir rocks at depths of less than 3,300 meters. In fact, Heald has stated that, ''If history is a dependable guide, the existence of producible oil or gas at a depth of not more than 2,000 meters, is almost a prerequisite if a major field is to be opened.'' Nevertheless, about 25 percent of present world production is derived from depths greater than 2,000 meters.

The depth versus production of world-wide oil fields is

Depth Meters	Production (percent of total volume)
<500	6.9
500–1000	32.3
1000–1500	26.1
>1500	34.7

Constituents, Types, and Properties

Petroleum is composed of many compounds of carbon and hydrogen, with minor oxygen, nitrogen, and a little sulfur. The numerous members of the different CH series each have different properties; at normal temperatures some are gases, some are liquids, and some are solid waxes. Since the proportion of these varies in different oils, no two oils are alike in their properties or constituents. A crude oil may contain CH members that give the oil a high gasoline content, or certain ones may be absent and the oil will contain little or no lubricants. Thus oils may be referred to as paraffin base, or asphaltic and naphthene, or mixed base.

In general, paraffin-base oils are light and yield good lubricants, and asphaltic-base oils are heavy, unsuited for good lubricants, and may be usable only for fuel oil. The chemical composition conveys little practical information, but the yield does as shown in Table 23-1.

The terms *light* and *heavy* refer to the gravity, measured in arbitrary units called API (American Petroleum Institute), which replaced Baumé units. A specific gravity of 1.00 equals 10 API; 0.9 equals 25.7 API (a medium oil); and 0.8 equals 45.4 (a light oil). By ''cracking'' these crude oils, more distillate gasoline is produced as about one-half of crude oil is now used for gasoline.

ORIGIN OF OIL AND GAS

It is now universally believed that oil and gas are of organic origin. The inorganic theories of volcanic origin, which involve the reaction of carbides within the earth to form acetylene and ultimately hydrocarbons, have been discredited because of the overwhelming accumulation of evidence favoring an origin from low forms of marine or brackish-water life. There exist, however, differences of opinion as to the details of the processes of conversion into petroleum constituents.

In brief, it is held that organic material buried in

Table 23-1 Yields of Contrasting Crude Oils in Percentage

Product	Pennsylvania Light	California Heavy	Mid-Continent Cushing
Benzenes (gasoline)	23	30	31
Kerosene	42	7	20
Gas oil	0	20	15
Lubricants	31	15	0
Residuals	4	25	27

marine muds underwent changes to produce natural hydrocarbons, and these subsequently moved into porous reservoir rocks and accumulated to form commercial oil pools.

The Problem

McCulloh has succinctly presented the questions that need answers in the present energy crisis faced by the industrial nations of the world in regard to oil and gas:

What are the ultimate potential United States and world resources of petroleum and natural gas, both those discovered and those waiting to be found? Where are they? What part of them can be found and exploited economically? What fraction of the total was originally within the territorial limits of the United States? How much of that fraction lies beneath the continental shelves—and where? To what extent will the United States be dependent in the future on imported petroleum, and how can we be assured of an economical supply? What should be the national goal and policy toward encouraging domestic exploration on land and beneath the continental shelves? What part of our national research and investment capability should be devoted to programs aimed at utilizing coal resources or nuclear "fuels" more fully in the future in place of petroleum and natural gas? Should special incentives be provided to support conservative but economically marginal operation of "stripper wells," of "tight gas reservoirs," of secondary recovery operations?

The answers to these questions, and myriads like them, are of the utmost concern or importance today to business leaders, industrialists, economists, politicians, international strategists, and military planners. To an important extent, the answers stem from geological, geophysical, and geostatistical research and investigations. Whether such questions are answered correctly or incorrectly, the life of every citizen living in the United States in the year 2000 will be affected profoundly by the actions that are being taken now on the basis of the answers available.

Production and Reserves

Since 1859, neither the United States nor the world has had proven reserves sufficient to last for more than about 20 to 30 years. Prior to about 1955, however, liquid petroleum discoveries within the United States amounted to about 2 barrels of new reserves discovered for every barrel consumed. Even at the present time, U.S. reserves are about 40 to 50 billion barrels with our consumption nearing 7 billion barrels, or enough to last about 7 years (Fig. 23-1), the lowest number of years in the history of oil production in the United States.

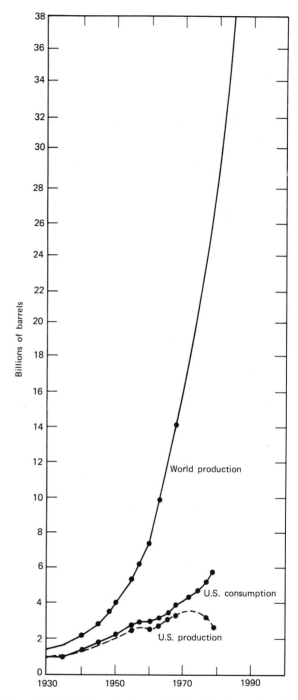

Figure 23-1 World production and United States production and consumption of crude petroleum, 1930–1971. (Modified From U.S. Bur. of Mines and Commodity Data Summaries.)

Only two fields have been found in the United States with reserves of more than 1 billion barrels, East Texas and northern Alaska. In the Middle East more than 25 fields containing in excess of 1 billion barrels each exist; there are about 55 in the world. Figure 23-2 shows the proven reserves of crude oil in several nations. This chart, ignoring the USSR

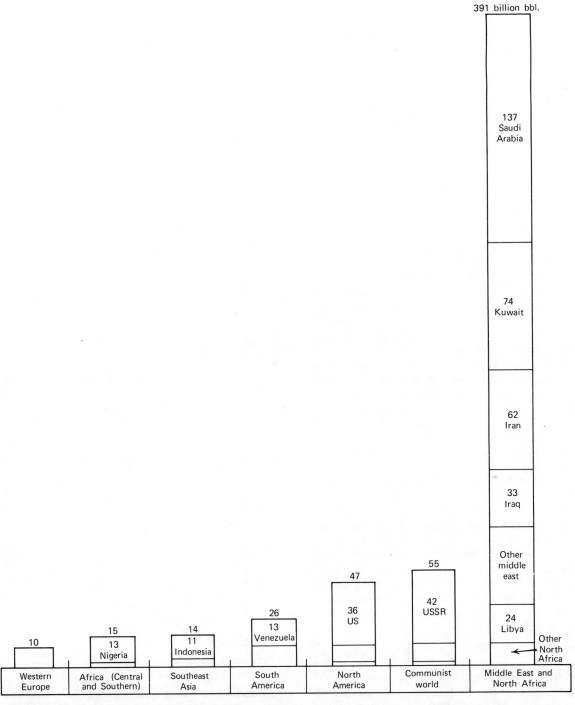

Figure 23-2 Proven reserves of crude oil, 1972. A potentially large discovery made in southern Mexico in 1974 may add significantly to North American reserves. (From *Time*, 1973; and Walkins, et al., 1975, *Our Geological Environments*, W. S. Sanders and Co., Philadelphia.)

reserves, shows that the Middle East and North Africa have 62 percent of the world's crude oil reserves; Saudi Arabia alone has 22 percent of the reserves. The location and concentration of some of the oil fields in the Middle East is shown in Fig. 23-3.

In regard to consumption, Fig. 23-1 shows the situation in detail. As recently as 1950, the United States produced about 60 percent of the world's crude oil and consumed practically all that it produced. At the present time, U.S. consumption is approaching 7 billion barrels per year of which the

Figure 23-3 Map of the Middle East showing the location of some of the well-known oil and gas seepages and of the oil pools. (From Leverson, *Petroleum Geology*, W. H. Freeman and Company, Publishers, San Francisco.)

United States produces less than 3.5 billion barrels. Obviously, the bell-shaped curve predicted by Hubbard (Fig. 1-3) is evident as U.S. reserves reach the zenith of the curve and turn down. But the world production has skyrocketed to about 40 billion barrels per year. That is about equal to the total U.S. reserves. Incidentally, this world production divided into the world's reserves of 632 billion barrels is enough to last the world for only 16 years at the present rate of consumption. Of course, consumption will continue to increase dramatically, and new reserves will be found. By the turn of the century, however, the entire world, including the Middle East, may be having a petroleum energy crisis, and at that time after 25 years of development within the United States of new energy systems, U.S.

technology, hopefully, may be of greater assistance to other nations in regard to their energy problems.

Distribution of Oil and Gas

It is obvious to even the casual observer that producible hydrocarbons are distributed extremely unevenly, not only as to area but also in the geological column. Pipe lines are major evidence of the localization of oil and gas and the need to transport them to the locations of the consumers. Figure 23-4 shows the network of gas pipe lines in the United States.

Source Materials and Source Places

The slow, oxygen-free decomposition of the remains of marine organisms is considered to be the

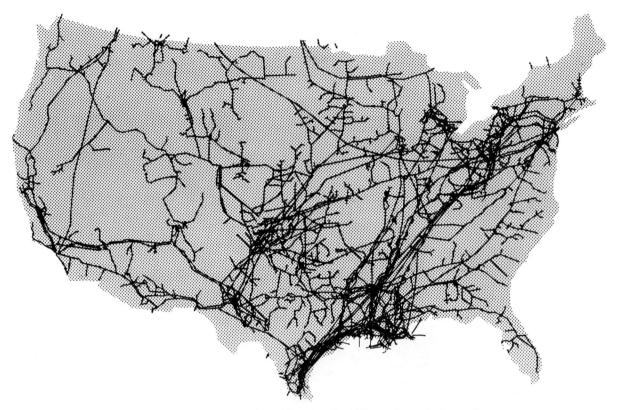

Figure 23-4 Built at a cost of $50 billion and irreplaceable today, the U.S. gas transmission and distribution network of 1 million miles is one of the most efficient methods to transmit energy, consuming less energy per unit shipped than, for instance, electricity.

source of petroleum hydrocarbons, a possibility that has been demonstrated by laboratory experiments. As slow as the process of conversion of organic material to hydrocarbons is, ^{14}C studies indicate that the process can occur in less than 10,000 years.

The higher forms of life that grow and live on the land are unimportant, but the lower planktonic organisms, such as diatoms and algae, that thrive abundantly near the surface of the sea, are now considered the most probable and important source materials. Some geologists think that the differences in oils may be due to differences in source material, but this seems quite unimportant, since the complicated processes that follow could yield vastly greater differences.

The organic remains accumulate in the bottom muds of lagoons or in depressions on the floor of shallow seas and there become incorporated in accumulating sediments. Trask has tested thousands of bottom mud samples and has found the organic content to be fairly constant for about 160 kilometers offshore and to range from 0.3 percent in deep sea oozes to 7 percent off the coast of California.

The average organic content is 2.5 percent in recent sediments and only 1.5 percent in ancient sediments. A high organic matter is no longer considered by Trask to be a reliable indication of a source rock.

Conversion to Petroleum. The bacteria that thrive in the upper mud of the sea floor are thought to change the organic matter into the mother material of oil by removing oxygen and nitrogen and producing other changes. Trask states that the composition of average plankton is 24 percent protein, 3 percent fat, and 73 percent carbohydrates, and thinks that it is not the fats, cellulose, and simple proteins that yield the oil but rather the complex proteins and carbohydrates. Bacterial processes on the sea floor yield oil droplets, prior to burial, as is evident from ^{14}C in recent bottom muds. With deeper burial, bacterial action presumably ceases, and then other, as yet unknown, chemical changes occur to yield oil before consolidation. Further post-consolidation changes probably take place during migration. Pressure probably exerts some influence, but temperatures could not have been high because certain

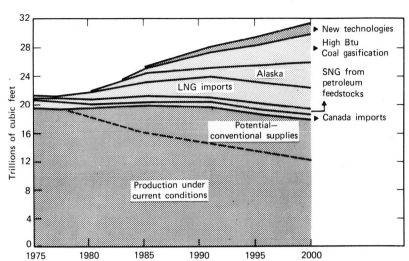

Figure 23-5 Future total potential supplies of gases.

compounds could not exist at 140 to 300°C. The liquid hydrocarbons that are formed are believed to be capable of dissolving other organic substances such as pigments, waxes, and fatty acids. During migration, other organic compounds might also be dissolved, thus continually changing the composition of the petroleum, and perhaps giving rise to the differences in oils.

During the conversion of organic matter into petroleum, natural gas is also formed. This is dominantly methane, much of which is lost to the surface. Natural gas reserves are also dwindling, but there are hopes of more discoveries (Fig. 23-5).

FORMATION OF AN OIL POOL

An oil pool is not a subterranean pond but an accumulation of petroleum in rock pores. Dispersed droplets generated in the source muds do not constitute an oil pool. The conditions necessary for this are (1) migration and accumulation, (2) suitable reservoir and caprocks, (3) suitable traps, and (4) retention.

Migration and Accumulation of Oil

Since many oil pools are found in sandstones, it follows that the dispersed droplets of oil generated within the source muds must have migrated out of the source rocks into the sands. The forces that cause migration are (1) compaction of the muds, (2) capillarity, (3) buoyancy, (4) gravity, and (5) currents.

Compaction. Source muds may contain up to 80 percent water. As covering sediments are depos-

ited, the accumulating weight gradually compacts the lower beds and the enclosed fluids are squeezed out into places of less pressure, such as pore spaces in sands but predominantly to the surface. The fluids may move up, or down, or even laterally. Athy estimates that compaction results in removal of half the water when burial has reached 300 meters and 85 percent at 1200 meters. Compaction is seemingly the most important factor causing migration of oil.

Capillarity. If oil-wet shales are in contact with water-wet sandstones, the water, owing to its higher surface tension, will move from the coarse sandstone pores into the fine, capillary, shale pores, and displace the oil therefrom into adjacent sandstones.

Buoyancy. Oil generally being lighter than water tends to rise upon a water surface. Free gas similarly rises above the oil. This "secondary migration" takes place within reservoir rocks. It is most effective where the pore spaces are large and where large volumes of fluids are involved; it gives rise to the common stratiform arrangement of salt water, oil, and perhaps also gas. If the host beds are horizontal, the oil will tend to rise to the tops of permeable beds and no pronounced accumulation may occur; if inclined, the oil tends to migrate up-dip and, as is shown later, accumulation into oil pools may occur.

Gravity. If water is present, the differences in specific gravity between oil and water give rise to buoyancy, but when water is absent, gravity causes the oil to migrate down-dip until arrested by impervious beds; this gives rise to rare synclinal accumulation.

Accumulation. Migration of oil generally leads to accumulation, which is the collection of oil droplets into pools. Oil may migrate without accumulating, or it may accumulate in noncommercial bodies, as at the top of horizontal beds, but concentrated accumulation is essential to produce commercial oil pools, and this in turn is dependent upon requisite reservoir rocks and traps.

Reservoir and Caprocks

Reservoir Rocks. Accumulation takes place only in porous and permeable rocks. A rocklike clay has high porosity but small permeability and, therefore, is not a suitable reservoir rock. Its numerous minute pores create frictional resistance to flow and cause fluids in them to be held firmly by capillarity. The most suitable reservoir rocks are sands and sandstones. Others are porous or cavernous limestone and dolomite, rarely fissured shale, and rarely jointed igneous rocks. Cavernous limestones, especially biotherms, may yield sensational and prolific flows of oil, as in Mexico. The total yield of oil from sandstone is less than half of the total production.

Unconsolidated sands of California average about 25 percent porosity. In sandstone, the cementing material decreases the porosity from 25 to 12 percent; an average porosity of many oil sandstones is 17 percent. The capacity of sand with 20 percent porosity is 1550 barrels per acre-foot, or 155,000 barrels for a sandstone 30 meters thick. The size, shape, and sorting of the sand grains is important, and the highest porosity results with well-sorted angular grains.

The greater the porosity, the greater the amount of oil that a reservoir rock can contain; and the larger the pore size, the greater the amount of oil it will yield, since with small grains and small pores more oil clings to the rock grains and is not recovered. The percentage of oil recovered from reservoir rocks was surprisingly small. Powers states that the Bradford sand of Pennsylvania yielded only 8 percent of pumping and an additional 20 percent by flushing; Oklahoma sands yielded 10 to 60 percent, and unconsolidated sands of California, with 25 to 40 percent porosity, yielded only 10 to 30 percent of the contained oil. The part of the original pore space that will yield its oil is called the *effective porosity*. Secondary recovery, with newer tech-

niques, has greatly increased the recovery of oil, sometimes to the extent of more than 90 percent.

Caprocks. A confining impervious caprock is necessary to retain oil in a reservoir rock. Shale and clay are the most common caprocks, but dense limestone and dolomite and gypsum also serve, and even well-cemented, fine-grained sandstone or shaly sandstones are effective. Water-wet impervious shales, clays, and limestones even retain gas. Good caprocks form effective seals to underlying oil and gas for long periods of geologic time, but poor ones permit slow escape of mobile hydrocarbons.

Suitable "Traps" or Structures

In inclined sedimentary beds the up-dip migration of oil will continue until it escapes at the surface or is arrested in some trap where it accumulates to form an oil pool. If the strata are folded into an anticline, up-dip migration ceases when the oil reaches the top of the arch (Fig. 23-6), where it accumulates to form an anticlinal oil pool. This is the most common kind of trap.

In the earlier period of oil exploration most of the Appalachian oil discovered was found in domes and anticlines. Consequently, it was assumed that structures were the controlling loci of oil accumulation and there arose such terms as *anticlinal theory of oil accumulation* (proposed by I. C. White, 1885), now in disuse, and *oil structure;* a dry hole was said to be *off structure*. Now, however, it is realized that there are also many other kinds of oil and gas *traps* and these may be divided into structural and stratigraphic traps, as shown in Table 23-2.

Folds. Up-folds, giving rise to *anticlines* and *domes*, have been the source of the greater part of the oil

Table 23-2 Kinds of Oil and Gas Traps

Structural Traps	Stratigraphic Traps
Anticline[a]	Unconformities[a]
Dome[a]	Ancient shore lines[a]
Monocline[a]	Sandstone lenses[a]
Terraces	Shoestring sands[a]
Synclines	Up-dip wedging of sands[a]
Faults[a]	Up-dip porosity diminution
Fissures	Overlaps[a]
Salt domes[a]	Reflected buried hills[a]
Igneous intrusions	Buried coral reefs[a]

[a] The more important types of oil reservoirs.

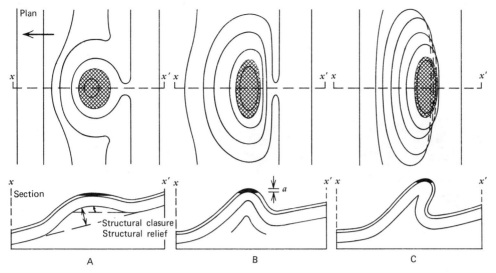

Figure 23-6 Idealized structural maps and sections of typical anticlinal dome folds; such folds are characteristic of many traps containing oil and gas pools. Oil shaded and black; arrow shows direction of dip. (From Leverson, *Petroleum Geology*, W. H. Freeman and Company, Publishers, San Francisco.)

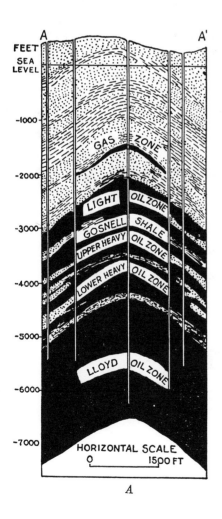

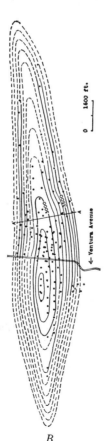

Figure 23-7 Ventura Avenue oil field, California. A, Section illustrating thick producing zones; B, subsurface structural contour map of the oil field. (Drawn by Decius; reproduced by Powers, AIME.)

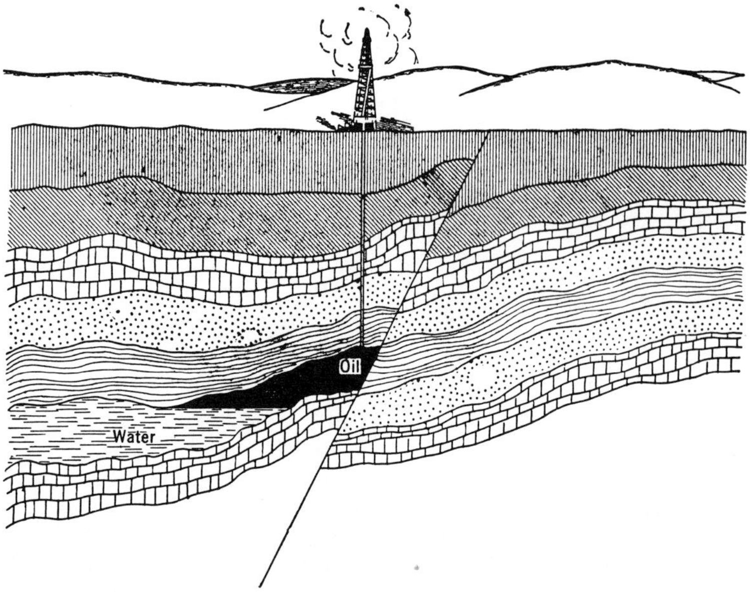

Figure 23-8 Oil trapped by fault
seal.

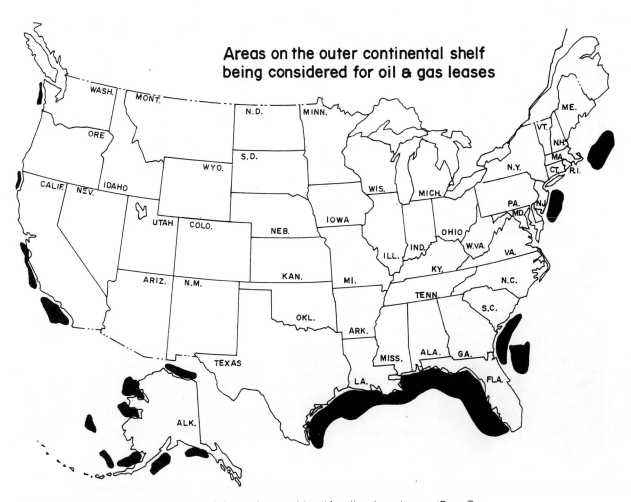

Figure 23-9 Ares on the outer continental shelf being considered for oil and gas leases. (From Geo.
times, October, 1975.)

492

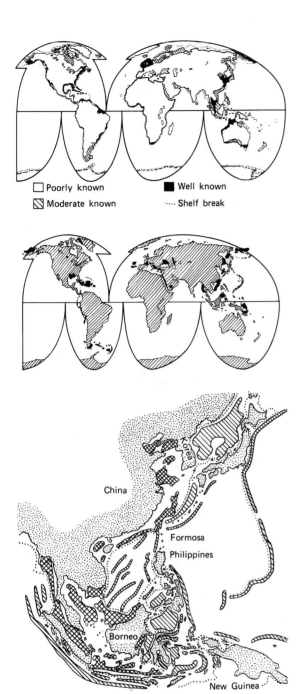

□ Poorly known   ■ Well known

▨ Moderate known   ···· Shelf break

Korea

China

Formosa

Philippines

Borneo

New Guinea

Figure 23-10 Top: Only parts of the world's continental shelves are known even moderately well, and fewer have been explored sufficiently to yield sediments, stratigraphy, and structure. Such detailed knowledge is a prerequisite for the development of offshore oil. Center: Deep marginal basins (solid color) and marginal trenches (dots) are associated with the convergence of ocean-floor and continental plates. Many of these basins may become oil provinces of the future, but both extensive exploration and new technology will be required for their development. Bottom: Many marginal basins off eastern Asia are promising for petroleum resources; some are only partly filled with sediments (single hatched), whereas others underlying the continental shelves have been filled to overflowing (double-hatched). Political—rather than geological—issues are likely to

so far produced. An anticline (Fig. 23-6) is an ideal petroleum "trap," since its extended flanks facilitate migration, its arch permits accumulation, and its crest or "center" generally contains domelike areas in which oil collects from all sides and is retained beneath the caprock. It is customary to draw structure contours on such a fold, and where the contours close it is called a "closed fold"; the amount of "closure," expressed in meters, is the height to the crest above the lowest contour that closes. The oil ordinarily accumulates within the closure. Folds with gentle dips offer larger areas for accumulation of oil; generally, the dips are less than 30° but they may reach as much as 60°, as at Ventura, California (Fig. 23-7). In asymmetric folds, the gentler-dipping limb often contains the most oil.

Anticlinal pools vary greatly in size. The highly productive Long Beach pool of California has an area of only 1305 acres; others may attain tens or even hundreds of square kilometers. Large anticlines are rarely productive throughout.

In ideal arrangement in an anticline, free gas, if present, is at the top, oil lies beneath, and the water below. Commonly, however, in deep wells where pressure is high, the gas is contained within the oil and is liberated only upon drilling. An individual productive sand may range from a few meters to a few hundred, or even to a thousand meters or more in thickness. Also, many productive sands may lie below each other within the same anticline. They may be separated by large intervening thicknesses of barren beds, or be essentially contiguous, as at Long Beach, California, where drills intersected productive horizons between depths of 870 and 2000 meters.

Anticlinal oil structures have been formed by compressive force, deposition of beds over ancient hills, and by differential compaction where shales have undergone greater compaction than sands. Many details of anticlines, beyond the scope of this chapter, are to be found in numerous books on petroleum geology.

Anticlines and domes form important oil traps in practically all the large oil fields of the world.

Synclines, in a few cases, serve as oil traps where water is absent, and *monoclines* and *terraces* constitute important oil structures in several places.

Faults. Faulting has given rise to many important

dictate the way in which these promising formations are explored and exploited. (From Technology Review, March/April 1975.)

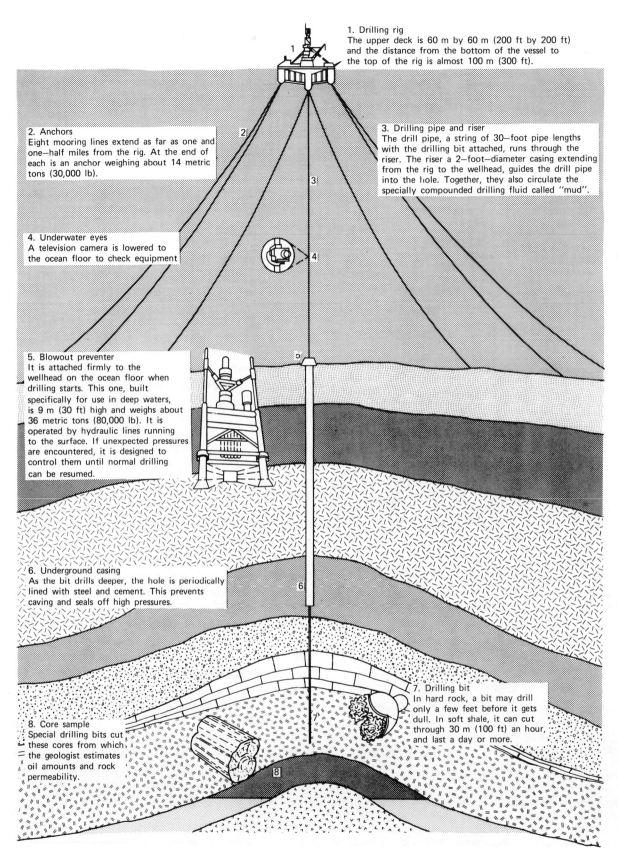

1. Drilling rig
The upper deck is 60 m by 60 m (200 ft by 200 ft) and the distance from the bottom of the vessel to the top of the rig is almost 100 m (300 ft).

2. Anchors
Eight mooring lines extend as far as one and one-half miles from the rig. At the end of each is an anchor weighing about 14 metric tons (30,000 lb).

3. Drilling pipe and riser
The drill pipe, a string of 30-foot pipe lengths with the drilling bit attached, runs through the riser. The riser a 2-foot-diameter casing extending from the rig to the wellhead, guides the drill pipe into the hole. Together, they also circulate the specially compounded drilling fluid called "mud".

4. Underwater eyes
A television camera is lowered to the ocean floor to check equipment

5. Blowout preventer
It is attached firmly to the wellhead on the ocean floor when drilling starts. This one, built specifically for use in deep waters, is 9 m (30 ft) high and weighs about 36 metric tons (80,000 lb). It is operated by hydraulic lines running to the surface. If unexpected pressures are encountered, it is designed to control them until normal drilling can be resumed.

6. Underground casing
As the bit drills deeper, the hole is periodically lined with steel and cement. This prevents caving and seals off high pressures.

7. Drilling bit
In hard rock, a bit may drill only a few feet before it gets dull. In soft shale, it can cut through 30 m (100 ft) an hour, and last a day or more.

8. Core sample
Special drilling bits cut these cores from which the geologist estimates oil amounts and rock permeability.

Figure 23-11 Diagrammatic view of an off-shore drilling rig. With the exception of the special equipment in the water, the drill hole is drilled in much the same way as in land-based drilling. (Courtesy, Exxon Corporation.)

oil pools as, for example, the Mexia-Powell field, Texas. In inclined beds as impervious shale may be faulted against the up-dip continuation of an oil sand, causing an effective seal and permitting oil accumulation beneath it (Fig. 23-8). Rarely, a thick fault gouge may serve as an effective upward seal to an oil sand. Open faults may permit the escape of oil to upper beds or to the surface.

Stratigraphic Traps. Stratigraphic traps, as distinguished from structural traps, are those formed by conditions of sedimentation in which lateral and vertical variations in thickness, texture, and porosity of beds result. These may result from intercalation of beds, interruptions to sedimentation, and sites of deposition with respect to shore lines. A sandstone bed may wedge out or undergo decrease in porosity and thus make a suitable trap. With the discovery of most of the obvious structural traps, increasing amounts of oil are being found in stratigraphic traps, and these will probably constitute the most important sources of future oil especially off the eastern coastline of the United States. More study will have to be given to them and new methods devised for locating them.

Unconformities produce an important class of oil reservoirs of which East Texas is the prime example. They mask tilted strata, overlays, variable porosity, and folding. Underlying, tilted sand beds may be sealed at the unconformity by overlying beds, and accumulation can take place. Angular unconformities are more effective than parallel unconformities. The underlying beds are commonly deeply weathered and therefore porous, and constitute oil containers. If the erosion surface is developed on limestone, solution caverns may be present. The basal beds above an unconformity are also commonly quite porous and may serve as reservoir rocks. Deeper drilling has disclosed oil associated with two and even three underlying unconformities, giving rise to what Levorsen has aptly termed "layer-cake" geology.

Shoestring sands are long, narrow bodies of sand enclosed within shaly beds. These sands may have been deposited in offshore bars by tidal currents in lagoons or by meandering streams. Many small pools have been found in them in the Mid-Continent field.

Shore-line sands formed on low-lying submerged coastal plains, tapering in shore and extending deeper offshore may be covered by clays and form suitable oil containers. This is spoken of as "lensing out" of sands. It generally gives rise to small pools, but it accounts for the eastern portion of the East Texas field.

Buried hills and structures have accounted for several important fields, such as the Oklahoma City, Seminole, and Amarillo pools along the buried granite ridge of Kansas and Oklahoma. Here, projecting granite ridges received a mantle of sediments, thin on top and thicker on the flanks, simulating anticlines, and giving rise to overlap. These stratigraphic features made excellent oil traps.

Other stratigraphic traps caused by porosity variation are *sandstone lenses* enclosed in shales, shore-line *wedges of sand* that become fine grained and impervious offshore (Burbank pool, Oklahoma), sandstone overlaps against impervious beds, up-dip seals by asphalt, and buried coral reef rocks (Alberta and Snyder, Texas).

Summary

An over-simplification of petroleum exploration is to (1) find a suitable source bed from which the hydrocarbons could have migrated which in the future will be predominantly offshore (Fig. 23-9 and 23-10), (2) determine if suitable reservoir rocks exist in the area, and (3) locate a suitable structure for the entrapment of petroleum. The only method used to determine if oil does exist in the trap is to drill (Fig. 23-11). In the United States alone, this has resulted in more than 2,222,300 wells totaling more than 2 billion meters of hole or almost 156 thousand times the diameter of the earth.

24

CERAMIC MATERIALS

China is the home of chinaware and the source of the word Kaolin—from Kaoling

The ceramic industry utilizes common minerals and rocks and produces lowly brick or tile, porcelains for utility, or beautiful chinaware. Its chief raw material is clay, but other substances are also necessary. In the making of brick, terra cotta, tile, or stoneware, clay alone is needed, but for whiteware or porcelain, feldspar and quartz are necessary. For certain porcelains, bauxite is essential, and some even require highly refractory substances, such as andalusite and kyanite. Bentonite, fuller's earth, and other aluminous substances, and pyrophyllite, zircon, and fluorspar are also drawn upon, but the backbone of the industry is clay.

CLAY

Clay is one of the most widespread and earliest mineral substances utilized by man. It carries the records of ancient races inscribed upon tablets, in brick buildings, in monuments, and in pottery. Its products portray the history of man, and by its beautiful wares we trace the development of the delicate artistry of the Chinese, the utility of the Romans, or the humor of the Incas. A wealth of artistic wares culminated in the eighteenth century, but today utility holds sway in the multitudinous utilizations of the varied clay products.

The term *clay* is applied to earthy substances consisting chiefly of hydrous aluminum silicates with colloidal material and specks of rock fragments, which generally become plastic when wet and stonelike under fire. These properties give clays their usefulness, since they can be molded into almost any form, which they retain after firing. Widespread accessibility, ease of extraction, and adaptability to so many uses has resulted in the products of clay entering the wide ramifications of modern industrial civilization. Clay has many other uses than in ceramics, particularly in building and manufacturing, but, for purposes of unity of treatment, all types of clay are considered together in this chapter.

Composition and Properties

Clay is not a mineral but an aggregate of minerals and colloidal substances. The constituents are so fine that, until the use of the X-ray for mineral determination, its exact composition was unknown. Some clay minerals can be observed in detail only by the electron microscope at magnifications greater than 5000 times. Residual clay is often called kaolin (after a hill in China where firing of ceramic ware was done), but this is now recognized as just one variety of clay. It was formerly thought that kaolin was composed of the mineral kaolinite. It is now known that, although kaolin contains considerable kaolinite, other clay minerals are also constituents.

The *clay minerals* are flakelike, lathlike, fiberlike, or hollow-tube-shaped, and they are recognized by the microscope, X-ray, and differential thermal analyses curves. They have replaceable bases; the one formed is determined by the mode of origin and may change in response to changes in environment. The clay minerals are

Group	Composition	Origin Occurrence	
A. Kaolinite			
1. Kaolinite	$Al_2Si_2O_5(OH)_4$	H, W	China clays,
		H	underclays, soils, wallrocks
2. Dickite	$Al_2Si_2O_5(OH)_4$	H	Wallrocks. U

496

	3. Nacrite	$Al_2Si_2O_5(OH)_4$		H	Wallrocks. U
	4. Anauxite	$Al_2Si_2O_5(OH)_4$		W	Soils. U
	5. Halloysite	$Al_2Si_2O_5(OH)_4$		H, W	Soils
	6. Endellite	$Al_2Si_2O_5(OH)_4 \cdot 2H_2O$		W	Soils
B.	Montmorillonite				
	1. Montmorillonite	$Mg_2Al_{10}Si_{24}O_{60}(OH)_{12}[Na_2,Ca]$		H, W	Soils, bentonite, fuller's earth,
				H	wallrocks
	2. Nontronite	$Fe'''Si_{22}Al_2O_{60}(OH)_{12}[Na_2,Ca]$		H	Veins
	3. Saponite	$Mg_{18}Si_{22}Al_2O_{60}(OH)_{12}[Na_2]$		H	Veins
	4. Beidellite	$Al_{13}Si_{19}Al_5O_{60}(OH)_{12}[Na_2]$		H	Vein gouge
	5. Hectorite	$Li_2Mg_{16}Si_{24}O_{60}(OH)_{12}[Na_2]$		W	Clays
C.	Hydrous micas (illite)	$(OH)_4K_2(Si_6 \cdot Al_2)Al_4O_{20}$		W	Soils, marine clays, underclays
D.	Miscellaneous				
	1. Palangorite (Attapulgite)	$Mg_5Si_8O_{20}(OH_2)_4 \cdot 4H_2O$		W	Fuller's earth
	2. Sepiolite-like	$Mg_6Si_8O_{20}(OH)_4 \cdot nH_2O$			
	3. Allophane	$Al + SiO_2 + H_2O$		W	Clays, soils

W = weathering; H = hydrothermal; U = uncommon; square brackets surround exhangeable bases.

The chemical equations are not indicative of the structural composition of clay minerals. Kaolinite has, for example, a composition expressed as $(OH)_8Si_4Al_4O_{18}$, which indicates that kaolinite consists of an Si_4O_{10} sheet structure (Fig. 24-1) with Al^{+3} substituting for some Si^{+4}, and with (OH) molecules oriented within and above the pockets formed by four tetrahedrons in an octahedral structure. As silicon and aluminum have different ionic charges of +4 and +3, respectively, the ionic charges of the colloidal particles are different; this enables them to absorb water to balance their charges and provide the plasticity of clays.

Clays also contain other substances such as rock fragments, hydrous oxides, and colloidal materials, and their nature often determines the value and use of a particular clay. Obviously, a different clay is required to make delicate porcelain from that usable for sewer pipes. The composition of the purest kaolins, from which particles of quartz have been washed out, resembles that of kaolinite. *Quartz* decreases plasticity and shrinkage and helps make clay refractory; if coarse, it must be removed. *Silica* in colloidal form increases the plasticity. *Alumina* makes clay refractory. *Iron oxide*, as well as feldspar, lowers the temperature of fusion and acts as a flux; it is also a strong coloring agent—a little of it makes burned clay buff-colored, much of it (over 5 percent) makes a red product; white-burning clays should have 1 percent or less. *Lime, magnesia,* and *alkalies* act as a flux; lime is a bleacher, but it also forms lumps of undesirable quicklime. *Titanium* acts as a flux at high temperatures. *Carbon* and *water* are driven off during burning.

Physical properties of importance are (1) plasticity, which permits raw clay to be shaped before

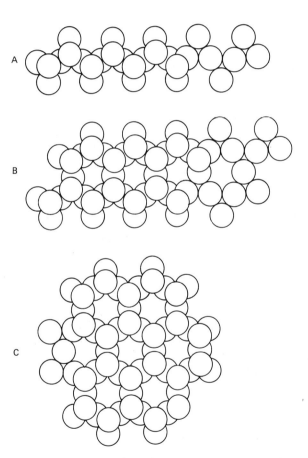

Figure 24-1 Single chains (A), double chains (B), and sheets (C), each illustrating a sharing of oxygens by adjacent tetrahedra. In (A), each tetrahedra shares two oxygens; in (B), some share two and some share three; and in (C), each shares three oxygens. Single chains are joined to other single chains and double chains are joined to other double chains by positively charged ions (such as iron, magnesium, sodium, calcium, and potassium). Similarly, sheets are joined to other sheets by such ions, making stacks of sheets. (From Ojakanges and Darby, 1976, *The Earth*, McGraw-Hill Book Company, New York.)

burning—"fat" or highly plastic clays may be mixed with "lean" clays to produce desired plasticity; (2) transverse strength; (3) shrinkage, both during drying and during burning—if high, the clay is useless for firing; (4) fusibility, which starts at 1000° C with low-grade clays and reaches 1300 to 1400° C for refractory clays. Fusible clays may be vitrified below 1300° C and refractory clays not below 1600° C. The size of clay particles is of the order of fineness of 0.002 mm diameter.

Uses of Clay

The uses of clay and clay products are too numerous to list completely. In domestic life clay is used in pottery, earthenware, china, cooking ware, vases, ornaments, plumbing fixtures, porcelain stoves, tiles, fire kindlers, oilcloths, linoleum, wallpaper, scouring soaps, and polishing bricks. It even finds a place as an adulterant in foods and medicine. In buildings it is used for building bricks, vitrified and enameled brick, building and conduit tile, tiles for floors, walls, and drains, copings, flues, chimney pots, sewer pipes, and foundation blocks. In the electrical industry it is used for conduits, cleats, sockets, insulators, and switches. In refractory ware it is used for fire brick, furnace linings, chemical stoneware, crucibles, retorts, glass-melting equipment, and saggers. Other important uses are for fulling cloth, foundry sands, terra cotta, emery wheels, rubber crucibles, water conduits, paving brick, septic tanks, railroad ballast, portland cement, filtering oils, paper making, and innumerable minor purposes.

Clay was used as a building material in Mexico and Southwestern United States in the forming of bricks or blocks of adobe, a sun-dried mixture of clay, straw, and water. According to Hosterman, an improvement in the construction of bricks was first made in the United States in 1585 at the Raleigh settlement on Roansoke Island, North Carolina. Dr. Daniel Coxe may have been the first to make pottery in the colonies.

The potter's wheel, which has been with man since his early beginning, allowed him to shape clay with his hands as the wheel turned. Improvements in the construction of ceramic ware supposedly culminated with the Chinese but their work was an art and not a science. Ancient pottery sites in China were littered with broken and cracked ceramic ware indicating that a very small percentage of their ceramic ware was successfully produced.

Wedgwood, in England, created a science of the ceramic industry by the careful selection of materials, machine-formed plastic wares, proper drying, kiln firing, and the development of clear glazes.

In addition, certain types of clays carry names designating their use, such as *slip clay,* used for glazing, *pottery clay, retort clay, earthenware clay, terra cotta clay, pipe clay, bleaching clay, bonding clay, foundry clay, sagger clay,* and *rubber clay.*

Kaolins or *china clays* are the purest, whitest, and most expensive clays. They are residual clays of limited occurrence. The higher-grade washed clays are used for fine coated printing papers, and in rubber, refractories, and pottery. They are also used for whiteware, for porcelains of all kinds, and for paper filler. Some so-called sedimentary, nonplastic, Cretaceous kaolins of Georgia and South Carolina are used as paper clays and for some grades of whiteware and are sometimes grouped under "whiteware clays."

Ball clays are good quality, sedimentary, plastic, refractory clays of restricted distribution. They are used chiefly in pottery and stoneware, and they are added to supply plasticity and high bonding qualities.

Fire clays include all refractory clays exclusive of kaolins and ball clays. They are mostly of sedimentary origin. There are four types of fire clay: (1) plastic clay, composed of poorly crystallized kaolinite and contains illite as the major impurity; (2) semiflint clay composed of comparatively well-crystallized kaolinite with impurities; (3) flint clay composed of well-crystallized kaolinite and containing no impurities; and (4) nodular flint clay, composed of well-crystallized kaolinite and containing diaspore or boemite.

Illite is the dominant impurity mineral in clays but is most common in marine shales. It may be the result of weathered debris being eroded, transported by streams, deposited in near-shore marine environments, and subject to diagenetic processes and lithification. It is sometimes referred to as extremely fine-grained mica.

Bentonite is a product of the devitrification and alteration of volcanic ash or tuff. The alteration probably began as the fine-grained ash settled through the seawater but continued during diagenesis. Most bentonite consists chiefly of montmorillmite, but some montmorillonite clays are still classed as bentonite clays under commercial usage.

Fuller's earth is composed predominantly of attapulgite (now known as palygorskite) occurring in the Attapulgus, Georgia-Quincy, Florida, district. Elsewhere, Fuller's earth is mainly composed of montmorillonite clay, and some of these deposits are actually bentonitic clays.

Production

About 90 percent of all clay produced is common brick clay, for which world statistics are not kept. The annual production of clays in the United States ranges from 60 to 65 million tons, valued at about $1,500 million.

Major difficulties are met in statistical records of different types of clay because different products are made from the same clay minerals. Kaolinite, for example, occurs in kaolin, ball clay, and fire clay. Nevertheless, Hosterman reports for 1969 that the United States produced 58,694,000 short tons of clay; exported 1,574,000 tons; and imported 82,000 tons. The only clays imported in any large quantity were kaolin and ball clay. Of the total clay production in 1969, Hosterman lists miscellaneous clay as 72 percent; fire clay, 12 percent; kaolin, 8 percent; bentonite, 5 percent; fuller's earth, 2 percent; and ball clay, 1 percent.

The value of clay varies considerably, but the average value per ton has increased about fivefold from about $5.75 per ton in 1933 to about $30/ton at the present time.

Technology

Common clays are mined in bulk and undergo no processing except for the removal of stones. Even ball clays are not purified. Kaolins, however, are

Type of Clay	Foremost Localities
High-grade kaolins	England, Czechoslovakia, Germany, France, China
Paper clays	Same as above
Ball clays	England, Germany
Fire clays	Germany, England, Belgium, United States
Bentonite	United States

Kaolins are not common. Where they occur, ceramic centers of world renown have sprung up, and fine chinaware made from them has international markets. China is the largest producer, the principal center being in Kiangsi Province, but important de-

washed, screened, settled in water, and filtered. Fire and burning clays are generally processed to remove rock particles. The present tendency is to process all good clays by centrifuging, oil flotation, and bleaching to obtain fine grain, to remove deleterious minerals, and to remove minerals that give undesired colors upon firing.

The clay is then mixed with water in a pug mill, de-aired, and molded into forms. After drying it goes to the kiln, where the temperature is raised gradually to the desired point. Most wares are fired just to the incipient fusion point. Before firing, special surfaces may be given to the clay. If salt is added to the fire, the exterior of wares fuse to give "salt-glazed" ware. Glazes and enamels are added after firing, by dipping, spraying, or dusting, and the ware is then refired to fuse and set them. A glaze is transparent, showing the under body color or design; an enamel is opaque, hiding the under color.

For pottery, plastic and nonplastic clays are carefully blended, stored for "curing," then skillfully molded for firing. Decorations and limited colors are added, followed by a second firing; the wares are then glazed. Lusterware is made by adding silver or copper to the glaze, firing at low temperature, and bringing out the metallic iridescent film by smoke from rosemary or gorse wood.

Distribution

Common clays are found everywhere, but high-grade clays are more restricted. Their chief distribution is as follows:

U.S. Distribution

Minor—North Carolina, Georgia, Delaware

Georgia, South Carolina

Kentucky, Tennessee, Mississippi

Pennsylvania, Ohio, Kentucky, Missouri, Indiana, Illinois, Maryland, Alabama, New Jersey, California, Colorado, Texas, and West Virginia, Utah, Arizona and Washington.

Wyoming, Mississippi, and Texas

posits occur in other provinces. Fine chinaware has been made there since A.D. 220, and 1 million people have been employed in the industry. The most important deposits, however, are those of Cornwall and Devon, from which kaolin is shipped all over the world for the making of chinaware and paper. The kaolins of France have made that country famous for its fine Limoges and Sèvres porcelains, but supplies are sufficient only for its own needs. They occur chiefly in Brittany and in the Departments of Allier and Drôme. Large German deposits have given rise to the famed Dresden ware and other high-quality wares and supply paper clay in quantity. Czechoslovakia also has extensive deposits from which its own famous chinawares have been made and which, along with Bavaria, Cornwall, and Georgia and North and South Carolina, supply most of the fine paper clays. In the United States, North Carolina is the most important producer of residual kaolin, but deposits occur also in Pennsylvania, Virginia, Maryland, Alabama, Connecticut, and Washington.

Five tons of kaolin was mined near Franklin, North Carolina, and shipped to Isiah Wedgewood in England in 1767 to 1768, which led to the Wedgewood products. Since that time, according to Hosterman, the demand for kaolin has increased so greatly that in 1970, production reached almost 5 million short tons valued at more than $100 million.

Refractory or *fire clays* are chiefly of sedimentary origin and in general underlie coal seams. Fire clays occur in Pennsylvania, Maryland, Ohio, Kentucky, Missouri, Indiana, Illinois, Georgia, central Europe, Japan, and many other localities. *Ball clay* of the best quality comes from England and Germany, but a considerable amount is supplied by Kentucky and Tennessee. *Slip clay* is mined in the United States only in New York and Pennsylvania, but it is found in the United Kingdom and central Europe. *Pottery* and *stoneware clays* are also chiefly sedimentary clays associated with fire clays. *Brick* and *tile clays* are the common variety found where the local market exists, such as metropolitan areas.

Bentonite and Fuller's Earth. Both of these commodities have a wide variety of physical properties that make them suitable for many uses. Both are used in drilling muds, bentonite when freshwater is present and palygorskite-type fuller's earth where saltwater occurs. The "Wyoming"-type bentonite has exchangeable sodium ions and when wetted with water will increase in volume 15 to 20 times. Other uses of the dry swelling type of bentonite is for bonding foundry sand and iron ore pellets. Bentonite is also used in petroleum refining as a catalyst, as a carrier for insecticides, and many other uses. Fuller's earth is sold for clarifying and processing mineral and vegetable oil. This is also used

in a specific form as a cleansing absorbing oil, removing grease and chemicals from floors, as animal litter, and as a soil conditioner.

The annual value of bentonite and fuller's earth exceeds $60 million. The major producing states are Wyoming, Mississippi, and Texas.

The fuller's earth deposits in southern Georgia and northern Florida consist mainly of palygorskite and montmorillinite and occur in the Hawthorne Formation of Miocene age.

Hosterman estimates resources of bentonite at almost 1 billion tons, much of which is suitable for binding iron ore pellets, but only limited resources of high-grade colloidal bentonite, which is required by the petroleum industry, is available.

Examples of Deposits of High-Grade Clays

England. The residual china clay deposits of Cornwall and Devon are the most important of the world because of their purity, standard shipping quality, and access to cheap transportation. They are the products of alteration of a porphyritic granite, in part sericitized. The china clay is absent from hilly granite and is restricted to low-lying areas, where it is abruptly delimited from thin, swampy soil covers. Its depth is unknown, since the economic limit of the open-pit operation is about 100 meters. Lilley states that the raw product averages 20 to 25 percent clay substance, and the remaining "sand" or other granite disintegration products are removed by processing. The "sand" also includes such minerals as cordierite, topaz, andalusite, tourmaline, and, locally, considerable fluorspar. The Stannon Marsh clay belt is about two kilometers long and a kilometer wide.

The depth, extent, and mineral assemblage have led most field investigators to conclude that the clay deposits were formed by hydrothermal alteration of the granite. Such is also Lilley's conclusion. Others maintain that it was the normal weathering of hydrothermally altered granite, because some deposits can be traced into partly altered granite and weathering is a common mode of origin. Hydrothermal action in the formation of clay minerals has been more widely recognized of late. A prime example is the halloysite deposit in the Tintic mining district, Utah.

Czechoslovakia. The extensive kaolins of Czechoslovakia are in great demand throughout Europe for paper clay. Lilley has described two interesting oc-

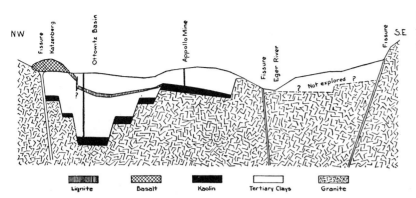

currences, the celebrated deposits of Zettlitz and the large areas of Pilsen.

The *Zettlitz* kaolin deposits lie in a graben south of the Erzgebirge (Fig. 24-2), forming a belt about 12 kilometers long and up to 4 kilometers wide. The kaolin lies on the granite sides and bottom of the graben at depths up to 60 meters beneath Tertiary sediments that contain lignites and basalt flows. The deposit is about 14 meters thick, and the upper part is the best; the lower part grades into granite. It is high-quality kaolin and contains 25 to 40 percent clay substance, the remainder being largely quartz. It has resulted from normal weathering although the proposal has been made that the thermal Carlsbad spring waters have modified it. The best areas of kaolinization correspond to overlying lignite; therefore, acids derived from the lignite are thought to be the important agencies of kaolinization.

In the *Pilsen* area the kaolin has been derived from the weathering of highly feldspathic, carboniferous, arkosic sandstones. The product, which is a very high-grade paper kaolin that is much sought throughout Europe and even in America, consists of 25 to 40 percent clay minerals, the remainder

being almost entirely quartz. The impurities are low. The beds are about 20 meters thick but attain 70 meters. Above the sandstones are thin remnants of Tertiary clays and sands with more or less lignite. The kaolinization is thought to have been produced during the Tertiary by solutions in percolating surface waters, derived largely from the decomposition of overlying peat.

Germany. The most important deposits occur in Saxony, north of the Erzgebirge, in the vicinity of the famous ceramic centers of Dresden, Halle, and Kemmlitz. According to Lilley, in the *Kemmlitz* area, the kaolin attains thicknesses of 50 meters in basinlike areas in lower Permian porphyry and is covered by an overburden of glacial drift and alluvium. It is white and contains 30 to 40 percent of kaolin substance. The deposits were formed by normal weathering, but it is thought that former overlying lignite beds contributed acid waters that helped purify the kaolin.

The deposits of *Halle* (Fig. 24-3) are mostly covered by Tertiary sands and clays that now contain or did contain beds of lignite. In places the deposits crop out. The high-quality kaolin has resulted from the alteration chiefly of the older and younger porphyry but also of other rocks. German geologists believe that lignite decomposition has been effective in its formation.

In *Bavaria*, near Arnberg, a group of kaolin deposits, like the Czechoslovakian deposits, was derived from arkosic sandstones. Here, however, the kaolin is of Triassic age and crops out as a belt 50 to 150 meters wide and 9 kilometers long. The kaolin content ranges from 10 to 25 percent. Permian arkosic sandstones have been less extensively kaolinized. Somewhat similar but relatively unimportant deposits occur in Thuringia. The Bavarian clays are also valuable paper clays.

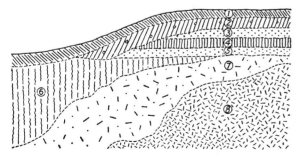

Figure 24-3 Section of part of Halle kaolin area. 1, Soil; 2, alluvium; 3 to 5, Tertiary sand and clay; 6, kaolinized Lower Permian beds; 7, kaolinized porphyry; and 8, granite porphyry. (From: Lilley-Behr, AIME.)

China. China is the largest producer, but not exporter, of china clay, and is the home of chinaware. Several varieties are produced, but the best material is that of the Chimen district, Anhwei, derived from feldspathic rocks.

United States. Residual deposits of low-grade clays are numerous south of the glaciated areas and east of the Rockies, and glacial clays are found in the north. Deposits of high-grade white clays, however, occur mostly in North Carolina, Virginia, Georgia, and Alabama, where they are derived mostly from pegmatite dikes. Those of North Carolina yield good-quality kaolin and halloysite but they are generally less desirable than European kaolins. The deposits are as much as 50 meters wide and 30 meters deep. The content of kaolin substance averages about 25 percent but reaches 40 percent. The Virginia deposit is a white, nonplastic clay developed on syenite. An unusual deposit occurs in Litchfield County, Connecticut, where a bed of feldspathic quartzite 300 meters long, lying between harder quartzites, has been weathered to kaolin and was preserved from glacial erosion. The white clays or so-called sedimentary kaolins of Cretaceous age in South Carolina and Georgia have been utilized for paper clays. Hydrothermal halloysite occurs in Utah and Nevada.

Selected References on Clays

General Reference 13.

The kaolin minerals. 1930. C. S. Ross and P. F. Kerr. U.S. Geol. Survey Prof. Paper 165-E.; also Prof. Paper 185-G, 1934. *Detailed mineralogy.*

High grade clays of the eastern United States. 1922. H. Ries and W. S. Bayley. U.S. Geol. Survey Bull. 708. *Detailed geographical descriptions.*

Modern concepts of clay minerals. 1942. R. E. Grim. Jour. Geol. v. 50, p. 225–275. *Clay concept and descriptions of clay minerals and crystal structures.*

Absorbent clays. 1943. P. G. Nutting. U.S. Geol. Survey Bull. 928-C, p. 127–221. *Summary of knowledge of absorbent and bleaching clays.*

Minerals of the montmorillonite group, their origin and relation to clays. 1945. C. S. Ross and S. B. Hendricks. U.S. Geol. Survey Prof. Paper 250-B. *Comprehensive discussion of the montmorillonite and related clay minerals.*

Diagnostic criteria for clay minerals. 1945. W. F. Bradley, Amer. Miner., v. 30, p. 704–713. *X-ray criteria.*

Clays. 1971. J. F. Wells, U.S. Bur. Mines Yearbook. v. 1–2, p. 287–308.

High-alumina kaolinitic clay in the United States. 1963. Helen Mark. U.S. Geol. Survey Mineral Inv. Resource Map MR-37.

Clays. 1970. J. D. Cooper. In Mineral facts and problems, U.S. Bur. Mine Bull. 650, p. 923–938.

Applied Clay Mineralogy. 1962. R. E. Grim, McGraw-Hill Book Company. 422p. *Outstanding textbook on clay.*

Clays and clay technology. 1955. J. D. Pask and M. D. Turner, ed. Dept. of Mines, Dept. of Natural Resources, State of California, Bull. 169. *Numerous papers presented at the First National Conference of Clays and Clay Technology.*

Clays. J. W. Hosterman. General Reference 19. *Latest summary information on clay.*

Relation of clay mineralogy to origin and recovery of petroleum. 1947. R. E. Grim. A. A. P. G., v. 31, p. 1491–1499. *Crystal structure reasons.*

Kaolins of North Carolina. 1947. J. L. Stuckey. AIME Tech. Pub. 2219. *Distribution, composition, and occurrences.*

Glossary of clay mineral names. 1949. Paul F. Kerr and P. K. Hamilton; Reference clay localities—United States. Paul F. Kerr and J. L. Kulp. Differential thermal analyses of reference clay mineral specimens. Paul F. Kerr, J. L. Kulp, and P. K. Hamilton. Am. Petrol. Inst. Proj. 49, Clay Mineral Standards Prel. Repts. 1, 2, and 3, Columbia University, New York. *Clay minerals, history, origin, formulas, properties, occurrences.*

FELDSPAR

Feldspar is the general name of a group of aluminum silicate minerals containing varying amounts of potassium, calcium, and sodium. They make up about 50 percent of many igneous rocks.

Feldspar is used in ceramics for the making of glass and pottery, both in the body of the ware and in the glaze. It is also used in enamels for household utensils, tile, porcelain sanitary ware, and other minor ceramic uses. About 35 percent of the feldspar produced in the United States is used for such purposes and 55 percent to supply alumina, potash, and soda for glass manufacture. Nepheline syenite, with about 24 percent alumina, however, is now a strong competitor of feldspar in the glass industry. Some feldspar is also an ingredient in scouring soaps, abrasives, roofing materials, and false teeth.

Kinds and Properties of Commercial Feldspars

Both potash and soda feldspars are the commercial varieties, the former being the more important; the plagioclases high in lime are undesirable. The potash feldspars (orthoclase and microcline) always contain some soda from included albite, and the soda feldspar always contains a little lime. Iron, manganese, and sericite are deleterious. Feldspars high in potash or soda, upon cooling from fusion, yield a solid glass, but lime-rich varieties become partially crystallized.

Producers and consumers know feldspar as a beneficiated mixture composed of feldspar minerals with ideally less than 5 percent but sometimes as much as 20 percent quartz, and minor quantities of other minerals. Perthitic microcline, albite, and oligoclase are the principal types of feldspar mined in the United States.

Occurrence

Although feldspars constitute about 50 percent of igneous rocks, commercial varieties are derived chiefly from pegmatite dikes, mainly granitic ones. They occur in areas of granitic, syenitic, and metamorphic rocks and, as are common with pegmatites, they are generally irregular in size, continuity, and distribution of the feldspar content. Commercial pegmatites are mostly lens-shaped and are as much as a hundred meters or so across and a kilometer or more in length. Most of the feldspar crystals range from a few centimeters to a meter or so in diameter but may attain 7 meters. Quartz, the most abundant associate, and other pegmatitic minerals have to be removed by processing. The cleaned product is finely ground for market.

Production and Distribution

Prior to 1946, according to Lesure, most of the feldspar produced in the United States consisted of hand-cobbed perthite and oligoclase from pegmatites. In 1946, flotation milling of finer grained pegmatite material expanded the potential sources of feldspar and quartz. By 1970, about 57 percent of the feldspar used in the United States was recovered by flotation, 8 percent was hand-sorted, and 35 percent was obtained from feldspar-silica mixtures. It is surprising, incidentally, that graphic-granite is not used more in the ceramic industry because it usually contains a uniform amount of quartz and feldspar. The industry, however, is slow to change and many concerns believe that it is best to mix feldspar and quartz derived from the reliable same deposits rather than switch to different raw material sources that may not produce as uniform a product.

Feldspar sells at a price varying from $11.00 to $26.00 per long ton depending on the flotation mesh size and the producer. The average price was $14.87 per long ton in 1970.

The United States is, or could be, self-sufficient in feldspar production.

Selected References on Feldspar
General Reference 13.
Spruce Pine district, North Carolina. 1940. C. S. Maurice. Econ. Geol., v. 35, p. 49–78; 158–187. *Detailed descriptions but difficult reading.*
Feldspar. F. G. Lesure. General Reference 19. *Excellent review and selected references.*
Feldspar. 1932. H. S. Spence. Canada Dept. Mines, Mines Branch Pub. 731, Ottawa, 1932. A *general treatise.*
General Reference 7.
New Jersey potential feldspar resources. 1948. J. M. Parker, III. N.J. Bur. of Min. Res. Bull. 5, Pt. 1. *Geological occurrence and resources.*
Internal Structure of Granitic Pegmatites. 1949. E. N. Cameron, et al. Monograph No. 2. Economic Geology Publishing Co.
Feldspar. 1949. B. C. Burgess. *Industrial Minerals and Rocks,* 2nd ed. N.Y. AIME, p. 345–373. *Excellent bibliography.*
The Study of Pegmatites. 1955. R. H. Jahns. Econ. Geol. 50th Ann. Vol., p. 1025–1130.

OTHER CERAMIC MATERIALS

Other materials are utilized in minor amounts in the ceramic industry either to supply certain desired ingredients of clay or to make special ceramic products.

Bauxite. Bauxite with its high alumina content is required for certain porcelains to add strength and resistance to heat, corrosion, abrasion, and spalling. Aluminous refractory brick, made of baked bauxite and a binder, are desired for certain furnace conditions. Fused aluminous refractories of the metallic type are cast directly from the melt.

Sillimanite Minerals. This group of minerals includes sillimanite, andalusite, kyanite, and dumortierite, which are considered as a group because of similar ceramic uses. These minerals are furnaced at 1545° C to give silica and müllite. This material, stable up to 1810° C, makes excellent refractories. Porcelains produced from it are strong, withstand high temperature, have small expansion, are excellent insulators, and resist corrosion. They are used for making spark plugs, electrical and chemical porcelains, enamelware, saggers, and hotel ware.

Refractories containing mullite are used for glass tanks, crucibles, furnace linings, fireboxes, and high-temperature cements. According to Kerr, the Champion Spark Plug Company uses dumortierite from Oreana, Nevada, for its spark plugs and markets Champion "sillimanite" made from dumortierite and andalusite. Dumortierite is more favored than the others because of its higher Al_2O_3 content. Kyanite is the least favored because it increases in volume upon changing to mullite. It is, therefore, generally precalcined before use. The mineralogy, occurrence, origin, and examples of deposits are given in Chapters 15 and 26.

Borax. Borax, although required in minor quantities, is an essential constituent of porcelain enamels used to coat iron and steel household and kitchen utensils, such as cooking utensils, stoves, refrigerators, washing machines, bathtubs, sinks, table tops, tiles, pipes, and many similar products that require an attractive, durable, and sanitary finish and for articles used in hospitals and doctors' offices. Borax also permits addition of pigments to enamels. Since borax lowers expansion and makes durable products, it is indispensable for many ceramic products, such as certain pottery, china, colored wares, glazes, and various heat-resisting glasses used in laboratory, kitchen, signal, thermometer, and optical glass. It imparts brilliance to glass as well as strength and durability.

Magnesite. Magnesium oxide or silicate obtained from calcining magnesite (or dolomite) is used in making ceramic bodies for electric stoves. Pure magnesia with 2 percent talc is used to make extruded insulators for radios, and pressed magnesia is made into insulators for high-conductivity copper conductors. Laboratory crucibles made from magnesia are used for refining metals because they are chemically inert and stand high temperature. Fused magnesia is used to make molds for certain glass objects, such as electric light bulbs.

Lithium Minerals. There has been increasing use of lithium compounds in ceramics. They are introduced directly into the batch as lepidolite, spodumene, or as chemically prepared salts, particularly lithium carbonate. Betz states that lithium (1) is a powerful flux along with feldspar, (2) permits the use of much less alkali, (3) has a mineralizing effect on ceramic bodies, (4) increases the fluidity and gloss of enamels and glazes, (5) reduces vaporization of the glaze, and (6) permits the manufacture of glass with high electric resistance and the power to transmit ultraviolet light. Lepidolite, with its fluorine and lithium, decreases expansion and increases the strength of ceramic bodies. It also is a good opacifier for opal and white opaque glasses, and is used for nonshattering glass.

Cornish Stone. Cornish stone, or china stone, is a kaolinized granite rich in orthoclase and albite that is used largely by English potters in place of feldspar. It contains from 6 to 15 percent kaolin, 55 to 77 percent feldspar, and 16 to 31 percent quartz. It is obtained in Cornwall, England, near the china clay deposits, and because it occurs in large bodies it can be mined cheaply. It is the principal source of the alumina used in British pottery.

Diaspore. Diaspore (see Refractories, Chapter 26) is used mainly for refractory fire brick, although some of it is added to certain clays to increase the alumina content and to give a hard porcelain.

Bentonite. Bentonite is used mainly for purposes other than ceramic, but it is also added to other clays and ceramic materials as an additional plasticizing and bonding agent to improve the products after burning.

Fluorspar. Fluorspar (Chapter 26) is used in the ceramic industry in the manufacture of enamels, in opalescent, opaque, and colored glass, and in facings for bricks and vitrolite.

Barite. Barite, converted to barium carbonate, is utilized in making optical glass and enamels and "granite ware" for coating metal dishes.

Potash Minerals. Potash minerals are employed for jewelry-type enamels to give high brilliancy and luster. The potassium-lead-silicate type use as much as 36 percent K_2O.

Talc. Talc is used in ceramics for making calcined talc (lava), which is harder than steel and can be tooled and threaded for use in gas tips, refractories and electrical insulators. For occurrence and origin see Chapter 15; for uses, distribution, production, and examples of deposits see Chapter 26.

Pyrophyllite. Pyrophyllite is used as a ceramic body

material in wall tile, tablewear, and electrical porcelains (see also Chap. 15).

Diatomite. Diatomite is used in the manufacture of glazes and enamels.

Zirconia. Zirconia (see Chapter 26) is used for certain porcelains and high refractory materials, such as crucibles for melting platinum and other metals, which withstand chemicals and high temperatures.

Selected Refernces
General Reference 7.
General Reference 19. Chapter on feldspar.

25

STRUCTURAL AND BUILDING MATERIALS

Present-day structural and building operations create large demands for bulk materials of the mineral kingdom. A bridge is built of steel, stone, and concrete, all mineral products; a highway of stone, sand, and cement forming concrete; and a modern city building from its concrete and steel foundation to its slate or metal roof employs steel for its structure, stone, cement or brick (clay) and mortar for its walls, metals for its plumbing, heating, wiring, elevators, and finishings, plaster (gypsum) for its walls and ceilings, cement for floors, roofing stones, or asbestos shingles, mineral wool for insulation, glass (glass sands) for windows, and mineral paints for finishing. Except for a minor amount of wood, it is entirely composed of mineral products. The materials are common substances occurring in large bulk. Mostly they are the stone, clay, and sand used by early man. Some are used directly; some are dressed; and others undergo considerable preparation. The demand for these products has increased so greatly that huge industries have sprung up around them. The group, some materials of which have already been considered, includes:

Building, roofing, and crushed stones	Pigments and fillers
Hydraulic cements	Heat and sound insulators—mineral wool
Sand, gravel and lightweight aggregates	Asphalt and bitumens
Gypsum and anhydrite	Clay products
Lime	Miscellaneous
Magnesite	

BUILDING, ROOFING, AND CRUSHED STONES

The stone industry is widespread, since rock is the most abundant of all material things. Except for a thin mantle of soil and water, it is the earth itself. Making use of it is one of the oldest human activities, and today in the United States there are about 2,800 quarries with an annual production of around 2,000 million tons, worth about 1,500 million dollars.

The term *rock* differs from *stone* in that the latter is applied to blocks or pieces broken for use. Not all rock, however, makes commercial stone; certain geological and physical properties are necessary, such as strength, durability, and ease of processing and quarrying. Most stone is used crude and untreated. It may be cut and dressed and is then termed *dimension* stone; it may be broken into aggregates and called *crushed* stone; or it may be treated and made into cement, lime, glass, or refractories. The common rocks used for commercial stone are granite and related igneous rocks, limestone, marble, slate, sandstone, and soapstone, all of which serve more than one purpose.

Building and Structural Stone

For 12,000 years or more man has used stone for building shelters. In the mighty pyramids of Egypt and in the great buildings of the Greeks, Incas, and Mayas he early overcame the problems of transportation of huge blocks, weighing up to 90 tons. Later, beauty followed grandeur, and the magnificent and exquisite cathedrals of Europe reflect the artistic use of building stone.

Today utility follows beauty. Seldom, however, are buildings now constructed of massive masonry, but thin slabs of stone or aluminum plate sheathe most steel skyscrapers and more modest buildings. In brick and concrete buildings, sills, trim, facings, and steps may be of stone. Stone is used for interiors in steps, fireplaces, wainscoting, baseboards, and other ways. If stone were used more in America, the fire hazard of buildings would be less.

Most building stone is dimension stone and is used as cut or finished stone, according to drawings supplied; as ashlar or rectangular blocks; as rough building stone; and as rubble, with one flat face. Such stone is sawed or chipped into shape. Dimension stone is also used for other structural purposes, such as bridges, abutments, fences, retaining walls, monuments, flagstones, paving stones, cobblestones, curb stones, switchboards, and blackboards.

Despite the abundance of rock outcrops, few rocks satisfy the requirements for dimension stone. The important ones are ease of quarrying, strength, color, hardness and workability, texture and porosity, and durability. Transportation is also an economic factor.

The varieties of building stone used in North America and their principal use and geographical source are listed in Table 25-1.

Quarrying. The rock exposure must be free from closely spaced joints, cracks, or other lines of

Table 25-1 North American Building Stones

Stone	Designation	Chief use	Important localities
Granite	Barre	Monuments	Vermont
	Quincy	Monuments	Massachusetts
	New Hampshire	Building	New Hampshire
	Maine	Paving blocks	Maine
	Westerly	Building and monuments	Rhode Island
	St. Cloud	Ornamental and building	Minnesota
	Wisconsin	Building	Wisconsin
	Winnipeg	Building	Manitoba
	Norwegian Pearl Gray (Laurvikite)	Decorative	Norway
	Rapakivi	Building	Finland
Limestone	(Oölitic)	Decorative	Indiana, Utah
	Coquina	Trim and building	Indiana
	Missouri	Building	Florida
	Kasota	Building	Missouri
	Texas	Decorative	Minnesota
		Building	Texas
	Tyndall	Building	Manitoba
Travertine	Salida	Decorative	Colorado
	Montana	Decorative	Montana
	Tivoli	Decorative	Italy
Marble	Vermont	Building, monument	Vermont
	Rutland	Decorative	Vermont
	Georgia	Building, monument	Georgia
	Carrara	Monument, decorative	Italy
	Siena Parian	Monument, decorative	Greece
	Savoie	Decorative	France
	Onyx	Decorative	Mexico
Serpentine	Verde antique	Decorative	
Sandstone	Berea or Ohio	Building, trim	Ohio
	Portland	Building, trim	Connecticut
	Blue stone	Flags	Pennsylvania, Kentucky, New York
Dolerite	Trap	Curbstones	Connecticut, New Jersey, Massachusetts
Soapstone	Soapstone	Switchboards, sinks, trim	Virginia

weakness or sizable sound blocks cannot be obtained. Some lines of weakness, such as well-spaced bedding and joint planes, are necessary to assist in quarrying and to permit breaking to one or more flat surfaces, else the quarried rock would have to be dressed on all sides. Deep and irregular weathering is undesirable.

Production. Reed reports that 926 quarries produced $100 million worth of dimension stone in the United States in 1977; however production was highest 50 years ago although the value was much less. All stone was 9.16 million tons valued at 2.3 billion in 1977.

The production of stone in 1969 according to type of rock is shown in Table 25-3.

Hardness and Workability. Hardness ranges from that of soft coquina or limestone to that of granite, which exceeds steel. Workability depends in part upon hardness. Limestone is easy and cheap to dress, but granite is expensive. Hardness is generally unimportant except where wear is concerned, such as for steps, flooring, or paving stones.

Color and Fabric. Color is very important. Many architects are inclined to let color predominate over other essential qualities. Color, however, is a matter of taste and even of fashion; the prized brownstone buildings of yesteryear are less desired today. For buildings, reds, buffs, grays, or white are preferred. White is undesirable for smoky cities, and blacks and dark grays are not popular. Some rock colors are permanent and are not affected by weathering, but stones containing minerals, such as pyrite, that oxidize and produce unsightly stains are undesirable as are limestone facings where indus-

Table 25-2 Comparison of Dimension and Crushed Stone Production from 1929 to 1970 (all figures in millions)

	Dimension stone	Crushed stone	Total (rounded)
1929 tonnage	4.7	93	98
value	$70	$94	$164
1939 tonnage	2.3	145	147
value	$25	$133	$158
1949 tonnage	1.8	222	224
value	$52	$289	$341
1959 tonnage	2.3	145	147
value	$25	$133	$158
1970 tonnage	1.6	873	875
value	$95	$1380	$1475

Table 25-3 Production of Stone in 1969, by Types of Stone, in the United States

Type	Percentage of total Tonnage	Percentage of total Value
Crushed stone		
Limestone (including dolomite)	73	$67
Granite	9	9
Marble	.2	2
Sandstone (including quartzite)	3	4
Traprock	10	11
Other	4.8	7
	100	100
Total millions—	861	$1326
Dimension stone		
Granite	37	$46
Limestone (including dolomite)	31	18
Marble	4	13
Sandstone (including quartzite)	17	11
Slate	8	9
Other	3	3
	100	100
Total millions—	1873	$99

Source: From Laurence, U.S. Geol. Surv. Paper 820, p. 167, 1973.

trial SO_2 + H_2O form gypsum and result in exfoliation of the limestone.

A pleasing fabric fetches high prices for ornamental purposes, such as columns or wainscoting. Some limestone breccias of varicolored fragments, hard agglomerates, porphyries, or certain banded or schistose stones give pleasing effects.

Porosity and Texture. Porosity affects freezing expansion and solution. Small pores give up water slowly and are, therefore, affected most by frost action. High permeability aids solvent action and chemical reactions. Texture affects workability, therefore cost; fine-textured rocks split and dress more readily than coarse ones. In many ornamental and monumental stones, texture is also vital, but the architectural trend in building stone is toward uneven texture.

Durability. Durability should be, but is not always, a determining factor in the choice of a building stone subject to weathering, climate, frost, heat, and fire. Cleopatra's Needle survived 3000 years of Egyptian climate but succumbed to a quarter century of New York climate. Weathering is caused by rain containing CO_2 and SO_2 and by frost action and solar heat. This produces crumbling, spalling, and exfoliation. In some of the English colleges, almost every stone has had to be replaced, and the Wren churches of England have suffered demolition of sculptures. Granite and allied rocks resist climate best; silica-cemented sandstone is good; close-textured, low-porosity limestone resists well; but coarse marble and limestone, poorly cemented sandstones, and much arkose succumbs fairly quickly. Fire and heat resistance is high with fine-grained, compact stones, such as limestone and fine sandstone, and less so in coarse crystalline rocks.

The life of building stones in New York ranges from 15 to 50 years for coarse brownstone and 50 to 200 years for granite.

Those rocks that combine several of the qualities listed above have attained widespread eminence as building stone; such are the well-known Indiana or Bedford limestone, Ohio sandstone, Minnesota granite, Vermont marble, Carrara marble (Italy), Mexican onyx marble, Italian travertine, Scottish granite, and Virginia soapstone.

Rockefeller Center in New York City is faced with Indiana oölitic limestone, a light gray colored rock. It is now dark gray in color as the result of fly ash, polluted air, and smoke. With the burning of fuels, especially high sulfur-containing coal, SO_2 forms and then reacts with water and the impregnated limestone to produce gypsum, which results in spalling of the limestone as the gypsum expands upon forming.

Extraction Methods. Building and other dimension stone is obtained by open quarrying, rarely from underground. Explosives are used sparingly so that the blocks will not be harmed. With hard rocks, like granite, wedges are driven into drilled holes to spring out the blocks. This is supplemented by light blasting, advantage being taken of the rift and grain. Smaller blocks may be extracted by wedging or by drilling small holes and using plug and feathers. With limestones, sandstones and marble narrow, parallel channel cuts are made with a channeling machine, and the blocks are then removed and subdivided by plug and feather or by cutting machines. Wire saws are also used now in many quarries to cut out blocks. For cut stone, saws, planes, and rubbing and polishing machines are used.

Crushed and Broken Stone
Owing to increased use of machinery and concrete and accelerated highway construction, the crushed

stone industry now vastly exceeds that of dimension stone in tonnage and value.

Raw materials are widespread, the requisite qualities are few, and quarries are widely scattered. Because crushed stone can be produced on a large scale by inexpensive operations, the product is relatively low cost. Since the cost of transportation is the chief limiting factor, the industry must depend upon local markets.

The rocks used are few, mainly alluvial gravels and limestone (including marble), with lesser amounts of trap rock (including dolerite, basalt, and andesite), granite, sandstone, and quartzite.

The particular rocks employed depend largely upon the desired uses. These fall into two main groups: (1) for use in the crude state, and (2) for chemical purposes. The former include highway construction, concrete aggregate, railroad ballast, and riprap or broken stone for retaining walls and sea walls. Minor uses are for sewage filter floors, shingle aggregates, and road ballast. For highway construction, fill, toughness, hardness, and strength are desired; and trap rock fits these requirements best, with granitic rocks next. Sandstone is used for aggregates and riprap.

The chemical purposes for which some limestone serves almost exclusively are in the manufacture of portland cement, and to an important but less extent, for fluxstone, lime, alkali works, agriculture, sugar factories, asphalt filler, refractory stone, glass and paper, calcium carbide, refractories, and rock wool.

The rocks are extracted from huge quarries by churn drilling and large-scale blasting, loaded by power shovel, crushed, and screened for proper sizing.

Roofing Stones (Slate)
Roofing stones are largely confined to slate, but a little sandstone flag is also used. *Slate* is a very durable rock with a pronounced parallel cleavage, differing thereby from phyllites and schists. It is the cleavage, strength, and durability that make it so excellent a roofing stone; it is noninflammable; also it is found in a variety of pleasing colors of gray, blue-black, red, green, purple, and mottled. The colors are mostly permanent. The cleavage of slate rarely coincides with bedding, and thin beds of different compaction can often be discerned as "ribbons" across the cleavage, which makes second-quality stone. Slate is found in metamorphic regions, and most of the commercial product of North

America comes from Pennsylvania and Vermont, with lesser amounts from the other Appalachian states.

In addition to roofing, slate is also used as mill stock for baseboards, steps, walks, mantles, shower-bath floors, blackboards, switchboards, school slate, as granules for surfacing shingles and roofing, and as pulverized slate or flour for fillers. The enormous waste attendant upon cleaving shingles and blocks has led to the increasing use of granules.

A slate quarry should have little overburden and should be free from siliceous or carbonaceous beds. Extraction involves greater care than with other rocks. Slate is generally sawn by wire saws into blocks that are wedged off, removed, and split by hand into shingles or blocks.

Manufactured substitutes. A large industry now exists that manufactures crushed rock such as slag or expanded lightweight aggregate. A large array of man-made facing stone slabs made of cemented mixtures of marble chips, ceramic material, glass, colored rock chips, and so forth, are being used in increasing abundance. These products are cemented in slabs or brick shape and sometimes ground smooth and polished or split to show the rough cleavage faces of the enclosed minerals. Prestressed concrete beams are formed in casts in which steel cable is stretched before the cast is filled with concrete, which, when hardened, has a much higher compressive strength and modules of elasticity. Prestressed cement beams are commonly used in highway underpasses and overpasses.

Selected References on Stone
The Stone Industries. 1934. Oliver Bowles. McGraw-Hill Book Company, New York.
General Reference 19. Chapter on construction stone. *Excellent review of U.S. resources and construction stone properties.*
The shape of things to come. 1971. D. R. Bliss. Stone Mag., v. 91, p. 147.
Stone. 1970. H. J. Drake. U.S. Bureau of Mines Yearbook, p. 19.
The Stone Industries, 2nd ed., 1939. Oliver Bowles. McGraw-Hill Book Company, New York. 519 p.
Limestone and dolomite. 1956. U.S. Bur. Mines Inf. Circ. 7738, 29 p.
Dimension stone. 1960. In J. L. Gillson, ed. *Industrial Minerals and Rocks,* 3rd ed.: Am. Inst. Mining Metall. and Petroleum Engineers, p. 321–337.
Slate. 1960. In *Industrial Minerals and Rocks,* 3rd ed. J. L. Gillson, ed. Am. Inst. Mining Metall. and Petroleum Engineers, p. 791–798.
Stone, 1970. J. D. Cooper. In Mineral facts and problems. U.S. Bur. Mines Bull, 650, p. 1219–1235.
Geologic appraisal of dimension-stone despoits. 1960. L. W. Currier, U.S. Geol. Survey Bull. 1109, 78 p.

HYDRAULIC CEMENTS

The vast expansion of highway and building construction has created great demand for cement, and the industry is now the sixth largest in value of output among American mineral industries. Cement is a manufactured material which when mixed with water sets or becomes hard, either in air or under water. Essentially, it is a mixture of about four parts of limestone and one of clay or shale, calcined to near fusion and ground to a powder.

One kind of cement was known to the Carthaginians, and the Romans discovered that quicklime added to the volcanic ash of Puzzuoli gave a cement that set under water. They used it to build aqueducts, baths, the Pantheon, and other structures, and it became known as *Puzzuolan* cement. It is still made in Europe.

In 1756 John Smeaton in England made "lime" by burning argillaceous limestone and discovered that it set under water. This was the start of the manufacture of *natural hydraulic cements* and inaugurated searches for similar limestone in Europe and America. It was first discovered in America during the building of the Erie Canal and supplied the hydraulic cement for making locks. A big natural cement industry sprang up around Rosendale, N.Y., and Lehigh, Pa. Since such limestones are not common, it became the practice to add either limestone or shale to give the correct proportion. Naturally, the composition was variable, which caused natural cement to lose favor. Natural cement from different localities varies considerably in composition, some containing from 15 to 18 percent MgO, about 2.5 percent Al_2O_3, from 18 to 25 percent SiO_2 and the remainder 75 percent $CaCO_3$.

Portland cement was discovered by Aspdin in England in 1824 and named by him because it resembled the famous Portland stone. It has now almost entirely displaced natural cement.

Ingredients and Manufacture. Portland cement is made by burning to a clinker a finely ground mixture containing about 75 percent CaO, and 25 percent clayey minerals, the latter consisting of 20 percent SiO_2, Al_2O_3, and Fe_2O_3, and 5 percent magnesia, alkalies, and so forth; MgO should not exceed 5 percent of the finished product. Three percent of gypsum is added before final grinding to prevent too rapid setting.

The calcining releases the CO_2, and the remaining constituents combine to form complex silicates, aluminates, and ferrates of calcium, which in turn break down to form other compounds. Tricalcium and dicalcium silicate and tricalcium aluminate are the chief constituents. The addition of water gives rise to a gel of hydrous compounds, which later crystallize and interlock, giving the hard set. Mixed with crushed rock and sand, the cement binds all together into a rocklike mass of great strength, forming *concrete*.

Limestone is the chief rock used to supply the calcium oxide, and the nearer it approaches the composition of the cement mixture the better. Pure limestone is neither necessary nor desirable. MgO should not exceed 10 percent, and pyrite should be absent; CaO is also supplied in part by marl, furnace slag, and oyster shells; and SiO_2 and Al_2O_3 are supplied by clay or shale. Sandstone or sand, bauxite, or iron ore may be added to supply deficiencies in SiO_2 and Al_2O_3 or Fe_2O_3. The combinations of raw materials used are (:1) limestone with clay or shale, (2) cement rock, alone or with high-calcium limestone, (3) blast furnace slag and limestone, (4) marl and clay, and (5) oyster shells and clay. In many plants the raw materials are now subjected to flotation to discard objectionable minerals. About 612 pounds of raw materials are used to a barrel of finished cement (376 pounds).

There is a tendency to make a variety of specialized cements adapted to particular uses, such as high-early-strength, masonry, low-heat, oil well, and high-alumina cements.

Production and Distribution. The raw materials for cement are so widely distributed that few states or countries lack them. Consequently, cement plants may be located wherever favorable associations of raw materials occur, where the need exists, and where ready transportation is available.

The world production normally amounts to 750 to 800 millions tons, and the main producing countries in order are Russia (17 percent), Japan (9 percent), United States (8 percent), Germany, United Kingdom, and Italy. The United States production is 60 million tons.

Selected References on Cements

The Stone Industries. 1934. O. Bowles. McGraw-Hill Book Company, New York. *Comprehensive book on industrial stones,* including cement rocks.

The solidification of cement. July 1977. D. D. Double and A. Hellawell. Sci. Am., v. 237, no. 1, p. 82–90.

General Reference 13. *Excellent treatise on cements.*

General Reference 19. 1928. *Chapter on light weight aggregates.*

Cements, Limes, and Plasters. E. C. Eckel. John Wiley & Sons, Inc., New York. *Good material on industrial limestones used for cement rock.*

SAND, GRAVEL, AND LIGHTWEIGHT AGGREGATES

Humdrum sand and gravel is the largest nonmetallic industry in volume in the United States. They are essential in modern construction, particularly in paving and building. They are widely distributed the world over and are generally used locally because their low value permits only very low-cost transport. Despite their low-unit cost, $1.2 to $2.50 per ton, the total sand and gravel produced in the United States amounts to more than a billion tons, worth from $1.2 to $2.5 billion per year (f.o.b.).

Sand is a broad term used to cover almost any comminuted rock or mineral, but technically it is restricted to quartz sand with minor impurities of feldspar, mica, and iron oxides. There are also black sands (magnetite), coral sands, gypsum sands, and others. Sand grains fall between 0.06 and 2 millimeters, and gravel from 2 to 100 mm. Gravel consists dominantly of quartz pebbles and grains but also includes pebbles of other rocks and minerals. Dry sand ranges from 90 to 110 pounds er cubic foot, and dry gravel from 90 to 107 pounds per cubic foot. Both are of detrital origin and sorted, cleansed, sculptured, and transported by water.

Occurrence. Sand and gravel occur as sedimentary beds, lenses, and pockets lying at or near the surface or interbedded with other sedimentary beds. They occur as fluvioglacial deposits, stream-channel and floodplain deposits, as present and elevated seashore deposits, as windblown deposits along and near bodies of water, as desert sand dunes, as marine and freshwater sedimentary beds, continental lakes along their shorelines, and even solid rock that is crushed. The deposits may be well sorted and may consist of almost pure sand grains or they

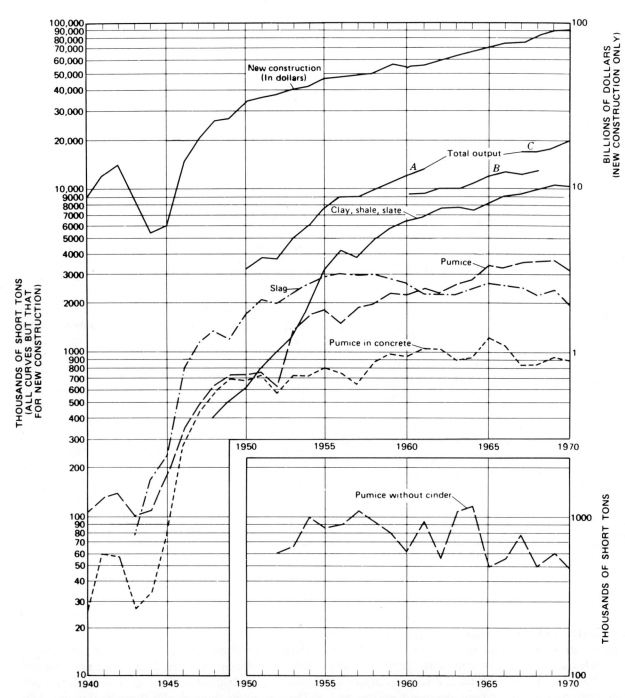

Figure 25-1 Production of materials used as structural lightweight aggregates in the United States, 1940–1970. New construction: Value of total new construction, exclusive of maintenance and repair; data from statistical abstracts of the United States, 1940–1970. Total output: Tonnage of all materials used as structural lightweight aggregates: *A*, Estimate in Rock Products, v. 65, no. 1, p. 83, 1962. *B*, Estimate by R. A. Grancher in Rock Products, v. 72, no. 12, p. 62, 1969 (lacks data for slate, 1960–1963). *C*, Estimate by Harvey Smith in Mining Engineering, v. 24, no. 1, p. 53, 1972. (From Bush, 1975, U.S. Geol. Survey Prof. Paper 820, p. 336.)

may consist of irregular sizes and impurities necessitating screening to size them or washing to purify them.

Uses. Sand and gravel are the normal weight aggregate used with cement to form concrete. Lightweight aggregates are providing stiff competition (Fig. 25-1) to sand and gravel. These materials consist of scoria, volcanic cinder, pumice and pumicite, expandable clays, shales and slates, diatomite, expanded blast furnace slag, and fly and bottom ash. Ultralightweight aggregates consist of perlite, expanded pumicite, and exfoliated vermiculite, and are commonly used for insulation and to form sound-absorbing or acoustical walls and ceilings. Expanded clays and shales make up more than 70 percent by weight of the structural lightweight aggregates used. The total value of all lightweight aggregates is about $150 million per year.

Expandable clays, shales, and slates average about 55 to 70 percent SiO_2, 12 to 20 percent Al_2O_3, and a few percent each of Fe_2O_3, MgO, LaO, Na_2O, and K_2O. The bloating range of these materials according to their composition is shown in Figure 25-2.

The uses and demand for sand and gravel fluctuate nationally and locally, as a function of large public work projects, such as dams, interstate highways, and the building industry.

California leads all other states in sand and gravel production. Nevertheless, one authority on sand and gravel has stated that "there are no undiscovered deposits near the metropolitan areas (of Los Angeles) that can be developed to meet demand." California is not unique in this regard, and it is quite certain that crushed rock will be developed as a source of coarse aggregate in many areas as crushed trap rock or dolerite has been used in the Connecticut Valley for some time.

For *building and paving,* clean, angular, sharp sand is desired for concrete and mortar; specifications for paving are less rigid. Gravel for concrete is largely replaced by superior crushed stone. For secondary roads, paving gravel is generally used as extracted. About one-half of the sand and gravel produced goes into concrete.

Railroad ballast requires only bulk and packing ability. Crushed stone is now largely replacing gravel for this purpose.

Molding sands are used in foundries for making molds to receive molten metal. Such sands should be refractory to withstand melting; cohesive to take and hold shape (therefore, some clay is necessary);

porous to allow the escape of gases; and low in iron.

Engine sands used for the prevention of wheel slipping should be fine and even-grained, sharp, and dry.

Abrasive sands should be sharp, clean, sized, and free from clay.

Filter sand for filtering water supplies should be clean; sized; quartzy; and free from clay, lime, and organic matter.

Fire and furnace sand should be refractory, clean, and quartzy. To this a binding clay is added.

Glass sands are treated in Chapter 20.

Selected References on Sand and Gravel

Non-Metallic Minerals. 1925. R. B. Ladoo. McGraw-Hill Book COmpany, New York. *Types, specifications, uses.*

General References 7, 13, and especially 19, p. 561–565.

Sand and Gravel. 1970. J. P. Cooper. In Mineral facts and problems. U.S. Bur. Mines Bull. 650, p. 1185–1199.

GYPSUM AND ANHYDRITE

These two common evaporite minerals, gypsum ($CaSO_4 \cdot 2H_2O$), and anhydrite ($CaSO_4$), are widely used in the construction industry and agriculture. Gypsum, by far the more important of the two, is used to retard the setting time in portland cement, and as an agricultural soil conditioner and fertilizer for which anhydrite is also used. Gypsum is calcined to form "plaster of paris" ($CaSO_4 \cdot \frac{1}{2}H_2O$) for use as plaster for construction and industrial purposes, and for manufacturing wallboard and other prefabricated products now widely used in the building industries.

Although nature slowly transforms the surficial surfaces of anhydrite to gypsum, man has not been successful in forming gypsum from anhydrite, a needed process because anhydrite is far more plentiful than gypsum.

Gypsum was known and prized by the Assyrians and Egyptians for making containers and for sculpturing. The Egyptians used it for making plaster for the building of the pyramids. The Greeks and Romans also used it for plaster, and the soft alabaster was desired for sculpturing. In the United States it was first used for "land plaster" in 1808.

Properties and Uses. Gypsum occurs in five varieties: (1) rock gypsum and (2) gypsite, an impure earthy form, both of which are used commercially;

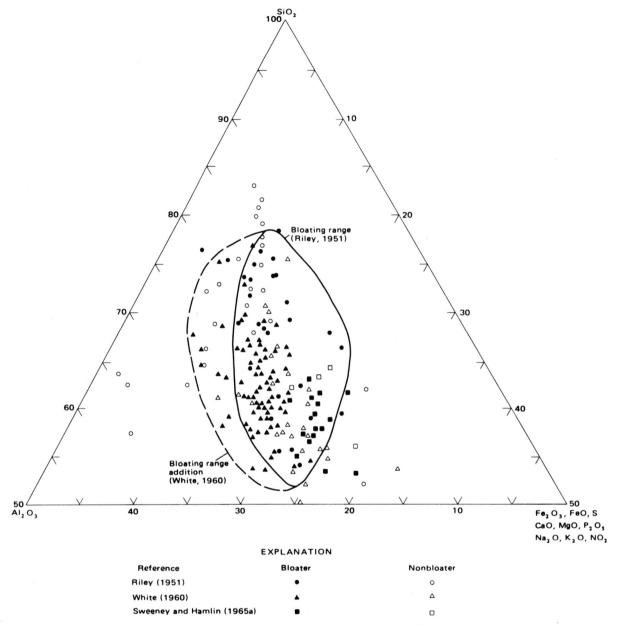

EXPLANATION

Reference	Bloater	Nonbloater
Riley (1951)	●	○
White (1960)	▲	△
Sweeney and Hamlin (1965a)	■	□

Figure 25-2 Comparative composition of bloating and nonbloating argillaceous rocks. (From: A. L. Bush, 1973, U.S. Geol. Survey Prof. Paper 820, p. 340.)

(3) alabaster, a massive, fine-grained, translucent variety; (4) satin spar, a fibrous silky form; and (5) selenite, a transparent crystal form.

The various plasters set quickly with smooth finish and are mainly cement plaster, plaster of paris, stucco, neat plaster, flow plaster, and finishing plaster. The laths, plasterboard, and wallboard consist of thin sheets of paper, felt, or thin wood, separated by compressed plaster and are used for walls, ceilings, partitions, roofing, and for fireproofing and sound-deadening. The other uses are for embedding plate glass and decorative stone for polishing; for various types of cements and casts; and for fillers, paints, crayons, insecticides, and other minor purposes.

Production and Distribution. The normal world production amounts to about 80 million tons annually, with the United States producing about 18 percent.

The U.S. annual production ranges from 12 to 15 million tons and is obtained chiefly from Michigan, Texas, Iowa, and California, in the order named. The most extensive deposits are in the western states, where it occurs in Pennsylvanian, Permian, Triassic, and Jurassic formations. It is, however, found with rocks of all ages from Silurian to Pleistocene.

Gypsum is widely distributed, but large production is confined to those countries of large local demand or cheap extraction and transportation. The extensive occurrences in the Paris Basin have given rise to the term *plaster of paris*.

Occurrence and Origin. Gypsum occurs as an evaporite in regular beds or lenses, in various states of purity, and in a great range of thickness from a few feet to hundreds of feet. Gypsum may occur singly or with anhydrite as a primary deposit, or as a surficial hydration product of anhydrite.

Extraction and Preparation. Since gypsum occurs chiefly in flat or gently inclined beds, most of it is mined by open pit methods. It is crushed and ground to a powder. That used for portland-cement retarder and agricultural purposes is sold as raw gypsum. The rest is calcined in kilns at temperatures of 300 to 350° F to remove part of the water of crystallization.This gives calcined gypsum or, if pure, plaster of paris. With the addition of water, the plaster sets in 6 to 8 minutes, so a retarder is necessary to delay the set 1 to 2 hours. This consists of glue, fiber, lime, or other materials. The impurities of gypsite commonly serve as effective retarders.

Selected References on Gypsum
Gypsum and Anhdyrite. 1939. F. T. Moyer. U.S. Bur. Mines Inf. Circ. 7049. *Occurrence, mining, and processing.*
Cements, Limes, and Plasters, 3rd ed. 1928. E. C. Eckel. John Wiley & Sons, Inc., New York. *Comprehensive discussion.*
General Reference 7, 13, and especially 19.
Gypsum, Texas. 1934. In Edwards limestone. Virgil E. Barnes; In Hockley Dome. H. B. Stenzel. Texas Min. Res. Univ. Texas Pub. 4301: 35–46; 205–226. *Evaporite beds and salt-dome occurrences.*
Gypsum. 1970. R. V. Ashizawa. In U.S. Bur. Mines Minerals Yearbook, v. 1–2, p. 547–553.
Gypsum and anhydrite in the United States, exclusion of Alaska and Hawaii. 1962. C. F. Withington. U.S. Geol. Survey Mineral Inv. Resource Map MR-23.
The Gypsum-Anhydrite equilibrium at one atmosphere. 1967. L. A. Hardie. Am. Mineral., v. 52, p. 171–200. *Stability and equilibrium temperature for the reaction $CaSO_4 \cdot 2H_2O = CaSO_4 + 2H_2O$ (liq. soln) as a function of activity of $H_2O(aH_2O)$ of the solution.*

LIME

Since the time of the ancients, lime has been used for mortar. It is simple and cheap to prepare. Limestone or other calcareous rock is heated in kilns to 903° C, when the CO_2 is driven off and quicklime (CaO) remains. This slakes with water; mixed with sand, it makes mortar or plaster. Commonly, the lime is prepared as hydrated lime, $Ca(OH_2)$, by adding the necessary water. One hundred pounds of pure limestone yield 56 pounds of lime. Dolomite may also be used, giving $CaO \cdot MgO$, which slakes more slowly and gives less heat than CaO lime.

Uses and Production. The utility of lime depends upon the stone from which it is made. With the addition of water it sets in air and finds wide use as mortar and plaster. Much of it, therefore, is consumed in the building trade, but it also has many uses in the chemical industries and is an important fertilizer. The chief uses are in metallurgy for smelting and for concentration of metalliferous ores, which consumes one-third of manufactured lime; another one-third is used by the chemical industry; with the remaining one-third used by construction industries, paper mills, agriculture, and others. Magnesia lime makes a strong, hard, elastic stucco.

Since the raw materials for lime are world-wide in distribution, it is produced almost everywhere according to need. The United States produced 20 million tons in 1979, valued around $830 million, from 42 states, with Ohio and Pennsylvania leading.

Selected References on Lime
General Reference 13 and 19.
Cements, Limes and Plasters. 1928. E. C. Eckel. John Wiley & Sons, Inc., New York. *Materials, properties, manufacture.*
Lime. O. Bowles and D. M. Banks. 1936. U.S. Bur. Mines Inf. Circ. 6884. *Uses, requirements, and technique.*
Chemistry and Technology of Lime. 1967. R. S. Boynton. Wiley-Interscience, New York. 449p.

MAGNESITE

Magnesite is desired chiefly as a raw material for magnesium compounds and metallic magnesium

(Chapter 21) and is used to some extent in its natural state. Its use expands with activity in the building and metallurgical industries. After calcination it yields material for refractory bricks, cements, and flooring. Dolomite is used in place of magnesite, when possible, because it is cheaper.

Properties and Uses. Magnesite occurs as both the crystalline and amorphous (crystocrystalline) varieties, the latter being generally the purer. It loses its carbon dioxide content upon heating, forming magnesia (MgO), which upon further heating develops into periclase. This resists hydration and carbonation at ordinary temperatures. The magnesite of commerce refers not only to $MgCO_3$ but also to the sintered products, magnesia, and to breunnerite, the carbonate with over 5 percent ferrous carbonate. It is burned to *caustic magnesite,* containing 2 to 7 percent CO_2, at 700 to 1200° C, and to *dead-burned magnesite,* with less than 0.5 percent CO_2, at 1450 to 1500° C. Most of the magnesite consumed in the United States goes into these products.

Caustic magnesite is used for oxychloride or Sorel cements, which are employed mainly for flooring and stucco. As a stucco it has many advantages over portland cement, except for weathering qualities. Since this difficulty has been overcome, its use for this purpose has increased. Its use for flooring and wallboard has also increased greatly. Mixed with magnesium chloride and fillers, it makes an inexpensive, hard, nonshrinking, dustless flooring material, which is fireproof, flexible, durable, and takes wax or polish. It is suitable for bathrooms, hospitals, and public buildings. An important use is as a chemical accelerator in rubber. It is molded into pipe coverings for heat insulation and has some ceramic use.

Dead-burned magnesite is a high-grade refractory in the metallurgical industry.

Magnesite is extensively used for metallic magnesium. Other minor applications are in the paper, ceramic, and glass industries, as an abrasive, as a source of CO_2, and for magnesium chemicals.

Production and Distribution. The annual world production is about 12 million tons, and the principal producing countries are Russia and Manchuria.

The United States production, amounting to about 600 thousand tons, with additional amounts from sea water and brines.

Occurrence and Origin. Magnesite has three modes of occurrence:

1. Replacement of dolomite or limestone, for example, Washington, Austria, Manchuria, Czechoslovakia, and Quebec.
2. Veins, for example, California, Greece, India, Russia, Yugoslavia, and Nevada.
3. Sedimentary beds, for example, Nevada.

The *replacement* deposits yield the "crystalline" variety and have resulted from progressive replacement (rarely complete) of limestone or dolomite by $MgCO_3$ through hydrothermal solutions. This forms bedded deposits, lenslike or irregular in shape, and of large size. They generally contain some ferrous iron.

The *veins* contain the hard amorphous variety and occupy fractures or crush zones in serpentine or ultrabasic rocks. They result from the breakdown of serpentine by hydrothermal carbonate solutions, accompanied by the release of silica, which forms opal or chalcedony.

The *sedimentary* deposits probably formed by evaporation but they are so interbedded with dolomite and other rocks to make their extracting unprofitable.

Extraction and Preparation. Magnesite is quarried, hand-sorted, crushed, and then calcined in kilns. For caustic magnesite, the kiln temperature is about 1200° C, and for dead-burned magnesite, to which iron ore is added for sintering, the temperature is about 1560° to 2,000° C.

Examples of Deposits

United States. Only two districts in the United States contain large deposits of crystalline magnesite, one at Gabbs, Nevada (Chapter 21), and the other in northeastern Stevens County of Washington. Small occurrences are known in New Mexico, Texas, and California.

The Gabbs deposits are large, covering an area of about 2 kilometers. It is suggested that the deposit is sedimentary but modified by a nearby granodiorite stock. All of the Gabbs magnesite and some brucite was sent to Henderson, Nevada, during World War II until 1966, to be recovered as magnesium. Of course, U.S. production of magnesium chloride provides a much greater source of magnesium from seawater, deep brine wells, and especially from the Great Salt Lake (Fig. 21-4), which

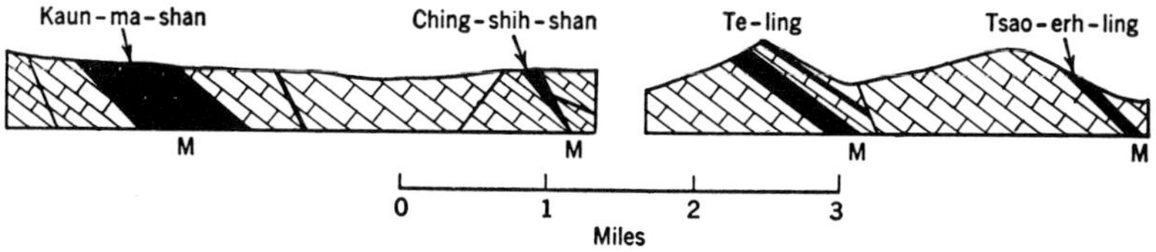

Kaun-ma-shan   Ching-shih-shan   Te-ling   Tsao-erh-ling

M   M   M   M

0   1   2   3
Miles

Figure 25-3 Sections of Manchuria magnesite deposits. (From: Niinomy, Econ. Geol.)

has recoverable reserves of magnesium chloride that exceed 600 million tons.

The large deposits of crystalline magnesite near *Chewelah, Washington,* occur as a hydrothermal replacement of Carboniferous dolomite. The deposits are huge, bedded lenses up to 330 meters long and 100 meters thick. The material varies in grain size, is white to red in color, and is low in iron. The reserves within 30 meters of the surface are large. The *California* deposits of the Coast Ranges occur as pure amorphous magnesite in serpentine in veins, lenses, or masses up to 8 meters wide. One deposit consists of boulders or nodules in serpentine breccia. A deposit in *Kern County* is apparently sedimentary. *Nevada* contains a sedimentary deposit up to 70 meters thick along the Muddy River. Brucite deposits, covered by hydromagnesite and formed as an alteration of dolomite, are worked in Nye County. This is made into "thomasite," a ferrite-bonded magnesia refractory.

Russia. Russia contains four large deposits of crystalline magnesite in the Urals; the largest, at Satka, has reserves of 145 million tons. The deposits extend for 5 miles in two parallel series of lenses up to 270 feet wide within a series of sediments. New deposits have been found at Savinsk, in eastern Siberia, where 2000 million tons of talc-rich magnesite exist.

Manchuria. The largest deposits of magnesite in the world are worked in Manchuria over a 15 kilometer belt (Fig. 25-3). They occur as replacements associated with dolomite in a series of metamorphosed Cambrian sediments. Individual deposits, accord-

ing to Niinomy, attain a length of 1900 meters and a thickness of 900 meters. The deposits of Korea are generally similar.

Others. The principal deposits of *Greece* are in Euboea, where veins and lenses of magnesite occur in serpentine (Fig. 25-4). One deposit is a series of lenses almost 2 kilometers long, of which individual lenses attain 220 meters in length and 60 meters in width. The largest Mandoudi lens is 1000 by 78 meters. Similar but smaller deposits occur in *Yugoslavia. Czechoslovakia* has a 120 kilometer belt containing numerous deposits, apparently of replacement origin in limestones. *Brazil* has a 300-million-ton deposit of crystalline magnesite and a more than 4 billion-ton-deposit of the evaporite mineral tachyhydrote.

Selected References on Magnesite
General References 7 and 19. *Chapter on magnesian refractory.*
Types of magnesite deposits and their origin. 1924. G. W. Bain. Econ. Geol., v. 19, p. 412–433. *Good general paper.*
Magnesite. 1931. P. M. Tyler. U.S. Bur. Mines Inf. Circ. 6437. *Occurrence, distribution, uses, and preparation.*
Preliminary report on magnesite deposits of Stevens Co., Washington. 1941. W. A. G. Bennett. Washington Dept. of Cohs. and Dev. Repr. Inv. 5, 22 pp. *Description of deposits.*
Production and properties of commercial magnesias. 1942. M. Y. Seaton. AIME Tech. Pub. 1496, 21 pp. *Deposits and description of production processes.*
Magnesium resources of the United States. 1957. R. E. Davis. U.S. Geol. Survey Bull. 1019-E, p. 373–515.

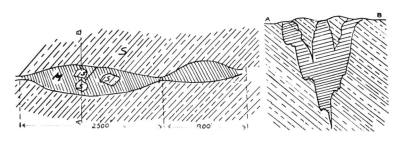

Figure 25-4 Sections of magnesite deposits, Mandoudi, Euboea, Greece. (From Sagui, Econ. Geol.)

Unusual marine evaporites with salts of calcium and magnesium chloride in Cretaceous Basins of Sergipe, Brazil. 1972. N. C. Wardlow. Econ. Geol., v. 67, p. 000.

Magnesium Compounds. 1971. R. Chin., U.S. Bur. Mines Minerals textbook.

Magnesite and brucite. 1936. W. W. Rudey and E. Callighan. U.S. Geol. Survey Bull. 871, p. 113–114.

PIGMENTS AND FILLERS

Mineral fillers, as described by Brown, are finely ground, generally white, mineral or rock additives that impart necessary or desirable physical properties to a product. Mineral pigments are mineral-derived powders added for the sole purpose of giving color. The functions of fillers and pigments overlap, inasmuch as one of the common purposes of a filler is coloring.

Mineral pigments are utilized directly as paints or to give color, body, or opacity to paints, stucco, plaster, cement, mortar, linoleum, rubber, plastics, or other materials. There are three classes: (1) natural mineral pigments, (2) pigments made by burning or subliming natural minerals, and (3) manufactured paints. Combinations of the three are common.

Natural Mineral Pigments

Natural mineral paints contain as their essential color constituents either limonite, hematite (or rarely magnetite), with or without mixtures of clay and manganese oxides. They form ochers, umbers, and siennas. Most of these were used by early man for decoration and drawings and were extensively employed in America in colonial days. Their chief use is for painting steel and ironwork, barns, and freight cars, since they consist of unfading minerals. The various colors and their brilliancy depend upon the proportions of the essential constituents. These natural mixtures are called ochers and are formed mostly by residual weathering.

Mineral red or Indian red is composed of hematite and is formed from the weathering of hematite or iron-bearing minerals. *Vermilion* has long been a New England barn paint. *Persian red*, the original Indian red, contains 65 to 72 percent Fe_2O_3 and comes from the Persian Gulf. *Spanish red*, containing 82 to 87 percent Fe_2O_3, is a nearly pure red hematite from Spain. *Venetian red* originally came from Italy.

Mineral yellow-brown is limonite, generally with some manganese oxide.

Ochers are mixtures of hematite, limonite, and clay, with 15 to 80 percent iron oxide, and yield yellow and brown colors. When roasted they yield reddish browns. *Umber* is an ocher with 11 to 25 percent manganese oxide, which gives it the typical brown color. When roasted it takes on a rich brown color of *burnt umber. Sienna* is a yellowish-brown ocher with less manganese oxide and more limonite, named from Sienna, Italy. There are several rich yellows, such as mineral yellow, Chinese yellow, and Roman earth. Roasting gives a browner tint, known as *burnt sienna.*

Greens, such as green earth, green ocher, terreverte, and Celadon and Verona greens are all low-tinting pigments derived from ferromagnesian silicates, greenstones, and rocks rich in chlorite.

Whites are obtained from gypsum, barite, talc, white clay, and other substances and are used for plaster, mortars, and cold-water paints.

Ground red and black slate and shale and other colored rocks also serve as natural pigments.

Natural mineral pigments are obtained from several states, led by Pennsylvania, Georgia, and New York, and from Spain, the Caspian Sea, Cyprus, Italy, France, South Africa, and India.

Manufactured Pigments

Iron oxide red and brown pigments are made by roasting ochers, umbers, and siennas and also iron ore and copper ore. Various shades of red, yellow, brown, and black are made from iron salts, such as ferrous sulfate or ferric chloride. Carbon black, for black paints and ink, is made by impinging flaming natural gas upon cold steel plates.

Most chemical paints are made from lead, zinc, titanium, barium, chromium, and carbon.

Lead paints were made chemically and by smelting. The lead compounds used are white lead (basic carbonate), sublimed lead (basic sulfate), red minium oxide, red, orange, and yellow chromates, yellow massicot, and orange litharge. Little lead paint is now made because of the toxicity of lead.

Zinc paints are made from zinc ore and from metallic zinc, giving zinc oxide and leaded zinc oxide. Many white paints are made of white lead and zinc oxide, which consumes 18 percent of the annual production of zinc.

Barium is used to make *lithopone*, a brilliant white paint composed of 70 percent barium sulfate and 30 percent zinc sulfide and oxide. After a phenomenal rise, lithopone has suffered severe competition from titanium paints.

Titanium yields white paints of such outstanding opacity and covering ability that they are growing competitors to lead and zinc paints. It is used as titanium oxide, or blended with barium sulfate or calcium compounds, or added to lead and zinc paints. The manufacture of TiO₂ pigment from ilmenite and rutile is the main use of these minerals; the industry consumed 96 percent of titanium ore production in 1969 and the use is increasing.

Other mineral compounds and the colors obtained follow.

Metal	Color	Compound Used
Chromium	Chrome green	Chromic oxide
	Chrome yellow	Lead chromate
	Guignet green	Lead chromate and Prussian blue
	Chrome red	Basic lead chromate
	Zinc yellow	Zinc and potassium chromates
Cadmium	Cadmium red	Cadmium sulfide and selenium
	Cadmium yellow	Cadmium sulfide
	Cadmium orange	Cadmium sulfide
Cobalt	Cobalt blue	Cobalt oxide and silicate
Mercury	Vermilion	Mercuric sulfide
Calcium	Venetian red	Iron oxide and calcium sulfate
Carbon	Carbon black	Natural gas soot
	Black	Natural graphite
	Balck roofing	Asphalt
	Black rouge	Precipitated or ground magnetite

All pigment minerals must be finely ground and must have opacity or hiding power and ability to absorb oil. The chemical composition is of minor importance. Various combinations of the materials listed may be used to produce intermediate colors.

Occurrence and Distribution. Most natural pigments are formed by residual concentration (see Chapter 13). Certain localities have become noted for certain colors. For example, yellow ocher, France and United States; sienna, Italy; umber, Cyprus; red oxides, Spain, Canada, and United States; Persian red, Persian Gulf; and various metallic paints, United States. Brown has compiled a table (Table 25-4) of natural iron oxide pigment colors and sources that illustrates the variety of colors produced from this ore mineral source.

Production. More than 15 rocks or minerals are commonly used as fillers, and in 1969, more than 7 million short tons of fillers valued at more than $350 million were used in the United States.

Selected References on Mineral Pigments
Non-Metallic Minerals. 1925. R. B. Ladoo. McGraw-Hill Book Company, New York. *Best articles; comprehensive; and good bibliography.*

Iron oxide mineral pigments. 1933. H. Wilson. U.S. Bur. Mines Bull. 370.
General Reference 5, 13, and especially 19. *Chapter on pigments and fillers.*
Iron oxide mineral pigments of the United States. 1933. H. Wilson. U.S. Bur. Mines Bull. 370, p. 198.
Nonmetallic Minerals, 2nd ed. 1951. R. B. Ladoo and W. M. Myers. McGraw-Hill Book Company, p. 605.

HEAT AND SOUND INSULATORS—MINERAL WOOL

There has been a rapidly increasing demand for heat and sound insulators in building construction, particularly in the United States. Heat insulation in buildings to conserve heat in winter and to render them cooler in summer is now even more important with the energy crisis. Various nonmetallic materials are used for this purpose, such as mineral wool, diatomite, exfoliated vermiculite, expanded gypsum, magnesia, mica, pumice, asbestos, and aluminum foil. Except for mineral wool and vermiculite, these materials have other more important uses and are described elsewhere. All the substances mentioned above have the property of holding small pockets of air between nonconducting fibers or plates, and nonmoving air makes a good

Table 25-4 Properties, Value, and Industrial Utilization of the Principal Mineral Fillers Used in the United States in 1969. (X, percentage unknown)

Mineral filler	Distribution by end use (in percent, if known)								Percentage of total commodity consumption	Approximate value (in thousands of dollars)	Distinguishing properties	Approximate tonnage used in United States for filler (in thousands of tons)	Remarks
	Paint	Rubber	Paper	Fertilizer	Pesticides	Plastics	Roofing	Other unidentified uses					
Asbestos	—	—	—	—	—	>75	X	X	25	20,000	Nearly all uses based on fibrous nature	200	Filler uses take mainly "fines" and "shorts"
Barite	75	20	—	—	—	—	—	5	4	>4,000	Highest density filler; inert, light colored	66	About 80 percent of the world's annual production is used for heavy drilling mud
Bentonite and other clays, excluding fuller's earth and kaolin	—	—	—	—	7	—	—	93	0.5	500	Absorbent	28	Not an important filler; mainly an insecticide diluent
Carbon black	X [b]	90±	—	—	—	5	—	X	95	200,000	Extremely fine particle size, only black filler	1,480	Excellent rubber reinforcing agent. Also a pigment.
Diatomite	X	—	X	X	X	X	—	X	20	7,200	Inert or reactive opaline silica depending upon whether calcined, skeletal silica particles	120	Used in raw state, milled or calcined for fillers. Large reserves
Fuller's earth	—	—	—	—	95	—	—	5	20	5,000	Absorbent, cost low	187	Mainly used for floor sweeping compound, rotary drilling muds, and decolorizing agent
Gypsum	X	—	X	—	X	—	—	—	<5	3,000±	Low cost, soft	100±	Minor filler mineral; used raw or calcined
Kaolin	4	12	75	3	<1	2	—	4	66	67,500	White, low oil absorption; abundant, low cost	3,164	Most important filler mineral. principal paper filler, important rubber filler

Material										Physical properties		Remarks
Limestone	X	X	—	X	X	—	X	5	22,000	High brightness, low oil absorption, availability good and cost low	1,300	Cheapest filler; has low oil absorption and good color. Main filler for putty, caulking, and sealants
Mica, ground	37	9	X[b]	—	52	<1	X	62	5,100	Micaceous structure, high dielectric strength, spangled sheen to wallpaper	79	Scrap mica, including some phlogopite and biotite. Mica and sericite schists are resources
Perlite	X	—	—	—	—	X	X	1	220	Lightweight; vesicular structure	5	Not an important filler; mainly used for plaster and concrete aggregate and filtering aid
Pyrophyllite	X	X	—	X	—	—	X	65±	1,000±	Inert, soft, high brightness	111	Statistics reported with talc in U.S. Bur. Mines Yearbooks; about 27,000 tons used for ceramics
Silica, ground	X	X	—	—	—	X	X	Small	?	Hardness; low oil absorption	?	For paint extender in deck paints, for high abrasive use
Talc	47	7	16	16	8	X	X	40	14,000	Inert, soft, high brightness; some varieties fibrous or platy	350	Some grades contain much tremolite, anthophyllite or serpentine
Wollastonite	X	X	—	—	—	X	—	25±	200	White, acicular-fibrous; low oil absorption; alkaline suspension	5	Mainly used in ceramics; newest important filler mineral
Total									350,000		7,200	

Source: C. E. Brown, U.S. Geol. Survey Prof. Paper 820, p. 336, 1972.

[a] Used as a pigment.

[b] Used in wallpaper.

heat insulator. Sound insulation, for the improvement of acoustics and reducing noise, depends upon continuous channels connected with many small pores; the channels admit the sound waves and the pores trap them. Also, sound is reflected by means of smooth-surfaced tile, brick, or plaster.

Mineral Wool

Mineral wool exhibits the lowest heat conductivity of any material except still air, and ranks below cork. It is also light, weighing only 8 to 12 pounds per cubic foot. It is an artificial product composed of extremely thin silicate fibers, 1 to 10 microns in diameter, and resembles wool. It includes innumerable minute air pockets and hence is a good insulator. The term includes rock wool, slag wool, glass wool, glass silk, and silicate cotton.

Manufacture. According to Lamar and Machin, mineral wool is made by subjecting molten silicate to shearing forces while cooling rapidly enough to prevent crystallization. This is accomplished by (1) blowing steam or air through molten silicate, making droplets that elongate into threads in air; (2) centrifuging on a large rotating disc; (3) allowing droplets to fall from an elevation; (4) spinning a thread upon a rotating drum from extended silicate, or from the molten glass to a steam of air blast so that the glass stream is not broken. The addition of fluorspar gives finer fibers.

Rock Wool. Rock wool is made from (1) "natural woolrocks," and (2) rock mixtures of "composite woolrocks." These are impure carbonate rocks that consist mainly of lime, magnesia, alumina, silica, and carbon dioxide; and contain about 40 to 65 percent carbonate. The natural woolrocks may consist of a single rock or a series of interbedded strata (limestone and shale or sand) that together constitute the necessary composition. The groups include impure limestone and dolomite high in silica or silt, cherty carbonate, calcareous or dolomite sandstone and shale, some glacial clays, and some gravels. Suitable mixtures may be compounded from high calcareous rocks and siliceous or argillaceous rocks. Wollastonite rock is used in California. High iron, sulfur, and calcium sulfide are objectionable. The materials are melted in cupola furnaces at as low a temperature as possible (800° C); CO_2 is eliminated; and molten silicate is formed.

Slag Wool. This is made from the slag of iron blast furnaces, which is high in CaO because limestone is used for fluxing iron ore. Copper and lead slags may also be used. A slag with 30 to 50 percent CaO or CaO + MgO would yield mineral wool; others need added ingredients to correct the composition. Much of the mineral wool or "rock wool" of today is made from slag instead of rock.

Glass Wool. Glass wool or glass silk is made from commercial "soft" glass or "soda-lime glass," which is made of the customary raw materials for glass making. Glass wools serve the same purpose as other mineral wools. The fibers are strong and flexible; because of this they are also used for weaving into glass fabrics that are strong, flexible, lustrous, nonfading, and noninflammable. Curtains, upholstery, and clothing are made of spun glass.

Distribution, Production, and Uses. Mineral wools are made wherever raw materials are available. In the United States much is manufactured in Illinois, Indiana, Wisconsin, New Jersey, Nevada, Maryland, Alabama, and California.

Mineral wool is used chiefly for house insulation. Minor uses are for insulating pipes, boilers, tanks, stoves, refrigerators, refrigerator cars, air-conditioned trains, and for insulating boards, blankets, protecting water plants (mulching wool), filters, and sound insulation.

Other Insulators

Vermiculite. Vermiculite is a mica that exfoliates under strong heat into very thin sheets. It is used for heat insulation in loose form or, with a binder added, is fabricated into forms. It is supposed to represent a hydrothermal alteration of biotite or phlogopite. Deposits occur at Libby, Montana, and in Colorado, Wyoming, North Carolina, Georgia, Newfoundland, and South Africa.

Diatomite. Diatomite or diatomaceous earth, an accumulation of myriad microscopic siliceous shells of diatoms, is very porous and so light that the dry powder weighs only 7 to 16 pounds per cubic foot. Its lightness and its minute cells, filled with immovable air, make it an ideal insulator; it is used in the form of powder, bricks, and blocks. Its chief use, however, is as a filter.

In 1971, the United States produced more diatomite than any other nation, about 650,000 short tons, worth about $50 million. The USSR is second in the production of diatomite.

Gypsum. Calcined gypsum in an expanded form (Insulex) is also used for heat insulation. Chemicals that produce gas are added to calcined gypsum, and when water is added the gases expand the plaster of paris to lightweight, cellular powder or flakes. Perforated gypsum board is also used for sound insulation.

Asbestos. Asbestos has other more important uses than for heat insulation, but the short fibers and refuse are utilized in asbestos cement and in covers for boiler and pipe insulation or are made into fireproof shingles, sheet roofing, and wallboards. The fabrics made from the longer fibers are also utilized for fireproof curtains, awnings, warship mats, gloves, aprons, and firemen's clothes. The long fibers, providing air channels between, are suitable for soundproofing.

Basic Magnesium Carbonate. This substance, made from calcined dolomite, is used with 15 percent asbestos to make pipe and boiler coverings for heat insulation.

Pumice. Pumice, a volcanic frothy glass, because of its large cellular space and light weight is an ideal natural insulator. It rarely occurs, however, in a form suitable for insulating use. A massive bed in California yields large blocks that can be sliced directly into sheets suitable for refrigerator insulation. Less massive material and granules are used in stucco and plaster for sound insulation and as a cement admixture. The United States produces about 70,000 tons annually, most of which is used as an abrasive.

Perlite. Perlite, an acid volcanic glass, when heated becomes porous and serves for insulation and light aggregate. Not all perlite expands sufficiently to be of commercial value. Value in 1979 was $24.85/T.

Selected References on Heat and Sound Insulators
General Reference 7 and 13
General Reference 19. *Excellent detailed data on insulation, including trade name products.*

BITUMEN-BEARING ROCKS

Bitumen-bearing rocks occur in many portions of the earth, but few exist in commercial amounts. Of course, until recently, few efforts have been made to exploit or evaluate such deposits.

Two notable exceptions are the Athabaska tar sands of Canada and the extensive deposits rimming the Uinta Basin in Utah, and the Piceance Basin in western Colorado.

The bitumen is a viscous to semisolid hydrocarbon that fills the interstices of consolidated or unconsolidated slightly porous and permeable rocks. Numerous names are attached to such deposits; for example, tar sands, oil-impregnated rocks, asphalt-bearing rocks, oil sands, bitumen-bearing rocks, and petroleum impregnated rocks. "Bitumen-bearing rocks" is preferred by Cashion and many others even though "tar sands" is the favored term in Canada and Utah.

Mineralogy. *Asphalt* is a common term even though it does not have a given mineral composition. Some of the specific minerals are gilsonite, wurtzitite, ozokerite, and others.

Production. The Athabaska tar sands are being mined and shipped as a slurry to processing plants by GCOS (Great Canadian Oil Sands, Ltd.) a subsidiary of Sun Oil Co. where it is converted to crude oil. Great Canadian Oil Sands is producing about 50,000 barrels per day, now at a profit, after 10 years of an annual loss in excess of $1 million.

Vertical gilsonite veins have been mined in Utah since the nineteenth century. The minable veins are a meter or two wide, hundreds of meters deep, and extend along the strike for kilometers as "straight as an arrow." Some of these veins are still being mined, as the asphalt zones in the "Tar-Sand Triangle" are extensive and are as much as 75 meters thick. A major research study is being made at the University of Utah on these deposits with the belief that these deposits may produce oil before the oil-shale properties are in production.

Other deposits. Asphalt Lake in Trinidad covers 114 acres. A similar lake occurs in Venezuela. In 1967, tar-sand deposits in Albania, Rumania, and Russia, and the asphalt-lake deposit in Trinidad were being exploited. In 1964, the LaBrea, Trinidad, operation produced 184,246 tons of material.

Reserves. Hydrocarbon reserves in these deposits are enormous. World reserves are estimated at 915 billion barrels—and this includes only those tar-sand deposits with more than 15 million barrels in reserves.

Selected References on Bitumen-Bearing Rocks
General Reference 19. *Excellent summary reviews bitumen-bearing rocks.*
Major tar-sand deposits of the world. 1967. P. H. Phizackerly and L. O. Scott. In Drilling and production. World Petroleum Cong. 7th, Mexico, Proc., v. 3, p. 551–557.
Selected bibliography on asphalt-bearing rocks of the United States and Canada to 1970. 1972. U.S. Geol. Survey Bull. 1352, 218 p.

MISCELLANEOUS MATERIALS

Clays and Clay Products. Clays and clay products are extensively used for building and structural materials. They are also widely used for a great variety of other materials and, therefore, are treated together as a unit under "Ceramic Materials" (Chapter 24).

Building materials utilize mainly the common, lower-grade, widely distributed types of clays and shales of sedimentary, residual, or glacial origin. The clay products that are employed in building operations are: various types of bricks; tiles in the form of hollow building tile, conduits, roofing, floor, fireproofing, drain, sewer, and mosaic; flue linings; wall coping; chimney pots; toilet fixtures; laundry tubs; enamel stoves; doorknobs; and other porcelain and whiteware. In addition, innumerable products of clay are used about the home.

Mica. Since mica is used mainly for purposes other than structural, it is treated separately and in detail in Chapter 27. A growing minor proportion is utilized for building and structural materials.

In the preparation of industrial sheet mica there is a large amount of scrap and waste. This and ground mica, which is ground-up scrap and pulverized mica-rich schists, are utilized in part for backing for roofing rolls and shingles, and in concrete and some cements. The better grades of wet ground mica are used in some paints, and mica is the luster-imparting ingredient in wallpaper.

26

METALLURGICAL AND REFRACTORY MATERIALS

A group of unrelated minerals finds common use in the large metallurgical industry. Some are used directly for making steel and other metallurgical products; others serve as refractory materials to withstand the high temperatures employed in metallurgical and related furnaces. The group includes fluorspar; cryolite; graphite; and refractory substances, such as magnesite and dolomite, aluminous compounds, refractory clays, diaspore, bauxite, silica, and zirconia.

FLUORSPAR

Fluorite or fluorspar (CaF_2) is a commercial mineral of critical importance because of its necessity in the steel industry. Since the discovery of this use in 1889 its production has increased rapidly. See Fig. 26-1.

Uses. About 33 percent of the fluorite produced is consumed in the basic open-hearth process of steel making. It facilitates fusion and the transfer of objectionable impurities such as sulfur and phosphorus into the slag and gives fluidity to the slag. About 18 percent is used for making hydrofluoric acid, used chiefly in the aluminum industry for making synthetic cryolite. The glass and enamel industry consumes about 15 percent. Other uses are listed in Table 26-2.

Occurrence and Origin. Worl, Van Alstine, and Shawe classify fluorine deposits into four groups,

1. Fluorine deposits associated with igneous rocks. These include disseminated and replacement deposits, such as the Thomas Range, Utah, topaz-bearing rhyolite; deposits in pegmatite, and carbonatite such as Mountain Pass, California; Okarusu, Southwest Africa; and Amba Dongar, India; and deposits in

contact aureoles, such as Spor Mountain, Utah.

2. Fluorine deposits associated with sedimentary rocks. These include those in volcanoclastic and lacustrine sedimentary rocks and those in evaporite, marine carbonate, and phosphorite deposits. Examples are the fluorspar deposits near Rome, Italy, and Rome, Oregon, where the deposits are associated with volcanic tuff; lacustrine brines and evaporite deposits and the brines in the rift valleys of Africa, where lake waters contain as much as 1627 ppm fluorine; and the fluorine in marine carbonates and evaporites of the Permian Preurals of the USSR. Certainly, the Phosphoria formation is also a prime example in the western United States.

3. Fluorine deposits associated with regionally metamorphosed rocks. Topaz and tourmaline occur in a topaz-quartz-sillimanite gneiss of Precambrian age in the central part of the Front Range, Colorado.

4. Fluorine in hydrothermal (magmatic or not) deposits. This group includes those in veins and mantos, pipes and stockworks, and in zones or alteration. There are numerous examples in India, Eastern Mongolia, in the western United States, and the North Pennine district, England. Others occur in Mexico, Spain, Brazil, South Africa, the USSR, Korea, Canada, Italy, France, Pakistan, and Thailand.

Production and Distribution. U.S. production and consumption of fluorspar has decreased recently, probably because of the decreased steel and aluminum production. World production has increased.

In 1970, the United States accounted for only 6 percent of the world's production of fluorspar but consumed 30 percent; thus the United States pro-

Table 26-1 Estimated Fluorspar Resources of the World.
(Millions of short tons of crude ore. The ratio of crude fluorspar ore to processed fluorspar recovered is about 3:1.)

Area	Identified resources[a]	Hypothetical resources[b]
United States	25	45
Mexico	25	50
Canada	10	15
Europe[c]	62	84
Africa[d]	39	310
Asia[e]	25	62
South America[f]	3	6
Australia	1	1
Total	190	573

Source: R. G. Work, et al. U.S. Geol. Survey Prof. Payner 820, 230, 1973.

[a] Chiefly exploitable economic ore containing 35 percent or more CaF_2 or equivalent in combined fluorspar and metal values. The average CaF_2 content is about 50 percent. Includes some subeconomic material.

[b] Some material of ore grade, but much rock is exploitable only under more favorable economic or technologic conditions. Deposits incompletely known because few subeconomic fluorspar deposits have been explored.

[c] Chiefly France, Germany, Italy, Spain, United Kingdom, and the U.S.S.R.

[d] Chiefly Kenya, Morocco, Mozambique, Rhodesia, Southwest Africa, Tunisia, and the Union of South Africa.

[e] Chiefly China, India, Korea, and Thailand.

[f] Chiefly Argentina and Brazil.

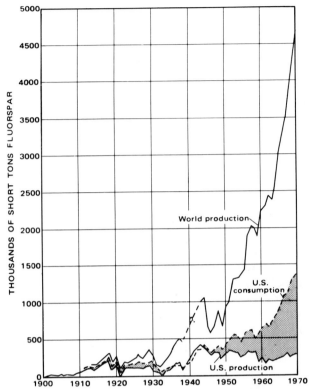

Figure 26-1 World production, 1913–1970, U.S. production, 1900–1970, and U.S. consumption, 1911–1970, of fluorspar, (CaF_2). (From R. G. Worl, et al., 1973, U.S. Geol. Survey Prof. Paper 820, p. 225.)

duced only 20 percent of what it consumed. The other 80 percent was imported from Mexico, Spain, Italy and the Republic of Africa. Figure 26-1 indicates the rapid rate of world production of CaF_2.

Mineralogy. Fluorine consists of a single stable isotope mostly as a single charged anion $(F)^-$, but also as a complex, such as $(BF_4)^-$. It is a highly active gaseous element and is prevalent in active volcanic areas as F_2, SiF_4, and H_2SiF_6. The fluorine ion has a radius of 1.33 Ångstroms, and with oxygen at 01.32 Ångstroms; and with proper valence conditions, diadochic substitution of F^- for $(O)^-$ or O^{2-} is common. Examples are fluorapatite, $Ca_5(PO_4, CO_3)_{.3}F$ and topaz $Al_2Si_4(F,O)_2$.

Fluorite (CaF_2) is by far the most common mineral of fluorine providing 44 percent of the fluorine consumed in the United States. Cryolite, Na_3AlF_6, formerly from Greenland, is now produced syn-

thetically by combining caustic soda, alumina, and hydrofluoric acid. It is then the major flux used to reduce alumina to aluminum in electrolytic cells, which accounts for 18 percent of the U.S. annual consumption of fluorine.

Table 26-2 Uses of Fluorspar

Metallurgical	Chemical and Miscellaneous
Iron foundry flux	Extraction of potash from feldspar and portland-cement dust
Electric furnace ferroalloys and alloy steels	
Producing nickel, Monel metal, brass	Making calcium carbide and cyanamid
Iron and manganese fluorides	Fluorides and silicofluorides
Smelting Au, Ag, Pb, and Cu	Making portland cement
Refining Cu, Pb, and Sb	Bond for emery wheels and electrodes
	Insecticides, preservatives, and dyestuffs
	Refrigerating fluids
	Ceramic purposes
	Rock-wool manufacture
	Microscope lenses

Other fluorine-bearing minerals are sellaite, MgF_2; villiaumite, NaF; and bastnaesite, $(Ce, Ca)(CO_3)F$.

Resources. The identified resources of the United States are about 25 million tons, largely in Illinois, Kentucky, Texas, New Mexico, Nevada, Utah, Colorado, Idaho, Montana, and Alaska. Other portions of the world have resources as indicated by Table 26-1.

Sources from marine phosphate rocks are enormous although the grade is low, about 3 percent fluorine. These deposits are chiefly in Florida; North Caroline; Tennessee; and in the phosphoria formation in Utah, Wyoming, Idaho, and Montana.

Extraction and Preparation. Fluorite is mined in the same manner as underground metalliferous deposits; rarely, it is obtained from open pits.

In some deposits fluorite occurs in such purity that a merchantable product is obtained by hand cobbing and sorting. Generally, however, it must be concentrated and cleaned by jigs or tables or by flotation. The last yields a product up to 98 percent pure and enables low-grade deposits to be worked.

The highest-grade product, known as "acid" fluorspar, is used for making hydrofluoric acid and requires 98 percent of calcium fluoride; for glasswork, 97 percent purity is required, and for steel furnaces 85 percent is satisfactory. The products are marketed in gravel, lump, and ground form.

Examples of Deposits

Illinois-Kentucky District. This district of 100 square kilometers spans the Ohio River and has been one of the largest and most productive fluorspar districts in the world.

The area is underlain by horizontal Mississippian sediments cut by mica-peridotite dikes. It is arched into a broad dome 50 kilometers across that has collapsed by intricate faulting, some of faults having displacements up to 400 meters.

The deposits are of two types: (1) steep replacement veins occupying faults of moderate displacement, and (2) horizontal, bedded or blanket replacement deposits. The former are the more important.

The vein deposits are typified by those near Rosiclare, Illinois. They are as much as 10 meters wide, and the Rosiclare vein has been mined for 2700 meters along the strike. The fault in which it occurs is 7 kilometers long. The veins contain chiefly fluorite and calcite but also some quartz, barite, and base-metal sulfides. Masses of pure fluorite several meters across are not uncommon. Calcite was deposited first, followed by fluorite, which replaced calcite and limestone; in the Daisy mine, minor calcite was seen to replace fluorite. The veins represent combined replacement and open-space filling; vein matter cements and also replaces limestone, fault breccia, and wall rock. Vugs lined with crystals of calcite and fluorite are common. They are typical hydrothermal deposits.

The bedded deposits near Cave In Rock. Illinois,

Figure 26-2 Polished surface of high-grade fluorite specimen (90 percent fluorite) showing dark and light banding. (From: Currier, Econ. Geol.)

occur in the flattish Fredonia limestone beneath shale. These interesting ore beds are from several centimeters to 4 meters thick, average 1 meter, and contain fluorite of high purity. The same minerals of the vein deposits, with the addition of marcasite, are present in minor amounts. Vugs yield handsome aggregates of fluorite crystals. These deposits exhibit remarkable parallel banding of fluorite (Figs. 26-2 and 26-3), interpreted by Bastin as rhythmic replacement and by Currier as preservation of bedding and cross-bedding in replaced limestone. Currier believes that the fluorite resulted from reaction of hydrothermal solutions containing hydrofluoric acid upon limestone, giving calcium fluoride.

Views on the origin of these deposits and those farther northeast in Ohio and Michigan have changed from the classical magmatic hydrothermal source to a connate sodium-calcium-fluorine brine solution, with low to moderate temperatures that migrated slowly through the carbonate beds. The source of the fluorine is unknown but could have been leached from evaporite beds, leached from sediments, or derived from a deep-seated origin.

Foreign Deposits. Important deposits of fluorspar, with ample reserves, occur in *Germany* in the Harz Mountains, Anhalt, Baden, Bavaria, Saxony, and Thuringia, where the spar is an important constituent of base-metal fissure veins. *French* fluorspar of high purity is obtained from Var, Puy-de-Dôme, Saône-et-Loire, and other places. It occurs in fissure veins associated with chalcedony, barite, and quartz. In *Great Britain* important deposits occur in Derbyshire and Durham in long fissure veins carrying also lead, zinc, barite, and calcite. Less important deposits occur in Cornwall and Devon in fissure veins carrying ores of lead, copper, tin, and tungsten. A large production comes from epithermal veins at St. Lawrence, *Newfoundland*. In *Russia* deposits occur at the Kara Sea and beyond Lake Baikal, and in *China* in Fengtien, Shantung, and Chekiana. Important fluorite deposits are also found in Bolzano, Italy; Mexico; Spain; the Transvaal; New South Wales; British Columbia; and Ontario.

Selected References on Fluorspar

General Reference 19. *Excellent review chapter.*
Origin of the bedding replacement deposits of fluorspar in the Illinois field. 1937. L. W. Currier, Econ. Geol., v. 32, p. 364–386. *Reviews theories of origin and proposes replacement.*
Fluorspar. 1972. H. B. Wood. Eng. and Mining Jour. v. 173, p. 80.
Fluorspar—A world review. 1971. B. L. Hodge. Ind. Minerals, No. 8, p. 9–29.
On the geochemical cycle of fluorine. 1947. T. F. W. Barth. Jour. Geol., v. 55, p. 420–426.
Fluorite in Mineral Facts and Problems. 1970. R. T. MacMillan. U.S. Bur. Mines Bull 650, p. 989–1000.
Geologic characteristics of fluorspar deposits in the western United States. 1958. W. C. Peters, Econ. Geol., v. 53, p. 553–588.
Fluorspar Deposits of the Western United States. 1945. J. L. Gillson. AIME Tech. Pub. 1783. *Survey of western deposits; bibliography.*
Fluorspar Resources of New Mexico. 1946. H. E. Rothrock, C. H. Johnson, and A. D. Hahn, New Mex. Bur. Min and Min. Resources Bull. 21, p. 245. *Veins and breccia deposits.*
General References 7 and 8.

CRYOLITE

Cryolite is a rare mineral occurring in only two places in the world, namely, *Miask, Russia, and Greenland* although small amounts of it do occur at Pikes Peak, Colorado.

This aluminum and sodium fluoride is desired for flux and solvent in the electrolytic bath used for the

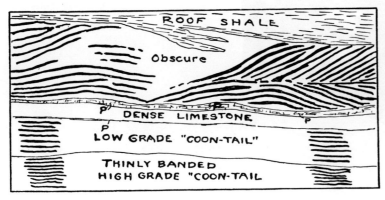

Figure 26-3 Sketch of mine face showing banding in fluorite. Black, impure bands; white, pure bands; *p*, shaly partings. (From Currier, Econ. Geol.)

manufacture of aluminum. Its scarcity has led to the manufacture of artificial cryolite. A small amount also finds use in the making of enamels, glass, glazes, insulating material, and insecticides.

At *Ivigtut, Greenland,* on the shore of Arsuk Fiord, cryolite, associated with pegmatite was quarried from a small mass of porphyritic granite. The deposit is irregular in shape, about 30 by 150 meters in area. The cryolite contains base-metal sulfides, siderite, and fluorite. The Greenland cryolite has been exhausted and almost all cryolite is now produced synthetically.

GRAPHITE

Graphite receives its name from *grapho,* "write." Once mistaken for lead, it was called "black lead" (also plumbago), and pencils made from it are still called "lead pencils." Although chemically the same as diamond and charcoal, it is a crystalline modification of carbon. It ranges in purity from 30 to 98 percent. In commerce it is classified into natural and artificial; the natural varieties are divided into "crystalline" and "amorphous," the only difference between them being that of grain size. Dis-

seminated graphite deposits contain from 2½ to 7 percent carbon.

The consumption by uses and the dollar values of each during 1969 is shown in Table 26-3.

The value of graphite products was $27,494,673 in 1979 which places the graphite industry as one of the smallest but growing mineral industries. The extent of U.S. imports of natural graphite is given in Table 26-4 for 1979.

Production and Consumption. Synthetic graphite is crystalline, not flaky, and is produced in high-temperature furnaces from petroleum, coke, anthracite, or other forms of amorphous carbon. The amount produced in 1969 was seven times the consumption of natural carbon during that year. Even though synthetic graphite is more costly than the natural carbon, the former is the only one pure enough to be used as moderator rods in nuclear reactors. It is also used in electrodes for electrosmelting and in anodes for the electrochemical industry.

Uses. The high melting temperature of graphite (3000°C) and its insolubility in acid create many uses for it. The oldest use is in making pencils, which still persists. The softest graphite is finely ground, mixed with clay and baked, the amount of clay and the time of baking giving the desired hardness. Other uses are listed in Table 26-3.

Table 26-3 Consumption, by Uses, of Natural Graphite in the United States in 1969

Use	Short tons	Value (thousands)
Batteries	(*a*)	(*a*)
Bearings	104	$51,307
Brake linings	1,066	368,998
Carbon brushes	588	285,869
Crucibles, retorts, stoppers, sleeves, and nozzles	4,639	780,793
Foundry facings	8,028	1,017,221
Lubricants	2,702	492,316
Packings	409	182,983
Paints and polishes	182	27,997
Pencils	2,190	666,095
Refractories	5,554	735,419
Rubber	225	73,054
Steelmaking	6,624	578,386
Other*b*	4,853	1,093,359
Total	37,164	6,353,797

Source: P. L. Weis, U.S. Geol. Survey Prof. Paper 820, p. 278, 1973.
a Withheld to avoid disclosing individual company confidential data; included in "Other."
b Includes adhesives, chemical equipment and processes, electronic products, gray iron castings, powdered-metal parts, small packages, specialities, and batteries. Data from U.S. Bur. Mines.

Table 26-4 Tonnage and Value of 1978 U.S. imports for consumption of natural graphite, by countries

Country	Total Tons	Total Value
Mexico	49,954	$1,723,000
Madagascar	4,865	1,490,000
Korea (Republic)	26,082	1,475,000
Sri Lanka	2,302	1,031,000
Germany (Fed. Rep.)	2,064	1,007,000
Switzerland	917	989,000
China (Mainland)	3,046	896,000
USSR	3,659	708,000
Others		2,381,000
Total		11,700,000

Source: Mineral Industry Surveys, Graphite in 1978, 1979.

Occurrence and Origin. Graphite occurs in metamorphic, igneous, and sedimentary rocks in flakes, lumps, and dust. It originates by:

1. Magmatic concentration—Irkutsk, Siberia.
2. Contact metasomatism—Calabogie, Ontario; Ticonderoga, New York.
3. Hydrothermal deposition in veins—Ceylon; San Gabriel, California; Dillon, Montana.
4. Metamorphism—Alabama; Raton, New Mexico; Bavaria; Madagascar; Sonora, Mexico.

Most graphite originates through metamorphic processes, and these, as well as other processes of origin, are discussed in Chapter 15.

Extraction and Preparation. Vein graphite is mined no differently from other vein deposits; the included lump graphite is hand cobbed or screened. Disseminated flake graphite makes up only a few percent of the enclosing rock and is generally mined from open cuts and concentrated to a product of 80 to 85 percent carbon. The concentrates may be further refined to about 95 percent carbon by finer grinding and screening and rarely by chemical removal of impurities.

Examples of Deposits

Russia. The deposits near Irkutsk, Siberia, are among the most important of the world and for long were the chief supply of lump pencil graphite. The graphite occurs in nepheline syenite and adjacent schist. Other extensive deposits have been developed in the Turukhansk and Yenisei districts, Siberia. Russian reserves are hundreds of millions of tons of flake graphite.

Korea. Korea in 1944 attained the largest production of graphite ever attained by any country but it has been surpassed by several nations recently. Ninety-six percent of it was amorphous. The deposits occur in schists in layers and lenses, many of which are near the contacts of Cretaceous granites. Several deposits are metamorphosed coal seams; others are metamorphosed graphitic argilites.

Sri Lanka. The deposits of Sri Lanka are the most productive of high-grade graphite. They are unusual in that the graphite occurs in fissure veins up to 25 meters wide, filled almost exclusively with large, thick plates and fibrous graphite. Subordinate quartz, pyrite, and calcite are present. The veins cut gneiss and marble and are intruded by granite and pegmatites. Some veins are worked to depths of 500 meters. Disseminated graphite also occurs in the gneisses and marble and some occurs in pegmatites.

Mexico. The largest producer at the present time is Mexico, which produces 75 percent of the U.S. imports of natural graphite. It has hundreds of million of tons of amorphous graphite with tenors between 80 and 85 percent graphite. The ore occurs in upper Triassic coal beds metamorphised by granite dikes. The beds are as much as 7.5 meters thick.

Canada. Commercial graphite deposits occur in Ontario and Quebec, where flake graphite is disseminated in graphic gneisses, schists, and crystalline limestone. It also occurs in masses enclosed in limestone near igneous intrusions, in narrow veins, and in pegmatite dikes. An unusual deposit is the Black Donald mine, Calabogie, Ontario, where an intrusive cuts Grenville limestones, forming a contact-metasomatic deposit of graphite, calcite and silicates 5 to 20 meters wide and known to a depth of more than 60 meters. The rock contains up to 60 percent graphite, of which one-fourth is flake and three-fourths is amorphous.

The potential of a major graphite deposit was recognized in 1974 by J. A. McGregor near the village of Southend at Reindeer Lake in northern Saskatchewan.

Open pit reserves are estimated at 1.8 million tons containing 10.32 percent carbon to a 60 meters depth. The ore zone is up to 40 meters wide and at least 550 meters long. Production of 9000 tons of flake graphite per year is projected.

United States. Although graphite has been mined in 27 states, the chief areas of production were Alabama, New York, Pennsylvania, and Texas. Production is now confined to Alabama and New York. The Alabama belt, 100 kilometers long and 8 kilometers wide, consists of mica schist and included pegmatites with graphitic bands from 6 to 30 meters wide that average from 2.5 to 3.5 percent flake graphite. Only weathered parts, to depths of 10 to 15 meters, are mined. Prouty thinks that the graphite was derived from former petroleum or petroleum products in the rock.

The New York occurrences lie in the Adirondack region, mostly near Ticonderoga. The important deposits are graphitic quartz-schists carrying 5 to

7 percent graphite and Grenville limestone intruded by granite. Nearby, granite pegmatites have intruded limestone and schist, producing a contact-metamorphic zone containing contact silicates and graphite.

In Chester County, Pennsylvania, graphite occurs in graphitic schist and crystalline limestone cut by pegmatites. Similar deposits occur in the Llano-Burnet region, Texas, where schists contain 10 to 14 percent flake graphite. Ceylon-type graphite has been produced from veins near Dillon, Montana. Near Raton, New Mexico, coal has been metamorphosed to graphite; similarly, amorphous graphite has been formed from coal at Cranston, Rhode Island. Beverly describes deposits in Los Angeles County, California, in shear zones in schists cut by pegmatites in which the graphite particles replace rock minerals.

The identified graphite resources of the United States are less than 1 million tons of flake graphite and a few thousand tons of 80 to 85 percent amorphous graphite, which is a by-product from the mining of flake carbon. Alaska has the largest potential reserves in the Kigluaik Range of the Steward Peninsula, where graphite occurs in gneiss and schist. These deposits are higher grade than those of the eastern United States but less accessible.

Selected References on Graphite
General Reference 7 and 19. Chapter on graphite.
Graphite (natural) 1970. R. W. Lewis. In Mineral Facts and Problems. U.S. Bur. Mines Bull. 650, p. 1025–1038.
The Adirondack graphite deposits. 1917. H. L. Alling, N.Y. State Mus. Bull. 199, 150p. *Excellent detailed descriptions of individual deposits.*
Occurrence and origin of the graphite deposits near Dillon, Mont. 1954. R. B. Ford, Econ. Geol., v. 49.
Graphite deposits of Ceylon. 1912. E. S. Bastin. Econ. Geol., v. 7, p. 419–445. Excellent geological description.
Graphite deposits of Ashland, Ala. 1925. J. S. Brown. Econ. Geol., v. 20, p. 208–248. *Disseminated type in schists.*
Origin of the graphite of Ceylon. 1943. D. N. Wadia. Deylon Rec. Dept. Miner. From. Pap. 1, p, 15–24. *Veins, disseminations, and pegmatite occurrences.*
Alabama flake graphite in World War II. 1945. H. D. Pallister and R. W. Smith. AIME Tech. Pub. 1909. *Occurrence and developments.*
Graphite for manufacture of crucibles. 1945. G. R. Gwinn. Econ. Geol., v. 40, p. 86.
Strategic graphite—A survey. 1960. E. N. Cameron and P. L. Weis. U.S. Geol. Survey Bull. 1982-E, p. 201–321. *Excellent paper on deposits and good bibliography.*

REFRACTORIES

Modern high-temperature metallurgical processes require refractory materials for furnace linings that will withstand high temperatures; if the materials show signs of fusion below 1500°C, they are not classed as refractories. Other requirements are that the refractories will not react with the materials being melted, that they are strong enough to withstand the weight and wear of molten metals and slag, that they resist cracking and spalling under temperature changes, and that they can be molded into bricks or other forms. Some refractories are required for high-temperature purposes other than furnace linings, such as for retorts, for electrical purposes, and for ceramics.

Purposes and Uses. In the iron and steel industries refractories are required for iron ore blast furnaces, steel furnaces, open hearths, soaking pits, reheating furnaces, boilers for generating steam, and coke ovens. Public utilities use them for electrical power and gas plants, locomotives, incinerators, and other purposes. Refractory linings are necessary for the smelting and refining of metals, making alloys, cement kilns, glass furnaces, oil stills, holders or saggers for firing ceramic materials, making spark plugs, and a host of minor purposes in which high temperatures are involved.

The typical products supplied are bricks and other preformed standard shapes and numerous special shapes, such as retorts, muffles, tiles, saggers, cupolas, and glass-tank blocks.

The chief refractory materials used in the United States and important sources are listed in Table 26-5.

Principal Varieties of Refractories
The common refractories are generally grouped under fire clay, silica, high alumina, magnesite, and chrome. All the materials, except chrome, are nonmetallic products. Most of them have already been discussed in other sections of this book.

Fire Clays. Fire clays make up the bulk of modern refractories. Most of them are combinations of nonplastic refractory flint clays, moderately refractory plastic clays, and high-alumina clays. Their heat

Table 26-5 Chief Refractory Materials Used in United States

Material	Chief constituents	Upper temperature limit, °C	Important sources
Clays:			England, Germany, Belgium,
kaolin	Al_2O_3,SiO_2	1785	Pennsylvania, Georgia, Alabama,
fire clay	Al_2O_3,SiO_2	1745	Maryland, Louisiana, Ohio, and
			Missouri
Silica:			
quartz	SiO_2	1700	Pennsylvania, Ohio, and Alabama
cristobalite	SiO_2	1700	California
diatomite	SiO_2	1615	California, Nevada
High alumina:			
bauxite	Al_2O_3,H_2O	2020	Arkansas, Guianas, and the Gold Coast
diaspore	Al_2O_3,H_2O	2000	Missouri
corundum	Al_2O_3	2030	Canada, South Africa, artificial
Sillimanite group	Al_2O_3,SiO_2	1810	California, Nevada, North Carolina, Virginia, India, and Kenya
Magnesia:			
magnesite	MgO,CO_2	2800	Washington, California, Manchuria Greece, and Austria
dolomite	MgO,CaO,CO_2	2485	Numerous places
periclase	MgO	2800	Rare, artificial
spinel	MgO,Al_2O_3	2100	Artificial
brucite	MgO,H_2O		Nevada
Chrome	Cr_2O_3,FeO	2050	Africa, Cuba, Turkey, and New Caledonia
Miscellaneous:			
beryllia	BeO	2400	Brazil, Argentina
graphite	C	Infusible	United States, Ceylon, Madagascar, and Mexico
limestone	CaO,CO_2	2485	Everywhere
rutile	TiO_2	1630	Virginia, Brazil, and India
thoria	ThO_2	3050	Artificial
yttria	Y_2O_3	2410	Artificial
zircon and baddeleyite	ZrO_2,SiO_2	2300	India, Brazil, and Australia

resistance ranges from 1500 to 1650°C, and they are utilized for furnaces, boilers, oil stills, and miscellaneous furnace linings. High-grade fire brick made from fire clays and the flint clays of Missouri yield superior brick with antispalling properties. Refractory bricks of fire clay are furnished for every purpose needed. Refractory cements are also made for these bricks, some of which contain silica or high-alumina compounds.

The occurrence, origin, and distribution of the clays used are discussed under "Ceramics," Chapter 24.

Silica Refractories. Silica in its numerous natural forms is a widely used refractory (Fig. 26-4). The most common forms are quartzite (ganister) and crushed quartz, which, bonded with 2 percent of lime, yield silica brick, and important acid refractory product utilized in many metallurgical processes. Common sand and diatomite are also used, and the Berea sandstone of Ohio is sawn into bricks and blocks. The chief ganister deposits in the United States are the Tuscarora and Baraboo quartzites of Pennsylvania and Wisconsin, respectively.

Silica bricks are highly refractory and do not soften appreciably before their melting point, but they do spall when suddenly heated or cooled.

High-Alumina Refractories. The *sillimanite group* of minerals sillimanite, andalusite, kyanite, dumortierite all convert to mullite with a melting point of 1800°C. This yields a high-cost but much desired refractory, especially for highly refractory and

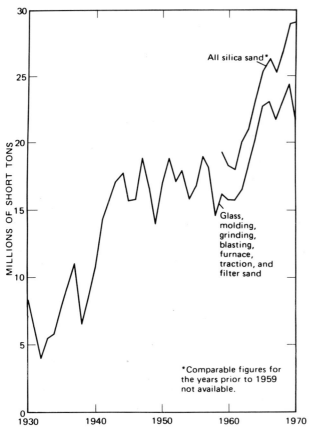

Figure 26-4 Production of silica sand in the United States. 1930–1970. (From Bur. of Mines and K. B. Ketner, 1973 U.S. Geol. Survey Prof. Paper 820, p. 578.)

tough porcelain bodies used for spark-plug cores, electrical porcelains, saggers, and laboratory porcelains. The geology of these minerals is discussed in Chapter 15 and their ceramic uses in Chapter 24.

Bauxite (Chapter 13) with the addition of a clay binder is utilized for making high-alumina refractory brick. Because of impurities present, such bricks fuse at a lower temperature than the melting point of alumina. They are used chiefly for linings for open-hearth steel furnaces. Bauxite brick is now largely displaced by diaspore brick.

High-alumina refractory bricks are made from *diaspore* and combinations of plastic and flint clays. "Sixty percent" brick is substituted for fire-clay brick in boilers, and "70 percent" brick is used for boilers, furnaces, and portland-cement kilns. The diaspore used contains 85 percent alumina and is obtained from Missouri, where it occurs in porous masses or as oölites in clay.

Magnesia Refractories. Dead-burned *magnesite* is desired chiefly as a basic refractory for basic slags

in metallurgical furnaces, for kilns, and for use with corrosive materials. The successful development of the basic converter, basic open-hearth, and electric furnaces is due in no small part to its use. The carbonate is calcined at 1500°C; the carbon dioxide is driven off; and the magnesia is converted to *periclase*. This is made into magnesite brick or used as granules, which form the hearth beds of basic steel furnaces, but not the roof because of a tendency to expand and spall. Magnesite is also used for "unburned magnesite brick," which resists spalling better than the ordinary magnesite brick. The sources and uses of magnesium are shown in Fig. 26-5. The occurrence of magnesite is given in Chapter 25.

Dolomite, when dead burned, is used for magnesia brick for basic open-hearth furnaces, being preferred to magnesite because it is cheaper.

Spinel is made artificially for refractory purposes by fusing calcined magnesite and alumina in an electric furnace. The product is then mixed with some ball clay and made into refractory brick, which has the formula of spinel. It is used as a special furnace brick and also for refractory cement.

Brucite, in minor quantities from large deposits in Nye County, Nevada, is calcined for a basic refractory for uses similar to those of magnesite brick.

Chrome. This mineral makes one of the best refractory substances known because of its high melting point and remarkable physical and chemical stability. It is made into high-duty brick that is used for side walls at the slag line in steel, copper, nickel, and chemical furnaces. Because of its lower cost it is displacing magnesite brick in certain fields. About 2.5 pounds of chromite is used per ton of steel produced. It is considered essential in steel furnaces. Chrome-magnesite bricks with about 75 percent chrome are also used. Chrome ore suitable as a refractory contains from 33 to 48 percent Cr_2O_3, 12 to 30 percent Al_2O_3, and about 17 percent MgO.

Zirconia Refractories. Zirconia (ZrO_2) is one of the most refractory substances known. It is strong and hard, resists spalling, and is relatively inert to slags and corrosive substances. Zirconia crucibles withstand temperatures up to 2500°C. They are used for refining precious metals and in electric furnaces. Zirconia and zircon brick are also made; zircon cement is used to coat other refractories.

The zirconium minerals used are zircon, the silicate; and baddeleyite, the oxide. They are obtained

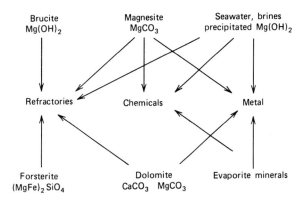

Figure 26-5 Use of principal raw materials of magnesium. (From A. J. Bodenlos and T. P. Thayer, 1973, U.S. Geol. Survey Prof. Paper 820, p. 378.)

from beach sands in India, Brazil, Florida, and New South Wales, along with monazite and ilmenite. For their geologic occurrence see Chapter 13.

Miscellaneous Refractory Materials *Graphite* and *carbon* are utilized to make refractory carbon brick. The graphite used is both natural and electric furnace product. Carbon is obtained from coke and charcoal. Carbon bricks are used for certain types of furnaces, boilers, and acid vats. They are, however, consumed in an oxidizing atmosphere at high temperatures, or in contact with fused oxides of metals.

Beryllia, with a melting point of 2300°C, forms a porcelainlike material suitable for crucibles and heat shields on satellites for protection as they reenter the earth's atmosphere at high velocities. It displaces alumina for higher-temperature work. *Rutile,* with lime as a binder, is made into brick for use as an acid refractory. *Thoria* yields a high-temperature refractory, more basic than zirconia, for extra heavy duty. *Olivine* has been found to be a desirable refractory material to replace chromite in steel furnaces, and for use above the slag line in nonferrous furnaces. Olivine is also mixed with magnesite to form the low-cost "forsterite" brick, which has excellent refractory qualities. Olivine, the chief mineral of dunite rock, is produced in North Carolina and Washington. *Soapstone* is sawed to desired shapes and used as a refractory for alkali recovery furnaces. *Pyrophyllite* blocks are also used for refractory purposes. *Talc* makes a very hard refractory for small shapes, such as gas tips; and *vermiculite* finds some uses. *Silicon carbide* (trade name, Carborundum) is an artificial product of excellent refractory qualities up to 2245°C. It is strong, acid proof, has high heat conductivity, and is used gen-

erally for the same purposes as high-alumina brick. *Fused alumina* (alundum, aloxite) is also widely used as a special artificial refractory for muffles, brick for certain types of furnaces, and for small-scale high-temperature work.

Selected References on Refractories

General Reference 19.
Refractories. 1931. F. H. Norton. McGraw-Hill Book Company, New York.
General Reference 7 and 13.
Magnesium. 1970. J. Paone. In Minerals facts and problems. U.S. Bur. Mines Bull. 650, p. 621–638.
Applied Clay Mineralogy, 1962. R. E. Grim. McGraw-Hill Book Company. 422p.
General Reference 11. 1968. The Gabbs magnesite brucite deposit, Nye County, Nevada. J. H. Schilling, p. 1608–1622.

FOUNDRY SANDS

Foundry sands are siliceous sands utilized to make forms for casting metal, and molding sands are those used for making molds. Some are used for the cores of castings. Since a bond is necessary to make foundry sands adhere, a certain proportion of clay is essential. Core sands are bonded with oil, resin, pitch, or other substances. The specifications for foundry sands are rigid. According to Ries the important properties are fineness, bonding strength, permeability, sintering point, and durability. The grains are partly angular and range in size from 20 microns to 3 millimeters. The fineness affects permeability, strength, and smoothness of casting. Permeability is necessary to permit escape of gases. If the sintering point is low, the sand will "burn" on to the casting.

Most foundry sands are of marine or lacustrine sedimentary origin. Dune sands or sandstones may also be used. Foundry sands are rather widely distributed. Prominent United States occurrences are the Hudson Valley; the Paleozoic beds of New Jersey, Ohio, Indiana, Illinois, and California; the St. Peter sandstone extending from wisconsin and Michigan to Oklahoma; and the Eureka quartzite in the Basin and Range Province.

The annual United States production is about 30 million tons.

Selected Reference on Foundry Sands

General Reference 19.
Silicon. 1970. F. E. Brantley. *In Mineral Facts and Problems.* U.S. Bur. Mines Bull. 650,p. 369–384.

LIMESTONE AND LIME

Limestone and lime are variously employed in the metallurgical industry for fluxstone in metallurgical furnaces, particularly those of iron and nonferrous metals. It serves to offset silica, to yield a thin slag, and to produce a basic slag that collects and retains impurities separating from the metal.

Lime is finding increasing metallurgical uses. Of the 120 million tons of lime made anually in the United States, about 83 percent is employed metallurgically for fluxes in furnaces. In basic open-hearth steel furnaces lime is added (with limestone) to flux off impurities. With the increasing use of scrap iron more lime is added. It is also used as a flux in electric steel furnaces and nonferrous smelters. In the flotation process of ore dressing, lime is added to make an alkaline circuit. In cyaniding, lime is added to neutralize soluble acid salts, to coagulate slimes, and to protect against carbon dioxide.

OTHER METALLURGICAL MATERIALS

Furnace Sand. Furnace sand is used to line bottoms and walls of open hearth and other furnaces. Such sands are high in silica: either they contain some bonding clay or else some fine clay is added. Mixed grain sizes are desirable so that the voids will be filled. When sintering occurs, a hard, resistant lining is formed.

Bauxite. In addition to its many other uses (Chapter 13), bauxite serves as a slag corrective in iron smelting.

Borax. Borax makes a good flux in brazing, welding, and soldering metals and is also utilized in smelting copper, refining gold and silver, and in assaying. Some borides act as deoxidizers to prevent scale and scum in making brass and bronze, and ferroboron is used in small quantities as a deoxidizer for certain steels. Boron steels of remarkable strength are made with as much as 2 percent boron.

Strontium Minerals. Strontium is alloyed with copper to increase hardness and decrease blow holes. Strontium and tin added to lead make a harder and more lasting metal for storage batteries. With tin it serves as a scavenger for metals containing copper, lead, zinc, or tin. In Germany strontianite is employed to desulfurize and dephosphorize iron and steel by making a thin slag.

Dolomite is employed as a flux in blast furnaces, particularly in the making of ferrosilicon and ferromanganese because it carries very little of the silica or the manganese into the slag.

Phosphate is employed metallurgically as ferrophosphorus, produced by smelting phosphate rock and iron ore, as an addition to special steels. The phosphorus content of pig iron is adjusted by the addition of phosphate rock.

27

INDUSTRIAL AND MANUFACTURING MATERIALS

A large number of unrelated nonmetallic materials are used for industrial and manufacturing purposes. Wares are made wholly of them, or they form important constituents of manufactured products, or they are employed in the production of other industrial products. This group does not include minerals utilized in the chemical industries, as they are treated in the following chapter. The group includes;

Asbestos	Mineral fillers
Mica	Mineral filters and purifiers
Talc	
Barite and witherite	Optical crystals
Glass sands	Miscellaneous materials

ASBESTOS

Asbestos is a term applied in commerce to naturally fibrous silicates that are amenable to mechanical separation into fine filaments of considerable tensile strength and flexibility; these fibers have, by virtue of unique combinations of physical and chemical properties, a great variety of industrial uses. Man-made fibers and naturally occurring organic fibers do not possess the full range of properties inherent in asbestos, and thus, at best, can be substituted only partially for the mineral fibers. These properties make asbestos of considerable value. Although used by the Romans for making lampwick that would not burn out, and for cremation cloths that became cleansed by fire, its real utilization did not start until 1878, after the discovery of the large Quebec deposits. During the twentieth century it has become one of the principal industrial minerals, and for many purposes, such as brake linings, it has no adequate substitute.

Uses. The uses of asbestos are so numerous that a listing given by Ross occupies four pages. There are two principal use-classes, spinning and nonspinning fiber. Spinning fiber comprises the longer grades of flexible fibers, and nonspinning comprises the shorter lengths of the same, and other short, matted, or harsh fibers. Spinning fiber is the most valuable and is spun into threads and yarns and woven into textiles. The fibers must possess strength and flexibility; individual filaments may be only 1/63,000 centimeter in diameter. The largest use of asbestos is for vehicular brake band linings and clutch facings, and an important one is for gaskets for steam fittings. The nonspinning fibers are chiefly valuable for various types of fireproofing and heat insulation materials. Some of the important uses of the 3000 known uses, are listed in Table 27-1.

Production and Distribution. Since 1940, the increase in the demand for absestos has been exceptional, exceeding 4 million short tons as illustrated by Fig. 27-1. This figure also indicates the reliance of U.S. demands upon foreign sources. Canada has increased its production by 400 percent since 1940 while Russia, in the same period of time, has increased its production by 15 times. Canada, Russia, and southern African deposits in Rhodesia, the Transvaal, and the Cape Provinces with a significant deposit in Swaziland, are the leading world producers and rank as listed. Prior to 1950, 90 to 95 percent of the world's annual production was derived from these three areas but dropped to 84 percent in 1968. The United States, Italy, and mainland China provided another 10 percent with the remaining percent derived from 17 other nations.

The use of a given lot of asbestos is governed by acceptable properties that are determined largely

Table 27-1 Important Uses of Asbestos

Spinning fiber	Nonspinning fiber

Asbestos cloth:
  brake lining and clutch
    facings
  theater curtains and
    scenery
  gaskets and sheet
    packings
  wall linings
  firemen's clothing
  fireproof gloves
  floor rugs and linings
  blankets and bags
  welding and acid
    equipment
  heat insulators
  steam and chemical
    equipment
  filter pads
  ovens
Asbestos rope, tape, yarn,
  and cord:
  fire-department equipment
  fire mats
  deck mats for war vessels
  all kinds of pipe and joint
    packing
  covering for electrical
    wires
  gaskets
  string asbestos cloth
  laboratory uses
  fire, acid, and electrical
    equipment
  wicks
	Asbestos millborad:
	  housing, household, and
	    factory wallboard
	  gaskets
	  stoves, ovens, and boilers
	  kilns, furnaces
	  garages
	  various insulators
	Asbestos paper:
	  floor and roofing paper
	  pipe covering
	  numerous heat insulators
	  electrical and steam
	    equipment
	Asbestos cements:
	  coverings for furnaces, and
	    pipes
	  floors, walls
	  heat tables
	  acid and chemical
	    equipment
	  burners
	Asbestos shingles
	Composition materials:
	  Various binders for
	    porcelains, plasters,
	    hard rubber,
	    phonograph records,
	    and electrical insulators

by fiber length. The longest fibers demand the highest prices and the shorter grades progressively are lower in value. (In recent years soft chrysotile fibers 2 centimeters long have sold for $1500 to $1600 per ton, whereas the shortest grades have brought $50 to $1,600 per ton). Thus asbestos must be separated from the parent rock, fiberized, and classified by length.

Asbestos Minerals. Asbestos is a commercial term applied to a group of minerals that separate readily into fibers. The minerals differ in chemical composition and in the strength, flexibility, and usefulness of their fibers. Broadly, they fall into two groups—serpentine and amphibole; the former includes the mineral chrysotile and the noncommercial picrolite, and the latter includes anthophyllite, crocidolite, amosite, tremolite, and actinolite.

Mountain leather, mountain wood, and mountain cork are varieties of amphibole.

Chrysotile is the most valuable variety. Its fibers are fine, silky, and strong; 4,350 meters of thread can be spun from a kilogram of it. Some varieties withstand temperatures up to 2750° C.

Crocidolite, or "blue asbestos" from South Africa, is a long, coarse, flexible spinning fiber with low fusibility and high resistance to acids.

Anthophyllite occurs as mass fiber, which is short, brittle, and nonspinning and is used chiefly for insulation. *Amosite* is an iron-rich variety of anthophyllite that occurs in unusually long, splin-

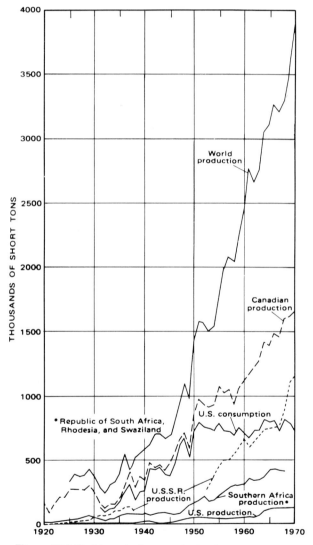

Figure 27-1 World production, production from major sources, U.S. production and consumption of asbestos, 1920–1970. (From A. F. Shrite, 1973, U.S. Geol. Survey Prof. Paper 820, p. 66.)

tery, coarse fibers, some of which can be spun; however, amosite is used chiefly as a binder for heat insulators.

Tremolite and *actinolite,* except for the Italian tremolite, have little commercial value; some actinolite is ground up for insulating purposes.

The length of fiber is important in determining the utility and value. Long fibers give strong yarns and threads; short fibers cannot be spun and are suitable for lower-cost products, such as boards, shingles, and cements. The fiber is graded on the basis of length, and identical quality ranges in price from $50 to $1,600 a ton according to length of fibers.

Occurrence and Origin. The mode of occurrence and origin of the different kinds of asbestos are stated in Chapter 15.

Extraction and Preparation. Asbestos deposits are mined by large- and small-scale methods, both in open cuts and underground. In Quebec, the fiber rock is extracted from large open cuts and loaded by power shovels. Some of the pits are so deep that underground methods are employed for the deeper rock.

In milling, distinction is made between "crude," "mill fibers," and "shorts." No. 1 Crude is spinning fiber over ¾ inch in length, and No. 2 Crude is from ⅜ to ¾ inch; both are hand-cobbed on a picking belt. Mill fibers are freed by crushing and beating the fiber rock, and shorts are the lowest-grade mill products. If fiber ⅜ inch long is worth $150 a ton, ¾-inch length is worth $1500 a ton; therefore, in milling, if long fiber is broken its value may be reduced three-quarters. Consequently, milling methods aim to separate fibers with minimum breakage. Crude is generally pounded by hand to free attached serpentine. Mill rock is crushed and screened in stages, the fiber being drawn off by suction. The crudes are prepared by mechanical beating and combing to separate and fluff the fiber for threads or yarns.

Examples of Asbestos Deposits

Examples of chrysotile in serpentinized ultrabasic rocks are: Thetford-Black Lake, Quebec; South Africa; Shabani, Rhodesia; Barberton, Transvaal; Bajenova, Saja and Uss, Russia; and Cyprus. Examples in serpentinized limestone are: Arizona; Laiyuau, China; Carolina, Transvaal; Madras, India; and Minusinsk, Russia. Crocidolite and amo-

site are confined to South Africa and Australia, and anthophyllite to the United States, Newfoundland, and Finland. Some tremolite is mined in Italy.

Canada. The zone of ultramafic bodies in the Appalachian belt extends from Alabama to Newfoundland. Hundreds of small to large such bodies occur in this belt. Certain rock types, such as peridotites and some dunite and pyroxenites, that exist in the ultramafic zone, have been altered to serpentinites. Until recently, the only mineable deposits were restricted to a 88 kilometer zone in the Eastern Townships of Southeastern Quebec. Farther northeast, abundant deposits have been found but only one, in Newfoundland, near Baie Vorte, has entered production—and that within the last decade.

Another Canadian belt of ophiolite terrain exists in Ontario and extends eastward into Quebec. Large-scale production of asbestos was initiated in Ontario with the opening of the Munro mine in 1950.

An ultramafic belt of Mesozoic age extends sporadically through California to Oregon and Washington, and through British Columbia to the Yukon Territory and Alaska. The Cassair deposit in British Columbia has reserves of 25 million tons of ore with a content of 8 to 10 percent asbestos that mills free of magnetite and includes a high proportion of long fibers. The development of this deposit was hampered, to say the least, by climate, permafrost, and transportation logistics.

Russia. The asbestos of Russia occurs chiefly in the Bajenova district, Urals, with minor amounts in Minusinsk, Yenisei River, and Maikop and Laba River in the Caucasus. In the Bajenova district, according to Keyser, there are about 20 pits developed in ellipsoidal masses, 1,000 by 300 meters, of a serpentine ultrabasic rock, which disclose cross-fiber chrysotile veins. The fiber contains sufficient ferrous iron to cause discoloration upon weathering. The quality is lower than Canadian asbestos, but, although the percentage of total fiber is about the same, the proportion of spinning fiber is higher. The reserves are large. Similar but smaller deposits have been found in the Sayan Range, west of the Altai Mountains, and in the Aktovrak and Uss districts. Little is known about the large, more recent finds in Russia.

Southern Rhodesia. This country contains much good spinning chrysotile in three districts, the Bulawayo (including the Shabani mine), Victoria, and Lomagundi.

The Shabani deposit occurs in the central part of a serpentine mass 17 by 2 to 5 kilometers, which has been altered from dunite. perhaps as the result of a granite intrusion. The fiber zone is up to 5 kilometers long and 180 meters wide, but the best fiber occurs in the lower 6 to 60 meters of the inclined serpentine body. Veins of chrysotile are up to 15 centimeters wide, but because of partings they rarely yield more than 3-inch fiber. The fiber seams parallel the contact. About 25 to 30 percent of the fiber produced is of spinning grade, and the shorts are not utilized. The depth is unknown, but reserves are stated to be 7 million tons. Similar deposits are mined at Victoria and Lomagundi along the Great Dyke.

South Africa. This country first produced the unique crocidolite and amosite. It also yields high-grade chrysotile from two types of occurrence: (1) cross-fiber ribbon rock in serpentinized dunite, in the Barberton district, and (2) cross fiber in a bed of serpentinized dolomite, in the Carolina district.

The *Barberton* occurrence exhibits remarkable and puzzling geological features. Near Kaapsche Hoop is a gently dipping (10 to 20°) serpentinized peridotite sill that is fiber-bearing for 5 kilometers in length and which, in the New Amianthus mine, has been followed down-dip for over 600 meters. There is an unusual zone or "ribbon line" or rhythmic banding, from 1.5 to 6 meters high, with cross-fiber veins parallel to the sill contact. The bottom 2 to 2.5 meters contain widely spaced fiber seams from 5 to 22 centimeters thick that make up about 10 percent of the entire rock. Above these are thinner (1 to 2 centimeters), more closely spaced seams, and then a zone of closely spaced

ribbon rock with glistening seams of silky chrysotile, averaging 15 to 30 seams to the foot, that aggregate 30 percent of the rock. Above this is a dolerite sill. The fiber recovery is about 20 percent of the rock mined. About 8 percent of the output is of spinning grade and 5 percent is over 2.5 centimeters in length. The remarkable parallelism and rhythmic alteration of the ribbon rock is baffling.

The *Carolina* district contains long, spinning chrysotile, within a 1.5 meter zone of dolomite, altered to serpentine, where it overlies a dolerite sill. Serpentine and fiber are of replacement origin.

The amosite deposits (Fig. 27-2) occur near Penge, Transvaal, as cross-fiber veins interbedded with banded siliceous-ferruginous slates. The fiber attains the remarkable length of 28 centimeters and averages 15 centimeters, but it is harsh and splintery. Its origin is an enigma.

The blue crocidolite is found near the amosite, under similar conditions, in a belt 400 kilometers long. The fiber length is up to 10 centimeters, about 12 to 20 percent of the total fiber being of spinning grade; the percentage of fiber is high. The reserves are large.

Cyprus. Chrysotile occurs at Amiandos, where a dunite plug is peripherally altered to serpentine, which contains irregular veinlets of "shingle-stock" fiber too short for spinning. The asbestos content runs only 1 to 2 percent.

United States. *Vermont* contains a few small deposits of the Quebec type but with negligible spinning fiber. Most of it is mill fiber that is used for molded brake linings, shingles, paper, and pipe covering.

In the Sierra Ancha of *Arizona* are small deposits of spinning chrysotile, free from iron. The deposits contain one or more horizontal seams of cross fiber

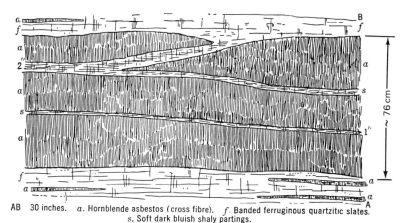

AB 30 inches.   *a*. Hornblende asbestos (cross fibre).   *f*. Banded ferruginous quartzitic slates.
       s. Soft dark bluish shaly partings.

Figure 27-2 Amosite asbestos occurrence, Penge mine, Transvaal, Republic of South Africa. Section, 30 m; *a*, hornblende asbestos; *f*, ferruginous quartzite slates; and *s*, shaly partings.

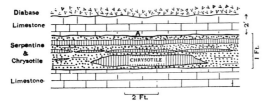

Figure 27-3 Asbestos bands in serpentine bands in dolomite, Sierra Ancha, Arizona.

within dolomite limestone beds of which certain bands have been altered to serpentine (Fig. 27-3). The fibers, which are as much as 15 centimeters long, occur only where the limestone has been cut by or overlies dolerite sills. Discontinuous seams occur en echelon. A replacement origin seems indicated. Somewhat similar occurrences are found in the Grand Canyon region.

In *California,* small quantities of mill fiber with a little spinning fiber occur in serpentinized dunite.

Of more than 90 asbestos prospects in California, at least half are chrysotile occurrences, about 30 are tremolite or anthophyllite occurrences. Intermittet mining has occurred. In 1962, a 2500-tons-of-ore-per-day operation began near Copperopoli, Calaveras County, one of the larger asbestos operations that California has seen.

Anthophyllite is produced commercially at Sall Mountain and Hollywood, Georgia, where 90 to 95 percent of the rock quarried is short fiber that is used for fireproofing and insulating, and in paint.

Australia. Large resources of crocidolite have been known to exist in the Hammersley Range in Western Australia. Mining was initiated in 1947 with continued exploration in an area of dimensions 320 kilometers long and 120 kilometers wide. The asbestos seams are thin and continuity is difficult

to predict. Production has ceased, but there are hopes of beginning again as access and other circumstances increase in favorability.

Selected References on Asbestos

World Asbestos Industry. 1970. Industrial minerals, London, no. 28, p. 17–29.

Asbestos. 1957. P. H. Riordon, and others. In Geol. of Canadian Ind. Mineral Deposits. 6th Comm. Min and Metall. Cong., p. 3–53.

General reference 19. Chapter on asbestos resources.

Asbestos, a mineral of unparalled properties. 1951. M. S. Badollet, Can. Inst. Min. and Metall. Trans., v. 54, p. 151–160.

The Asbestos Industry. 1955. O. Bowles, U.S. Bur. Mines Bull. 552, p. 122.

Asbestos in the United States. 1962. A. H. Chidester and A. F. Shride. U.S. Geol. Survey Mineral Inv. Resource Map MR-17.

General Reference 19. Chapter on asbestos.

MICA

Mica is a group name for several minerals familiar to the layman, because of its uncommon cleavage, which allows the minerals to split into thin sheets. The minerals are all hydrous aluminum silicates varying in amounts of iron, potassium, lithium, and other cations.

The amount of mica consumed annually is relatively small, but it is essential to the electrical industries because of its form and physical properties. No other mineral combines in itself so many desirable properties for the qualifications imposed by the electrical industry.

Mica Minerals. The chief mica minerals are

Mineral Name	Commercial Name	Chief Constituents in Addition to Al and SiO_2
Muscovite	White mica	Potash
Phlogopite	Amber mica	Magnesia, potash
Biotite	Black mica	Magnesia, iron, potash
Vermiculite	Jeffersite	Magnesia, iron, potash
Lepidolite	Lithia mica	Lithium, fluorine, potash
Zinnwaldite	Lithia-iron mica	Lithium, iron, potash, fluorine
Roscoelite	Vanadium mica	Vanadium, magnesium, iron
Fushsite	Chrome mica	Chrome

The first three are the only micas employed commercially as mica, and the first two are the principal ones; the lithium and vanadium micas are used as sources of those elements, and some vermiculite is used chiefly for sound and heat insulation.

Properties and Uses. An important property of mica is its perfect basal cleavage, which permits it to be split into sheets or films. Muscovite and phlogopite are split into sheets 0.025 millimeters in thickness; biotite does not split as well. This property, combined with low heat conductivity, high dielectric constant, flexibility, resilience, and toughness, make the light-colored micas outstanding electrical insulators. Muscovite withstands temperatures up to 550° C and phlogopite up to 1000° C. At higher temperatures they lose their water content. Crystallographic imperfections of twinning, intergrowth, or distortion, and foreign mineral inclusions spoil their splitting qualities, and much of the mica mined can be used only for scrap and grinding mica, to make mica powder.

Most of the sheet mica produced, that is pieces of at least a square centimeter, is used for electrical insulation, chiefly in condensers, tubes, radio, radar, and television. Large sheets go into electrical heater elements, such as toasters, and smaller sizes into segments, discs, and washers for electric-light sockets and fuses. Films are bonded with shellac into plates that are cut to desired shapes. Airplane spark plugs are built of bonded phlogopite. Clear sheets are used for heat windows and gas masks.

Wet ground mica gives luster to wallpaper and some paints; it is used as a filler in rubber and some plastics. Dry ground mica is used for roofing materials, stucco finish, lubricants, rubber dusting powders, and molded electric insulation. Biotite, obtained from biotite schists, is little used in sheets, and most of it goes into ground mica.

Most mica is obtained by quarrying with carefully controlled blasting to prevent damage to the crystals. It is hand-cobbed, sorted, freed of adhering rock, and broken into sections. Skilled trimmers hand-split and trim the sheets into marketable sizes. The larger the sheet the more valuable the mica. Much waste and scrap result from the production of sheet mica. Mica is marketed as bloc, sheet, splittings, punch, and wet or dry ground.

Production and Distribution. Since 1970, no sheet mica has been produced in the United States. This is the result of several factors, such as the result of improved technology, the availability of substitute and artifically grown mica, and the high cost of labor needed for hand-sorting of the sheet mica. The consumption of sheet mica in the United States, however, is almost 22 million kilograms, which is imported primarily from Bihar and Madras, India; Brazil; and the Malagasy Republic.

Scrap mica production is increasing in the United States as indicated by Fig. 27-4. This material can be mined by bulk methods, washed, screened, and pulverized for use as a lubricant, rolled roofing or asphalt shingles, paint, rubber goods, and welding rods.

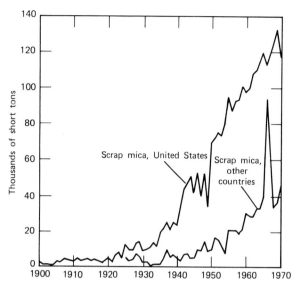

Figure 27-4 Production of scrap mica in the United States, 1900–1970, compared with that in the rest of the world, 1925–1970. Data from U.S. Geological Survey and U.S. Bureau of Mines, Mineral Resources of the United States and Minerals Yearbook, 1900–1970. (From F. G. Lesure, 1973, U.S. Geol. Survey Prof. Paper 820, p. 18.)

Prices. Lesure reports that the selling price of mica ranges from only a few cents per pound for punch or scrap to many dollars a pound for large sheets of the best quality. In 1970, prices for sheet mica ranged from $0.85 a pound for sheets 1½ inches across to $8.00 a pound for sheets 8 inches or more across. Punch mica brought $0.07 to $0.12 a pound. Scrap mica generally brought $30 to $40 per short ton. Ground mica prices were $0.02 to $0.04 a pound for dry-ground mica and $0.075 to $0.095 a pound for wet-ground mica.

Occurrence and Origin. Commercial muscovite occurs only in silicic pegmatites, mostly in association with granitic intrusives. Commercial phlogopite comes from quartz-free, basic pegmatites in Canada and Madagascar. Biotite is obtained from metamorphic biotite schists. Lithia mica is confined to granitic pegmatites of limited areas. Vermiculite is regarded as hydrothermally altered biotite, which at Libby, Montana, represented originally rich concentrations in a basic intrusive. Roscoelite occurs in sandstones in Colorado, Utah, and Arizona. Commercial mica occurs as zonally distributed "books" in the pegmatites. A single crystal weighing about 2 metric tons was found at Spruce Pine, North Carolina; a crystal 4 meters across was uncovered in South Africa, and a crystal 2 by 2.8 meters weighing almost 7 metric tons was found at Eau Claire, Ontario. Mica rock carries about 10 percent of mica, of which only 1 percent is of sheet size.

States that produced scrap mica in 1970 were Alabama, Arizona, Colorado, Connecticut, Georgia, New Hampshire, New Mexico, South Carolina, and South Dakota. North Carolina is the principal producer of scrap mica.

Selected References on Mica

Mica. 1929. H. S. Spence. Can. Dept. Mines, Mines Branch, No. 701, Ottawa. *Canadian and worldwide occurrences.*
Mica. 1942. J. A. Dunn. Rec. Geol. Survey, India, Vol. 76, Bull. 10, p. 80. *Indian mica; uses, occurrences, mining, preparation, and statistics.*
Muscovite mica in Brazil. 1946. D. D. Smythe. AIME Tech. Pub. 1972, p. 1-24. *Occurrence, types, quality, and evaluation.*
The New England mica industry. 1946. H. M. Bannerman and E. N. Cameron. AIME Tech. Pub. 2024, p. 1-8. *Occurrence, production, yield, and future outlook.*
Internal Structure of Granitic Pegmatites. 1949. E. N. Cameron, et al. Mon. 2. Ec. Geol. Publ. Co.
General Reference 8, p. 139-142 and General Reference 19, p. 415-423.
General Reference 7.
Economic geology of the Spruce Pine pegmatite district, North Carolina. 1944. J. C. Olson. North Carolina Dir. Mineral Resources Bull. 43, p. 67.
Mica deposits of the Blue Ridge in North Carolina. 1968. F. G. Lesure. U.S. Geol. Survey Prof. Paper 577, p. 124.
Pegmatite investigations, 1942-45. Black Hills, South Dakota. 1953. L. Page. U.S. Geol. Survey Prof. Paper 247, p. 228.
The study of pegmatites. 1955. R. H. Jahns. Econ. Geol. 50th Ann. Vol., pt. 2, p. 1925-1130.

TALC

Talc, the softest of all minerals, is known to everyone because of talcum powder. The pure, soft mineral in trade is called talc, but the term is also used to include *steatite,* a massive compact variety, and *soapstone,* which consists essentially of talc but includes other minerals. Pyrophyllite is often improperly included with talc because it is used for some of the same purposes.

Properties and Uses. The softness, flakiness, and stability of talc make it a desirable substance for many purposes. In industry, it is roughly classified into hard and soft, and fibrous and flaky, and is marketed as crude, ground, and sawed; about 90 percent of the United States talc is marketed as ground talc. The fibrous variety contains tremolite. In the United States talc is utilized for various purposes as shown in Fig. 27-5.

In paint, talc ("asbestine") serves as an extender and is the white pigment of cold-water paints. In ceramics, ground talc is used for wall tile, electrical porcelain, and dinnerware. Calcined talc or "lava," formerly used for gas tips, is harder than steel and is made into blocks that can be intricately tooled and threaded for use in electrical insulators, radio tubes, and refractories. Only the finest grades of talc are used in toilet powders, lotions, and face creams.

Production. The world production of talc pyrophyllite, and soapstone amounts annually to seven million tons, of which the United States supplied in 1979, 245 million tons, the remainder coming chiefly from Korea, France, Italy, Japan, Canada, Spain, China, India, Norway, and USSR.

The United States production comes mainly from Vermont, North Carolina, California, Montana, Texas, Alabama, Virginia, New York and Georgia. The New York talc goes to the paint, ceramic, and paper industries; North Carolina to the rubber industry; and Vermont to roofing materials. The soapstone comes from Virginia, and pyrophyllite from North Carolina. Imports come mostly from Italy, France, and Canada. Steatite also comes from India and Sardinia.

Occurrence and Origin. Talc and soapstone occur as lenses in metamorphosed dolomites, schists, or gneisses, or as large bodies in altered ultrabasic intrusives. The occurrences and origin are treated in Chapter 15.

Extraction and Preparation. Talc is mined by underground and open cut methods, and soapstone is quarried in large blocks ready for sawing. The talc is dry ground and air separated and is carried through a fine grinding stage. Cosmetic talc is hand-sorted, screened, ground very fine, and bolted through silk cloth.

Distribution and Examples of Deposits

Commercial talc deposits are confined to regions of metamorphism and occur along the Appalachian axis from Canada to the Carolinas; in the Canadian Shield; in the folded rocks of the western Americas;

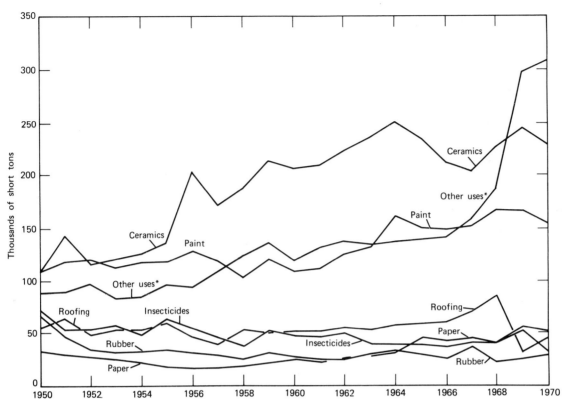

Figure 27-5 Talc and soapstone produced and used by major consuming industries in the United States. 1950–1970. Other uses include asphalt filler, crayons, exports, fertilizer, floor and wall tile, foundry facings, electrical insulation, joint cement, plastics, rice polishing, textile coating, toilet preparations, and miscellany. (From C. E. Brown, 1973, U.S. Geol. Survey Prof. Paper 820, p. 621.)

in the metamorphosed masses of the Alps and Pyrenees; in the Italian Piedmont; in the shield areas of Russia, Scandinavia, South Africa, and India; and in Manchuria, China, and Japan.

United States. The deposits of St. Lawrence County, *New York,* occur in contact-metamorphosed Grenville limestone cut by granite, in a zone of tremolite and enstatite; the tremolite has been altered to talc. About 20 lenticular bodies are known, ranging in width from 6 to 45 meters, one of which, at Talcville, has reached a depth of 180 meters. Shearing has played an important role in developing talc schist zones.

The *Vermont* deposits occur in serpentine or in adjacent schists or gneisses. Associated minerals are dolomite, actinolite, chlorite, biotite, apatite, magnetite, and pyrrhotite. Gillson believes the chlorite replaced the host rocks and was later altered to talc by hydrothermal solutions.

Virginia supports the largest soapstone industry. The deposits occur in a belt 50 kilometers long, centering at Schuyler. There the soapstone bodies

occur in altered ultrabasic intrusions in the Loudon schists, are sheetlike in form, and are 50 to 90 meters thick and 450 to 600 meters long. The soapstone consists of talc with minor quantities of chlorite, carbonate, amphibole, pyroxene, and magnetite. The original rock was altered progressively to hornblende, actinolite, chlorite, and then to talc, by hydrothermal solutions. The product is sawed into slabs and blocks.

In *California,* irregular bodies of white talc and associated tremolite and serpentine occur at a diorite-limestone contact, and in altered dolomites or magnesites in *Washington.* In *North Carolina* and *Georgia* a belt of Cambrian marble, in which are large tremolite crystals, contains minable lenses of talc, and the pyrophyllite deposits of *North Carolina* replace a volcanic-sedimentary belt.

Canada. In the Madoc district, Ontario, massive, white talc occurs in lenses in contact-metamorphosed Grenville dolomite near a granite intrusion. The deposits are lenslike masses 7.5 to 12 meters wide and up to 150 meters long.

France and Italy. The finest grades of toilet talc come from the Pyrenees and the Alps. The deposits occur in two beds intercalated with schists and slates, and Gillson suggests they may represent replacements of included dolomite or serpentine. The beds are 15 to 60 meters thick, have been traced several kilometers, and contain blocks of dolomite and granite. Some Italian talc occurs in serpentine.

Selected References on Talc and Soapstone
General Reference 19, p. 619-626. *Excellent review and bibliography.*
Origin of talc and soapstone. 1933. H. H. Hess. Econ. Geol., v. 28, p. 634-657.
Talc, steatite, and soapstone; pyrophyllite. 1940. H. S. Spence. Can. Dept. Mines, Mines Branch, No. 803, Ottawa. *Comprehensive treatise of all phases.*
Block Talc. 1947. J. E. Eagle. Am. Ceramic Soc., v. 26, p. 272-274. *Use and occurrences of block talc.*
Selected bibliography of talc in the U. S. 1963. U.S. Geol. Survey Bull. 1182-C, 26 p.
Talc resources of the United States. 1964. A. H. Chidester, et. al. U.S. Geol. Survey Bull. 1167, 61 p.

BARITE AND WITHERITE

Barite (BaSO$_4$) and to a minor extent witherite (BaCO$_3$), also known as barytes and heavy spar, are heavy, inert, and stable. These properties make them valuable. Barite formerly was used chiefly as a filler and adulterant, but it is now used in glass and paint, but the oil-well drilling industries consume 80 percent of the world's production.

Properties and Uses. Barite is the chief constituent of lithopone paint (70 percent BaSO$_4$, 30 percent ZnS), an interior white or light-colored paint. Ground barite is used mostly as an inert volume and weight filler in well-drilling mud, rubber, glass, linoleum, oilcloth, paper, plastics and resins, and as a paint extender. It is used in tanning, making highly surfaced papers, such as playing cards, for buttons, poker chips, printer's ink, and face powder. It is also employed in making glass, glazes, and enamels.

Barite yields most barium chemicals, of which there are many, although witherite is also used. The chief one is blanc fixe or precipitated barium sulfate, used for paints. Mixed with TiO$_2$ it makes a white paint of outstanding opacity. Precipitated barium carbonate is used in ceramics, in making optical glass, "granite ware" enamel, and for flat velvety paints. Barium peroxide is used to make hydrogen peroxide and other chemicals.

In the United States, almost 80 percent of the 1.5 million tons of barite consumed annually is ground finely (10 percent, minus 325 mesh) and is then used as drilling mud, the circulation of which lubricates the drill stem, cools the drill bit, seals the walls of the hole, removes cuttings, and confines the high oil and gas pressures at depth.

Production and Distribution. The annual world production of barite is about 8 million tons. The chief sources in order of production are the United States, Germany, United Kingdom, Canada, France, India, Italy, Greece, Korea, Cuba, Spain, and Eire. Important Russian production is unreported. In the United States over three-quarters of the production comes from Texas, Louisiana, and Nevada, the remaining quarter comes from Georgia, Tennessee, Missouri, Arizona, California, North and South Carolina, Arkansas, and Virginia.

Occurrence and Origin. Commercial barite deposits are formed as (1) fissure and cavity fillings—the Appalachians, California, Nevada, Germany, and Great Britain; (2) breccia fillings—Nova Scotia and Virginia; (3) bedded deposits—California, Arkansas, and Nevada; and (4) residual deposits—Missouri, Georgia, and Virginia. The residual deposits have been derived from the other three types.

Barite is also a gangue mineral of many metalliferous veins, but they yield only a little by-product mineral. Veins and replacement deposits are the important sources in Germany and Great Britain, but residual and bedded deposits yield most of the barite of the United States. These are accumulations of barite in residual clays that resulted from the weathering of bedrock with included barite deposits or sedimentary deposits, respectively.

Examples of Deposits
Some of the barite deposits of the *western United States* are associated with intrusions of Tertiary age, but these are notable exceptions of vein deposits associated with Precambrian igneous rocks such as Mountain Pass, California, and the Wet Mountains, Custer County, Colorado.

The notable discovery by the U.S. Geological Survey geologists of an extensive, undoubtedly sedimentary deposit of barite east of Monitor Valley in the Toquima Range, Nye County, Nevada, has strengthened the doubts of hydrothermal origin of many barite deposits, especially those in the eastern and midwestern states. With the exception of Magnet Cove, Arkansas, an igneous association with these deposits is not obvious. In addition, there is no associated evidence of base metal or hydro- thermal alteration or mineralization.

The Northumberland Canyon deposit west of Monitor Valley is being mined by open pit techniques and trucked to Mina, Nevada, the closest railroad line. This deposit is typical of black bedded barite deposits, which are believed to be the result of inorganic and organic chemical processes of deposition with the source of the barite suggested to be submarine exhalative origin.

Residual barite deposits are evident in central Missouri, Georgia (Fig. 27-6) and in some of the Appalachian States where the barite is usually white in color. In contrast, most sedimentary deposits exhibit black barite, the source of which is enigmatic. Suggestions range from telethermal and hydrothermal sources to more likely barium-rich brine solutions reaching sulfate-rich marine or brackish water that could mix locally and form barite deposits.

In *Germany* large barite vein deposits occur particularly in Thuringenwald and the Harz Mountains, where witherite is also mined. In *Great Britain* important barite vein deposits are mined in Ayrshire, Scotland, and in Shropshire, Devonshire, and Derbyshire; witherite in quantity is obtained from the Settlingstones mines in Northumberland. Vein deposits also occur in Italy, Greece, France, Spain, Canada, and other countries.

Selected References on Barite

Barite deposits of Virginia. 1937. R. S. Edmundson. AIME Tech. Pub. 725. *Detailed descriptions and sources.*

Barite and barium products. 1930. R. M. Santmyers. U.S. Bur. Mines Inf. Circ. 6221. *Distribution, production, uses.*

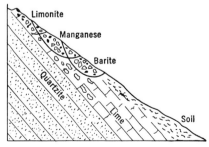

Figure 27-6 Residual barite in Georgia. (From Hull, Geol. Survey Ga.)

Barite deposits of central Missouri. 1947. W. B. Mather. AIME Tech. Pub. 2246, p. 1–15. *Fissure, breccia, circle, solution channel, replacement, and residual deposits.*

General Reference 7 and 19, p. 75–84.

Newly discovered bedded barite deposits in Northumberland Canyon, Nye County, Nevada. 1969. D. R. Shawe, et al., Econ. Geol., v. 64, p. 245–254.

Barite—World Production, resources, and future prospects. 1970. D. A. Probst. U.S. Geol. Survey Bull. 1321, 46 p.

GLASS SANDS

The chief constituent of glass is silica sand and minor constituents are soda and lime. Sands have been used for glass making for over 5000 years.

Composition. Glass sands have a silica content of 95 to 99.8 percent, but the impurities are important; the kind of glass produced depends upon them. *Alumina* is not harmful for common glass, but it should be less than 4 percent, and for optical glass less than 0.1 percent; a little aids in preventing devitrification. *Iron oxide* imparts green and yellow tints; only for amber glass can it reach 1 percent; small amounts can be neutralized by decolorizing agents. Manganese, after long exposure to the sun, imparts delicate amethystine or lavender colors, much prized by glass collectors. *Lime, magnesia,* and *alkalies* are desired; they are constituents of all common glasses. The grains of glass sands should be even-sized of less than 20 and more than 100 mesh. Coarse sand produces batch scum.

Occurrence and Distribution. Glass sands are obtained chiefly from cemented sandstones but also from unconsolidated deposits. The sandstone must be friable and break readily around the grains. Glass sands range in age from Devonian to Recent.

Glass sands are found in the United States in the Devonian Oriskany sandstone of West Virginia and Pennsylvania and in the Ordovician St. Peter sandstone of Illinois and Missouri, which furnish three-fourths of the supply. They range in thickness from 18 to 60 meters and in places are very pure and so friable that the sand can be hydraulicked and pumped out. In other places they are quarried. New Jersey Tertiary sands are less pure and are utilized mainly for bottle and window glass. About 10 million tons of glass sands are produced annually in the United States.

Preparation and Glass Manufacture. The prepared sand and the other proper ingredients to make the type of glass desired are mixed into a "batch." Common window glass consists of about 72 percent SiO_2, 14 percent $CaO + MgO$, 12 percent Na_2O, and 1 to 2 percent $Al_2O_3 + Fe_2O_3 + SO_3$. This is fused to a melt; CO_2 escapes, and the soda and lime unite with silica to form silicates. The batch is cooled and poured, pressed, rolled, or blown into desired shapes. Salt cake and soda ash (Chapter 21) are used for bottle, sheet, and plate glass; magnesite is used as an opacifier in glazes.

Pyrex laboratory and cooking glass contains 80 percent SiO_2, 12 percent B_2O_3, and 8 percent Na_2O and Al_2O_3. *Flint glass* contains lead oxide, which gives it brilliance for cut glass. *Quartz glass* is desired for crucibles, tubes, dishes, quartz lights, and ultraviolet windows. *Optical glass* is the highest-quality glass made; in it the potash content is twice that of soda. Best-quality *crystal glass,* with high brilliance and good musical tone, is a potash, lead-silica glass with 7 to 8 percent K_2O. Potash also imparts chemical durability and resistance and is used for *bulb glass.* In X-ray *lead glass* and *church-window glass* practically all the alkali is potash. *Colored glasses* contain coloring agents; metallic gold gives gold ruby glass; selenium gives red glass; chromium and copper give greens; cobalt gives blue; cadmium and uranium give yellows; manganese gives violet; iron oxide gives browns; and calcium fluoride and tin oxide give opal glass.

Selected References on Glass Sands

Introductory Economic Geology, 2nd ed. 1938. W. A. Tarr. McGraw-Hill Book Company, New York, pp. 491-496.
Glass sands and glass making materials in Georgia. 1940. Georgia Div. Mines, Mining, and Geol., Atlanta.
Some glass sands from New Jersey. 1949. A. S. Wilderson and J. E. Comeforo. Econ. Geol., v. 44, p. 63-67. *Occurrences and analyses.*
Silica 1970. In Mineral facts and problems. U.S. Bur. Mines Bull. 650, p. 368-384.
Selected annotated bibliography of high-grade silica of the United States and Canada. 1957. M. C. Jaster, U.S. Geol. Survey Bull. 1019-H, p. 609-673.
General Reference 13 and 19. Excellent chapter on silica sands.

MINERAL FILLERS

Mineral fillers are finely ground and generally inert, cheap mineral substances that are added to such manufactured products as paint, paper, rubber, lin-oleum, and other materials to give body, weight, opacity, wear, toughness, or other useful properties. They generally change the properties of the product but are themselves unchanged. They are not necessarily adulterants, but some are.

Many fillers are waste products; others are used solely or partly as fillers. All are produced cheaply because generally they need only to be ground and sized; a few are prepared. Their number is large, and their applications are even larger. In addition to natural fillers there are several manufactured fillers. A list of the more important fillers, modified from Ladoo, is given in Table 27-2.

The natural fillers are described in Chapter 25.

Products and processes in which fillers are used (after Ladoo) are listed in Table 27-3.

Some of the mineral fillers most used in the more important products are shown in Table 27-4.

Selected References on Mineral Fillers

Nonmetallic Minerals. 1925. R. B. Ladoo. McGraw-Hill Book Company, New York, p. 364-367. *Gives details of fillers.*
General References, 7, 13 and 19.

MINERAL FILTERS AND PURIFIERS

Certain mineral substances possess properties of base exchange and selective adsorption that make them highly efficient filters and purifiers of mineral, animal, and vegetable oils, sugar, fats, water, chemicals, and other materials. The oil and sugar industries, in particular, consume large quantities of them. The ones used in considerable volume are: fuller's earth or bleaching clays, diatomite, bentonite, and filter sand; those less commonly used are bauxite, tripoli, and alunite. Some of these substances have been described in other chapters of this book.

Diatomite

Diatomite, a sedimentary rock, consists of microscopic siliceous tests of diatoms and rarely of other silica-secreting organisms. It resembles chalk or clay but contains chiefly silica and 3 to 10 percent water with a little alumina, iron oxides, and alkalies. It is also known as diatomaceous earth, diatomaceous silica, kieselguhr, and (incorrectly) infusorial tripoli earth. It has gone under 22 different names.

Properties and Uses. Diatomite is friable, porous, and so light that when dry it will float on water. Its specific gravity is about 2.1, and the dried powder

Table 27-2 Natural and Manufactured Mineral Fillers

Natural Rocks		
Anhydrite	Flint	Rottenstone
Bentonite	Fuller's earth	Sand
Chalk	Gypsum	Serpentine
Clay	Limestone	Shale
Coal	Magnesite	Slate
Diatomaceous earth	Marble dust	Soapstone
Dolomite	Pumice	Talcstone
Natural Minerals		
Apatite	Graphite	Sulfur
Asbestos	Iron oxides	Talc
Barite	Mica	Tripolite
Calcite	Ocher	Umber
Celestite	Pyrophyllite	Vermiculite
Feldspar	Quartz	Witherite
Manufactured Mineral Products		
Blanc fixe	Magnesia	Sublimed white lead
Calcium sulfate	Pearl Filler	(sulfate)
Carbon black	Pearl hardening	Whiting
Gypsum (artificial)	Portland cement	Zinc oxide
Lime (hydrated)	Satin white	
Lithopone	White lead (carbonate)	

is only 0.45. Its absorptive power is such that it may carry 25 to 45 percent of water. It is insoluble in acids but soluble in alkalies. The important commercial properties of diatomite are its porosity, fineness of pores, absorptive power, light weight, and low heat conductivity.

It is used as a filter and filler, for heat and sound insulation, for abrasives, in building materials, in ceramics, and for miscellaneous chemical purposes.

Its chief use, according to Cummins and Mulryan, is as filter of the materials listed.

Various sugars and glucose	Gas purification
Mineral oils and products	Varnishes and lacquers
	Starches and pastes
Molasses	Trade waste and effluents
Fruit juices and beverages	Vinegar and flavoring extracts
Beer, wine, and liquor	Dyestuffs
Vegetable and animal oils and fats	Medicines
Malt products and extracts	Cosmetics and perfumes
Liquid soap	Vitamin extracts
Gelatin glue and adhesives	
Metallurgical slimes and solutions	
Water and sewage	

Production. The United States leads the world in

Table 27-3 Uses of Mineral Fillers

Artificial stone	Insulators	Plastics
Asphalt surfacing[a]	Leather	Polishes
Cements[a]	Linoleum and oilcloth	Putty
Ceramics	Lubricants	Roofing materials[a]
Cordage	Paint[a]	Rubber[a]
Dry batteries	Paper[a]	Soap
Dynamite[a]	Phonograph records	Textiles
Fertilizers[a]	Pipe coverings	Tooth powder and paste
Insecticides	Plasters[a]	Wood finishing

[a] Indicates more important consuming industries.

Table 27-4 Important Mineral Fillers

Sheet roofing		Fertilizers	Paints	Paper
Mica		Sand	Diatomite	Clay
Talc		Limestone	Calcium carbonate	Talc
Slate dust		Dolomite	Whiting	Asbestos
Silica dust		Phosphate rock	Ground limestone	Gypsum
Dolomite			Mica	Barite
Cement			Barite	Blanc fixe
Diatomite			Clay	Calcite
			Amorphous silica	Magnesite
Plastics		*Rubber*	*Cosmetics*	
China clay		Carbon black	Talc	
Gypsum		Clay	Lithopone	
Asbestos		Barite	Barite	
Mica		Chalk	Chalk	
Diatomite		Diatomite	Kaolin	
Barite		Magnesite	Fuller's earth	
		Whiting	Diatomite	

diatomite production, followed by Russia, Denmark, France, West Germany, and Italy, and is the dominant producer of filtration quality material. The United States exported 176,000 short tons of diatomite in 1969.

Occurrence and Distribution. Diatomite occurs in marine and freshwater sedimentary beds and in existing lake and swamp bottoms. The beds are mostly a few centimeters to about a meter thick. One of the largest deposits is in the marine Monterey series near Lompoc, California, where beds up to 420 meters thick are worked on a large scale. Other thick deposits occur in Tertiary beds of the California Coast Ranges. They range in age from Cretaceous to Recent but are most abundant in the Tertiary. The diatom remains must have accumulated under conditions where the deposition of other sediments was temporarily inhibited. The silica is thought to have been extracted by the microscopic plants from silicates, such as clay, suspended in the water. The United States production, the largest in the world, comes mainly from California, Washington, and Oregon, with some from Idaho and Nevada. The Florida, New York, and New Hampshire deposits occur in modern bogs, lakes, or marshes.

Extraction and Preparation. Diatomite is dredged or quarried and removed by mechanical shovels or scrapers. After drying, it is milled and converted into powder, aggregates, or bric, depending upon its use. For filtering, powder is desired; for this the important specifications are microscopic structure, particle size, chemical purity and inertness, low density, and filter performance.

Selected References on Diatomite
General Reference 7, 13, and 19, p. 191–195.
Diatomite, its occurrence, preparation and uses. 1928. V. L. Eardley-Wilmot. Can. Mines Branch Bull. 691. *Comprehensive.*
Diatomites of the Pacific Northwest as filter-aids. 1944. K. C. Skinner, et. al. U.S. Bur. Mines Bull. 460. *Occurrences, properties, and tests.*
Accumulation of diatomaccons deposits. 1942. P. S. Conger. Am. Jour. Sci., v. 82, p. 55–66.
Diatomites in Metals, Minerals, and Fuels. 1971. U.S. Bur. Mines Mineral Yearbook 1969, p. 483–485.
The Diatoms of the United States, Exclusive of Alaska and Hawaii. 1966. R. Patric and C. W. Reimer. Nat'l Sci., Philadelphia, Mon. 13, 688 p.
The ubiquitous diatom—A brief survey of the present state of knowledge. 1960. K. K. Lohman. V258A, p. 180–191.
Diatoms. 1951. P. S. Conger. Sci. Monthly, v. 73, p. 315–323.
Dissolution and precipitation of silica at low temperatures. 1956. K. B. Krauskopf. Geochem et Cosmochim Acta, v. 10, p. 1–26.
General Reference 19. Chapter on diatomite.

Fuller's Earth or Bleaching Clay

Fuller's earth is a variety of clay, so named because it has been used by fullers to full, or remove grease from cloth. Its marked absorptive powers, however, have caused it to be more widely used to filter and decolor oils, fats, and greases. The term has come to be applied to certain natural bleaching clays that possess high absorptive capacity for oil without activation, and which are mostly found in the southeastern United States. It is one of the varieties of bleaching clay. Fuller's earth is chemically little different from other clays and its composition is no indication of absorptive power with the exception of attapulgite. The composition of clays is given in Chapter 24.

Properties and Uses. Bleaching clays fall into two broad groups: (1) those naturally active and (2) those that become active after artificial activation by acid treatment. Fuller's earth belongs to the former class and does not respond satisfactorily to activation. Bleaching clays are rarely plastic, and most of them are nonslaking and generally disintegrate in water. Used fuller's earth can generally be treated by heat and reused about 20 times. It is characterized by a large water content, foliated structure, and a tendency to adhere strongly to the tongue when dry. Not much is known of the seat of decolorizing power; a test is necessary. Nutting thinks that the replacement of a base in the surface of a mineral by the hydrogen of water (or of acid), followed by removal of the hydrogen to leave an open bond, appears to be essential to produce an active absorbing surface. Charged particles unite, positive with negative, and this union of acidoid particles with the basic colored ions in oil is thought to be the basis of the bleaching action in active clays. If an active clay is heated higher than the temperature necessary to drive off the combined water, the absorbing power is lost.

The dominant use (about 70 percent) of fuller's earth is in petroleum refining for filtering and clarifying petroleum products, mainly lubricants, and about 10 percent is utilized in the refining of vegetable oils and animal fats. Its use is growing in domestic water purification. It also removes putrescence, odors, and even coliform bacteria from oily waste waters.

In oil refining, colored oils passed through fuller's earth come out colorless. In addition, it removes naphtha gum, and improves the sludge and carbon content, acidity, and viscosity of lubricating oils. It removes color, odor, and taste from vegetable and animal oils. Fuller's earth cannot be reused after filtering vegetable oils.

Minor uses include printing, abrasives, and detection of coloring agents in food products, fillers,

and cosmetics. In oil refining fuller's earth is being displaced by activated bentonite.

Production and Distribution. The United States is the leading producer of fuller's earth (about 1200 thousand tons annually), of which one-half is from Florida and Georgia, and a third from Texas. Considerable quantities are produced in England, Germany, Russia, France, and Japan.

Occurrence and Origin. Fuller's earth is mainly of sedimentary origin but some is formed in hydrothermally altered volcanics. Exceptions are the residual earth in Arkansas and glacial silt in Massachusetts. It occurs interstratified with geologically late sands and clays. The large deposits of Georgia and northern Florida occur in the Hawthorne formation of Miocene age at shallow depth, and were formed in coastal embayments; some deposits occur in lake beds.

It is thought that most fuller's earth originated from volcanic ash. Montmorillonite and attapulgite are the principal constituents, and montmorillonite is characteristic of bentonite. Nutting believes that it was derived from bentonite by natural leaching in surface water, assisted perhaps by plant acids and bacteria. This is further suggested by the occurrence of bentonite at the base of thick beds in Georgia and Florida. "Ash deposited in situ becomes bentonite; that which is gradually washed in, with leaching of grain surfaces, became fuller's earth." Extended weathering of fuller's earth may yield kaolin. Bentonite is mined in California, Texas, and elsewhere, and is sold as fuller's earth.

Extraction and Preparation. Fuller's earth is mined in open cuts by power shovel or scraper. Tunneling may be used for deep clays. It is dried to about 15 percent moisture, crushed, ground, and screened. If the fuller's earth is used for percolation, absorption granules are used; if by contacting method, powder is agitated with the oil and later filtered off.

Bentonite

Bentonite is a clay composed dominantly of montmorillonite and minor beidellite, with small amounts of igneous rock minerals. It contains about 5 to 10 percent of alkaline earths and alkalies and 3 percent ferric iron.

According to Hosterman there are two classes of bentonites: (1) the "sodium" types that swell and increase in volume 15 to 20 times when wetted in water, remain in suspension in thin water dispersions, and are nonbleaching; and (2) the "calcium" types that do not swell, that settle in water, and that have absorptive properties. Type 2 is analogous to fuller's earth, is used as a bleaching clay after activation, and is displacing fuller's earth as an oil clarifier. Activated bentonitic clays are not excelled in activity by any other type of bleaching clay.

Properties and Uses. The activable group of bleaching clays are mostly of bentonitic origin. Those that respond to artificial activation with sulfuric acid are several times more efficient absorbents than natural active clays. Apparently, the chemical composition does not determine if a clay will respond to activation.

The bentonitic bleaching clays are characterized by a waxy appearance and by rapid slaking in water without swelling. They are heavier and denser than fuller's earth. Some of these clays, according to Hosterman, are naturally active to start with and also show increased activity with acid treatment. Activation consists simply in the removal of sodium ions loosely held at the surface so that in filtration other chosen ions may be absorbed and thus removed from suspension.

The uses of activated bentonite are the same as for fuller's earth; smaller quantities are required, and it has a lower oil retention. Its use has more than trebled in the last 10 years in contrast with that of other bleaching clays.

Type 1, or the swelling type of bentonite, is used primarily for rotary oil drilling mud, but also for medicinal, cosmetic, and pharmaceutical preparations; for leakage prevention; in insecticides and concrete admixture; as an emulsifying agent in asphalt; as a gelatinizing agent; for dewatering wood pulp; and as a wine clarifier and paste-forming agent.

Production and Distribution. Figures on world production of bentonite are not available, but that of the United States is around 1,200,000 tons, of which about one-quarter is used as a bleaching clay. The main production of bleaching clay comes from Texas, Oklahoma, Alabama, California, Arizona, Mississippi, Kentucky, and Tennessee; smaller deposits occur in all of the western states. The swelling nonactivable type bentonite comes mainly from Wyoming and South Dakota. The best grade "Wyoming" or "sodium" bentonite occurs in the Mowry shale of Cretaceous age in Wyoming and South Dakota. The Clay Spur bentonite bed, near the top of this formation in a belt surrounding the Black Hills, has been a major source of bentonite, and it still contains large reserves. Bleaching bentonites also occur in Germany, Russia, France, and Japan, and nonbleaching bentonites occur in Canada, Mexico, Italy, and China.

Extraction and Preparation. The following steps are involved for bleaching clay: (1) preparation of raw clay for treaters, (2) activating, (3) washing out impurities, (4) mechanical dewatering, (5) drying, (6) grinding, and (7) packaging. For activation, water is added to make a slurry; then sulfuric acid (rarely hydrochloric) is added. The mixture is kept agitated at a temperature of 95 to 100° C for 2 to 12 hours. The material is then dried and ground to a powder.

Other Mineral Filters

Filter Sand. Sands are widely used for the filtering of municipal and industrial water supplies to remove sediment and certain bacteria. Filter sands should be even-grained, high in silica, with little soluble material. A certain ratio of fine to coarse grains is desirable. Such sands are obtained from the same deposits that yield glass sands.

Bauxite. Bauxite (Chapter 19) since 1937 has been growing in use as a mineral filter. It is effective in decolorizing, deodorizing, and desulfurizing mineral oils and other liquids, and is particularly desired for filtering kerosene, lubricants, and paraffinic oils. It has to be activated, and, despite its high cost, it competes with cheaper fuller's earth because it can be revivified indefinitely.

Alunite from Marysville, Utah, is used to a minor extent for decolorizing and deodorizing oils.

Activated magnesia is a new compound used as a decolorizing, neutralizing, or absorbing agent in industrial processes. It has an absorptive power many times that of bentonite.

Selected References on Bleaching Clays

Bleaching Clay. 1933. A. D. Rich. In Ind. Min and Rocks, AIME Dupl. *Good for technology and bibliography.* Chap. 5, Bentonite, by P. Bechtner. *Occurrences and uses.*

Geologic Features of Some Bleaching Clays. Jan. 1940. G. A. Schroter and Ian Campbell. AIME Min. Tech, 1-31. *Excellent articles on geology, properties, and mineralogy; an extensive bibliography.*

Bleaching clays. 1933. P. G. Nutting. U.S. Geol. Survey Circ. 3. *Properties and investigations of absorption properties.*

Clays. 1971. J. R. Wells. U.S. Bureau of Mines Yearbook, v. 1-2, p. 287-308.

Applied Clay Mineralogy. 1962. R. E. Grim. McGraw-Hill Book Company. Excellent text on clays.

Absorbent Clays. 1943. P. G. Nutting. U.S. Geol. Survey Bull. 928-C, p. 127-221. *Physical and chemical properties, tests, and distribution.*

Geology of Bentonite Deposits, Caspar, Wyo. 1946. G. Dengo. Geol. Survey Wyo. Bull. 37. *Occurrence, origin, and properties.*

General References 7 and especially 19, p. 123-139.

OPTICAL CRYSTALS

Various mineral crystals are in demand because of certain optical properties they possess. These are used mainly in microscopes, spectroscopes, and other scientific instruments. Although demand for them is small, they are essential for specific purposes.

Iceland Spar. This is calcite of uncommon transparency, purity, and perfection of crystallization. It possesses the property of making a dot viewed through it appear as two dots; that is, it is doubly refracting.

This property is utilized in making Nicol prisms, which constitute the analyzer and polarizer of polarizing microscopes used for the examination of rock slices and polished ore surfaces; Nicol prisms are also employed in scientific instruments for measuring the sugar content of a solution, comparative colors, intensity of light, dichroism of crystals, in polariscopes and polarimeters for studying polarized light, determining index of refraction, and examining spectra.

Iceland spar for such optical purposes must be free from flaws, bubble inclusions, and twinning; only a very small percentage of the spar recovered is suitable. Most of it is obtained from clay pockets in cavities in basic igneous rocks. Iceland was long a source of the best optical material. The Cape Province of South Africa has also been supplying optical grade spar; there it occurs embedded in soft clay filling cavities in weathered diabase. A small quantity comes from the United States and Mexico. Polaroid is replacing the Nicol prism for some uses.

Fluorite. Fluorite has a low index of refraction, disperses light faintly, and normally displays no double refraction. These properties make clear fluorite crystals desired for high-power microscope lenses to correct spherical abberation and chromatic errors. Obviously, optical fluorite must be free from flaws and inclusions and preferably should be water-clear. Fluorite is also used in the spectrograph and other instruments, where transparency of both ends of the spectrum is desired.

Quartz. Quartz crystals of fine quality and true optical character are indispensable in radio, radar, and telephone and telegraphic operations. For radio transmission, very thin plates of absolute uniformity control the transmission frequency. For microscopic examination of minerals and rocks in transmitted, polarized light, quartz wedges, with the flat sides cut parallel to the optic axis and the long edge inclined 45° to it, are essential. Crystals must be flawless and transparent. Much quartz is now being produced synthetically, 49 percent in 1979.

Tourmaline. Tourmaline is also a double refracting mineral, and plates of it cut parallel to the principal axis, and set normal to each other obliterate all light. It is used for the study of polarized light, and also for "crystals" in radio transmitters to give a definite frequency of sending waves.

Mica. Basal plates of clear, flawless mica are made for use with the polarizing microscope.

Selenite. Cleavage plates of clear, flawless selenite, known as the sensitive plate, are accessories of polarizing microscopes to reveal the slightest double refraction in mineral sections.

Polaroid. This is a recent synthetic material that polarizes light in one plane, and, as with the Nicol prism, when two plates are crossed no light is transmitted. It is beginning to replace the Nicol prism for some purposes in the optical field, but its greatest use so far is for polaroid sun glasses, and on automobile headlights and windshields to cut down automobile light glare.

Selected References on Optical Crystals

Deposits of Quartz Crystal, Brazil. F. L. Knouse. General Reference 16, 173-184. *Occurrences in vugs, veins, replacement bodies, and pegmatites.*

Quartz Crystals in Brazil. 1946. W. D. Johnston, Jr., and R. D. Butler. Geol. Soc. Amer. Bull. 57, p. 601-650. *Occurrence and origin of main source.*

MISCELLANEOUS MATERIALS

Lime. (See also Chapter 25) Nearly one-fourth of the 4 million tons of quick and hydrated lime manufactured annually in the United States is consumed industrially; this, together with chemical and metallurgical uses, accounts for over 52 percent of the lime made. Paper mills and glass works take most of it, but it finds innumerable miscellaneous applications.

In paper mills the sulfite process utilizes milk of lime along with sulfur dioxide for its calcium and magnesium bisulfites to obtain the acid-digesting liquor. It is also used in the soda and sulfate process of paper manufacture.

In glass works, it is the alkaline-earth constituent of the batch for union with silica. In the tanning industry lime is a dehairer and retards putrefaction. Among the many other industrial uses of lime, some of the most important are: sugar refining, brick, gas purification, insecticides, fungicides and disinfectants, magnesia, paints, petroleum refining, gelatin, glue, baking powder, grease, bleaching liquids and powder, soap and fat, salt refining, rubber, and polishers and buffers.

Magnesite. (See Chapter 25.) In the caustic form magnesite is used extensively as an accelerator in rubber; it also increases resilience and tensile strength.

Aluminous Cement. A rapid hardening cement, aluminous cement is resistant to chemicals and heat, and is made from bauxite (Chapter 19).

Carbon Dioxide. Carbon dioxide finds considerable industrial use as dry ice for a refrigerant for ice cream and perishable goods, for candy and varnish making, for golf-ball centers, and freezing water pipes for repairs. The gas is used for soda and various carbonated waters and beverages and in refrigerating machines. The liquid is used as a "safe" explosive. Some of the gas used for making both liquid and solid carbon dioxide is obtained from natural gas wells and springs, but it is more often produced by burning carbon in an excess of oxygen, by heating carbonates or by the action of acids on these, as for example, $M_2CO_3 + 2HCl \rightarrow 2MCl + H_2O + CO_2$. Some is obtained by a by-product from coke and from chemical and metallurgical works. Dry ice is obtained directly by the expansion of carbon dioxide emitted from carbon dioxide wells in Mexico and Colorado.

Nitrates. (See Chapter 29.) Nitrates are used for refrigerants, explosives, fireworks, and flares. Many other items listed under chemical uses (Chapter 28) also have industrial application.

Phosphates. Phosphates find a wide application in minor industrial products, but their principal use is in fertilizer.

28

CHEMICAL MINERALS

There are a number of more or less unrelated nonmetallic substances that are used chemically. Some are utilized in the raw state, chemically or medically; others must be prepared; and others supply desired ingredients. Of course, many minerals are sought for their chemical composition rather than for their physical properties, which are utilized dominantly or entirely for purposes other than chemical, such as potash and fluorite. Some of the chemical minerals are also used in other industries; many other minerals find minor applications in the chemical industry.

Those treated in this chapter as chemical minerals are

Salt and brines	Potash
Borax and borates	Sulfur
Sodium carbonate and sodium sulfate	Nitrates and nitrogen
Calcium and magnesium chloride	Strontium minerals
Bromine and iodine	Miscellaneous chemical minerals

The identified and potential resources of some of these minerals are given in Table 28-1.

Salt is the most familiar of all minerals. Its production started with the beginning of man, and each person consumes about 12 pounds of it a year. Because it is a primary human need its acquisition entailed exploration, warfare, conquest, and barter. Salt became a precious substance in those countries that lacked it. For a long time it was a basis for government taxation and monopoly, with ensuing power and inevitable political unrest and upheaval. Now, however, its value as a food and food preservative is overshadowed by an industrial demand that makes it one of the most essential raw materials in modern chemical industry.

Properties and Uses. Natural salt and salt brines generally contain impurities. For food salt, even small quantities of calcium sulfate, and calcium and magnesium chlorides, are objectionable; they also make the salt deliquesce. The chemical control of its preparation is, therefore, important.

Salt is so widespread in its uses that the materials made with it or by it are continuously encountered in everyday routine. The chief fields of most of the consumption are:

In Industry:
Raw products for chemicals and acids—soda, sodium bicarbonate, caustic soda, caustic ash, sal soda, and acids.
As a chemical in manufacturing:
Metallurgical industries—treating, smelting, and refining of ores and metals.
Chemical industries—packing-house uses, soaps, dyes, emulsions, tanning, food and wood preservative, cements, blasting, water purification, lacquers, cotton and paper bleaching and road surfacing.
Ceramics—glazes, shrinkage prevention, dental, and vitrifiers.
Refrigeration agent—ice, ice cream, chemical works, oil refining, cold storage, and household refrigerators.

In Agriculture: cattle food, fertilizers, hay preservative, weed eradicators, insecticides, soil amenders, and dairying.

In Medicine: drugs, medicines, cleansers, both internal and external.

In the Home: food preservative, refrigerant, cleansers, whiteners, stain removers, and miscellaneous uses.

Occurrence and Origin Commerical salt is obtained from five sources: (1) sedimentary bedded deposits, (2) brines, (3) seawater, (4) surface playa deposits, and (5) salt domes. Salt is generally referred to as rock salt or brine. The first includes solid salt deposits. Brines include ocean water; salt

Table 28-1 Evaporative and Brine Resources of the United States (Expressed as years supply at current rates of domestic consumption)

Commodity	Identified resources[a] (reserves[b] and subeconomic deposits)	Undiscovered resources (hypothetical[c] and speculative[d] resources)
Potassium compounds	100 years	Virtually inexhaustible
Salt	1000 + years	Do
Gypsum and anhydrite	500 + years	Do
Sodium carbonate	6000 years	5000 years
Sodium sulfate	700 years	2000 years
Borates	300 years	1000 years
Nitrates	Unlimited (air)	Unlimited (air)
Strontium	500 years	2000 years
Bromine	Unlimited (seawater)	Unlimited (seawater)
Iodine	100 years	500 years
Calcium chloride	100 + years	100 + years
Magnesium	Unlimited (seawater)	Unlimited (seawater)

Source: U.S. Geol. Survey Prof. Paper 820, p. 213, 1973.

[a] Identified resources: Specific, identified mineral deposits that may or may not be evaluated as to extent and grade, whose contained minerals may or may not be profitably recoverable with existing technology and economic conditions.

[b] Reserves: Identified deposits from which minerals can be extracted profitably with existing technology and under present economic conditions.

[c] Hypothetical resources: Undiscovered mineral deposits, whether of recoverable or subeconomic grade, that are geologically predictable as existing in known districts.

[d] Speculative resources: Undiscovered mineral deposits, whether of recoverable or subeconomic grade, that may exist in unknown districts or in unrecognized or unconventional form.

lake waters; and subsurface brines, both natural and artificial. The brines are of all degrees of saturation and include bitterns with other chlorides, bromides, iodides, and sulfates. Of the United States production, 70 percent of the salt is extracted from natural or synthetic brines or seawater; 30 percent is mined as a solid. Subsurface deposits of rock salt are the principal sources of salt in the United States. About half of the production is from salt domes, and most of the balance from stratified marine evaporites.

Bedded Deposits. The sedimentary beds occur intercalated with common strata, and with gypsum, anhydrite, or potash minerals. The beds range from a few inches to several hundred feet in thickness. Generally, several beds lie above each other. One

well in Kansas, for example, penetrated 32 salt beds; New York state has 7 beds aggregating over 120 meters, and Phalen states that at one place in Michigan there is a thickness of 270 meters. The Southwestern Permian Basin, in Texas, according to Sellards and Baker, contains 300 meters of salt, and near Carlsbad, New Mexico, one drill hole disclosed over 360 meters of salt and included potash minerals.

Most bedded rock salt is associated with gypsum or anhydrite and in places contains potash minerals.

Brines. Ocean and salt lake brines are less used in America than elsewhere. Subsurface brines in sandstones and other porous rocks are regarded as connate or buried seawater. Some brines form locally by solution of rock salt beds. The most important

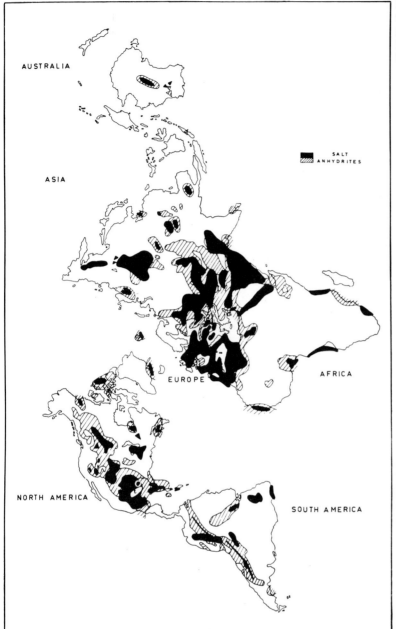

Figure 28-1 Composite distribution of evaporites of all ages. (From Kozary, et al., 1962, in Saline deposits, Geol. Soc. Amer. Special Paper 88, p. 57.)

subsurface brines of the United States are those in the Mississippian and Pennsylvanian beds in Michigan, Ohio, Pennsylvania, New York, and West Virginia. They supply a considerable part of the salt produced and also yield bromine and calcium chloride. Artificial brines are made by introducing water into salt beds and recovering the resulting brine.

Salt Domes. Only 8 of the more than 250 salt domes (Fig. 28-3) in the Gulf Coast region of Texas, Louisiana, and Mississippi are salt mines; the largest is

the Grand Salina salt dome in east Texas. This dome is roughly a cylinder in shape, approximately 3 kilometers in diameter and 2500 to 3300 kilometers in length or depth.

Playas. Salt, along with borax, potash, and other chemicals, occurs as surficial deposits from desiccated salt lakes, such as the numerous playas of the Great Basin, western part of the United States including the deposits of the Great Salt Lake Desert,

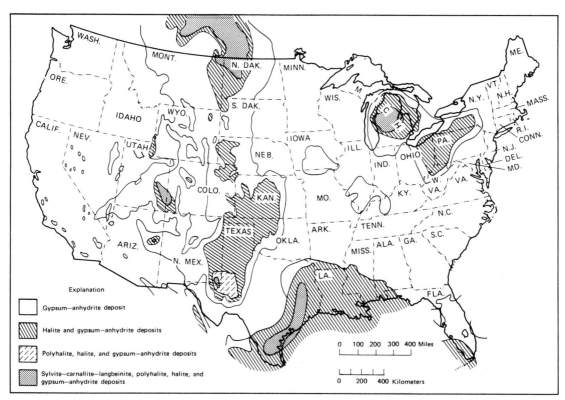

Figure 28-2 Marine evaporite deposits of the United States. (From U.S. Geol. Survey Prof. Paper 820, p. 199, 1973.)

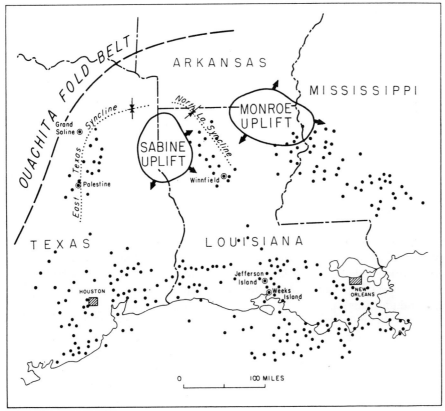

Figure 28-3 Location map of the Gulf Coast salt dome area. Dots represent salt domes. (From G. I. Atwood, 1962, in Saline deposits, Geol. Soc. Amer. Special Paper 88, p. 30.)

Utah. It also occurs in the sands and clays of playa marshes.

Production and Distribution. Among the marine evaporites, salt (NaCl) is by far the most valuable mineral resource, ranking with coal, limestone, iron and sulfur as a basic industrial raw material.

About 50 million tons of salt with a value of more than $538 million was produced in 1970 by 54 companies that operated 99 plants in 17 states, according to Smith, et al. China, Russia, Germany, and the United kingdom are other major producers. Evaporite and brine resources of the United States are given in Table 28-1. World production is 184 million tons.

Origin of Salt Deposits. Salt beds originate by the evaporation of saline waters. The process is simple; reduction in the volume of water results in saturation, with early deposition of gypsum or anhydrite, next of salt, and later of potash and other minerals, as described in detail in Chapter 12.

Extraction and Preparation. Salt is marketed as brine, evaporated salt, or rock salt; pressed blocks of salt may be made from the last two. Most of the salt consumed in the chemical industry is used as brine, either natural or artificial. It contains impurities of hydrogen sulfide, iron compounds, and calcium and magnesium salts that have to be removed.

Evaporated salt is obtained from brines by artificial and solar evaporation. For pure salt, the impurities of the brines are removed by aeration and by carefully contolled chemical treatment. Solar salt is produced by progressive evaporation in ponds such as those of the Great Salt Lake. Rock and evaporated salt are compressed under great pressure into blocks of about 25 kilograms for salting cattle.

Examples of Deposits

Examples of salt deposits occur in practically every country in the world; some have rock salt deposits, some playa deposits, and others brines or seawater. The rock deposits, except for salt domes, are generally similar the world over. The beds display variable attitudes, and because of the solubility of salt, they rarely outcrop.

In the *United States,* the bedded deposits of Michigan occur in the synclinal Michigan basin between Lakes Huron and Michigan, and near Detroit in the Silurian Saline formation (Fig. 28-2). Natural brines occur in salt beds, higher beds, and even in glacial till. The salt beds of New York and Ohio, of the same age, parallel the Great Lakes and dip southward beneath Pennsylvania. Mississippian salt beds occur in Michigan, Pennsylvania, and Virginia; and Pennsylvanian beds in southern Ohio and West Virginia (Fig. 28-3). The salt domes occur in Texas, Mississippi, and Louisiana. Playa deposits are important in Utah, Nevada, and California. Extensive flat-lying beds of Permian salt occur throughout the Permian basin. No commercial salt occurs in New England or in the southeastern or northwestern states.

In *Canada,* salt beds similar to those of Michigan and New York occur in Ontario, Nova Scotia, New Brunswick, and the Prairie Provinces. In *Europe* the chief deposits are Triassic salt beds in England and the Permian deposits of central Europe. Oligocene beds yield salt from the Rhine trench in Alsace. The considerable salt industry of *India* is a government monopoly. The salt comes from the salt lake brines of Rajputana and rock salt of the Punjab. Salt is also a monopoly industry in *China,* which is one of the large producers of the world (Fig. 28-1).

Selected References on Salt

Salt deposits. 1949. W. C. Phalen. Ind. Min. and Rocks, AIME Publ. *Good general account, particularly on industrial phases.*

Salt resources of the United States. 1919. W. C. Phalen. U.S. Geol. Survey Bull. 669. *Broadly descriptive.*

Recent developments in the use of salt. 1936. C. D. Tooker. AIME Tech. Pub. 723. *Brief discussion.*

Marketing of salt. 1939. F. E. Harris. U.S. Bur. Mines Inf. Circ. 7062.

General Reference 7, and especially 19, p.197–216.

Saline deposits. 1968. R. B. Matez, ed. Geol. Soc. Am. Special Paper 89, 701 p.

BORAX AND BORATES

Borax, one of the most important chemical minerals, and boric acid are the two principal commercial boron compounds. Borax is obtained in part from the natural borax and in part from borates, which also are the source of boric acid. Schaller points out that the borate industry has been featured by the successive discarding of one mineral for another more usable. First, borax from lake muds was used; then ulexite and borax were obtained from playas and borax marshes. These impure products were displaced by bedded deposits of purer colemanite and ulexite from Death Valley—made known by "It burns blue Katy," and "Twenty Mule Team Borax." In 1925 a new borate, kernite, was discovered in high-grade deposits along with borax, and these, along with boron compounds extracted from lakes, are the present source of refined borax.

Boron Minerals. Of the 60 known boron minerals, only 7 are used commercially, namely, the water-soluble sodium borates, borax ($Na_2B_4O_7 \cdot 10H_2O$) and kernite ($Na_2B_4O_8 \cdot 4H_2O$); the water-insoluble calcium borates, colemanite, ulexite, and priceite; the magnesium borate, boracite; and sassolite (boric acid). The first four, along with brine, have been the supply minerals of the United States; priceite supplies the Turkish, sassolite the Italian, and boracite, the German production. The solubility of borax and kernite makes them preferred. The noteworthy feature about kernite is that it dissolves readily in water and upon evaporation yields normal borax. These two minerals today supply most of the borax of the world.

Uses. Chemically refined borax and the many compounds made from it are used in many articles of everyday life. Borax is a well-known household commodity but is of greater importance in industry. According to Schaller, "the cleanser, the pharmacist, the paper and textile maker, the metallurgist, the brazier, and the jeweler all use borax. Hardly any other substance enters into so many diversified lines of manufacture."

It is also used by the manufacturers of glass, glass fiber, enamel, chemical and agricultural products, gasoline additives, fire retardants, and metal alloys.

It is a good cleanser, either directly or in soaps. It is an ingredient in baking powder, food preservatives, flavoring extracts, syrups and pickles, and insecticides. Medicinally, it is a mild antiseptic. It prevents rancidity in cosmetics, pastes, and glues; mold in fruits, leather, textiles, paper, and lumber; and disease in sugar beets and celery. It aids tanning and gives a smooth, soft finish to leather. It fireproofs wood and textiles. It gives glaze to glazed papers and is a flux for glass. It is used in manufacturing paper, candles, implements, carpets, drugs, shoes, hats, dyes, ink, jewelry, oil, paints, polishes, tobacco, and tools. Other boron compounds are used in oils, varnishes, inks, linoleum, and electric equipment.

Production and Distribution. The world production of borax amounts to about 1.6 million tons, of which the United States produces about 60 percent containing 560,000 tons of B_2O_3 with a value of $85 million; the remainder comes from Argentina, Italy, and Turkey, with a little from Chile, Bolivia, Germany, and Tibet. The United States source materials are obtained from California, Nevada, and Oregon. The famed Death Valley deposit now produces colemanite.

Occurrence and Origin. Boron compounds are obtained commercially from: (1) bedded deposits beneath old playas from which brines are pumped from the interstices of the beds below the surface, (2) brines of saline lakes and marshes, (3) encrustations around playas and lakes, and (4) hot springs and fumaroles. The bedded deposits consist of ulexite, colemanite, borax, or kernite, along with other minor boron minerals, in Tertiary and later clays. The kernite deposits consist of about 25 percent clay and 75 percent of equal amount of kernite and borax. Details of occurrence are given in the individual examples. These have been, and are, the main sources of the United States boron. Lake brines are next in importance. They represent concentrations of former larger lakes and contain borates along with soluble chlorides, carbonates, and sulfates, such as occur in Searles Lake (described in Chapter 12).

Encrustations around saline lakes and playas form impure surficial deposits a foot or so thick. These formerly were important in Death Valley and around Searles and Borax Lakes and still are in Argentina, Chile, Bolivia, and China. Hot springs and fumaroles carrying boric acid yield boron compounds in Italy.

The origin of the various types of borate deposits involves simple concentration and evaporation followed by many chemical mineralogical transformations before final fixation in the borate minerals now found.

The original source of the the boron of the United States deposits was presumably spring waters and sediments containing above-average amounts of boron. Boric acid was probably emitted, which upon reacting with lime and soda, formed ulexite in clay. Colemanite was formed from ulexite by leaching out the soluble sodium borate, which accumulated in lakes where evaporation either yielded borax brines or encrustations. Some borax became buried under later sediments; part of this was transformed from borax with 10H$_2$O to kernite with 4H$_2$O by partial dehydration, perhaps brought about by burial. Some kernite was later hydrated to tincalconite with 5H$_2$O and back again to borax. Kernite exposed during mining was hydrated to borax.

Extraction and Preparation. Bedded borate deposits are extracted by underground mining methods, and kernite and borax are crushed together, roasted to remove the water, separated from the clay, and refined to borax.

At Searles Lake California, brines are pumped out and the various constituents are separated by complicated chemical treatment, which is essentially evaporation followed by fractional crystallization with careful control of temperature and concentration. During evaporation, the sodium carbonate, sulfide, and chloride are precipitated; then, when saturation with potassium chloride occurs, rapid cooling causes it to be precipitated, and further cooling gives borax and other salts, which are then refined to pure borax.

Examples of Deposits

The *United States* production comes almost entirely from California-bedded deposits near Kramer, Searles Lake, and Boron, California. The Kramer area in the Mojave Desert contains borax and kernite in a basin 1.5 kilometers wide by 6.5 kilometers long. The borates occur in clay and shale in a basin of folded Tertiary sediments and volcanic tuffs at depths between 100 to 400 meters and in bodies from 25 to 35 meters thick. Borates constitute about 75 percent of the deposits; in places there are large masses of pure kernite or pure borax or mixtures of the two. Some tincalconite and probertite are present, and ulexite and colemanite occur in the wall rocks. Three meter layers of borax lie above and below kernite. The kernite occurs only in the disturbed portions of the deposit; where borax occurs without kernite, the bedding is more regular. Veinlets of borax cut kernite, but Schaller and Hoyt Gale think that the kernite is a transformation from borax.

Searles Lake, in the Mojave Desert, yielding borax and also sodium carbonate, sodium sulfate, and potassium chloride, is described in Chapter 12.

Impure borate encrustations, colemanite, and ulexite, occur in several localities in California and Nevada and sparsely in Oregon.

Argentina, second in world production, contains layers and nodules of ulexite in playas. Similar deposits occur in the Atacama Desert of *Chile*, at Las Salinas, *Peru*, and in *Bolivia*. In *Turkey* borax is obtained from deposits of priceite (pandermite) in buried playa beds in Anatolia. The unique *Italian* occurrences, in the province of Pisa, are volcanic steam vents, carrying boric acid, emitted from fissures that cut Cretaceous and Eocene sediments. The superheated steam is condensed, yielding a weak solution of boric acid, which is evaporated by natural steam and fumarolic vapors to yield borax.

Selected References on Borates
Mineral resources of the region around Boulder Dam. Heavy chemical minerals. 1936. U.S. Geol. Surv. Bull. 871, 92–113. *Good brief sections on California and Nevada borates.*
Borates. 1932. Imp. Inst. London. *Brief world review and bibliography.*
Boron. 1931. R. M. SANTMYERS. U.S. Bur. Mines Inf. Circ. 6499. *Statistical survey.*
Borate Minerals of Kramer, Mojave Desert, Calif. 1930. W. T. SCHALLER. U.S. Geol. Survey Prof. Paper 158-I. *Descriptive mineralogy of borates.*
General Reference 1, 7, and especially 19.
Boron. 1972. R. B. KISTLER. Min. Eng., v. 24, p. 36–37.

SODIUM CARBONATE AND SODIUM SULFATE

The carbonate and sulfate of soda are two mainstays of the chemical industry, and the production of their compounds is referred to as the "alkali" industry. The bicarbonate (NaHCO$_3$) is known to all households as soda or baking soda. In industry the anhydrous carbonate (Na$_2$CO$_3$) is known as *soda ash* and the sulfate as *salt cake*, and *soda* refers to either the carbonate or oxide of sodium. The purified carbonated salts are marketed as soda ash, bicarbonate, natron (Na$_2$CO$_3$·10H$_2$O), and trona (Na$_2$CO$_3$·NaHCO$_3$·2H$_2$O), and the sulfate as salt cake, Glauber's salt (Na$_2$SO$_4$·10H$_2$O), and niter cake.

Properties and Uses. Sodium sulfate is one of the most important chemical minerals. Soda ash is used in the glass industry (50 percent), used to make other chemicals (40 percent), in the paper and pulp industry (8 percent) and miscellaneous uses such as making soap detergents, and water softenors (2 percent).

It is also used in fire extinguishers, oil refining, making synthetic rubber, and explosives. Glauber salt, is used for dyes and as a medicine. Glauber salt is dehydrated for shipment as "crude salt cake," which dissolved in water and recrystallized below 30° C gives Glauber salt again, or if recrystallized above 35° C gives soda ash.

Production and Distribution. The world production of sodium compounds is not known, but the United States produces annually about 8 million tons of sodium carbonate. The other chief producing countries are Germany, Canada, Chile, Peru, Russia, Persia, India, Egypt, Kenya, and South Africa. The United States production of trona is now primarily from the new deposits in Wyoming. Some production of carbonate is from Searles, Montana, and Owens Lakes, California; of sulfate, from California, Texas, Wyoming, North Dakota, Arizona, Utah, and Nevada. Canadian supplies come from Saskatchewan, Alberta, and British Columbia lakes, which are estimated to contain 115 million tons of hydrous salts, mainly sodium sulfate, and are the largest known reserves.

Occurrence and Origin. Sodium is released during the weathering of igneous rocks and in certain regions reaches local basins, where it is precipitated as the carbonate or sulfate. Sulfur is contributed by the oxidation of pyrite in igneous rocks, by volcanic sources, hot springs and, locally, from gypsum beds. If a solution of gypsum is mixed with a solution of sodium carbonate, evaporation will give calcium carbonate and sodium sulfate. Apparently, this is how the trona deposits of Wyoming formed in the Eocene Green River Lake.

The natural sodium compounds are obtained from the alkali and bitter lakes. In Searles Lake an area of some 50 square kilometers in the midst of a larger playa is underlain by spongy salt. It was formerly 200 meters deep but is now only a marsh and at times is dry. Brine from the upper salt crust is withdrawn by wells 3 to 6 meters deep. Beneath this to depths of 25 to 60 meters is a relatively solid mass of intermingled saline minerals. At greater depths bicarbonate increases and there is a thin bed rich in borax. Drill holes show hard bottom reefs of almost pure bicarbonate with some burkeite. For many years it yielded large quantities of borax and now yields sodium sulfate, potash, borax, and soda.

Owens Lake, 250 square kilometers in area, has a salinity of over 200 parts per thousand and yields sodium carbonate, of which it is estimated to contain 8 million tons. Mono Lake contains 300 million tons of salts. Soda Lake and Columbus marsh in Nevada were former sources of sodium carbonate and borax. A drought in North Dakota disclosed eight dried-up lake bottoms that aggregate 25 million tons of sodium sulfate. One playa in Kenya, Africa, contains 200 million tons of soda. Lake Magadi in Kenya generates new trona at about the same rate as the present annual production of 150,000 to 200,000 tons. Huge deposits of trona are now being mined in Sweetwater County, Wyoming. Natural soda ash, 60 percent pure, occurs in the Green River formation at depths of 120 to 1050 meters over an area of more than 3,500 square kilometers. Between 30 and 40 billion tons occur in beds more than 2 meters thick. In 1970 this source provided more than one-third of the U.S. consumption, and with the completion of a fourth plant in 1975 as much as 80 to 90 percent of the U.S. consumption of soda ash was derived from this source.

Prior to the Wyoming trona deposits development, soda ash was produced by the Solvay Process of mixing $CaCO_3$ and $NaCl$ with ammonia. With the first production of trona from Wyoming, four of the existing 10 plants closed. The 6 remaining plants have subsequently been shut down.

One of the larger trona operations is the Texasgulf, Inc., mine and plant near Green River, Wyoming. The plant has a capacity of 1 million tons of soda ash per year and is readily expandable to 2 million tons, which would be almost 30 percent of the U.S. consumption.

The trona deposits are in two horizontal beds at depths of about 410 and 425 meters. They are consistent in grade and average more than 90 percent pure trona. The reserves total about 176 million tons.

Two shafts, 4.8 and 6.6 meters in diameter, have been bottomed at a depth of 470 meters and 13 kilometers of underground development have been completed.

Extraction Preparation. Searles Lake brine contains the following percentages of important constituents, $NaCl$ 16.3, Na_2SO_4 6 to 9, KCl 4.7, Na_2CO_3 3.4, $NaHCO_3$ 0.77, $Na_2B_4O_7$ 1.39, H_2O 65.6. The sodium, potassium, and boron compounds are recovered. The brine is pumped to multiple evaporators in which $NaCl$, Na_2SO_4, and Na_2CO_3 separate. The remaining solution is quickly cooled to deposit KCl, and the residue yields borax.

Solid salts are harvested on the surface, mined, or later leached. The crude salts are generally refined.

Manufactured salt cake is a by-product in making hydrochloric acid; NaCl treated with H_2SO_4 yields HCl and Na_2SO_4 (salt cake). By another process NaCl is treated with SO_2, O, and H_2O, yielding Na_2SO_4 and HCl. Manufactured Glauber salt is made from salt cake, as previously described. Niter cake ($NaHSO_4$) is a residual product in the manufacture of nitric acid from saltpeter (sodium nitrate) and sulfuric acid.

Selected References on Sodium Sulfate and Carbonate

General References 7,13, and especially 19.
Stratigraphy of the trona deposits in the Green River formation, southwest Wyoming. 1971. W. C. Culbertson. Contrib. to Geol., v. 10, p. 15-23. Wyoming University.

CALCIUM AND MAGNESIUM CHLORIDE

Calcium and magnesium chloride are obtained from natural brines during the extraction of salt, and calcium chloride is also a by-product in the manufacture of sodium carbonate. Bromine is extracted with the chlorides.

Calcium chloride obtained from brines generally contains magnesium chloride and other salts and in this form is used directly for many purposes. That obtained from sodium carbonate manufacture receives its chlorine from salt and its calcium from added limestone. Some is also recovered in Germany from carnallite.

Calcium chloride is a mineral, hydrophyllite, which occurs at Vesuvius; hydrous calcium chloride is artificial. The anhydrous form has great avidity for water, which property determines its chief use.

Calcium chloride is used chiefly for laying dust on highways, stabilizing roads, and for ice control on highways and railway tracks. It forms a refrigerating brine and an antifreeze in cooling condensers in oil fields. It dustproofs coal and coke, cures concrete, and preserves wood. It is a dehydrant for laboratory uses, pipe lines, gas, and air conditioning. It has some use in curing and canning vegetables, as a deodorizer, and in many chemicals.

Magnesium chloride is a source of metallic magnesium and it also allays dust.

The chief sources of these chlorides are from brines wells in Michigan, Ohio, West Virginia, California, and the oceans. Sources in the world are nearly unexhaustible.

Selected References on Chlorides

Calcium Chloride. 1934. P. M. Tyler. U.S. Bur. Mines Inf. Circ. 6781.
General References 7, 18, and 19.

BROMINE AND IODINE

Bromine is a brownish corrosive liquid that occurs as bromide, along with salt and magnesium chlorides. Formerly, it was obtained chiefly from the potash deposits of Germany and France, but now it is extracted entirely from deep brine wells, of which five are located in Michigan, five in Arkansas, and one in California. Of the world production of 330 million metric tons, the United States produces 65 percent. It is marketed chiefly for gasoline additives such as ethylene bromide, but also as potassium and sodium bromide and other chemicals.

The inexhaustible oceanic supply contains about 60 ppm of bromine from which it could be liberated by acidification and oxidation with chlorine and blown out by compressed air. Seawater extraction is not being done even though it is almost economic today. Well brines in Michigan contain 1300 to 2900 ppm bromine; California brines contain about 850 ppm bromine as a by-product. Seawater contains only about 65 ppm, but this can be concentrated to 1750 ppm by solar evaporation before processing.

Bromine is used chiefly to make ethyl or antiknock gasoline. Bromides are used in the dye, photographic, and motion picture industries, and for many medicinal purposes. Bromine compounds are also used for chemical, metallurgical, ore dressing, tear gas, and hand-grenade purposes.

Iodine is the heaviest nonmetallic element, and its vapor is one of the heaviest known gases. It is widely distributed in minute quantities in the mineral, vegetable, and animal kingdoms and is obtained as a by-product of Chilean nitrates, from natural brines, and from kelp (seaweed). The United States production, formerly came from oil-well brines in California and Louisiana. Present production of iodine comes chiefly from gas-well brines in the Chiba Pennsula of Japan (65 percent) and from nitrate areas of northern Chile (33 percent). Two percent comes from brines at Midland, Michigan.

Iodine is used for biologic and industrial purposes. Its lack in diet is a cause of goiter, and a preventive is the addition of potassium iodide to table salt and water supplies. It is also added to animal feedstuffs and fertilizers. Medicinally, it is an antiseptic, and is also used internally. It finds use as a laboratory reagent, in many chemicals, and its main use is for sensitizing solutions for photographic plates, films, and papers. Iodine is also used in the dye, tanning, and other industries.

Selected References on Bromine and Iodine
Strategic Mineral Supplies. 1939. G. A. Roush. Chap. 13, Iodine. McGraw-Hill Book Company, New York.
General Refernces 7, 13, and 19.
Mineral Industries Surveys, Bromine in 1979.

POTASH

In the United States, in 1970, potassium compounds were produced by 13 companies that operated plants in five states. About 87 percent of production comes from crystalline deposits in southeastern New Mexico and east central Utah.

Texasgulf is operating a potash mine near Moab, Utah, that was formerly mined by underground methods until a disastrous explosion occurred because of natural gas seeping into the mine. Only solution mining is now being conducted. The evaporites occur in the Paradox formation. Salt beds underlie over 30,000 square kilometers in southeastern Utah, southwestern Colorado, and northwestern New Mexico (Fig. 28-2). The salt beds contain potash over a large part of this area. Individual beds range in thickness as the result of thinning and swelling from 6 to 240 meters at the basin center to zero on the flanks. Soluble potassium minerals are dissolved along with halite, which occasionally makes up more than 80 percent of the saline beds. Nevertheless, this deposit is a major U.S. producer of potash.

Concentrated brines are also a source of potash, which is recovered from wells or saline lakes, and lacustrine sediments of continental origin in arid regions. Potassium is an essential plant nutrient needed to maintain and expand food production for man and beast. About 95 percent of the potassium compounds used in the world are used as fertilizers.

The world production of potash is slightly less than 20 million tons of K₂O. Eastern Europe and the United States each produce about 35 percent; western Europe about 27 percent; with the remainder provided by South America, Africa, and Asia.

Potassium carbonate and bicarbonate, formerly obtained from wood ashes, are now manufactured from potash salts, coal, and limestone. The carbonate is employed to make hydroxide and other potassium compounds. The *hydroxide* or *caustic potash* is one of the strongest bases known and is employed for making soap, dyes, laundry and pharmaceutical preparations, disinfectants, laboratory reagents, and for other chemicals, such as potassium chromate and oxalic acid.

Potassium iodide is obtained from the hydroxide or carbonate, and is employed in medicine, photography, and reagents. *Potassium ferricyanide* is employed to make blueprint paper and for blue pigments. *Potassium chlorate,* a strong oxidizing agent, is made from the chloride and is used in making explosives, caps, fireworks, toothpastes, medicines, and dyes. *Perchlorate,* a stronger oxidizing agent, is used mainly for railroad fuses. *Potassium permanganate,* another strong oxidizer, is also made from the hydroxide, and is used as an antiseptic, in gas mask canisters, and has other industrial uses. *Potassium bitartrate,* another derivative is used for baking powder and cream of tartar.

About 75,000 tons of potash salts are used annually for chemical purposes.

SULFUR

Sulfur is perhaps the most important chemical mineral and one of the most widespread native elements. It occurs as native sulfur and in sulfides and sulfates. Native sulfur is the chief commercial source, and pyrite is the only other mineral mined for its sulfur. Substantial amounts of sulfur gases are also recovered from smelters, industrial plants, and sour crude oil and natural gas.

In approximate order of the increasing cost of mining or recovery, the following raw materials provide most of the world's sulfur and sulfuric acid supply, as compiled by Bodenles:

1. Elemental sulfur deposits in evaporite rocks.
2. Hydrogen sulfide contained in sour natural gas.
3. Organic sulfur compounds contained in petroleum.
4. Massive deposits of pyrite.
5. Elemental sulfur deposits in volcanic rocks.
6. Ores of metallic sulfide minerals (copper, zinc, lead, nickel, and molybdenum).

Additional large resources of sulfur occur in the

following minerals or compounds:

7. Beds of the calcium sulfate minerals, anhydrite and gypsum.
8. Organic sulfur compounds in tar sands.
9. Organic sulfur compounds and pyrite in coal beds.
10. Organic sulfur compounds and pyrite in oil shale and shale-rich organic matter.

Uses. Sulfur and its chemical compounds are used in virtually every industry and agricultural effort in the world. Principal uses in the United States are primarily for soluble fertilizers, synthetic fibers, plastics, papers, pigments, explosives, petroleum products, drugs, and insecticides. Smelter and refining industries in the Rocky Mountains, including the Wasatch Mountains, and extending into Alberta and British Columbia, recover SO_2 and produce phosphoric acid for fertilizer by reacting produced sulfuric acid with phosphate rock from the Phosphoria formation.

In the paper industry sulfur is burned to SO_2,

which is absorbed by milk of lime solution to make calcium bisulfite and sulfurous acid for digesting wood pulp. Hard rubber for storage batteries contains 30 percent sulfur, and tires about 1.5 percent.

Sulfuric acid is also employed in petroleum refining to remove gums, tars, and corrosives, to scrub coking gases, and to purify benzol and tuluol. Numerous chemicals are made from it. Metallurgically, it is used for pickling iron and steel before galvanizing, and in making electrolytic zinc and copper. In the paint industry it is used for making titanium oxide. In explosive manufacture, sulfur is used for black powder, and nitrocotton is made in a bath of sulfuric and nitric acids.

Production and Distribution. The total sulfur production in the United States, including exports and imports, is shown in Fig. 28-4. Since the turn of the century, when sulfur domes began production, almost 90 percent of U.S. sulfur was derived from the Frasch process. Beginning in about 1950, other sources of sulfur had to be developed; in 1953 the United States began to import sulfur even though

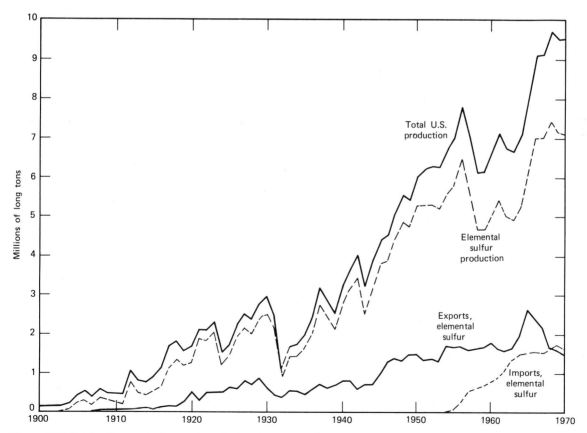

Figure 28-4 Total sulfur production and production, imports, and exports of elemental sulfur of the United States, 1900–1970. (From Bur. of Mines, and A. J. Bodenlos, 1973, U.S. Geol. Survey Prof. Paper 820, p. 607.)

exports reached new highs. Sulfur production in the United States today is about 12 million metric tons which 63 percent is from the Frasch process. The U.S. production is 38 percent of the world production, followed by Canada (20 percent), Poland (13 percent), and France (6.5 percent). The remainder is provided by Russia, Italy, Japan, Chile, Mexico, and Spain.

World sulfur is derived from petroleum and natural gases (50 percent), the Frasch process (42 percent), and ores (8 percent).

Figure 28-5 indicates the sulfur production from pyrites, base metal ores, and petroleum and natural gas. The hydrocarbon source has become an important source but still remains less than 10 percent of the elemental sulfur sources.

Occurrence and Origin. Native sulfur occurs (1) as depositions around volcanoes—Japan, Mexico, and Chile; (2) in salt-dome caprocks—Gulf Coast, United States and the Middle East; and (3) as sedimentary beds—Russia and Sicily.

Pyrite (sulfur-ore type) occurs as massive pyritic replacement bodies with associated pyrrhotite, chalcopyrite, and some sphalerite. Examples are Rio Tinto in Spain, Portugal, Cyprus, Norway, the Urals, Noranda, Quebec and Ducktown, Tennes-

see. Some flotation by-product pyrite is obtained from other types of lode deposits.

Native sulfur originates by several different means. Deposits associated with volcanism may be formed by (1) condensation of sulfur vapors; and (2) bacterial production of H_2S and oxidation to $S°$. Sulfur may be deposited by oxidation and by sulfur bacteria from thermal springs containing H_2S. One or the other of these processes has probably given rise to most of the sulfur of volcanic regions and also contributed the sulfur of sedimentary beds, such as those of Sicily, Russia, and some Japanese lakes (Chapter 9).

The sulfur of salt domes is associated with limestone, gypsum or anhydrite, and sulfur water (as are also the Sicilian deposits).

Those salt dome deposits containing native sulfur are associated with, or contain, petroleum and are located at relatively shallow depth where the temperatures do not exceed about 90°C. These conditions permit the viability of sulfate-reducing anaerobic bacteria, specifically *Desulfovibrio desulfuricans* and *Clostridium nigrificans*. Their energy source is from hydrocarbons; their oxygen source is from the sulfate in the cap rock. These bacteria liberate H_2S and CO_2; the former is exothermically oxidized to sulfur, not SO_2, because of

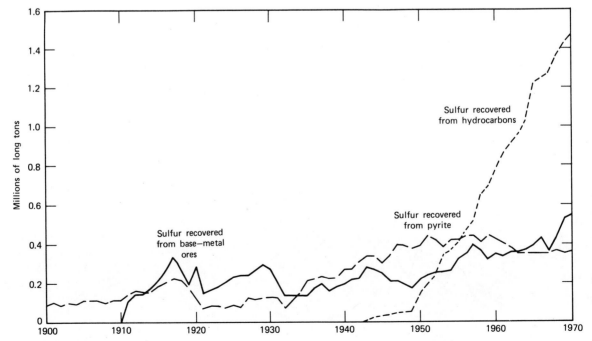

Figure 28-5 Production of sulfur contained in pyrites, recovered from petroleum and natural gas, and recovered from base metal ores, United States 1900–1970. (From Bur. of Mines, and A. J. Bodenlos, 1973, U.S. Geol. Survey Prof. Paper 820, p. 608.)

the low partial pressure of oxygen, and the latter combines with the Ca released from the anhydrite to form white $CaCO_3$.

Extraction and Preparation. The Gulf Coast salt-dome sulfur is extracted ingeniously and simply by the Frasch process (Fig. 28-6). A well with 10-inch casing is driven through the overlying formations to the sulfur deposit. Inside of it is an 8-inch hot water pipe, and inside of this is a return sulfur pipe containing an air pipe. Water heated to 140°C is discharged into the sulfur bed under pressure and melts the sulfur at 134°C; then the thin molten sulfur collecting at the bottom is forced by hot compressed air up the discharge pipe. The red liquid sulfur is piped to a stock pile, where it solidifies to pure yellow sulfur.

Sicilian sulfur is mined by underground methods to a depth of 240 meters. It is placed in a battery of connected furnaces where hot gases volatilize the sulfur out of the ore, which is then collected and condensed to marketable sulfur. The extraction is 80 to 85 percent. These deposits were once the sole

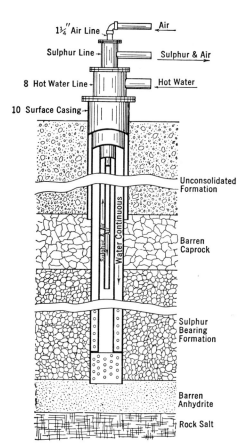

Figure 28-6 Frasch process of extracting sulfur. (From Lundy, Ind. Min. and Rocks.)

source of the world's sulfur. By taking advantage of this fact and raising the price of the sulfur, Great Britain protested and sailed her battleships around the island, "while back home", British scientists developed new sulfur production from roasting Cornwall sulfides. Sicily never really recovered from this experience and the deposits, which are structurally complex, now provide a very small amount of the world sulfur demand.

Pyrite is generally roasted in furnaces to sulfur dioxide, which is recovered directly for sulfuric acid manufacture, or for sulfite paper pulp, or is compressed into a liquid for shipment. Lump ore or concentrates containing from 40 to 50 percent sulfur are used. The Westcott process recovers elemental sulfur from pyrite concentrates by contacting pyrite with chloridizing gases, forming ferrous chloride and gaseous sulfur that is condensed to liquid sulfur. The ferrous chloride is oxidized to Fe_2O_3, and the chlorine is recovered. Two tons of pyrite yield about 1 ton of sulfur and 1.5 tons of iron oxide.

Examples of Deposits

In the *United States* the sulfur deposits of the Gulf Coast area (Fig. 28-3) are most numerous. Their occurrence is unique, as they lie in the caprock of some salt domes (Chapter 9), which have been intruded upward from thousands of meters in depth.

The core of salt, containing some anhydrite, is overlain directly by anhydrite or gypsum (Fig. 28-7). The anhydrite is the residual, less soluble accumulation of material originally present in the upper part of the soluble salt plug; it has been altered in part to gypsum.

The sulfur occurs in the lower part of the cavernous limestone or calcite rock and also in the upper part of the gypsum, but only that in the calcite rock is recovered. It occurs in bunches, seams, specks, and crystals and, stochiometrically makes up 30 percent of the volume. Some barite, celestite, and strontianite, and rarely galena, sphalerite, and manganese sulfides occur in the caprock series. The sulfur zone ranges from 8 to 100 meters in thickness; a thickness of 30 meters is common. The mineralized caprock is overlain by 2 to 70 meters of barren caprock, and this by unconsolidated sediments up to the surface, which may exceed several thousands of meters or more above the caprock.

The sedimentary deposits of Sicily (Fig. 9-5) and Russia (Fig. 28-8) are described in Chapter 9. Most of the other deposits of the world are associated with fumaroles or hot springs connected with vol-

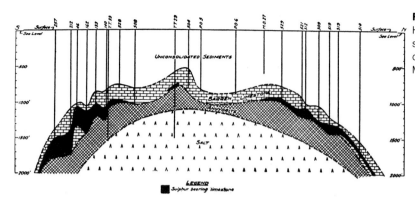

Figure 28-7 N-S section across Hoskins Mound salt dome with sulfur cap (black) mostly in white calcite above anhydrite. (From: Marx, Am. Assoc. Petr. Geol.)

canism, where sulfur fills fissures, vugs, and rock pores. Such deposits are found in Japan, Java, Mexico, and the Andes; Watanabe recorded remarkable eruptions of molten sulfur from Siretoko-Iôsan volcano in Japan in 1936 during which time pure sulfur flows with a temperature of 120°C, cascaded down the valley to form a deposit 1500 meters long, 20 to 25 meters wide, and 5 meters thick.

Selected References on Sulfur

General Reference 11. Chap. 47, by W. T. Lundy. *Good résumé and exhaustive bibliography from which adequate references may be selected.*

Sulfur. 1930. R. H. Ridgeway. U.S. Bur. Mines Inf. Circ. 6329. Pyrites. Inf. Circ. 6523, 1930. *Occurrence, technology, and statistics.*

Genesis of the sulfur deposits of the U.S.S.R. 1935. P. M. Murzaiev. Econ. Geol., v. 32, p. 69–103. *Sedimentary and volcanic types.*

The sulfur mines of Sicily. 1923. C. Sagul. Econ. Geol., v. 18, p. 278–287.

The Boling dome, Texas. 1933. A. G. Wolf. XVI Int. Geol. Cong. Guidebook, v. 6, p. 86–91. *A sulfur-bearing salt dome.*

Eruptions of molten sulfur from Siretoko-Iôsan volcano, Japan. 1940. T. Watanabe. Jap. Jour. Geol. and Geog., v. 17, p. 289–310. *An amazing cascading flow of molten volcanic sulfur.*

General Reference 7 and 19, p. 605–618.

Origin of Gulf Coast salt dome sulfur deposits. 1957. H. W. Feely and J. L. Kulp. Am. Assoc. Petrol. Geol. Bull., v. 41, p. 1802–1853.

Sulfur. 1970. R. W. Lewis. In Minerals facts and problems. U.S. Bureau of Mines Bull 650, p. 1247–1265.

Caprock development and salt-stock movement. 1970. A. J. Bodenles. In *Geology and technology of Gulf Coast salt - A symposium. Baton Rouge, School of Geoscience, Louisiana State University, p. 73–86c.*

NITRATES AND NITROGEN

Nitrogen compounds are dominantly used for fertilizer. However, the 15 percent of the supply that is used chemically plays an important role in the chemical industry.

The sources of the nitrogen compounds used chemically are from by-product ammonia and ammonium sulfate from coking ovens, synthetic ammonia and nitrogen firing from the air, and synthetic cyanamide and cyanogen from the air. The Chilian deposits, which were once the sole source of nitrates now provide only a few percent of the world's demand.

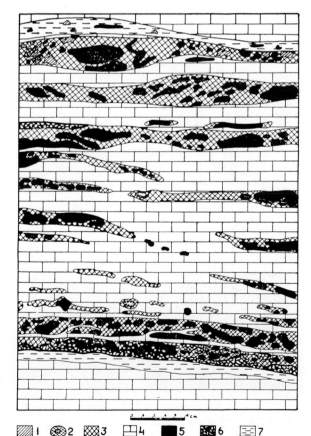

Figure 28-8 Sedimentary sulfur deposit at Knibyshev, Russia. 1, Lutecite; 2, calcite vugs; 3, calcite; 4, bituminous limestone; 5, pure sulfur; and 6, bituminous sulfur. (From Murzaiev, Econ. Geol.)

Tabel 28-2. Chief Uses of Important Nitrogen Compounds

Compound	Chief Chemical Uses	Important Minor Uses
Sodium nitrate	Explosives, nitric acid, and chemicals	Glass and minor chemicals
Potassium nitrate	Explosives, glass, and pickling meat	Acids, chemicals, and medicine
Nitric acid	Nitration of hydrocarbons; aniline dyes, explosives, plastics, rayon, leather solvent, textiles, and sulfuric acid	Photography, lacquers, enamels, varnishes, many chemicals, and textiles
Various nitrates	Explosives, hydrogen peroxide, inks, dyes, and chemicals	Chemicals, lamp filaments, photography, mirrors, and medicines
Ammonia	Refrigerant, household, chemical reagent, silks, dyes, and steel hardening	Hydrogen, welding, water purification, and neutralizer
Ammonium compounds	Medicines, tanning, dyeing, printing, galvanizing iron, chemicals, oxidizer, celluloid, hydrogen peroxide, and sodium carbonate	Batteries, soldering, photography, textiles, baking, scouring, smelling salts, and "laughing gas"
Cyanamide	Chemicals, dyes, and explosives	
Cyanogen	Various cyanides, gold extraction, indigo, photography, electroplating	Dyeing, painting, printing, blueprint paper, and fumigating.

The important nitrogen compounds made or used chemically and their chief uses are listed in Table 28-2.

LITHIUM MINERALS

Lithium, the lightest metal known and with a melting point of 180°C, is finding increasing chemical and industrial uses.

Properties and Uses. Lithium minerals are the source materials for numerous lithium compounds, many of which are used in pharmaceutics, particularly in mineral waters and lithia tablets. Lithium is employed for making Edison storage batteries, and it supplies the red color in flares, fireworks, and signal rockets. Lithium compounds are used for photography, welding aluminum, dental cement, curing meat, making ammonia, making rayon, reducing agents, and purifying helium and other gases. The hydroxide is one of the most important commercial compounds of the metal and is the one used in heavy-duty storage batteries. The chloride is finding growing use in dehumidifying air for air conditioning and industrial dyeing.

The metal finds use as an alloy of aluminum, magnesium, and zinc, for light airplane metals. Bearing-metal alloys, copper electrodes, and lead sheathings are benefited by a little lithium. It serves as a deoxidizer and purifying agent for refining nickel, iron, and copper metals and alloys.

Occurrence and Distribution. The commercial lithium minerals occur in pegmatite dikes, but considerable amounts of lithium compounds have also been obtained lately as a by-product from the brines of Searles Lake, California, the Great Salt Lake, and drilling in Playas, such as Silver Peak, Nevada.

The brine wells at Silver Peak, Nevada, have become the world's largest source of lithium and comprise about 50 percent of the U.S. reserves.

The former source of lithium was entirely from pegmatites, where the minerals spodumene, lepidolite, and amblygonite were hand-cobbed to provide the limited source of lithium. At present, however, lithium is by far the most economically important of the rare alkalies. Two of the more frequently mentioned potential uses of lithium are for anodes and as a constituent of the electrolyte in storage batteries to power electric automobiles. The other is as an energy source to breed tritium for fusion reactors.

Although the brines at Silver Peak, Nevada, and Searles Lake, California, provide the dominant demand for lithium, some lithium is recovered from the pegmatites of the Black Hills of South Dakota, notably the Etta mine, where spodumene crystals up to 14 meters long and almost 2 meters in diameter occur; one yielded 90 tons of spodumene, and crystals 2 to 3 meters long are common. Amblygonite and some lepidolite occur in the same deposits. Much lepidolite and some amblygonite is obtained from pegmatites in California and from the Harding mine in New Mexico. Important spodumene deposits occur in Knox County, Maine, and scattered localities in North Carolina. Hess describes some spodumene dikes in North Carolina up to 300 meters long and 30 meters wide with spodumene that carries 6½ percent lithia. The only important occurrence outside of the United States is a Bernick Lake, Manitoba, and Barreutc, Quebec, where, at the former, a pegmatite dike 25 meters wide is said to contain 95,000 tons of lithium. Some of the material contains substantial quantities of caesium and rubidium. Other foreign occurrences are Southwest Africa and France, with minor amounts in Czechoslovakia, Sweden, Germany, Spain, and Western Australia. The supply of lithium minerals seems more than adequate to take care of the demand.

Selected References on Lithium
Lithium. 1932. Imp. Inst. London. *Good summary and bibliography.*
General Reference 11. Chap. 23, by R. W. Mumford. Good review of minerals and occurrences.
Lithium in New England. 1938. F. L. Hess and O. C. Ralston. Eng. and Min. Jour., v. 139, p. 48–49.
Spodumene pegmatites of North Carolina. 1940. F. L. Hess. Econ. Geol., v. 35, p. 942–966. *Good detailed description of occurrence.*
Exploration of Harding tantalum-lithium deposits, Taos, N. Mex. 1946. John H. Soulé. U.S. Bur. Mines Rept. Inv. 3986. *Spodumene in large pegmatite.*
Lithium. 1970. A. M. Cummings. In Mineral facts and problems. U.S. Bur. Mines Bull. 650, p. 531–536.
General Reference 7 and 19.

STRONTIUM MINERALS

The two commercial strontium minerals are celestite (sulfate) and strontianite (carbonate), the former being the common ore and the latter the more valuable because of its lower manufacturing cost. Strontianite is named from Strontian, Scotland, its type locality.

Properties and Uses. Strontium is desired for its salts, used chiefly to desaccharize beet-sugar mo-

lasses. The nitrate is in constant demand for pyrotechnics—the red color of fireworks, rockets, railroad signals and tracer bullets. It is used for electric batteries, paints, rubber, waxes, purifying caustic soda, for making glasses, glazes, and enamels, for refrigerant solutions, and for medical preparations. Celestite is a substitute for barite as a rubber filler, and the carbonate is used in steel furnaces.

Occurrence and Distribution. Celestite occurs as a gangue mineral in vein deposits with galena, barite, and calcite; as sedimentary beds associated with gypsum; as disseminations in limestone; and in solution cavities in limestone. Strontianite occurs as a secondary alteration product of celestite and as replacements of limestone. Strontium is given off in solfataras and is a constituent of seawater. Most deposits of strontium minerals are enclosed within sedimentary rocks; a few occur in igneous and metamorphic rocks.

The most important deposits are the celestite deposits of *Gloucester* and *Somerset, England,* where celestite occurs as irregular masses and lenticular bodies in Triassic marl associated with gypsum. It also occupies flow openings under the marl. In *Westphalia, Germany,* are bedded deposits in Zechstein dolomite, underlying which are irregular masses in the Kulm. Near Münster are small strontianite deposits in veins in Cretaceous marls. In *Sicily* fairly large deposits of celestite are associated with sulfur and gypsum that occur in fissures underlying upper Miocene beds. Several celestite vein deposits occur in limestone in *Ontario.* Other deposits occur in *Mexico* and *Spain.* In the *United States* a number of deposits are known, many of which are large, but they cannot compete except during wartime, with the imported product from England. Strontianite deposits are few and small. The total world production seldom exceeds 15,000 tons a year, all of which normally comes from England and Germany.

Selected References on Strontium
Strontium. 1929. R. M. SANTMYERS. U.S. Bur. Mines Econ. Paper 41. *Brief survey;* and Inf. Circ. 7200, by C. L. Harness, 1942.
Strontium. Nov. 1940. L. SANDERSON. Can. Min. Jour. v. 61, p. 726–728. *Brief summary.*
General References 7 and 19.

MISCELLANIOUS CHEMICAL MINERALS

Alum Minerals. The alums are a series of double sulfate isomorphs with potash alum. Some are natural minerals; others are artificial chemicals made from natural raw materials. Some of the commercially used natural alums are common alum or kalinite (K, Al), soda alum or mendozite (Na, Al), ammonium alum or tschermigite (NH_4 Al), alum rock or alunite (Al, K), alunogen (Al). All contain water in addition to SO_4^-. They are formed by the evaporation of alum waters.

Manufactured alums are made by treating bauxite with acid or from clay, cryolite, pyritic rocks, lignite, alunite, and leucite. About 400 to 500 thousand tons of aluminum salts are produced annually in the United States.

Potash alum is used for treating furs, and for medicinal and special uses. Ammonium alum is used for stucco, cast stone, statuettes, and for sundry purposes. Sodium-aluminum sulfate is used for baking powder. Aluminum sulfate (not true alum) is used for dyeing, pigments, tanning, water purification, paper, fire-fighting compounds, and for decolorizers and deodorizers.

Barium. (See Chapter 27.) About 22 percent of the barite mined is used for barium chemicals. Barite is furnace reduced with carbon to "black ash" or barium sulfide, or to barium chloride. The principal substance made from this is precipitated barium sulfate or blanc fixe (Chapter 20). Precipitated barium carbonate is used in ceramics or for barium peroxide. Other important barium chemicals are chloride, chlorate, chromate, dioxide, hydrate, nitrate, and peroxide.

Bauxite. (See Chapter 19.) About 175,000 to 200,000 tons of bauxite are used annually in the United States for many chemicals. The most important of these are aluminum sulfate, aluminum hydrate, aluminum chloride, and sodium aluminate. These substances have many chemical uses.

Carbon Dioxide. (See Chapter 27.) Carbon dioxide is obtained as a natural gas, manufactured, and as a by-product. In addition to its industrial uses, it has many chemical uses.

Dolomite. Utilized in the sulfite process of paper making dolomite supplies the acid liquor, which is a solution of magnesium and calcium bisulfites prepared from dolomite or dolomitic limestone. Roasted dolomite is used to neutralize acid water, and it is also used in the preparation of certain magnesium salts.

Epsomite. Epsomite ($MgSO_4 \cdot 7H_2O$), or epsom salts, is obtained from spring waters and brines to make commercial epsom salts.

Kieserite. Found with potash deposits kieserite ($MgSO_4 \cdot H_2O$) is also used for the same purpose, as are *dolomite* and *magnesite*.

Fluorite. (See Chapter 26.) Fluorite that is utilized for making hydrofluoric acid must contain not less than 98 percent calcium fluoride. It has many other chemical applications. It is utilized in the extraction of potassium from feldspar and portland-cement flue dust; in making calcium carbide and cyanamide; inorganic and organic fluorides and silicofluorides, including insecticides, preservatives, and dyestuffs; and in making "Freon," an important noninflammable, nonexplosive, and nontoxic refrigerator fluid.

Greensand. Greensand (marl) is extensively used chemically as a water softener. Because of its high rate of base exchange, water is softened as rapidly as it can be forced through the greensand, which regenerates equally speedily.

Magnesite. Magnesite is one of the principal raw products from which various magnesium salts used in chemistry and pharmacy are produced. Magnesium citrate is used for saline beverages and by the dispenser. The oxide and carbonate are used for cosmetics, toothpaste, antiacid tablets, the hydroxide in water gives milk of magnesia. Other pharmaceutical salts include epsom salts, salicylate, lactate, phosphate, chloride, and iodide.

Monazite. In small quantities monazite is used to obtain thorium, which converted into nitrate is used to manufacture incandescent mantles for gas, gasoline, or kerosene lamps, for arc light carbons, and for rare-earth oxides. The accompanying mesothorium is a substitute for radium. The residual salts of cerium and lanthanum are made into an alloy of iron to make cigarette sparkers.

Lime. Lime is widely employed in the chemical industry, which uses about one-eighth of all that produced in the United States. The largest consumption is in water purification. When lime is added to "hard" water, it combines with the excess CO_2, forming $CaCO_3$, which, with the bicarbonate already present, is precipitated and removed by filtration. Lime and soda ash are added to remove "permanent" hardness caused by magnesium and calcium sulfates. Lime also serves as a coagulating agent and as a disinfectant. Another growing use is for sewage and trade-waste purification.

The next most important use is for making calcium carbide and cyanamide. Among some of the minor chemical uses are: for making ammonium, potassium, and sodium chemicals, calcium carbonate, baking powder, salt refining, pigments, and varnish constituents.

Phosphate rock. Phosphate rock is utilized chemically for numerous products, of which the most important are phosphorus and phosphoric acids. These are used to make many acids, phosphides, and phosphates that have a wide use in metallurgy, photography, medicine, sugar refining, soft drinks, preserved foods, food products, ceramics, textiles, and matches. Elemental phosphorus is used in safety matches, medicines, shells, tracer bullets, grenades, smoke screens, and fireworks.

A large phosphate deposit was discovered in 1961 below the Abu Tartur Plateau in the western desert of Egypt. Initial investigations have put estimated phosphorite reserves at nearly 1000 million tons of good-grade material possibly amenable to underground mining. No portion of the property is yet in production nor has it thoroughly been prospected. One area that is under study measures about 10 kilometers along its horizontal extent.

The Phosphate formation is Cretaceous in age, probably more specifically Lower Campanian to Upper Maestrichtian in age. The formation ranges from 17 to 57 meters in thickness occurring at a depth of 170 to 288 meters below the surface.

The P_2O_5 content of the more-productive layer. which averages 3.86 meters thick, varies from 23.82 to 30.19 percent averaging 25.6 percent, slightly less than the minimum of 28.2 percent P_2O_5 of marketable raw material. It will be some time, however, before production begins and means of transport from this remote area are available.

Helium. An inert gas, helium is extracted from natural gas in Kansas and the panhandle of Oklahoma and Texas. Helium has been used mainly for lighter-than-air craft, having a lifting force of 66 pounds per 1000 cubic feet compared with 71 for hydrogen. Its uses are to prevent deep-sea divers from getting the "bends," for certain metallurgical operations (such as welding in an inert atmosphere), in radio tubes and electric signs, and for mixing with oxygen for treating pneumonia.

Its use mushroomed during World War II but will increase further if its potential as a coolant in power transmission, is realized. Certain metals at the temperature of liquid helium become superconductors. Also, as described by Ward and Pierce, in power generation by nuclear fusion, thermonuclear reactions may be confined by magnetic fields produced by high-intensity superconducting magnets that could use liquid helium as a refrigerant. Unfortunately, helium is being wasted at the rate of billions of cubic feet per year.

29

ABRASIVES AND ABRASIVE MINERALS

Many minerals and rocks of diverse composition but with one thing in common—hardness—are employed as natural abrasives. (Table 29-1) These are used in the natural state, except for processing and bonding. Natural abrasives, however, are being supplanted by artificial abrasives, which in turn are made from mineral products. Most of the abrasive materials are dealt with incidentally in different parts of this book.

Abrasives are defined by Thoden as substances that are used to clean or dress the surfaces of other minerals, or to comminute other materials by abrasion and percussion.

ABRASIVE MATERIALS.

Natural abrasives are divided into three groups: (1) high-grade natural abrasives, which include, in the order of hardness, diamond, corundum, emery, and garnet; (2) siliceous abrasives, consisting of various forms of silica; and (3) miscellaneous abrasives, including buffing and polishing powders. The materials utilized are listed in Table 29-2.

Use and Preparation. Natural abrasives may be used (1) in the natural form, for example, sand and pumice; (2) after shaping, such as millstones; and (3) after being ground into grains or powders and made up into wheels or papers.

Abrasives have a wide application in all phases of industry, and their use rises and falls with industrial trends. Because the automobile industry is the largest consumer, the use of abrasives parallels the fluctuations of automobile and steel output. Industries like the airplane industry increase the call for certain types of abrasives, such as industrial diamonds.

The chief uses of some of the natural abrasives are given in Table 29-3 (after Eardley-Wilmot).

Abrasives used in the soap industry (including scouring soap) include pumice, pumicite, feldspar, diatomite, bentonite, talc, silica, chalk, and clays. Metal polishes use pumice, emery, diatomite, silica, tripoli, chalk, fuller's earth, clay, bauxite, magnesite, and oxides of metals. Many of the materials that serve as abrasives have greater use for other purposes. Manufactured abrasives have replaced many of the natural abrasives for certain applications, particularly in industrial metalwork.

INDUSTRIAL DIAMONDS

One thinks of the diamond primarily as a gemstone, which it is as $200 million of cut diamonds are imported to the United states annually, along with $300 million of uncut diamonds. However, about 12 million carats of industrial diamonds are imported into the United states annually. The difference between this and the demand of 20 million carats is provided by synthetic diamonds. There are two types of industrial stone: the carbonado or carbon, the hard black diamond; and the bort, which includes small stones, fragments, and badly colored or flawed stones.

Carbonados. These are hard, tough, black diamonds that come from Bahia, Brazil, and are utilized for dies in wire drawing, for diamond-drill bits used in exploring for ores and oil, and for tools employed for dressing and truing abrasive wheels. Diamond-tipped tools are used for boring metals and other hard substances, such as Bakelite, in the auto and electrical industries, where precision work is necessary.

Bort. Bort has come to replace carbonados for many industrial purposes because of its lower cost. It is extensively employed in the precision manufacture of airplane and motor car engines, both for boring and for abrasion of surfaces. War activity

Table 29-1 Resources, in Short Tons, of Abrasive Materials in the United States

Commodity	Identified resources (including reserves)	Reserves hypothetical
Diamond	0 (Synthetic diamonds are being made)	
Corundum	125,000	> 2,000,000
Emery	3,000,000	> 4,000,000
Garnet	14,000,000	> 100,000,000
Tripoli (and related materials)	4,500,000,000	> 10,000,000,000
Silica sand	Large	Very large

Source: From U.S. Geol. Surv. Prof. Paper 820, 1973

greatly stimulates the consumption of industrial diamonds.

Diamond rock-drill bits are now largely mechanically set with many small borts, which makes them cheaper than carbon bits and greatly lowers the cost of diamond drilling. Borts are even employed extensively for drilling blasting holes. A typical drilling bit of this type contains an average of 178 small stones weighing 7 carats.

Diamond dust made by grinding bort is used as an abrasive for the cutting of gemstones, minerals, and carbides. Crushed diamonds are now impregnated in a molten metal bond for fast cutting wheels, and similar cutting wheels are made with diamond fragments in a powder-metal bond. bort comes mainly from Zaire, Gold Coast, Angola, Sierra Leone, and South Africa.

Industrial diamonds are replacing the natural borts at a rapid rate.

CORUNDUM

Corundum (Al_2O_3) has a hardness of 9 as compared with 10, or 40 on a linear scale, for the diamond. There are three varieties: gems (ruby, sapphire), corundum, and emery, the last being a mixture of corundum and magnetite. Its hardness makes it a natural abrasive, but it is being supplanted by Carborundum. It is used as loose grain in optical grinding, on paper and cloth, and in the form of abrasive wheels.

Occurrence and Origin. Corundum is formed as (1) magmatic segregations in quartz-free igneous rocks, such as nepheline syenite (Chapter 12), (2) as a reaction casing between pegmatite dikes and intruded basic igneous rocks by a process of desilication of the dike.

The desilication type is the most common occur-

Table 29.2

High Grade	Siliceous	Miscellaneous	
Diamond	Sandstone	Bauxite	Siliceous carbide
Corundum	Quartzite	Magnesite	(Carborundum, etc.)
Emery	Novaculite	Magnesium oxide	Fused alumina
Garnet	Flint	Ground feldspar	(Alundum, aloxite,
	Chert	Chalk	etc.)
	Silicified limestone	Lime	Boron carbide
	Quartz	China clay	Metallic oxides
	Quartz mica schist	Talc	Lampblack
	Sand	Oxides of tin, iron,	Carbon black
	Tripoli	chromium, and	
	Pumice	manganese	
	Diatomite		

Table 29-3 Chief Natural Abrasives and Their Uses

Material	Abrasive	Chief Uses
High-Grade Natural Abrasives		
Diamond	Crystal	Cutting, boring, and wheel truing
	Dust	Airplane engines, and gem and rock cutting
Corundum	Wheels	Metal cutting and lens polishing
	Papers and cloths	Metals and hardwood; optical
Emery	Wheels	Snagging metals
	Papers and cloths	Metals and hardwood
	Loose grains	Finishing and polishing metals; glass grinding
Garnet	Papers and cloths	Hardwoods; paint and varnished surfaces
	Loose grains	Glass grinding
Siliceous Abrasives		
Sandstone	Grindstone	Grinding saws, knives, metals, etc.
	Pulpstones	Grinding wood for paper pulp
	Sharpening stones	Hand sharpening
	Oilstones	Fine sharpening of steels
Quartz, flint	Burrstones	Grinding flour, pigments, etc.
	Pebbles	Grinding ores in mills
	Slabs	Fine hand sharpening
	Crushed	Soft wood (sandpapers)
Sand	Grains	Sand blasting and glass grinding
	Papers and cloths	Woods and metals
Pumice	Blocks	Rubbing plant and varnish
	Grains	Glass and scouring powders
Diatomite	Powder	Metal polish and dental powder
Tripoli	Powder	Metal buffing
Volcanic dust	Grains	Scouring powders and cleansers
Rottenstone	Powder	Scouring powders
Soft Abrasives		
Feldspar	Powder	Scouring and cleansing
Clays	Powder	Buffing of metals
Dolomite	Calcined	Buffing of metals
Lime	Powder	Buffing of metals
Bauxite	Powder	Buffing of metals
Chalk	Powder	Buffing of silver and metals
Lampblack	Powder	Buffing silverware
Black rouge	Powder	Buffing metals and mineral surfaces
Red rouge	Powder	Buffing metals, optical glass, and mineral surfaces
Green rouge	Powder	Buffing hard metals and mineral surfaces
Tin oxide	Powder	Buffing metals and mineral surfaces

rence and is the origin ascribed by Hall to the South African deposits, and by others to the peridotite occurrences.

Distribution. Commercial deposits of corundum are known in Canada, South Africa, United States, India, Madagascar, and Russia.

In eastern Ontario, *Canada,* are three intrusions of nepheline syenite that contain corundum. The important one is in Renfrew County, near Craigmont, where nepheline syenite pegmatite carries crystals of corundum in commercial quantities. In other places, disseminated grains of corundum occur in the syenite and larger grains flank intrusive pegmatite dikes.

South Africa is the chief corundum area of the

world. Brown and ruby corundum occur in deposits flanking pegmatite dikes that intrude basic igneous rocks. The border zone is a highly altered mass, called plumasite, with much dark mica and corundum. Corundum also occurs as a placer mineral. In *India,* the home of the ruby and sapphire, isolated crystals of corundum are picked up over weathered syenite and in schists intruded by pegmatites. In *Madagascar* the mineral is found in the weathered portions of a metamorphic rock containing graphite, mica, and corundum. In *Russia,* corundum is obtained from an anorthositic rock called kyschtynnite and from dikes of corundum syenite. These occurrences are probably of magmatic origin. In the *United States* corundum occurs in the southern Appalachian belt at the contact of dunite intrusions with gneiss in North Carolina, and in Georgia where feldspar dikes intersect peridotite masses. This type is probably formed by desilication.

Selected References on Corundum

Corundum, emery, and diamond. 1927. V. L. Eardley-Wilmot. Part II, Abrasives, Can. Dept. Mines, Mines Branch, No. 675, Ottawa. *General survey.*

Corundum in the Union of South Africa. 1935. W. Kupferbinger. Geol. Survey, S. Africa, Geol. Ser., Bull. 6. *Occurrence and origin of desilication type.*

Corundum in South Africa. 1947. R. W. Metcalf. U.S. Bur. Mines Min. Trade Notes, Vol. 24, No. 2. *General survey.*

Corundum in North Carolina. 1943. W. A. White. N. Car. Div. Mines and Res. Rep. Inv. *Associated with basic igneous rocks.*

General Reference 13 and 18.

Corundum and Emery. 1970. J. D. Cooper U.S. Bur. Mines Bull. 650, p. 935–951.

EMERY

Emergy is a natural mixture of corundum, magnetite, and some hematite and spinel, named from Cape Emeri, Greece. Spinel emery contains considerable spinel, and corundum may be lacking (American). Feldspathic emery contains much plagioclase. Three commercial grades of emery are recognized: Greek, Turkish, and American.

The hardness and cutting quality of emery depend upon the amount of corundum present. It is coarse or fine grained and tough, and it withstands intense heat. It is used chiefly as grinding wheels, as emery cloth, and as grains and flour for glass polishing. The Grecian is generally the hardest, and

the Turkish next, but the American varieties are soft and are used mainly in pastes and composition.

Most of the commerical emery comes from Naxos in Greece, Turkey, and United States.

Occurrence and Origin. Emery is formed mainly by contact metasomatism and occurs in irregularly shaped bodies in crystalline limestone, altered basic igneous rocks, and chlorite and hornblende schists. In New York some also occurs in veins.

Distribution. On the island of *Naxos, Greece,* are lenticular masses of emery 100 meters long and up to 50 meters wide, enclosed in crystalline limestone and formed by contact metasomatism. In *Aidin, Turkey,* are irregular masses of emery, 70 by 100 meters, enclosed in crystalline limestone interfabricated with schists and gneisses. The main supply comes from boulders in residual clay. Similar deposits are worked in the *Urals, Russia.* The *United States* contains deposits of emery in New York, Massachusetts, and Virginia. Near Peekskill, New York, spinel emery occurs in the Cortlandt basic igneous complex near mica schist inclusions, in the form of sharply defined veins contained in sillimanite-cordierite-garnet-quartz rocks. One vein, according to Zodac, is 15 meters wide and a 1.6 kilometers long. At Chester, Massachusetts, emery occurs in pockets in a band of sericite schist. In Virginia, spinel emery occurs in lenticular bodies in (1) schist and quartzite, and (2) in granite cut by pegmatites. Watson considers that these were formed by high-temperature replacement akin to contact metasomatism.

Selected References on Emery

Corundum, emery, and diamond. 1927. V. L. Eardley-Wilmot. P. II Abrasives, Can. Dept. Mines, Mines Branch, No. 675, Ottawa. *Comprehensive treatment.*

Peekskill, N. Y., emery deposits. 1930. J. L. Gillson and J. E. A. Kania. Econ. Geol., v. 25, p. 506–527.

Mineralogy, composition, and origin of the emery of Turkey. 1946. J. De Lapparent. Jour. Amer. Cer. Soc. v. 29, no. 11, p. 200. *Abstract of occurrences.*

General Reference 13 and 19.

GARNET

Garnet is a group name applied to about 15 different complex silicate species with generally similar characteristics but of different composition, and with metals that are replaceable with each other. Some of the varieties are listed below.

Varieties	Composition	Geologic Environment
Almandite	Fe-Al-garnet	Mica schists and gneisses—metamorphic origin
Andradite	Ca-Fe	Contact-metasomatic limestone
Grossularite	Ca-Al	Contact-metasomatic impure limestone
Pyrope	Mg-Al	Eclogite—deep-seated igneous origin
Spessartite	Mn-Al	Granitic rocks and pegmatites
Uvarovite	Ca-Cr	Serpentine
Rhodolite	Fe-Mg-Al	

Almandite, or common garnet, is the one generally used as an abrasive, but andradite and rhodolite are also utilized. The production of garnet is relatively small.

Qualifications and Uses. For abrasive purposes, the important properties are hardness, toughness, and fracture. Sharp angular fractures with brittle edges that form new cutting edges are necessary. Rounded fragments are not desired. The grains should be at least the size of a pea, and the garnet content of the rock should not be less than 10 percent.

Abrasive garnet is used as garnet cloth or paper, and as loose grains, and is employed particularly for hard woods. Its cutting power is 2 to 6 times that of sandpaper. Garnet paper is also employed for finishing hard rubber, celluloid, leather, felt and silk hats and for rubbing down varnished and painted surfaces, automobile bodies, and copper and brass. Garnet grains are used for surfacing plate glass and ornamental stones and for gang-sawing marble and slate.

Occurrence and Distribution. Garnets are constituents of igneous rocks and are formed by metamorphism in schists and by contact metasomatism in calcareous rocks. They are widely distributed, but commercial deposits are few. Ninety-five percent of the production comes from the United States, the largest mine in the world being in New York, with smaller ones in New Hampshire, North Carolina, and Massachusetts. Negligible amounts come from Spain, India, Canada, and Madagascar.

Examples of Deposits.
New York includes garnet deposits in Warren, Essex, and Hamilton Counties; the important deposits are those of the Barton mines and the North River Garnet Company. The garnet is almandite and occurs in garnet gneiss in grains and crystals mostly between bean size and a few centimeters or more in diameter. Crystals up to a third of a meter are common, and a few are a meter in diameter. The garnet content of the Barton quarry averages 13 percent. At the North River quarry on Thirteenth Lake similar gneiss carries 4 to 8 percent garnet in crystals up to 7 or 8 centimeters in diameter. In *New Hampshire,* near North Wilmot, almandite garnet constitutes 40 to 60 percent of the enclosing garnet schist. The deposit is 30 by 60 meters and 8 meters deep. The garnets are thought to have been formed by granitic emanations. In *North Carolina,* at Sugar Loaf Mountain, is a deposit of rhodolite garnet, which averages 20 to 25 percent of the enclosing schist. In *Idaho,* garnet production comes from Fernwood.

In *Canada* many small deposits are known, but the only production has come from Lennox and Addington Counties, Ontario, where almandite occurs in gneiss. In *Spain* pink garnets have been concentrated into alluvial deposits in Almeira. Some alluvial garnet is known in Madras, Deccan, and Mysore, India; and in Madagascar, Ceylon, and Czechoslovakia.

Selected References on Garnet
General Reference 13 and 19.
Abrasives, Part III. 1927. V. L. Eardley-Wilmot, Can. Dept. Mines, Mines Branch, No. 677, Ottawa.
Classification and occurrence of garnets. 1938. W. I. Wright. Amer. Min., v. 23, p. 436–445. *Largely mineralogical.*
Garnet, its mining, milling, and utilization. 1935. U.S. Bur. Mines Bull. v. 265, p. 54.

NATURAL SILICA STONE ABRASIVES

Siliceous rocks are quarried and shaped into forms suitable for use, such as grindstones, pulpstones, millstones, and hand stones of various types. The rock used is dominantly sandstone, but quartzite, quartz-mica schist, silicified limestone, flint, chert, and novaculite are also used. One half million tons are used for grindstones and 1 million tons are used as crushed sand in the United States annually.

Grindstones. Grindstones are made mostly from sandstones. The sandstone must be of uniform hardness, possess sharp and even grain, and be sufficiently cemented to insure tenacity but at the same time permit crumbling away to prevent glaze. The larger the grain the faster the cutting power, but with corresponding roughness. Sound blocks must be available in sizes from 1 to 2 meters across. Relatively few sandstones have all of these qualifications.

The main sources are Carboniferous sandstones of the United States, Great Britain, Canada, and Germany. In the *United States* most of the grindstones come from the Dunkard and Berea sandstones of Ohio, and to a lesser extent of West Virginia and Michigan. In *England* the celebrated Newcastle stones come from Coal Measures sandstones near Newcastle, where excellent fine-grained stones suitable for edge tools are quarried. Other similar stones are known as Derbyshire, or Peak, and Yorkshire stones, which are obtained from the Millstone Grits. In *Canada* an old industry obtained grindstones in Nova Scotia and New Brunswick from Carboniferous and Permo-Carboniferous sandstone beds. Production of grind stones in the United States is waning because of artificial abrasives.

Pulpstones. Pulpstones are used for the grinding of pulp logs to make paper pulp. They are made of sandstone, of much the same type as grindstones, but the beds must be capable of yielding blocks 5 feet in diameter, and with a face from 1 to 3 meters wide. The qualifications are generally similar to those of grindstones, but the cementing should be weak enough to permit the bond to wear and the harder quartz grains to protrude. Coarse grain yields too coarse a wood fiber. After quarrying, the stones are seasoned for 1 to 2 years.

Pulpstones may come from the same beds as grindstones. The Millstone Grits of England yield the best, and West Virginia and Ohio yield the next best grade; Canada produces good stones in British Columbia and New Brunswick.

Millstones. Millstones are large circular stones run horizontally or on edge and include burrstone and chaser stone. They are made from any hard suitable sandstone, quartzite, quartz conglomerate, granite, or basalt. True burrstone is chalcedonic silica, originally employed for grinding grain but now used for grinding paints, fertilizers, and graphite. Chaser stones are large heavy stones desired mainly for grinding feldspar, quartz, barite, and other minerals. Most of these stones are produced in Italy, Great Britain, United States, Canada, and Germany, and yearly production has amounted to over 800,000 tons.

Grinding Pebbles and Liners. Pebbles of selected flint and quartzite are used in large rotating ball mills for the grinding of metallic ores, paints, cement, gypsum, clays, ceramic minerals, fillers, and powder abrasives. Steel balls have supplanted stone for many uses where steel cuttings do no harm, but stone must be employed for ceramic and certain other materials. The best-known flint pebbles come from deposits in Denmark, others from the shores of Belgium, France, and England; the United States and Canada yield small amounts.

Sharpening Stones. This group includes the familiar scythestones, whetstones, waterstones, honestones, oilstones, razor hones, and holystones. They are made from fine sandstones, siliceous argillite, and schist. Natural Belgian honestone is desired for hones and fine sandstone and novaculite for oilstones. Hindostan stone from Indiana and Queen Creek stone from Ohio serve for waterstones. Holystones or rubbing stones, made of coarse sandstone, are employed for rubbing auto bodies, furniture, and concrete.

Selected References on Silica Abrasive Stones
General Reference 13 and 19.
Abrasives, Part I. 1927. V. L. Eardley-Wilmot. Can. Dept. Mines, Mines Branch, No. 673, Ottawa.

NATURAL SILICA ABRASIVES

Natural silica abrasives are silica in various natural forms. Some of them are used as mined, but mostly they are processed and ground to grains or powder.

Most of them serve purposes other than abrasives and have been considered elsewhere in this book. This group has a wider industrial application than natural silica stones; it includes diatomite, tripoli, abrasive sand, ground sand and sandstone, ground quartz, pumice and pumicite.

Diatomite. (See Chapter 27.) Diatomaceous earth, dominantly a filler, serves as an abrasive principally in metal (silver) polishes, powders and paste, automobile polishes, dental powders and paste, and as an abrading agent in match heads and box sides.

Tripoli. (See Chapter 13.) Tripoli, a porous, earthy substance that is nearly pure silica, and rottenstone, a siliceous argillaceous limestone, are classed in trade as "soft silicas." About 30,000 tons are produced annually, of which 30 to 50 percent is used as abrasives. It is used chiefly for buffing blocks and powders, for the buffing of metals and plated wares, but also for rubbing down auto bodies and painted surfaces and in scouring and cleaning soaps and powders. Tripoli results from residual weathering of chert and cherty and siliceous rocks (see Chapter 12). The largest deposits are in Missouri-Oklahoma, Tennessee Valley, and Illinois. Near Seneca, Missouri, the deposits occur in beds 1 to 7 meters thick under thin overburden and were derived from cherty Boone limestone. Those in the western Tennessee Valley are described by E. L. Spain as a well-compacted 8 meter bed of incoherent tripoli that crops out in bluffs under 30 meters of cover. The Butler deposit contains 98 percent or more silica and was derived from the weathering of Mississippian cherty limestones from which the calcareous material has been leached. The southern Illinois tripoli is in compact beds, contains much chert, and has to be ground.

Abrasive Sand. Abrasive sands are employed for sandblasting, sandpaper, grinding or surfacing plate glass, and for scouring stone. United States production comes mainly from Cape May, New Jersey, Illinois, Ohio, and the Virginias. These sands must be sharp, clean, and fairly uniform in grain. Coarse varieties are used for heavy cast iron and steel work.

Ground Sand and Sandstone. Friable sandstone and sand are ground to finer sizes for plate glass grinding, sandblasting, sandpaper, and other finer abrasive products. About 3 tons of sand are required to surface 1 ton of plate glass. The Ottawa sandstone of Illinois furnishes much of the raw product, and Illinois, New Jersey, Ohio, and Pennsylvania are the chief United States sources. Most other countries supply all their needs. The ceramic industry consumes about 40 percent of the million tons or so of ground sand and sandstone produced. Abrasive use also includes cleansing and cleaning powders and pastes for household and industrial use.

Ground Quartz. Clean quartz, crushed and graded, is used for "flint" sandpaper, harsher metal polishes, in metallurgical works, and various scouring compounds. About $1\frac{1}{2}$ milliontons a year are produced in the United States. The quartz is derived from veins, pegmatite dikes, and pure quartzite beds. California, Virginia, North Carolina, Maine, Maryland, and New York are important producers.

Amorphous Silica. Natural amorphous silica from southwestern Illinois, Tennessee, and Georgia is sometimes listed with tripoli. It is used in buffing and polishing compounds. Chemically precipitated, amorphous silica serves the same purposes.

Pumice and Pumicite. These substances are silicates and not silica but are generally included under silica or siliceous abrasives. Pumice is the natural volcanic product or frothy glass of steam-expanded siliceous lava, and pumicite is volcanic ash. About 16 percent of an annual production of about 3 million tons in the United States is devoted to abrasive purposes.

Lump pumice is used for dressing furniture and musical instruments, preparing metal surfaces for silver plating, cleaning, lithographic stones, and rubbing and polishing fine tools and instruments. Ground pumice and pumicite make excellent cleansers because of the thin sharp glass shards contained in them. Much of this material is made up into scouring soaps. It is also an ingredient in various polishing compounds for celluloid, hard rubber, and bone, and is included in rubber erasers.

The United States supply of pumice comes from California, Kansas, Nebraska, New Mexico, Oklahoma, and Oregon; pumicite comes from Kansas, Nebraska, Nevada, and Oklahoma. Canada, Italy, Japan, New Zealand, and other countries are also producers.

Selected References on Siliceous Abrasives
Abrasives, Part I, Siliceous abrasives. 1927. V. L. Eardley-Wilmot. Can. Dept. Mines, Mines Branch, No. 673. *Complete survey.*

Tripoli. R. W. Metcalf. 1946. U. S. Bur. Mines Inf. Circ. 7371. *General survey.*
General References 13 and 18.

MISCELLANEOUS "SOFT" ABRASIVES

Calcite ground to a fine powder is used to a small extent as a nonscratch metal polish.

Chalk or whiting is a widely used soft polisher for silverware, gold, nickel, and chromium plate, brass, buttons, and so forth.

China clay and pipe clay are used as polishing powders for soft metals. Pipe clay was once the standard polish for military and naval uniform buttons and belts.

Dolomite calcined to unhydrated calcium and magnesium oxides, called *Vienna lime,* is a common buffer for various metals, pearl, and celluloid. It is of particular value for nickel plate, to give the deep "under-surface" blue color in highly polished nickel articles.

Sheffield lime is a similar product.

Feldspar ground to powder makes a desirable scouring powder and soap for cleansing porcelain, glass, and enameled surfaces. It is fairly hard, but, being softer than these surfaces, it does not scratch them. It is said to be the chief abrasive constituent of "Bon Ami" cleanser. It is also a constituent of bonds for cementing vitrified abrasive wheels.

Fuller's earth is a soft abrasive for grease removal and high-grade polish for silver and chromium wares.

Lampblack, made from the burning of oil, is a fine, soft buffer and polish for burnishing silverware, black celluloid, and buttons.

Lime is a constituent of Vienna lime.

Magnesia is used as a soft polisher for metal surfaces and also a final polisher for polished surfaces of metals and minerals prepared for microscopic study.

Magnesite serves as a bonding agent and abrasive for making emery wheels.

Metallic oxides are widely employed as final buffing and polishing agents for metals, mineral surfaces, and optical glass. *Crocus* is a hard, purple-red hydrated iron oxide made up as a paste for buffing tin, steel, cutlery, and other metal surfaces requiring a high finish. *Rouge* is a hydrated oxide of iron used as a soft metal polisher and for polishing lenses of eyeglasses, microscope lenses, and other optical products.

Green rouge is an oxide of chromium and is a very fine, fast buffer and polisher for platinum and gives the high polish on stainless-steel wares. It is also a polisher of hard minerals in polished ore sections. *Black rouge,* or precipitated magnetite, is used as a buffer on cloth for the final finish on plate and cut glass. It is also an excellent polisher to remove final scratches from soft minerals in polished ore surfaces.

Tin oxide is employed to give a high finish to certain metals, polished ore surfaces, and stone glazing.

Manganese dioxide is a good polishing medium, but it is dirty.

Talc is used as a polisher for rice grains, soft metals, and leather.

MANUFACTURED ABRASIVES

Artificial abrasives are supplanting many natural abrasives. Some of them are extremely hard and sharp and are of uniform quality.

The chief mineral products involved in their manufacture are bauxite, silica, carbon, lime, magnesia, clay, salt, boron and tungsten minerals, iron and steel, rubber, natural gas and oil, and some natural abrasives. The manufactured products are divided into silicon carbide, fused alumina, boron carbide, and metallic abrasives. In Mohs' scale of hardness, quartz is 7, corundum 9, and diamond 10, but really 40, the carbides rank between 10 and 14, boron carbide being 14, the next hardest substance to the diamond. Silicon carbide and fused alumina are the most used.

Silicon Carbide. Silicon carbide (SiC) is known under such trade names as Carborundum, Crystalon, and Carbolon and is made by the fusion of crushed petroleum coke and silica sand in an electric furnace for 36 hours at 2,200°C. It is used mainly in cutting wheels and papers and cloths. Its hardness is 13, Mohs' scale.

Fused Alumina. Fused alumina is sold under the trade names of Alundum, Aloxite, and so forth, and is made of bauxite fused in an electric furnace with coke and iron for 24 hours at 1,000°C. Corundum may be used in place of bauxite. The fused product is used in the form of vitrified wheels and powders.

Boron Nitride. The next hardest substance to the diamond, cubic boron nitride is replacing the diamond and diamond dust for many abrasive purposes. It is made into molded products for extrusion dies, thread guides in textile machinery, and sandblast nozzles. The powder is suitable for many grinding and lapping operations where extreme hardness is important. It is an electric furnace product made from coke and dehydrated boric acid.

Metallic Abrasives. Included in this group of abrasives are crushed steel, steel shot, angular steel, and steel wool, all of which are made from special irons and steels. They are used for metal blasting, for cleaning castings, forgings, and metals, core drilling, and stone sawing. These abrasives can be reclaimed and used 200 or 300 times. Steel wool is a nonclogging abrasive for woodwork, paint and varnish, and aluminum ware.

Diamonds. The diamond is by far the most important manufactured abrasive. It is manufactured from a mixture of graphite and a catalyst of iron, manganese, chromium, cobalt, nickel, or other metal under conditions of high temperature and pressure, typically 3000°C and 70,000 bars of pressure for several hours. Artificial diamonds may be challenged as an abrasive by artificial cubic boron nitride, invented in 1957 and manufactured in the same way as diamonds. It can abrade diamond and remains more stable at high temperatures than does diamond.

General References on Abrasives
Abrasives. Parts I-IV. 1927-1929. V. L. Eardley-Wilmot. Can. Dept. Mines Branch, Nos. 673, 677, 699, Ottawa. *Exhaustive treatment and complete bibliography to those* dates.
Boron carbide, the new abrasive. 1947. N. Barkeris. Jour. Am. Ceram. Soc., v. 30, no. 1.
General References 13 and 19.

General References on Economic Geology
The list below is a group of general references with which the student of economic geology should have familiarity. It is not intended to be a comprehensive bibliography as some references for each subject are given at the end of each section or chapter; rather, it covers those general references that embrace many subjects, to avoid repeating them in the Selected References given throughout the volume.
1. *Journal of Economic Geology.* Economic Geology Publishing Co. New Haven, Conn. Eight numbers per year, v. 1, 1905, to present. The outstanding journal of the world relating to economic geology and containing original articles on all phases of economic geology.

2. *Mineralium Deposita.* Springer Verlag, Berlin. Four issues per year, v. 1, 1965, to the present. An international journal for geology, mineralogy, and geochemistry of ore deposits and the official organ of the Society for Geology Applied to Mineral Deposits.
3. *Geochemica et Cosmochimica Acta.* Pergamon Press, Elmsford, N. Y. Twelve numbers per year, v. 1, 1946. An international journal of the Geochemical Society with articles pertaining to all aspects of geochemistry.
4. *Annotated Bibliography of Economic Geology.* Economic Geology Publishing Co., Urbana, Ill. Two numbers per year, from v. 1, 1933, to the present. Gives abstracts of all important papers in the world relating to economic geology.
5. *Bibliography and Index of North American Geology.* U.S. Geological Survey, Washington, D.C. Biyearly, from 1785 to present; consolidated each 10 years. Gives authors and titles of all articles on geology in North America.
6. *Bibliography and Index of Geology Exclusive of North America.* Geological Society of America, New York. Yearly, from 1933 to present. Covers foreign geology, by authors and titles, with brief annotations.
7. *Minerals Yearbook.* Annual volumes. U.S. Bureau of Mines, Washington, D.C. Annual volumes giving statistics and reviews for each metal and nonmetal, covering production, consumption, trade prices, and reviews by states and counties.
8. *Mineral Resources of the United States.* By staffs of the U.S. Bureau of Mines and U.S. Geological Survey. 212 p. Public Affairs Press, Washington, D.C., 1948. Summary of domestic reserves and sufficiency and separate treatment of 39 minerals.
9. *Mineral Deposits.* 4th ed. 1933. Waldemar Lindgren. 930 p. McGraw-Hill Book Company, New York. A leading text and reference book on advanced economic geology; covers metallic and several nonmetallic deposits.
10. *Ore Deposits of the Western United States—Lindgren Volume.* 1933. 797 p. American Institute of Mining and Metallurgical Engineers, New York. Contributions by 44 authors relating to the various groups of mineral deposits in the western United States.
11. *Ore Deposits in the United States, 1933–1967, Graton Sales Volumes.* 1968. J. D. Ridge, ed. American Institute of Mining and Metallurgical Engineers, New York, Contributions on descriptions of new mines developed since the Lindgren volume. Consists of two volumes with 82 chapters. Excellent reference source with contributions from 150 geologists on the diversity, genesis, and problems relating to the origin of metalliferous deposits in the United States.

12. *Ore Deposits*. C. F. Park, Jr., and R. A. McDiarmid., 522 p. 1975, 3rd Ed. McGraw-Hill Book Co. One of the later up-to-date texts on facts and theories of how ore deposits form and where.

13. *Minerals for the Chemical and Allied Industries,* 2nd ed. 1961, 2nd Edition. S. J. Johnstone and Margery G. Johnstone. 788 p. John Wiley & Sons, Inc., New York, A concise account of the sources, nature, modes of occurrence, and methods of treatment of all industrial minerals, including their uses and the products derived from them.

14. *Geochemistry of Hydrothermal Ore Deposits*. 1967. H. L. Barnes, ed. 670 p. Holt, Rinehart, and Winston, New York, A comprehensive review of hydrothermal and geochemical processes and ore deposits on specific types by 18 different authors.

15. *Ore Deposits as Related to Structural Features*. 1942. W. H. Newhouse, ed. 280 p. Princeton University, Press, Princeton, N.J. Brief résumés of the most important mineral deposits that exhibit localization of ore by structural features.

16. *Structural Geology of Canadian Ore Deposits—A Symposium*. 1948. 948 p. Geology Division of the Canadian Institute of Mining and Metallurgy, Montreal. A comprehensive assemblage by belts or areas of the geology and structure of all Canadian mining districts or mines with emphasis on their structural controls.

17. *Minerals in World Affairs*. 1943. T. S. Lovering. 394 p. Prentice-Hall, Inc., Englewood Cliffs, N.J. The uses, technology, distribution, geology, country, occurrences, and political features of 16 important metals and minerals.

18. *Seventy-five Years of Progress in the Mineral Industry—Anniversary Volume*. 1947. B. Parsons, ed. American Institute of Mining and Metallurgical Engineers, New York. Authoritative contributions by group of authors on various phases of the mineral industry.

19. *United States Mineral Resources*. 1973. Edited by D. A. Brobet and W. P. Pratt. 722 p. U.S. Geological Survey Professional Paper 820. Brief but succinct papers on all minerals, metals, and energy fuels of importance to the United States and their reserves.

20. *Geology and Mineral Deposits of Canada*. 1970. R. J. W. Douglas ed. 838 p. Economic Geology Report No. 1. Excellent comprehension report on mineral and energy resources, reserves and deposits of Canada.

21. *Ore Petrology*. 1972. R. L. Stanton. 713 p. McGraw-Hill Book Company, New York. Excellent source of information on the study of ores as rocks, or petrology.

22. *Mining Companies of the World*. 1975. Mining Journal Books, Ltd. London, England. Links all of the mining companies to their mines.

23. *Exploration Mining and Geology*. 1978. W. C. Peters. 696 p. John Wiley and Sons. Excellent and most recent text on mineral deposits, prospecting, and the exploration geologist.